Springer Collected Works in Mathematics

More information about this series at http://www.springer.com/series/11104

1963

Herbert Robbins

Selected Papers

Editors

Tze Leung Lai · David Siegmund

Reprint of the 1985 Edition

Springer

Author
Herbert Robbins (1915–2001)
Department of Statistics
Columbia University
New York, NY
USA

Editors
Tze Leung Lai
Department of Statistics
Stanford University
Stanford, CA
USA

David Siegmund
Department of Statistics
Stanford University
Stanford, CA
USA

ISSN 2194-9875
Springer Collected Works in Mathematics
ISBN 978-1-4939-7133-6 (Softcover)

Library of Congress Control Number: 2012954381

Mathematics Subject Classification (2010): 01-A75, 60-xx, 60-2xx

Printed on acid-free paper

This Springer imprint is published by Springer Nature
The registered company is Springer Science+Business Media LLC
The registered company address is: 233 Spring Street, New York, NY 10013, U.S.A.

Herbert Robbins
Selected Papers

Edited by T. L. Lai and D. Siegmund

Springer-Verlag
New York Berlin Heidelberg Tokyo

Herbert Robbins
Department of Statistics
Columbia University
New York, NY 10027
U.S.A.

Editors

T. L. Lai
Department of Statistics
Columbia University
New York, NY 10027
U.S.A.

D. Siegmund
Department of Statistics
Stanford University
Stanford, CA 94305
U.S.A.

AMS Classification: 01-A75, 60-XX, 60-2XX

Library of Congress Cataloging in Publication Data
Robbins, Herbert.
Selected papers.
Bibliography: p.
1. Mathematical statistic—Collected works.
I. Lai, T. L. II. Siegmund, David, 1941–
III. Title.
QA276.A12R6325 1985 519.5 85-4724

With 20 Illustrations

ISBN 0-387-96137-2 Springer-Verlag New York Berlin Heidelberg Tokyo

9 8 7 6 5 4 3 2 1

ISBN-13: 978-1-4612-9568-6 e-ISBN-13: 978-1-4612-5110-1
DOI:10.1007/ 978-1-4612-5110-1

PREFACE

Herbert Robbins is widely recognized as one of the most creative and original mathematical statisticians of our time. The purpose of this book is to reprint, on the occasion of his seventieth birthday, some of his most outstanding research.

In making selections for reprinting we have tried to keep in mind three potential audiences: (1) the historian who would like to know Robbins' seminal role in stimulating a substantial proportion of current research in mathematical statistics; (2) the novice who would like a readable, conceptually oriented introduction to these subjects; and (3) the expert who would like to have useful reference material in a single collection.

In many cases the needs of the first two groups can be met simultaneously. A distinguishing feature of Robbins' research is its daring originality, which literally creates new specialties for subsequent generations of statisticians to explore. Often these seminal papers are also models of exposition serving to introduce the reader, in the simplest possible context, to ideas that are important for contemporary research in the field. An example is the paper of Robbins and Monro which initiated the subject of stochastic approximation.

We have also attempted to provide some useful guidance to the literature in various subjects by supplying additional references, particularly to books and survey articles, with some remarks about important developments in these areas.

The reprinted articles have been divided into three groups with a brief introductory essay and road map for each group. The first is Empirical Bayes Methodology and Compound Decision Theory. This is a well-defined sequence of papers beginning in 1951, which is probably Robbins' most original and is arguably his most important contribution to statistics. The second group is Sequential Experimentation and Analysis. This is a large, but less well-defined set of papers which includes Robbins' research on tests of power one and boundary crossing probabilities, sequential allocation rules, and fixed precision estimation, in addition to his work in stochastic approximation. The third group, Probability and Inference, could alternatively be labeled "Miscellaneous." This group includes Robbins' research in the 40's and 50's in geometric probability, quadratic forms, ergodic theory, and stochastic processes, as well as his later research on optimal stopping theory, probability, estimation, and legal applications of statistical methods.

We would like to thank the following publishers for permission to reprint articles: Academic Press; Almqvist and Wiksell; American Mathematical Society; Columbia Law Review Association; Institute of Mathematical Statistics; Institute of Mathematics, Academia Sinica, Taiwan; Mathematical Association of America; Mathematical Institute of the Hungarian Academy of Sciences; North-Holland; Union of Japanese Scientists and Engineers; University of California Press; Weitzman Science Press of Israel.

We particularly thank Warren Page and the Mathematical Association of America for their permission to use Page's "An interview with Herbert Robbins."

New York and Stanford
November 6, 1984

T. L. LAI AND D. SIEGMUND

CONTENTS

B. Adaptive Allocation of Treatments in Sequential Experiments

C. Sequential Estimation and Testing

D. Power-One Tests and Related Boundary Crossing Probabilities

Part Three: Probability and Inference

CONTENTS

PUBLICATIONS OF HERBERT ROBBINS

[1935] [1] On a class of recurrent sequences. *Bull. Amer. Math. Soc.* **43**, 413–417.

[1939] [2] A theorem on graphs. *Amer. Math. Monthly* **46**, 281–283.

[1941] [3] On the classification of the mappings of a 2-complex. *Trans. Amer. Math. Soc.* **49**, 308–324.

[4] *What is Mathematics*? (with R. Courant). Oxford Univ. Press, New York.

[1943] [5] A note on the Riemann integral. *Amer. Math. Monthly* **50**, 617–618.

[1944] [6] Two properties of the function cos *x*. *Bull. Amer. Math. Soc.* **50**, 750–752.

[7] On the measure of a random set. *Ann. Math. Statist.* **15**, 70–74; also **18**, 297.

[8] On distribution-free tolerance limits in random sampling. *Ann. Math. Statist.* **15**, 214–216.

[9] On the expected values of two statistics. *Ann. Math. Statist.* **15**, 321–323.

[1945] [10] On the measure of a random set II. *Ann. Math. Statist.* **16**, 342–347.

[1946] [11] On the (C,1) summability of certain random sequences. *Bull. Amer. Math. Soc.* **52**, 699–703.

[1947] [12] Complete convergence and the law of large numbers (with P. L. Hsu). *Proc. Natl. Acad. Sci. USA* **33**, 25–31.

[1948] [13] Some remarks on the inequality of Tchebychef. In *Studies and Essays* (ed. K. O. Friedrichs et al.), 345–350. Interscience, New York.

[14] Convergence of distributions. *Ann. Math. Statist.* **19**, 72–76.

[15] On the asymptotic distribution of the sum of a random number of random variables. *Proc. Natl. Acad. Sci. USA* **34**, 162–163.

[16] The central limit theorem for dependent random variables (with Wassily Hoeffding). *Duke Math. J.* **15**, 773–780.

[17] The asymptotic distribution of the sum of a random number of

random variables. *Bull. Amer. Math. Soc.* **54**, 1151–1161.
[18] The distribution of a definite quadratic form. *Ann. Math. Statist.* **19**, 266–270.
[19] Mixture of distributions. *Ann. Math. Statist.* **19**, 360–369.
[20] The distribution of Student's *t* when the population means are unequal. *Ann. Math. Statist.* **19**, 406–410.
[1949] [21] Application of the method of mixtures to quadratic forms in normal variates (with E. J. G. Pitman). *Ann. Math. Statist.* **20**, 552–560.
[1950] [22] Competitive estimation. *Ann. Math. Statist.* **21**, 311.
[23] A generalization of the method of maximum likelihood: estimating a mixing distribution. *Ann. Math. Statist.* **21**, 314.
[24] The problem of the greater mean (with R. R. Bahadur). *Ann. Math. Statist.* **21**, 469–487; also **22**, 310.
[1951] [25] Asymptotically subminimax solutions of compound statistical decision problems. *Proc. Second Berkeley Symposium Math. Statist. Probab.* **1** Univ. of Calif. Press, 131–148.
[26] A stochastic approximation method (with Sutton Monro). *Ann. Math. Statist.* **22**, 400–407.
[27] Minimum variance estimation without regularity assumptions (with D. G. Chapman). *Ann. Math. Statist.* **22**, 581–586.
[1952] [28] Some aspects of the sequential design of experiments. *Bull. Amer. Math. Soc.* **58**, 527–535.
[29] A note on gambling systems and birth statistics. *Amer. Math. Monthly* **59**, 685–686.
[1953] [30] Ergodic property of the Brownian motion process (with G. Kallianpur). *Proc. Natl. Acad. Sci. USA* **39**, 525–533.
[31] On the equidistribution of sums of independent random variables. *Proc. Amer. Math. Soc.* **4**, 786–799.
[32] Ergodic theory of Markov chains admitting an infinite invariant measure (with T. E. Harris). *Proc. Natl. Acad. Sci. USA* **39**, 860–864.
[1954] [33] The sequence of sums of independent random variables (with G. Kallianpur). *Duke Math. J.* **21**, 285–308.
[34] Two-stage procedures for estimating the difference between means (with S. G. Ghurye). *Biometrika* **41**, 146–152.
[35] A note on information theory. *Proc. Inst. Radio Eng.* **42**, 1193.
[36] A remark on the joint distribution of cumulative sums. *Ann. Math. Statist.* **25**, 614–616.
[37] A one-sided confidence interval for an unknown distribution function. *Ann. Math. Statist.* **25**, 409.
[1955] [38] A remark on Stirling's formula. *Amer. Math. Monthly* **62**, 26–29.
[39] Asymptotic solutions of the compound decision problem for two completely specified distributions (with J. F. Hannan). *Ann. Math. Statist.* **26**, 37–51.
[40] The strong law of large numbers when the first moment does not exist (with C. Derman). *Proc. Natl. Acad. Sci. USA* **41**, 586–587.
[1956] [41] An empirical Bayes approach to statistics. *Proc. Third Berkeley Symposium Math. Statist. Probab.* **1**, 157–163. Univ. of Calif. Press.
[42] A sequential decision problem with a finite memory. *Proc. Natl. Acad. Sci. USA* **42**, 920–923.
[1957] [43] The theory of probability. In *Insights into Modern Mathematics*

(ed. F. L. Wren), 336–371. Natl. Council of Teachers of Mathematics, Wash., D.C.

[1959] [44] Sequential estimation of the mean of a normal population. In *Probability and Statistics* (ed. U. Grenander), 235–245. Almqvist and Wicksells, Stockholm.

[45] Probability. In *Spectrum* (ed. R. Ginger), 100–114. Henry Holt, New York.

[46] Comments on a paper by James Albertson. *J. Biblical Literature* **78**, 347–350.

[1960] [47] A statistical screening problem. In *Contributions to Probability and Statistics* (ed. I. Olkin et al.), 352–357. Stanford Univ. Press.

[1961] [48] On sums of independent random variables with infinite moments and 'fair' games (with Y. S. Chow). *Proc. Natl. Acad. Sci. USA* **47**, 330–335.

[49] A martingale system theorem and applications (with Y. S. Chow). *Proc. Fourth Berkeley Symposium Math. Statist. Probab.* **1**, 93–104, Univ. of Calif. Press.

[50] Recurrent games and the Petersburg paradox. *Ann. Math. Statist.* **32**, 187–194.

[1962] [51] A Bayes test of $p \leq \frac{1}{2}$ versus $p > \frac{1}{2}$ (with S. Moriguti). *Rep. Statist. Appl. Res. JUSE* **9**, 39–60.

[52] Some numerical results on a compound decision problem. In *Recent Developments in Information Theory and Decision Processes* (eds. R. E. Machol and P. Gray), 56–62. Macmillan, New York.

[53] Testing statistical hypotheses; the compound approach (with E. Samuel). In *Recent Developments in Information Theory and Decision Processes* (eds. R. E. Machol and P. Gray), 63–70. Macmillan, New York.

[1963] [54] A new approach to a classical statistical decision problem. In *Induction: Some Current Issues* (eds. H. E. Kyburg, Jr. and E. Nagel), 101–110. Wesleyan Univ. Press.

[55] On optimal stopping rules (with Y. S. Chow). *Z. Wahrsch. Verw. Gebiete* **2**, 33–49.

[56] A renewal theorem for random variables which are dependent or non-identically distributed (with Y. S. Chow). *Ann. Math. Statist.* **34**, 390–395.

[57] Some problems of optimal sampling strategy (with C. L. Mallows). *J. Math. Anal. and Appl.* **6**, 1–14.

[58] The empirical Bayes approach to testing statistical hypotheses. *Rev. Internat. Statist. Inst.* **31**, 182–195.

[1964] [59] The empirical Bayes approach to statistical decision problems. *Ann. Math. Statist.* **35**, 1–20.

[60] On the 'parking' problem (with A. Dvoretzky). *Publ. Math. Inst. Hung. Acad. Sci. Ser. A* **9**, 209–225.

[61] Optimal selection based on relative rank—the 'Secretary Problem' (with Y. S. Chow, S. Moriguti, and S. M. Samuels). *Israel J. Math.* **2**, 81–90.

[1965] [62] On optimal stopping rules for S_n/n (with Y. S. Chow). *Illinois J. Math.* **9**, 444–454.

[63] Moments of randomly stopped sums (with Y. S. Chow and H.

Teicher). *Ann. Math. Statist.* **36**, 789–799.

[64] On the asymptotic theory of fixed-width sequential confidence intervals for the mean (with Y. S. Chow). *Ann. Math. Statist.* **36**, 457–462.

[1966] [65] An extension of a lemma of Wald (with E. Samuel). *J. Applied Probab.* **3**, 272–273.

[1967] [66] A class of optimal stopping problems (with Y. S. Chow). *Proc. Fifth Berkeley Symp. Math. Statist. Probab.* **1**, 419–426. Univ. of Calif. Press.

[67] On values associated with a stochastic sequence (with Y. S. Chow). *Proc. Fifth Berkeley Symposium Math. Statist. Probab.* **1**, 427–440. Univ. of Calif. Press.

[68] Finding the size of a finite population (with D. A. Darling). *Ann. Math. Statist.* **38**, 1392–1398.

[69] Some complements to Brouwer's fixed point theorem. *Israel J. Math.* **5**, 225–226.

[70] Iterated logarithm inequalities (with D. A. Darling). *Proc. Natl. Acad. Sci. USA* **57**, 1188–1192.

[71] Inequalities for the sequence of sample means (with D. A. Darling). *Proc. Natl. Acad. Sci. USA* **57**, 1577–1580.

[72] Confidence sequences for mean, variance, and median (with D. A. Darling). *Proc. Natl. Acad. Sci. USA* **58**, 66–68.

[73] A sequential analogue of the Behrens–Fisher problem (with G. Simons and N. Starr). *Ann. Math. Statist.* **38**, 1384–1391.

[1968] [74] A sequential procedure for selecting the largest of k means (with M. Sobel and N. Starr). *Ann. Math. Statist.* **39**, 88–92.

[75] Estimating the total probability of the unobserved outcomes of an experiment. *Ann. Math. Statist.* **39**, 256–257.

[76] The limiting distribution of the last time $s_n \geq n\varepsilon$ (with D. Siegmund and J. Wendel). *Proc. Natl. Acad. Sci. USA* **61**, 1228–1230.

[77] Some further remarks on inequalities for sample sums (with D. A. Darling). *Proc. Natl. Acad. Sci. USA* **60**, 1175–1182.

[78] Some nonparametric tests with power one (with D. A. Darling). *Proc. Natl. Acad. Sci. USA* **61**, 804–809.

[79] Iterated logarithm inequalities and related statistical procedures (with D. Siegmund). In *Mathematics of the Decision Sciences* **2** (eds. G. B. Dantzig and A. F. Veinott), 267–279. Amer. Math. Soc., Providence.

[1969] [80] Probability distributions related to the law of the iterated logarithm (with D. Siegmund). *Proc. Natl. Acad. Sci. USA* **62**, 11–13.

[81] Confidence sequences and interminable tests (with D. Siegmund). *Bull. Internat. Statist. Inst.* **43**, 379–387.

[1970] [82] Statistical methods related to the law of the iterated logarithm. *Ann. Math. Statist.* **41**, 1397–1409.

[83] Boundary crossing probabilities for the Wiener process and sample sums (with D. Siegmund). *Ann. Math. Statist.* **41**, 1410–1429.

[84] Sequential estimation of an integer mean. In *Scientists at Work* (ed. T. Dalenius et al.) 205–210. Almqvist and Wicksells, Uppsala.

[85] Optimal stopping. *Amer. Math. Monthly* **77**, 333–343.

[1971] [86] *Great Expectations: The Theory of Optimal Stopping* (with Y. S. Chow and D. Siegmund). Houghton Mifflin, Boston.

[87] A convergence theorem for non-negative almost supermartingales and

some applications (with D. Siegmund). In *Optimizing Methods in Statistics* (ed. J. Rustagi), 233–257. Academic Press, New York.

[88] Simultaneous estimation of large numbers of extreme quantiles in simulation experiments (with A. S. Goodman and P. A. W. Lewis). *I.B.M. Research* RC3621. Yorktown Heights, New York.

[1972] [89] Reducing the number of inferior treatments in clinical trials (with B. Flehinger, T. Louis, and B. Singer). *Proc. Natl. Acad. Sci. USA* **69**, 2993–2994.

[90] On the law of the iterated logarithm for maxima and minima (with D. Siegmund). *Proc. Sixth Berkeley Symposium Math. Statist. Probab.* **3**, 51–70. Univ. of Calif. Press.

[91] A class of stopping rules for testing parametric hypotheses (with D. Siegmund). *Proc. Sixth Berkeley Symposium Math. Statist. Probab.* **4**, 37–41. Univ. of Calif. Press.

[1973] [92] Statistical tests of power one and the integral representation of solutions of certain partial differential equations (with D. Siegmund). *Bull. Inst. Math., Acad. Sinica* **1**, 93–120.

[93] Mathematical probability in election challenges (with M. O. Finkelstein). *Columbia Law Rev.* **73**, 241–248.

[1974] [94] The statistical mode of thought. In *The Heritage of Copernicus* (ed. J. Neyman), 419–432. M.I.T. Press.

[95] The expected sample size of some tests of power one (with D. Siegmund). *Ann. Statist.* **2**, 415–436.

[96] Sequential tests involving two populations (with D. Siegmund). *J. Amer. Statist. Assoc.* **69**, 132–139.

[97] A sequential test for two binomial populations. *Proc. Natl. Acad. Sci. USA* **71**, 4435–4436.

[98] Sequential estimation of p in Bernoulli trials (with D. Siegmund). In *Studies in Probability and Statistics* (ed. E. J. Williams), 103–107. Jerusalem Acad. Press.

[1975] [99] Sequential estimation of p with squared relative error loss (with P. Cabilio). *Proc. Natl. Acad. Sci. USA* **72**, 191–193.

[100] *Introduction to Statistics* (with J. Van Ryzin). Science Research Assoc., Palo Alto.

[101] Wither mathematical statistics? *Supplement to Advances in Applied Probab.* **7**, 116–121.

[1976] [102] Maximally dependent random variables (with T. L. Lai). *Proc. Natl. Acad. Sci. USA* **73**, 286–288.

[1977] [103] Prediction and estimation for the compound Poisson distribution. *Proc. Natl. Acad. Sci. USA* **74**, 2670–2671.

[104] Sequential decision about a normal mean (with T. L. Lai and D. Siegmund). In *Statistical Decision Theory and Related Topics II* (ed. S. S. Gupta), 213–221. Academic Press, New York.

[105] Strong consistency of least-squares estimates in regression models (with T. L. Lai). *Proc. Natl. Acad. Sci. USA* **74**, 2667–2669.

[1978] [106] A class of dependent random variables and their maxima (with T. L. Lai). *Z. Wahrsch. Verw. Gebiete* **42**, 89–111.

[107] Adaptive design in regression and control (with T. L. Lai). *Proc. Natl. Acad. Sci. USA* **75**, 586–587.

[108] Limit theorems for weighted sums and stochastic approximation

processes (with T. L. Lai). *Proc. Natl. Acad. Sci. USA* **75**, 1068–1070.

[109] Review of "Adventures of a Mathematician" by S. M. Ulam. *Bull. Amer. Math. Soc.* **84**, 107–109.

[110] Strong consistency of least squares estimates in multiple regression (with T. L. Lai and C. Z. Wei). *Proc. Natl. Acad. Sci. USA* **75**, 3034–3036.

[1979] [111] Strong consistency of least squares estimates in multiple regression II (with T. L. Lai and C. Z. Wei). *J. Multivariate Anal.* **9**, 343–361.

[112] Sequential rank and the Polya urn (with J. Whitehead). *J. Applied Probab.* **16**, 213–219.

[113] Adaptive design and stochastic approximation (with T. L. Lai). *Ann. Statist.* **7**, 1196–1221.

[114] Local convergence theorems for adaptive stochastic approximation schemes (with T. L. Lai). *Proc. Natl. Acad. Sci. USA* **76**, 3065–3067.

[1980] [115] Estimation and prediction for mixtures of the exponential distribution. *Proc. Natl. Acad. Sci. USA* **77**, 2382–2383.

[116] Sequential medical trials (with T. L. Lai, B. Levin, and D. Siegmund). *Proc. Natl. Acad. Sci. USA* **77**, 3135–3138.

[117] Some estimation problems for the compound Poisson distribution. In *Asymptotic Theory of Statistical Tests and Estimation* (ed. I. M. Chakravarti), 251–257. Academic Press, New York.

[118] An empirical Bayes estimation problem. *Proc. Natl. Acad. Sci. USA* **77**, 6988–6989.

[1981] [119] Consistency and asymptotic efficiency of slope estimates in stochastic approximation schemes (with T. L. Lai). *Z. Wahrsch. Verw. Gebiete* **56**, 329–360.

[120] Selecting the highest probability in binomial or multinomial trials (with B. Levin). *Proc. Natl. Acad. Sci. USA* **78**, 4663–4666.

[1982] [121] Iterated least squares in multiperiod control (with T. L. Lai). *Advances in Applied Math.* **3**, 50–73.

[122] Adaptive design and the multi-period control problem (with T. L. Lai). In *Statistical Decision Theory and Related Topics III* **2** (ed. S. S. Gupta), 103–120. Academic Press, New York.

[123] Estimating many variances. In *Statistical Decision Theory and Related Topics III* **2** (ed. S. S. Gupta), 251–261. Academic Press, New York.

[1983] [124] Some thoughts on empirical Bayes estimation. *Ann. Statist.* **11**, 713–723.

[125] A note on the underadjustment phenomenon (with B. Levin). *Statist. Probab. Letters* **1**, 137–139.

[126] Sequential design of comparative clinical trials (with T. L. Lai, and D. Siegmund). In *Recent Advances in Statistics* (ed. J. Rustagi et al.), 51–68. Academic Press, New York.

[127] Adaptive choice of mean or median in estimating the center of a symmetric distribution (with T. L. Lai and K. F. Yu). *Proc. Natl. Acad. Sci. USA* **80**, 5803–5806.

[1984] [128] Some breakthroughs in statistical methodology. *College Math. J.* **15**, 25–29.

[129] Optimal sequential sampling from two populations (with T. L. Lai). *Proc. Natl. Acad. Sci. USA* **81**, 1284–1286.

[130] Asymptotically efficient adaptive allocation rules (with T. L. Lai). *Advances in Applied Math*. **6**, in press.

[131] Asymptotically optimal allocation of treatments in sequential experiments (with T. L. Lai). In *Design of Experiments: Ranking and Selection* (eds. T. Santner and A. Tamhane). Marcel Dekker, New York.

[132] A probabilistic approach to tracing funds in the law of restitution (with M. O. Finkelstein). *Jurimetrics* **24**, 65–77.

[133] Urn models for regression analysis, with applications to employment discrimination studies (with B. Levin). *J. Amer. Statist. Assoc.*, in press.

Reprinted from the College Mathematics Journal
Vol. 15, No. 1, January 1984

An Interview with Herbert Robbins

by
Warren Page

Herbert Robbins

Herbert Robbins is best known to most mathematicians for his collaboration with Richard Courant in writing the classic *What is Mathematics?*, but in the following forty years, Robbins' work has earned him the reputation as one of the world's leading statisticians. Jerzy Neyman once indicated how rare it was for a professor to place on a high pedestal one of his colleagues—of necessity, one of his "competitors." And yet, in Neyman's two survey papers on major advances in statistics during the second half of this century, the main breakthroughs cited by Neyman were all obtained by Robbins. (See "Breakthroughs in Statistical Methodologies" following this interview.)

Robbins is the Higgins Professor of Mathematical Statistics at Columbia University, and is known among friends for combining a cynical humor with a deep involvement in humanitarian causes. Among his well-known sayings are:

"No good deed shall go unpunished,"

and (when asked by a university administrator if they might be taking a risk in recommending a junior faculty member for tenure):

"If he washes out in his research, we can always make him a dean."

In 1978 Jack Kiefer of Cornell wrote that "Robbins (then 63 years old) does not seem to have slowed down at all, and he is as lively and original now as in the past." That's still true. In 1982, I interviewed him at his home in East Setauket, New York, and it was immediately clear that he loves people and conversation. His broad interests, his many friendships, and meaningful involvements—like good vintage wine—seem to grow fuller and richer with the passing time. All this, of course, made our meeting rewarding and memorable. It also made my job very difficult: trying to confine him to a few specific issues was like trying to pin down an enthusiastic, animated, intellectual octopus. The following is only a sample of what was retrieved from the occasion.

* * *

Page: *You first became nationally known in* 1941 *as the co-author, with Richard Courant, of* "*What is Mathematics?*". *So let's begin there.* "*What is Mathematics?*" *has been translated into several languages, and more than* 100,000 *copies have been sold thus far. What has made this such a mathematical best seller*?

Robbins: A Russian translation of *What Is Mathematics?* was published shortly after World War II, and it provoked the Russians to produce their own version—some of them call it the Anti-Courant & Robbins—because aspects of Courant and Robbins are not in harmony with certain Soviet mathematical tendencies. This Russian work presumably reflected the correct line on mathematics—its history, the contributions of Russian mathematicians, its current state, its importance, and its Marxist foundations. When I looked at the English translation, for the first time I realized what a good book *What Is Mathematics?* really is. It wasn't written by a committee, and it seemed to have no great practical importance to governments or their defense departments. It was written by two people who collaborated in an intense manner on a subject that concerned them very deeply all their lives. When I started working on *What Is Mathematics?* I was twenty-four, one year out from my

Ph.D. It represented what I had learned about mathematics, what I hoped it would become, and where I wanted to find my place. For Courant, it represented a summation of what he had already done during his rich life as a mathematician and, in his later years, as an administrator and promoter of mathematical institutions. Courant and Robbins spoke for mathematics at a particular time (1939–1941) that can never be repeated. It would be impossible to write that book now; mathematics has changed so much, and the nature of the mathematical enterprise is so different. I would classify *What Is Mathematics?* as more a literary than a scientific work. It belongs to the tradition that started in French intellectual circles when Newton and Leibnitz discovered the calculus. People without formal scientific education wanted to understand this new notion. Salons were held, and philosophers gave lectures to crowned heads and rich bourgeois about the calculus. *What Is Mathematics?* belongs to that tradition of high vulgarization, as the French call it.

Page: *How did a young topologist like yourself come to work with Courant*?

Robbins: (*Laughter*) That's like asking, "What's a nice girl like you doing in a place like this?" Well, I had just earned my Ph.D. and I was beginning a one-year appointment at the Institute for Advanced Study in Princeton as Marston Morse's assistant. I needed a permanent job. When Courant came by looking for someone to work at New York University, Morse suggested me and Courant offered me a job. He had the book in mind, but I didn't know that at the time. When I began teaching at NYU in 1939, I was supporting my mother and young sister. For me, money was sine qua non. My salary as an instructor at NYU remained fixed at $2,500 a year during 1939–1942. That was my sole support; there were no NSF grants then. Some time during the beginning of my first year at NYU, Courant said to me: "I've been given a little money to work up some old course material into a book on mathematics for the general public. Would you like to help me with it? I can pay you $700–$800 for your assistance." I was in no position to turn down extra money for a legitimate enterprise, and the idea of communicating my ideas about mathematics to the educated layman appealed to me.

Page: *What was it like working with Courant*?

Robbins: I thought of him as an accomplished and worldly person who, in the fullness of his career, decided to devote some time to explaining to the world what mathematics was, rather than to write another research paper. I felt that I could help him in this. As work on the book progressed, however, the amount of time I had to devote to it became larger and larger, and I soon came to feel that it was interfering with any future career I might have in research. I used to commute regularly to his house in New Rochelle and, in a way, I became a member of the family for a while. In fact, I actually lived nearby for some time so that I could work with him when he wasn't busy. For about two years we worked very closely together exchanging drafts of chapters. But, as you probably know, the whole thing came to an abrupt and grinding halt in a rather dramatic confrontation described in Constance Reid's book *Courant in Göttingen and New York*.

Page: *Did your confrontation with Courant come right after the book had been written?*

Robbins: That's right. As Reid indicated, Courant felt that my collaboration was so helpful that he came to me early in our arrangement to propose joint authorship. He wasn't going to pay me any more, however, because as the joint author I'd probably want to spend even more time on the book. I agreed, since I had already become engrossed in writing the book. My first indication of what was really going on came when I went to the printers, to go over the final page proofs, and the last page I saw was the title page '*What Is Mathematics?* by *Richard Courant*.' This was like being doused with a bucket of ice water. "My God," I thought, "what's going on here? The man's a crook!" By then the book had been written, except that Courant never showed me the preface in which he thanked me for my collaboration. The dedication page to his children was also written without my collaboration.

You mentioned earlier that more than 100,000 copies of *What Is Mathematics?* had been sold. That may be, but when I recently asked Oxford University Press how many copies have been sold they told me I had no right to know. Courant copyrighted the book in his own name without my knowledge. He had a wealthy friend who paid for having the plates made, and he got Oxford to agree to distribute the book. It was a unique arrangement in which he retained the copyright and received a much larger portion of royalties. After this had been done, Courant informed me that he completely controlled the book and he would remit to me, from time to time, a portion of his royalties. And so every year, for a number of years, I used to get a note from Courant saying, "Dear Herbert, enclosed is a check for such and such an amount representing your share of royalties from *What Is Mathematics?*." I never knew how many copies were sold or how much he got, and I still don't. This arrangement continued up to the time of his death a few years ago, when his son Ernest became his legatee.

Page: *What happened then?*

Robbins: Three or four years ago, *What Is Mathematics?* appeared in paperback. Just prior to that, in order to simplify matters, Ernest offered to buy out my share, and I agreed to renounce all further claims on the book if we could set some reasonable figure for my doing so. But he never went through with this, although the sum had been agreed on—and when the book came out in paperback, I stopped getting anything at all. In fact, on the jacket of the hard cover edition (Robbins taking out his edition: copyright renewed, 16th printing, 1977) here you see something about the late Richard Courant and here's something about the present Herbert Robbins. But on the paperback edition, one finds that the mysterious Herbert Robbins appears only on the title page as co-author; on the back cover it looks as though it's entirely Courant's book. So, even after his death, there has been an intensification of the campaign not merely to deny my financial rights in the book, but even to conceal the fact that I was its co-author.

Page: *Were you ever given an explanation why Courant didn't treat you as might have been expected?*

Robbins: Some of Courant's friends came to me and said, "You see, in Europe, it's quite customary for a younger man to do the work while the older man is credited with being the formal author. This has happened before with many people and, in particular, with Courant. Don't be upset, etc." As a non-European, not acquainted with this tradition, I refused. "It wasn't fair! I had taken his word; I wouldn't have put the effort I did into this unless it was going to be a joint book." The drama continued, with more visits by Courant's emissaries, including some distinguished European mathematicians whom he had brought here. But I was adamant and wouldn't agree to be quiet. I threatened to make a fuss if Courant didn't include my name on the title page. Courant finally agreed to do so.

Page: *Were there any mathematicians who gave you guidance and encouragement during critical periods of your professional development?*

Robbins: No. What they gave me was something perhaps more important. The leading mathematicians I encountered made me want to tell them: "You son-of-a-bitch, you think that you're smart and I'm dumb. I'll show you that I can do it too!" It was like being the new kid in the neighborhood. You go out into the street and the first guy you meet walks up to you and knocks you down. Well, that's not exactly guidance or encouragement. But it has an effect.

Page: *Who have been the most impressive mathematicians you've known?*

Robbins: The first mathematician who impressed me was William Fogg Osgood, author of *Funktionentheorie*, because he had a beautiful white beard. I was a freshman at Harvard and, being from a little town, I had never seen anyone like him. I was also impressed by Julian Lowell Coolidge because he spoke with a lisp that sounded very upper class. There are many ways to be impressive. These people impressed me as personae; I thought it must have taken several generations to produce people like them. The first mathematician I met who impressed me as a mathematician was Marston Morse, and I regard him as one of the two or three most powerful mathematicians America has produced. Morse was not a wide-ranging mathematician of the Hilbert type, but he created the theory of the calculus of variations in the large, and whatever he needed he learned, borrowed, or created for himself.

Page: *How did you come to meet Morse?*

Robbins: In the 1931 Harvard–Army football game, Harvard was losing at half-time. During the intermission, Harvard's President A. Lawrence Lowell said to the cadets' commandant: "Your boys may be able to beat us in football, but I'll bet we can beat you in mathematics." The commandant accepted the challenge, and it was agreed that Army and Harvard would have a mathematics competition the following year. Since cadets had only two years of mathematics at West Point, Harvard limited its team membership to sophomores. Lowell's relative, William Lowell Putnam, agreed to put up a prize—the forerunner of today's Putnam Prize in mathematics. In 1931, I was taking freshman calculus. Having just entered Harvard with practically no high school mathematics, I knew calculus would be useful if I ever wanted to study any of the sciences. At the end of my freshman year, much to my surprise I was asked by the mathematics department to join the Harvard math team. Marston Morse was our coach. We met with him on several occasions to prepare for the competition, and that's how I first met Marston. As it happened, incidentally, Army won that mathematics competition.

Page: *How and when did you make the decision to become a mathematician?*

Robbins: Morse, G. D. Birkhoff, and Whitney were the three mathematicians who most influenced me because I got to know them quite well for short periods of time and, in very different ways, they formed my image of what a mathematician was. Meeting these three early in my education turned my thoughts to mathematics as a possible career.

Page: *I'm sure you also had great teachers in other subjects. What, in particular, did these three convey to you about mathematics?*

Robbins: One of my professors at Harvard, a famous literary critic, used to walk in with a briefcase full of books and lecture on the Romantic Poets. He'd take out a book, read a poem, and then comment on it. Now this represented real scholarship that left me totally cold. To my mind, this wasn't being creative. He was talking about what others had done. I would rather have done these things. He talked about Coleridge; I would like to have written "The Rime of the Ancient Mariner." On the other hand, Marston Morse impressed me deeply. Even though what he was talking about meant nothing to me—I didn't know the first thing about the Betti numbers of a complex and the number of critical points of a function defined on it —I could see that he was on fire with creation. There was something going on in his mind of a totally different nature from anything I'd seen before. That's what appealed to me.

H. Robbins and Marston Morse 7

Page: *Morse seems to have played a pretty prominent role in your life.*

Robbins: At that time, Marston's life was pretty much at low ebb. His wife had left him, and he was living as a bachelor at Harvard. I pitied him almost . . . but in a way, I didn't pity him; I was scared stiff of him intellectually. I was an undergraduate and, although I had never taken a course with him, I got to know him since he was living at the college. One day he said to me, "I'm leaving Harvard and going to the Institute for Advanced Study. You stay here and when you get your Ph.D. in mathematics, come to the Institute to be my assistant." Six years later —I hadn't seen him since—I sent him a telegram: "HAVE PH.D. IN MATHEMATICS." He immediately wired back: "YOU ARE MY ASSISTANT STARTING SEPTEMBER 1." Marston was, in a way, the type of person I would like to have been. He was a father figure to me—my own father died when I was thirteen. Marston and I were about as different as two people could be; we disagreed on practically everything. And yet, there was something that attracted me to Marston that transcended anything I knew. I suppose it was his creative, driving impulse—this feeling that your house could be on fire, but if there was something you had to complete, then you had to keep at it no matter what.

Page: *What was it that originally attracted you to topology?*

Robbins: My affair with topology was rather accidental. Hassler Whitney had come back from a topology conference in Moscow around 1936, and in a talk at Harvard on some of the topics discussed at the conference, he mentioned an unsolved problem that seemed to be important. Since I was then a graduate student looking for a special field to work in—not particularly topology, since I hadn't even taken a course in the subject—I asked Whitney to let me work on it. That's how I got started. I had set myself a time limit from the beginning: if I didn't get my Ph.D. within three years after starting graduate work, I would leave the field of mathematics. Midway through my third year, when they asked me whether I wanted to continue my fellowship for another year, I told them that I wouldn't be coming back next year. Although I did manage to complete my thesis that year, I didn't feel that I had become a topologist; I thought I had become a Ph.D.—a kind of generalized mathematician.

Page: *How did you become a statistician?*

Robbins: My first contact with statistics came when I was teaching at NYU. Courant had invited Willy Feller to give a course in probability and statistics, but at the last minute Feller couldn't come. The course had been advertised, but now there was no one at NYU with any interest in probability or statistics. As the youngest and most defenseless person in the department, I was assigned to teach the course. It must have been a pretty terrible course because I knew nothing about either subject. This was just before I joined the Navy in World War II, not as a mathematician but as a reasonably able-bodied person.

It was in the Navy, in a rather strange way, that my future career in statistics originated. I was reading in a room, close to two naval officers who were discussing the problem of bombing accuracy. In no way could I keep from overhearing their conversation: "We're dropping lots of bombs on an airstrip in order to knock it out, but the bomb impacts overlap in a random manner, and it doesn't do any good to obliterate the same area seventeen times. Once is enough." They were trying to decide how many bombs were necessary to knock out maybe 90% of an area, taking into account the randomness of impact patterns. The two officers suspected that some research groups working on the problem were probably dropping poker chips on the floor in order to trace them out and measure the total area they covered. Anyway, I finally stopped trying to read and asked myself what really does happen when you do that? Having scribbled something on a piece of paper, I walked over to the officers and offered them a suggestion for attacking the problem. Since I wasn't engaged in war research, they were not empowered to discuss it with me. So I wrote up a short note and sent it off to one of the two officers. In due course, it

Robbins and Courant, 1942

9

came to the attention of some mathematical research group working on the problem. However, I had no clearance to discuss classified matters, so there was a real communications problem: how were they going to find out my ideas without telling me something I shouldn't know? (What I shouldn't know was, in fact, the Normandy invasion plans.) Well, in some mysterious way, what I had done came to the attention of Marston Morse, and he saw to it that my note reached the right people. Shortly afterward, S. S. Wilks, then editor of the *Annals of Mathematical Statistics*, asked me to referee a paper by Jerzy Neyman and Jacob Bronowski (author of *The Ascent of Man*) on this very same problem. I recommended rejecting their paper as: "a rather unsuccessful attempt at solving a problem that is easily solved if it's done the right way, and here's how to do it." Wilks wrote back that he had to publish the paper because Neyman was one of the authors. But he also wanted me to publish a paper on what I'd written to him. So, after the war in Europe ended, there's an issue of the *Annals* containing the paper by Neyman and Bronowski, followed immediately by my paper which, so to speak, says "Please disregard the preceding paper. Here's the solution to the problem that they can't solve." That was my first publication in the field of statistics. But even then I had no idea that I would become a statistician. What I had been doing was not statistics, but some rather elementary probability theory.

J. Neyman and H. Robbins

Page: *What did you do after four years in the Navy?*

Robbins: I had a career crisis. My pre-war job had been as an instructor at NYU, and I had already burned my bridges there. Jobs were scarce, so with my back pay

10

from the Navy I bought a farm in Vermont. I went there with my wife—I had gotten married during the war—to figure out what to do next. I thought I was going to leave mathematics and the academic profession completely. Then fate struck again with a telephone call from Harold Hotelling of Columbia University's Economics Department. Hotelling's primary interest was in mathematical statistics. Since Columbia had not allowed him to create a department of mathematical statistics, Hotelling had just accepted an offer to do so at Chapel Hill. The idea of such a department was being promoted at the University of North Carolina by a very energetic statistician, Gertrude Cox. Hotelling offered me an associate professorship in this newly created department. I thought he'd telephoned the wrong Robbins, and I offered to get out my AMS directory to find the Robbins he'd intended to call. Hotelling insisted that there was no mistake, even though I told him that I knew nothing about statistics. He didn't need me as a statistician; he wanted me to teach measure theory, probability, analytic methods, etc. to the department's graduate students. Having read my paper in the *Annals of Mathematical Statistics*, Hotelling felt that I was just the sort of person he was looking for. "Don't question it any further," he insisted. "The salary will be $5,000 a year." That was in 1946, and it was double my salary at NYU four years earlier. It was a very good salary at that time. So, with some trepidation, I agreed. At Chapel Hill, I attended seminars and got to know several very eminent statisticians. Soon I began to get some idea about what was going on in that subject and finally, at age thirty-two, I became really interested in statistics.

Chapel Hill, 1946
Front row, left to right: W. Hoeffding, H. Robbins, R. C. Bose, H. Hotelling, S. N. Roy

Chapel Hill, 1950 H. Robbins and R. A. Fisher

The Creative Process

Page: *Herman Chernoff characterizes your innovations as having been based mainly on extra-mathematical insight, and Mark Kac describes your contributions as marked by power, great originality, and equally great elegance. Is there anything you can share with us about the creative process—your feelings and experiences—during the germination of some new insight or breakthrough*?

Robbins: I'm always pleased to hear my work praised, but the things I've been associated with that are really important have not been done by me at all. I've merely been the vehicle by which something has done them. When something significant is happening, I have a feeling of being used—my fingers are writing, but there's a lot of noise and it's hard for me to get the message. Most of the time I'm just sitting there, in an almost detached manner, thinking: "Well, here's another day's wastebasket full of paper. Nothing's come through. Maybe another day. Maybe I should stay up tonight and try some more." I stay up nights when my wife and children have gone to sleep. Over and over again I keep working at it, trying to understand something which after months or even years turns out to be so simple that I should have seen it in the first ten minutes. Why does it take so long? Why haven't I done ten times as much as I have? Why do I bother over and over again trying the wrong way when the right way was staring me in the face all the time? I don't know.

12

Page: *How do you feel after having made a discovery?*

Robbins: I feel like someone who has climbed a little mountain the wrong way. Once I get the message, so to speak, I try to write it up as clearly as possible. Then I want to get away from it. I didn't do it. I don't want to see it again; I had enough trouble with it. I want to push this onto the rest of the world: "Look, there's lots more that has to be done, but don't expect me to do it. I've done my duty. I've contributed to the Community Chest. Now let somebody else carry on."

Page: *The fact is that you reached the summit, you made the discoveries.*

Robbins: Yes, I take some pride in that. Had I not lived, certain things would not have been done. But a world consisting of lots of me's would be intolerable. One is enough.

I always look for something terribly simple, because very simple things are often overlooked. In a way, my strengths are due to my weaknesses. Others are technically much better than I, but it never occurs to them to do the dumb kinds of things that occur to me. A good example is stochastic approximation. Lots of people said: "My God, we can generalize that; we can do it under much weaker restrictions; etc." And I thought, "Yes, that's true. But why didn't somebody do it sixty years ago?"

Page: *Let's stay with the creative process. What do you do when you're blocked or stymied?*

Robbins: There's nothing I can do except try not to get panicky about it. If I live long enough, one of these days I will stop and never get another idea. I have no idea when that will come, but it will come . . . or maybe it has already.

Page: *Why? I want an existence proof.*

Robbins: Look at Einstein. He seems not to have had an inspiration in physics during the last thirty or so years of his life—at least none that could be compared with the great ones he had from 1905 to 1920. Here was a man who had perhaps the greatest intellect that God ever created and, in the last years of his life, nothing much came of it. It wasn't because Einstein was frivolous or dissipating his energy; he had done what he could and there came an end. As I watched him at the Institute one year, he never complained about it and no one mentioned it, but everyone knew that he was essentially finished as a scientist. And Newton? The same thing. From about age thirty on, he did absolutely nothing in science. He had a career as Master of the Mint, he carried on a great deal of activity with friends—controversies over who invented the calculus, and so on—but the last half of his life was totally sterile from a scientific viewpoint.

Page: *Isn't this fear of "drying up" something every researcher and, in fact, every creative person has?*

Robbins: Yes, but it doesn't end here. Take the guy who cracked the genetic code, for instance. "What's he done recently?" In this country, the question is always: "What have you done recently?" . . . "Oh yes, you did such–and–such, but how about last year?" It's not so much that others are asking this question, we're taught here to ask it of ourselves. I constantly find myself asking what have I done during the last year or so, and how does it compare with what I did thirty years ago?

Page: *That sounds like sequential doom. Are you saying that we've come to expect an ever-increasing sequence of better and better encores of ourselves?*

Robbins: Right. And this creates a lot of frustration and anxiety. If I were a promising young tennis player, I'd hope to get better and better, and finally to win at Wimbledon. And then I'd become a teacher—one can't go on forever playing competitively with twenty-year-olds. No one would think less of me if I didn't enter Wimbledon at age 65. But what I'm doing is not tennis. As a mathematician, I'm using my brain, and there's no reason why it shouldn't be as good, if not better, than it was thirty years ago. "So why is it not?" I ask myself. I've been in statistics now for thirty-five years and I would like to try something else. Why don't I try going into molecular biology? Or sociology, or economics? Am I incapable of the mental effort, or am I just too weary? These complicated questions, raised by increasing expectations, I can't answer.

Page: *Although physical prowess—say reaction time—is crucial in sports, it makes no significant difference if one's insight into a mathematics problem takes a second or a year. Perhaps we should consider the mathematician's personal drive to succeed and the price he's willing to pay for success.*

Robbins: Younger mathematicians have a greater desire to become known and make a reputation. This weakens with age, either through frustration if they don't succeed or through satiation if they do. And even if one does make it to the top, was it worthwhile? Is it worth continuing to strive for more? The really successful mathematician, if he's honest, must assess his life in terms of having foregone meaningful relations with others—wife, children, colleagues, friends, etc. As one becomes older, he becomes less likely to want to pay the price for new successes. The theorems that I've proved aren't going to be much good or as comforting to me as would be close friends when I'm old and perhaps infirm.

Page: *Having been both a topologist and a statistician, have you perceived any difference between those involved in these two fields or, more generally, between those in pure mathematics and those in applied areas?*

Robbins: I don't think I can distinguish any behavioral or personality differences. However, I recall that when I started out, applied mathematicians were looked down on by pure mathematicians. If you got a Ph.D. in mathematics and your professors thought you weren't really very good, then they'd suggest that you would do well to get an actuarial job in an insurance company, or an applied job with an industrial firm. If you weren't a pure mathematician, you weren't top drawer.

I remember a well-known mathematician, alive today, who started out in pure mathematics and then became interested in probability and statistics. While talking with him one day—I was quite young at the time—I asked what he thought was the most important work I might do during the next few years in the field of probability and statistics. To my amazement, he turned red with emotion and almost pleaded: "Robbins, the most important thing you can do is to show mathematicians that probability theory and mathematical statistics are really part of mathematics." I was absolutely dumbfounded. Evidently his former colleagues had made him feel that he was no longer a member of the elite when he became involved with probability and statistics. He had never been able to survive the blow to his ego that his defection from the realm of pure mathematics had caused.

Page: *Are there feelings of jealousy and/or competitiveness among individuals working on the frontiers of developments in mathematics?*

Robbins: Competitiveness and jealousy seem to belong more to my generation than to the current one. When I was young, there was a great deal of it. Young people now don't have as much. Although there's a reasonably well-defined pecking order, I don't see them motivated by the same burning desire to be Number One and to cast discredit on all their competitors. That was quite common when I was in my 20s and 30s. Maybe because it's easier now. Mathematics is a way of making a living, like selling insurance. When I started out, to be mathematician was a rare choice: there weren't many jobs, and one had to be prepared to give up certain things for the enjoyment of doing mathematics. Today, of course, mathematicians work everywhere.

Page: *Is a little bit of competitiveness healthy for those engaged in research?*

Robbins: Competitiveness, as far as I'm concerned, has an ambiguous quality. Sometimes I think that I'm the best in the world, since I'm the only one who looks at things the way I do. So, in this sense, I'm beyond competition. The other feeling is one of total ineptness. There are many fields of mathematics in which I don't even know the elements. I've tried to learn them, but I can't remember things from one day to the next. These are fields in which I just fall on my face every time I try. I can't help being anxious about not really knowing what others are talking about. I should know these things because my students have to. Thank God I'm not being examined!

Mathematical Reflections and Projections

Page: *In what directions is the field of statistics evolving?*

Robbins: Let's take just the field, called biostatistics, that deals with the application of statistical methods to human health and disease. The demand for trained biostatisticians is enormous, but there's absolutely no supply. If I were given ten million dollars to spend for advancing science, I could spend it trying to produce one or two good biostatisticians. Statistical methods that are currently being used were mostly developed in England for analyzing such things as agricultural experiments and industrial processes. Many of these techniques are being blindly applied to situations for which they are not adapted. The methodology for handling important problems in biostatistics does not exist. It's just beginning now; its Newton or Einstein has yet to appear.

Page: *How and where can one become trained as a biostatistician?*

Robbins: A mathematically capable student who wants to become directly involved with problems of human welfare should be doing biostatistics. Unfortunately, there's very little encouragement to do this, and there are very few places now to learn biostatistics. A mathematics department would never think of advising anyone to study it. I would like to see a distinguished mathematics department in this country tell its students: "You are very capable and you could have a career in algebraic geometry, or whatever, but we would like to encourage you to go into biostatistics."

Page: *Since it's so demanding just to keep abreast of one's own field of specialization, is it possible to stay mathematically literate in general?*

Robbins: It's harder and harder. I am not totally illiterate in mathematics. If you're in a university, about the best you can do now is to go to as many seminars and listen to as many one-hour lectures as you can, and just hope that some of it will sink in. But to really keep up with the literature now is impossible. Even when I was a graduate student, in the 1930s, it was just barely possible to have a fairly good idea of most of what was going on. Maybe then somebody could still have said, "Anything that's going on in mathematics is of interest to me and with a little effort, if necessary, I'll read the latest paper on it." No more, it's not possible.

Page: *Is it better to be a mathematical specialist or a generalist, and how difficult is it to be either in a meaningful manner?*

Robbins: That's like asking if it's better to be a decathlon athlete or a high jumper. You do what's best for you. You do what God has given you the wherewithal to do it with. If you are pretty good in a lot of things without being world class in any one of them, you'll find some field or activity which requires exactly that, and nobody else will do it as well. No high jumper could win the decathlon. The person finds the problem. You can't decide what kind of mathematician to be.

Page: *How do today's mathematicians compare with those of earlier generations?*

Robbins: I once enunciated a law of human development: *The total amount of intelligence remains constant while the population increases exponentially*. If you ask who the great mathematicians of the present day are, and how they compare with those of fifty years ago, people will tell you that we've got so many bright people now who can do things which nobody could do fifty years ago. I take that with a grain of salt. I don't believe we've got all these greatly gifted mathematicians and all these young geniuses. Hilbert is Hilbert, and there won't be another one like him for some time.

Page: *Is there anything society can do to help produce future Hilberts?*

Robbins: There's not much difference between creativity in music and in mathematics. We have not seen, nor been able to create, a modern music that compares with the Baroque, even though millions of dollars are spent annually on music instruction in the high schools, and seventy-eight Americans have won the international competition in this, that, and the other thing. What comes out is pedestrian and not of much interest. During the Sputnik era, the country became concerned with its technical capabilities and we decided to strengthen our scientific establishment. Mathematics became a national priority. Everyone was running around reforming mathematics instruction, creating the new math, rewriting textbooks. More student scholarships and faculty research grants were awarded. There's no doubt that the effect of all that was to produce more mathematics, but I don't know that anything significant came out of it. It may have produced a lot of utility-grade mathematicians who have written lots of mediocre stuff. We viewed the problem in the same sense as our annual output of steel. Maybe we're only number three in the world in annual steel output. Would it do us any good to be number one? Who cares? The point is that nobody knows how to produce a Bach or a Newton.

Page: *There seems to be a greater publish-or-perish pressure today than ever before. Has this resulted in an increased tendency for academicians to jump on new mathematical bandwagons in order to take advantage of greater opportunities for publication?*

Robbins: Journals have proliferated to the extent that there's no real problem in getting published somewhere, although there's still a distinction between publishing in refereed and nonrefereed journals. Many published papers are of no real interest or value. I've often thought that when I become enormously wealthy, I'll establish the very prestigious Herbert Robbins Prize in Mathematics. It would have one condition: the recipient shall never publish another paper. As to mathematicians jumping onto new bandwagons, I believe that most people place self-aggrandizement and obvious rewards ahead of duty to the truth, so to speak. I like to think that when I was young, one did something because that was what one wanted to do, regardless of whether anyone paid for it or listened to it. But this is probably an illusion of age and selective recall.

Page: *Is it easier to become better known in some fields of mathematics than in others?*

Robbins: In number theory, there are a number of classical conjectures—Goldbach's, Fermat's, etc. Anyone who makes a contribution to them gets instant fame because these are such famous problems—even though they seem to be somewhat outside the general domain of mathematics. An affirmative solution of Goldbach's conjecture would have no obvious consequences in any other domain of mathematics, or even in number theory itself. It's just a glorified champion Rubik's Cube puzzle. One could, of course, say: "I proved Goldbach's conjecture and that makes me the greatest mathematician of our time." That can be justified in the sense that some very powerful mathematicians have tried and failed. It's like saying that you're the first person to climb Mt. Everest. There are many conjectures hanging around—problems that nobody's been able to prove or disprove—and they represent standing challenges for young mathematicians to try their muscles on.

Page: *But what about new breakthroughs or discoveries not rooted in historical precedent?*

Robbins: I wasn't able to prove Goldbach's conjecture, but I did invent empirical Bayes, stochastic approximation, and tests of power one. That's something like saying, "I failed to win Wimbledon, but I invented a new game called clinker ball, and I was the local club champion when there were only a few others who played it." Personally, I would rather have done some of the things I've done than some of those other things which I haven't. In a sense, I'm simply saying that I love my wife, and I'd rather have married her than somebody else who might be more famous. To have proved that π and e are transcendental were great accomplishments. These were outstanding problems, and everybody knew that anyone who could solve them would be famous. But, in a sense, nothing much came of it; nothing was created that wasn't there before. To have created some important new fields that didn't exist before doesn't make me a great mathematician, but it does contribute significantly to the general progress of the mathematical sciences.

Page: *So, in mathematics, there seem to be two different types of activities?*

Robbins: Yes. One is to find the answers to problems that have been raised earlier by others, and the other is to create techniques which will then find the problems to which they can be applied.

Page: *Let's broaden our focus. How are mathematicians regarded by society at large, and how does this compare with the public's view of physical and social scientists?*

Robbins: The public has a terrible fear of mathematics. I think it's quite real, and it's not going to be overcome by restructuring the curriculum or anything else—say, like painless dentistry. The ability and the desire to think abstractly and rigorously is not generally fostered in our society. Most people haven't the faintest idea of what mathematicians do, how they think, or what they contribute to society. Mathematicians are regarded with a sort of awe that attaches to any scientist—although we're not really scientists—because we are engaged in a very elusive form of activity.

Page: *How do you feel about being a mathematician?*

Robbins: Let me answer your question this way. Most people acquire a certain expertise, and they work in fields where their expertise can be used. I don't have any expertise. If I were Picasso, I could wake up in the morning and say: "Well, I think I'll paint a Picasso today." And by the end of the day I would have painted a real, genuine Picasso. Although it may not be one of my best, it would be another Picasso and it would be discussed by art critics and sold to collectors, and so on. Another day, another painting. Now if I get up in the morning and say, "I think I'll do something in mathematical statistics," at the end of the day I've got a wastebasket full of paper and nothing to show for it. And likewise the next day, and the next. I cannot do something by willing myself to do it, and what I finally produce is usually complete junk. I've probably wasted more paper than any mathematician in the world. I have no idea whether I'll ever do anything worth talking about for the rest of my life. I'm not even like a dentist who comes home and can tell his wife: "Today I did three fillings and two root-canals, and I saved several people from serious tooth decay. Now let's have dinner." What did I do today? I talked to a few people. I tried to think about something and it came to nothing. Finally, I found that I was just repeating what some other researcher had already done. The day's been a total loss.

Page: *Doesn't this place a pretty severe burden on one's self esteem and character?*

Robbins: Most mathematicians are unable to cope with this. I see so many who have stopped working, or are just repeating themselves and basking in former glory. There are so many ways this emotional deprivation can get to you—the fact that you're just looking at the interior of your skull as though you were inside an egg, and there's no world except what you see inside. In most cases, there's no real contact with humanity, history, or culture in general.

Teaching and Learning Mathematics

Page: *Statements have been made to the effect that the good researcher who is a good teacher (undergraduate, or even graduate) is the exception. What has been your experience: are good researchers usually poor teachers?*

Robbins: Good researchers are often poor teachers; bad researchers are almost always poor teachers. The reason that you have poor teachers is that you have poor persons: undeveloped, ignorant, intellectually poverty-stricken individuals who have nothing to offer their students except the subject matter itself. They have no joie de vivre, enthusiasm, or curiosity for learning. They'd be poor in any profession.

Page: *Do you enjoy teaching? What, in particular, do you like and/or dislike about teaching?*

Robbins: I like to think that I'm a teacher by profession; research is what I do for fun. I want to show people what I've seen, that no one else has seen, so that they can share it with me. My teaching is like a man struggling with a bear. You don't know how it's going to come out, the result is not preordained. But that can be very painful too. Teaching should be like a competition between two antagonists with the outcome really in doubt. And yet you don't want it to be a clumsy job. Things are never settled: every answer raises new questions and begins a new cycle in the subject.

Page: *Have students changed much during your forty-five year teaching career?*

Robbins: There seems to be a regression toward mediocrity: lots of fairly good students, but not as many really bright or as many really dumb students. I don't see many outstanding, dedicated, obsessed, self-motivated freaks. Right now everyone wants to get an MBA, or to get into medical school, or into computer science, or some other highly remunerative graduate field. When I went to college in the 30s, not that many people went on to graduate study to prepare for a job in some special field. If your parents could afford it, or if you were very bright and got a scholarship, you went to college to get an education. I don't want to set myself up as a critic of today's youth, comparing them to a utopia that I envisage in the past. Nevertheless, I have a feeling now of teaching in a trade school; I didn't have the feeling of being a student in a trade school.

Page: *People do learn to learn differently, and today's youth seem to be getting an increasing diet of television, videogames, computers, and other interactive modes of learning. Will future students find classroom lectures uninspiring, and textbook or informational reading unbearably dull?*

Robbins: I have three children who spend a lot of time watching television and damn little time reading books. I don't know if any of them is going to get into college, or what kind of college it will be if they do get in. All I know is that they're a lot different from what I was like. When I was their age, I used to go down to the public library after school, and come home with an armful of books. I'd read them all before the next day—I must have read every book in the library. I don't believe that expertise at computer programming and interactive this, that, and the other thing is any substitute for the written word and the human voice. I don't have a home computer myself, and I'm not anxious for children to learn programming at an early age. I'm still hoping they will learn to read, think, and interact with people rather than machines. Anyone who reads a newspaper will see that parents are now being told that computers are the secret of success. "Send your children to computer school on weekends so that they'll get that edge in the race for success." We could all be replaced by computers, I'm sure. This would be advantageous to the efficiency of computations, but it's not the kind of world that interests me.

Page: *Will any important subjects become much easier or much more difficult to teach in the future because of changing technology, student intellect, or societal values?*

Robbins: Roughly speaking, this is the same as asking what will be the world's record for the 100-meter dash in the year 2500. I think we've gone about as far as we can go (maybe someone can shave half a second or so from the record) unless we mutate into a strikingly different breed. There's just so much energy and so much time for training, and that's it. The subjects of mathematics? I do not see them, as a result of efforts by some future Bourbaki, becoming simple and within the grasp of young children, so to speak, without painstaking introduction, slow step-by-step increments, and historical approaches. I don't see any reason to believe that there's going to be any great simplification or greater accessibility of mathematical knowledge in the future, no matter what amount of technology, training, or machinery is used. Nobody is going to run 100-meters in five seconds, no matter how much is invested in training and machines. The same can be said about using the brain. The human mind is no different now from what it was five thousand years ago. And when it comes to mathematics, you must realize that this is the human mind at an extreme limit of its capacity.

Page: *What about rapidly expanding scholarly disciplines such as mathematics: will more education and graduate study or training be required of future students who want to begin a meaningful career in mathematics?*

Robbins: I think less is being demanded now than used to be. You don't have to know any foreign languages, for example. If someone can write a doctoral thesis, all other deficiencies will be forgiven, and he'll get his Ph.D. even if he's never taught and has never convinced anybody that he knows the difference between mathematics and computer programming. There's a job market out there eager to swallow up such novitiates. They don't have to earn Ph.D's; an M.A. is fine, even a B.A. in mathematics is fine.

Knowledge and Power

Page: *We know that knowledge is power. But power is also knowledge insofar as prevailing political systems mold and determine what knowledge is created and utilized. Should scientists promote the creation and development of all kinds of knowledge for the sake of knowledge and the enlightenment of mankind, or should there be limitations —external or self-imposed—to the quest for knowledge?*

Robbins: Well, let me mention a remark that J. Robert Oppenheimer once made. As you know, his attitude toward the H-bomb changed from being opposed to being in favor of it. Oppenheimer said that originally he was opposed to the H-bomb because it served no useful purpose. But once a really clever way of making it had been proposed, it was so "sweet"—from the point of view of physics —that it was impossible not to try it. My blood ran cold when I read that. What kind of enterprise were we engaged in when something can be so technologically attractive that, even though it may involve the death of millions of people, a scientist must do it because of its scientific sweetness? One of the things I'm happy about is that I didn't work on nuclear weapons. I know many mathematicians who contributed to producing fission and fusion weapons. I'm glad I didn't. But then, nobody asked me to.

Page: *What about those who teach mathematical techniques that may be used for destructive purposes?*

Robbins: I hope that nothing I do will be used for purposes I don't approve of, but I know perfectly well that it will. It's inevitable. There's nothing I can do about it. As a teacher, I have become increasingly alienated from teaching because it gives me so little opportunity to explain to students that technique is not what it's all about: the desire to prove theorems is not what made me go into mathematics; there's more to life than learning how to get a Ph.D. I'm trying to tell them that the world we live in is not what it should be, and that they should spend most of their time not directly learning techniques, but rather learning what the world ought to be like, and how they should act to help make it so. What worries me is that, in my own student days, I had the benefit of contact with a very small number of people whose lives—not written words—influenced me profoundly. My students know nothing about me outside the classroom. They have no idea of how I live, why I'm doing what I'm doing, or what I think about the world. I feel frustrated. I can't turn my classroom into a pulpit; I'm in the wrong profession for that. But I damn well don't want to teach arc-welding to a bunch of robots who'll go out and arc-weld everything in sight for whoever's paying for it. So, in a sense, I might just as well admit that I'm not all that different from those who worked on nuclear weapons, because I'm teaching techniques to young people without knowing what use they'll make of them, or whether they understand what it's really all about. And I'm afraid they don't in most cases. How can I tell them?

Page: *Do you feel that scientists have a responsibility to become involved in issues of social concern*?

Robbins: Nobody has the responsibility to extend himself into a field beyond his competence. But if you feel this is something that concerns you, it's your duty to become involved. Otherwise, you'll be frustrated and bitter, and the world will be all the poorer. Proving theorems should be permitted to anyone. But if you're not an idiot in the Greek sense of being a private person, then you'll want to talk to others —scientists and nonscientists—about issues that concern everybody. And you will do so. Mathematicians have a very poor track record in this respect; the one exception being their participation in issues of human rights. The proportion of mathematicians defending human rights probably exceeds their proportion in the sciences as a whole. Perhaps I should mention that the chairman of the National Academy of Science's Committee on Human Rights is Lipman Bers. In Russia, many mathematicians try to support the Helsinki Accords.

Page: *To what do you attribute this*?

Robbins: Part of the reason may be that their original concern for issues of human rights led them into careers as mathematicians. People who have a predilection for resisting abusive social policies will tend to prefer activities which are not directly useful to their government's enactment of these policies. In Russia, for example, if you wanted to work in the sciences and not be controlled by the Party apparatus, you'd choose a science as far removed from practicality as possible. Pure mathematics would be a good career choice—in which case, the State would be more likely to place you at an institute not directly concerned with military matters. It seems quite unlikely that algebraic geometry can be used for military or political purposes. The Soviet Union can afford to allow very capable mathematicians to do pure research since they'll bring credit to the State in an indirect manner. Most other scientists are directly involved with something that can be used by the State.

Page: *Have you ever used your mathematical expertise in matters of civil importance?*

Robbins: During the last ten years, I've become interested in the applications of probability and statistics to legal proceedings. Recent developments now make statistical evidence not only admissible in court, but preponderant in certain cases. In the past, someone would file a discrimination suit as an individual, citing direct anecdotal evidence of being denied fair treatment. Today, however, one files a legal suit as a member of a class, and the evidence is the data on how people are being hired or rewarded as a class. Thus, the evidence is statistical: although no single individual can be said to have been maltreated, the class as a whole may have been found to be treated unfairly.

A really serious problem that emerges is due to the fact our purely scientific statistical apparatus is not really well adapted for legal proceedings: statistical tools created for quality control in the chemical industry must not be misapplied in deciding issues of discrimination. This problem is exacerbated by the use of computer programs, since one can feed numbers into a computer and then interpret the output any way one wishes. Calling something evidence of discrimination doesn't really make it such. Imagine how difficult it must be for a judge and jury to interpret this type of evidence. As a consultant in legal matters, I find over and over again that "evidence"—results based on putting numbers into formulas and computer programs—is being misrepresented and totally perverted by statistical "experts."

New Horizons

Page: *Is there anything you still want to accomplish?*

Robbins: I'd give up my next five papers to write a good string quartet, but I've never been able to. In fact, I don't seem to have any choice in the matter. Sometimes, maybe at 2 a.m., I'm awakened by a feeling of someone knocking: "Hey Herb, you've been to the movies, you've taken your kids to the beach, and you've socialized with the neighbors. Now let's get back to business. You never really figured out what went on in this problem, and you don't even remember where you left it. But I remember, so please get up. We've got some work to do." Finally, I get up and start working, feeling as if I've been Cinderella at the ball and, now that midnight has struck, I've got to go back to cleaning up my mathematical house. After all, that's what I'm for. The last few years of my career have been unusual—I've actually gone back to working on empirical Bayes and stochastic approximation after a thirty-year hiatus. I feel that I didn't do quite as much as I should have, and no one else has done them justice. Once I give them another push in the right direction, I'll be able to relax and not worry about them.

Page: *Albert Schweitzer once said that the great secret of success is to go through life "as a man who never gets used up." At age 67 and still going strong (more than eighteen publications in the last five years), what's your secret for not getting used up?*

Robbins: The one thing I must express is my great fortune in having found some wonderful young people to collaborate with. They've helped prolong my mathematical work far beyond what it would have been in isolation. Of course I'm not doing pure, abstract, postulate-theorem-proof mathematics. I'm involved with the mathematics that relates to the mysterious and fascinating phenomena of chance that I see around me. I'm trying to create methods for looking at the real world, and I'm better able to do this now than when I was young because I know more about the world and its problems. I'm trying now to solve some mathematical problems that I've been thinking about for fifty years.

Perhaps the real difficulty in answering your question stems from the fact that you're interviewing a sixteen-year old kid who happens to be inhabiting the body of a sixty-seven year old man. You're looking at the body, but I'm afraid you're listening to the kid.

Herbert Robbin's parents

Part One

EMPIRICAL BAYES METHODOLOGY AND COMPOUND DECISION THEORY

Robbins' pioneering paper [41]* on empirical Bayes methodology and his earlier paper [25]* on the related subject of compound decision theory were acclaimed by Neyman (1962) as "two breakthroughs in the theory of statistical decision making."

Consider the simple problem of estimating a parameter θ based on the observed datum X, whose probability density function $p(x|\theta)$ depends on θ. The Bayesian approach assumes the existence of a prior distribution B for θ; and the Bayes estimate minimizing squared error loss,

$$E[t(X)-\theta]^2 = \int E_\theta[t(X)-\theta]^2 \, dB(\theta),$$

is given by the posterior mean

$$t(X) = E(\theta|X) = \int \theta p(x|\theta) \, dB(\theta)/f(X), \tag{1}$$

where $f(x) = \int p(x|\theta) \, dB(\theta)$ denotes the marginal density of X. The difficulty with implementing the Bayesian approach is specification of the prior distribution, which may be impossible in practice.

In many applications, however, one is faced with n structurally similar problems of estimating θ_i from X_i $(i=1,2,\ldots,n)$, where X_i has probability density function $p(x|\theta_i)$ and the θ_i can be assumed to have the same distribution as B (cf. Neyman, 1962; Cover, 1968; Copas, 1972; Carter and Rolph, 1974; Efron and Morris, 1977; Hoadley, 1981; Rubin, 1981; Morris, 1983). As a concrete example suppose that the ith automobile driver in a certain sample is observed to have X_i accidents in a given year. Assuming

*Papers reprinted in this book are marked here by an asterisk when they first appear.

that X_i has a Poisson distribution

$$p(x|\theta_i)=\exp(\theta_i)\theta_i^x/x!, \qquad x=0,1,\ldots, \tag{2}$$

we want to estimate the "accident-proneness" parameter θ_i for each of the n drivers. If the distribution B of accident proneness in the population of drivers were known, the Bayes estimate of θ_i is the $t(X_i)$ given by (1), which in the present case reduces to

$$t(x)=(x+1)f(x+1)/f(x). \tag{3}$$

Of course B is essentially never known; and Robbins' idea is first to estimate $f(x)$ on the basis of the n observations $X_1,\ldots, X_n$ and then to use this estimate as a substitute for the unknown f in (3), leading to an empirical Bayes estimate of the form $t(x; X_1,\ldots, X_n)$.

The same ideas are applicable to the so-called "compound decision problems," in which the θ_i are regarded as unknown constants rather than an unobservable random sample from a distribution B, and the overall loss function is the sum of the loss functions of the individual problems. Following the seminal paper [25], Robbins made further contributions to compound decision theory in the papers [39]*, [52], and [53]. Other important developments along this line include the work of Hannan (1957), Samuel (1963a, 1964, 1965), Van Ryzin (1966a,b), and Vardeman (1978) on sequential compound decision problems, the papers of Gilliland *et al.* (1969, 1976) on Bayes procedures in compound problems, and of Hannan and Van Ryzin (1965) on convergence rates. Parallel to Robbins' paper [25], another pioneering landmark in compound decision theory is Stein's (1956) inadmissibility result concerning the usual estimator of a multivariate normal mean. This work, and the subsequent James–Stein (1961) estimator, initiated an important and fruitful area of research in multivariate estimation and admissibility theory.

After Robbins' pioneering paper [41] on empirical Bayes methodology, his students Johns (1957, 1961) and Samuel (1963b) undertook further developments of the subject. In the early 60's Robbins returned to empirical Bayes in his ISI article [58] on testing and in his Rietz lecture [59]* on general statistical decision problems. Other important developments in empirical Bayes methodology in the 1960's and early 1970's include the work of Miyasawa (1961) and Kagan (1962) on estimation, Teicher's (1961, 1963) papers on the identifiability of mixtures, Cogburn's (1965, 1967) stringent solutions to empirical Bayes decision problems, Meeden's (1972) work on admissible empirical Bayes procedures, the papers of Martz and Krutchkoff (1969) and of Wind (1973) on empirical Bayes methods in regression, the work of Rutherford and Krutchkoff (1968) on estimating a prior distribution, the papers of Johns and Van Ryzin (1971, 1972) on convergence rates, Copas' (1969) survey paper and Maritz's (1970) monograph. A particularly distinguished contribution is the series of papers by

Efron and Morris (1972a,b; 1973a,b) that provide an empirical Bayes context to Stein-type estimators and study their properties from this point of view.

After a long hiatus Robbins returned to empirical Bayes in the late 1970's. In [103]*, [115] and [117] he extended the empirical Bayes methodology to prediction problems. Suppose, for example, that one is interested in predicting the number of accidents in a future year incurred by all those drivers who were accident-free in the observed year. Under the Poisson model (2) for each driver, Robbins [103] showed that a natural empirical Bayes predictor is the number of drivers who had exactly one accident during the observed year. He also obtained the asymptotic distribution of this predictor. In [118]* he considered combining Stein's linear (parametric) empirical Bayes with his general (nonparametric) empirical Bayes approach in estimation problems. This idea of an appropriate adaptive choice between competing estimators is developed further in [127] in the problem of choosing the mean or median as an estimate of the center of a symmetric distribution. In [123]* he developed empirical Bayes methods for the simultaneous estimation of several variances. His Neyman lecture [124]* presents his recent research on linear and general empirical Bayes approaches to estimation, prediction, and testing the null hypothesis that a treatment has no effect.

The last decade has witnessed a growing interest in empirical Bayes methodology in a variety of applications (cf. Efron and Morris, 1977, Morris, 1983; Susarla and Van Ryzin, 1977, 1978). There is a wealth of beautiful ideas, interesting problems, and useful applications in empirical Bayes and compound decision theory, as was envisioned by Robbins some thirty-five years ago.

References

Carter, G. and Rolph, J. (1974). Empirical Bayes methods applied to estimating fire alarm probabilities. *J. Amer. Statist. Assoc.* **69**, 880–885.

Cogburn, R. (1965). On the estimation of a multivariate location parameter with squared error loss. In ***Bernoulli*, *Bayes*, *and Laplace Anniversary Volume*** (eds. J. Neyman and L. LeCam), 24–29. Springer-Verlag, Berlin.

Cogburn, R. (1967). Stringent solutions to decision problems. *Ann. Math. Statist.* **38**, 447–463.

Copas, J. B. (1969). Compound decisions and empirical Bayes. *J. Roy. Statist. Soc. Ser. B* **31**, 397–425.

Copas, J. B. (1972). Empirical Bayes methods and the repeated use of a standard. *Biometrika* **59**, 349–360.

Cover, T. (1968). Learning in pattern recognition. In *Methodologies of Pattern Recognition* (ed. S. Watanabe), 111–132. Academic Press, New York.

Efron, B. and Morris, C. (1972a). Empirical Bayes on vector observations: an extension of Stein's method. *Biometrika* **59**, 335–347.

Efron, B. and Morris, C. (1972b). Limiting the risk of Bayes and empirical Bayes estimators II—the Empirical Bayes case. *J. Amer. Statist. Assoc.* **67**, 130–139.

Efron, B. and Morris, C. (1973a). Stein's estimation rule and its competitors—an empirical Bayes approach. *J. Amer. Statist. Assoc.* **68**, 117–130.

Efron, B. and Morris, C. (1973b). Combining possibly related estimation problems. *J. Roy. Statist. Soc. Ser. B* **35**, 379–421.

Efron, B. and Morris, C. (1977). Stein's paradox in statistics. *Scientific American* **236**, 119–237.

Gilliland, D. C. and Hannan, J. F. (1969). On the extended compound decision problem. *Ann. Math. Statist.* **40**, 1536–1541.

Gilliland, D. C., Hannan, J. F., and Huang, J. S. (1976). Asymptotic solutions to the two state component compound decision problems, Bayes versus diffuse priors on proportions. *Ann. Statist.* **4**, 1101–1112.

Hannan, J. F. (1957). Approximation to Bayes risk in repeated play. In *Contributions to the Theory of Games* **3**, 97–139 (eds. M. Dresher and A. W. Tucker). Princeton Univ. Press.

Hoadley, B. (1981). Quality management plan (QMP). *Bell System Tech. J.* **60**, 215–273.

James, W. and Stein, C. (1961). Estimation with quadratic loss. *Proc. Fourth Berkeley Symp. Math. Statist. Probab.* **1**, 361–379. Univ. Calif. Press.

Johns, M. V. (1957). Non-parametric empirical Bayes procedures. *Ann. Math. Statist.* **28**, 649–669.

Johns, M. V. (1961). An empirical Bayes approach to non-parametric two-way classification. In *Studies in Item Analysis and Prediction* (ed. H. Solomon), 221–232. Stanford Univ. Press.

Johns, M. V. and Van Ryzin, J. R. (1971). Convergence rates for empirical Bayes two-action problems I. Discrete case. *Ann. Math. Statist.* **42**, 1521–1539.

Johns, M. V. and Van Ryzin, J. R. (1972). Convergence rates for empirical Bayes two-action problems II. Continuous case. *Ann. Math. Statist.* **43**, 934–947.

Kagan, A. M. (1962). An empirical Bayes approach to the estimation problem. *Doklady Acad. Sci. USSR* **147**, 1020–1021.

Maritz, J. S. (1970). *Empirical Bayes Methods*. Methuen, London.

Martz, H. and Krutchkoff, R. G. (1969). Empirical Bayes estimators in a multiple linear regression model. *Biometrika* **56**, 367–374.

Meeden, G. (1972). Some admissible empirical Bayes procedures. *Ann. Math. Statist.* **43**, 96–101.

Miyasawa, K. (1961). An empirical Bayes estimator of the mean of a normal population. *Bull. Internat. Statist. Inst.* **38**, 181–188.

Morris, C. N. (1983). Parametric empirical Bayes inference: theory and applications. *J. Amer. Statist. Assoc.* **78**, 47–55.

Neyman, J. (1962). Two breakthroughs in the theory of statistical decision-making. *Rev. Internat. Statist. Inst.* **30** (1), 11–27.

Rubin, D. (1981). Using empirical Bayes techniques in the Law School Validity studies. *J. Amer. Statist. Assoc.* **75**, 801–827.

Rutherford, J. R. and Krutchkoff, R. G. (1967). The empirical Bayes approach: estimating the prior distribution. *Biometrika* **54**, 672–675.

Samuel, E. (1963a). Asymptotic solutions of the sequential compound decision problem. *Ann. Math. Statist.* **34**, 1079–1094.

Samuel, E. (1963b). An empirical Bayes approach to the testing of certain parametric hypotheses. *Ann. Math. Statist.* **34**, 1370–1385.

Samuel, E. (1964). Convergence of the losses of certain rules for the sequential compound decision problem. *Ann. Math. Statist.* **35**, 1606–1621.

Samuel, E. (1965). Sequential compound estimators. *Ann. Math. Statist.* **36**, 879–889.

Stein, C. (1956). Inadmissibility of the usual estimator for the mean of a multivariate normal distribution. *Proc. Third Berkeley Symp. Math. Statist. Probab.* **1**, 197–206. Univ. Calif. Press.

Susarla, V. and Van Ryzin, J. R. (1977). On the empirical Bayes approach to multiple decision problems. *Ann. Statist.* **5**, 172–181.

Susarla, V. and Van Ryzin, J. R. (1978). Empirical Bayes estimation of a distribution (survival) function from right censored observations. *Ann. Statist.* **6**, 740–754.

Teicher, H. (1961). Identifiability of mixtures. *Ann. Math. Statist.* **32**, 244–248.

Teicher, H. (1963). Identifiability of finite mixtures. *Ann. Math. Statist.* **34**, 1265–1269.

Van Ryzin, J. R. (1966a). The sequential compound problem with $m \times n$ finite loss matrix. *Ann. Math. Statist.* **37**, 954–975.

Van Ryzin, J. R. (1966b). Repetitive play in finite statistical games with unknown distributions. *Ann. Math. Statist.* **37**, 976–994.

Vardeman, S. B. (1978). Admissible solutions of finite state sequence compound decision problems. *Ann. Statist.* **6**, 673–679.

Wind, S. L. (1973). An empirical Bayes approach to multiple linear regression. *Ann. Statist.* **1**, 93–103.

ASYMPTOTICALLY SUBMINIMAX SOLUTIONS OF COMPOUND STATISTICAL DECISION PROBLEMS

HERBERT ROBBINS
UNIVERSITY OF NORTH CAROLINA

1. Summary

When statistical decision problems of the same type are considered in large groups the minimax solution may not be the "best," since there may exist solutions which are "asymptotically subminimax." This is shown in detail for a classical problem in the theory of testing hypotheses.

2. Introduction

Consider the following *simple statistical decision problem.* The random variable x is normally distributed with variance 1 and mean θ, where θ is known to have one of the two values ± 1. It is required to decide, on the basis of a single observation on x, whether the true value of θ is 1 or -1, in such a way as to minimize the probability of error.

For any decision rule R the probability of error will depend on the true value of θ. Let

$$(1) \qquad \eta(R) = P[\text{error} \mid R, \theta = -1], \quad \delta(R) = P[\text{error} \mid R, \theta = 1].$$

By a suitable choice of R we can give to $\eta(R)$ any desired value between 0 and 1; unfortunately, if R is chosen so that $\eta(R)$ is near 0 then $\delta(R)$ will be near 1, and in this circumstance lies the problem.

For any constant c let R_c be the decision rule which asserts "$\theta = \text{sgn}(x - c)$"; thus in using R_c we assert "$\theta = 1$" if $x > c$ and "$\theta = -1$" if $x < c$. Then

$$(2) \qquad \eta(R_c) = \int_c^{\infty} f(x+1)\,dx = F(-1-c),$$

$$\delta(R_c) = \int_{-\infty}^{c} f(x-1)\,dx = F(-1+c),$$

where we have set

$$(3) \qquad f(x) = \frac{1}{\sqrt{2\pi}}\, e^{-x^2/2}, \qquad F(x) = \int_{-\infty}^{x} f(y)\,dy = 1 - F(-x).$$

It is clear from (2) that

(4) for any number η between 0 and 1 there exists a number

$$c = c(\eta) \text{ such that } \eta(R_c) = \eta.$$

Moreover, using the fundamental lemma of Neyman and Pearson it can be shown that

(5) for any c and any decision rule R such that $\eta(R) \leqq \eta(R_c)$,
$$\delta(R) \geqq \delta(R_c).$$

It follows from (4) and (5) that we need only admit into competition decision rules of the form R_c, but it remains to choose the proper value for c.

An examination of (2) shows that the value $c = 0$ is of particular interest. Let us denote by $\bar{R}$ the rule R_c with $c = 0$; thus in using $\bar{R}$ we assert "$\theta = \text{sgn}(x)$." Now for any c,

$$(6) \quad \max[\eta(R_c), \delta(R_c)] = \max[F(-1-c), F(-1+c)] = F(-1+|c|),$$

and this attains its minimum value $F(-1) = .1587$ for $c = 0$. It follows from (4)–(6) that for any decision rule R (not necessarily of the form R_c),

$$(7) \quad \max[\eta(R), \delta(R)] \geqq \max[\eta(\bar{R}), \delta(\bar{R})] = \eta(\bar{R}) = \delta(\bar{R}) = F(-1),$$

and it can be shown that this inequality is strict unless $R = \bar{R}$, where we regard two decision rules as equal if and only if they arrive at the same decision with probability 1 for all values of θ. Thus $\bar{R}$ is the unique decision rule which minimizes the maximum possible probability of error, or, in Wald's terminology, $\bar{R}$ is the unique *minimax* decision rule. As is often the case with minimax solutions, $\bar{R}$ has the agreeable property that the probability of error is independent of the value of θ.

The unique minimax property of $\bar{R}$ is a strong argument in favor of $\bar{R}$ but is not in itself a compelling reason for regarding $\bar{R}$ as the "best" solution of the decision problem. Suppose, for the sake of argument, that there existed another rule R for which

$$(8) \quad \eta(R) = F(-1) + \epsilon_1, \quad \delta(R) = \epsilon_2,$$

where both ϵ_1 and ϵ_2 are small positive numbers, say less than .001. This would not contradict the minimax property (7) of $\bar{R}$. Still, there would be little doubt that R is preferable to $\bar{R}$, for in using R we would achieve a *much smaller* probability of error when $\theta = 1$ at the cost of only a *slightly greater* probability of error when $\theta = -1$.

Of course, there is no such rule R. In fact, it follows easily from the previous discussion that for any rule R such that (8) holds,

$$(9) \quad \epsilon_1 + \epsilon_2 \geqq F(-1),$$

equality holding only for $\epsilon_1 = 0$, $\epsilon_2 = F(-1)$, $R = \bar{R}$. Hence ϵ_1 and ϵ_2 cannot *both* be made small, and the gain, $F(-1) - \epsilon_2$, of any rule R over $\bar{R}$ when $\theta = 1$ is more than balanced by the loss, ϵ_1, when $\theta = -1$. The fact that any improvement over $\bar{R}$ when $\theta = 1$ must be accompanied by an even greater deterioration when $\theta = -1$, goes beyond the minimax property of $\bar{R}$ and greatly strengthens the view that $\bar{R}$ is in fact the "best" decision rule.

Statistical decision problems often occur, or can be considered, in large groups. Thus let $x_1, \ldots, x_n$ be independent random variables, each normally distributed with variance 1, and with respective means $\theta_1, \ldots, \theta_n$, where $\theta_i = \pm 1$, $i = 1$,

$\ldots, n$. *No relation whatever is assumed to hold among the unknown parameters* θ_i. To emphasize this point, x_1 could be an observation on a butterfly in Ecuador, x_2 on an oyster in Maryland, x_3 the temperature of a star, and so on, all observations being taken at different times. Let it be required to decide, on the basis of the observed values $x_1, \ldots, x_n$, for every $i = 1, \ldots, n$ whether $\theta_i = 1$ or -1, in such a way as to minimize the expected total number of errors. The parameter space Ω of this *compound statistical decision problem* consists of the 2^n points $\boldsymbol{\theta} = (\theta_1, \ldots, \theta_n)$, $\theta_i = \pm 1$. It is natural to suppose that the "best" solution of the compound problem consists in applying to each of the x_i the "best" solution of the original simple problem and therefore in asserting "$\theta_i = \operatorname{sgn}(x_i)$, $i = 1, \ldots, n$." Let us again call this (compound) decision rule $\bar{R}$. It is indeed true that $\bar{R}$ remains for every n the unique minimax solution, in that for any rule $R \neq \bar{R}$ which may be applied in the compound problem,

$$\max_{\boldsymbol{\theta} \in \Omega} [\text{exp. no. of errors} \mid R, \boldsymbol{\theta}] > \max_{\boldsymbol{\theta} \in \Omega} [\text{exp. no. of errors} \mid \bar{R}, \boldsymbol{\theta}] .$$

We shall see, however, that *for large* n, $\bar{R}$ *can no longer be regarded as the "best" decision rule in the compound problem.* Nor is this due to any special property of the simple decision problem with which we began; it lies rather in the fundamental operation of "compounding" and will occur in a large class of compound decision problems.

3. Statement of the compound decision problem. The rule $\bar{R}$

Let

$$x_1, \ldots, x_n \tag{10}$$

be independent random variables, each normally distributed with variance 1, and with respective means

$$\theta_1, \ldots, \theta_n , \qquad \theta_i = \pm 1 . \tag{11}$$

On the basis of the observed sample (10) we are to decide for every $i = 1, \ldots, n$ whether the true value of θ_i is 1 or -1, in such a way as to minimize the expected total number of errors.

Denote by Ω the set of all 2^n possible parameter points $\boldsymbol{\theta} = (\theta_1, \ldots, \theta_n)$, $\theta_i = \pm 1$. For any $\boldsymbol{\theta}$ in Ω the density function of the sample vector $\boldsymbol{x} = (x_1, \ldots, x_n)$ is, by hypothesis,

$$\phi(\boldsymbol{x}, \boldsymbol{\theta}) = \frac{1}{(2\pi)^{n/2}} e^{-\sum_{i=1}^{n} (x_i - \theta_i)^2/2} = \frac{1}{(2\pi)^{n/2}} e^{-(\boldsymbol{x}^2 + n)/2} \cdot e^{\boldsymbol{\theta}\boldsymbol{x}} . \tag{12}$$

If $\boldsymbol{\theta}$ and $\boldsymbol{\theta}'$ are any two points of Ω take

$$w(\boldsymbol{\theta}', \boldsymbol{\theta}) = \frac{1}{n} (\text{no. of } i \text{ for which } \theta'_i \neq \theta_i) = \frac{1}{2n} \sum_{i=1}^{n} \left| \theta'_i - \theta_i \right| \tag{13}$$

as a measure of the loss involved when the true parameter point is $\boldsymbol{\theta}$ and the decision "$\boldsymbol{\theta} = \boldsymbol{\theta}'$" is taken. [The factor $1/n$ in (13) is used in order to stabilize certain later formulas as n varies.] Order the points of Ω arbitrarily as $\boldsymbol{\theta}^{(1)}, \ldots, \boldsymbol{\theta}^{(2^n)}$.

The most general (randomized) decision rule R amounts to specifying as a function of $\mathbf{x}$ a probability distribution $p_j(\mathbf{x}), j = 1, \ldots, 2^n$, on Ω:

$$\text{(14)}\quad R:\ p_j(\mathbf{x}),\qquad j = 1, \ldots, 2^n,\qquad p_j(\mathbf{x}) \geqq 0,\qquad \sum_{j=1}^{2^n} p_j(\mathbf{x}) \equiv 1 .$$

For given $\mathbf{x}$ the rule R asserts "$\boldsymbol{\theta} = \boldsymbol{\theta}^{(j)}$" with probability $p_j(\mathbf{x})$. When $\boldsymbol{\theta}$ is the true parameter point the expected loss in using R is given by the ***risk function***

$$\text{(15)}\quad L(R, \boldsymbol{\theta}) = \int \Big[\sum_{j=1}^{2^n} p_j(\mathbf{x})\, w(\boldsymbol{\theta}^{(j)}, \boldsymbol{\theta}) \Big] \phi(\mathbf{x}, \boldsymbol{\theta})\, d\mathbf{x}$$

$$= \frac{1}{2n} \sum_{i=1}^{n} \int \Big[\sum_{j=1}^{2^n} p_j(\mathbf{x}) \left| \theta_i^{(j)} - \theta_i \right| \Big] \phi(\mathbf{x}, \boldsymbol{\theta})\, d\mathbf{x} .$$

This can be put into a more convenient form as follows. Let

$$\text{(16)}\quad u_i(\mathbf{x}) = \frac{1}{2} \sum_{j=1}^{2^n} p_j(\mathbf{x}) (1 + \theta_i^{(j)}),\qquad i = 1, \ldots, n,$$

$= \text{conditional probability, given } \mathbf{x}, \text{ of deciding that } \theta_i = 1,\ 0 \leqq u_i(\mathbf{x}) \leqq 1 .$

Then

$$\text{(17)}\quad \sum_{j=1}^{2^n} p_j(\mathbf{x}) \left| \theta_i^{(j)} - \theta_i \right| = \begin{cases} 2u_i(\mathbf{x}) & \text{if } \theta_i = -1, \\ 2[1 - u_i(\mathbf{x})] & \text{if } \theta_i = 1, \end{cases}$$

$$= 1 + \theta_i - 2\operatorname{sgn}(\theta_i)\, u_i(\mathbf{x}) \text{ for } \theta_i = \pm 1 .$$

For any $\boldsymbol{\theta}$ in Ω let

$$\text{(18)}\quad p(\boldsymbol{\theta}) = \frac{1}{2n} \sum_{i=1}^{n} (1 + \theta_i) = \frac{1}{n} (\text{no. of } i \text{ for which } \theta_i = 1),\ 0 \leqq p(\boldsymbol{\theta}) \leqq 1 .$$

Then from (15)–(18) we have

$$\text{(19)}\quad L(R, \boldsymbol{\theta}) = p(\boldsymbol{\theta}) - \frac{1}{n} \sum_{i=1}^{n} \operatorname{sgn}(\theta_i) \int \phi(\mathbf{x}, \boldsymbol{\theta})\, u_i(\mathbf{x})\, d\mathbf{x} .$$

This shows that $L(R, \boldsymbol{\theta})$ (although in general not R itself) depends only on the n functions (16).

The maximum likelihood estimate of the true parameter point $\boldsymbol{\theta}$ is, by (12),

$$\hat{\boldsymbol{\theta}} = [\operatorname{sgn}(x_1), \ldots, \operatorname{sgn}(x_n)] .$$

The corresponding decision rule will be denoted by $\bar{R}$:

$$\text{(20)}\quad \bar{R}: \text{"}\theta_i = \operatorname{sgn}(x_i),\qquad i = 1, \ldots, n.\text{"}$$

For the rule $\bar{R}$, $u_i(\mathbf{x}) = 1$ or 0 according as $x_i > 0$ or $x_i < 0$, so that from (19),

$$\text{(21)}\quad L(\bar{R}, \boldsymbol{\theta}) = \frac{1}{n} \sum_{i=1}^{n} \Big[\frac{1 + \theta_i}{2} - \operatorname{sgn}(\theta_i) \int_{x_i > 0} \phi(\mathbf{x}, \boldsymbol{\theta})\, d\mathbf{x} \Big] .$$

Using the notation of (3), (21) becomes

$$(22)\qquad L(\tilde{R}, \boldsymbol{\theta}) = \frac{1}{n}\sum_{i=1}^{n}\left[\frac{1+\theta_i}{2} - \operatorname{sgn}(\theta_i)\int_0^\infty f(x_i - \theta_i)\,dx_i\right]$$

$$= \frac{1}{n}\sum_{i=1}^{n}\left[\frac{1+\theta_i}{2} - \operatorname{sgn}(\theta_i)\,F(\theta_i)\right] = \frac{1}{n}\sum_{i=1}^{n} F(-1) \equiv F(-1) = .1587$$

for every $\boldsymbol{\theta}$ in Ω. Thus $\tilde{R}$ has the constant risk $F(-1)$ no matter what the true parameter point $\boldsymbol{\theta}$.

Returning to the general case where R is any decision rule with associated functions (16) we shall consider certain weighted sums of $L(R, \boldsymbol{\theta})$ taken over all $\boldsymbol{\theta}$ in Ω. For any $k = 0, 1, \ldots, n$ let Ω_k denote the set of all $\boldsymbol{\theta}$ in Ω for which $p(\boldsymbol{\theta}) = k/n$; thus $\boldsymbol{\theta} \in \Omega_k$ if exactly k of its components are 1. Let a function $h(\boldsymbol{\theta}) \geqq 0, \not\equiv 0$ be defined on Ω such that $h(\boldsymbol{\theta}) = \text{constant} = b_k$ for $\boldsymbol{\theta} \in \Omega_k$, $k = 0, 1, \ldots, n$. Then from (19) we have

$$(23)\qquad \sum_{\Omega} h(\boldsymbol{\theta})\,L(R, \boldsymbol{\theta}) = \sum_{\Omega} h(\boldsymbol{\theta})\,p(\boldsymbol{\theta}) - \frac{1}{n}\sum_{i=1}^{n}\int\left[\sum_{\Omega} h(\boldsymbol{\theta})\operatorname{sgn}(\theta_i)\,\phi(\boldsymbol{x}, \boldsymbol{\theta})\right]u_i(\boldsymbol{x})\,d\boldsymbol{x}\,.$$

This will be a minimum with respect to R for given $h(\boldsymbol{\theta})$ [in Wald's terminology, R will be a "Bayes solution" corresponding to $h(\boldsymbol{\theta})$] if and only if for a.e. $\boldsymbol{x}$,

$$(24)\qquad u_i(\boldsymbol{x}) = \begin{cases} 1 & \text{if } \sum_{\Omega} h(\boldsymbol{\theta})\operatorname{sgn}(\theta_i)\,\phi(\boldsymbol{x}, \boldsymbol{\theta}) > 0\,, \\ 0 & \text{otherwise}\,. \end{cases}$$

Let

$$(25)\qquad \begin{aligned} \Omega_{k,i}^{+} &= \text{all } \boldsymbol{\theta} \text{ in } \Omega_k \text{ for which } \theta_i = 1\,, \\ \Omega_{k,i}^{-} &= \text{all } \boldsymbol{\theta} \text{ in } \Omega_k \text{ for which } \theta_i = -1\,, \end{aligned}$$

so that

$$\Omega_k = \Omega_{k,i}^{+} + \Omega_{k,i}^{-}\,, \quad k = 0, 1, \ldots, n\,; \quad i = 1, \ldots, n\,.$$

Then (24) asserts that $u_i(\boldsymbol{x}) = 1$ when

$$(26)\qquad \sum_{k=0}^{n} b_k\left[\sum_{\Omega_{k,i}^{+}}\phi(\boldsymbol{x}, \boldsymbol{\theta}) - \sum_{\Omega_{k,i}^{-}}\phi(\boldsymbol{x}, \boldsymbol{\theta})\right] > 0\,.$$

Multiplying by the positive factor $(2\pi)^{n/2}e^{(\boldsymbol{x}^2+n)/2}\cdot e^{\mathbf{1}\boldsymbol{x}}$, where $\mathbf{1} = (1, \ldots, 1)$, (26) is seen to be equivalent to

$$(27)\qquad \sum_{k=0}^{n} b_k\left[\sum_{\Omega_{k,i}^{+}} e^{(\mathbf{1}+\boldsymbol{\theta})\boldsymbol{x}} - \sum_{\Omega_{k,i}^{-}} e^{(\mathbf{1}+\boldsymbol{\theta})\boldsymbol{x}}\right] > 0\,.$$

Let

$$(28)\qquad S_k^{(i)} = \sum e^{2(x_{j_1} + \cdots + x_{j_k})}\,, \qquad k = 1, \ldots, n-1\,,$$

$$S_{-1}^{(i)} = S_n^{(i)} = 0\,, \qquad S_0^{(i)} = 1$$

where the summation is over all $\binom{n-1}{k}$ combinations of the integers 1, . . . , $i-1, i+1, \ldots, n$ taken k at a time. Then (27) may be written as

$$\sum_{k=0}^{n} b_k \left[e^{2x_i} S_{k-1}^{(i)} - S_k^{(i)} \right] > 0 ,$$

or, finally, as

$$x_i > \tfrac{1}{2} \ln \frac{\sum_{k=0}^{n-1} b_k S_k^{(i)}}{\sum_{k=0}^{n-1} b_{k+1} S_k^{(i)}} . \tag{29}$$

It follows that $\sum_{\Omega} h(\theta) L(R, \theta) =$ *min. for the (nonrandomized) rule*

$$R: \text{“}\theta_i = \operatorname{sgn}\left(x_i - \tfrac{1}{2} \ln \frac{\sum_{k=0}^{n-1} b_k S_k^{(i)}}{\sum_{k=0}^{n-1} b_{k+1} S_k^{(i)}} \right), \qquad i = 1, \ldots, n .\text{”} \tag{30}$$

If we regard two rules as equal if and only if they give the same decision with probability 1 for all θ in Ω, then the minimizing rule (30) is unique.

Example 1. $b_k = 1, k = 0, 1, \ldots, n$. In this case (30) shows that $\sum_{\Omega} L(R, \theta) =$ min. for $R = \bar{R}$ defined by (20). Since

$$\sum_{\Omega} L(R, \theta) > \sum_{\Omega} L(R, \theta) \qquad \text{for } R \neq \bar{R} , \tag{31}$$

$$L(\bar{R}, \theta) \equiv F(-1) ,$$

it follows that $\bar{R}$ is the ***unique minimax decision rule:***

$$\max_{\theta \in \Omega} L(R, \theta) > \max_{\theta \in \Omega} L(\bar{R}, \theta) = F(-1) \qquad \text{for every } R \neq \bar{R} . \tag{32}$$

The result (32) for $n = 1$ is well known. Our purpose in proving it for arbitrary n is to show that $\bar{R}$ remains the unique minimax solution even when we admit rules of the general type (14) which make the decision on each θ_i depend on the whole sample (10). For $\bar{R}$, the decision on θ_i depends only on x_i. This agrees with the fact that, since the components (10) are independent, and since no relation is assumed to hold among the components (11) (that is, since the true parameter point θ can be any point in Ω), it is x_i alone which contains “information” about the value of θ_i.

From the minimax point of view our decision problem is completely solved in favor of $\bar{R}$ by (32). Nevertheless, we shall consider two other examples of decision rules obtained by minimizing weighted sums.

We shall call a decision rule R *symmetric* if $L(R, \theta) =$ constant $= c_k$ for $\theta \in \Omega_k$, $k = 0, 1, \ldots, n$. [Any rule of the form (30) is easily seen to be symmetric.] In the class of symmetric rules it is of interest to minimize the sum $\sum_{k=0}^{n} c_k$. This will be done in the following example.

Example 2. $b_k = \binom{n}{k}^{-1}$, $k = 0, 1, \ldots, n$. (30) shows that

$$\sum_{k=0}^{n} \binom{n}{k}^{-1} \sum_{\Omega_k} L(R, \boldsymbol{\theta}) = \min.$$

in the class of all decision rules, symmetric or not, for the symmetric rule

$$\text{(33)} \qquad \bar{R}: \text{``}\theta_i = \operatorname{sgn}\left(x_i - \tfrac{1}{2}\ln \frac{\sum_{k=0}^{n-1}\binom{n}{k}^{-1} S_k^{(i)}}{\sum_{k=0}^{n-1}\binom{n}{k+1}^{-1} S_k^{(i)}} \right), \qquad i = 1, \ldots, n.\text{''}$$

Let $L(\bar{R}, \boldsymbol{\theta}) = \bar{c}_k$ for $\boldsymbol{\theta}$ in Ω_k, $k = 0, 1, \ldots, n$. If R is *any* symmetric rule with $L(R, \boldsymbol{\theta}) = c_k$ for $\boldsymbol{\theta}$ in Ω_k then

$$\sum_{k=0}^{n} c_k = \sum_{k=0}^{n} \binom{n}{k}^{-1} \sum_{\Omega_k} L(R, \boldsymbol{\theta}),$$

since there are $\binom{n}{k}$ points in Ω_k. *Hence* $\bar{R}$ *defined by* (33) *minimizes* $\sum_{k=0}^{n} c_k$ *in the class of all symmetric decision rules.* Since $\bar{R} \neq \bar{R}$ it follows incidentally that

$$\text{(34)} \qquad \frac{1}{n+1} \sum_{k=0}^{n} \bar{c}_k < F(-1).$$

As a final example we shall consider the decision problem when the parameter space is some fixed Ω_k. This corresponds to the case when the *number* k (but not the positions) of the values 1 (and hence also of the values -1) in the sequence (11) is known.

Example 3. $b_k = 1$ for some fixed k, $0 \leqq k \leqq n$, and $b_j = 0$ for $j \neq k$. Then $\sum_{\Omega_k} L(R, \boldsymbol{\theta}) = \min.$ uniquely for the symmetric rule

$$\text{(35)} \qquad \bar{R}_k: \text{``}\theta_i = \bar{R}_k \operatorname{sgn}\left(x_i - \tfrac{1}{2}\ln \frac{S_k^{(i)}}{S_{k-1}^{(i)}} \right), \qquad i = 1, \ldots, n.\text{''}$$

We shall not attempt to determine numerically the constant $L(\bar{R}_k, \boldsymbol{\theta})$, $\boldsymbol{\theta} \in \Omega_k$ [it is, of course, $< F(-1)$]. As in example 1 it follows that $\bar{R}_k$ is the unique minimax decision rule when $\boldsymbol{\theta}$ is restricted to Ω_k:

$$\text{(36)} \qquad \max_{\boldsymbol{\theta} \in \Omega_k} L(R, \boldsymbol{\theta}) > \max_{\boldsymbol{\theta} \in \Omega_k} L(\bar{R}_k, \boldsymbol{\theta}) \qquad \text{for every } R \neq \bar{R}_k.$$

This result is somewhat surprising, as the following considerations show. For definiteness take $k = 1$, so that $\bar{R}_k$ becomes

$$\bar{R}_1: \text{“}\theta_i = \operatorname{sgn}\left(e^{2x_i} - \sum_{j\neq i} e^{2x_j}\right), \qquad i = 1, \ldots, n \text{.”}$$

For $n > 2$ the probability of the decision “$\boldsymbol{\theta} = (-1, -1, \ldots, -1)$” is positive (since this decision will be taken when all the x_i are nearly equal) even though it is known to involve exactly one error!

A more plausible rule than $\bar{R}_1$ when $\boldsymbol{\theta}$ is known to lie in Ω_1 would be to assign the value $\theta_i = 1$ to that i for which $x_i = \max(x_1, \ldots, x_n)$ and $\theta_j = -1$ for $j \neq i$. This rule, call it R, always assigns to $\boldsymbol{\theta}$ a value in Ω_1 and has a constant risk in Ω_1, as does $\bar{R}_1$, but from (36) it follows that $L(R, \boldsymbol{\theta}) > L(\bar{R}_1, \boldsymbol{\theta})$ for $\boldsymbol{\theta} \in \Omega_1$, so that $\bar{R}_1$ is uniformly better than R in Ω_1. Of course, if one is restricted to decision rules which assign to $\boldsymbol{\theta}$ a value in Ω_1 then R is presumably minimax. Corresponding remarks hold for $k = 2, \ldots, n-1$.

4. R^*, a competitor of $\bar{R}$

We have proved (32) that for the unrestricted compound decision problem where $\boldsymbol{\theta}$ is known only to lie in Ω, the rule $\bar{R}$ defined by (20) is the unique minimax solution. We now make the, perhaps surprising, statement that *for large values of n there are strong reasons for regarding $\bar{R}$ as a relatively poor decision rule.* In support of this assertion we propose the following rule R^* as a competitor of $\bar{R}$. Let

(37
$$\bar{x} = \frac{1}{n}\sum_{i=1}^{n} x_i ,$$

$$x^* = \begin{cases} \infty & \text{if } \bar{x} \leqq -1 , \\ \frac{1}{2}\ln\dfrac{1-\bar{x}}{1+\bar{x}} & \text{if } -1 < x < 1 , \\ -\infty & \text{if } \bar{x} \geqq 1 , \end{cases}$$

$$R^*: \text{“}\theta_i = \operatorname{sgn}(x_i - x^*), \qquad i = 1, \ldots, n \text{.”}$$

Observe that R^* makes the decision on each θ_i depend on all the components (10) and not solely on x_i. Now it may be that the n populations from which the x_i are drawn are entirely different and completely unrelated, as in the last paragraph of section 2. The use of the “hybrid” mean $\bar{x}$ might then seem to be meaningless physically and pointless statistically. Furthermore, the rule R^* is not “admissible” in Wald's sense; that is, there exists a rule R such that $L(R, \boldsymbol{\theta}) \leqq L(R^*, \boldsymbol{\theta})$ for every $\boldsymbol{\theta}$ in Ω, the strict inequality holding for at least one, and possibly all, $\boldsymbol{\theta}$. This is known to follow from the fact that R^* is not of the form (30) of the “Bayes solutions” of the present problem. Nevertheless, on the principle that the proof of the pudding lies in the eating, let us compare R^* with $\bar{R}$ by computing the risk function $L(R^*, \boldsymbol{\theta})$.

For any $0 \leqq p \leqq 1$ and any $n = 1, 2, \ldots$, let [see (3) for notation]

$$(38)\quad h(p, n) = pF(-2p\sqrt{n}) + (1-p)F[-2(1-p)\sqrt{n}]$$
$$+\int_{-2(1-p)\sqrt{n}}^{2p\sqrt{n}} \left\{ pF\left[\sqrt{\frac{n}{n-1}}\left(-1-\frac{x}{\sqrt{n}}+\tfrac{1}{2}\ln\frac{1-p+\frac{x}{2\sqrt{n}}}{p-\frac{x}{2\sqrt{n}}}\right)\right]\right.$$
$$\left. + (1-p)F\left[\sqrt{\frac{n}{n-1}}\left(-1-\frac{x}{\sqrt{n}}-\tfrac{1}{2}\ln\frac{1-p+\frac{x}{2\sqrt{n}}}{p-\frac{x}{2\sqrt{n}}}\right)\right]\right\} f(x)\,dx\,.$$

It is plausible from inspection of (38), and it can be proved rigorously, that

$$(39)\quad \lim_{n\to\infty} h(p, n) = h(p) = pF\left(-1+\tfrac{1}{2}\ln\frac{1-p}{p}\right) + (1-p)F\left(-1-\tfrac{1}{2}\ln\frac{1-p}{p}\right)$$

uniformly for all $0 \leqq p \leqq 1$. We note also that

$$(40)\quad h(p, n) = h(1-p, n),\quad h(p) = h(1-p),\quad h(0) = h(1) = 0\,,$$
$$h(.5, n) > F(-1),\quad h(p) < F(-1),\ \text{for } p \neq .5,\quad h(.5) = F(-1)\,.$$

By elementary calculation which we omit here it can be shown that

$$(41)\quad L(R^*, \theta) = h[p(\theta), n]\,,$$

from which it follows that

$$(42)\quad \lim_{n\to\infty} \{L(R^*, \theta) - h[p(\theta)]\} = 0$$

uniformly for all θ in Ω.

A few values of $h(p)$ and $h(p, 100)$ are given in table I, computed by Mr. J. F. Hannan. [The entries for $h(p, 100)$ are averages of strict upper and lower bounds and are not guaranteed beyond two significant figures.] From the table we see that

TABLE I

p	$F(-1)$	$h(p)$	$h(p, 100)$	$h(p, 1000)$
0.0 or 1.0	.1587	0	.0041	
.1 or .9	.1587	.0691	.0763	
.2 or .8	.1587	.1121	.1174	
.3 or .7	.1587	.1387	.1439	
.4 or .6	.1587	.1538	.1591	
.5	.1587	.1587	.1628	.1591

for $n = 100$, R^* has a *slightly higher* risk than $\bar{R}$ for p near .5 and a *much lower* risk for p near 0 or 1. As $n \to \infty$, this phenomenon becomes more pronounced. Since (39) and the last two relations in (40) hold, we call R^* an *asymptotically subminimax* decision rule as $n \to \infty$.

A statistical decision problem is sometimes regarded as a game between the statistician S and Nature [1]. In the present problem if S should use the decision rule R^* then Nature could counter by seeing to it that $p(\theta) \simeq .5$. Since $L(R^*, \theta) > L(\bar{R}, \theta) = F(-1)$ for $p(\theta) \simeq .5$, S would do better, as far as expectations are concerned, to use $\tilde{R}$. But if Nature is not an opponent but a neutral observer of the game then $p(\theta)$ may not be $\simeq .5$, and in using R^* rather than $\tilde{R}$, S would be balancing the possibility of a slightly higher risk in return for that of a much lower one. As $n \to \infty$, the set of values of $p(\theta)$ for which $L(R^*, \theta) > L(\tilde{R}, \theta)$ converges to the single point .5 and the excess of $L(R^*, \theta)$ over $L(\tilde{R}, \theta)$ in the neighborhood of this point tends to 0, while the excess of $L(\tilde{R}, \theta)$ over $L(R^*, \theta)$ near $p(\theta) = 0$ or 1 tends to $F(-1)$. Even for large n this is not, of course, a compelling reason for preferring R^* to $\tilde{R}$, especially if there is reason to believe that $p(\theta)$ *is* near .5, but we shall not labor this point here.

The reader will have observed that R^* can only be used in applications in which all the values (10) are at hand before any of the individual decisions concerning the θ_i are to be made. This will often be the case in practice. Even when it is not, R^* can be used, after all the values (10) are known, to supersede preliminary decisions based, say, on $\tilde{R}$, or perhaps on some rule which uses the values $x_1, \ldots, x_i$ to decide the value of θ_i.

We emphasize that R^* is by no means advanced as in any sense a "best" rule. Its chief virtue as an asymptotically subminimax rule is its comparative simplicity, both in application and in the computation of its risk function (38). A possible candidate for a rule superior to R^* in every respect save simplicity is the rule $\bar{R}$ defined by (33); unfortunately, the risk function $L(\bar{R}, \theta)$ seems difficult to compute. It is possible that $\bar{R}$ is uniformly better than R^*. On the other hand, it may be that the limiting value of $L(\bar{R}, \theta)$ as $n \to \infty$ and $p(\theta) \to p$ is $h(p)$, in which case $\bar{R}$ and R^* would be asymptotically equivalent in performance. Finally, it is possible that no rule has a limiting risk function uniformly below $h(p)$, in which case R^* would be "asymptotically admissible."

In the preceding discussion the rule R^* was introduced without motivation. In what follows we shall show how R^* came to be considered, in a way which indicates that the existence of asymptotically subminimax decision functions is to be expected in a wide class of problems.

5. Heuristic motivation for R^*

A decision rule R with corresponding functions (16) will be called *simple* if for some function $u(x)$,

$$(43) \qquad u_i(\mathbf{x}) = u(x_i), \qquad i = 1, \ldots, n.$$

(For $n = 1$ any rule is simple. For $n > 1$, of the specific rules $\tilde{R}$, $\bar{R}$, $\tilde{R}_k$, R^* considered thus far, only $\tilde{R}$ is simple.) For any simple rule R (19) becomes

$$(44) \quad L(R, \theta) = p(\theta) - \frac{1}{n}\sum_{i=1}^{n} \operatorname{sgn}(\theta_i) \int \phi(\mathbf{x}, \theta)\, u(x_i)\, d\mathbf{x}$$
$$= p(\theta) - \int \{ p(\theta) f(x-1) - [1 - p(\theta)] f(x+1) \} u(x)\, dx.$$

This shows incidentally that every simple rule is "symmetric" (definition in section 3, preceding example 2) and that for fixed R, $L(R, \theta)$ is a linear function of $p(\theta)$.

Now let λ be any constant, $0 \leqq \lambda \leqq 1$, and choose the function $u(x)$, $0 \leqq u(x) \leqq 1$, so as to maximize the integral

$$(45) \qquad \int [\lambda f(x+1) - (1-\lambda) f(x-1)]\, u(x)\, dx\,.$$

This occurs if and only if for a.e. x,

$$(46) \qquad u(x) = u_\lambda(x) = \begin{cases} 1 & \text{if } \lambda f(x+1) - (1-\lambda) f(x-1) > 0\,, \\ 0 & \text{otherwise}\,, \end{cases}$$

which determines the simple rule

$$(47) \qquad R_\lambda\text{: } \text{``}\theta_i = \operatorname{sgn}\left(x_i - \tfrac{1}{2}\ln\frac{1-\lambda}{\lambda}\right), \qquad i = 1, \ldots, n\,.\text{''}$$

It follows that *when* $p(\boldsymbol{\theta}) = \lambda$, R_λ *minimizes* $L(R, \boldsymbol{\theta})$ *in the class of all simple rules.* The risk function of R_λ is, by (44), for any λ and any $\boldsymbol{\theta}$,

$$(48) \quad L(R_\lambda, \boldsymbol{\theta}) = p(\boldsymbol{\theta}) F\left(-1 + \tfrac{1}{2}\ln\frac{1-\lambda}{\lambda}\right) + [1 - p(\boldsymbol{\theta})] F\left(-1 - \tfrac{1}{2}\ln\frac{1-\lambda}{\lambda}\right).$$

The family of simple rules R_λ, $0 \leqq \lambda \leqq 1$, is "complete" in Wald's sense: if R is any simple rule then there exists a λ such that

$$(49) \qquad L(R_\lambda, \boldsymbol{\theta}) \leqq L(R, \boldsymbol{\theta}) \quad \text{for every } \boldsymbol{\theta} \text{ in } \Omega\,.$$

To show this directly in the present case, take any simple rule R with associated function $u(x)$ and choose that λ for which

$$(50) \quad \int f(x+1)\, u_\lambda(x)\, dx = F\left(-1 + \tfrac{1}{2}\ln\frac{1-\lambda}{\lambda}\right) = \int f(x+1)\, u(x)\, dx\,;$$

then necessarily

$$(51) \quad \int f(x-1)\, u_\lambda(x)\, dx = F\left(-1 - \tfrac{1}{2}\ln\frac{1-\lambda}{\lambda}\right) \leqq \int f(x-1)\, u(x)\, dx\,,$$

since otherwise (45) would not be maximized by $u_\lambda(x)$. Now (49) follows directly from (44), (50), (51). We note that $R_\lambda = \tilde{R}$ for $\lambda = .5$, and that $L(R_{.5}, \boldsymbol{\theta}) \equiv F(-1)$.

It follows from (49) that in the class of simple rules we may confine ourselves to the family R_λ. It remains to choose λ. From (48) we see that for fixed λ, $L(R_\lambda, \boldsymbol{\theta})$ is a linear function of $p(\boldsymbol{\theta})$ with extreme values $F\left(-1 - \tfrac{1}{2}\ln\dfrac{1-\lambda}{\lambda}\right)$, $F\left(-1 + \tfrac{1}{2}\ln\dfrac{1-\lambda}{\lambda}\right)$ assumed respectively when $p(\boldsymbol{\theta}) = 0, 1$. It follows that for $\lambda \neq .5$,

$$(52) \quad \max_{\boldsymbol{\theta} \in \Omega} L(R_\lambda, \boldsymbol{\theta}) = \max\left[F\left(-1 - \tfrac{1}{2}\ln\frac{1-\lambda}{\lambda}\right),\ F\left(-1 + \tfrac{1}{2}\ln\frac{1-\lambda}{\lambda}\right)\right]$$
$$> \tfrac{1}{2}\left[F\left(-1 - \tfrac{1}{2}\ln\frac{1-\lambda}{\lambda}\right) + F\left(-1 + \tfrac{1}{2}\ln\frac{1-\lambda}{\lambda}\right)\right] > F(-1)\,,$$

which is a stronger inequality than the minimax character of $R_{.5} = \tilde{R}$, previously shown to hold in the class of all decision rules. Since by (52) the linear risk function $L(R_\lambda, \theta)$ for $\lambda \neq .5$ lies above $F(-1)$ for more than half the interval $0 \leqq p(\theta) \leqq 1$, and rises above it at one end by more than it falls below it at the other, it seems reasonable (when nothing is known about θ) to regard $\tilde{R} = R_{.5}$ as the "best" of the rules R_λ and hence of all simple rules. Thus when we are restricted to *simple* rules the minimax rule $\tilde{R}$ *is* the "best," and asymptotically subminimax rules do not exist. On the other hand, if $p(\theta)$ is known then by the previous discussion culminating in (48),

$$(53) \qquad L(R_{p(\theta)}, \theta) = h[p(\theta)] \quad [\text{see } (39)] < F(-1) \text{ for } p(\theta) \neq .5\,.$$

Thus if $p(\theta)$ were known we could, by using the simple rule $R_{p(\theta)}$, achieve the risk function $h[p(\theta)]$ which lies below the risk function $F(-1)$ of $\tilde{R}$. {Of course, by using the nonsimple rule $\tilde{R}_k$ with $k = np(\theta)$ [see (35)] we could still further reduce the risk function.} In fact, the curve $y = h(p)$ is the envelope of the one parameter family of straight lines

$$y = y(p, \lambda) = pF\left(-1 + \tfrac{1}{2}\ln\frac{1-\lambda}{\lambda}\right) + (1-p)F\left(-1 - \tfrac{1}{2}\ln\frac{1-\lambda}{\lambda}\right)$$

and lies below each of them, including the line $y = y(p, .5) = F(-1)$.

In practice, of course, $p(\theta)$ will rarely be known. However, and this is the key to the matter, *we can estimate $p(\theta)$ from the sample* (10) *and then use the rule R_λ with λ replaced by our estimate of $p(\theta)$*. Let us see how this attempt to lift ourselves by our own bootstraps works out.

We must first choose some estimator of $p(\theta)$. One's first thought is to use the method of maximum likelihood. As was pointed out in section 3, the maximum likelihood estimate of θ is

$$\hat{\theta} = \hat{\theta}(x) = [\text{sgn}(x_1), \ldots, \text{sgn}(x_n)]\,,$$

so that (presumably)

$$p(\hat{\theta}) = \frac{1}{n}(\text{no. of } i \text{ for which } x_i > 0)$$

is the maximum likelihood estimate of $p(\theta)$. It is easily seen that

$$E[p(\hat{\theta})|\theta] = [1 - 2F(-1)]\,p(\theta) + F(-1)\,,$$

so that for $p(\theta) \neq .5$, $p(\hat{\theta})$ is a biased estimator whose bias does not tend to 0 as $n \to \infty$ unless $p(\theta) \to .5$. We can correct for bias by using the unbiased estimator

$$z = \frac{p(\hat{\theta}) - F(-1)}{1 - 2F(-1)}$$

which has variance

$$\text{var}[z|\theta] = \frac{1}{n}\frac{F(-1)[1-F(-1)]}{[1-2F(-1)]^2} \simeq \frac{.29}{n}$$

and is nearly normal for large n.

Consider instead, however, the unbiased estimator

$$v = \tfrac{1}{2}(1 + \bar{x})\,, \qquad \bar{x} = \frac{1}{n}\sum_{i=1}^{n} x_i\,,$$

which is normal with variance

$$\text{var}\,[v \mid \boldsymbol{\theta}] = \frac{.25}{n}$$

less than that of z. Since $0 \leqq p(\boldsymbol{\theta}) \leqq 1$ it seems natural to truncate v at 0 and 1 and to use the modified estimator

$$v' = \begin{cases} 0 \text{ if } v \leqq 0\,, \\ v \text{ if } 0 < v < 1\,, \\ 1 \text{ if } v \geqq 1\,, \end{cases}$$

which, though biased, seems "better" than v. We shall adopt, arbitrarily, v' as our estimator for $p(\boldsymbol{\theta})$. v' is clearly consistent as $n \to \infty$; in fact

$$\lim_{n \to \infty} E\{[v' - p(\boldsymbol{\theta})]^2 \mid \boldsymbol{\theta}\} = 0$$

uniformly for all $\boldsymbol{\theta}$ in Ω. Now the rule R_λ with λ replaced by the estimate v' of $p(\boldsymbol{\theta})$ is simply R^* defined by (37), and this provides the motivation for considering R^*.

Since v' is a consistent estimator of $p(\boldsymbol{\theta})$ it is clear from (53) that (42) must hold. The performance of R^* for finite n must, of course, be worked out by computation on the basis of (41) and (38), as was done in the table of section 4.

There are, of course, other ways in which we might obtain decision rules to compete with $\bar{R}$. For example, we might use an integer-valued estimate of $k = np(\boldsymbol{\theta})$ in the rule $\bar{R}_k$. Again, we might use an iterative process: first estimating $p(\boldsymbol{\theta})$ then using R_λ with λ replaced by its estimate to obtain the decision "$\boldsymbol{\theta} = \boldsymbol{\theta}^{(k)}$" for some k, and finally using R_λ again with λ replaced by $p(\boldsymbol{\theta}^{(k)})$. There is also the rule $\bar{R}$ derived in section 3 as the solution of a minimum problem. Compared with any of these, R^* has at least the advantage of simplicity.

Because of the demonstrated properties of R^* it seems safe to say that for "large" n there exist, among the nonsimple rules, worthy competitors of the minimax rule $\bar{R}$. If this be admitted one then has the problem of finding the "best" decision rule. The definition of "best" is an open question at the moment, but at least it appears that "best" does not equal "minimax."

The existence of asymptotically subminimax decision functions is not confined to problems of the "compound" type. For example, let x have a binomial distribution (n, θ) and let it be required to estimate θ by some function $t = t(x)$ so as to minimize the quantity

$$L(t, \theta) = nE[(t-\theta)^2 \mid \theta] = n \sum_{x=0}^{n} \binom{n}{x} \theta^x (1-\theta)^{n-x} [t(x) - \theta]^2 .$$

For the conventional estimator $t_1 = x/n$ we have

$$L(t_1, \theta) = \theta(1-\theta)\,, \tag{54}$$

while the minimax estimator is

$$t_2 = \frac{x + \dfrac{\sqrt{n}}{2}}{n + \sqrt{n}}\,,$$

for which

$$L(t_2, \theta) = \frac{1}{4\left(1+\frac{1}{\sqrt{n}}\right)^2}. \tag{55}$$

As $n \to \infty$, (55) $\to \frac{1}{4}$ which is greater than (54) except for $\theta = .5$. Thus t_1 is asymptotically subminimax, although in this case it is the minimax risk function (55) which varies with n. The question of whether t_1 is "better" than t_2 has been raised by Hodges and Lehmann [2].

6. General remarks on compound decision problems

A wide class of statistical decision problems can be brought under a general scheme, due to Wald, in which there is given (1) a sample space X of points x, (2) a parameter space ω of points θ such that for every θ in ω there corresponds a probability distribution P_θ on X, (3) a class $\mathfrak{D}$ of decisions D, (4) a loss function $w(D, \theta) \geqq 0$ representing the cost of taking the decision D when the true value of the parameter is θ. Any function $u = u(x)$ with values in $\mathfrak{D}$ is called a decision function, and the function

$$L(u, \theta) = \int w[u(x), \theta]\, dP_\theta(x) \tag{56}$$

is called the risk function. The statistical decision problem ρ is to find the decision function u which in some sense minimizes the risk function (56) over ω. For example, we may seek the u which minimizes the quantity

$$\int_\omega L(u, \theta)\, dG(\theta), \tag{57}$$

where $G(\theta)$ is a given distribution on ω, and which is called the *Bayes solution* of ρ corresponding to $G(\theta)$. Or we may require that u be a *minimax* solution for which the quantity

$$\max_{\theta \in \omega} L(u, \theta)$$

is a minimum.

It often happens that one deals with a set of n independent and, in general, unrelated, decision problems of the same mathematical form. Thus, let $x_1, \ldots, x_n$ be independent random variables such that each x_i presents the same problem ρ. Each x_i will be distributed in X with a distribution P_{θ_i}, $\theta_i \in \omega$, but no relation is assumed to hold among the n parameter values $\theta_1, \ldots, \theta_n$. For each x_i a decision $D_i \in \mathfrak{D}$ must be taken. We shall take the quantity

$$\frac{1}{n}\sum_{i=1}^{n} w(D_i, \theta_i)$$

as a measure of the loss incurred by any set of decisions $D_1, \ldots, D_n$ when the true parameter values are respectively $\theta_1, \ldots, \theta_n$. If a decision function u_i depending on x_i alone is used for the i-th decision then, setting $\boldsymbol{\theta} = (\theta_1, \ldots, \theta_n)$ and $\boldsymbol{u} = (u_1, \ldots, u_n)$, the risk function will be

$$L(\boldsymbol{u}, \boldsymbol{\theta}) = \int \cdots \int \frac{1}{n}\sum_{i=1}^{n} w[u_i(x_i), \theta_i]\, dP_{\theta_1}(x_1) \ldots dP_{\theta_n}(x_n) \tag{58}$$

$$= \frac{1}{n}\sum_{i=1}^{n} \int w[u_i(x), \theta_i]\, dP_{\theta_i}(x).$$

The problem of minimizing (58) by proper choice of the $u_i(x)$ is essentially the same as that of minimizing (56). However, if the whole set of values $x_1, \ldots, x_n$ is known before the individual decisions are to be made, then we can permit u_i to depend on *all* the values $x_1, \ldots, x_n$, so that the risk function will be

$$(59)\quad L(\boldsymbol{u}, \boldsymbol{\theta}) = \frac{1}{n}\sum_{i=1}^{n}\int \ldots \int w\,[u_i(x_1, \ldots, x_n), \theta_i]\, dP_{\theta_1}(x_1) \ldots dP_{\theta_n}(x_n).$$

The problem of minimizing (59) over the n-fold Cartesian product Ω of ω with itself, consisting of all ordered n-tuples $\boldsymbol{\theta} = (\theta_1, \ldots, \theta_n)$, $\theta_i \in \omega$, is quite different from the original problem $\mathcal{P}$ involving (56). We shall denote the problem of minimizing (59) by $\mathcal{P}^{(n)}$ and call it the *compound decision problem* corresponding to the simple problem $\mathcal{P}$.

At first sight it may seem that the use of decision functions of the general form $u_i(x_1, \ldots, x_n)$ is pointless, since the values x_j for $j \neq i$ can contribute no information concerning θ_i; this because the distribution P_{θ_j} of x_j depends only on θ_j which was not assumed to be in any way related to θ_i. From this point of view we should stick to *simple* decision functions of the form $u_i(x_1, \ldots, x_n) = u(x_i)$ where $u(x)$ is the "best" solution of $\mathcal{P}$. The example of section 4, however, shows that there may be great advantages in using nonsimple decision functions of the general form $u_i(x_1, \ldots, x_n)$. In that example the minimax solution of $\mathcal{P}^{(n)}$ is afforded by the simple decision functions $\bar{u}_i(x_1, \ldots, x_n) = \bar{u}(x_i)$, where $\bar{u}(x)$ is the minimax solution of $\mathcal{P}$, but although $\bar{u}$ was seen to be the "best" solution of $\mathcal{P}$, the existence as $n \to \infty$ of an asymptotically subminimax solution of $\mathcal{P}^{(n)}$ showed that the minimax solution of $\mathcal{P}^{(n)}$ was not the "best" for large n. This phenomenon is to be expected in many cases, as we shall see.

The most interesting Bayes solutions of $\mathcal{P}^{(n)}$ are those obtained by minimizing the integral of (59) over Ω with respect to some distribution $G(\boldsymbol{\theta})$ which is invariant under all permutations of the components $\theta_1, \ldots, \theta_n$ of $\boldsymbol{\theta}$; the corresponding Bayes solutions $\boldsymbol{u}$ will then be *symmetric* in the sense that the risk function (59) will be invariant under all permutations of $\theta_1, \ldots, \theta_n$. In general the Bayes solutions may be expected to be complicated in structure and difficult to evaluate in performance.

We shall now give a heuristic rule for constructing certain nonsimple solutions of $\mathcal{P}^{(n)}$. For any $\boldsymbol{\theta} = (\theta_1, \ldots, \theta_n)$ in Ω let $G_{\boldsymbol{\theta}}(\theta)$ be the cumulative distribution function of the probability distribution of a random variable θ for which $P[\theta = \theta_i] = 1/n$, $i = 1, \ldots, n$; that is, if ω is the real line,

$$(60)\qquad G_{\boldsymbol{\theta}}(\theta) = \frac{1}{n}(\text{no. of } i \text{ for which } \theta_i \leqq \theta),\ -\infty < \theta < \infty.$$

Suppose that the distribution P_θ on X has a density function $f(x, \theta)$; then for any simple rule $\boldsymbol{u}$ such that $u_i(x_1, \ldots, x_n) = u(x_i)$, (58) becomes

$$(61)\qquad L(\boldsymbol{u}, \boldsymbol{\theta}) = \int \Big[\frac{1}{n}\sum_{i=1}^{n} w\,[u(x), \theta_i]\, f(x, \theta_i)\Big]\, dx$$

$$= \int \Big[\int w\,[u(x), \theta]\, f(x, \theta)\, dG_{\boldsymbol{\theta}}(\theta)\Big]\, dx.$$

Now if $G_{\boldsymbol{\theta}}(\theta)$ is known (that is, if the components θ_i of $\boldsymbol{\theta}$ are known apart from their order) then (61) will be a minimum for that function $u(x) = u[x; G_{\boldsymbol{\theta}}(\theta)]$ such that for every fixed x, $u(x) = t$, where t is that number for which

$$\int w(t, \theta) f(x, \theta) dG_{\boldsymbol{\theta}}(\theta) = \min. \tag{62}$$

Denote the simple decision rule for which $u_i(x_1, \ldots, x_n) = u[x_i; G_{\boldsymbol{\theta}}(\theta)]$ by $u[G_{\boldsymbol{\theta}}(\theta)]$; this will clearly be better than any simple rule which depends on $x_1, \ldots, x_n$ alone. Of course, in order to use the rule $u[G_{\boldsymbol{\theta}}(\theta)]$ we must know $G_{\boldsymbol{\theta}}(\theta)$, which depends on the unknown $\boldsymbol{\theta}$. *Thus we must devise a method for estimating $G_{\boldsymbol{\theta}}(\theta)$ from the observed values $x_1, \ldots, x_n$.* [Actually, we need only be able to estimate the left hand side of (62) for every x and t.] This involves finding a solution to the following problem:

(I). Let $x_1, \ldots, x_n$ be independent random variables such that the density function of x_i is $f(x, \theta_i)$, where $\theta_1, \ldots, \theta_n$ are n arbitrary unknown elements of a parameter set ω. The joint density function of the x_i is therefore $\prod_{i=1}^{n} f(x_i, \theta_i)$. Let $\boldsymbol{\theta} = (\theta_1, \ldots, \theta_n)$. From the observed values $x_1, \ldots, x_n$ we are to form an estimator $G(\theta; x_1, \ldots, x_n)$ of the cumulative distribution function (60) which for large n shall be "close" to (60) with probability near 1 for all possible values of $\boldsymbol{\theta}$.

Assuming problem (I) to be solved we can apply in the compound decision problem $\rho^{(n)}$ the nonsimple decision rule

$$u^*(x_1, \ldots, x_n) = \{u[x_1, G(\theta; x_1, \ldots, x_n)], \ldots, u[x_n, G(\theta; x_1, \ldots, x_n)]\}; \tag{63}$$

that is, $u[G_{\boldsymbol{\theta}}(\theta)]$ with $G_{\boldsymbol{\theta}}(\theta)$ replaced by its estimate $G(\theta; x_1, \ldots, x_n)$. If our solution of problem (I) is a good one then for large n (63) will be better than any simple rule. In particular, if the minimax solution (assumed unique) of ρ is denoted by $u(x)$, then (63) will be better than the simple rule

$$u(x_1, \ldots, x_n) = [u(x_1), \ldots, u(x_n)]. \tag{64}$$

If (64) is the minimax solution of $\rho^{(n)}$ in the class of all decision rules, nonsimple included, then (63) will be *asymptotically subminimax.*

We have seen in section 5 that problem (I) can be solved in the very simple case in which ω consists of only two elements, ± 1, and $f(x, \theta)$ is the normal density function with mean θ and variance 1. The function $G_{\boldsymbol{\theta}}(\theta)$ is then completely determined by the number $p(\boldsymbol{\theta}) =$ (no. of i for which $\theta_i = 1)/n$, of which a consistent estimator is $(1 + \bar{x})/2$, where $\bar{x} = \sum_{i=1}^{n} x_i/n$.

Before proceeding further with problem (I) let us consider a different but analogous problem:

(II). Let $x_1, \ldots, x_n$ be independent random variables each with a common density function

$$h_G(x) = \int f(x, \theta) dG(\theta)$$

where $f(x, \theta)$ is the same as in problem (I) and $G(\theta)$ is an unknown distribution on ω. The joint density function of the x_i is therefore $\prod_{i=1}^{n} h_G(x_i)$. From the observed values $x_1, \ldots, x_n$ we are to form an estimator $G(\theta; x_1, \ldots, x_n)$ of the unknown $G(\theta)$ which for large n will be "close" to $G(\theta)$ with probability near 1 for all $G(\theta)$ in some class $\mathcal{G}$.

Problem (II) is a generalization of a classical problem in the theory of estimation. Let $\mathcal{G}$ be the class of distributions concentrated at some *single point* of ω; then the joint density function of the x_i is simply $\prod_{i=1}^{n} f(x_i, \theta)$, with θ unknown, and we require a consistent estimator of θ. Under certain conditions on $f(x, \theta)$ and ω, the *method of maximum likelihood* provides a solution:

$$(65) \qquad \hat{\theta}(x_1, \ldots, x_n) = \text{that } \theta \text{ in } \omega \text{ for which } \prod_{i=1}^{n} f(x_i, \theta) = \max .$$

More generally, it has been announced in an abstract [3] that under certain conditions the "generalized method of maximum likelihood" provides a solution of problem (II):

$$(66) \quad \hat{G}(\theta; x_1, \ldots, x_n) = \text{that } G(\theta) \text{ in } \mathcal{G} \text{ for which } \prod_{i=1}^{n} h_G(x_i) = \max .$$

Problem (II) is itself of interest in statistical decision problems in which there is a *prior distribution of parameters*. Returning to the problem $\mathcal{P}$ stated at the beginning of this section, if θ is itself a random variable with known distribution $G(\theta)$ on ω then the best solution of $\mathcal{P}$ is that u which minimizes the integral (57). However, if $G(\theta)$ exists but the statistician knows only that it belongs to some class $\mathcal{G}$, then in the problem $\mathcal{P}^{(n)}$ he can estimate $G(\theta)$ by solving problem (II) and then determine $u(x)$ by minimizing (57) with $G(\theta)$ replaced by the estimate $G(\theta; x_1, \ldots, x_n)$ [3]. However, even the assumption of an existing but unknown prior distribution $G(\theta)$ will be questionable in most applications of statistics, and we merely mention the matter here.

We have stated that under certain conditions problem (II) can be solved by the generalized method of maximum likelihood. Problem (I) is more difficult, and it is easily seen that a solution of problem (I) would in general provide a solution of problem (II). Conversely, however, as a heuristic principle *we can in some cases solve problem* (I) *by acting "as though" it were problem* (II); in fact, any solution $G(\theta; x_1, \ldots, x_n)$ of problem (II) which is a symmetric function of $x_1, \ldots, x_n$ [as, for example, (66)] will at the same time provide a possible solution of problem (I). In justification of this principle we point out that if $\theta_1, \ldots, \theta_n$ form a random sample from a distribution $G(\theta)$, then for large n the empirical cumulative distribution function of $\theta_1, \ldots, \theta_n$ will tend uniformly to $G(\theta)$ with probability 1 as $n \to \infty$. Hence the random variables $x_1, \ldots, x_n$ of problem (II) will act much like those of problem (I), insofar as symmetric functions of the x_i are concerned. Questions of uniformity, of course, have to be considered before any precise theorem can be stated, and the whole subject of problems (I) and (II) requires and will repay a careful treatment. In particular, the generalized method of maximum likelihood, even if in theory it provides a solution to these problems, will in practice be extremely difficult to apply.

REFERENCES

[1] A. Wald, "Statistical decision functions," *Annals of Math. Stat.*, Vol. 20 (1949), pp. 165–205.

[2] J. L. Hodges, Jr. and E. L. Lehmann, "Some problems in minimax point estimation," *Annals of Math. Stat.*, Vol. 21 (1950), pp. 182–197.

[3] H. Robbins, "A generalization of the method of maximum likelihood: estimating a mixing distribution," abstract, *Annals of Math. Stat.*, Vol. 21 (1950), pp. 314–315.

Reprinted from
Proc. Second Berkeley Symposium Math. Statist. Prob.
1, 131–148 (1951)

Reprinted from THE ANNALS OF MATHEMATICAL STATISTICS
Vol. 26, No. 1, March, 1955

ASYMPTOTIC SOLUTIONS OF THE COMPOUND DECISION PROBLEM FOR TWO COMPLETELY SPECIFIED DISTRIBUTIONS[1]

BY JAMES F. HANNAN AND HERBERT ROBBINS
Michigan State College and *Columbia University*

1. Summary. A compound decision problem consists of the simultaneous consideration of n decision problems having identical formal structure. Decision functions are allowed to depend on the data from all n components. The risk is taken to be the average of the resulting risks in the component problems. A heuristic argument for the existence of good asymptotic solutions was given by Robbins ([1] Sec. 6) and was preceded by an example (component decisions between $N(-1,1)$ and $N(1,1)$) exhibiting, for sufficiently large n, a decision function whose risk was uniformly close to the envelope risk function of "simple" decision functions.

The present paper considers the class of problems where the components involve decision between any two completely specified distributions, with the risk taken to be the weighted probability of wrong decision. For all sufficiently large n, decision functions are found whose risks are uniformly close to the envelope risk function of "invariant" decision functions.

2. Statement and reduction of the problem. The problem of testing a simple statistical hypothesis against a simple alternative can be formulated as follows. Let x be a random variable (of arbitrary dimensionality) which is known to have one of the two distinct distribution functions $F(x, \theta)$ for $\theta = 0$ or 1. On the basis of a single observation on x (we consider only the nonsequential case) it is require to decide whether the true value of the unknown parameter θ is 0 or 1.

Statistical decision problems of the same formal structure often occur, or can be considered, in large groups. We shall, therefore, take as a single entity the following compound decision problem. Let n be a fixed positive integer and let $x_1, \cdots, x_n$ be independent random variables, each of which has the distribution function $F(x, \theta)$ with respective parameter values $\theta_1, \cdots, \theta_n$, with $\theta_i =$ 0 or 1. Let $\mathbf{x} = (x_1, \cdots, x_n)$ denote the vector of observations and $\boldsymbol{\theta} = (\theta_1, \cdots, \theta_n)$ the unknown vector of parameters; $\boldsymbol{\theta}$ is known to belong to the set Ω consisting of all 2^n possible vectors of n components, each 0 or 1. On the basis of $\mathbf{x}$ it is required to decide the true value of $\boldsymbol{\theta}$, which amounts to deciding for every $i = 1, \cdots, n$ whether $\theta_i = 0$ or 1.

Any vector of n functions $\mathbf{t} = (t_1(\mathbf{x}), \cdots, t_n(\mathbf{x})$ is a (randomized) decision function for the compound decision problem if for $i = 1, \cdots, n$, $0 \leq t_i(\mathbf{x}) \leq 1$, and if the conditional probabilities, given $\mathbf{x}$, of deciding that $\theta_i = 0$

Received July 20, 1954.

[1] This research was completed at the University of North Carolina and was supported in part by the United States Air Force under Contract AF 18(600)-83 monitored by the Office of Scientific Research.

or 1 are respectively $1 - t_i(\mathbf{x})$ and $t_i(\mathbf{x})$. If for some function $t(x)$, $t_i(\mathbf{x}) = t(x_i)$ for $i = 1, \cdots, n$, then $\mathbf{t}$ will be called *simple* and will be denoted by t.

We assume the practical background of the problem provides two positive constants,

a, the loss incurred in deciding "$\theta_i = 0$" when the true value of $\theta_i = 1$,

b, the loss incurred in deciding "$\theta_i = 1$" when the true value of $\theta_i = 0$.

For any Borel set B we define

$$\mu_\theta(B) = \int_B dF(x, \theta), \qquad \theta = 0, 1; \tag{2.1}$$

$$\mu(B) = \mu_0(B) + \mu_1(B). \tag{2.2}$$

Since μ_0 and μ_1 are absolutely continuous with respect to the finite measure μ, generalized probability density functions $f(x, 0)$ and $f(x, 1)$ exist such that for any Borel set B

$$\mu_\theta(B) = \int_B f(x, \theta)\, d\mu, \qquad \text{for } \theta = 0, 1.$$

We note that the relation

$$f(x, 0) + f(x, 1) = 1, \qquad \text{a.e. } (\mu), \tag{2.3}$$

is obtainable from the identity in Borel B,

$$\int_B 1\, d\mu = \mu(B) = \mu_0(B) + \mu_1(B) = \int_B (f(x, 0) + f(x, 1))\, d\mu.$$

The joint generalized probability density function of $\mathbf{x}$ with respect to the product measure μ^n, when the parameter vector is $\boldsymbol{\theta} = (\theta_1, \cdots \theta_n)$, is $f(\mathbf{x}, \boldsymbol{\theta}) = \prod_1^n f(x_i, \theta_i)$. The expected loss on the ith decision in using a decision function $\mathbf{t} = (t_1(\mathbf{x}), \cdots, t_n(\mathbf{x}))$ is, for $i = 1, \cdots, n$,

$$R_i(\mathbf{t}, \boldsymbol{\theta}) = \int [a\theta_i(1 - t_i(\mathbf{x})) + b(1 - \theta_i)t_i(\mathbf{x})] f(\mathbf{x}, \boldsymbol{\theta})\, d\mu^n.$$

The average expected loss on all n decisions, which we define to be the risk of $\mathbf{t}$, is therefore

$$R(\mathbf{t}, \boldsymbol{\theta}) = \frac{1}{n} \sum_1^n R_i(\mathbf{t}, \boldsymbol{\theta}) = \int \frac{1}{n} \sum_1^n \{a\theta_i[1 - t_i(\mathbf{x})] + b(1 - \theta_i)t_i(\mathbf{x})\} f(\mathbf{x}, \boldsymbol{\theta})\, d\mu^n. \tag{2.4}$$

This is equivalent to defining the loss of the decision $\mathbf{d} = (d_1, \cdots, d_n)$ in Ω, given $\boldsymbol{\theta}$ in Ω, to be

$$W(\mathbf{d}, \boldsymbol{\theta}) = \frac{1}{n} \sum_1^n [a\theta_i(1 - d_i) + b(1 - \theta_i)\, d_i],$$

since this definition implies that the decision function $\mathbf{t}$ induces the conditional (for fixed $\mathbf{x}$) expected loss

$$W(\mathbf{t}(\mathbf{x}), \boldsymbol{\theta}) = \frac{1}{n} \sum_1^n \{a\theta_i[1 - t_i(\mathbf{x})] + b(1 - \theta_i)t_i(\mathbf{x})\}. \tag{2.5}$$

By the remark (2.3), $f(\mathbf{x}, \boldsymbol{\theta})$ is expressible as

$$(2.6)\quad f(\mathbf{x}, \boldsymbol{\theta}) = \prod_1^n f(x_i, \theta_i) = \prod_1^n [\theta_i f(x_i, 1) + (1 - \theta_i)(1 - f(x_i, 1))].$$

The Halmos-Savage [2] form of the theorem linking sufficient statistics and density factorization shows that $(f(x_1, 1), \cdots, f(x_n, 1))$ is a sufficient statistic for $\boldsymbol{\theta}$ in Ω. Let

$$(2.7)\quad \begin{cases} \mathbf{z} = (z_1, \cdots, z_n), & z_i = f(x_i, 1); \\ \nu_\theta(I) = \mu_\theta(x \mid f(x, 1) \text{ in } I), \quad \text{for all Borel sets } I; & \theta = 0, 1; \\ \nu(I) = \nu_0(I) + \nu_1(I). \end{cases}$$

It is easily verified (see [3] Sec. 32) that the sufficient statistic $\mathbf{z}$ (and, consequently, the measures ν_0, ν_1, and ν) is independent of the choice (2.2) of an underlying measure μ relative to which μ_0 and μ_1 are absolutely continuous. We note that $\mathbf{z}$ has, with respect to the product measure ν^n, the generalized probability density

$$(2.8)\quad d(\mathbf{z}, \boldsymbol{\theta}) = \prod_1^n [\theta_i z_i + (1 - \theta_i)(1 - z_i)].$$

Returning to (2.4) we see that the vector of conditional expectations

$$E[\mathbf{t}(\mathbf{x}) \mid \mathbf{z}] = (E[t_1(\mathbf{x}) \mid \mathbf{z}], \cdots, E[t_n(\mathbf{x}) \mid \mathbf{z}])$$

is a decision rule having the same risk as $\mathbf{t}$. We denote this rule by $\mathbf{t}(\mathbf{z}) = (t_1(\mathbf{z}), \cdots, t_n(\mathbf{z}))$ and, using (2.5) express its risk,

$$R(\mathbf{t}, \boldsymbol{\theta}) = \int W(\mathbf{t}(\mathbf{z}), \boldsymbol{\theta})\, d(\mathbf{z}, \boldsymbol{\theta})\, d\nu^n$$

$$(2.9)\quad = \frac{a}{n} \sum_1^n \theta_i \int (1 - t_i(\mathbf{z}))\, d(\mathbf{z}, \boldsymbol{\theta})\, d\nu^n + \frac{b}{n} \sum_1^n (1 - \theta_i) \cdot \int t_i(\mathbf{z})\, d(\mathbf{z}, \boldsymbol{\theta})\, d\nu^n.$$

3. Simple decision functions. If $\mathbf{t} = t$ is a simple decision function, then (2.9) simplifies to

$$(3.1)\quad R(t, \boldsymbol{\theta}) = \frac{a}{n} \sum_1^n \theta_i \int (1 - t(z)) z\, d\nu + \frac{b}{n} \sum_1^n (1 - \theta_i) \int t(z)(1 - z)\, d\nu.$$

Setting $\bar{\theta} = (1/n)\sum \theta_i$ = proportion of 1's among the n components of θ, we can write (3.1) in the form

$$(3.2)\quad R(t, \boldsymbol{\theta}) = \int a\bar{\theta}(1 - t(z))z + b(1 - \bar{\theta})t(z)(1 - z)\, d\nu.$$

The value $\bar{\theta}$ is necessarily rational, of the form k/n for $k = 0, \cdots, n$. We now define for any real number p such that $0 \leqq p \leqq 1$, and any ν-measurable function $t(z)$ such that $0 \leqq t(z) \leqq 1$, the expression

$$B(t, p) = \int ap(1 - t(z))z + b(1 - p)t(z)(1 - z)\, d\nu. \tag{3.3}$$

Thus (3.2) becomes

$$R(t, \boldsymbol{\theta}) = B(t, \bar{\theta}). \tag{3.4}$$

From (3.3) it is clear that for any fixed p, $B(t, p)$ is a minimum if and only if for almost every $(\nu)z$, $t(z)$ is of the form

$$t_p(z) = \begin{cases} 1, & \text{if } apz > b(1 - p)(1 - z); \\ 0, & \text{if } apz < b(1 - p)(1 - z); \\ \text{arbitrary}, & \text{if } apz = b(1 - p)(1 - z). \end{cases} \tag{3.5}$$

The minimum value of $B(t, p)$ is

$$\begin{aligned} \phi(p) = B(t_p, p) &= \int \min\,[apz, b(1 - p)(1 - z)]\, d\nu \\ &= ap \int_0^{C(p)} d\nu_1 + b(1 - p)\left(1 - \int_0^{C(p)} d\nu_0\right) \end{aligned} \tag{3.6}$$

where $C(p) = b(1 - p) / [ap + b(1 - p)]$ for $0 \leqq p \leqq 1$. It follows from (3.4) that, for any simple decision function t and any $\boldsymbol{\theta}$ in Ω,

$$R(t, \boldsymbol{\theta}) \geqq \phi(\bar{\theta}), \tag{3.7}$$

equality holding if and only if $t(z)$ is of the form (3.5) with $p = \bar{\theta}$.

We shall now establish some properties of the function $\phi(p)$ defined for $0 \leqq p \leqq 1$ by (3.6). For $0 \leqq p_1 < p_2 \leqq 1$, and $0 < s < 1$,

$$\begin{aligned} s \min[ap_1z,\ \ b(1 - p_1)(1 - z)] &+ (1 - s) \min[ap_2z,\ \ b(1 - p_2)(1 - z)] \\ &= \min[asp_1z,\ \ bs(1 - p_1)(1 - z)] \\ &\qquad + \min[a(1 - s)p_2z,\ \ b(1 - s)(1 - p_2)(1 - z)] \\ &\leqq \min[a\{sp_1 + (1 - s)p_2\}z,\ \ b\{1 - sp_1 - (1 - s)p_2\}(1 - z)], \end{aligned}$$

with strict inequality if and only if $p_1 < C(z) < p_2$ (or alternatively $C(p_2) < z < C(p_1)$). Integrating with respect to ν we obtain from (3.6) that ϕ is a concave function of p,

$$s\phi(p_1) + (1 - s)\, \phi(p_2) \leqq \phi(sp_1 + (1 - s)p_2), \tag{3.8}$$

with equality if and only if $\nu[z \mid C(p_2) < z < C(p_1)] = 0$. Since $\phi(0) = \phi(1) = 0$, this implies

$$\phi(p) > 0, \qquad 0 < p < 1, \qquad \text{unless } \nu[z \mid 0 < z < 1] = 0. \tag{3.9}$$

The exception here is the trivial case where ν_0 and ν_1 (and hence μ_0 and μ_1) are disjoint measures.

The continuity of $\phi(p)$ can be established in the following form. If $0 \leqq p < p' \leqq 1$, then $0 \leqq C(p') < C(p) \leqq 1$ and

$$\min[azp,\ \ b(1-z)(1-p)] - \min[azp',\ \ b(1-z)(1-p')]$$
$$= \begin{cases} -az(p'-p), & 0 \leqq z < C(p'); \\ b(1-z)(1-p) - azp', & C(p') \leqq z \leqq C(p); \\ b(1-z)(p'-p), & C(p) < z \leqq 1. \end{cases}$$

Hence it follows that, for $0 \leqq z \leqq 1$,

$$-az(p'-p) \leqq \min[azp,\ \ b(1-z)(1-p)] - \min[azp',\ \ b(1-z)(1-p')] \leqq b(1-z)(p'-p).$$

Integrating with respect to ν, we obtain

$$(3.10) \qquad -a(p'-p) \leqq \phi(p) - \phi(p') \leqq b(p'-p), \qquad p' > p.$$

Interchanging p and p', multiplying by -1 achieves

$$(3.11) \qquad -b(p-p') \leqq \phi(p) - \phi(p') \leqq a(p-p'), \qquad p' < p.$$

From (3.3) we have, for any $0 \leqq p, p' \leqq 1$,

$$B(t_p, p') - \phi(p) = \int (p'-p)\{a(1-t_p(z))z - bt_p(z)(1-z)\}\, d\nu$$

$$(3.12) \qquad = (p'-p)\int \{a(1-t_p(z))z - bt_p(z)(1-z)\}\, d\nu,$$

$$\leqq \begin{cases} (p'-p)a, & p' > p; \\ (p-p')b, & p' < p. \end{cases}$$

The last three inequalities, (3.10) to (3.12), and the definition of ϕ imply

$$0 \leqq B(t_p, p') - \phi(p') \leqq (a+b)\,|\,p - p'\,|.$$

From (3.4) it follows that for any $0 \leqq p \leqq 1$ and any $\boldsymbol{\theta}$ in Ω,

$$0 \leqq R(t_p, \boldsymbol{\theta}) - \phi(\bar{\theta}) \leqq (a+b)\,|\,p - \bar{\theta}\,|.$$

Thus, we have proved

THEOREM 1. *Suppose that in some manner an approximate value p of the true proportion $\bar{\theta}$ is known. Then the statistician who uses the simple decision function $\mathbf{t} = t_p$ will achieve a risk $R(t_p, \boldsymbol{\theta})$ which is within $(a+b)\,|\,p-\bar{\theta}\,|$ of the minimum attainable risk $\phi(\bar{\theta})$ in the class of all simple decision functions.*

4. "Consistent" estimation of a proportion. We use (without loss of generality) the canonical form of the problem (see (2.6) to (2.9)), assuming only that the original measures, μ_0 and μ_1, are not identical. Thus we are concerned with the random vector $\mathbf{z} = (z_1, \cdots, z_n)$ having the density $d(\mathbf{z}, \boldsymbol{\theta})$ defined by (2.8). We consider the problem of finding an estimator $p_n(\mathbf{z})$ for the proportion $\bar{\theta}$ of 1's in the first n coordinates of $\boldsymbol{\theta}$.

Since our principal interest is in the asymptotic estimation problem, we will consider the sequence of problems defined for $n = 1, 2, \cdots$ as embedded in the probability space of infinite sequences $\mathbf{z} = (z_1, z_2, \cdots)$ with p-measure induced by the density $d(\mathbf{z}, \boldsymbol{\theta})$ defined on the first n coordinates of $\mathbf{z}$ and $\boldsymbol{\theta}$, respectively. We emphasize this aspect in this and the following section by referring to $\boldsymbol{\theta}$ in Ω_∞.

Avoiding a discussion of "optimum" estimation, we devote this section to the consideration of a subclass H of the class U of all unbiased estimators of $\bar{\theta}$. This subclass H is to be the class of all estimators of the form

$$\bar{h}(\mathbf{z}) = \frac{1}{n}\sum_{1}^{n} h(z_i), \tag{4.1}$$

where $h(z)$ is an unbiased estimator of θ.

As a measure of the risk of an estimator in H we use its variance,

$$n \operatorname{Var} \bar{h}(\mathbf{z}) = \bar{\theta}\, V_1(h) + (1 - \bar{\theta})\, V_0(h), \tag{4.2}$$

where $V_1(h) = \int (h(z) - 1)^2\, d\nu_1$ and $V_0(h) = \int (h(z))^2\, d\nu_0$. We single out an interesting subclass of H by investigating the existence and representation of elements h minimizing, for fixed p, with $0 < p < 1$,

$$pV_1(h) + (1 - p)V_0(h). \tag{4.3}$$

For any pair of real numbers $\lambda = (\lambda_0, \lambda_1)$ we define an extension of (4.3) to the set of all $g(z)$ such that $\int |g(z)|\, d\nu < \infty$:

$$F_\lambda(g) = p\left[\int g^2\, d\nu_1 - 1\right] + (1 - p)\int g^2\, d\nu_0 - 2\lambda_1\left[\int g\, d\nu_1 - 1\right] - 2\lambda_0 \int g\, d\nu_0. \tag{4.4}$$

Using the density representation of these integrals and the restriction on the domain of g, we have

$$F_\lambda(g) = \int g^2(z)[pz + (1 - p)(1 - z)] - 2g(z)[\lambda_1 z + \lambda_0(1 - z)] \cdot d\nu - p + 2\lambda_1. \tag{4.5}$$

For fixed λ_0 and λ_1 the integrand here is a minimum, for each fixed z, if and only if

$$g(z) = g_p(z \mid \lambda) = \frac{\lambda_1 z + \lambda_0(1 - z)}{pz + (1 - p)(1 - z)}, \qquad \text{a. e. } (\nu). \tag{4.6}$$

Since $g_p(z \mid \lambda) \leqq \max [|\lambda_1|, |\lambda_0|] / \min [p, 1 - p]$, we find that $g_p(z \mid \lambda)$ is a unique (ν) minimum of $F_\lambda(g)$ over its domain. If there exists a determination of λ_0 and λ_1 such that $g_p(z \mid \lambda)$ is an unbiased estimator of θ, we will denote it by $h_p(z)$. Then we have for all unbiased h

$$(4.7) \quad pV_1(h) + (1 - p)V_0(h) = F_\lambda(h) \geqq F_\lambda(h_p) = pV_1(h_p) + (1 - p)V_0(h_p),$$

where equality holds only if $h(z) \equiv h_p(z) \quad (\nu)$.

The estimator $g_\lambda(z \mid p)$ will be unbiased if λ_0 and λ_1 satisfy

$$(4.8) \quad \lambda_1 \int \frac{z}{pz + (1 - p)(1 - z)}\, d\nu_i + \lambda_0 \int \frac{1 - z}{pz + (1 - p)(1 - z)}\, d\nu_i = i, \qquad i = 0, 1.$$

The determinant of these equations is

$$\left[\int \frac{z(1 - z)}{pz + (1 - p)(1 - z)}\, d\nu\right]^2 - \int \frac{(1 - z)^2}{pz + (1 - p)(1 - z)}\, d\nu \cdot \int \frac{z^2}{pz + (1 - p)(1 - z)}\, d\nu.$$

It is nonpositive by the Schwarz inequality. If it is equal to zero, z and $1 - z$ are linearly dependent a.e. (ν). Since z and $1 - z$ are densities, linear dependence would imply that ν_0 and ν_1 are identical; this possibility has been excluded here.

Thus, an explicit representation of estimators in H, minimizing (4.2) for $\boldsymbol{\theta}$ such that $\bar{\theta} = p$, is obtained as

$$\bar{h}_p(\mathbf{z}) = \frac{1}{n} \sum_1^n h_p(z_i), \qquad 0 < p < 1,$$

where $h_p(z) = g_p(z \mid \lambda)$, and λ is a solution of (4.8). This class of estimators merits our interest since each is admissible relative to H and each satisfies the following theorem.

Theorem 2. *Let $\bar{h}(\mathbf{z})$ be any estimator in H such that $|h(z)| + 1 < M < \infty$. Define*

$$p_n(\mathbf{z}) = \begin{cases} 0, & \bar{h}(\mathbf{z}) < 0; \\ \bar{h}(\mathbf{z}), & 0 \leqq \bar{h}(\mathbf{z}) \leqq 1; \\ 1, & 1 < \bar{h}(\mathbf{z}). \end{cases}$$

Then, (a) $0 \leqq p_n(z) \leqq 1$, and (b) for any $\epsilon > 0$ there exists an $N(\epsilon)$ such that, for any $\boldsymbol{\theta}$ in Ω_∞,

$$\Pr[|p_n(z) - \bar{\theta}| > \epsilon \quad \text{for some } n \geqq N(\epsilon)] \leqq \epsilon.$$

Proof. It will clearly be sufficient to show that $\bar{h}(\mathbf{z})$ satisfies part (b). For this purpose we introduce $\bar{h}_i(k)$ for $i = 0$ or 1 and $k = 1, 2, \cdots$ to denote the arithmetic mean of k independent random variables, each having the probability measure $\nu_i h^{-1}$. We express

$$\bar{h}(\mathbf{z}) = (1/n) \sum [\theta_j h(z_j) + (1 - \theta_j) h(z_j)]$$

$$(4.9)\qquad = \bar\theta \sum \theta_j h(z_j) / \sum \theta_j + (1 - \bar\theta)\sum(1 - \theta_j)h(z_j) / \sum(1 - \theta_j)$$
$$= \bar\theta \bar h_1(n\bar\theta) + (1 - \bar\theta)\bar h_0(n - n\bar\theta),$$

$$(4.10)\qquad \bar h(\mathbf{z}) - \bar\theta = \bar\theta[\bar h_1(n\bar\theta) - 1] + (1 - \bar\theta)\bar h_0(n - n\bar\theta).$$

The strong law of large numbers yields the existence of functions $N_1(\eta)$ and $N_0(\eta)$, defined for $\eta > 0$ and such that

$$(4.11)\qquad \Pr[|\bar h_i(k) - i| > \eta \quad \text{for some } k \geqq N_i(\eta)] \leqq \eta, \qquad i = 0, 1.$$

We fix $\epsilon > 0$ and consider the term $\bar\theta[\bar h_1(n\bar\theta) - 1]$ of (4.10). Since $\bar\theta \leqq 1$ and $|\bar h_1(n\bar\theta) - 1| \leqq M$, we have

$$(4.12)\qquad |\bar\theta[\bar h_1(n\bar\theta) - 1]| \leqq \min\ \{\bar\theta M,\ |\bar h_1(n\bar\theta) - 1|\}.$$

Hence, for any fixed integer N and any fixed point $\boldsymbol\theta$ in Ω_∞, if $\mathbf{z}$ is such that there exist $n_{\mathbf{z}} \geqq N$ with $\bar\theta[\bar h_1(n_{\mathbf{z}}\bar\theta) - 1] > \epsilon/2$, then

$$n_{\mathbf{z}}\bar\theta > n_{\mathbf{z}}M^{-1}\epsilon/2, \qquad |\bar h_1(n_{\mathbf{z}}\bar\theta) - 1| > \epsilon/2.$$

Hence there exists $k > NM^{-1}\epsilon/2$ such that $|\bar h_1(k) - 1| > \epsilon/2$. Consequently

$$(4.13)\qquad \begin{aligned} &\Pr\{\bar\theta[\bar h_1(n\bar\theta) - 1] > \epsilon/2 \text{ for some } n \geqq N\} \\ &\qquad \leqq \Pr\{|\bar h_1(k) - 1| > \epsilon/2 \text{ for some } k > NM^{-1}\epsilon/2\}. \end{aligned}$$

Thus, if $N \geqq 2M\epsilon^{-1}N_1(\epsilon/2)$, it follows from (4.11) for $i = 1$ that the right side of (4.13) is less than $\epsilon/2$ uniformly for all $\boldsymbol\theta$ in Ω_∞.

We can deal in a similar fashion with the term $(1 - \bar\theta)\bar h_0(n - n\bar\theta)$ of (4.10). Thus we obtain that part (*b*) is satisfied by

$$(4.14)\qquad N(\epsilon) = 2M\epsilon^{-1} \max\,[N_1(\epsilon/2), N_0(\epsilon/2)].$$

5. Nonsimple decision functions t. We have seen in Section 3 that if p is a good approximation to $\bar\theta$, then the simple decision function $\mathbf{t} = t_p$ (see (3.5)) does about as well as is possible for any simple decision function. Although a good approximation p to $\bar\theta$ is not generally available to the statistician, we have seen in Section 4 that for large n a good estimator $p_n(\mathbf{z})$ of $\bar\theta$ is always available.

It is natural to combine these two results by using the decision function t_p with the constant p replaced by the random variable $p_n(z)$. This amounts to using the nonsimple decision function $t^* = (t_1^*(\mathbf{z}), \cdots, t_n^*(\mathbf{z}))$ such that for $i = 1, \cdots, n$,

$$(5.1)\qquad t_i^*(\mathbf{z}) = \begin{cases} 1, & \text{if } z_i > C(p_n(\mathbf{z})), \\ 0, & \text{otherwise.} \end{cases}$$

(We have chosen, arbitrarily, one way of resolving the ambiguity in the definition (3.5) when $z = C(p)$.)

In practice, $\mathbf{t}^*$ can be used only when all the values $z_1, \cdots, z_n$ are known before the individual decisions on the values of $\theta_1, \cdots, \theta_n$ have to be made.

We shall now investigate the behavior, for large n, of the risk function $R(\mathbf{t}^*, \boldsymbol\theta)$.

We begin by considering the loss of the decision rule t_p (determined by (3.5) and by $t_p(C(p)) = 0$), where $\mathbf{z}$ is observed and $\boldsymbol{\theta}$ is the true parameter point. From (2.5) we obtain

$$W(t_p(\mathbf{z}), \boldsymbol{\theta}) = a\bar{\theta}[\sum \theta_i(1 - t_p(z_i)) / \sum \theta_i] \tag{5.2}$$
$$+ b(1 - \bar{\theta})[\sum (1 - \theta_i)t_p(z_i) / \sum (1 - \theta_i)]$$
$$= a\bar{\theta}S_1(C(p) \mid n\bar{\theta}) + b(1 - \bar{\theta})[1 - S_0(C(p) \mid n - n\bar{\theta})],$$

where $S_i(v \mid k)$ is the sample distribution function of k independent random variables, each distributed with probability measure ν_i for $i = 0$ or 1 and $k = 1, 2, \cdots$. For future use we define

$$D_1^+(k) = \sup_{0 \leq v \leq 1} \left[S_1(v \mid k) - \int_0^v d\nu_1 \right], \quad D_0^-(k) = \sup_{0 \leq v \leq 1} \left[\int_0^v d\nu_0 - S_0(v \mid k) \right]. \tag{5.3}$$

Using an alternative form of (3.3), it follows from (5.2) that

$$W(t_p(\mathbf{z}), \boldsymbol{\theta}) = a\bar{\theta} \left\{ S_1(C(p) \mid n\bar{\theta}) - \int_0^{C(p)} d\nu_1 \right\}$$
$$+ b(1 - \bar{\theta}) \left\{ \int_0^{C(p)} d\nu_0 - S_0(C(p) \mid n - n\bar{\theta}) \right\} + B(t_p, \boldsymbol{\theta}).$$

Hence, from (5.3) and Theorem 1,

$$W(t_p(\mathbf{z}), \boldsymbol{\theta}) \leq a\bar{\theta}D_1^+(n\bar{\theta}) + b(1 - \bar{\theta})D_0^-(n - n\bar{\theta}) + \phi(\bar{\theta}) + (a + b)\,|p - \bar{\theta}|.$$

Since this holds for each $0 \leq p \leq 1$ we have, for any estimate $p_n(\mathbf{z})$ satisfying (*a*) of Theorem 1,

$$W(t_{p_n(\mathbf{z})}(\mathbf{z}), \boldsymbol{\theta}) \leq a\bar{\theta}D_1^+(n\bar{\theta}) \tag{5.4}$$
$$+ b(1 - \bar{\theta})D_0^-(n - n\bar{\theta}) + \phi(\bar{\theta}) + (a + b)\,|p_n(\mathbf{z}) - \bar{\theta}|.$$

By the Glivenko-Cantelli Theorem [4]

$$\Pr[\lim_{k\to\infty} D_1^+(k) = 0] = \Pr[\lim_{k\to\infty} D_0^-(k) = 0] = 1.$$

This is equivalent to the existence of functions $N_1^*(\eta)$ and $N_0^*(\eta)$, defined for $\eta > 0$, such that

$$\Pr[D_1^+(k) > \eta \text{ for some } k \geq N_1^*(\eta)] \leq \eta,$$
$$\Pr[D_0^-(k) > \eta \text{ for some } k \geq N_0^*(\eta)] \leq \eta.$$

As in (4.11) to (4.14), we have that if

$$N^*(\epsilon) = 2\epsilon^{-1} \max[aN_1^*(\epsilon/2), bN_0^*(\epsilon/2)], \qquad \epsilon > 0,$$

then, for any $\boldsymbol{\theta}$ in Ω_∞,

$$\Pr[a\bar{\theta}D_1^+(n\bar{\theta}) + b(1 - \bar{\theta})D_0^-(n - n\bar{\theta}) > \epsilon \text{ for some } n \geq N^*(\epsilon)] \leq \epsilon. \tag{5.5}$$

Returning to (5.4) we see that, for any $p_n(z)$ satisfying the conclusion of Theorem 2, (5.5) and Theorem 2 combine to furnish for any $\epsilon > 0$

$$n_0(\epsilon) = \max[N^*(\epsilon/2), \quad N(\epsilon/2(a+b))]$$

such that, for any $\boldsymbol{\theta}$ in Ω_∞,

$$\Pr\{W(t_{p_n(\mathbf{z})}(\mathbf{z}), \boldsymbol{\theta}) - \phi(\bar{\theta}) > \epsilon \text{ for some } n \geqq n_0(\epsilon)\} \leqq \epsilon. \tag{5.6}$$

The argument from (5.2) onward proves

THEOREM 3. *If $p_n(\mathbf{z})$ satifies the conclusions (a) and (b) of Theorem 2, then the decision function $\mathbf{t}^*$ defined by* (5.1) *is such that to any $\epsilon > 0$ there corresponds an $n_0(\epsilon)$ such that for any $\boldsymbol{\theta}$ in Ω_∞*

$$\Pr\{W(\mathbf{t}^*(\mathbf{z}), \boldsymbol{\theta}) \leqq \phi(\bar{\theta}) + \epsilon \text{ for all } n \geqq n_0(\epsilon)\} \geqq 1 - \epsilon.$$

Since $W(\mathbf{t}^*(\mathbf{z}), \boldsymbol{\theta}) \leqq \max\,[a, b]$, Theorem 3 implies

THEOREM 4. *Under the assumptions of Theorem 3, $\mathbf{t}^*$ is such that to any $\epsilon > 0$ there corresponds an $n_1(\epsilon)$ such that for any $n \geqq n_1(\epsilon)$ and any $\boldsymbol{\theta}$ in Ω,*

$$R(\mathbf{t}^*, \boldsymbol{\theta}) < \phi(\bar{\theta}) + \epsilon.$$

Thus, $\mathbf{t}^*$ is a nonsimple decision function which for large n does about as well as could be done by any simple decision function even if $\bar{\theta}$ were known exactly.

6. Invariant and R-invariant decision functions. Let $(P(1), \cdots, P(n))$ be an arbitrary permutation of $(1, \cdots, n)$. Define, for any real vector $\boldsymbol{\xi} = (\xi_1, \cdots, \xi_n)$,

$$P\boldsymbol{\xi} = (\xi_{P(1)}, \cdots, \xi_{P(n)}). \tag{6.1}$$

From (2.5) and (2.8) we note that $W(P\mathbf{t}, P\boldsymbol{\theta}) = W(\mathbf{t}, \boldsymbol{\theta})$ and $d(P\mathbf{z}, P\boldsymbol{\theta}) = d(\mathbf{z}, \boldsymbol{\theta})$. From (2.9) we obtain for any decision function $\mathbf{t}$, by a change of variable of integration,

$$\begin{aligned} R(\mathbf{t}, P\boldsymbol{\theta}) &= \int W(\mathbf{t}(\mathbf{z}), P\boldsymbol{\theta})\, d(\mathbf{z}, P\boldsymbol{\theta})\, d\nu^n \\ &= \int W(\mathbf{t}(P\mathbf{z}), P\boldsymbol{\theta})\, d(P\mathbf{z}, P\boldsymbol{\theta})\, d\nu^n = R(P^{-1}\mathbf{t}P, \boldsymbol{\theta}). \end{aligned} \tag{6.2}$$

We call a decision function $\mathbf{t}$ *invariant* if $P^{-1}\mathbf{t}P = \mathbf{t}$ for all P. We denote by *R-invariant* a decision function $\mathbf{t}$ such that

$$R(P^{-1}\mathbf{t}P, \boldsymbol{\theta}) = R(\mathbf{t}, \boldsymbol{\theta}), \qquad \text{for all } P \text{ and } \boldsymbol{\theta}. \tag{6.3}$$

The risk of an R-invariant decision function can be expressed as an explicit function of $\bar{\theta}$ by formally averaging the representation (2.9) over all P. From (2.9), (6.2), and (6.3),

$$R(\mathbf{t}, \boldsymbol{\theta}) = \frac{1}{n!}\sum_P R(\mathbf{t}, P\boldsymbol{\theta}) = \frac{1}{n!}\sum_P \int W(\mathbf{t}(\mathbf{z}), P\boldsymbol{\theta})\, d(\mathbf{z}, P\boldsymbol{\theta})\, d\nu^n. \tag{6.4}$$

Changing the order of summations and integration, we find that if $\mathbf{t}$ is R-invariant its risk, $R(\mathbf{t}, \boldsymbol{\theta})$, is expressible in the form

$$R(\mathbf{t}, \boldsymbol{\theta}) = \frac{1}{n} \sum_1^n \int \left\{ a \frac{1 - t_i(\mathbf{z})}{n!} \sum_P \theta_{P(i)}\, d(\mathbf{z}, P\boldsymbol{\theta}) + b \frac{t_i(\mathbf{z})}{n!} \sum_P (1 - \theta_{P(i)})\, d(\mathbf{z}, P\boldsymbol{\theta}) \right\} d\nu^n. \tag{6.5}$$

For any real vector $\mathbf{r} = (r_1, \cdots, r_n)$ we define for every integer k

$$L(k, n, \mathbf{r}) = \begin{cases} \dfrac{1}{n!} \displaystyle\sum_P \prod_1^k r_{P(j)} \prod_{k+1}^n (1 - r_{P(j)}), & k = 0, 1, \cdots, n; \\ 0, & \text{otherwise}; \end{cases}$$

$L(k, n - 1, \check{r}_i = L[k, n - 1, (r_1, \cdots, r_{i-1}, r_{i+1}, \cdots, r_n)], \quad i = 1, 2, \cdots, n.$

Then

$$\sum_P \theta_{P(i)}\, d(\mathbf{z}, P\boldsymbol{\theta}) = z_i \sum_{\theta_{P(i)}=1} \prod_{j \neq i} [\theta_{P(j)} z_j + (1 - \theta_{P(j)})(1 - z_j)] = z_i \left(\sum_1^n \theta_k \right) (n - 1)!\, L\left(\sum_1^n \theta_k - 1,\ n - 1,\ \check{z}_i \right); \tag{6.6}$$

$$\sum_P (1 - \theta_{P(i)})\, d(\mathbf{z}, P\boldsymbol{\theta}) = (1 - z_i) \sum_1^n (1 - \theta_k)(n - 1)! \cdot L\left(\sum_1^n \theta_k,\ n - 1,\ \check{z}_i \right). \tag{6.7}$$

It follows from (6.5) that for R-invariant $\mathbf{t}$

$$R(\mathbf{t}, \boldsymbol{\theta}) = \frac{1}{n} \sum_1^n \int \{ a[1 - t_i(\mathbf{z})] z_i \bar{\theta} L(n\bar{\theta} - 1,\ n - 1,\ \check{z}_i) + b t_i(\mathbf{z})(1 - z_i)(1 - \bar{\theta}) L(n\bar{\theta},\ n - 1, \check{z}_i) \}\, d\nu^n. \tag{6.8}$$

Denoting the ith summand of (6.8) by $R_i(\mathbf{t} \mid \bar{\theta})$, we note for later use that, if $\mathbf{t}$ is actually invariant,

$$R_i(\mathbf{t} \mid \bar{\theta}) = R_1(\mathbf{t} \mid \bar{\theta}) = R(\mathbf{t}, \boldsymbol{\theta}), \qquad i = 1, \cdots, n. \tag{6.9}$$

This follows from the invariance and a permutation of the variables of integration in the representation of $R_i(\mathbf{t} \mid \bar{\theta})$ in (6.8). It further follows from this representation that, for any fixed $\boldsymbol{\theta}$, the integrand of $R_i(\mathbf{t} \mid \bar{\theta})$ is, for each fixed $\mathbf{z}$, at a minimum with respect to t_i if and only if t_i is of the form defined, for $p = 0$, $1/n, \cdots, (n - 1)/n, 1$, by

$$t_{ip}(\mathbf{z}) = \begin{Bmatrix} 1, \\ 0, \\ \text{arbitrary}, \end{Bmatrix} a z_i p L(np - 1,\ n - 1,\ \check{z}_i) \begin{Bmatrix} > \\ < \\ = \end{Bmatrix} b(1 - z_i)(1 - p) \cdot L(np,\ n - 1,\ \check{z}_i), \tag{6.10}$$

with p equal to $\bar{\theta}$. We denote $(t_{1p}(\mathbf{z}), \cdots, t_{np}(\mathbf{z}))$ by $\mathbf{t}_p$. With a proper determination of the arbitrary part, any decision function of this form (6.10) is invariant. Hence, the representation of its risk in the form (6.8) is valid and $R(\mathbf{t}, \boldsymbol{\theta})$ for each fixed $\boldsymbol{\theta}$ is minimized over the class of R-invariant decision functions by an invariant decision function of the form (6.10).

Let

$$\phi_n(\bar{\theta}) = \inf_{R\text{-invariant } \mathbf{t}} R(\mathbf{t}, \boldsymbol{\theta}). \tag{6.11}$$

From (6.8), (6.9), and (6.10) we obtain

$$\phi_n(\bar{\theta}) = R_1(t_{\bar{\theta}} \mid \bar{\theta})$$

$$= \int \min [az_1\bar{\theta}L(n\bar{\theta} - 1, \ n - 1, \check{z}_1), \quad b(1 - z_1)(1 - \bar{\theta})L(n\bar{\theta}, \ n - 1, \ \check{z}_1)] \, d\nu^n. \tag{6.12}$$

Since $L(k, \ n - 1, \ \check{z}_1)$ is a symmetric function of $z_2, \cdots, z_n$, we have for any symmetric measurable set S

$$\int_S L(k, \ n - 1, \ \check{z}_1) \, d\nu^{n-1} = \nu_1^k \nu_0^{n-1-k}(S). \tag{6.13}$$

It follows that the integral with respect to $z_2, \cdots, z_n$ of a z_1-section of the integrand of (6.12) is

$$az_1\bar{\theta}[1 - \nu_1^{n\bar{\theta}-1}\nu_0^{n-n\bar{\theta}}(S_{z_1})] + b(1 - z_1)(1 - \bar{\theta}) \, \nu_1^{n\bar{\theta}}\nu_0^{n-1-n\bar{\theta}}(S_{z_1})$$

where S_{z_1} is the symmetric set

$$S_{z_1} = [(z_2, \cdots, z_n) \mid az_1\bar{\theta}L(n\bar{\theta} - 1, \ n - 1, \ \check{z}_1) > b(1 - z_1)(1 - \bar{\theta})L(n\bar{\theta}, \ n - 1, \ \check{z}_1)].$$

The developments from (6.4) onward set the stage for

THEOREM 5. *If* $\phi(\bar{\theta})$ *and* $\phi_n(\bar{\theta})$ *are defined by* (3.6) *and* (6.12), *respectively, then for any* $\epsilon > 0$ *there exists* $N(\epsilon)$ *such that for any* $n \geqq N(\epsilon)$ *and any* $\boldsymbol{\theta}$

$$\phi(\bar{\theta}) - \epsilon < \phi_n(\bar{\theta}) \leqq \phi(\bar{\theta}).$$

PROOF. That $\phi_n \leqq \phi$ for all $\boldsymbol{\theta}$ follows from the fact that every simple decision function is invariant.

For the nontrivial part of the proof we fix $\epsilon > 0$. From the continuity of $\phi(p)$ ((3.10) and (3.11)) we obtain a $\delta(\epsilon)$ such that

$$\phi(p) - \epsilon < 0 \quad \text{if } p < \delta \ \text{ or } p > 1 - \delta. \tag{6.14}$$

Thus, it will be sufficient to show the existence of an $N(\epsilon)$ which suffices for the theorem when $0 < \delta \leqq \bar{\theta} \leqq 1 - \delta$. We will obtain this from the following measure theoretic lemma, a slight generalization of which is proved in [5].

LEMMA. *If m_0 and m_1 are nondisjoint p-measures on a σ-algebra of subsets χ of a set X, then for every ϵ, $\delta > 0$ there exists $N(\epsilon, \delta)$ such that for any pair of positive integers r, s with $r + s \geqq N(\epsilon, \delta)$ and $\delta \leqq r/(r + s) \leqq 1 - \delta$,*

$$|m_0^r m_1^s(S) - m_0^{r-1} m_1^{s+1}(S)| < \epsilon \textit{ uniformly for all symmetric } S \textit{ in } \chi^{r+s}.$$

From this lemma we obtain that if $n - 1 \geqq N(\eta, \delta^*)$ and $\delta^* \leqq n\bar{\theta}/(n - 1) \leqq 1 - \delta^*$,

$$[1 - \nu_1^{n\bar{\theta}-1}\nu_0^{n-n\bar{\theta}}(S_{z_1})] + \nu_1^{n\bar{\theta}}\nu_0^{n-1-n\bar{\theta}}(S_{z_1}) > 1 - \eta$$

uniformly for all z_1. From this it follows that $\phi_n(\bar{\theta}) > (1 - \eta)\phi(\bar{\theta}) = \phi(\bar{\theta}) - \eta\phi(\bar{\theta})$. Since the $\max_p \phi(p) \leqq c = ab / (a + b)$, we see that in view of (6.14) the choice $N(\epsilon) = N(c^{-1}\epsilon, 2^{-1}\delta(\epsilon))$ will complete the proof of Theorem 5.

Theorem 5 can be combined with Theorem 3 or Theorem 4 to give a strengthened endorsement of $\mathbf{t}^*$. For large n, $\mathbf{t}^*$ does about as well as could be done by any R-invariant decision function even if $\bar{\theta}$ were known exactly.

7. Bayes' and minimax solutions. Let $\mathbf{t} = (t_1(\mathbf{z}), \cdots, t_n(\mathbf{z}))$ be any decision function in the compound decision problem. For any $\boldsymbol{\theta}$ in Ω we can write the risk (2.9) in the form

$$R(\mathbf{t}, \boldsymbol{\theta}) = \int \frac{1}{n} \sum_1^n \{a\theta_i[1 - t_i(\mathbf{z})]\, d(\mathbf{z}, \boldsymbol{\theta}) + b(1 - \theta_i)t_i(\mathbf{z})\, d(\mathbf{z}, \boldsymbol{\theta})\}\, d\nu^n.$$

By a weight function we mean any function $\beta(\boldsymbol{\theta}) \geqq 0$ defined on Ω and not identically 0. For any weight function $\beta(\boldsymbol{\theta})$ and any decision function $\mathbf{t}$ we define the weighted risk of $\mathbf{t}$ relative to $\beta(\boldsymbol{\theta})$ as

$$\begin{aligned} B(\mathbf{t}, \beta) = \sum_{\boldsymbol{\omega}} \beta(\boldsymbol{\omega}) \cdot R(\mathbf{t}, \boldsymbol{\omega}) = \int \frac{1}{n} \sum_1^n \{&(1 - t_i(\mathbf{z}))\, a \sum_{\boldsymbol{\omega}} \beta(\boldsymbol{\omega})\omega_i\, d(\mathbf{z}, \boldsymbol{\omega}) \\ &+ t_i(\mathbf{z}) b \sum_{\boldsymbol{\omega}} \beta(\boldsymbol{\omega})(1 - \omega_i)\, d(\mathbf{z}, \boldsymbol{\omega})\}\, d\nu^n. \end{aligned}$$

For fixed $\beta(\boldsymbol{\theta})$ this will be a minimum if and only if for almost every $(\nu^n)\,\mathbf{z}$ and for every $i = 1, \cdots, n$, $t_i(\mathbf{z})$ is of the form

$$(7.1) \qquad t_i(\mathbf{z} \mid \beta) = \begin{Bmatrix} 1, \\ 0, \\ \text{arbitrary,} \end{Bmatrix} a \sum_{\boldsymbol{\omega}} \beta(\boldsymbol{\omega})\omega_i\, d(\mathbf{z}, \boldsymbol{\omega}) \begin{Bmatrix} > \\ < \\ = \end{Bmatrix} b \sum_{\boldsymbol{\omega}} \beta(\boldsymbol{\omega})(1 - \omega_i)\, d(\mathbf{z}, \boldsymbol{\omega}).$$

Any decision function $\mathbf{t}$ of the form (7.1) is called a *Bayes' solution* relative to the weight function $\beta(\boldsymbol{\theta})$.

For the remainder of this section we restrict our attention to symmetric $\beta(\boldsymbol{\theta})$, that is, such that

$$(7.2) \qquad \binom{n}{k} \beta(\boldsymbol{\theta}) = \beta_k \qquad \text{for all } \boldsymbol{\theta} \text{ with } \sum_1^n \theta_j = k, \quad k = 0, 1, \cdots, n.$$

For symmetric β we obtain the representation (see (6.8) to (6.11))

$$a \sum_{\omega} \beta(\omega)\omega_i\, d(\mathbf{z}, \omega) = az_i \sum_{k=0}^{n} \beta_k \frac{k}{n} L(k-1,\ \ n-1,\ \ \check{z}_i),$$

$$b \sum_{\omega} \beta(\omega)(1-\omega_i)\, d(\mathbf{z}, \omega) = b(1-z_i) \sum_{k=0}^{n} \beta_k (1-k/n) L(k,\ \ n-1,\ \ \check{z}_i).$$

Several particular cases of (7.2) hold special interest for us. If for some integer k_0 with $0 \leqq k_0 \leqq n$, (7.2) takes the form

$$\beta_{k_0} = 1, \qquad \beta_k = 0 \qquad\qquad k = 0, \cdots k_0 - 1,\ k_0 + 1, \cdots, n,$$

then the corresponding Bayes' solution (7.1) is a decision function of the form (6.10).

If for some constant p with $0 \leqq p \leqq 1$, expression (7.2) takes the form

$$\beta_k = \binom{n}{k} p^k (1-p)^{n-k},$$

then the corresponding Bayes' solution (7.1) is a decision function of the form (3.5) provided the Bayes' solution is required to be simple on the arbitrary part of its definition.

Referring back to Section 3, we have by the previous paragraph that each of the simple decision functions $\mathbf{t} = t_p$ for $0 \leqq p \leqq 1$, defined by (3.5), is a Bayes' solution for the compound decision problem. It can be shown (we omit the simple proof) that there always exists a value $0 < p' < 1$ and a determination of $t_{p'}$ for which the coefficient of p in (3.3) vanishes. Letting r be the constant value of $B(t_{p'}, p)$ it follows that $R(t_{p'}, \boldsymbol{\theta}) \equiv r$ for every $\boldsymbol{\theta}$ in Ω.

Since $t_{p'}$ has constant risk and is a Bayes' solution relative to a weight function which is positive for every $\boldsymbol{\theta}$ in Ω, it follows that $t_{p'}$ is the admissible minimax decision function, unique in the sense of risk. That is

$$\sup_{\boldsymbol{\theta} \text{ in } \Omega} R(\mathbf{t}, \boldsymbol{\theta}) = \text{minimum for } \mathbf{t} = t_{p'},$$

and if $\mathbf{t}$ is any other decision function such that $\sup R(\mathbf{t}, \boldsymbol{\theta}) = r$, then $R(\mathbf{t}, \boldsymbol{\theta}) \equiv r$.

8. Admissibility. Since the minimax decison function $t_{p'}$ is simple and has constant risk r independent of n, it follows that

$$r = R(t_{p'}, \boldsymbol{\theta}) \geqq \inf_t R(t, \boldsymbol{\theta}) = \phi(\bar{\theta}) \qquad\qquad \text{for all } \boldsymbol{\theta} \text{ in } \Omega,$$

and hence $r = \phi(p') = \max_p \phi(p)$.

If μ_0 and μ_1 are nondisjoint, $\phi(0) = \phi(1) = 0 < r$ and we conclude from Theorem 4 that, for any $0 < \epsilon < r$, the minimax decision function is ϵ-inadmissible (See [6] for definitions) for all sufficiently large n, since decision functions of the type $\mathbf{t}^*$ are ϵ-better.

The present paper has failed to exhibit a $\mathbf{t}^*$ which is admissible (or even to show the existence of such). This deficiency is remedied, at least asymptotically, by the fact that, for any $\epsilon > 0$ Theorems 4 and 5 together imply that any decision function of the type t^* is ϵ-uniformly-best (relative to the class of R-invariant decision functions) for all sufficiently large n.

REFERENCES

[1] HERBERT ROBBINS, "Asymptotically subminimax solutions of compound statistical decision problems," *Proceedings of the Second Berkeley Symposium on Mathematical Statistics and Probability*, University of California Press, 1951.

[2] P. R. HALMOS AND L. J. SAVAGE, "Application of the Radon-Nikodym theorem to the theory of sufficient statistics," *Ann. Math. Stat.*, Vol. 20 (1949), pp. 225–242.

[3] P. R. HALMOS, *Measure Theory*, D. Van Nostrand Company, 1950.

[4] M. FRÉCHET, *Recherches Théoriques Modernes sur le Calcul des Probabilités*, Vol. 1. Gauthier-Villars, Paris, 1937.

[5] J. F. HANNAN, "Asymptotic solutions of compound decision problems," *Institute of Statistics Mimeograph Series*, University of North Carolina, No. 68 (1953), pp. 1–77.

[6] J. WOLFOWITZ, "On ϵ-complete classes of decision functions," *Ann. Math. Stat.*, Vol. 22 (1951), pp. 461–465.

AN EMPIRICAL BAYES APPROACH TO STATISTICS

HERBERT ROBBINS
COLUMBIA UNIVERSITY

Let X be a random variable which for simplicity we shall assume to have discrete values x and which has a probability distribution depending in a known way on an unknown real parameter Λ,

$$(1) \qquad p(x \mid \lambda) = Pr[X = x \mid \Lambda = \lambda],$$

Λ *itself being a random variable* with *a priori* distribution function

$$(2) \qquad G(\lambda) = Pr[\Lambda \leqq \lambda].$$

The unconditional probability distribution of X is then given by

$$(3) \qquad p_G(x) = Pr[X = x] = \int p(x \mid \lambda)\, dG(\lambda),$$

and the expected squared deviation of any estimator of Λ of the form $\varphi(X)$ is

$$(4) \qquad \begin{aligned} E[\varphi(X) - \Lambda]^2 &= E\{E[(\varphi(X) - \Lambda)^2 \mid \Lambda = \lambda]\} \\ &= \int \sum_x p(x \mid \lambda)[\varphi(x) - \lambda]^2 dG(\lambda) \\ &= \sum_x \int p(x \mid \lambda)[\varphi(x) - \lambda]^2 dG(\lambda), \end{aligned}$$

which is a minimum when $\varphi(x)$ is defined for each x as that value $y = y(x)$ for which

$$(5) \qquad I(x) = \int p(x \mid \lambda)(y - \lambda)^2 dG(\lambda) = \text{minimum}.$$

But for any fixed x the quantity

$$(6) \qquad \begin{aligned} I(x) &= y^2 \int p\, dG - 2y \int p\lambda\, dG + \int p\lambda^2 dG \\ &= \int p\, dG \left(y - \frac{\int p\lambda\, dG}{\int p\, dG}\right)^2 + \left[\int p\lambda^2 dG - \frac{(\int p\lambda\, dG)^2}{\int p\, dG}\right] \end{aligned}$$

is a minimum with respect to y when

$$(7) \qquad y = \frac{\int p\lambda\, dG}{\int p\, dG},$$

the minimum value of $I(x)$ being

$$(8) \qquad I_G(x) = \int p(x \mid \lambda)\lambda^2 dG(\lambda) - \frac{[\int p(x \mid \lambda)\lambda\, dG(\lambda)]^2}{\int p(x \mid \lambda)\, dG(\lambda)}.$$

Research supported by the United States Air Force through the Office of Scientific Research of the Air Research and Development Command, and by the Office of Ordnance Research, U.S. Army, under Contract DA-04-200-ORD-355.

Hence the *Bayes estimator* of Λ corresponding to the *a priori* distribution function G of Λ [in the sense of minimizing the expression (4)] is the random variable $\varphi_G(X)$ defined by the function

$$\varphi_G(x) = \frac{\int p(x|\lambda)\,\lambda dG(\lambda)}{\int p(x|\lambda)\,dG(\lambda)}, \tag{9}$$

the corresponding minimum value of (4) being

$$E[\varphi_G(X) - \Lambda]^2 = \sum_x I_G(x). \tag{10}$$

The expression (9) is, of course, the expected value of the *a posteriori* distribution of Λ given $X = x$.

If the *a priori* distribution function G is known to the experimenter then φ_G defined by (9) is a computable function, but if G is unknown, as is usually the case, then φ_G is not computable. This trouble is not eliminated by the adoption of arbitrary rules prescribing forms for G (as is done, for example, by H. Jeffreys [1] in his theory of statistical inference). It is partly for this reason—that even when G may be assumed to exist it is generally unknown to the experimenter—that various other criteria for estimators (unbiasedness, minimax, etc.) have been proposed which have the advantage of not requiring a knowledge of G.

Suppose now that the problem of estimating Λ from an observed value of X is going to occur repeatedly with a fixed and known $p(x|\lambda)$ and a fixed but unknown $G(\lambda)$, and let

$$(\Lambda_1, X_1), (\Lambda_2, X_2), \cdots, (\Lambda_n, X_n), \cdots \tag{11}$$

denote the sequence so generated. [The Λ_n are independent random variables with common distribution function G, and the distribution of X_n depends only on Λ_n and for $\Lambda_n = \lambda$ is given by $p(x|\lambda)$.] If we want to estimate an unknown Λ_n from an observed X_n and if the previous values $\Lambda_1, \cdots, \Lambda_{n-1}$ are by now known, then we can form the empirical distribution function of the random variable Λ,

$$G_{n-1}(\lambda) = \frac{\text{number of terms } \Lambda_1, \cdots, \Lambda_{n-1} \text{ which are} \leqq \lambda}{n-1}, \tag{12}$$

and take as our estimate of Λ_n the quantity $\psi_n(X_n)$, where by definition

$$\psi_n(x) = \frac{\int p(x|\lambda)\,\lambda dG_{n-1}(\lambda)}{\int p(x|\lambda)\,dG_{n-1}(\lambda)}, \tag{13}$$

which is obtained from (9) by replacing the unknown *a priori* $G(\lambda)$ by the empirical $G_{n-1}(\lambda)$. Since $G_{n-1}(\lambda) \to G(\lambda)$ with probability 1 as $n \to \infty$, the ratio (13) will, under suitable regularity conditions on the kernel $p(x|\lambda)$, tend for any fixed x to the Bayes function $\varphi_G(x)$ defined by (9) and hence, again under suitable conditions, the expected squared deviation of $\psi_n(X_n)$ from Λ_n will tend to the Bayes value (10).

In practice, of course, it will be unusual for the previous values $\Lambda_1, \cdots, \Lambda_{n-1}$ to be known, and hence the function (13) will be no more computable than the true Bayes function (9). *However, in many cases the previous values $X_1, \cdots, X_{n-1}$ will be available to the experimenter at the moment when Λ_n is to be estimated*, and the question then arises whether it is possible to infer from the set of values $X_1, \cdots, X_n$ the approximate form of the unknown G, or at least, in the present case of quadratic estimation, to approximate the value of the functional of G defined by (9). To this end we observe that for any fixed x the empirical frequency

$$(14) \qquad p_n(x) = \frac{\text{number of terms } X_1, \cdots, X_n \text{ which equal } x}{n}$$

tends with probability 1 as $n \to \infty$ to the function $p_G(x)$ defined by (3), no matter what the *a priori* distribution function G. Hence there arises the following mathematical problem: from an approximate value (14) of the integral (3), where $p(x \mid \lambda)$ is a known kernel, to obtain an approximation to the unknown distribution function G, or at least, in the present case, to the value of the Bayes function (9) which depends on G. (This problem was posed in [4].) The possibility of doing this will depend on the nature of the kernel $p(x \mid \lambda)$ and on the class, say $\mathcal{G}$, to which the unknown G is assumed to belong. In order to fix the ideas we shall consider several special cases, the first being that of the *Poisson* kernel

$$(15) \qquad p(x \mid \lambda) = e^{-\lambda} \frac{\lambda^x}{x!}; \qquad x = 0, 1, \cdots; \lambda > 0;$$

$\mathcal{G}$ being the class of all distribution functions on the positive real axis.

In this case we have

$$(16) \qquad p_G(x) = \int p(x \mid \lambda)\, dG(\lambda) = \int_0^\infty e^{-\lambda} \lambda^x dG(\lambda) / x!$$

and

$$(17) \qquad \varphi_G(x) = \frac{\int_0^\infty e^{-\lambda} \lambda^{x+1} dG(\lambda)}{\int_0^\infty e^{-\lambda} \lambda^x dG(\lambda)},$$

and we can write the fundamental relation

$$(18) \qquad \varphi_G(x) = (x+1) \cdot \frac{p_G(x+1)}{p_G(x)}.$$

If we now define the function

$$(19) \quad \varphi_n(x) = (x+1) \frac{p_n(x+1)}{p_n(x)} = (x+1) \cdot \frac{\begin{array}{c}\text{number of terms } X_1, \cdots, X_n \\ \text{which are equal to } x+1\end{array}}{\begin{array}{c}\text{number of terms } X_1, \cdots, X_n \\ \text{which are equal to } x\end{array}}$$

then no matter what the unknown G we shall have for any fixed x

$$(20) \qquad \varphi_n(x) \to \varphi_G(x) \text{ with probability 1 as } n \to \infty.$$

This suggests using as an estimate of the unknown Λ_n in the sequence (11) the computable quantity

$$(21) \qquad \varphi_n(X_n),$$

in the hope that as $n \to \infty$,

$$(22) \qquad E[\varphi_n(X_n) - \Lambda_n]^2 \to E[\varphi_G(X) - \Lambda]^2.$$

We shall not investigate here the question of whether (22) does actually hold for the particular function (19) or whether (19) represents the best possible choice for minimizing in some sense the expected squared deviation. (See [8].)

It is of interest to compute the value of (10) for various *a priori* distribution functions G in order to compare its value with the expected squared deviation of the usual (maximum likelihood, minimum variance unbiased) Poisson estimator, X itself, for which

$$E(X-\Lambda)^2 = E\Lambda = \int_0^\infty \lambda dG(\lambda). \tag{23}$$

Suppose, for example, that G is a gamma type distribution function with density

$$G'(\lambda) = C\lambda^{b-1}e^{-h\lambda}; \quad \lambda, b, h > 0; C = h^b/\Gamma(b). \tag{24}$$

By elementary computation we find that

$$E\Lambda = \frac{b}{h}, \quad \text{Var}\,\Lambda = \frac{b}{h^2} \tag{25}$$

and

$$\varphi_G(x) = \frac{x+b}{1+h}, \quad E[\varphi_G(X) - \Lambda]^2 = \frac{b}{h(1+h)}; \tag{26}$$

hence

$$\frac{E[\varphi_G(X) - \Lambda]^2}{E(X-\Lambda)^2} = \frac{1}{1+h}. \tag{27}$$

For example, if $b = 100$, $h = 10$ then

$$E\Lambda = 10, \text{Var}\,\Lambda = 1, \varphi_G(x) = \frac{x+100}{11}, \frac{E[\varphi_G(X) - \Lambda]^2}{E(X-\Lambda)^2} = \frac{1}{11}. \tag{28}$$

An even simpler case occurs when Λ has all its probability concentrated at a single value λ. In this case, of course, the Bayes function is

$$\varphi_G(x) = \lambda, \tag{29}$$

not involving x at all, and

$$E[\varphi_G(X) - \Lambda]^2 = 0, \tag{30}$$

while as before

$$E(X-\Lambda)^2 = E\Lambda = \lambda. \tag{31}$$

Here the sequence (11) consists of observations $X_1, \cdots, X_n, \cdots$ from the same Poisson population (although this fact may not be apparent to the experimenter at the beginning); the traditional estimator $\varphi(x) = x$ does not take advantage of this favorable circumstance and continues to have the expected squared deviation λ after any number n of trials.

As a second example we take the *geometric* kernel

$$p(x|\lambda) = (1-\lambda)\lambda^x; \quad x = 0, 1, \cdots; 0 < \lambda < 1; \tag{32}$$

for which

$$p_G(x) = \int_0^1 (1-\lambda)\lambda^x dG(\lambda), \varphi_G(x) \tag{33}$$

$$= \frac{\int_0^1 (1-\lambda)\lambda^{x+1} dG(\lambda)}{\int_0^1 (1-\lambda)\lambda^x dG(\lambda)} = \frac{p_G(x+1)}{p_G(x)}.$$

Here it is natural to estimate Λ_n by (21) with the definition

$$(34)\qquad \varphi_n(x) = \frac{\text{number of terms } X_1,\cdots,X_n \text{ which are equal to } x+1}{\text{number of terms } X_1,\cdots,X_n \text{ which are equal to } x}.$$

Our third example will be the *binomial* kernel

$$(35)\qquad p_r(x\mid\lambda) = \binom{r}{x}\lambda^x(1-\lambda)^{r-x}; \qquad x=0,1,\cdots,r;\ 0\leq\lambda\leq 1.$$

Here r is a fixed positive integer representing the number of trials, X the number of successes, and Λ the unknown probability of success in each trial. G may be taken as the class of all distribution functions on the interval (0, 1). In this case

$$(36)\qquad \begin{cases} p_{G,r}(x) = \int p_r(x\mid\lambda)\,dG(\lambda) = \binom{r}{x}\int_0^1 \lambda^x(1-\lambda)^{r-x}dG(\lambda), \\ \varphi_{G,r}(x) = \dfrac{\int_0^1 \lambda^{x+1}(1-\lambda)^{r-x}dG(\lambda)}{\int_0^1 \lambda^x(1-\lambda)^{r-x}dG(\lambda)}, \end{cases}$$

so that we can write the fundamental relation

$$(37)\qquad \varphi_{G,r}(x) = \frac{x+1}{r+1}\cdot\frac{p_{G,r+1}(x+1)}{p_{G,r}(x)}; \qquad x=0,1,\cdots,r.$$

Let

$$(38)\qquad p_{n,r}(x) = \frac{\text{number of terms } X_1,\cdots,X_n \text{ which are equal to } x}{n}$$

then $p_{n,r}(x)\to p_{G,r}(x)$ with probability 1 as $n\to\infty$. Now consider the sequence of random variables

$$(39)\qquad X_1', X_2', \cdots, X_n', \cdots$$

where X_n' denotes the number of successes in, say, the first $r-1$ out of the r trials which produced X_n successes, and let

$$(40)\qquad p_{n,r-1}(x) = \frac{\text{number of terms } X_1',\cdots,X_n' \text{ which are equal to } x}{n};$$

then $p_{n,r-1}(x)\to p_{G,r-1}(x)$ with probability 1 as $n\to\infty$. Thus if we set

$$(41)\qquad \varphi_{n,r}(x) = \frac{x+1}{r}\cdot\frac{p_{n,r}(x+1)}{p_{n,r-1}(x)},$$

then

$$(42)\qquad \varphi_{n,r}(x)\to\frac{x+1}{r}\cdot\frac{p_{G,r}(x+1)}{p_{G,r-1}(x)} = \varphi_{G,r-1}(x)$$

with probability 1 as $n\to\infty$. If we take as our estimate of Λ_n the value

$$(43)\qquad \varphi_{n,r}(X_n')$$

then for large n we will do about as well as if we knew the *a priori* G but confined ourselves to the first $r-1$ out of each set of r trials. For large r this does not sacrifice much information, but it is by no means clear that (43) is the "best" estimator of Λ_n that could be devised in the spirit of our discussion.

As a final example consider any kernel of the "Laplacian" type

$$p(x \mid \lambda) = e^{\lambda x} f(x)\, h(\lambda). \tag{44}$$

We have

$$\begin{aligned} p_G(x) &= f(x) \int e^{\lambda x} h(\lambda)\, dG(\lambda), \\ \varphi_G(x)\, p_G(x) &= f(x) \int \lambda e^{\lambda x} h(\lambda)\, dG(\lambda) = f(x) \frac{d}{dx} \left\{ \frac{p_G(x)}{f(x)} \right\}, \end{aligned} \tag{45}$$

provided the differentiation under the integral sign is justified. Hence

$$\varphi_G(x) = \frac{d}{dx} \log \left\{ \frac{p_G(x)}{f(x)} \right\}. \tag{46}$$

Perhaps a satisfactory approximation to $\varphi_G(x)$ might be obtained by replacing $p_G(x)$ in (46) by a smoothed interpolation based on $p_n(x)$. The kernel (44) has been considered by M. C. K. Tweedie [6] and I am indebted to him for this example.

Until now we have been concerned only with approximating to the Bayes function $\varphi_G(x)$ defined by (9). In many cases we shall want an approximation to some other functional of the unknown *a priori* distribution function G; in particular to G itself. We shall make a few remarks about this problem in the general case in which X is not restricted to discrete values but has a distribution function

$$F(x \mid \lambda) = Pr\,[X \leq x \mid \Lambda = \lambda] \tag{47}$$

depending on the random variable Λ whose distribution function G is unknown. The unconditional distribution function of X is then

$$F_G(x) = Pr\,[X \leq x] = \int F(x \mid \lambda)\, dG(\lambda), \tag{48}$$

and there is assumed to be available an infinite sequence $X_1, X_2, \cdots$ of independent random variables with the common distribution function F_G. The empirical distribution function

$$F_n(x) = \frac{\text{number of terms } X_1, \cdots, X_n \text{ which are} \leq x}{n} \tag{49}$$

is known to converge uniformly to $F_G(x)$ with probability 1 as $n \to \infty$.

Problem: to find in terms of $F_n(x)$ a distribution function $G_n(\lambda)$ which will converge as $n \to \infty$ to the unknown $G(\lambda)$.

Let $\mathcal{G}$ denote some class of distribution functions to which the unknown G is assumed to belong. ($\mathcal{G}$ might, for example, be the class of all distribution functions, or all those with total mass distributed on some fixed finite interval.) The correspondence

$$F_G(x) = \int F(x \mid \lambda)\, dG(\lambda) \tag{50}$$

maps $\mathcal{G}$ onto some class of distribution functions which we shall denote by $\mathcal{F}$. We shall assume that the kernel $F(x \mid \lambda)$ is such that this mapping is one-to-one. Now, since we know an approximation F_n to F_G, it would be natural to seek an approximation to G by solving the functional equation (50) for G with F_G replaced by F_n. Unfortunately, in general this will be impossible since F_n will not belong to the class $\mathcal{F}$. [For example, if $F(x \mid \lambda)$ is continuous in x then all elements of $\mathcal{F}$ will be continuous, whereas F_n is a step function.] However, we may proceed as follows. Let F_n^* be any element of $\mathcal{F}$ whose distance (in the sense of maximum absolute value of the difference for all x) from F_n is within $\epsilon_n \to 0$ of the minimum distance of F_n from $\mathcal{G}$ (this is the "minimum distance" method of Wolfowitz), and let G_n be defined by the relation

$$F_n^*(x) = \int F(x \mid \lambda)\, dG_n(\lambda). \tag{51}$$

Then $F_n^* \to F_G$ in the maximum difference metric, and under suitable conditions on the kernel $F(x \mid \lambda)$ it will follow that $G_n \to G$. We shall go into this question in more detail elsewhere, but at least it indicates one possible way of obtaining an empirical approximation to a "mixing" distribution G from observations on the "mixed" distribution F_G. (See [5] also.) This problem, special cases of which have occurred several times in the statistical literature (see, for example, [2], [4], [7] and pp. 84–102 in [3]), awaits a satisfactory solution and seems to be of considerable importance.

I should like to express my appreciation to A. Dvoretzky, J. Neyman, and H. Raiffa for helpful discussions and suggestions.

REFERENCES

[1] H. Jeffreys, *Theory of Probability*, 2d ed., Oxford, Clarendon Press, 1948.

[2] P. F. Lazarsfeld, "A conceptual introduction to latent structure analysis," *Mathematical Thinking in the Social Sciences*, Glencoe, Ill., Free Press, 1954, pp. 349–387.

[3] J. Neyman, *Lectures and Conferences on Mathematical Statistics and Probability*, 2d ed., Washington, U.S. Department of Agriculture Graduate School, 1952.

[4] H. Robbins, "Asymptotically subminimax solutions of compound statistical decision problems," *Proceedings of the Second Berkeley Symposium on Mathematical Statistics and Probability*, Berkeley and Los Angeles, University of California Press, 1951, pp. 131–148.

[5] ———, "A generalization of the method of maximum likelihood: estimating a mixing distribution," (abstract), *Annals of Math. Stat.*, Vol. 21 (1950), pp. 314–315.

[6] M. C. K. Tweedie, "Functions of a statistical variate with given means, with special reference to Laplacian distributions," *Proc. Camb. Phil. Soc.*, Vol. 43 (1947), pp. 41–49.

[7] R. von Mises, "On the current use of Bayes' formula," *Annals of Math. Stat.*, Vol. 13 (1942), pp. 156–165.

[8] M. V. Johns, Jr., "Contributions to the theory of empirical Bayes procedures in statistics," doctoral thesis at Columbia University (unpublished), 1956.

Reprinted from
Proc. Third Berkeley Symposium Math. Statist. Prob.
1, 157–163 (1956)

Reprinted from THE ANNALS OF MATHEMATICAL STATISTICS
Vol. 35, No. 1, March, 1964

THE EMPIRICAL BAYES APPROACH TO STATISTICAL DECISION PROBLEMS[1]

BY HERBERT ROBBINS

Columbia University

1. Introduction. The empirical Bayes approach to statistical decision problems is applicable when the same decision problem presents itself repeatedly and independently with a fixed but unknown *a priori* distribution of the parameter. Not all decision problems in practice come to us imbedded in such a sequence, but when they do the empirical Bayes approach offers certain advantages over any approach which ignores the fact that the parameter is itself a random variable, as well as over any approach which assumes a personal or a conventional distribution of the parameter not subject to change with experience. My own interest in the empirical Bayes approach was renewed by recent work of E. Samuel [10], [11] and J. Neyman [6], to both of whom I am very much indebted. In keeping with the purpose of the Rietz Lecture I shall not confine myself to presenting new results and shall try to make the argument explicit at the risk of being tedious. In the current controversy between the Bayesian school and their opponents it is obvious that any theory of statistical inference will find itself in and out of fashion as the winds of doctrine blow. Here, then, are some remarks and references for further reading which I hope will interest my audience in thinking the matter through for themselves. Considerations of space have confined mention of the non-parametric case, and of the closely related "compound" approach in which no *a priori* distribution of the parameter is assumed, to the references at the end of the article.

2. The empirical Bayes decision problem. We begin by stating the kind of statistical decision problem with which we shall be concerned. This comprises

(a) A *parameter space* Λ with generic element λ. λ is the "state of nature" which is unknown to us.

(b) An *action space* A with generic element a.

(c) A *loss function* $L(a, \lambda) \geqq 0$ representing the loss we incur in taking action a when the parameter is λ.

(d) An *a priori distribution* G of λ on Λ. G may or may not be known to us.

(e) An *observable random variable* x belonging to a space X on which a σ-finite measure μ is defined. When the parameter is λ, x has a specified probability density f_λ with respect to μ.

The problem is to choose a *decision function* t, defined on X and with values in

Received 22 October 1963.

[1] Presented as the Rietz Lecture at the 26th Annual Meeting of the Institute of Mathematical Statistics, August 29, 1963, at Ottawa, Canada. Research sponsored by National Science Foundation Grant NSF-GP-316 at Columbia University.

A, such that when we observe x we shall take the action $t(x)$ and thereby incur the loss $L(t(x), \lambda)$. For any t the expected loss when λ is the parameter is

$$(1) \qquad R(t, \lambda) = \int_X L(t(x), \lambda) f_\lambda(x) \, d\mu(x),$$

and hence the overall expected loss when the *a priori* distribution of λ is G is

$$(2) \qquad R(t, G) = \int_\Lambda R(t, \lambda) \, dG(\lambda),$$

called the *Bayes risk* of t relative to G. We can write

$$(3) \qquad R(t, G) = \int_X \phi_G(t(x), x) \, d\mu(x),$$

where we have set

$$(4) \qquad \phi_G(a, x) = \int_\Lambda L(a, \lambda) f_\lambda(x) \, dG(\lambda).$$

To avoid needless complication we shall assume that there exists a decision function (d.f.) t_G such that for a.e. (μ) x,

$$(5) \qquad \phi_G(t_G(x), x) = \min_a \phi_G(a, x).$$

Then for any d.f. t

$$(6) \qquad R(t_G, G) = \int_X \min_a \phi_G(a, x) \, d\mu(x) \leq R(t, G),$$

so that, defining

$$(7) \qquad R(G) = R(t_G, G) = \int_X \phi_G(t_G(x), x) \, d\mu(x),$$

we have

$$(8) \qquad R(G) = \min_t R(t, G).$$

Any d.f. t_G satisfying (5) minimizes the Bayes risk relative to G, and is called a *Bayes d.f.* relative to G. The functional R defined by (7) is called the *Bayes envelope functional* of G. When G is known we can use t_G and thereby incur the minimum possible Bayes risk $R(G)$.

There remains the problem of what to do when G is not known. To extreme Bayesians and extreme non-Bayesians this question will be empty, since to the former G will always be known, by introspection or otherwise, and to the latter G will not even exist. We shall, however, take the position that G exists, so that $R(t, G)$ is an appropriate criterion for the performance of any d.f. t, but that G is not known, and therefore t_G is not directly available to us.

Suppose now that the decision problem just described occurs repeatedly and independently, with the same unknown G throughout (for examples see [6]). Thus let

$$(9) \qquad (\lambda_1, x_1), \quad (\lambda_2, x_2), \quad \cdots$$

be a sequence of pairs of random variables, each pair being independent of all the other pairs, the λ_n having a common *a priori* distribution G on Λ, and the conditional distribution of x_n given that $\lambda_n = \lambda$ being specified by the probability density f_λ. At the time when the decision about λ_{n+1} is to be made we have observed $x_1, \cdots, x_{n+1}$ (the values $\lambda_1, \lambda_2, \cdots$ remaining always unknown). We can therefore use for the decision about λ_{n+1} a function of x_{n+1} *whose form depends upon* $x_1, \cdots, x_n$; i.e. a function

$$t_n(\cdot) = t_n(x_1, \cdots, x_n; \cdot), \tag{10}$$

so that we shall take the action $t_n(x_{n+1}) \varepsilon A$ and thereby incur the loss $L(t_n(x_{n+1}), \lambda_{n+1})$. Our reason for doing this instead of using a fixed d.f. $t(\cdot)$ for each n (as might seem reasonable, since the successive problems are independent and have the same structure) is that we hope for large n to be able to extract some information about G from the values $x_1, \cdots, x_n$ which have been observed, hopefully in such a way that $t_n(\cdot)$ will be close to the optimal but unknown $t_G(\cdot)$ which we would use throughout if we knew G.

We therefore define an "empirical" or "adaptive" *decision procedure* to be a sequence $T = \{t_n\}$ of functions of the form (10) with values in A. For a given T, the expected loss on the decision about λ_{n+1}, *given* $x_1, \cdots, x_n$, will be (cf. (3))

$$\int_X \phi_G(t_n(x), x)\, d\mu(x), \tag{11}$$

and hence the overall expected loss will be

$$R_n(T, G) = \int_X E\phi_G(t_n(x), x)\, d\mu(x), \tag{12}$$

where E denotes expectation with respect to the n independent random variables $x_1, \cdots, x_n$ which have the common density with respect to μ on X given by

$$f_G(x) = \int_\Lambda f_\lambda(x)\, dG(\lambda), \tag{13}$$

the symbol x in (12) playing the role of a dummy variable of integration and not a random variable. From (5) and (12) it follows that always

$$R_n(T, G) \geqq R(G). \tag{14}$$

DEFINITION. If

$$\lim_{n\to\infty} R_n(T, G) = R(G) \tag{15}$$

we say that T is *asymptotically optimal* (*a.o.*) relative to G.

We now ask whether we can find a T which is in some sense "as good as possible" for large n relative to some class $\mathcal{G}$ of *a priori* distributions which we are willing to assume contains the true G. In particular, *can we find a T which is a.o. relative to every G in* $\mathcal{G}$? ($\mathcal{G}$ may be the class of all possible distributions on Λ.)

3. Some generalities on asymptotic optimality. Comparing (2.7) and (2.12) we see by Lebesgue's theorem on dominated convergence that for $T = \{t_n\}$ to be a.o. relative to a G it suffices that

$$\text{(A)} \qquad \lim_{n\to\infty} E\phi_G(t_n(x), x) = \phi_G(t_G(x), x) \qquad \text{(a.e. } (\mu)\ x),$$

and

$$\text{(B)} \quad E\phi_G(t_n(x), x) \leqq H(x) (\text{all } n), \qquad \text{where} \int_X H(x)\, d\mu(x) < \infty.$$

The main problem is (A); we shall summarily dispose of (B) by assuming that

$$\text{(C)} \qquad \int_\Lambda L(\lambda)\, dG(\lambda) < \infty,$$

where we have set

$$\text{(1)} \qquad 0 \leqq L(\lambda) = \sup_a L(a, \lambda) \leqq \infty.$$

Then setting

$$\text{(2)} \qquad H(x) = \int_\Lambda L(\lambda) f_\lambda(x)\, dG(\lambda) \geqq 0,$$

we have by (2.4) for any T,

$$\text{(3)} \qquad \phi_G(t_n(x), x) \leqq H(x) \qquad (\text{all } n),$$

and by (C),

$$\text{(4)} \quad \int_X H(x)\, d\mu(x) = \int_\Lambda L(\lambda) \int_X f_\lambda(x)\, d\mu(x)\, dG(\lambda) = \int_\Lambda L(\lambda)\, dG(\lambda) < \infty,$$

and (3) and (4) imply that (B) holds. Moreover, from (4) it follows that

$$\text{(5)} \qquad H(x) < \infty \qquad \text{(a.e. } (\mu)\ x)$$

and hence to prove that (A) holds it will suffice to prove that

$$\text{(D)} \qquad \text{p} \lim_{n\to\infty} \phi_G(t_n(x), x) = \phi_G(t_G(x), x) \qquad \text{(a.e. } (\mu)\ x),$$

where by p lim we mean limit in probability. Hence (C) and (D) suffice to ensure that T is a.o. relative to G.

Let a_0 be an arbitrary fixed element of A and define

$$\text{(6)} \qquad \Delta_G(a, x) = \int_\Lambda [L(a, \lambda) - L(a_0, \lambda)] f_\lambda(x)\, dG(\lambda)$$

and

$$\text{(7)} \qquad L_0(x) = \int_\Lambda L(a_0, \lambda) f_\lambda(x)\, dG(\lambda),$$

so that under (C) we have, for a.e. (μ) x,

$$\phi_G(a, x) = L_0(x) + \Delta_G(a, x). \tag{8}$$

Suppose we can find a sequence of functions

$$\Delta_n(a, x) = \Delta_n(x_1, \cdots, x_n; a, x) \tag{9}$$

such that for a.e. (μ) x,

$$\text{p}\lim_{n\to\infty} \sup_a |\Delta_n(a, x) - \Delta_G(a, x)| = 0. \tag{10}$$

Let ϵ_n be any sequence of constants tending to 0 and set (subject to measurability conditions)

$$t_n(x) = t_n(x_1, \cdots, x_n; x) = \text{any element } \bar{a} \varepsilon A \tag{11}$$

such that $\Delta_n(\bar{a}, x) \leqq \inf_a \Delta_n(a, x) + \epsilon_n$. Then by (2.5) and (8),

$$\begin{aligned} 0 &\leqq \Delta_G(t_n(x), x) - \Delta_G(t_G(x), x) \\ &= [\Delta_G(t_n(x), x) - \Delta_n(t_n(x), x)] + [\Delta_n(t_n(x), x) - \Delta_n(t_G(x), x)] \\ &\qquad + [\Delta_n(t_G(x), x) - \Delta_G(t_G(x), x)]. \end{aligned} \tag{12}$$

Given any $\epsilon > 0$ we have by (10) that for large n with probability as near 1 as we please the right hand side of (12) will be $\leqq \epsilon + \epsilon_n + \epsilon$; thus

$$\text{p}\lim_{n\to\infty} \Delta_G(t_n(x), x) = \Delta_G(t_G(x), x) \qquad (\text{a.e. } (\mu)\ x), \tag{13}$$

which by (8) implies (D). We have therefore proved

THEOREM 1. *Let G be such that* (C) *holds, let $\Delta_n(a, x)$ be a sequence of functions of the form* (9) *and such that* (10) *holds, and define $T = \{t_n\}$ by* (11). *Then T is a.o. relative to G.*

When A is *finite* this yields

COROLLARY 1. *Let $A = \{a_0, \cdots, a_m\}$ be a finite set, let G be such that*

$$\int_\Lambda L(a_i, \lambda)\, dG(\lambda) < \infty \qquad (i = 0, \cdots, m), \tag{14}$$

and let $\Delta_{i,n}(x) = \Delta_{i,n}(x_1, \cdots, x_n; x)$ for $i = 1, \cdots, m$ and $n = 1, 2, \cdots$ be such that for a.e. (μ) x,

$$\text{p}\lim_{n\to\infty}\Delta_{i,n}(x) = \int_\Lambda [L(a_i, \lambda) - L(a_0, \lambda)] f_\lambda(x)\, dG(\lambda). \tag{15}$$

Set $\Delta_{0,n}(x) = 0$ and define

$$t_n(x) = a_k, \quad \text{where } k \text{ is any integer } 0 \leqq k \leqq m \text{ such that} \quad \Delta_{k,n}(x) = \min[0, \Delta_{1,n}(x), \cdots, \Delta_{m,n}(x)]. \tag{16}$$

Then $T = \{t_n\}$ is a.o. relative to G.

In the important case $m = 1$ (hypothesis testing) this becomes

COROLLARY 2. *Let* $A = \{a_0, a_1\}$, *let* G *be such that*

$$\int_\Lambda L(a_i, \lambda)\, dG(\lambda) < \infty \qquad (i = 0, 1), \tag{17}$$

and let $\Delta_n(x) = \Delta_n(x_1, \cdots, x_n; x)$ *be such that for a.e.* (μ) x,

$$\text{p}\lim_{n\to\infty} \Delta_n(x) = \Delta_G(x) = \int_\Lambda [L(a_1, \lambda) - L(a_0, \lambda)] f_\lambda(x)\, dG(\lambda). \tag{18}$$

Define

$$\begin{aligned} t_n(x) &= a_0, & \text{if } \Delta_n(x) \geqq 0, \\ &= a_1, & \text{if } \Delta_n(x) < 0. \end{aligned} \tag{19}$$

Then $T = \{t_n\}$ *is a.o. relative to* G.

We proceed to give an example in which a sequence $\Delta_n(x)$ satisfying (18) can be constructed.

4. The Poisson case. We consider the problem of testing a one-sided null hypothesis $H_0 : \lambda \leqq \lambda^*$ concerning the value of a Poisson parameter λ. Thus let $\Lambda = \{0 < \lambda < \infty\}$, $A = \{a_0, a_1\}$, where a_0 = "accept H_0", a_1 = "reject H_0" and

$$\begin{aligned} X &= \{0, 1, 2, \cdots\}, \qquad \mu = \text{counting measure on } X, \\ f_\lambda(x) &= e^{-\lambda}\lambda^x/x!. \end{aligned} \tag{1}$$

It remains to specify the loss functions $L(a_i, \lambda)$; we shall take them to be

$$\begin{aligned} L(a_0, \lambda) &= 0 & \text{if } \lambda \leqq \lambda^*, \\ &= \lambda - \lambda^* & \text{if } \lambda \geqq \lambda^*, \\ L(a_1, \lambda) &= \lambda^* - \lambda & \text{if } \lambda \leqq \lambda^*, \\ &= 0 & \text{if } \lambda \geqq \lambda^*. \end{aligned} \tag{2}$$

Thus (very conveniently)

$$L(a_1, \lambda) - L(a_0, \lambda) = \lambda^* - \lambda \qquad (0 < \lambda < \infty), \tag{3}$$

and

$$\Delta_G(x) = \int_\Lambda [L(a_1, \lambda) - L(a_0, \lambda)] f_\lambda(x)\, dG(\lambda) = \int_0^\infty (\lambda^* - \lambda) \frac{e^{-\lambda}\lambda^x}{x!}\, dG(\lambda). \tag{4}$$

Now by (2.13),

$$f_G(x) = P(x_j = x) = \int_0^\infty \frac{e^{-\lambda}\lambda^x}{x!}\, dG(\lambda), \tag{5}$$

so that we can write

$$\Delta_G(x) = \lambda^* f_G(x) - (x + 1) f_G(x + 1). \tag{6}$$

Define

$$\text{(7)}\qquad \begin{aligned}\delta(x, y) &= 1 \qquad \text{if } x = y,\\ &= 0 \qquad \text{if } x \neq y,\end{aligned}$$

and consider the expression

$$\text{(8)}\qquad u_n(x) = u_n(x_1, \cdots, x_n; x) = n^{-1} \sum_{j=1}^{n} \delta(x, x_j).$$

Noting that

$$\text{(9)}\qquad E\delta(x, x_j) = P(x_j = x) = f_G(x),$$

it follows from the law of large numbers that

$$\text{(10)}\qquad \text{p}\lim_{n\to\infty} u_n(x) = f_G(x) \qquad (x = 0, 1, \cdots),$$

and hence that, setting

$$\text{(11)}\qquad \Delta_n(x) = \lambda^* u_n(x) - (x+1)u_n(x+1),$$

we have for $x = 0, 1, \cdots$

$$\text{(12)}\qquad \text{p}\lim_{n\to\infty} \Delta_n(x) = \lambda^* f_G(x) - (x+1)f_G(x+1) = \Delta_G(x).$$

Setting

$$\text{(13)}\qquad \begin{aligned}t_n(x) &= a_0 \qquad \text{if } \lambda^* u_n(x) - (x+1)u_n(x+1) \geqq 0,\\ &= a_1 \qquad \text{otherwise,}\end{aligned}$$

it follows from Corollary 2 of Section 3 that T *is a.o. relative to every* G *such that*

$$\text{(14)}\qquad \int_0^\infty \lambda \, dG(\lambda) < \infty.$$

We remark that we could equally well have defined $u_n(x)$ to be

$$\text{(15)}\qquad u_n(x) = [1/(n+1)] \sum_{j=1}^{n+1} \delta(x, x_j),$$

since (10) holds as well for (15) as for (8). Using (15) the corresponding T would require us to take action a_0 on λ_n if and only if

$$\text{(16)}\qquad \lambda^* \geqq \frac{(x_n + 1)(\text{number of terms } x_1, \cdots, x_n \text{ equal to } x_n + 1)}{(\text{number of terms } x_1, \cdots, x_n \text{ equal to } x_n)}.$$

For the problem of point estimation with squared error loss it was shown by M. V. Johns, Jr. [2] that the right hand side of (16) is in fact an a.o. point estimator of λ_n for every G such that

$$\text{(17)}\qquad \int_0^\infty \lambda^2 \, dG(\lambda) < \infty.$$

The much easier result of the present section on hypothesis testing uses a loss structure (2) suggested by Johns in [4].

The relation (6) was basic to our construction (11) of a sequence $\Delta_n(x)$ satisfying (3.18); now, (6) is a special property of the Poisson distribution (1) and the loss structure (2), and therefore it might seem that the application of Corollary 2 to empirical Bayes hypothesis testing would be very limited. Such is not the case. The application of Corollary 2 to more general loss structures, and to many of the most common discrete and continuous parametric distributions of statistics is discussed in [4], [9], [11]. Instead of giving a review of these results here, we shall consider a case in which no asymptotically optimal T exists but the empirical Bayes approach is still useful.

5. An example in which asymptotic optimality does not exist but all is not lost. Let x be a random variable with only two values, 0 and 1, with respective probabilities $1 - \lambda$ and λ, the unknown parameter λ lying in the interval $\Lambda = \{0 \leqq \lambda \leqq 1\}$. On the basis of a single observation of x we want to estimate λ; if our estimate is $a \in A = \{0 \leqq a \leqq 1\}$ the loss will be taken to be $L(a, \lambda) = (\lambda - a)^2$. A d.f. t is determined by the two constants $t(0), t(1)$ which are at our disposal on the unit interval A; the expected loss in using t for a given λ is

$$\begin{aligned} R(t, \lambda) &= (1 - \lambda)(\lambda - t(0))^2 + \lambda(\lambda - t(1))^2 \\ &= t^2(0) + [t^2(1) - 2t(0) - t^2(0)]\lambda + [1 - 2t(1) + 2t(0)]\lambda^2. \end{aligned} \tag{1}$$

Consider the particular family of d.f.'s t_α defined for $0 < \alpha < 1$ by setting

$$t_\alpha(0) = \tfrac{1}{2}\alpha, \qquad t_\alpha(1) = \tfrac{1}{2}(1 + \alpha). \tag{2}$$

It is easily seen from (1) that

$$R(t_\alpha, \lambda) = \tfrac{1}{4}[\alpha^2 + (1 - 2\alpha)\lambda]. \tag{3}$$

For $\alpha = \frac{1}{2}$ we shall denote t_α by t^*, so that

$$t^*(0) = \tfrac{1}{4}, \qquad t^*(1) = \tfrac{3}{4}, \qquad R(t^*, \lambda) \equiv \tfrac{1}{16}. \tag{4}$$

For any *a priori* distribution G of λ let

$$\nu_i = \nu_i(G) = \int_0^1 \lambda^i \, dG(\lambda) \qquad (i = 1, 2). \tag{5}$$

Then from (1) it follows that for any d.f. t,

$$\begin{aligned} R(t, G) &= \int_0^1 R(t, \lambda)\, dG(\lambda) \\ &= t^2(0) + \nu_1[t^2(1) - 2t(0) - t^2(0)] + \nu_2[1 - 2t(1) + 2t(0)]. \end{aligned} \tag{6}$$

Apart from the trivial cases in which $\nu_1 = 0$ or 1 we have after some simple algebra the formula

$$\begin{aligned} R(t, G) &= (\nu_1 - \nu_2)(\nu_2 - \nu_1^2)/\nu_1(1 - \nu_1) \\ &\quad + (1 - \nu_1)[t(0) - (\nu_1 - \nu_2)/(1 - \nu_1)]^2 + \nu_1[t(1) - \nu_2/\nu_1]^2, \end{aligned} \tag{7}$$

from which it follows that for a given G, $R(t, G)$ is minimized uniquely by the Bayes d.f. t_G for which

$$(8)\qquad t_G(0) = (\nu_1 - \nu_2)/(1 - \nu_1), \qquad t_G(1) = \nu_2/\nu_1,$$

with

$$(9)\qquad R(G) = R(t_G, G) = (\nu_1 - \nu_2)(\nu_2 - \nu_1^2)/\nu_1(1 - \nu_1).$$

Each t_α (and in particular t^*) is a Bayes d.f.; it suffices to find a distribution of λ such that

$$(10)\qquad (\nu_1 - \nu_2)/(1 - \nu_1) = \tfrac{1}{2}\alpha, \qquad \nu_2/\nu_1 = (1 + \alpha)/\alpha,$$

and this is provided e.g. by the distribution G_α with the density

$$(11)\qquad [B(\alpha, 1 - \alpha)]^{-1}\lambda^{\alpha-1}(1 - \lambda)^{(1-\alpha)-1}$$

for which

$$(12)\qquad \nu_1 = \alpha, \qquad \nu_2 = \tfrac{1}{2}\alpha(1 + \alpha), \qquad R(G_\alpha) = \tfrac{1}{4}\alpha(1 - \alpha).$$

The fact that t^* is the Bayes d.f. relative to $G_{\frac{1}{2}}$ and that for any G,

$$(13)\qquad R(t^*, G) = \int_0^1 R(t^*, \lambda)\, dG(\lambda) = \tfrac{1}{16}$$

has the important consequence that

$$(14)\qquad \sup_G R(t, G) > \tfrac{1}{16} \qquad \text{for every} \quad t \neq t^*.$$

For if for some t', $\sup_G R(t', G) \leqq \frac{1}{16}$, then in particular

$$(15)\qquad \tfrac{1}{16} = R(t^*, G_{\frac{1}{2}}) \leqq R(t', G_{\frac{1}{2}}) \leqq \tfrac{1}{16},$$

so that

$$(16)\qquad R(t', G_{\frac{1}{2}}) = R(t^*, G_{\frac{1}{2}}) = \tfrac{1}{16}$$

and therefore $t' = t^*$. Thus t^* is the unique "minimax" d.f. in the sense that *it minimizes the maximum Bayes risk relative to the class of all a priori distributions* G. When nothing is known about G it is therefore not unreasonable (the avoidance of a direct endorsement of any decision function is customary in the literature) to use t^*; the Bayes risk will then be $\frac{1}{16}$ irrespective of G, while for any $t \neq t^*$ the Bayes risk will be $> \frac{1}{16}$ for some G (in particular for any G with $\nu_1 = \frac{1}{2}$, $\nu_2 = \frac{3}{8}$; e.g. $G_{\frac{1}{2}}$).

For any $0 < \alpha < 1$ let $\mathcal{G}_\alpha$ denote the class of all G such that $\nu_1(G) = \alpha$. For any G in $\mathcal{G}_\alpha$ (in particular for G_α) we see from (2) and (7) after a little algebra that

$$(17)\qquad R(t_\alpha, G) = \tfrac{1}{4}\alpha(1 - \alpha) \qquad (G \in \mathcal{G}_\alpha)$$

irrespective of the value of $\nu_2(G)$. It therefore follows as above that

$$(18)\qquad \sup_{G \in \mathcal{G}_\alpha} R(t, G) > \tfrac{1}{4}\alpha(1 - \alpha) \qquad \text{for every} \quad t \neq t_\alpha,$$

so that relative to the class $\mathcal{G}_\alpha$, t_α is the unique minimax d.f. in the sense that *it minimizes the maximum Bayes risk relative to the class* $\mathcal{G}_\alpha$. If nothing is known about G except that $\nu_1(G) = \alpha$ it is therefore not unreasonable to use t_α; the Bayes risk will then be $\frac{1}{4}\alpha(1 - \alpha)$, while for any other d.f. the Bayes risk will be $>\frac{1}{4}\alpha(1 - \alpha)$ for some G in $\mathcal{G}_\alpha$ (in particular for any G with $\nu_1 = \alpha$, $\nu_2 = \frac{1}{2}\alpha(1 + \alpha)$; e.g. G_α).

It follows from the above (or can be verified directly) that

$$(\nu_1 - \nu_2)(\nu_2 - \nu_1^2)/\nu_1(1 - \nu_1) \leqq \tfrac{1}{4}\nu_1(1 - \nu_1) \leqq \tfrac{1}{16}; \tag{19}$$

equality holding respectively when $\nu_2 = \frac{1}{2}\nu_1(1 + \nu_1)$ and when $\nu_1 = \frac{1}{2}$.

Suppose now that we confront this decision problem repeatedly with an unknown G. The sequence $x_1, x_2, \cdots$ is an independent and identically distributed sequence of 0's and 1's with

$$P(x_i = 1) = \int_0^1 \lambda \, dG(\lambda) = \nu_1(G), \qquad P(x_i = 0) = 1 - \nu_1(G); \tag{20}$$

thus the distribution of the x_i depends only on $\nu_1(G)$. Since t_G, defined by (8), involves $\nu_2(G)$ as well, it follows that no T can be a.o. relative to every G in a class $\mathcal{G}$ unless ν_2 is a function of ν_1 in $\mathcal{G}$, which is not likely to be the case in practice.

On the other hand, let

$$u_n = (1/n) \sum_{i=1}^n x_i, \tag{21}$$

and consider the decision procedure $\tilde{T} = \{t_n\}$ with

$$t_n(0) = \tfrac{1}{2}u_n, \qquad t_n(1) = \tfrac{1}{2}(1 + u_n). \tag{22}$$

For any G in $\mathcal{G}_\alpha$, by the law of large numbers, $u_n \to \alpha$ and hence $t_n \to t_\alpha$ with probability 1 as $n \to \infty$. In fact, since

$$Ex_i = \alpha = Ex_i^2, \qquad \operatorname{Var} x_i = \alpha(1 - \alpha), \tag{23}$$

it follows that

$$Eu_n = \alpha, \qquad Eu_n^2 = \operatorname{Var} u_n + \alpha^2 = \alpha(1 - \alpha)/n + \alpha^2 \tag{24}$$

and hence from (6) that

$$\begin{aligned} &R_n(\tilde{T}, G) \\ &= E[\tfrac{1}{4}u_n^2 + \alpha\{\tfrac{1}{4}(1 + 2u_n + u_n^2) - u_n - \tfrac{1}{4}u_n^2\} + \nu_2(1 - 1 - u_n + u_n)] \\ &= \tfrac{1}{4}E[u_n^2 - 2u_n + \alpha] = \tfrac{1}{4}\alpha(1 - \alpha)[1 + 1/n] = R(t_\alpha, G)[1 + 1/n]. \end{aligned} \tag{25}$$

Thus for large n we will do almost as well by using $\tilde{T}$ as we could do if we knew $\nu_1(G) = \alpha$ and used t_α. We have in fact for $G \in \mathcal{G}_\alpha$,

$$R_n(\tilde{T}, G) - R(t_\alpha, G) = \alpha(1 - \alpha)/4n \leqq 1/16n, \tag{26}$$

while

$$R_n(\tilde{T}, G) - R(t^*, G) = \alpha(1-\alpha)(n+1)/4n - \tfrac{1}{16} \tag{27}$$
$$= -[\tfrac{1}{4}(1-2\alpha)]^2 + \alpha(1-\alpha)/4n.$$

This example illustrates the fact that even when an a.o. T does not exist, or when it does exist but $R_n(T, G)$ is too slowly convergent to $R(G)$, it may be worthwhile to use a T which is at least "asymptotically subminimax." R. Cogburn has work in progress in this direction for the case in which x has a general binomial distribution (for which see also [9] and [11]). The general problem of finding "good" T's has hardly been touched, and efforts along this line should yield interesting and useful results.

6. Estimating the *a priori* distribution: the general case. Returning to the general formulation of Sections 2 and 3 and confining ourselves for simplicity to the case $A = \{a_0, a_1\}$ and $\Lambda = \{-\infty < \lambda < \infty\}$, we recall that an a.o. T exists relative to the class $\mathcal{G}$ defined by (3.17) whenever we can find a sequence $\Delta_n(x) = \Delta_n(x_1, \cdots, x_n; x)$ such that for a.e. $(\mu)x$,

$$\text{(1)} \quad \text{p}\lim_{n\to\infty} \Delta_n(x) = \Delta_G(x) = \int_{-\infty}^{\infty} [L(a_1, \lambda) - L(a_0, \lambda)] f_\lambda(x)\, dG(\lambda)$$

for every G in $\mathcal{G}$. One way to construct such a sequence (not the one used in Section 4) is to find a sequence $G_n(\lambda) = G_n(x_1, \cdots x_n; \lambda)$ of random distribution functions in λ such that

$$\text{(2)} \quad P[\lim_{n\to\infty} G_n(\lambda) = G(\lambda) \text{ at every continuity point } \lambda \text{ of } G] = 1.$$

If we have such a sequence G_n of random estimators of G then we can set

$$\Delta_n(x) = \int_{-\infty}^{\infty} [L(a_1, \lambda) - L(a_0, \lambda)] f_\lambda(x)\, dG_n(\lambda); \tag{3}$$

if for a.e. (μ) fixed x the function

$$[L(a_1, \lambda) - L(a_0, \lambda)] f_\lambda(x) \tag{4}$$

is continuous and bounded in λ, then the Helly-Bray theorem guarantees (1).

We shall now describe one method—a special case of the "minimum distance" method of J. Wolfowitz—for constructing a particular sequence $G_n(\lambda)$ of random estimators of an unknown G, and then prove a theorem which ensures that under appropriate conditions on the family $f_\lambda(x)$ the relation (2) will hold for any G whatever.

In doing this we shall relax the condition that the distribution of x for given λ is given in terms of a density f_λ with respect to a measure μ, and instead assume only that for every $\lambda \varepsilon \Lambda = \{-\infty < \lambda < \infty\}$, $F_\lambda(x)$ is a specified distribution function in x, and for every fixed $x \varepsilon X = \{-\infty < x < \infty\}$, $F_\lambda(x)$ is a Borel measurable function of λ. (A function $F(x)$ defined for $-\infty < x < \infty$ will be called a distribution function if F is non-decreasing, continuous on the right, and $\lim_{x\to-\infty} F(x) = 0$, $\lim_{x\to\infty} F(x) = 1$.)

For any distribution function G of λ we define

$$(5) \qquad F_G(x) = \int_{-\infty}^{\infty} F_\lambda(x)\, dG(\lambda);$$

then F_G is a distribution function in x.

Let $x_1, x_2, \cdots$ be a sequence of independent random variables with F_G as their common distribution function, and define

$$(6) \quad B_n(x) = B_n(x_1, \cdots, x_n; x) = (\text{no. of terms } x_1, \cdots, x_n \text{ which are } \leqq x)/n.$$

For any two distribution functions F_1, F_2 define the distance

$$(7) \qquad \rho(F_1, F_2) = \sup_x |F_1(x) - F_2(x)|,$$

and let ϵ_n be any sequence of constants tending to 0.

Let $\mathcal{G}$ be any class of distribution functions in λ which contains G, and define

$$(8) \qquad d_n = \inf_{\bar{G}\in\mathcal{G}} \rho(B_n, F_{\bar{G}}).$$

Let $G_n(\lambda) = G_n(x_1, \cdots, x_n; \lambda)$ be any element of $\mathcal{G}$ such that

$$(9) \qquad \rho(B_n, F_{G_n}) \leqq d_n + \epsilon_n.$$

We say that the sequence G_n so defined is *effective* for $\mathcal{G}$ if (2) holds for every G in $\mathcal{G}$. We now state

THEOREM 2. *Assume that*

(A) *For every fixed x, $F_\lambda(x)$ is a continuous function of λ.*

(B) *The limits $F_{-\infty}(x) = \lim_{\lambda\to-\infty} F_\lambda(x)$, $F_\infty(x) = \lim_{\lambda\to\infty} F_\lambda(x)$ exist for every x.*

(C) *Neither $F_{-\infty}$ nor F_∞ is a distribution function.*

(D) *If G_1, G_2 are any two distribution functions in λ such that $F_{G_1} = F_{G_2}$, then $G_1 = G_2$.*

Then the sequence G_n defined by (9) is effective for the class $\mathcal{G}$ of all distribution functions in λ.

PROOF. By the Glivenko-Cantelli theorem,

$$(10) \qquad P[\lim_{n\to\infty} \rho(B_n, F_G) = 0] = 1.$$

Since

$$(11) \qquad \begin{aligned} \rho(F_{G_n}, F_G) &\leqq \rho(F_{G_n}, B_n) + \rho(B_n, F_G) \\ &\leqq d_n + \epsilon_n + \rho(B_n, F_G) \leqq \rho(B_n, F_G) + \epsilon_n + \rho(B_n, F_G), \end{aligned}$$

it follows from (10) that with probability 1 the sequence $x_1, x_2, \cdots$ is such that

$$(12) \qquad \lim_{n\to\infty} \int_{-\infty}^{\infty} F_\lambda(x)\, dG_n(\lambda) = \int_{-\infty}^{\infty} F_\lambda(x)\, dG(\lambda) \qquad \text{(uniformly in } x).$$

Now consider any fixed sequence $x_1, x_2, \cdots$ such that (12) holds and let G_{k_n} be any subsequence of G_n such that $G_{k_n}(\lambda) \to G^*(\lambda)$ at every continuity point λ of G^*, where G^* is a distribution function in the weak sense, $0 \leqq G^*(-\infty)$, $G^*(\infty) \leqq 1$. By a simple extension of the Helly-Bray theorem it follows from (A) and (B) that for every x,

$$\lim_{n\to\infty} \int_{-\infty}^{\infty} F_\lambda(x)\, dG_{k_n}(\lambda) = \int_{-\infty}^{\infty} F_\lambda(x)\, dG^*(\lambda) + G^*(-\infty)F_{-\infty}(x) + [1 - G^*(\infty)]F_\infty(x), \tag{13}$$

and hence from (12) that for every x,

$$\int_{-\infty}^{\infty} F_\lambda(x)\, dG(\lambda) = \int_{-\infty}^{\infty} F_\lambda(x)\, dG^*(\lambda) + G^*(-\infty)F_{-\infty}(x) + [1 - G^*(\infty)]F_\infty(x). \tag{14}$$

If we can show that $G^*(-\infty) = 0$ and $G^*(\infty) = 1$ then it will follow from (D) that $G = G^*$, and hence, since G^* denoted the weak limit of *any* convergent subsequence of G_n, that (2) holds. We shall complete the proof of the theorem by showing that (C) implies that $G^*(-\infty) = 0$ and $G^*(\infty) = 1$.

Since $F_{-\infty}$ is the limit as $\lambda \to -\infty$ of F_λ, it is a nondecreasing function of x such that

$$0 \leqq F_{-\infty}(-\infty), \qquad F_{-\infty}(\infty) \leqq 1, \tag{15}$$

and similarly for F_∞. Let $x \to -\infty$ in (14). By Lebesgue's theorem of bounded convergence,

$$0 = G^*(-\infty)F_{-\infty}(-\infty) + [1 - G^*(\infty)]F_\infty(-\infty). \tag{16}$$

Hence if $G^*(-\infty) \neq 1$ then $F_{-\infty}(-\infty) = 0$, and if $G^*(\infty) \neq 1$ then $F_\infty(-\infty) = 0$. Similarly, by letting $x \to \infty$ in (14) we see that if $G^*(-\infty) \neq 0$ then $F_{-\infty}(\infty) = 1$, and if $G^*(\infty) \neq 1$ then $F_\infty(\infty) = 1$. Suppose now that a_n is any sequence of constants converging to a limit a from the right. Then from (14), putting $x = a_n$, letting $n \to \infty$, and subtracting (14) for $x = a$, we see that

$$G^*(-\infty)[F_{-\infty}(a+0) - F_{-\infty}(a)] + [1 - G^*(\infty)][F_\infty(a+0) - F_\infty(a)] = 0. \tag{17}$$

Hence if $G^*(-\infty) \neq 0$ then $F_{-\infty}(a+0) = F_{-\infty}(a)$, and if $G^*(\infty) \neq 1$ then $F_\infty(a+0) = F_\infty(a)$. It follows that if $G^*(-\infty) \neq 0$ then $F_{-\infty}$ is a distribution function and if $G^*(\infty) \neq 1$ then F_∞ is a distribution function. Hence by (C), $G^*(-\infty) = 0$ and $G^*(\infty) = 1$. This completes the proof.

EXAMPLE 1. (location parameter) Let F be a continuous distribution function (e.g., the normal distribution function) with a characteristic function which never vanishes,

$$\phi_F(t) = \int_{-\infty}^{\infty} e^{ixt}\, dF(x) \neq 0 \qquad \text{(all } t\text{)}. \tag{18}$$

Set $F_\lambda(x) = F(x - \lambda)$; then (A), (B), (C) hold. If G_1, G_2 are any two distribution functions such that $F_{G_1} = F_{G_2}$; i.e., such that

$$\int_{-\infty}^{\infty} F(x-\lambda)\, dG_1(\lambda) = \int_{-\infty}^{\infty} F(x-\lambda)\, dG_2(\lambda) \qquad \text{(all } x\text{)}, \tag{19}$$

then

$$\phi_F(t)\phi_{G_1}(t) = \phi_F(t)\phi_{G_2}(t) \qquad \text{(all } t\text{)}, \tag{20}$$

and hence

$$\phi_{G_1}(t) = \phi_{G_2}(t) \qquad \text{(all } t\text{)}, \tag{21}$$

so that $G_1 = G_2$. Hence (D) holds, and by Theorem 1 the sequence G_n defined by (9) is effective for the class $\mathcal{G}$ of all distributions G of λ.

When the parameter space Λ is not the whole line, the statement and proof of Theorem 2 can be appropriately modified. As an example, suppose $\Lambda = \{0 \leqq \lambda < \infty\}$. Then we can prove in exactly the same way as for Theorem 2

THEOREM 3. *Assume that*

(A) *As in Theorem* 2.

(B) *The limit* $F_\infty(x) = \lim_{\lambda\to\infty} F_\lambda(x)$ *exists for every* x.

(C) F_∞ *is not a distribution function.*

(D) *If* G_1, G_2 *are any two distribution functions in* λ *which assign unit probability to* $\Lambda = \{0 \leqq \lambda < \infty\}$ *such that*

$$\int_\Lambda F_\lambda(x)\, dG_1(\lambda) = \int_\Lambda F_\lambda(x)\, dG_2(\lambda) \qquad \text{(all } x\text{)},$$

then $G_1 = G_2$.

Then the sequence G_n *defined by* (9) *is effective for the class* $\mathcal{G}$ *of all distributions which assign unit probability to* Λ.

EXAMPLE 2. (Poisson parameter) Let

$$\begin{aligned} F_0(x) &= 0 && \text{for } x < 0, \\ &= 1 && \text{for } x \geqq 0, \end{aligned} \tag{22}$$

and for $0 < \lambda < \infty$ let

$$F_\lambda(x) = \sum_{0 \leqq i \leqq x} e^{-\lambda}\lambda^i/i!. \tag{23}$$

Then (A), (B), and (C) hold.

Let $G \in \mathcal{G}$; then

$$F_G(x) = \int_\Lambda F_\lambda(x)\, dG(\lambda) = \sum_{0 \leqq i \leqq x} \int_\Lambda f_\lambda(i)\, dG(\lambda), \tag{24}$$

where

$$\begin{aligned} f_0(i) &= 1 && \text{for } i = 0, \\ &= 0 && \text{for } i = 1, 2, \cdots \\ f_\lambda(i) &= e^{-\lambda}\lambda^i/i! && \text{for } i = 0, 1, \cdots \text{ and } 0 < \lambda < \infty. \end{aligned} \tag{25}$$

Now

$$(26)\qquad \begin{aligned} F_G(0) &= \int_\Lambda f_\lambda(0)\, dG(\lambda) \\ F_G(n) - F_G(n-1) &= \int_\Lambda f_\lambda(n)\, dG(\lambda) \qquad (n = 1, 2, \cdots), \end{aligned}$$

so if $F_{G_1} = F_{G_2}$ then

$$(27)\qquad \int f_\lambda(n)\, dG_1(\lambda) = \int f_\lambda(n)\, dG_2(\lambda) \qquad (n = 0, 1, \cdots).$$

Define the set functions

$$(28)\qquad H_j(B) = \int_B e^{-\lambda}\, dG_j(\lambda) \Big/ \int_\Lambda e^{-\lambda}\, dG_j(\lambda) \qquad (j = 1, 2);$$

then H_j is a probability measure on the Borel sets. Since by (27),

$$(29)\qquad c = \int_\Lambda e^{-\lambda}\, dG_1(\lambda) = \int_\Lambda f_\lambda(0)\, dG_1(\lambda) = \int_\Lambda f_\lambda(0)\, dG_2(\lambda) = \int_\Lambda e^{-\lambda}\, dG_2(\lambda),$$

we can write

$$(30)\qquad H_j(B) = \frac{1}{c}\int_B e^{-\lambda}\, dG_j(\lambda) \qquad (j = 1, 2)$$

where $0 < c < \infty$. Since

$$(31)\qquad dH_j/dG_j = e^{-\lambda}/c$$

we have for $n = 1, 2, \cdots$ and $j = 1, 2$

$$(32)\qquad \int_\Lambda \lambda^n\, dH_j(\lambda) = \frac{1}{c}\int_\Lambda e^{-\lambda}\lambda^n\, dG_j(\lambda) = \frac{n!}{c}\int_\Lambda f_\lambda(n)\, dG_j(\lambda),$$

so that by (27)

$$(33)\qquad \alpha_n = \int_\Lambda \lambda^n\, dH_1(\lambda) = \int_\Lambda \lambda^n\, dH_2(\lambda) \qquad (n = 1, 2, \cdots);$$

moreover, since $0 \leqq e^{-\lambda}\lambda^n \leqq n!$ for $0 \leqq \lambda < \infty$, we have

$$(34)\qquad 0 \leqq \alpha_n = \frac{1}{c}\int_\Lambda e^{-\lambda}\lambda^n\, dG_j(\lambda) \leqq \frac{n!}{c} \qquad (n = 1, 2, \cdots),$$

so that the series $\sum_1^\infty (\alpha_n/n!)(\frac{1}{2})^n < \infty$. From a theorem of H. Cramér (*Mathematical Methods of Statistics*, p. 176) it follows that $H_1 = H_2$. Since

$$(35)\qquad G_j(B) = \int_B \frac{dG_j}{dH_j}\, dH_j(\lambda) = \int_B ce^{\lambda}\, dH_j(\lambda) \qquad (j = 1, 2)$$

it follows that $G_1 = G_2$, so that (D) holds.

We conclude with an example in which $\Lambda = \{0 < \lambda < \infty\}$, 0 here playing the role of $-\infty$ in Theorem 2.

EXAMPLE 3. (uniform distribution) Define for $\lambda \varepsilon \Lambda = \{0 < \lambda < \infty\}$

$$
(36) \qquad \begin{aligned} F_\lambda(x) &= 0 && \text{for } x \leqq 0, \\ &= x/\lambda && \text{for } 0 < x < \lambda, \\ &= 1. && \text{for } x \geqq \lambda. \end{aligned}
$$

Then

$$
(37) \qquad \begin{aligned} \lim_{\lambda \to 0} F_\lambda(x) &= 0 && \text{for } x \leqq 0, \\ &= 1 && \text{for } x > 0 \end{aligned}
$$

and

$$
(38) \qquad \lim_{\lambda \to \infty} F_\lambda(x) \equiv 0
$$

are not distribution functions, and (A), (B), (C) hold.

For any G which assigns unit probability to Λ we have for $x > 0$,

$$
(39) \qquad F_G(x) = \int_\Lambda F_\lambda(x)\, dG(\lambda) = \int_{\{0<\lambda\leqq x\}} 1 \cdot dG(\lambda) + x \int_{\{\lambda > x\}} \frac{dG(\lambda)}{\lambda}.
$$

Hence if $F_{G_1} = F_{G_2}$ then

$$
(40) \qquad G_1(x) + x \int_{\{\lambda > x\}} \frac{dG_1(\lambda)}{\lambda} = G_2(x) + x \int_{\{\lambda > x\}} \frac{dG_2(\lambda)}{\lambda}.
$$

If x is any common continuity point of G_1 and G_2 then

$$
(41) \qquad \int_{\{\lambda > x\}} \frac{dG_j(\lambda)}{\lambda} = \left[\frac{G_j(\lambda)}{\lambda}\right]_x^\infty + \int_{\{\lambda > x\}} \frac{G_j(\lambda)}{\lambda^2}\, d\lambda = -\frac{G_j(x)}{x} + \int_{\{\lambda > x\}} \frac{G_j(\lambda)}{\lambda^2}\, d\lambda
$$

so that

$$
(42) \qquad G_j(x) + x \int_{\{\lambda > x\}} \frac{dG_j(x)}{\lambda} = x \int_{\{\lambda > x\}} \frac{G_j(\lambda)}{\lambda^2}\, d\lambda,
$$

and hence from (40),

$$
(43) \qquad \int_{\{\lambda > x\}} \frac{G_1(\lambda)}{\lambda^2}\, d\lambda = \int_{\{\lambda > x\}} \frac{G_2(\lambda)}{\lambda^2}\, d\lambda
$$

at every continuity point $x > 0$ of G_1 and G_2. Differentiating with respect to x gives $G_1(x) = G_2(x)$ for every such x and hence $G_1 = G_2$. (*Cf.* [13] and references for the question of when the "identifiability" assumption (*D*) holds.)

The defect of the method of Theorem 2 for estimating G is that the choice of G_n satisfying (9) is non-constructive. In contrast, the method of Section 4, which bypasses the estimation of G and gives t_G directly, is quite explicit for the given parametric family and loss structure.

In the next section we shall indicate a method for estimating G which works in the case of a parameter space Λ consisting of a *finite* number of elements, and which may possibly be capable of generalization.

7. Estimating the *a priori* distribution: the finite case. Consider the case in which the observable random variable $x \varepsilon X$ is known to have one of a *finite* number of specified probability distributions $P_1, \cdots, P_r$, which one depending on the value of a random parameter $\lambda \varepsilon \Lambda = \{1, \cdots, r\}$ which has an unknown *a priori* probability vector $G = \{g_1, \cdots, g_r\}$, $g_i \geqq 0$, $\sum_{i=1}^r g_i = 1$, such that $P(\lambda = i) = g_i$. Observing a sequence $x_1, x_2, \cdots$ of independent random variables with the common distribution

(1) $$P_G(x \varepsilon B) = \sum_1^r g_i P_i(B),$$

our problem is to construct functions

(2) $$g_{i,n} = g_{i,n}(x_1, \cdots, x_n)$$

such that $g_{i,n} \geqq 0$, $\sum_{i=1}^r g_{i,n} = 1$, and whatever be G,

(3) $$P[\lim_{n\to\infty} g_{i,n} = g_i] = 1(i = 1, \cdots, r).$$

A *necessary* condition for the existence of such a sequence $g_{i,n}$ is clearly that

(A) If $G = \{g_1, \cdots, g_r\}$ and $\bar{G} = \{\bar{g}_1, \cdots, \bar{g}_r\}$

are any two probability vectors such that for every B,

(4) $$\sum_{i=1}^r g_i P_i(B) = \sum_{i=1}^r \bar{g}_i P_i(B)$$

then $G = \bar{G}$. We shall now show that (A) is also *sufficient*.

Denote by μ any σ-finite measure on X with respect to which all the P_i are absolutely continuous and such that their densities $f_i = dP_i/d\mu$ are square integrable:

(5) $$\int_X f_i^2(x)\, d\mu(x) < \infty \qquad (i = 1, \cdots, r).$$

(For example, we can always take $\mu = P_1 + \cdots + P_r$, since then $0 \leqq f_i(x) \leqq 1$ and hence

(6) $$\int_X f_i^2(x)\, d\mu(x) \leqq \int_X f_i(x)\, d\mu(x) = 1.)$$

The functions f_i are elements of the Hilbert space H over the measure space (X, μ). From (A) they are *linearly independent*. For if $c_1 f_1 + \cdots + c_r f_r = 0$ for some constants c_i not all 0, then by renumbering the f_i we can write

(7) $$c_1 f_1 + \cdots + c_k f_k = c_{k+1} f_{k+1} + \cdots + c_q f_q$$

with $c_1, \cdots, c_q$ all positive and $1 \leqq q \leqq r$. Integrating over X we obtain

(8) $$c_1 + \cdots + c_k = c_{k+1} + \cdots + c_q = c > 0,$$

and hence

$$(9)\quad G = \{c_1/c, \cdots, c_k/c, 0, \cdots, 0\} \neq \bar{G} = \{0, \cdots, 0, c_{k+1}/c, \cdots, c_q/c\}$$

are such that (4) holds, contradicting (A).

Now let H_j denote the linear manifold spanned by the $r - 1$ functions $f_1, \cdots, f_{j-1}, f_{j+1}, \cdots, f_r$. We can then write uniquely

$$(10)\qquad f_j = f_j' + f_j'' \qquad (j = 1, \cdots, r)$$

with

$$(11)\qquad f_j' \,\varepsilon\, H_j, \quad f_j'' \perp H_j, \quad f_j'' \neq 0.$$

Hence, setting

$$(12)\qquad \phi_j(x) = f_j''(x) \Big/ \int_X [f_j''(x)]^2 \, d\mu(x),$$

we have

$$(13)\qquad \int_X \phi_j(x) f_k(x) \, d\mu(x) = \begin{cases} 1 & \text{if } j = k, \\ 0 & \text{if } j \neq k. \end{cases}$$

Now define

$$(14)\qquad \bar{g}_{i,n} = n^{-1} \sum_{\nu=1}^{n} \phi_i(x_\nu), \quad g_{i,n} = [\bar{g}_{i,n}]^+ \Big/ \sum_{j=1}^{r} [\bar{g}_{j,n}]^+,$$

where $[a]^+$ denotes $\max(a, 0)$. If $x_1, x_2, \cdots$ are independent random variables with the common distribution (1), their common density with respect to μ is

$$(15)\qquad \sum_{j=1}^{r} g_j f_j(x),$$

so that by (13),

$$(16)\quad E\phi_i(x_\nu) = \int_X \phi_i(x) \sum_{j=1}^{r} g_j f_j(x) \, d\mu(x) = \sum_{j=1}^{r} g_j \int_X \phi_i(x) f_j(x) \, d\mu(x) = g_j.$$

The strong law of large numbers then implies (3).

Returning now to the problem of Section 2 in which we have an action space A and a loss structure $L(a, \lambda)$, here specified by functions $L(a, i)$ $(i = 1, \cdots, r)$, let us assume for simplicity that

$$(17)\qquad 0 \leqq L(a, i) \leqq L < \infty \quad (\text{all } a \,\varepsilon\, A \text{ and } i = 1, \cdots, r).$$

Then (2.6) becomes

$$(18)\qquad \Delta_G(a, x) = \sum_{i=1}^{r} [L(a, i) - L(a_0, i)] f_i(x) g_i,$$

and we can set

$$\Delta_n(a, x) = \sum_{i=1}^{r} [L(a, i) - L(a_0, i)]f_i(x)g_{i,n}, \tag{19}$$

so that

$$\sup_a |\Delta_n(a, x) - \Delta_G(a, x)| \leq L\sum_{i=1}^{n} f_i(x)|g_i - g_{i,n}|. \tag{20}$$

Since $f_i(x) < \infty$ for a.e. $(\mu)x$ it follows from (3) that with probability 1, (2.10) holds, so that $T = \{t_n\}$ defined by (2.11) is a.o. relative to every $G = (g_1, \cdots, g_r\}$.

It would be interesting to try to extend this method of estimating G to the case of a continuous parameter space, say $\Lambda = \{-\infty < \lambda < \infty\}$. One possible way is the following. Suppose for definiteness that λ is the location parameter of a normal distribution with unit variance, so that $x_1, x_2, \cdots$ have the common density function

$$f_G(x) = \int_{-\infty}^{\infty} f(x - \lambda)\, dG(\lambda); \qquad f(x) = (2\pi)^{-\frac{1}{2}}e^{-\frac{1}{2}x^2} \tag{21}$$

with respect to Lebesgue measure on $X = \{-\infty < x < \infty\}$.

For any $n \geqq 1$ let

$$\lambda_1^{(n)} < \lambda_2^{(n)} < \cdots < \lambda_{k_n}^{(n)} \tag{22}$$

be constants, and let $g_{i,n}$ $(i = 1, \cdots, k_n)$ be defined by (14) where the $f_j(x)$ of (5) are replaced by $f(x - \lambda_j^{(n)})$. Consider the random distribution function

$$G_n(\lambda) = \sum g_{i,n} \quad \text{(sum over all } i \text{ such that } \lambda_i^{(n)} \leqq \lambda). \tag{23}$$

Can we choose the values k_n and (22) for each n so that, whatever be G,

$$P[G_n \to G] = 1? \tag{24}$$

REFERENCES

[1] HANNAN, J. F. and ROBBINS, H. (1955). Asymptotic solutions of the compound decision problem for two completely specified distributions. *Ann. Math. Statist.* **26** 37–51.

[2] JOHNS, M. V., JR. (1956). Contributions to the theory of non-parametric empirical Bayes procedures in statistics. Columbia Univ. Dissertation.

[3] JOHNS, M. V., JR. (1957). Non-parametric empirical Bayes procedures. *Ann. Math. Statist.* **28** 649–669.

[4] JOHNS, M. V., JR. (1961). An empirical Bayes approach to non-parametric two-way classification. *Studies in Item Analysis and Prediction* (ed. by H. Solomon), pp. 221–232. Stanford Univ. Press.

[5] MIYASAWA, K. (1961). An empirical Bayes estimator of the mean of a normal population. *Bull. Inst. Internat. Statist.* **38** 181–188.

[6] NEYMAN, J. (1962). Two breakthroughs in the theory of statistical decision making. *Rev. Inst. Internat. Statist.* **30** 11–27.

[7] ROBBINS, H. (1950). Asymptotically subminimax solutions of compound statistical decision problems. *Proc. Second Berkeley Symp. Math. Statist. Prob.* 131–148.

[8] ROBBINS, H. (1955). An empirical Bayes approach to statistics. *Proc. Third Berkeley Symp. Math. Statist. Prob.* **1** 157–164.
[9] ROBBINS, H. (1963). The empirical Bayes approach to testing statistical hypotheses. To appear in *Rev. Inst. Internat. Statist.*
[10] SAMUEL, E. (1963). Asymptotic solutions of the sequential compound decision problem. *Ann. Math. Statist.* **34** 1079–1094.
[11] SAMUEL, E. (1963). An empirical Bayes approach to the testing of certain parametric hypotheses. *Ann. Math. Statist.* **34** 1370–1385.
[12] SAMUEL, E. Strong convergence of the losses of certain decision rules for the compound decision problem. Unpublished.
[13] TEICHER, H. (1961). Identifiability of mixtures. *Ann. Math. Statist.* **32** 244–248.

Reprinted from
Proc. Natl. Acad. Sci. USA
Vol. 74, No. 7, pp. 2670–2671, July 1977
Statistics

Prediction and estimation for the compound Poisson distribution

(empirical Bayes/accident proneness)

HERBERT ROBBINS

Department of Mathematical Statistics, Columbia University, New York, New York 10027

Contributed by Herbert Robbins, May 2, 1977

ABSTRACT An empirical Bayes method is proposed for predicting the future performance of certain subgroups of a compound Poisson population.

Suppose that last year a randomly selected group of n people experienced, respectively, $x_1, \ldots, x_n$ "accidents" of some sort. Under similar conditions during next year the same n people will experience some as yet unknown number of accidents $y_1, \ldots, y_n$. Let i be any positive integer. We are concerned with predicting, from the values $x_1, \ldots, x_n$ alone, the value of

$$S_{i,n} = \{\text{sum of the } y_j\text{s for which the corresponding } x_j\text{s are } <i\} \quad [1]$$

that is, *the total number of accidents that will be experienced next year by those people who experienced less than i accidents last year.*

The quantity

$$\{\text{sum of the } x_j\text{s that are } <i\} \quad [2]$$

is a poor predictor of [1], since it will usually underestimate [1] by a considerable amount. Thus, for $i = 1$ the value of [2] is 0, but the people who were lucky last year (perhaps only a small fraction of n) are not likely to remain so next year, and hence $S_{1,n}$ will be rather large if people in general are accident prone.

We shall show that, in contrast to [2], a good predictor of $S_{i,n}$ is the quantity

$$E_{i,n} = \{\text{sum of the } x_j\text{s that are } \le i\} \quad [3]$$

In particular, for $i = 1$, a good predictor of $S_{1,n} = \{$sum of the y_js for which the corresponding x_js are 0$\}$ is the quantity $E_{1,n} = \{$number of people who had exactly 1 accident last year$\}$.

The main theorem

Let (λ,x,y) be a random vector $(0 < \lambda < \infty;\ x, y = 0, 1, \ldots)$ such that

Given λ, x *and* y *are independent Poisson random variables with mean* λ. [4]

We denote the distribution function of λ by G, but make no assumptions about it.

The conditional probability function of x, given λ, is the Poisson

$$f(x|\lambda) = e^{-\lambda}\lambda^x/x! \quad [5]$$

with a similar expression for $f(y|\lambda)$. The unconditional probability function of x is the compound Poisson

$$f(x) = \int_0^\infty (e^{-\lambda}\lambda^x/x!)dG(\lambda) \quad [6]$$

with a similar expression for $f(y)$. It follows from Eq. 5 that

$$E(x|\lambda) = \lambda, \quad E(x^2|\lambda) = \lambda + \lambda^2, \quad \text{Var}(x|\lambda) = \lambda,$$
$$Ex = E\lambda, \quad E(x^2) = E\lambda + E(\lambda^2),$$
$$\text{Var } x = E\lambda + \text{Var } \lambda \quad [7]$$

and similarly for y.

We now define the random variables

$$v = \begin{cases} 1 \text{ if } x < i \\ 0 \text{ if } x \ge i \end{cases}, \quad w = vy = \begin{cases} y \text{ if } x < i \\ 0 \text{ if } x \ge i \end{cases} \quad [8]$$

and observe that by [4]

$$E(w|\lambda) = E(vy|\lambda) = E(v|\lambda)\cdot E(y|\lambda) = \lambda E(v|\lambda)$$
$$= \sum_{x=0}^{i-1} (e^{-\lambda}\lambda^{x+1}/x!) \quad [9]$$

so that

$$Ew = \sum_{x=0}^{i-1} \int_0^\infty (e^{-\lambda}\lambda^{x+1}/x!)dG(\lambda)$$
$$= \sum_{x=0}^{i-1} (x+1)f(x+1) = \sum_{x=0}^{i} xf(x) \quad [10]$$

Defining

$$u = \begin{cases} x \text{ if } x \le i \\ 0 \text{ if } x > i \end{cases} \quad [11]$$

we see that

$$Eu = Ew \quad [12]$$

Let (λ_j,x_j,y_j) for $j = 1, \ldots, n$ be independent random vectors with the same distribution as (λ,x,y), and define

$$S_{i,n} = \sum_1^n w_j = \text{sum of the } y_j\text{s for which the corresponding } x_j\text{s are } <i \quad [13]$$

$$E_{i,n} = \sum_1^n u_j = \text{sum of the } x_j\text{s that are } \le i \quad [14]$$

$$N_{i,n} = \sum_1^n v_j = \text{number of } x_j\text{s that are } <i \quad [15]$$

Then, as $n \to \infty$ the law of large numbers and the central limit theorem combine to show that

$$\sqrt{n}\left\{\frac{E_{i,n}}{N_{i,n}} - \frac{S_{i,n}}{N_{i,n}}\right\} = \frac{\sum_1^n (u_j - w_j)/\sqrt{n}}{\sum_1^n v_j/n} \to N(0,A_i^2) \quad [16]$$

in distribution, where

$$A_i^2 = E(u-w)^2/E^2v \quad [17]$$

In order to evaluate Eq. 17 in terms of Eq. 6 it is convenient to define

$$a_i = \sum_{x=0}^{i-1} f(x), \qquad b_i = \sum_{x=0}^{i} xf(x), \qquad c_i = \sum_{x=0}^{i} x^2 f(x) \quad [18]$$

so that

$$Eu = Ew = b_i, \qquad Ev = E(v^2) = a_i, \qquad E(u^2) = c_i \quad [19]$$

A little computation shows that

$$E(uw) = c_i - b_i, \qquad E(w^2) = b_i - b_{i+1} + c_{i+1} \quad [20]$$

Hence, from Eq. 17

$$A_i^2 = [c_i - 2(c_i - b_i) + b_i - b_{i+1} + c_{i+1}]/a_i^2 \quad [21]$$
$$= [3b_i - b_{i+1} + c_{i+1} - c_i]/a_i^2$$

We note that as $i \to \infty$, $A_i^2 \to 2Ex = 2E\lambda$.

The random variables

$$a_{i,n} = [\text{number of } x_1, \ldots, x_n \text{ that are } < i]/n$$
$$b_{i,n} = [\text{sum of those } x_1, \ldots, x_n \text{ that are } \le i]/n \quad [22]$$
$$c_{i,n} = [\text{sum of squares of those } x_1, \ldots, x_n \text{ that are } \le i]/n$$
$$A_{i,n}^2 = [3b_{i,n} - b_{i+1,n} + c_{i+1,n} - c_{i,n}]/a_{i,n}^2$$

converge as $n \to \infty$ to a_i, b_i, c_i, and A_i^2, respectively. Hence, from Eqs. 16 and 21 we have the

THEOREM. *For any fixed* i = 1,2, . . . , *as* n → ∞

$$\frac{\sqrt{n}(E_{i,n} - S_{i,n})}{A_{i,n} \cdot N_{i,n}} \to N(0,1) \text{ in distribution.} \quad [23]$$

As n → ∞, *the probability tends to 0.95 that*

$$\frac{E_{i,n}}{N_{i,n}} - \frac{(1.96)A_{i,n}}{\sqrt{n}} < \frac{S_{i,n}}{N_{i,n}} < \frac{E_{i,n}}{N_{i,n}} + \frac{(1.96)A_{i,n}}{\sqrt{n}} \quad [24]$$

The endpoints of the confidence interval [24] for the ratio $S_{i,n}/N_{i,n}$ are functions of $x_1, \ldots, x_n$ alone, and do not involve any knowledge of or assumption about the nature of the underlying mixing distribution function G. Of course, for finite n the exact probability that [24] will hold does depend on G.

We can also use $E_{i,n}$ to estimate the parameter (i.e., functional of G)

$$G_i = E(\lambda|x < i) = \frac{\sum_{x=0}^{i-1} f(x)E(\lambda|x)}{\sum_{x=0}^{i-1} f(x)} \quad [25]$$

Since

$$E(\lambda|x) = \frac{\int_0^\infty e^{-\lambda}\lambda^{x+1}\,dG(\lambda)}{\int_0^\infty e^{-\lambda}\lambda^{x}\,dG(\lambda)} = \frac{(x+1)f(x+1)}{f(x)} \quad [26]$$

we can write

$$G_i = \frac{\sum_{x=0}^{i-1}(x+1)f(x+1)}{\sum_{x=0}^{i-1} f(x)} = \frac{\sum_{x=0}^{i} xf(x)}{\sum_{x=0}^{i-1} f(x)} = \frac{b_i}{a_i} = \frac{Eu}{Ev} \quad [27]$$

Hence, as $n \to \infty$

$$\sqrt{n}\left[\frac{E_{i,n}}{N_{i,n}} - G_i\right] = \frac{\sum_1^n \left(u_j - \frac{Eu}{Ev}\cdot v_j\right)/\sqrt{n}}{\sum_1^n v_j/n} \to N(0,B_i^2) \quad [28]$$

in distribution, where by computation

$$B_i^2 = [a_i c_i - 2b_i b_{i-1} + b_i^2]/a_i^3 \quad [29]$$

As $i \to \infty$, $B_i^2 \to \text{Var } x = E\lambda + \text{Var } \lambda$. An approximately 95% confidence interval for G_i is

$$\frac{E_{i,n}}{N_{i,n}} \pm \frac{(1.96)B_{i,n}}{\sqrt{n}} \quad [30]$$

where $B_{i,n}^2$ is defined by Eq. 29 with a_i replaced by $a_{i,n}$, etc.

Remarks

(*A*) The parameter G_i defined by Eq. 25 is equal to $E(y|x < i)$. This follows from [4] and the fact that $E(y|\lambda) = \lambda$.

(*B*) It is easy to show (Schwarz inequality) that

$$G_1 \le G_2 \le \ldots \to E\lambda \quad [31]$$

Since the sample estimates $E_{i,n}/N_{i,n}$ of the G_i are not necessarily increasing in i, this suggests that for some purposes it should be possible to improve on them by smoothing (isotonic regression).

(*C*) It would be interesting to compare Eq. 28 with the behavior of the maximum likelihood estimator of G_i, either with respect to all possible Gs or within some parametric class such as those with the probability density function

$$g_{\alpha,\theta}(\lambda) = \frac{e^{-\lambda/\theta}\cdot\lambda^{\alpha-1}}{\theta^\alpha\Gamma(\alpha)} \quad [32]$$

with two unknown positive parameters α, θ.

(*D*) Many other quantities can be predicted or estimated from $x_1, \ldots, x_n$ by similar methods. For example, to estimate

$$E(\lambda|x \ge i) = E(y|x \ge i) = \frac{E\lambda - b_i}{1 - a_i} \quad [33]$$

we can use

$$\frac{\text{sum of the } x_j\text{s that are } > i}{\text{number of } x_j\text{s that are } \ge i} \quad [34]$$

As $n \to \infty$ it can be shown that

$$\sqrt{n}(\,(34) - (33)\,) \to N(0,C_i^2) \text{ in distribution} \quad [35]$$

where

$$C_i^2 = [(1 - a_i)(E(x^2) - c_i) - (Ex - b_i)^2]/(1 - a_i)^3 \quad [36]$$

The sequence $E(\lambda|x \ge i)$ is also increasing in i.

(*E*) Analogous results hold for mixtures of some important parametric families other than the Poisson; e.g., negative exponential, negative binomial, and normal. The binomial case is interesting as an example of nonidentifiability.

This work was supported by the Office of Naval Research Grant N00014-75-C-0560-P00001.

Proc. Natl. Acad. Sci. USA
Vol. 77, No. 12, pp. 6988–6989, December 1980
Statistics

An empirical Bayes estimation problem

(compound Poisson distribution)

HERBERT ROBBINS

Columbia University, New York, New York 10027

Contributed by Herbert Robbins, August 25, 1980

ABSTRACT Let x be a random variable such that, given θ, x is Poisson with mean θ, while θ has an unknown prior distribution G. In many statistical problems one wants to estimate as accurately as possible the parameter $E(\theta|x = a)$ for some given $a = 0,1,\ldots$. If one assumes that G is a Gamma prior with unknown parameters α and β, then the problem is straightforward, but the estimate may not be consistent if G is not Gamma. On the other hand, a more general empirical Bayes estimator will always be consistent but will be inefficient if in fact G is Gamma. It is shown that this dilemma can be more or less resolved for large samples by combining the two methods of estimation.

Let $x, x_1, x_2, \ldots, x_n$ be independent with probability function

$$f(x) = \int_0^\infty f(x|\theta)dG(\theta), \qquad [1]$$

where $f(x|\theta) = e^{-\theta}\theta^x/x!$ (Poisson with mean θ) and G is unknown. We are concerned with estimating from observed values of $x_1, \ldots, x_n$ the parameter

$$h = h(a,G) = E(\theta|x = a) = \frac{\int_0^\infty \theta f(a|\theta)dG(\theta)}{f(a)} = \frac{(a+1)f(a+1)}{f(a)}, \qquad [2]$$

where a is some fixed integer $0,1,\ldots$; h is in fact the minimum mean squared error estimator of θ when x is observed to have the value a.

Even in the absence of any information about G, a strongly consistent and asymptotically normal estimator T_n of h can be obtained as follows. Define

$$u = u(x) = \begin{cases} 1 \text{ if } x = a \\ 0 \text{ if } x \neq a \end{cases}, \quad v = \begin{cases} x \text{ if } x = a+1 \\ 0 \text{ if } x \neq a+1 \end{cases}, \qquad [3]$$

and set $u_i = u(x_i)$, $v_i = v(x_i)$. Then $h = Ev/Eu$, so that as $n \to \infty$

$$T_n = \frac{\sum_1^n v_i}{\sum_1^n u_i} \to h \quad \text{with probability 1}, \qquad [4]$$

and by the central limit theorem

$$\sqrt{n}(T_n - h) = \frac{\sum_1^n (v_i - hu_i)}{\left(\sum_1^n u_i/n\right)\sqrt{n}} \sim \frac{\sum_1^n (v_i - hu_i)}{Eu \cdot \sqrt{n}} \xrightarrow{\mathcal{D}} N(0,\sigma_1^2), \qquad [5]$$

where

$$\sigma_1^2 = \frac{E(v-hu)^2}{E^2u} = \frac{(a+1)^2 f(a+1)}{f^2(a)}\left[1 + \frac{f(a+1)}{f(a)}\right]$$
$$\sim n \cdot (a+1)^2\left[1 + \frac{n(a+1)}{n(a)}\right] n(a+1)/n^2(a), \qquad [6]$$

with $n(j)$ = number of values $x_1, \ldots, x_n$ that equal $j (j = 0,1,\ldots)$. [7]

From [5] and [6] it follows that

$$\frac{T_n - h}{\sqrt{\dfrac{(a+1)^2 n(a+1)}{n^2(a)}\left[1 + \dfrac{n(a+1)}{n(a)}\right]}} \xrightarrow{\mathcal{D}} N(0,1), \qquad [8]$$

so that for large n, an approximately 95% confidence interval for h is given by

$$(a+1)\cdot\frac{n(a+1)}{n(a)} \pm \frac{1.96(a+1)}{\sqrt{n(a)}} \times \sqrt{\frac{n(a+1)}{n(a)}\left[1 + \frac{n(a+1)}{n(a)}\right]}. \qquad [9]$$

Although T_n, a "general" empirical Bayes estimator of h, is always consistent, there are circumstances in which its use would be inefficient. For example, if we assume that G is a member of the two-parameter Gamma family of conjugate priors with probability density function

$$g(\theta) = G'(\theta) = \frac{\alpha^\beta}{\Gamma(\beta)}e^{-\alpha\theta}\theta^{\beta-1}, \qquad [10]$$

where α and β are unknown positive parameters, then an easy computation shows that

$$f(x) = \frac{\Gamma(x+\beta)}{x!\Gamma(\beta)}\left(\frac{\alpha}{1+\alpha}\right)^\beta\left(\frac{1}{1+\alpha}\right)^x, \qquad [11]$$

$$h = \frac{(a+1)f(a+1)}{f(a)} = \frac{a+\beta}{1+\alpha}. \qquad [12]$$

Now for [10]

$$E(\theta^r) = \frac{\Gamma(r+\beta)}{\Gamma(\beta)\alpha^r} \; (r \geq 0), \qquad [13]$$

so that

$$E\theta = \frac{\beta}{\alpha}, \quad \text{Var } \theta = \frac{\beta}{\alpha^2}. \qquad [14]$$

Moreover, from [1]

$$E(x|\theta) = \text{Var}(x|\theta) = \theta, \qquad [15]$$

so that

$$Ex = E\theta, \quad \text{Var } x = E\theta + \text{Var } \theta. \qquad [16]$$

From [16] and [14] we have

$$Ex = \beta/\alpha, \quad \text{Var } x = \beta(1+\alpha)/\alpha^2, \qquad [17]$$

The publication costs of this article were defrayed in part by page charge payment. This article must therefore be hereby marked "*advertisement*" in accordance with 18 U. S. C. §1734 solely to indicate this fact.

and hence

$$\alpha = \frac{Ex}{\mathrm{Var}\,x - Ex}, \quad \beta = \frac{E^2x}{\mathrm{Var}\,x - Ex}. \qquad [18]$$

It follows from [12] that

$$h = Ex + \left(1 - \frac{Ex}{\mathrm{Var}\,x}\right)(a - Ex), \qquad [19]$$

and there may be priors G other than those of the form [10] for which [19] also holds for the given value of a; in other words, for which $h = k$, where h is defined by [2] and k by

$$k = Ex + \left(1 - \frac{Ex}{\mathrm{Var}\,x}\right)(a - Ex). \qquad [20]$$

For any such G, one could estimate h by replacing Ex and Var x in [19] by the consistent moment estimators

$$\bar{x}_n = \frac{1}{n}\sum_1^n x_i, \quad s_n^2 = \frac{1}{n}\sum_1^n (x_i - \bar{x}_n)^2, \qquad [21]$$

obtaining

$$W_n = \bar{x}_n + \left(1 - \frac{\bar{x}_n}{s_n^2}\right)(a - \bar{x}_n). \qquad [22]$$

Whenever G has a finite fourth moment,

$$\sqrt{n}(W_n - k) \xrightarrow{\mathcal{D}} N(0,\sigma_2^2), \qquad [23]$$

where σ_2^2 is a function of a and moments of x that will be made explicit later. Now, if $h = k$ (as it is for [10]), it is plausible that the "restricted" empirical Bayes estimator W_n should be more efficient than T_n as an estimator of h. However, in using W_n we run the risk that if in fact $h \neq k$, then W_n will not even be a consistent estimator of h.

The foregoing remarks suggest combining T_n and W_n to obtain an estimator that will behave like T_n when $h \neq k$ and like W_n when $h = k$. One such estimator is

$$Z_n = \frac{1 \cdot W_n + n^{\epsilon}(T_n - W_n)^2 \cdot T_n}{1 + n^{\epsilon}(T_n - W_n)^2}, \qquad [24]$$

where ϵ is any constant such that $\frac{1}{2} < \epsilon < 1$. Then

$$\sqrt{n}(Z_n - h) = \frac{\sqrt{n}(W_n - h) + n^{\epsilon}(T_n - W_n)^2(T_n - h)\sqrt{n}}{1 + n^{\epsilon}(T_n - W_n)^2}. \qquad [25]$$

There are two cases to consider.

(a) Suppose that $h = k$. Then from [5] and [23], $|T_n - W_n| = 0(n^{-1/2})$ in probability, so that

$$n^{\epsilon}(T_n - W_n)^2 = 0(n^{\epsilon - 1}) \to 0, \qquad [26]$$

$$\sqrt{n}(Z_n - h) \sim \sqrt{n}(W_n - h) \xrightarrow{\mathcal{D}} N(0,\sigma_2^2). \qquad [27]$$

(b) Suppose $h \neq k$. Then $|T_n - W_n| \to |h - k| > 0$ in probability, so that

$$\sqrt{n}(Z_n - h) \sim \sqrt{n}(T_n - h) \xrightarrow{\mathcal{D}} N(0,\sigma_1^2). \qquad [28]$$

Thus, Z_n is asymptotically equivalent to W_n if $h = k$ and to T_n if $h \neq k$.

Z_n is, like T_n, a general empirical Bayes estimator of h, in that it is consistent no matter what the form of G, whereas W_n is only consistent when $h = k$. Thus Z_n dominates W_n. However, does Z_n dominate T_n? When $h \neq k$, they are equivalent. When $h = k$, Z_n is better than T_n iff $\sigma_2^2 < \sigma_1^2$. We have in [6] a formula for σ_1^2; we now exhibit one for σ_2^2, obtained by the usual asymptotics, as in ref. 1, chapter 28:

$$\sigma_2^2 = \frac{(2m - a)^2}{\mu_2} + \frac{m^2(m - a)^2}{\mu_2^4}(\mu_4 - \mu_2^2) - \frac{2m(m - a)(2m - a)}{\mu_2^3} \cdot \mu_3, \qquad [29]$$

where

$$m = Ex, \quad \mu_i = E(x - m)^i (i \geq 2). \qquad [30]$$

In the Gamma case [10] some tedious algebra gives

$$\sigma_2^2 = \frac{1}{\beta(1 + \alpha)^3}\{\alpha^2[2\alpha + 3 + 2\beta(1 + \alpha)]a^2 - 2\alpha\beta[\alpha + 2 + 2\beta(1 + \alpha)]a + \beta^2[\alpha^2 + 2\alpha + 2 + 2\beta(1 + \alpha)]\}, \qquad [31]$$

whereas

$$\sigma_1^2 = \frac{(\alpha + \beta)a!\Gamma(\beta)(1 + \alpha)^{a + \beta - 2}[(1 + \alpha)(1 + a) + a + \beta]}{\alpha^{\beta}\Gamma(a + \beta)}. \qquad [32]$$

In particular, for $a = 0$, $\sigma_2^2 < \sigma_1^2$ iff

$$\alpha^{\beta}[\alpha^2 + 2\alpha + 2 + 2\beta(1 + \alpha)] < (1 + \alpha)^{\beta + 1}(1 + \alpha + \beta). \qquad [33]$$

For large β this clearly holds, but not for β near 0, since as $\beta \to 0$

$$\frac{\sigma_2^2}{\sigma_1^2} \to \frac{\alpha^2 + 2\alpha + 2}{\alpha^2 + 2\alpha + 1} > 1, \qquad [34]$$

so that T_n is better than W_n or Z_n for sufficiently small β. When $\alpha = 1$, for example, the critical value of β is $\simeq 0.48$. That σ_2^2 is not always less than σ_1^2 in the case [10] may come as a surprise, but it is presumably due to the fact that the method of moments is not optimal for estimating α and β. In fact, if one is willing to assume a Gamma prior [10] (and not merely that $h = k$), then it would be reasonable to estimate h by replacing in [12] the unknown α and β by their maximum likelihood estimates $\hat{\alpha}_n$ and $\hat{\beta}_n$, obtaining the estimator

$$\hat{h}_n = \frac{a + \hat{\beta}_n}{1 + \hat{\alpha}_n}. \qquad [35]$$

[I have not attempted to verify that the asymptotic variance of $\sqrt{n}(\hat{h}_n - h)$ as a function of a, α and β is always less than [32].] One could then define $\hat{Z}_n$ by replacing W_n by $\hat{h}_n$ in [24], thus hopefully assuring asymptotic efficiency in the case [10] and consistency in all cases. (But what of efficiency if G is not of the form [10] but $\hat{h}_n \to k' = h$?) Many variations on this theme are possible. In fact, Z_n could be any combination $U_nW_n + (1 - U_n)T_n$ with $U_n \to 1$ if $h = k$ and $U_n\sqrt{n} \to 0$ if $h \neq k$ in probability; for example, $Z_n = W_n$ or T_n according as $|T_n - W_n| < n^{-1/4}$ or $\geq n^{-1/4}$.

Instead of h we may want to estimate $H = E(\theta | x \geq a)$ or to predict the future performance of that subgroup of persons whose x values are $= a$ or $\geq a$, etc. For an indication of the practical relevance of such problems, the reader may consult refs. 2–4. The idea embodied in [24] has many other applications in the direction of combining efficiency with robustness.

This research was supported by the National Science Foundation and the National Institute of General Medical Sciences.

1. Cramér, H. (1946) *Mathematical Methods of Statistics* (Princeton University Press, Princeton, NJ).
2. Robbins, H. (1977) *Proc. Natl. Acad. Sci. USA* **74**, 2670–2671.
3. Robbins, H. (1980) *Proc. Natl. Acad. Sci. USA* **77**, 2382–2383.
4. Robbins, H. (1980) *Asymptotic Theory of Statistical Tests and Estimation* (Academic, New York), pp. 251–257.

ESTIMATING MANY VARIANCES

Herbert Robbins[1]

Department of Mathematical Statistics
Columbia University
New York, New York, U.S.A.

Suppose that an unknown random parameter θ with distribution function G is such that given θ, an observable random variable x has conditional probability density $f(x|\theta)$ of known form. If a function $t = t(x)$ is used to estimate θ, then the expected squared error with respect to the random variation of both θ and x is

$$(1) \qquad E(t-\theta)^2 = \int \int (t(x)-\theta)^2 f(x|\theta)dx\, dG(\theta).$$

For fixed G we can seek to minimize (1) within any desired class of functions t, such as the class of all linear functions A + Bx, or the class of all Borel functions whatsoever. The minimum of (1) within the class of linear functions is achieved for the function $\tilde{t}$ defined by

$$(2) \qquad \tilde{t}(x) = E\theta + \frac{\mathrm{Cov}(\theta,x)}{\mathrm{Var}\, x} \cdot (x-Ex),$$

with

$$(3) \qquad E(\tilde{t}-\theta)^2 = \mathrm{Var}\, \theta - \frac{\mathrm{Cov}^2(\theta,x)}{\mathrm{Var}\, x},$$

while the unrestricted minimum of (1) is achieved for the function t* defined by

[1]Research supported by grants from the National Science Foundation and the National Institute of General Medical Sciences. Parts of this work were also done under the author's affiliation with Brookhaven National Laboratory.

Statistical Decision Theory and Related Topics III, Vol. 2

ISBN 0-12-307502-5

(4) $$t^*(x) = E(\theta|x) = \int \theta f(x|\theta)dG(\theta) \Big/ \int f(x|\theta)dG(\theta),$$

with

(5) $$E(t^*-\theta)^2 = E\ \mathrm{Var}(\theta|x).$$

When G is unknown to the statistician, the empirical Bayes approach is to approximate (2) or (4) by observing the x-components of a large number of independent pairs (θ_i, x_i) $(i = 1,\dots,n)$ with the same joint distribution as the pair (θ,x) of the preceding paragraph. It is not obvious how this approximation is to be effected in practice, nor whether one should try to approximate (2), (4), or perhaps something else, for a given n.

We begin by trying to approximate (2), and assume for simplicity that $f(x|\theta)$ is such that for known constants a,b,c

(6) $$E(x|\theta) = \theta, \quad \mathrm{Var}(x|\theta) = a + b\theta + c\theta^2.$$

From the first of these relations it follows that

(7) $$Ex=E\theta,\ \mathrm{Cov}(\theta,x)=\mathrm{Var}\ \theta,\ E(x-\theta)^2=\mathrm{Var}\ x-\mathrm{Var}\ \theta=E\mathrm{Var}(x|\theta),$$

and when the second is added that

(8) $$\mathrm{Var}\ \theta = [\mathrm{Var}\ x - a - bEx - cE^2x]/(1+c).$$

We can therefore write (2) in the form

(9) $$\tilde{t}(x) = Ex+\left[1- \frac{c(\mathrm{Var}\ x+E^2x)+a+b\ Ex}{(1+c)\mathrm{Var}\ x}\right] (x-Ex),$$

with

(10) $$E(\tilde{t}-\theta)^2 = \frac{\mathrm{Var}\ \theta\ [\mathrm{Var}\ x - \mathrm{Var}\ \theta]}{\mathrm{Var}\ x}$$

$$= \frac{\mathrm{Var}\ x - a - b\ Ex - cE^2x}{(1+c)\ \mathrm{Var}\ x}\ E(x-\theta)^2.$$

The usual estimators of Ex and Var x are

(11) $$\bar{x} = \frac{1}{n}\sum_1^n x_i, \quad s^2 = \frac{1}{n-1}\sum_1^n (x_i - \bar{x})^2,$$

and substituting them (or statistics asymptotically equivalent to them as $n \to \infty$) into (9) we obtain the l.e.B. (linear empirical Bayes) function $\tilde{t}_n$ defined by

(12) $$\tilde{t}_n(x) = \bar{x} + \left[1 - \frac{c(s^2 + \bar{x}^2) + a + b\bar{x}}{(1+c)s^2}\right]^+ (x - \bar{x})$$

as a consistent approximation to $\tilde{t}$. We then estimate θ_i by $\tilde{t}_n(x_i)$ $(i = 1,\ldots,n)$, and hope that, under mild assumptions on the nature of G,

(13) $$E(\tilde{t}_n(x_i) - \theta_i)^2 = \frac{1}{n}\sum_1^n E(\tilde{t}_n(x_i) - \theta_i)^2 \to E(\tilde{t} - \theta)^2$$

$$\sim \frac{s^2 - a - b\bar{x} - c\bar{x}^2}{(1+c)s^2} E(x-\theta)^2 \quad \text{as} \quad n \to \infty.$$

The usual example is that in which for some known $\sigma > 0$

(14) $$f(x|\theta) = \frac{1}{\sigma}\phi\left(\frac{x-\theta}{\sigma}\right),$$

where $\phi(x)$ is the standard normal p.d.f. (or more generally any p.d.f. with mean 0 and variance 1), so that

(15) $$E(x|\theta) = \theta, \quad Var(x|\theta) = \sigma^2,$$

and (12) and (10) become

(16) $$\tilde{t}_n(x) = \bar{x} + \left[1 - \frac{\sigma^2}{s^2}\right]^+ (x - \bar{x}), \quad \frac{E(x-\theta)^2}{E(\tilde{t}-\theta)^2} = \frac{Var\ x}{Var\ x - \sigma^2} \sim \frac{s^2}{s^2 - \sigma^2}.$$

Consider now the less familiar problem of estimating many *variances* (instead of means), in samples of the same size from normal populations. To define the basic pair (θ, x), for some $r = k/2$ $(k = 1,2,\ldots)$ let $y_1,\ldots,y_{2r+1}$ be i.i.d. normal with mean μ and variance θ, and let

(17) $$x = \frac{1}{2r} \sum_{1}^{2r+1} (y_i - \bar{y})^2 = \text{sample variance.}$$

Then $2rx/\theta$ has the chi-squared distribution with $2r$ degrees of freedom, and the conditional p.d.f. of x given θ is of the gamma family with known shape parameter r and unknown scale parameter θ,

(18) $$f(x|\theta) = \frac{r^r \theta^{-r}}{\Gamma(r)} e^{-\frac{rx}{\theta}} x^{r-1} \qquad (r, x, \theta > 0),$$

for which

(19) $$E(x|\theta) = \theta, \; \text{Var}\,(x|\theta) = \theta^2/r,$$

so that

(20) $$\tilde{t}_n(x) = \bar{x} + [1 - \frac{s^2 + \bar{x}^2}{(1+r)s^2}]^+ (x - \bar{x}),$$

$$\frac{E(x-\theta)^2}{E(\tilde{t}-\theta)^2} = 1 + \frac{1}{r}\,(1 + \frac{E^2\theta}{\text{Var}\,\theta}) \sim \frac{(r+1)s^2}{rs^2 - \bar{x}^2}\,.$$

Thus, for n independent samples of size $2r + 1$ from normal populations with unknown means μ_i and unknown variances θ_i, if x_i denotes the ith sample variance then a l.e.B. estimator of θ_i is

(21) $$\tilde{t}_n(x_i) = \bar{x} + [1 - \frac{s^2 + \bar{x}^2}{(1+r)s^2}]^+ (x_i - \bar{x}) \qquad (i = 1, \ldots, n).$$

For large n this will always be better, and in some cases much better, than the usual estimator x_i. For example, if $P(\theta = 99) = P(\theta = 101) = 1/2$, so that $E\theta = 100$, $\text{Var}\,\theta = 1$, then

(22) $$\frac{E(x-\theta)^2}{E(\tilde{t}-\theta)^2} = 1 + \frac{1}{r}\,(10{,}001),$$

which is $> 1{,}000$ for $r \leq 10$; i.e., sample size $= 2r+1 \leq 21$.

We now consider, for the same gamma family (18), the problem of approximating the unrestricted or general empirical Bayes (g.e.B.) estimator (4), of which the denominator is

(23) $$f(x) = \int f(x|\theta)dG(\theta) = \text{marginal p.d.f. of } x.$$

Now, $f(x)$ can be estimated from $x_1,\ldots,x_n$ by various techniques of density estimation. The numerator of (4), however, bears no immediate relation to $f(x)$, and it is not obvious how to approximate it from the observed $x_1,\ldots,x_n$. Happily, it can be shown for (18), by interchanging the order of integration, that for all $x > 0$ the identity

(24) $$t^*(x) = E(\theta|x) = rx^{r-1}\int_x^\infty y^{1-r}\, f(y)dy \Big/ f(x)$$

holds, and this gives t^* in terms of f without explicit reference to G, which is what we want. The numerator of $t^*(a)$ for any fixed $a > 0$ is the expectation of the random variable

(25) $$\psi(x) = \begin{cases} r\, a^{r-1} x^{1-r} & \text{if } x \geq a \\ 0 & \text{if } x < a, \end{cases}$$

and can be estimated by $\sum_1^n \psi(x_i)/n$.

For the gamma family (18), the l.e.B. and g.e.B. estimators (2) and (4) are equal when G is such that

(26) $$G'(\theta) = g(\theta) = \frac{\alpha^\beta}{\Gamma(\beta)}\, e^{-\frac{\alpha}{\theta}}\, \theta^{-\beta-1}$$

for some α, $\beta > 0$ with $\beta + r > 1$; their common value is the linear function

(27) $$\frac{\alpha + rx}{\beta + r - 1}.$$

Density estimation can be avoided by changing the problem from that of the simultaneous estimation of many unknown variances θ_i to that of estimating something else. We mention three closely related versions of a different problem.

I. For any fixed $a \geq 0$ define

$$(28) \qquad H = H(a) = E(\theta|x \geq a) = \frac{\int_a^\infty E(\theta|x)f(x)dx}{P(x \geq a)} = \frac{\int_a^\infty rx^{r-1}\int_x^\infty y^{1-r}f(y)dy\,dx}{P(x \geq a)}$$

$$= \frac{\int_a^\infty y^{1-r}f(y)\int_a^y rx^{r-1}dx\,dy}{P(x \geq a)} = \frac{\int_a^\infty x[1-(\frac{a}{x})^r]f(x)dx}{P(x \geq a)} = \frac{Ev}{Eu},$$

where by definition

$$(29) \qquad u = \begin{cases} 1 & \text{if } x \geq a \\ 0 & \text{if } x < a \end{cases}, \quad v = ux[1-(\frac{a}{x})^r].$$

We can estimate H consistently as $n \to \infty$ by the ratio T_n/N_n, where

$$(30) \qquad N_n = \sum_1^n u_i, \; T_n = \sum_1^n v_i = \sum_1^n u_i x_i [1-(\frac{a}{x_i})^r].$$

In fact, by the central limit theorem, as $n \to \infty$

$$(31) \qquad \sqrt{n}\,(\frac{T_n}{N_n} - H) = \frac{\sum_1^n (v_i - Hu_i)}{(\sum_1^n u_i/n)\sqrt{n}} \xrightarrow{D} N(0,\sigma_1^2),$$

where

$$(32) \qquad \sigma_1^2 = \frac{E(v-Hu)^2}{E^2 u} \sim \frac{n}{N_n}\left\{\frac{1}{N_n}\sum_1^n v_i^2 - (\frac{T_n}{N_n})^2\right\},$$

so that

$$(33) \qquad \frac{(T_n/N_n)-H}{\frac{1}{\sqrt{N_n}}\sqrt{\frac{1}{N_n}\sum_1^n v_i^2 - (T_n/N_n)^2}} \xrightarrow{D} N(0,1).$$

Thus, for large n, an approximately 95% confidence interval for H is given by

$$(34) \qquad \frac{T_n}{N_n} \pm \frac{1.96}{\sqrt{N_n}} \sqrt{\frac{1}{N_n} \sum_1^n v_i^2 - \left(\frac{T_n}{N_n}\right)^2}.$$

The width of this interval tends almost surely to 0 as $n \to \infty$.

An interesting question: in what sense, if any, is T_n/N_n an asymptotically optimal estimator of H? Another: find a confidence band for the function $H(a) = E(\theta|x \geq a)$ for all $0 \leq a_o < a < a_1 \leq \infty$.

II. Let

$$(35) \qquad R_n = \sum_1^n u_i\theta_i = \text{sum of all the } N_n \text{ values } \theta_i \text{ for which } x_i \geq a.$$

We note that

$$(36) \qquad Eu\theta = P(x \geq a)\, E(\theta|x \geq a) = Eu(Ev/Eu) = Ev,$$

so that as $n \to \infty$

$$(37) \qquad \frac{T_n - R_n}{\sqrt{n}} \xrightarrow{D} N(0, \sigma_2^2),$$

where

$$(38) \qquad \sigma_2^2 = E(v - u\theta)^2 = E(v^2 - 2v\theta + \theta^2).$$

III. Suppose that given θ_i, x_i and x_i' are independent, each with p.d.f. $f(x|\theta_i)$. For the case (18), from the ith of n normal populations we take two independent samples, each of size 2r+1, and denote the two sample variances by x_i and x_i'. Let

$$(39) \qquad S_n = \sum_1^n u_i x_i' = \text{sum of all the } N_n \text{ values } x_i'$$

for which $x_i \geq a$.

Since

(40) $E(ux'|\theta) = E(u|\theta)\cdot\theta = E(u\theta|\theta)$,

it follows from (36) that

(41) $Eux' = Eu\theta = Ev$,

so that as $n \to \infty$

(42) $$\frac{T_n - S_n}{\sqrt{n}} \xrightarrow{D} N(0, \sigma_3^2),$$

where

(43) $$\sigma_3^2 = E(v-ux')^2 = E(v^2 - 2v\, x' + ux'^2).$$

To complete the discussion of II and III above, we state without proof that $\sigma_2^2 = Ew$ and $\sigma_3^2 = Ez$, where

(44) $$w = u[\frac{1}{1+r}\cdot x^2 + (r-2)a^r x^{2-r} + a^{2r}x^{2-2r} - \frac{r^2}{1+r}a^{1+r}x^{1-r}],$$

$$z = u[\frac{2}{1+r}\cdot x^2 + (r-3)a^r x^{2-r} + a^{2r}x^{2-2r} - \frac{r(r-1)}{1+r}a^{1+r}x^{1-r}].$$

Thus, in analogy with (33),

(45) $$\frac{(T_n/N_n) - (R_n/N_n)}{\frac{1}{\sqrt{N_n}}\sqrt{\frac{1}{N_n}\sum_1^n w_i}} \xrightarrow{D} N(0,1), \quad \frac{(T_n/N_n) - (S_n/N_n)}{\frac{1}{\sqrt{N_n}}\sqrt{\frac{1}{N_n}\sum_1^n z_i}} \xrightarrow{D} N(0,1).$$

For $r = 1$, when (18) reduces to $f(x|\theta) = \theta^{-1}e^{-x/\theta}$, the negative exponential density with mean θ, (44) reduces to

(46) $z = 2w = u(x-a)^2$.

Version III is particularly interesting because the second relation of (45) can be used not only to provide a *prediction interval* for S_n/N_n if the x_i' have not yet been observed, but also to provide a *test* of the null hypothesis that the sample variances x_i and x_i' for which $x_i \geq a$ are in fact from normal populations with the same unknown population variance σ_i^2. To invent an example, suppose that 2r + 1 vital measurements of some sort are taken on each of n persons at different times. It is considered that large variability of the measurements is unhealthy, so to each of the N_n persons with sample variance $x_i \geq a$ (and only to such persons) a purportedly variance reducing treatment is administered. Then 2r+1 new measurements are taken on each of these N_n people, yielding sample variances x_i'. It is found that S_n/N_n is less than T_n/N_n, so that the treatment seems to reduce variance. It is objected that this is precisely what would be expected from a placebo : regression towards mediocrity. But if

$$(47) \qquad \frac{S_n}{N_n} \leq \frac{T_n}{N_n} - \frac{d}{\sqrt{N_n}} \sqrt{\frac{1}{N_n}\sum_1^n z_i}$$

for some d equal to 2 or more, say, it would indicate that the treatment does reduce variance when applied to people with high x_i values, *even though no control group was used*. An analogous result holds for other g.e.B. problems; e.g., for means instead of variances.

Remarks

A. Problems I - III, along with various generalizations of them, may be more important in practice than the original problem of simultaneous inference about many parameters with which e.B. and compound decision theory were introduced in 1950.

B. As indicated in Robbins [7] and Copas [1], there is no basic distinction in practice between the e.B. and the compound formulations, in the latter of which we regard $\theta_1,\ldots,\theta_n$ as unknown constants that are held fixed in repetitions of the experiment that gives rise to $x_1,\ldots,x_n$. The James-Stein estimator of (16) was originally devised to show the inadmissibility of the usual estimator of the multivariate normal mean in the compound sense, while Efron and Morris (see Efron [2] for references) have emphasized the e.B. aspect of (16).

C. Whether Neyman [3] was right in thinking that the e.B. and compound approaches will have some effect on statistical practice remains to be seen. The following quotation from Efron [2] emphasizes one oddity of the compound theory that may account for its neglect in practice: "If the different θ_i refer to obviously disjoint problems (e.g., θ_1 is the speed of light, θ_2 is the price of tea in China, θ_3 is the efficacy of a new treatment for psoriasis, etc.), combining the data can produce a definitely uncomfortable feeling in the statistician." (Compare Robbins [7], page 133, lines 1-4.) A most amusing example of combining different sorts of data is given in E.S. Pearson [4]. Everyone who is interested in Bayesianism should read this paper, together with K. Pearson [5], [6], keeping in mind the warning by D. V. Lindley (page 421 of Copas [1]) on the effect of even a seemingly slight deviation from orthdoxy: "there is no one less Bayesian than an empirical Bayesian".

D. We can combine the two problems that led to (16) and (20) respectively, in order to estimate simultaneously many means *and* variances. Adopting a new notation, let x_{ij} be independent and normal for $i = 1,\ldots,N$ and $j = 1,\ldots,n_i = 2r_i + 1 \geq 2$, with unknown $\mu_i = Ex_{ij}$ and $\sigma_i^2 = \operatorname{Var} x_{ij}$. Define

$$\text{(48)} \qquad \bar{x}_i = \frac{1}{n_i}\sum_j x_{ij}, \quad s_i^2 = \frac{1}{n_i - 1}\sum_j (x_{ij} - \bar{x}_i)^2, \quad q^2 = \frac{1}{N}\sum_i s_i^2,$$

$$\bar{x} = \frac{1}{N}\sum_i \bar{x}_i, \quad S^2 = \frac{1}{N}\sum_i (\bar{x}_i - \bar{x})^2, \quad d^4 = \Big[\frac{1}{N}\sum_i \Big(\frac{r_i}{1+r_i}\Big)s_i^4 - q^4\Big]^+.$$

Then one of the possible l.e.B. ways of estimating the σ_i^2 and μ_i is to use

$$\text{(49)} \qquad \tilde{\sigma}_i^2 = q^2 + \Big[1 - \frac{d^4 + q^4}{(1+r_i)d^4 + q^4}\Big](s_i^2 - q^2),$$

and

$$\tilde{\mu}_i = \bar{x} + \left[1 - \frac{(\tilde{\sigma}_i^2/n_i)}{(\tilde{\sigma}_i^2/n_i) - \frac{1}{N} \sum_i (\tilde{\sigma}_i^2/n_i) + S^2}\right]^+ (\bar{x}_i - \bar{x}). \tag{50}$$

REFERENCES

[1] Copas, J. B. (1969). Compound decisions and empirical Bayes. *J. Roy. Statist. Soc. Ser. B 31*, 397-423.

[2] Efron, B. (1975). Biased versus unbiased estimation. *Adv. in Math. 16*, 259-277.

[3] Neyman, J. (1962). Two breakthroughs in the theory of statistical decision making. *Rev. Int. Statist. Inst. 30*, 11-27.

[4] Pearson, E. S. (1925). Bayes' theorem examined in the light of experimental sampling. *Biometrika 17*, 388-442.

[5] Pearson, K. (1920). The fundamental problem of practical statistics. *Biometrika 13*, 1-16 and 300-301.

[6] Pearson, K. (1924). Note on Bayes' theorem. *Biometrika 16*, 190-193. (See also W. Burnside's note on page 189.)

[7] Robbins, H. (1950). Asymptotically subminimax solutions of compound statistical decision problems. *Proc. Second Berkeley Symposium Math. Stat. and Prob.*, 131-148. University of California Press, Berkeley.

[8] Robbins, H. (1955). An empirical Bayes approach to statistics. *Proc. Third Berkeley Symposium Math. Stat. and Prob. 1*, 157-164. University of California Press, Berkeley.

The Annals of Statistics
1983, Vol. 11, No. 3, 713–723

JERZY NEYMAN MEMORIAL LECTURE

SOME THOUGHTS ON EMPIRICAL BAYES ESTIMATION[1]

BY HERBERT ROBBINS

Columbia University

Examples are given to illustrate the "linear" and "general" empirical Bayes approaches to estimation. A final example concerns testing the null hypothesis that a treatment has had no effect.

1. The component problem. Let (θ, x) be a random vector such that (a) θ has a distribution function G, and (b) conditionally on θ, x has a probability density function $f(x \mid \theta)$ of known form with respect to a σ-finite measure m. We want to estimate θ by some function $t = t(x)$. The mean squared error of estimate with regard to the random variation of both θ and x is

$$E(t - \theta)^2 = \iint (t(x) - \theta)^2 f(x \mid \theta)\, dm(x)\, dG(\theta). \tag{1}$$

For given G we can try to minimize (1) within any desired class of functions t. The minimum of (1) within the class of *linear* functions $A + Bx$ is attained for

$$\tilde{t}(x) = E\theta + \frac{\operatorname{Cov}(\theta, x)}{\operatorname{Var} x}(x - Ex), \tag{2}$$

and is equal to

$$E(\tilde{t} - \theta)^2 = \operatorname{Var}\theta - \frac{\operatorname{Cov}^2(\theta, x)}{\operatorname{Var} x}. \tag{3}$$

The minimum of (1) within the class of *all* Borel functions is attained for

$$t^*(x) = E(\theta \mid x) = \frac{\int \theta f(x \mid \theta)\, dG(\theta)}{\int f(x \mid \theta)\, dG(\theta)}, \tag{4}$$

and is equal to

$$E(t^* - \theta)^2 = E \operatorname{Var}(\theta \mid x). \tag{5}$$

If, as we shall assume, G is unknown to the statistician, then neither $\tilde{t}$ nor t^* can be used directly to estimate θ.

2. The empirical Bayes approach. Suppose that we are dealing with a large number N of independent versions of the component problem: (θ_i, x_i) for $i = 1, \cdots, N$ are i.i.d. random vectors such that the θ_i have distribution function G, while the probability

Received October 1982; revised February 1983.

[1] Research supported by Department of Energy, National Science Foundation, and National Institutes of Health.

AMS 1970 *subject classifications.* Primary 62C12; secondary 62C25.

Key words and phrases. Empirical Bayes, prediction, testing hypotheses.

density function of x_i conditionally on θ_i is $f(x \mid \theta_i)$. By observing $x_1, \cdots, x_N$, which are i.i.d. random variables with marginal p.d.f.

$$f(x) = \int f(x \mid \theta) \, dG(\theta), \tag{6}$$

we can try to gather enough information about G to approximate (2), a modest aim, or even (4), a more grandiose one. It is not obvious how this is to be done, nor whether we should try to approximate (2), (4), or perhaps something else.

One reason that we might be content with approximating (2) rather than (4) is that (2) involves only the constants Ex, Var x, $E\theta$, and Cov(θ, x), which for some families $f(x \mid \theta)$ are easy to estimate from $x_1, \cdots, x_N$. In addition, (2) often provides a minimax value of (1) within the class of all G with given $E\theta$ and Var θ. We may even believe that G belongs, or is close to belonging, to some special parametric family of distribution functions for which (4) is in fact a linear function of x, so that (2) and (4) are equal.

We refer to methods that seek to approximate (2) as *linear empirical Bayes* (l.e.B.), and to those that seek to approximate (4) as *general empirical Bayes* (g.e.B.). For some families $f(x \mid \theta)$ it might be desirable to replace the class of linear functions by the class of functions of the form $(x + A)/(x + B)$, or by some other class. We would then try to approximate whatever function in that class minimizes (1), referring to this as a *restricted empirical Bayes* (r.e.B.) method. For $N \to \infty$ the g.e.B. approach, when feasible, would seem to be better than any restricted approach, such as the linear, but for moderate N a suitable restricted approach with rapid convergence as $N \to \infty$ may be better than a general one that converges slowly.

If the θ_i are regarded not as random variables but simply as N arbitrary unknown constants, and if the loss function is the sum of the N squared errors of estimation, then the problem of simultaneously estimating all the θ_i is called a *compound* estimation problem. The same methods that apply to the e.B. problem will also apply to the compound problem, as indicated in Robbins (1951). Stein's (1956) inadmissibility result and the James-Stein (1961) estimator belong to the compound theory, but were put into an e.B. context in a series of papers by Efron and Morris; e.g., (1973). Except in Section 7 we confine ourselves in what follows to the e.B. theory, starting with l.e.B.

3. Linear empirical Bayes estimation. We shall assume for simplicity that $f(x|\theta)$ is such that for all θ

$$E(x \mid \theta) = \theta. \tag{7}$$

Then

$$Ex = E\theta, \quad \mathrm{Cov}(\theta, x) = \mathrm{Var}\,\theta, \quad E(x - \theta)^2 = \mathrm{Var}\,x - \mathrm{Var}\,\theta = E\,\mathrm{Var}(x \mid \theta), \tag{8}$$

and (2) and (3) can be written as

$$\tilde{t}(x) = Ex + \frac{\mathrm{Var}\,\theta}{\mathrm{Var}\,x}(x - Ex), \quad E(\tilde{t} - \theta)^2 = \frac{\mathrm{Var}\,\theta \cdot E\,\mathrm{Var}(x \mid \theta)}{\mathrm{Var}\,\theta + E\,\mathrm{Var}(x \mid \theta)}. \tag{9}$$

We shall also assume, in addition to (7), that for some known constants a, b, c

$$\mathrm{Var}(x \mid \theta) = a + b\theta + c\theta^2. \tag{10}$$

Then

$$\mathrm{Var}\,x = E\,\mathrm{Var}(x \mid \theta) + \mathrm{Var}\,E(x \mid \theta) = a + bE\theta + cE^2\theta + (c + 1)\mathrm{Var}\,\theta, \tag{11}$$

so that

$$\mathrm{Var}\,\theta = \frac{\mathrm{Var}\,x - (a + bEx + cE^2x)}{c + 1}.$$

This suggests as an approximation to $\tilde{t}(x)$ the function

$$\tilde{\theta}(x) = \bar{x} + \left[1 - \frac{cs^2 + a + b\bar{x} + c\bar{x}^2}{(c+1)s^2}\right](x - \bar{x}), \tag{12}$$

where Ex and $\mathrm{Var}\, x$ in the first equation of (9) are conventionally estimated by

$$\bar{x} = \frac{1}{N}\sum_1^N x_i \quad \text{and} \quad s^2 = \frac{1}{N-1}\sum_1^N (x_i - \bar{x})^2. \tag{13}$$

(Of course, other estimators of Ex and $\mathrm{Var}\, x$ could be used. Moreover, the term in square brackets in (12) is an estimate of $\mathrm{Var}\,\theta/\mathrm{Var}\, x$, which by (8) lies beween 0 and 1, so that we should truncate this term accordingly if sampling fluctuations of $\bar{x}$ and s^2 so require. We shall not bother to indicate this notationally.)

We call

$$\tilde{\theta}_i = \tilde{\theta}(x_i) = \bar{x} + \left[1 - \frac{cs^2 + a + b\bar{x} + c\bar{x}^2}{(c+1)s^2}\right](x_i - \bar{x}) \tag{14}$$

an l.e.B. estimator of θ_i, and under some mild restrictions on the nature of G we hope that with reasonable rapidity as $N \to \infty$

$$\begin{aligned} E(\tilde{\theta}_i - \theta_i)^2 &= \frac{1}{N}\sum_1^N E(\tilde{\theta}_i - \theta_i)^2 \to E(\tilde{t} - \theta)^2 \\ &= \frac{\mathrm{Var}\,\theta \cdot E\,\mathrm{Var}(x \mid \theta)}{\mathrm{Var}\,\theta + E\,\mathrm{Var}(x \mid \theta)} = \frac{\mathrm{Var}\,\theta}{\mathrm{Var}\, x} E(x - \theta)^2, \end{aligned} \tag{15}$$

where for large N

$$\frac{\mathrm{Var}\,\theta}{\mathrm{Var}\, x} = \frac{\mathrm{Var}\,\theta}{a + bE\theta + cE^2\theta + (c+1)\mathrm{Var}\,\theta} \cong 1 - \frac{cs^2 + a + b\bar{x} + c\bar{x}^2}{(c+1)s^2}. \tag{16}$$

When the x_i are such that the last quantity is appreciably less than 1 it will seem advantageous to estimate θ_i by $\tilde{\theta}_i$ instead of by x_i, even if a decision to do so was not made in advance.

The most familiar case of (14) is that where for a known $\sigma > 0$, and for m = Lebesgue measure,

$$f(x \mid \theta) = \phi\left(\frac{x - \theta}{\sigma}\right)\Big/\sigma, \tag{17}$$

where ϕ is the p.d.f. of some random variable with mean 0 and variance 1, such as the standard normal. Then

$$E(x \mid \theta) = \theta, \quad \mathrm{Var}(x \mid \theta) = \sigma^2, \tag{18}$$

so that $a = \sigma^2$, $b = c = 0$ in (10), and (14) becomes

$$\tilde{\theta}_i = \bar{x} + \left[1 - \frac{\sigma^2}{s^2}\right](x_i - \bar{x}), \tag{19}$$

with

$$\frac{\mathrm{Var}\,\theta}{\mathrm{Var}\, x} = \frac{\mathrm{Var}\,\theta}{\sigma^2 + \mathrm{Var}\,\theta} \cong 1 - \frac{\sigma^2}{s^2}. \tag{20}$$

The estimator (19) is a variant of the James-Stein (1961) estimator of a multivariate normal mean referred to at the end of Section 2 above.

Before considering other examples, we shall generalize slightly the assumptions of Section 2 on the vectors (θ_i, x_i), which we now assume to be independent but not necessarily identically distributed, and such that

(21a) the θ_i have the same mean $E\theta$ and the same variance $\operatorname{Var}\theta$ for all i

(21b) $E(x_i \mid \theta_i) = \theta_i$,

(21c) for some known constants a_i, b_i, c_i,

$$\operatorname{Var}(x_i \mid \theta_i) = a_i + b_i\theta_i + c_i\theta_i^2.$$

Then with s^2 defined by (13),

$$Ex_i = E\bar{x} = E\theta, \quad Es^2 = 1/N \textstyle\sum_1^N \operatorname{Var} x_i = \bar{a} + \bar{b}E\theta + \bar{c}E^2\theta + (\bar{c}+1)\operatorname{Var}\theta, \tag{22}$$

where $\bar{a} = \sum_1^N a_i/N$, etc. If, without regard to efficiency, we simply estimate

$$E\theta \quad \text{by} \quad \bar{x},$$

$$\operatorname{Var}\theta \quad \text{by} \quad \frac{s^2 - (\bar{a} + \bar{b}\bar{x} + \bar{c}\bar{x}^2)}{\bar{c}+1},$$

and

$$\operatorname{Var} x_i = E \operatorname{Var}(x_i \mid \theta_i) + \operatorname{Var} E(x_i \mid \theta_i) = a_i + b_iE\theta + c_iE^2\theta + (c_i + 1)\operatorname{Var}\theta$$

$$\text{by} \quad a_i + b_i\bar{x} + c_i\bar{x}^2 + \frac{(c_i+1)}{(\bar{c}+1)}\{s^2 - (\bar{a} + \bar{b}\bar{x} + \bar{c}\bar{x}^2)\},$$

then we can approximate

$$\tilde{t}(x_i) = E\theta + \frac{\operatorname{Var}\theta}{\operatorname{Var} x_i}(x_i - E\theta) \tag{23}$$

by

$$\tilde{\theta}_i = \bar{x} + \left[1 - \frac{(\bar{c}+1)(a_i + b_i\bar{x} + c_i\bar{x}^2) + c_i\{s^2 - (\bar{a} + \bar{b}\bar{x} + \bar{c}\bar{x}^2)\}}{(\bar{c}+1)(a_i + b_i\bar{x} + c_i\bar{x}^2) + (c_i+1)\{s^2 - (\bar{a} + \bar{b}\bar{x} + \bar{c}\bar{x}^2)\}}\right](x_i - \bar{x}). \tag{24}$$

We continue to call (24) an l.e.B. estimate of θ_i under the assumptions (21). When $a_i = a$, $b_i = b$, $c_i = c$, (24) reduces to (14).

4. Examples of l.e.B. estimation.

EXAMPLE 1. For $i = 1, \cdots, N$ let y_{ij} for $j = 1, \cdots, n_i$ be independent, conditionally on $\theta_1, \cdots, \theta_N$, with mean θ_i and known variance σ_i^2, and let

$$x_i = \frac{1}{n_i}\textstyle\sum_{j=1}^{n_i} y_{ij} = i\text{th sample mean.} \tag{25}$$

Then

$$E(x_i \mid \theta_i) = \theta_i, \quad \operatorname{Var}(x_i \mid \theta_i) = \frac{\sigma_i^2}{n_i} = a_i. \tag{26}$$

Assuming that the θ_i are independent with a common mean and variance, (24) yields the l.e.B. estimator

$$\tilde{\theta}_i = \bar{x} + \left[1 - \frac{a_i}{(a_i - \bar{a}) + s^2}\right](x_i - \bar{x}). \tag{27}$$

Of course, it is inefficient to estimate $E\theta$ by $\bar{x}$ instead of by the weighted average

$$\textstyle\sum_1^N (n_ix_i/\sigma_i^2)/\sum_1^N (n_i/\sigma_i^2), \tag{28}$$

and in practice we would use (27) only when the a_i are nearly equal. Many other modifications of (27) could be made to improve its performance for moderate N. We shall not pursue these questions here, although they are important in the practical application of l.e.B. methods.

EXAMPLE 2. As in Example 1 but with the σ_i^2 unknown. Defining

(29) $$s_i^2 = \frac{1}{n_i - 1} \sum_{j=1}^{n_i} (y_{ij} - x_i)^2 = i\text{th sample variance},$$

and assuming that the n_i are large, we could replace the $a_i = \sigma_i^2/n_i$ in (27) by $a_i = s_i^2/n_i$ and still regard (27) as an l.e.B. estimate of θ_i. However, if the y_{ij} are *normal*, then setting $r_i = (n_i - 1)/2$ we have

(30) $$E(s_i^2 \mid \sigma_i^2) = \sigma_i^2, \quad \text{Var}(s_i^2 \mid \sigma_i^2) = \sigma_i^4/r_i,$$

so that (21b) holds with θ_i replaced by σ_i^2 and x_i by s_i^2, and (21c) holds with $a_i = b_i = 0$, $c_i = 1/r_i$. If we regard the σ_i^2 as independent random variables with a common mean and variance, we obtain from (24) the l.e.B. estimates (cf. Robbins (1982), formula (49))

(31) $$\tilde{\sigma}_i^2 = q^2 + \left[1 - \frac{R^4 + q^4}{R^4 + q^4 + r_i(R^4 - \bar{c}q^4)}\right](s_i^2 - q^2),$$

where by definition

(32) $$q^2 = \frac{1}{N} \sum_1^N s_i^2, \quad R^4 = \frac{1}{N-1} \sum_1^N (s_i^2 - q^2)^2, \quad \bar{c} = \frac{1}{N} \sum_1^N \frac{1}{r_i}.$$

For large N we could then use (27) with $a_i = \tilde{\sigma}_i^2/n_i$ to obtain "doubly" l.e.B. estimates of the population means θ_i. The assumption of normality for the y_{ij} can be dropped by a suitable generalization of (31).

Another l.e.B. approach that dispenses with normality to the simultaneous estimation of many population means $\theta_1, \cdots, \theta_N$ is the following. Let $F_1, \cdots, F_N$ be i.i.d. *random distribution functions*, and for given F_i $(i = 1, \cdots, N)$ let the y_{ij} $(j = 1, \cdots, n_i)$ be distributed according to F_i. Define x_i by (25), s_i^2 by (29), $\bar{x}$ and s^2 by (13), and let

$$\theta_i = \int t \, dF_i(t) = \text{random mean of population } i = E(y_{ij} \mid F_i),$$

$$\sigma_i^2 = \int (t - \theta_i)^2 \, dF_i(t) = \text{random variance of population } i = \text{Var}(y_{ij} \mid F_i).$$

Then $\theta_1, \cdots, \theta_N$ are i.i.d. random variables, as are $\sigma_1^2, \cdots, \sigma_N^2$. We observe that

$$E(x_i \mid F_i) = \theta_i, \quad Ex_i = E\theta_i = E\theta = E\bar{x}, \quad \text{Cov}(\theta_i, x_i) = \text{Var}\,\theta,$$

$$\text{Var}\,x_i = E\,\text{Var}(x_i \mid F_i) + \text{Var}\,E(x_i \mid F_i) = E\,\text{Var}(x_i \mid F_i) + \text{Var}\,\theta$$

$$Es^2 = (1/N) \sum_1^N \text{Var}\,x_i = (1/N) \sum_1^N E\,\text{Var}(x_i \mid F_i) + \text{Var}\,\theta,$$

so that (23) can be written as

(33) $$\tilde{t}(x_i) = E\theta + \left[\frac{Es^2 - (1/N) \sum_1^N E\,\text{Var}(x_i \mid F_i)}{Es^2 + E\,\text{Var}(x_i \mid F_i) - (1/N) \sum_1^N E\,\text{Var}(x_i \mid F_i)}\right](x_i - E\theta).$$

But, defining q^2 by (32), we have

$$\text{Var}(x_i \mid F_i) = \frac{\sigma_i^2}{n_i}, \quad E(s_i^2 \mid F_i) = \sigma_i^2, \quad Es_i^2 = E\sigma_i^2 = E\sigma^2 = Eq^2,$$

so that for large N,

$$E\,\mathrm{Var}(x_i\,|\,F_i) = \frac{E\sigma^2}{n_i} \cong \frac{q^2}{n_i}, \quad Es^2 \cong s^2, \quad E\theta \cong \bar{x}.$$

Hence, we can approximate $\tilde{t}(x_i)$ by (27) with

$$a_i = \frac{q^2}{n_i} = \frac{1}{n_i}\left(\frac{1}{N}\sum_1^N s_i^2\right), \tag{34}$$

in contrast to $a_i = \sigma_i^2/n_i$ known, $a_i = s_i^2/n$, or $a_i = \tilde{\sigma}_i^2/n_i$ with $\tilde{\sigma}_i^2$ given by (31) when the F_i are normal.

Example 3. Let y_i denote the number of successes in n_i Bernoulli trials with success probability p_i, and let $x_i = y_i/n_i$. Then

$$E(x_i\,|\,p_i) = p_i, \quad \mathrm{Var}(x_i\,|\,p_i) = p_i(1-p_i)/n_i,$$

so that in (21c) $a_i = 0$, $b_i = -c_i = 1/n_i$, and the l.e.B. estimate (24) of $\theta_i = p_i$ becomes

$$\tilde{p}_i = \bar{x} + \left[1 - \frac{\bar{x}(1-\bar{x}) - s^2}{\bar{x}(1-\bar{x}) - s^2 + n_i\{s^2 - \bar{b}\bar{x}(1-\bar{x})\}}\right](x_i - \bar{x}).$$

This estimate was introduced by Copas (1972) with a slightly different motivation.

Example 4. Returning to (14), let x denote the number of successes before the first failure in a sequence of Bernoulli trials with success probability $p = 1 - q$. Then

$$f(x\,|\,p) = qp^x \quad (x = 0, 1, \cdots) \tag{35}$$

and

$$E(x\,|\,p) = p/q, \quad \mathrm{Var}(x\,|\,p) = p/q^2. \tag{36}$$

To satisfy (7) and (10) we introduce $\theta = p/q$ as the parameter to be estimated, so that

$$E(x\,|\,\theta) = \theta, \quad \mathrm{Var}(x\,|\,\theta) = \theta(1+\theta), \tag{37}$$

with $a = 0$, $b = c = 1$ in (10). Then (14) becomes

$$\tilde{\theta}_i = \bar{x} + \left[1 - \frac{s^2 + \bar{x}(1+\bar{x})}{2s^2}\right](x_i - \bar{x}), \tag{38}$$

which by (15) and (16) improves on x_i for large N by the factor

$$\frac{\mathrm{Var}\,\theta}{\mathrm{Var}\,x} = \frac{\mathrm{Var}\,\theta}{E\theta(1+E\theta) + 2\,\mathrm{Var}\,\theta} \cong \left\{1 - \frac{\bar{x}(1+\bar{x})}{s^2}\right\}\Big/ 2. \tag{39}$$

5. G.e.B. estimation; an example. The problem of estimation for the examples $f(x\,|\,\theta)$ of the preceding section, along with many others, can also be considered from the g.e.B. aspect of approximating (4) instead of (2), when the component problems are identical and N is large. We shall treat here only Example 4, the geometric distribution with parameter $\theta = p/q$, for which (6) becomes

$$f(x) = \int_0^1 qp^x\,dG(p) \tag{40}$$

and the unrestrictedly best estimator of θ is

$$t^*(x) = E(\theta\,|\,x) = \int_0^1 p^{x+1}\,dG(p)/f(x). \tag{41}$$

Since

$$p^{x+1} = \sum_{i=0}^{\infty} qp^{x+i+1}, \tag{42}$$

we can write $t^*(x)$ in terms of $f(x)$ as

$$t^*(x) = [f(x+1) + f(x+2) + \cdots]/f(x). \tag{43}$$

This suggests defining a g.e.B. estimate of θ_i by

$$\theta_i^* = \frac{\text{number of terms } x_1, \cdots, x_N \text{ that are greater than } x_i}{\text{number of terms } x_1, \cdots, x_N \text{ that are equal to } x_i}. \tag{44}$$

We remark that, contrary to the situation in Example 4 of the preceding section, it is easy to obtain a g.e.B. estimator of p_i itself rather than of $\theta_i = p_i/q_i$. Since

$$E(p \mid x) = \int_0^1 qp^{x+1}\, dG(p)/f(x) = f(x+1)/f(x), \tag{45}$$

it is natural to define a g.e.B. estimate of p_i by

$$p_i^* = \frac{\text{number of terms } x_1, \cdots, x_N \text{ that are equal to } (x_i+1)}{\text{number of terms } x_1, \cdots, x_N \text{ that are equal to } x_i}. \tag{46}$$

We have in (38) and (44) two quite different e.B. estimators of $\theta_i = p_i/q_i$ for the case (35). But when G belongs to the Beta family, with

$$G'(p) = g(p) = \frac{1}{B(a,b)} p^{a-1} q^{b-1} \quad (a, b > 0), \tag{47}$$

then (2) and (4) are equal, so that (38) and (44) are approximations to the same thing. In fact, when (47) holds (40) becomes

$$f(x) = \frac{B(x+a, b+1)}{B(a,b)}, \tag{48}$$

and hence by (41)

$$t^*(x) = \frac{B(x+1+a, b)}{B(x+a, b+1)} = \frac{x+a}{b} = \tilde{t}(x). \tag{49}$$

From (7), (8), and (11) it follows that

$$Ex + \frac{\operatorname{Var} x - Ex(1+Ex)}{2 \operatorname{Var} x}(x - Ex) = \frac{x+a}{b}, \tag{50}$$

so that

$$b = \frac{2 \operatorname{Var} x}{\operatorname{Var} x - Ex(1+Ex)} \cong \frac{2s^2}{s^2 - \bar{x}(1+\bar{x})},$$

$$a = Ex \cdot \left[\frac{\operatorname{Var} x + Ex(1+Ex)}{\operatorname{Var} x - Ex(1+Ex)}\right] \cong \frac{\bar{x}\{s^2 + \bar{x}(1+\bar{x})\}}{s^2 - \bar{x}(1+\bar{x})}, \tag{51}$$

$$\frac{x+a}{b} \cong \bar{x} + \left[1 - \frac{s^2 + \bar{x}(1+\bar{x})}{2s^2}\right](x - \bar{x}) = (38) \text{ with } x_i = x.$$

Thus, for the family of Beta priors the l.e.B. estimator (38) of θ_i is an estimate by the method of moments of $t^*(x_i)$. (The g.e.B. estimator (44) is a more generally valid estimate of $t^*(x_i)$, but presumably converges more slowly to (41) as $N \to \infty$.) We could, of course, estimate a and b in (49) by maximum likelihood, finding for given $x_1, \cdots, x_N$ the maximum with respect to a, b of

$$\prod_{i=1}^{N} \frac{B(x_i + a, b + 1)}{B(a, b)},$$

but this is a dubious procedure if there is a possibility of a non-Beta prior. The virtue of (38) would seem to lie in the fact that for large N it is nearly equal to (2) for *any* G. (It is also true that

$$\tilde{t}(x) = E\theta + \frac{\operatorname{Var}\theta}{E\theta(1 + E\theta) + 2\operatorname{Var}\theta}(x - E\theta) \tag{52}$$

is minimax in the class of all priors with given $E\theta$ and Var θ, but no great importance attaches to this fact in itself.) For very large N, (44) would seem to be preferable to (38) unless there is some iron-clad reason to believe that G is Beta. A method of combining the best features of l.e.B. and g.e.B. is given in Robbins (1980), but finding a more or less optimal way of treating both moderate and large values of N will require considerable theoretical and computational study.

6. Prediction and testing. We have thus far considered only the usual e.B. problem of simultaneous estimation of many parameters. We consider now a problem (cf. Robbins, 1977) in which no parameters at all are to be estimated.

Let $\theta_1, \cdots, \theta_N$ be i.i.d. with unknown distribution function G, and conditionally on θ_i let x_i and y_i be independent random variables with same probability density function $f(\cdot \mid \theta_i)$. Thus, (x_i, y_i) for $i = 1, \cdots, N$ are i.i.d. random vectors with joint marginal p.d.f.

$$f(x, y) = \int f(x \mid \theta) f(y \mid \theta)\, dG(\theta). \tag{53}$$

We suppose that the x_i have been observed but not the y_i; the θ_i, in practice, will never be observed. Let A be some set of interest in the x-space, and let

$$S = \sum_{x_i \in A} y_i \tag{54}$$

be *the sum of all the not-yet-observed y_i values for which the corresponding observed x_i values belong to A*. Problem: find a good function $g(x_1, \cdots, x_N)$ to use as a predictor of S.

For example, N typists are chosen at random from some population about which nothing is known, and each types a text until he or she makes an error. Let x_i denote the number of symbols typed correctly by typist i, and let ν denote the number of typists whose x_i values are $\geq a$, where a is some fixed positive integer. These ν typists are to be set to typing again, until each makes an error. Let S denote the total number of symbols that will be typed correctly by the ν typists in this second round. From the values $x_1, \cdots, x_N$, predict S. (It is assumed that for typist i, the typing of a correct or incorrect symbol forms a sequence of Bernoulli trials with constant probability p_i of success throughout both rounds, p_i being characteristic of typist i but varying with i in some unknown manner, since nothing is known about the population from which the N typists were randomly selected.)

Clearly, ν times the average of $x_1, \cdots, x_N$ will generally underestimate S, while the sum of the ν values of x_i that are $\geq a$ will generally overestimate S. What, then, is a reasonable predictor for S, and how accurate is it likely to be? We shall treat this as a g.e.B. problem.

For the density (35), (53) becomes

$$f(x, y) = \int_0^1 q^2 p^{x+y}\, dG(p), \tag{55}$$

a function of $x + y$ *only*. It follows that

$$\text{(56)} \qquad \begin{array}{l}\text{conditionally on the value of } z = x + y, \\ \text{the distribution of } x \text{ is uniform on the set } \{0, 1, 2, \cdots, z\}.\end{array}$$

From now on we ignore the explicit form (55) of $f(x, y)$ and assume only (56), which is more general.

Let (x, y) be any random vector with $x, y = 0, 1, \cdots$ such that (56) holds, and let (x_i, y_i), $i = 1, \cdots, N$, be i.i.d. with the same distribution as (x, y). Let

$$\text{(57)} \qquad S = \sum_{x_i \ge a} y_i = \sum_1^N u(x_i)\, y_i,$$

where by definition

$$\text{(58)} \qquad u(x) = \begin{matrix} 1 & \text{if} & x \ge a \\ 0 & \text{if} & x < a. \end{matrix}$$

By (56),

$$\text{(59)} \qquad E[u(x)\,y \mid z] = \frac{1}{z+1} \sum_{x=a}^{z} (z - x) = \frac{1 + 2 + \cdots + (z - a)}{z + 1}$$

for $z \ge a$, and is 0 for $z < a$. Now define

$$\text{(60)} \qquad v(x) = (x - a)^+ = \begin{matrix} x - a & \text{if} & x \ge a \\ 0 & \text{if} & x < a \end{matrix},$$

and observe that

$$\text{(61)} \qquad E[v(x) \mid z] = \frac{1}{z+1} \sum_{x=a}^{z} (x - a) = \frac{1 + 2 + \cdots + (z - a)}{z + 1}$$

for $z \ge a$, and is 0 for $z < a$. Since z was arbitrary, it follows from (59) and (61) that $E[u(x)y] = E[v(x)]$, and hence that *the statistic*

$$\text{(62)} \qquad T = \sum_1^N v(x_i)$$

has the same expected value as S. Thus,

$$\text{(63)} \qquad S - T = \sum_1^N \left(u(x_i)\, y_i - v(x_i)\right)$$

is the sum of N i.i.d. random variables with mean 0, so that as $N \to \infty$

$$\text{(64)} \qquad \frac{S - T}{\sigma \sqrt{N}} \to N(0, 1) \quad \text{in distribution,}$$

where by definition

$$\text{(65)} \qquad \sigma^2 = E[u(x)\,y - v(x)]^2 = E[u(x)\{y - v(x)\}^2],$$

and

$$\text{(66)} \qquad \begin{aligned} E[u(x)\{y - v(x)\}^2 \mid z] &= E[u(x)\{z - x - (x - a)\}^2 \mid z] \\ &= E[u(x)\{z - 2x + a\}^2 \mid z] = \frac{1}{z+1} \sum_{x=a}^{z} (z - 2x + a)^2 \\ &= \frac{1}{z+1} \sum_{i=0}^{k} (k - 2i)^2 \quad \text{where} \quad k = z - a \\ &= \frac{k(k+1)(k+2)}{3(z+1)} \end{aligned}$$

for $z \ge a$, and is 0 for $z < a$.

We now define

$$w(x) = (x - a)^+(x - a + 1), \tag{67}$$

and observe that

$$E[w(x) \mid z] = \frac{1}{z+1} \sum_{x=a}^{z} (x - a)(x - a + 1) = \frac{1}{z+1} \sum_{i=0}^{k} i(i+1) = \frac{k(k+1)(k+2)}{3(z+1)}$$

for $z \geq a$, and is 0 for $z < a$. Hence $Ew(x) = \sigma^2$.

It follows that as $N \to \infty$

$$(1/N) \textstyle\sum_1^N w(x_i) \to Ew(x) = \sigma^2 \quad \text{in probability}, \tag{68}$$

and hence from (64) that as $N \to \infty$

$$\frac{S - T}{\sqrt{\sum_1^n w(x_i)}} \to N(0, 1) \quad \text{in distribution}, \tag{69}$$

so that *for large N an approximately* 95% *prediction interval for S is given by*

$$\textstyle\sum_1^N (x_i - a)^+ \pm 1.96 \sqrt{\sum_1^N (x_i - a)^+(x_i - a + 1)}. \tag{70}$$

This quantifies in the case of (56) the general phenomenon of "regression toward mediocrity."

If it should turn out that S does not lie in the interval (70), it would be reasonable to suspect that the ν typists with scores $x_i \geq a$ on the first round have changed their performance characteristics p_i on the second round. If some "treatment" has in fact been given to these typists between the first and second rounds, we can test the null hypothesis H_0 that the treatment has had no effect *even though no randomly chosen subset of the* ν *has been singled out as a control group*; cf. Robbins (1982).

7. Concluding remarks. Neyman (1962) brought the compound and e.B. approaches to the attention of the general statistical public, and gave great impetus to the subsequent development of the theory. According to a recent note of Forcina (1982), the e.B. approach was anticipated by Gini in 1911.

We conclude with the following estimation theorem (cf. Tsui and Press, 1982), the proof of which is an interesting elementary exercise. Let $x_1, \cdots, x_N$ be independent Poisson random variables with respective means $\theta_1, \cdots, \theta_N$, and let $\bar{x} = \sum x_i/N$, $\bar{\theta} = \sum \theta_i/N$, $\alpha = \bar{\theta}/(1 + \bar{\theta})$. Then for any $N \geq 2$

$$E\left\{\frac{1}{N} \sum_{i=1}^{N} \frac{\left(\frac{\bar{x} x_i}{1 + \bar{x}} - \theta_i\right)^2}{\theta_i}\right\} < \alpha + \frac{1}{N}(1 - \alpha^2) < 1. \tag{71}$$

As stated, (71) belongs to the compound theory; it becomes e.B. if the θ_i, as well as the x_i, are random variables, possibly dependent, but with finite expectations. Then the same inequality holds, with α equal to $\mu/(1 + \mu)$, $\mu = \sum E\theta_i/N$.

I think that Neyman would have appreciated this result.

REFERENCES

COPAS, J. B. (1972). Empirical Bayes methods and the repeated use of a standard. *Biometrika* **59** 349–360.

EFRON, B. and MORRIS, C. (1973). Stein's estimation rule and its competitors—an empirical Bayes approach. *J. Amer. Statist. Assoc.* **68** 117-130.

FORCINA, A. (1982). Gini's contributions to the theory of inference. *Int. Statist. Rev.* **50** 65-70.

JAMES, W. and STEIN, C. (1961). Estimation with quadratic loss. *Proc. Fourth Berkeley Symp. Math. Statist. Probab.* 361-379. Univ. of Calif. Press.

NEYMAN, J. (1962). Two breakthroughs in the theory of statistical decision making. *Rev. Intern. Statist. Inst.* **30** 11-27.

ROBBINS, H. (1951). Asymptotically subminimax solutions of compound statistical decision problems. *Proc. Second Berkeley Symp. Math Statist. Probab.* 131-148. Univ. of Calif. Press.

ROBBINS, H. (1956). An empirical Bayes approach to statistics. *Proc. Third Berkeley Symp. Math. Statist. Probab.* **1** 157-163. Univ. of Calif. Press.

ROBBINS, H. (1977). Prediction and estimation for the compound Poisson distribution. *Proc. Natl. Acad. Sci. U.S.A.* **74** 2670-2671. See also Robbins, H. (1979). Some estimation problems for the compound Poisson distribution. *Asymptotic Theory of Statistical Tests and Estimation* 251-257. Academic, New York.

ROBBINS, H. (1980). An empirical Bayes estimation problem. *Proc. Natl. Acad. Sci. USA* **77** 6988-6989.

ROBBINS, H. (1982). Estimating many variances. *Statistical Decision Theory and Related Topics* III, Vol. 2, 218-226. Academic, New York.

STEIN, C. (1956). Inadmissibility of the usual estimator for the mean of a multivariate normal distribution. *Proc. Third Berkeley Symp. Math. Statist. Probab.* **1** 197-206. Univ. of Calif. Press.

TSUI, K.-W. and PRESS, S. J. (1982). Simultaneous estimation of several Poisson parameters under *k*-normalized squared error loss. *Ann. Statist.* **10** 93-100.

DEPARTMENT OF STATISTICS
COLUMBIA UNIVERSITY
NEW YORK, NEW YORK 10027
and
APPLIED MATHEMATICS DEPARTMENT
BROOKHAVEN NATIONAL LABORATORY
UPTON, NEW YORK 11973

Part Two

SEQUENTIAL EXPERIMENTATION AND ANALYSIS

A. Stochastic Approximation

In 1951, Robbins and his student, Sutton Monro, founded the subject of stochastic approximation with the publication of their celebrated paper [26]*. Consider the problem of finding the root θ (assumed unique) of an equation $g(x) = 0$. In the classical Newton–Raphson method, for example, we start with some value x_0, and if at stage n our estimate of the root is x_n, we define the next approximation by

$$x_{n+1} = x_n - g(x_n)/g'(x_n). \tag{1}$$

Suppose that g is unknown and that at the level x we observe the output $y = g(x) + \varepsilon$, where ε represents some random error with mean 0 and variance $\sigma^2 > 0$. Suppose that in analogy with (1) we use the recursion

$$x_{n+1} = x_n - y_n/\beta, \tag{2}$$

where $\beta = g'(\theta)$ is momentarily assumed known, and $y_n = g(x_n) + \varepsilon_n$ is the observed output at the design level x_n. By (2), the convergence of x_n to θ would entail the convergence ε_n to 0, which does not hold for typical models of random noise (e.g., iid ε_n).

Assuming that $g(x) > 0$ if $x > 0$ and $g(x) < 0$ if $x < 0$, Robbins and Monro suggested using instead of (2) the recursion

$$x_{n+1} = x_n - a_n y_n, \tag{3}$$

where a_n are positive constants such that

$$\sum a_n^2 < \infty \quad \text{and} \quad \sum a_n = \infty. \tag{4}$$

*Papers reprinted in this book are marked here by an asterisk when they first appear.

Under certain assumptions on the random errors ε_n, they showed that x_n converges to θ in L_2. Later, Blum (1954) showed that x_n also converges a.s. to θ, while Chung (1954) and Sacks (1958) showed that an asymptotically optimal choice of a_n is $a_n \sim 1/(n\beta)$, for which

$$n^{1/2}(x_n - \theta) \xrightarrow{\mathscr{D}} N(0, \sigma^2/\beta^2). \tag{5}$$

The Robins–Monro stochastic approximation scheme can be readily modified to find the maximum (or minimum) of a regression function, as was first shown by Kiefer and Wolfowitz (1952). A more general class of recursive stochastic algorithms that include both the Robbins–Monro and the Kiefer–Wolfowitz schemes was subsequently proposed by Dvoretzky (1956) who also established their almost sure and L_2 consistency. This result is commonly called "Dvoretzky's approximation theorem." In 1971, Robbins and Siegmund [87]* proved a general convergence theorem (for nonnegative almost supermartingales) which includes the almost sure part of Dvoretzky's approximation theorem as a special case and which can be readily applied to establish the almost sure convergence of a broad class of recursive stochastic algorithms.

There is an extensive literature on convergence and asymptotic normality of the Robbins–Monro and Kiefer–Wolfowitz schemes and their multivariate extensions. Monographs and surveys of the field up to the early 1970's include the review papers of Schmetterer (1961, 1969) and of Fabian (1971), and the books by Albert and Gardner (1967) and by Nevel'son and Has'minskii (1972).

In the late 1970's Robbins returned to the subject of stochastic approximation. In a series of joint papers [107]*, [108], [113]*, [114], and [119] with Lai, adaptive stochastic approximation schemes in ignorance of $\beta = g'(\theta)$ are developed. Not only are these schemes asymptotically optimal, in the sense of (5), for estimating θ, but they are also asymptotically optimal for minimizing $\Sigma_{i=1}^n y_i^2$. The latter goal arises in adaptive control applications, as is discussed in [107] and is developed further in the papers [121], [122] on the so-called "multiperiod control problem" in econometrics. Thus, stochastic approximation is not only useful for estimating the root (or maximum) of a regression function but also provides asymptotically optimal solutions to adaptive control problems. Further extensions of the stochastic approximation idea to develop recursive feedback control schemes for linear dynamic systems are given by Goodwin, Ramadge, and Caines (1982) and by Lai and Wei (1982).

Other developments in stochastic approximation during the last decade include recursive estimation (cf. Fabian, 1978; Solo, 1981), constrained optimization in stochastic systems (cf. Kushner and Clark, 1978), and stability analysis of associated ordinary differential equations (cf. Ljung, 1977).

References

Albert, A. E. and Gardner, L. A. (1967). *Stochastic Approximation and Nonlinear Regression*. MIT Press, Cambridge, MA.

Blum, J. (1954). Approximation methods which converge with probability one. *Ann. Math. Statist.* **25**, 382–386.

Chung, K. L. (1954). On a stochastic approximation method. *Ann. Math. Statist.* **25**, 463–483.

Dvoretzky, A. (1956). On stochastic approximation. *Proc. Third Berkeley Symp. Math. Statist. Probab.* **1**, 39–56. Univ. Calif. Press.

Fabian, V. (1971). Stochastic approximation. In *Optimizing Methods in Statistics* (ed. J. S. Rustagi), 439–470. Academic Press, New York.

Fabian, V. (1978). On asymptotically efficient recursive estimation. *Ann. Statist.* **6**, 854–866.

Goodwin, G. C., Ramadge, P. J., and Caines, P. E. (1981). Discrete time stochastic adaptive control. *SIAM J. Control & Optimization* **19**, 829–853.

Kiefer, J. and Wolfowitz, J. (1952). Stochastic estimation of the maximum of a regression function. *Ann. Math. Statist.* **23**, 462–466.

Kushner, H. J. and Clark, D. S. (1978). *Stochastic Approximation Methods for Constrained and Unconstrained Systems*. Springer-Verlag, New York.

Lai, T. L. and Wei, C. Z. (1982). Least squares estimates in stochastic regression models with applications to identification and control of dynamic systems. *Ann. Statist.* **10**, 154–166.

Ljung, L. (1977). Analysis of recursive stochastic algorithms. *IEEE Trans. Autom. Contr.* **AC-22**, 551–575.

Nevel'son, M. B. and Has'minskii, R. Z. (1972). *Stochastic Approximation and Recursive Estimation*. Amer. Math. Soc. Translations, Providence.

Sacks, J. (1958). Asymptotic distribution of stochastic approximation procedures. *Ann. Math. Statist.* **29**, 373–405.

Schmetterer, L. (1961). Stochastic approximation. *Proc. Fourth Berkeley Symp. Math. Statist. Probab.* **1**, 597–609. Univ. Calif. Press.

Schmetterer, L. (1968). Multidimensional stochastic approximation. In *Multivariate Analysis II* (ed. P. R. Krishnaiah), 443–460. Academic Press, New York.

Solo, V. (1981). The second order properties of a time series recursion. *Ann. Statist.* **9**, 307–317.

A STOCHASTIC APPROXIMATION METHOD[1]

By Herbert Robbins and Sutton Monro

University of North Carolina

1. Summary. Let $M(x)$ denote the expected value at level x of the response to a certain experiment. $M(x)$ is assumed to be a monotone function of x but is unknown to the experimenter, and it is desired to find the solution $x = \theta$ of the equation $M(x) = \alpha$, where α is a given constant. We give a method for making successive experiments at levels $x_1, x_2, \cdots$ in such a way that x_n will tend to θ in probability.

2. Introduction. Let $M(x)$ be a given function and α a given constant such that the equation

$$M(x) = \alpha \tag{1}$$

has a unique root $x = \theta$. There are many methods for determining the value of θ by successive approximation. With any such method we begin by choosing one or more values $x_1, \cdots, x_r$ more or less arbitrarily, and then successively obtain new values x_n as certain functions of the previously obtained $x_1, \cdots, x_{n-1}$, the values $M(x_1), \cdots, M(x_{n-1})$, and possibly those of the derivatives $M'(x_1), \cdots, M'(x_{n-1})$, etc. If

$$\lim_{n\to\infty} x_n = \theta, \tag{2}$$

irrespective of the arbitrary initial values $x_1, \cdots, x_r$, then the method is effective for the particular function $M(x)$ and value α. The speed of the convergence in (2) and the ease with which the x_n can be computed determine the practical utility of the method.

We consider a stochastic generalization of the above problem in which the nature of the function $M(x)$ is unknown to the experimenter. Instead, we suppose that to each value x corresponds a random variable $Y = Y(x)$ with distribution function $Pr[Y(x) \le y] = H(y \mid x)$, such that

$$M(x) = \int_{-\infty}^{\infty} y \, dH(y \mid x) \tag{3}$$

is the expected value of Y for the given x. Neither the exact nature of $H(y \mid x)$ nor that of $M(x)$ is known to the experimenter, but it is assumed that equation (1) has a unique root θ, and it is desired to estimate θ by making successive observations on Y at levels $x_1, x_2, \cdots$ determined sequentially in accordance with some definite experimental procedure. If (2) holds *in probability* irrespective of any arbitrary initial values $x_1, \cdots, x_r$, we shall, in conformity with usual statistical terminology, call the procedure *consistent* for the given $H(y \mid x)$ and value α.

[1] This work was supported in part by the Office of Naval Research.

In what follows we shall give a particular procedure for estimating θ which is consistent under certain restrictions on the nature of $H(y \mid x)$. These restrictions are severe, and could no doubt be lightened considerably, but they are often satisfied in practice, as will be seen in Section 4. No claim is made that the procedure to be described has any optimum properties (i.e. that it is "efficient") but the results indicate at least that the subject of stochastic approximation is likely to be useful and is worthy of further study.

3. Convergence theorems. We suppose henceforth that $H(y \mid x)$ is, for every x, a distribution function in y, and that there exists a positive constant C such that

$$Pr[\,|\,Y(x)\,| \leq C] = \int_{-C}^{C} dH(y \mid x) = 1 \qquad \text{for all } x. \tag{4}$$

It follows in particular that for every x the expected value $M(x)$ defined by (3) exists and is finite. We suppose, moreover, that there exist finite constants α, θ such that

$$M(x) \leq \alpha \quad \text{for} \quad x < \theta, \qquad M(x) \geq \alpha \quad \text{for} \quad x > \theta. \tag{5}$$

Whether $M(\theta) = \alpha$ is, for the moment, immaterial.

Let $\{a_n\}$ be a fixed sequence of positive constants such that

$$0 < \sum_{1}^{\infty} a_n^2 = A < \infty. \tag{6}$$

We define a (nonstationary) Markov chain $\{x_n\}$ by taking x_1 to be an arbitrary constant and defining

$$x_{n+1} - x_n = a_n(\alpha - y_n), \tag{7}$$

where y_n is a random variable such that

$$Pr[y_n \leq y \mid x_n] = H(y \mid x_n). \tag{8}$$

Let

$$b_n = E(x_n - \theta)^2. \tag{9}$$

We shall find conditions under which

$$\lim_{n \to \infty} b_n = 0 \tag{10}$$

no matter what the initial value x_1. As is well known, (10) implies the convergence in probability of x_n to θ.

From (7) we have

$$\begin{aligned} b_{n+1} &= E(x_{n+1} - \theta)^2 = E[E[(x_{n+1} - \theta)^2 \mid x_n]] \\ &= E\left[\int_{-\infty}^{\infty} \{(x_n - \theta) - a_n(y - \alpha)\}^2 \, dH(y \mid x_n)\right] \\ &= b_n + a_n^2 E\left[\int_{-\infty}^{\infty} (y - \alpha)^2 \, dH(y \mid x_n)\right] - 2a_n E[(x_n - \theta)(M(x_n) - \alpha)]. \end{aligned} \tag{11}$$

Setting

(12) $$d_n = E[(x_n - \theta)(M(x_n) - \alpha)],$$

(13) $$e_n = E\left[\int_{-\infty}^{\infty} (y - \alpha)^2 \, dH(y \mid x_n)\right],$$

we can write

(14) $$b_{n+1} - b_n = a_n^2 e_n - 2a_n d_n .$$

Note that from (5)

$$d_n \geq 0,$$

while from (4)

$$0 \leq e_n \leq [C + |\alpha|]^2 < \infty.$$

Together with (6) this implies that the positive-term series $\Sigma\, a_n^2 e_n$ converges.

Summing (14) we obtain

(15) $$b_{n+1} = b_1 + \sum_{j=1}^{n} a_j^2 e_j - 2 \sum_{j=1}^{n} a_j d_j .$$

Since $b_{n+1} \geq 0$ it follows that

(16) $$\sum_{j=1}^{n} a_j d_j \leq \tfrac{1}{2}\left[b_1 + \sum_{1}^{\infty} a_n^2 e_n\right] < \infty.$$

Hence the positive-term series

(17) $$\sum_{1}^{\infty} a_n d_n$$

converges. It follows from (15) that

(18) $$\lim_{n\to\infty} b_n = b_1 + \sum_{1}^{\infty} a_n^2 e_n - 2 \sum_{1}^{\infty} a_n d_n = b$$

exists; $b \geq 0$.

Now suppose that there exists a sequence $\{k_n\}$ of nonnegative constants such that

(19) $$d_n \geq k_n b_n, \qquad \sum_{1}^{\infty} a_n k_n = \infty.$$

From the first part of (19) and the convergence of (17) it follows that

(20) $$\sum_{1}^{\infty} a_n k_n b_n < \infty.$$

From (20) and the second part of (19) it follows that for any $\epsilon > 0$ there must exist infinitely many values n such that $b_n < \epsilon$. Since we already know that $b = \lim_{n\to\infty} b_n$ *exists*, it follows that $b = 0$. Thus we have proved

LEMMA 1. *If a sequence* $\{k_n\}$ *of nonnegative constants exists satisfying* (19) *then* $b = 0$.

Let

(21) $$A_n = |x_1 - \theta| + [C + |\alpha|](a_1 + a_2 + \cdots + a_{n-1});$$

then from (4) and (7) it follows that

(22) $$Pr[|x_n - \theta| \leq A_n] = 1.$$

Now set

(23) $$\bar{k}_n = \inf\left[\frac{M(x) - \alpha}{x - \theta}\right] \quad \text{for} \quad 0 < |x - \theta| \leq A_n .$$

From (5) it follows that $\bar{k}_n \geq 0$. Moreover, denoting by $P_n(x)$ the probability distribution of x_n, we have

(24) $$d_n = \int_{|x-\theta| \leq A_n} (x - \theta)(M(x) - \alpha)\, dP_n(x) \geq \int_{|x-\theta| \leq A_n} \bar{k}_n |x - \theta|^2\, dP_n(x) = \bar{k}_n b_n .$$

It follows that the particular sequence $\{\bar{k}_n\}$ defined by (23) satisfies the first part of (19).

In order to establish the second part of (19) we shall make the following assumptions:

(25) $$\bar{k}_n \geq \frac{K}{A_n}$$

for some constant $K > 0$ and sufficiently large n, and

(26) $$\sum_{n=2}^{\infty} \frac{a_n}{(a_1 + \cdots + a_{n-1})} = \infty .$$

It follows from (26) that

(27) $$\sum_1^{\infty} a_n = \infty,$$

and hence for sufficiently large n

(28) $$2[C + |\alpha|](a_1 + \cdots + a_{n-1}) \geq A_n .$$

This implies by (25) that for sufficiently large n

(29) $$a_n \bar{k}_n \geq a_n \frac{K}{A_n} \geq \frac{a_n K}{2[C + |\alpha|](a_1 + \cdots + a_{n-1})},$$

and the second part of (19) follows from (29) and (26). This proves

LEMMA 2. *If* (25) *and* (26) *hold then* $b = 0$.

The hypotheses (6) and (26) concerning $\{a_n\}$ are satisfied by the sequence $a_n = 1/n$, since

$$\sum_1^\infty \frac{1}{n^2} = \frac{\pi^2}{6}, \qquad \sum_{n=2}^\infty \left[\frac{1}{n\left(1 + \frac{1}{2} + \cdots + \frac{1}{n-1}\right)}\right] = \infty.$$

More generally, any sequence $\{a_n\}$ such that there exist two positive constants c', c'' for which

$$\frac{c'}{n} \leq a_n \leq \frac{c''}{n} \tag{30}$$

will satisfy (6) and (26). We shall call any sequence $\{a_n\}$ which satisfies (6) and (26), whether or not it is of the form (30), a *sequence of type* $1/n$.

If $\{a_n\}$ is a sequence of type $1/n$ it is easy to find functions $M(x)$ which satisfy (5) and (25). Suppose, for example, that $M(x)$ satisfies the following strengthened form of (5): for some $\delta > 0$,

$$M(x) \leq \alpha - \delta \quad \text{for} \quad x < \theta, \qquad M(x) \geq \alpha + \delta \quad \text{for} \quad x > \theta. \tag{5'}$$

Then for $0 < |x - \theta| \leq A_n$ we have

$$\frac{M(x) - \alpha}{x - \theta} \geq \frac{\delta}{A_n}, \tag{31}$$

so that

$$\bar{k}_n \geq \frac{\delta}{A_n}, \tag{32}$$

which is (25) with $K = \delta$. From Lemma 2 we conclude

THEOREM 1. *If* $\{a_n\}$ *is of type* $1/n$, *if* (4) *holds, and if* $M(x)$ *satisfies* (5′) *then* $b = 0$.

A more interesting case occurs when $M(x)$ satisfies the following conditions:

$$M(x) \text{ is nondecreasing,} \tag{33}$$

$$M(\theta) = \alpha, \tag{34}$$

$$M'(\theta) > 0. \tag{35}$$

We shall prove that (25) holds in this case also. From (34) it follows that

$$M(x) - \alpha = (x - \theta)[M'(\theta) + \epsilon(x - \theta)], \tag{36}$$

where $\epsilon(t)$ is a function such that

$$\lim_{t\to 0} \epsilon(t) = 0. \tag{37}$$

Hence there exists a constant $\delta > 0$ such that

$$\epsilon(t) \geq -\tfrac{1}{2}M'(\theta) \quad \text{for} \quad |t| \leq \delta, \tag{38}$$

so that

(39) $$\frac{M(x) - \alpha}{x - \theta} \geq \frac{1}{2} M'(\theta) > 0 \quad \text{for} \quad |x - \theta| \leq \delta.$$

Hence, for $\theta + \delta \leq x \leq \theta + A_n$, since $M(x)$ is nondecreasing,

(40) $$\frac{M(x) - \alpha}{x - \theta} \geq \frac{M(\theta + \delta) - \alpha}{A_n} \geq \frac{\delta M'(\theta)}{2A_n},$$

while for $\theta - A_n \leq x \leq \theta - \delta$,

(41) $$\frac{M(x) - \alpha}{x - \theta} = \frac{\alpha - M(x)}{\theta - x} \geq \frac{\alpha - M(\theta - \delta)}{A_n} \geq \frac{\delta M'(\theta)}{2A_n}.$$

Thus, since we may assume without loss of generality that $\delta/A_n \leq 1$,

(42) $$\frac{M(x) - \alpha}{x - \theta} \geq \frac{\delta M'(\theta)}{2A_n} \quad \text{for} \quad 0 < |x - \theta| \leq A_n,$$

so that (25) holds with $K = \delta M'(\theta)/2 > 0$. This proves

THEOREM 2. *If* $\{a_n\}$ *is of type* $1/n$, *if* (4) *holds, and if* $M(x)$ *satisfies* (33), (34), *and* (35), *then* $b = 0$.

It is fairly obvious that condition (4) could be considerably weakened without affecting the validity of Theorems 1 and 2. A reasonable substitute for (4) would be the condition

(4′) $$|M(x)| \leq C, \qquad \int_{-\infty}^{\infty} (y - M(x))^2 \, dH(y \mid x) \leq \sigma^2 < \infty \qquad \text{for all } x.$$

We do not know whether Theorems 1 and 2 hold with (4) replaced by (4′). Likewise, the hypotheses (33), (34), and (35) of Theorem 2 could be weakened somewhat, perhaps being replaced by

(5″) $$M(x) < \alpha \quad \text{for} \quad x < \theta, \qquad M(x) > \alpha \quad \text{for} \quad x > \theta.$$

4. Estimation of a quantile using response, nonresponse data. Let $F(x)$ be an unknown distribution function such that

(43) $$F(\theta) = \alpha \; (0 < \alpha < 1), \qquad F'(\theta) > 0,$$

and let $\{z_n\}$ be a sequence of independent random variables each with the distribution function $Pr[z_n \leq x] = F(x)$. On the basis of $\{z_n\}$ we wish to estimate θ. However, as sometimes happens in practice (bioassay, sensitivity data), we are not allowed to know the values of z_n themselves. Instead, we are free to prescribe for each n a value x_n and are then given only the values $\{y_n\}$ where

(44) $$y_n = \begin{cases} 1 & \text{if } z_n \leq x_n \quad (\text{"response"}), \\ 0 & \text{otherwise} \quad (\text{"nonresponse"}). \end{cases}$$

How shall we choose the values $\{x_n\}$ and how shall we use the sequence $\{y_n\}$ to estimate θ?

Let us proceed as follows. Choose x_1 as our best guess of the value θ and let $\{a_n\}$ be any sequence of constants of type $1/n$. Then choose values $x_2, x_3, \cdots$ sequentially according to the rule

$$x_{n+1} - x_n = a_n(\alpha - y_n). \tag{45}$$

Since

$$Pr[y_n = 1 \mid x_n] = F(x_n), \qquad Pr[y_n = 0 \mid x_n] = 1 - F(x_n), \tag{46}$$

it follows that (4) holds and that

$$M(x) = F(x). \tag{47}$$

All the hypotheses of Theorem 4 are satisfied, so that

$$\lim_{n\to\infty} x_n = \theta \tag{48}$$

in quadratic mean and hence in probability. In other words, $\{x_n\}$ is a consistent estimator of θ.

The *efficiency* of $\{x_n\}$ will depend on x_1 and on the choice of the sequence $\{a_n\}$, as well as on the nature of $F(x)$. For any given $F(x)$ there doubtless exist more efficient estimators of θ than any of the type $\{x_n\}$ defined by (45), but $\{x_n\}$ has the advantage of being distribution-free.

In some applications it is more convenient to make a group of r observations at the same level before proceeding to the next level. The nth group of observations will then be

$$y_{(n-1)r+1}, \cdots, y_{nr}, \tag{49}$$

using the notation (44). Let $\bar{y}_n$ = arithmetic mean of the values (49). Then setting

$$x_{n+1} - x_n = a_n(\alpha - \bar{y}_n), \tag{50}$$

we have $M(x) = F(x)$ as before, and hence (48) continues to hold.

The possibility of using a convergent sequential process in this problem was first mentioned by T. W. Anderson, P. J. McCarthy, and J. W. Tukey in the Naval Ordnance Report No. 65-46(1946), p. 99.

5. A more general regression problem. It is clear that the problem of Section 4 is a special case of a more general regression problem. In fact, using the notation of Section 2, consider any random variable Y which is associated with an observable value x in such a way that the conditional distribution function of Y for fixed x is $H(y \mid x)$; the function $M(x)$ is then the regression of Y on x.

The usual regression analysis assumes that $M(x)$ is of known form with unknown parameters, say

$$M(x) = \beta_0 + \beta_1 x, \tag{51}$$

and deals with the estimation of one or both of the parameters β_i on the basis of observations $y_1, y_2, \cdots, y_n$ corresponding to observed values $x_1, x_2, \cdots, x_n$. The method of least squares, for example, yields the estimators b_i which minimize the expression

$$\sum_{i=1}^{n} (y_i - [\beta_0 + \beta_1 x_i])^2. \tag{52}$$

Instead of trying to estimate the parameters β_i of $M(x)$ under the assumption that $M(x)$ is a linear function of x, we may try to estimate the value θ such that $M(\theta) = \alpha$, where α is given, without any assumption about the form of $M(x)$. If we assume only that $H(y \mid x)$ satisfies the hypotheses of Theorem 2 then the sequence of estimators $\{x_n\}$ of θ defined by (7) will at least be consistent. This indicates that a distribution-free sequential system of making observations, such as that given by (7), is worth investigating from the practical point of view in regression problems.

One of us is investigating the properties of this and other sequential designs as a graduate student; the senior author is responsible for the convergence proof in Section 3.

Reprinted from
Ann. Math. Statist.
22, 400–407 (1951)

OPTIMIZING METHODS IN STATISTICS

A CONVERGENCE THEOREM FOR NON NEGATIVE ALMOST SUPERMARTINGALES AND SOME APPLICATIONS*

H. Robbins and D. Siegmund

Columbia University

1. Introduction and Summary

The purpose of this paper is to give a unified treatment of a number of almost sure convergence theorems by exploiting the fact that the processes involved possess a common "almost supermartingale" properties To be precise, let (Ω,F,P) be a probability space and $F_1 \subset F_2 \subset \ldots$ a sequence of sub-σ-algebras of F. For each $n = 1,2, \ldots$ let z_n, β_n, ξ_n, and ζ_n be non-negative F_n-measurable random variables such that

$$E(z_{n+1}|F_n) \leq z_n(1+\beta_n) + \xi_n - \zeta_n. \tag{1}$$

In Section 2 below we prove

Theorem 1. $\lim_{n\to\infty} z_n$ exists and is finite and $\sum_1^\infty \zeta_n < \infty$ a.s. on

$$\{\sum_1^\infty \beta_n < \infty, \quad \sum_1^\infty \xi_n < \infty\}.$$

In Sections 3-6 we apply this result (and the closely related Theorem 2) to obtain a number of convergence theorems, including those of Kolmogorov (strong law of large numbers for independent, non-identically distributed random variables), Blackwell (minimax theorem for vector- valued payoffs in game theory), MacQueen (cluster analysis), and some results in stochastic approximation.

*Research supported by NIH Grant 5-R01-GM-16895-03 and ONR Grant N00014-67-A-0108-0018.

The approach we use is similar to those in [7] and [8]. (We thank V. Fabian for pointing out the reference [7] to us.)

2. Proof of Theorem 1, Theorem 2, and Inequalities

To prove Theorem 1, let

$$A = \{ \sum_1^\infty \beta_n < \infty, \sum_1^\infty \xi_n < \infty \}$$

and define

$$z_n' = z_n \prod_1^{n-1} (1+\beta_k)^{-1}, \quad \xi_n' = \xi_n \prod_1^{n} (1+\beta_k)^{-1},$$

and

$$\zeta_n' = \zeta_n \prod_1^{n} (1+\beta_k)^{-1}.$$

Then from (1)

$$E(z_{n+1}' | F_n) \leq z_n' + \xi_n' - \zeta_n' \tag{2}$$

and

$$\Sigma \xi_n' \leq \Sigma \xi_n < \infty \text{ on } A. \tag{3}$$

Let

$$u_n = z_n' - \sum_1^{n-1} (\xi_k' - \zeta_k') \tag{4}$$

and for $a > 0$

$$t = \inf \{n: \sum_1^n \xi_k' > a\} . \tag{5}$$

On $\{t > n\}$, we have by (2)

$$E(u_{n+1} | F_n) = E(z_{n+1}' - \sum_1^n (\xi_k' - \zeta_k') | F_n) \leq u_n,$$

and hence

$$E(u_{t \wedge (n+1)} | F_n) = u_t I_{\{t \leq n\}} + E(u_{n+1} | F_n) I_{\{t > n\}} \leq u_{t \wedge n}$$

so $\{u_{t\wedge(n\ 1)}, F_n, 1 \leq n < \infty\}$ is a supermartingale. Since

$$u_{t\wedge n} \geq - \sum^{t\wedge n-1} \xi_k \geq -a$$

for all n it follows from the martingale convergence theorem that

$$\lim_{n \to \infty} u_{t\wedge n} \text{ exists}$$

and is finite a.s., i.e., $\lim_{n \to \infty} u_n$ exists and is finite on

$$\{t = \infty\} = \{ \sum_1^{\infty} \xi'_n \leq a\}.$$

Since a is arbitrary, we see that $\lim_{n \to \infty} u_n$ exists and is finite a.s. on $\{\Sigma\, \xi'_n < \infty\}$. Hence by (4) $\lim_{n \to \infty} z'_n$ exists and is finite and $\Sigma\, \zeta'_n < \infty$ on $\{ \Sigma\, \xi'_n < \infty\}$, and it follows from (3),

$$z_n = z'_n \prod_1^{n-1} (1 + \beta_k),$$

and the inequality

$$\zeta_n \leq \zeta'_n \prod_1^{\infty} (1 + \beta_k)$$

that $\lim_{n \to \infty} z_n$ exists and is finite and $\sum_1^{\infty} \zeta_n < \infty$ a.s. on A.

<u>Theorem 2.</u> Suppose (1) is satisfied, and for $b > 0$ let

$$\beta_n(b) = \beta_n I\{z_n \leq b\},\ \xi_n(b) = \xi_n I\{z_n \leq b\}, \quad (6)$$

$$\zeta_n(b) = \zeta_n I_{\{z_n \leq b\}}. \quad (6)$$

Then $\lim_{n \to \infty} z_n$ exists a.s. on $B = \bigcap_b \{\Sigma\beta_n(b) < \infty, \Sigma\xi_n(b) < \infty\}$, and $\sum_1^\infty \zeta_n < \infty$ a.s. on $B \cap \{\lim_n z_n < \infty\}$.

<u>Proof</u>: By (1)

$$E(z_{n+1} \wedge b|F_n) \leq [(1+\beta_n)z_n \quad \xi_n - \zeta_n] \wedge b. \tag{7}$$

Hence on $\{z_n \leq b\}$

$$E(z_{n+1} \wedge b|F_n) \leq (1+\beta_n)z_n + \xi_n - \zeta_n =$$

$$(1+\beta_n(b))\,(z_n \wedge b) + \xi_n(b) - \zeta_n(b)$$

by (6), while on $\{z_n > b\}$

$$E(z_{n+1} \wedge b|F_n) \leq b = (1+\beta_n(b))\,(z_n \wedge b) + \xi_n(b) - \zeta_n(b),$$

so that (1) holds for $z_n \wedge b$. It follows from Theorem 1 that $\lim_{n\to\infty} z_n \wedge b$ exists and

$$\sum_1^\infty \zeta_n(b) < \infty$$

a.s. on $\{\Sigma\, \beta_n(b) < \infty,\ \Sigma\, \xi_n(b) < \infty\}$. Since b is arbitrary, the first assertion of the theorem follows at once. If $\omega \in B$ and $\lim_{n\to\infty} z_n(\omega) < \infty$, then for some b, $\lim_{n\to\infty} z_n(\omega) < b$ and hence $\zeta_n(b)\,(\omega) = \zeta_n(\omega)$ for all $n = 1,2,\ldots$ from which the second assertion of the theorem follows.

A. Dvoretzky, who read the original manuscript of this paper, is responsible for shortening the proof of Theorem 1 and improving the form of Theorem 2. He has pointed out to us that the present method enables one to prove the following generalization of Theorem 1. Theorem 2 may be similarly generalized, and even more general versions of both theorems are possible.

<u>Theorem 1'.</u> For each $n = 1,2,\ldots$, let $\theta_n > 0$ and η_n be F_n-measurable random variables, and suppose that

$$E(z_{n+1}|F_n) \leq \theta_n z_n + \eta_n. \tag{8}$$

Then $\inf \sum_{k=1}^{n} \eta_k \prod_{i=1}^{k} \theta_i^{-1} > -\infty$ a.s. on

$$\{ \sup_n \sum_{k=1}^{n} \eta_k \prod_{i=1}^{k} \theta_i^{-1} < \infty\},$$

and $\lim_{n\to\infty} z_n$ exists and is finite and $\lim_{n\to\infty} \sum_{k=1}^{n} \eta_k \prod_{i=k+1}^{n} \theta_i^{-1}$ exists and is finite a.s. on $\{ \lim_{n\to\infty} \prod_{1}^{n} \theta_k$ exists and is finite,

and $\lim_{n\to\infty} \sum_{k=1}^{n} \eta_k \prod_{i=1}^{k} \theta_i^{-1}$ exists and is not $+\infty\}$.

The following inequalities are simple and useful generalizations of well-known results in martingale theory. Some applications are given in Sections 5 and 7. Moreover, Proposition 2 below yields an upcrossing inequality from which Theorem 1 follows easily (cf. problems 10 and 11 on p. 40 of [4]). Doob's submartingale convergence theorem may then be deduced directly from Theorem 1. (Suppose that $\{x_n, F_n, 1 \leq n < \infty\}$ is a submartingale and that

$$\sup_n Ex_n^+ < \infty.$$

To prove that $\lim_{n\to\infty} x_n$ exists it suffices to show that

$$\lim_{n\to\infty} \max(x_n, a)$$

exists for every $-\infty < a < \infty$, and since $\{\max(x_n, a) - a, F_n, 1 \leq n < \infty\}$ is a nonnegative submartingale, we may assume without loss of generality that $x_n \geq 0$ and $\sup_n Ex_n < \infty$. Now $E(x_{n+1}|F_n) = x_n + \xi_n$, where

$$\xi_n = E(x_{n+1}|F_n) - x_n \geq 0$$

and

$E\left(\sum_1^\infty \xi_n\right) \leq \sup_n E(x_n) < \infty$, so by Theorem 1, $\lim_n x_n$ exists a.s.)

Proposition 1. If (1) holds, then for any $m = 1,2, \ldots$ and finite stopping time $t \geq m$,

$$E(z_t | F_m) \leq z_m + E\left[\sum_{k=m}^{t-1} (\beta_k z_k + \xi_k) | F_m\right] \tag{9}$$

Proof Let $a > 0$, let s = first $n \geq m$ such that $z_n \geq a$,

$$\sum_m^n \beta_k \geq a, \quad \text{or} \quad \sum_m^n \xi_k \geq a \; ; \; s = \infty$$

if no such n exists, and let

$$u_n^{(m)} = z_n - \sum_{k=m}^{n-1} (\beta_k z_k + \xi_k). \tag{10}$$

As in the proof of Theorem 1, it may be verified that

$$\{u_{t\wedge s\wedge n}^{(m)}, F_n, m \leq n < \infty\}$$

is a supermartingale. Hence, for any $n \geq m$,

$$E(u_{t\wedge s\wedge n}^{(m)} | F_m) \leq u_m^{(m)},$$

or equivalently by (10),

$$E(z_{t\wedge s\wedge n} | F_m) \leq z_m + E\left[\sum_m^{t\wedge s\wedge n-1} (\beta_k z_k + \xi_k | F_m)\right]. \tag{11}$$

Letting $a \to \infty$, $n \to \infty$, we obtain (9) from (11) and the monotone convergence theorem and Fatou's lemma for conditional expectations.

Proposition 2.

If (1) holds, for any $a > 0$, $m = 1,2, \ldots$, and $n \geq m$

$$P\{\max_{m\le k\le n} z_k \ge a|F_m\} \le a^{-1}\left[z_m + E\left(\sum_m^{n-1} \xi_k|F_m\right)\right] + E\left(\sum_m^{n-1} \beta_k\,|F_m\right). \tag{12}$$

Proof. Let $t = \inf\{k: k \ge m, z_k \ge a\}$. Then $aP\{z_{t\wedge n} \ge a|F_m\} \le E(z_{t\wedge n}|F_m)$, which together with (9) and the definition of t proves (12).

3. The Strong Law of Large Numbers

The line of reasoning in the following application of Theorem 1 to the strong law of large numbers is typical of most of our applications. Application 1 has been obtained by Neveu [9], p. 148. Corollary 1 is Kolmogorov's classical result. Corollary 2, which is stated here for use in Section 6, was obtained independently of Neveu by Dubins and Freedman [5].

Application 1. Let $y_1, y_2, \ldots$ be a sequence of random variables such that $E|y_k| < \infty$ and $E(y_{n+1}|F_n) = 0$ $(n = 1, 2, \ldots)$. Assume that for each $n = 2,3,\ldots$, c_n is a positive F_{n-1}-measurable random variable with $c_1 \le c_2 \le \ldots$. Then

$$\lim_{n\to\infty} c_n^{-1} \left|\sum_1^n y_k\right|$$

exists and is finite a.s. on $\{\sum c_k^{-2} E(y_k^2|F_{k-1}) < \infty\}$; it equals 0 a.s. on

$$\{c_n \uparrow \infty, \sum c_k^{-2} E(y_k^2|F_{k-1}) < \infty\}.$$

Proof. Let

$$z_n = c_n^{-2}\left(\sum_1^n y_k\right)^2,$$

$$\zeta_n = \frac{c_{n+1}^2 - c_n^2}{c_{n+1}^2}\, z_n,$$

239

exists and is finite a.s. It is equal to 1 on $\{\Sigma p_k = \infty\}$.

Proof. We employ Application 1 with $y_n = r_n - p_n$, $F_n = B(r_1, \ldots, r_n)$, $c_n = p_1 + \cdots + p_n$ and hence

$$E(y_{n+1}^2 | F_n) = p_n(1 - p_n) \leq p_n.$$

Obviously

$$\Sigma \frac{p_n(1 - p_n)}{(p_1 + \cdots + p_n)^2} \leq \Sigma \frac{p_n}{(p_1 + \cdots + p_n)^2} < \infty$$

on $\{\Sigma p_n < \infty\}$, and by Lemma 1 below this series also converges on $\{\Sigma p_n = \infty\}$. Hence

$$\lim_{n \to \infty} \left(\frac{r_1 + \cdots + r_n}{p_1 + \cdots + p_n} - 1 \right) = \lim_{n \to \infty} \frac{\sum_1^n (r_k - p_k)}{\sum_1^n p_k}$$

exists and is finite a.s. and equals 0 a.s. on $\{\Sigma p_n = \infty\}$.

Lemma 1. Let $0 < a_n \uparrow \infty$. Then

$$\sum_1^\infty \frac{a_{n+1} - a_n}{a_{n+1}^2} < \infty \tag{15}$$

and

$$\sum_1^\infty \frac{a_{n+1} - a_n}{a_{n+1}} = \infty \quad . \tag{16}$$

Proof. To prove (15), note that

$$\sum_1^\infty \frac{a_{n+1} - a_n}{a_{n+1}^2} \leq \sum_1^\infty \int_{a_n}^{a_{n+1}} x^{-2} dx = \int_{a_1}^\infty x^{-2} dx < \infty.$$

To prove (16), first suppose $\sup_n a_{n+1}/a_n = \infty$. Then infinitely many of the terms

$(a_{n+1} - a_n)/a_{n+1}$ exceed ½ and hence (16)holds. On the other hand if $\sup_n a_{n+1}/a_n = a < \infty$, then

$$\sum_1^\infty \frac{a_{n+1} - a_n}{a_{n+1}} \geq \sum_1^\infty \frac{a_n}{a_{n+1}} \int_{a_n}^{a_{n+1}} x^{-1}dx \geq a^{-1} \int_{a_1}^{\infty} x^{-1}dx = \infty.$$

4. Stochastic Approximation

Let $\{y(x), -\infty < x < \infty\}$ be a family of random variables with $Ey(x) = M(x)$, $\operatorname{Var} y(x) = \sigma^2(x) < \infty$ $(-\infty < x < \infty)$. The Robbins-Monro process [10] is designed to find, under certain assumptions, the location of the root $M(x) = 0$ of the regression fuction M. Starting with an arbitrary random variable x_1, we define successively $x_2, x_3, \ldots$, by

$$x_{n+1} = x_n - a_n y_n, \tag{17}$$

where a_n is a non-negative (measurable) function of $x_1, y_1, \ldots, y_{n-1}$, and conditional on $x_1, y_1, \ldots, y_{n-1}$ the random variable y_n has the distribution of $y(x_n)$. The following application of Theorem 1 generalizes results of Blum [2] and Gladyshev [7].

<u>Application 2.</u> If σ and M are measurable,

$$\sigma(x) + |M(x)| \leq a + b|x| \quad \text{for some } a,b > 0, \tag{18}$$

there exists a real number θ such that for each $\epsilon > 0$

$$\inf_{\epsilon < x-\theta < \epsilon^{-1}} M(x) > 0, \qquad \sup_{\epsilon < x-\theta < \epsilon^{-1}} M(x) < 0, \tag{19}$$

$$\sum a_n^2 < \infty \text{ for every sequence } x_1, y_1, \ldots, \tag{20}$$

and

$$\sum a_n = \infty \text{ for every sequence } x_1, y_1, \ldots \text{ such that } \sup |x_n| < \infty, \tag{21}$$

then $\lim_{n\to\infty} x_n = \theta$ with probability one.

<u>Proof.</u> Let $F_n = B(x_1, y_1, \dots, y_{n-1})$, $z_n = (x_n - \theta)^2$. By (17), (18), (19) and the inequality

$$(a + b)^2 \le 2(a^2 + b^2)$$

$$E(z_{n+1}|F_n) = z_n - 2a_n(x_n - \theta)E(y_n|F_n) + a_n^2 E(y_n^2|F_n)$$

$$= z_n - 2a_n|x_n - \theta||M(x_n)| + a_n^2(\sigma^2(x_n) + M^2(x_n))$$

$$\le z_n(1 + 4b^2a_n^2) - 2a_n|x_n - \theta||M(x_n)| + a_n^2(2a^2 + 4b^2\theta^2).$$

Putting

$$\beta_n = 4b^2a_n^2, \quad \xi_n = 2a_n^2(a^2 + 2b^2\theta^2),$$

and

$$\zeta_n = 2a_n|x_n - \theta||M(x_n)|,$$

we see from (20) and Theorem 1 that $\lim_{n\to\infty} z_n$ exists and is finite and

$$\sum_1^\infty \zeta_n = \sum_1^\infty a_n\, |x_n - \theta||M(x_n)| < \infty$$

a.s. Hence by (19) and (21) $\lim_{n\to\infty} |x_n - \theta| = 0$ with probability one.

Dvoretzky [6] has given an example which shows that the growth condition on M given by (18) cannot be dropped completely. If $M(x) = x|x|$, $\sigma^2 = 0$, $a_n = 1/n$, and $x_1 = 6$, then it is easy to see that $\sup x_n = +\infty$, $\liminf x_n = -\infty$, and the sequence (x_n) has no finite point of accumulation. It is also easy to give examples which show that the growth condition on σ cannot be dropped completely.

The following simple modification of the Robbins-Monro process permits removal of the growth condition on M. (See also Remarks (i) and (ii) below.) Let $\underline{\theta} = \bar{\theta}$ and define

$$\tilde{x}_{n+1} = \begin{cases} (\tilde{x}_n - a_n y_n) I^{+}_{\{\tilde{x}_n - a_n y_n \geq \underline{\theta}\}} + \underline{\theta}\, I_{\{\tilde{x}_n - a_n y_n < \underline{\theta}\}} & \text{if } \tilde{x}_n > \bar{\theta} \\ \tilde{x}_n - a_n y_n & \text{if } \underline{\theta} \leq \tilde{x}_n \leq \bar{\theta} \\ (\tilde{x}_n - a_n y_n) I^{+}_{\{\tilde{x}_n - a_n y_n \leq \bar{\theta}\}} + \bar{\theta}\, I_{\{\tilde{x}_n - a_n y_n > \bar{\theta}\}} & \text{if } \tilde{x}_n \leq \underline{\theta}, \end{cases}$$

where $\tilde{x}_1$ is arbitrary, and as before a_n is a non-negative function of $\tilde{x}_1, y_1, \ldots, y_{n-1}$ and conditional on $\tilde{x}_1, y_1, \ldots, y_{n-1}$ the random variable y_n has the distribution of $y(\tilde{x}_n)$.

Application 3. If σ and M are measurable, M is locally bounded,

$$\sigma^2(x) \leq a + bx^2 \quad \text{for some } a, b > 0 \tag{23}$$

(19), (20), and (21) hold (with $\tilde{x}$'s instead of x's), then $\lim_{n \to \infty} \tilde{x}_n = \theta$ with probability one.

Proof. Without loss of generality we may assume that $\theta = 0$. By (22) and Chebyshev's inequality, on $\{\tilde{x}_n > \bar{\theta}\}$ we have

$$\begin{aligned} E(\tilde{x}^2_{n+1}|F_n) &\leq E((\tilde{x}_n - a_n y_n)^2 | F_n) + \underline{\theta}^2 P\{\tilde{x}_n - a_n y_n < \underline{\theta} | F_n\} \\ &\leq \tilde{x}^2_n - 2a_n \tilde{x}_n M(\tilde{x}_n) + a_n^2 E(y_n^2 | F_n) + \underline{\theta}^2 P\{a_n y_n > \bar{\theta} - \underline{\theta} | F_n\} \\ &\leq \tilde{x}^2_n - 2a_n \tilde{x}_n M(\tilde{x}_n) + a_n^2 (\sigma^2(\tilde{x}_n) + M^2(\tilde{x}_n)) \left(1 + \frac{\underline{\theta}^2}{(\bar{\theta} - \underline{\theta})^2} \right). \end{aligned}$$

Similar computations on $\{\underline{\theta} \leq \tilde{x}_n \leq \bar{\theta}\}$ and $\{\tilde{x}_n < \underline{\theta}\}$ show that in general

$$E(\tilde{x}^2_{n+1}|F_n) \leq \tilde{x}^2_n - 2a_n \tilde{x}_n M(\tilde{x}_n) + a_n^2 (\sigma^2(\tilde{x}_n) + M^2(\tilde{x}_n)) \left(1 + \frac{\bar{\theta}^2 + \underline{\theta}^2}{(\bar{\theta} - \underline{\theta})^2} \right).$$

Hence by Theorem 2 and the line of reasoning used in the

proofs of Applications 1 and 2,

$$\lim_{n \to \infty} \tilde{x}_n = 0$$

a.s. on $\{\lim\inf |\tilde{x}_n| < \infty\}$. To complete the proof it suffices to show that $P\{\lim\inf |\tilde{x}_n| = \infty\} = 0$. Let

$$\tilde{z}_n = [(\tilde{x}_n - \theta \vee \bar{\theta})^+]^2.$$

Using (19) and (23) it may be shown that

$$\begin{aligned} E(\tilde{z}_{n+1}|F_n) &\leq \tilde{z}_n(1 + 2ba_n^2) + a_n^2(a + 2b(\theta\vee\bar{\theta})^2) && \text{if } \tilde{z}_n > 0 \\ &\leq Ca_n^2 && \text{if } \tilde{z}_n = 0, \end{aligned}$$

where C depends on $\underline{\theta}$, $\bar{\theta}$, θ,

$$\sup_{\underline{\theta} \leq x \leq \theta\vee\bar{\theta}} |M(x)|, \quad \text{and} \quad \sup_{\underline{\theta} \leq x \leq \theta\vee\bar{\theta}} \sigma(x).$$

Hence in general

$$E(\tilde{z}_{n+1}|F_n) \leq \tilde{z}_n(1 + 2ba_n^2)$$
$$+ a_n^2(a + 2b(\theta\vee\bar{\theta})^2) + Ca_n^2 I_{\{\tilde{z}_n = 0\}},$$

and by Theorem 1, $P\{\lim \tilde{z}_n = \infty\} = P\{\lim \tilde{x}_n = \infty\} = 0$. Similarly, $P\{\lim \tilde{x}_n = -\infty\} = 0$. From (22) if $\lim\sup \tilde{x}_n = +\infty$ and $\lim\inf \tilde{x}_n = -\infty$, then $\tilde{x}_n \in [\underline{\theta}, \bar{\theta}]$ infinitely often and hence $\lim\inf |\tilde{x}_n| < \infty$. Thus $P\{\lim\inf |\tilde{x}_n| = \infty\} = 0$, which completes the proof.

Remarks. (i) The conditions (18) and (23) may be dropped if (20) is appropriately strengthened. For example, for the modified Robbins-Monro process defined by (22), if the there exists a known measurable function $\tilde{\sigma}$ (which may be assumed to vanish on an arbitrary bounded interval) such that

$$\sigma(x) = O(\tilde{\sigma}(x)) \quad \text{as} \quad |x| \to \infty,$$

then

$$\Sigma a_n^2 \; (1 + \tilde{\sigma}^2(\tilde{x}_n)) < \infty \quad \text{for every sequence } \tilde{x}_1, y_1, \ldots \tag{24}$$

together with (19), (21), and the local boundedness of M and σ implies the convergence of $\tilde{x}_n$ to θ with probability one.

(ii) The preceding modification of the Robbins-Monro process has practical as well as theoretical interest. The experimenter who has some prior knowledge of the value of θ may wish to use (22) with the hope that the speed of convergence will be increased if in fact $\theta \in [\underline{\theta},\bar{\theta}]$, while being protected against the possibility that $\theta \notin [\underline{\theta},\bar{\theta}]$ in the sense that convergence to the true value of θ still occurs with probability one.

(iii) Schmetterer [11] has shown that Dvoretzky's example for the unmodified Robbins-Monro process with σ^2 bounded depicts a general phenomenon, i.e., with probability one that either $\lim x_n = \theta$ or that $\lim\sup x_n = +\infty$, $\lim\inf x_n = -\infty$, and (x_n) has no finite point of accumulation. This result follows easily from the method of proof of Aplication 3.

In [6] Dvoretzky proved a general convergence theorem, which includes Blum's result for the Robbins-Monro process and the corresponding result for the Kiefer-Wolfowitz method for estimating the maximum of a regression function as special cases. The probability one part of Dvoretzky's theorem (he also proved mean square convergence) follows from Theorem 1, as the following generalization of his result shows.

<u>Application 4.</u> Let X be a Hilbert space. For each n = 1, 2, ... let $X_n = X \times X \times \ldots \times X$ (n factors) and suppose that $T_n\colon X_n \to X$ is a Borel-measurable transformation satisfying for some $\theta \in X$

$$\|T_n(x_1, \ldots, x_n) - \theta\| \quad \max(a_n, (1+b_n)\|x_n - \theta\| - c_n) \tag{25}$$

where a_n, b_n, and c_n are non-negative Borel measurable functions defined on X_n such that

$$\lim_{n \to \infty} a_n = 0 \qquad \text{for all } x_1, x_2, \ldots \tag{26}$$

$$\Sigma\, b_n < \infty \text{ for all } x_1, x_2, \ldots \tag{27}$$

and

$$\Sigma\, c_n = \infty \text{ for all } x_1, x_2, \ldots \text{ such that } \sup \|x_n\| < \infty. \tag{28}$$

Let $x_1, y_1, y_2, \ldots$ be random variables taking values in X and define

$$x_{n+1} = T_n(x_1, \ldots, x_n) + y_n \quad (n = 1,2, \ldots). \tag{29}$$

If

$$E(y_n | x_1, y_1, \ldots, y_{n-1}) = 0 \text{ a.s. } (n = 1,2, \ldots) \tag{30}$$

and

$$\Sigma\, E(\|y_n\|^2 | x_1, y_1, \ldots, y_{n-1}) < \infty \text{ a.s.} \tag{31}$$

then $\lim_{n \to \infty} \|x_n - \theta\| = 0$ with probability one.

Proof. The proof is a modification of the first part of Dvoretzky's argument. It may be assumed that $\theta = 0$. Let

$$F_n = B(x_1, y_1, \ldots, y_{n-1}),\ z_n = [(\|x_n\| - a_{n-1})^+]^2.$$

For any F_n-measurable random variable u_n such that $\|u_n\| \le a_n$ it follows from the triangle inequality that

$$z_{n+1} \le \|x_{n+1} - u_n\|^2$$

and hence by (29) and (30)

$$E(z_{n+1} | F_n) \le \|T_n - u_n\|^2 + E(\|y_n\|^2 | F_n). \tag{32}$$

Putting

$$u_n = T_n I_{\{\|T_n\| \le a_n\}} + a_n \frac{T_n}{\|T_n\|} I_{\{\|T_n\| > a_n\}},$$

we obtain from (32)

$$E(z_{n+1} | F_n) \le [(\|T_n\| - a_n)^+]^2 + E(\|y_n\|^2 | F_n). \tag{33}$$

Now by (25), if $\|x_n\| \le a_n$

$$(\|T_n\| - a_n)^+ \le (b_n a_n - c_n)^+$$

while if $\|x_n\| > a_n$

$$(\|T_n\| - a_n)^+ \le [(1 + b_n)(\|x_n\| - a_n) + b_n a_n - c_n]^+,$$

so in either case

$$[(\|T_n\| - a_n)^+]^2 \le [(1 + b_n)(\|x_n\| - a_n)^+ - c_n + a_n b_n]^2.$$

Using the inequality $(a + b)^2 \le (1 + b)a^2 + b(1 + b)$ which is valid for all $b \ge 0$ and all real a, we obtain

$$[(\|T_n\| - a_n)^+]^2 \le (1 + a_n b_n)(1 + b_n)^2 z_n - c_n z_n^{\frac{1}{2}}$$

$$+ \ (1 + a_n b_n) c_n^2 + a_n b_n (1 + a_n b_n).$$

Hence by (33), (1) holds with

$$\beta_n = b_n (1 + a_n b_n)(2 + b_n), \ \zeta_n = c_n z_n^{\frac{1}{2}}$$

and

$$\xi_n = (1 + a_n b_n) c_n^2 + a_n b_n (1 + a_n b_n) + E(\|y_n\|^2 | F_n). \tag{34}$$

Now if (25) holds for a given sequence (c_n), it holds a fortiori for any smaller sequence. Hence, replacing c_n by $\min(1, c_n)$, we may assume that $c_n \le 1$, since the series

$$\sum_1^\infty \min(1, c_n)$$

diverges if $\sum_1^\infty$ does. Moreover, defining

$$c_n' = \begin{cases} 0 & \text{if } c_1 + \dots + c_n = 0 \\ c_n/(c_1 + \dots + c_n) & \text{if } c_1 + \dots + c_n > 0, \end{cases}$$

we see that (25) continues to hold with c_n' in place of c_n, and by Lemma 1

$$\Sigma c'_n = \infty \text{ for all } x_1, x_2, \ldots \text{ such that } \sup \|x_n\| < \infty$$

$$\Sigma (c'_n)^2 \leq \Sigma \frac{c_n}{(c_1 + \cdots + c_n)^2} < \infty \text{ for all } x_1, x_2, \ldots$$

Hence it may be assumed that

$$c_n \leq 1, \ \Sigma c_n^2 < \infty \quad \text{for all } x_1, x_2, \ldots \ . \tag{35}$$

Then by (26), (27), (34) and (35) $P\{\Sigma \beta_n < \infty, \Sigma \xi_n < \infty\} = 1$, and it follows from Theorem 1 that

$$\lim_{n \to \infty} z_n$$

exists and is finite and $\Sigma \zeta_n = \Sigma c_n z_n^{\frac{1}{2}} < \infty$ a.s. By (28) $\lim_{n \to \infty} z_n = 0$ a.s. which by (26) implies that

$$\lim_{n \to \infty} \|x_n\| = 0$$

a.s.

5. Minimax Theorem for Vector Payoffs

In [1] Blackwell considers two person games in which each player has a finite number of possible actions, and the return to Player I if he takes action i ($1 \leq i \leq r$) and Player II takes action j ($1 \leq j \leq s$) is a random variable distributed according to m(i,j), a probability distribution over a closed, bounded, convex set X (not depending on (i,j)) in N space. The question Blackwell posed is whether in a sequence of repeated plays of such a game Player I can force the center of gravity $\bar{x}_n$ of his rewards $x_1, \ldots, x_n$ in the first n plays to converge to a given set S no matter what Player II does. Letting δ_n denote the squared distance from $\bar{x}_n$ to S, Blackwell shows that under certain hypotheses on S and the matrix (m(i,j)) there exists a particular sequence of strategies for Player I such that no matter what Player II does

$$E(\delta_n | \delta_1, \ldots, \delta_{n-1}) \leq (1 - 2/n)\delta_{n-1} + c/n^2 \text{ if } \delta_{n-1} > 0, \tag{36}$$

$$0 \leq \delta_n \leq a, \tag{37}$$

and $$|\delta_n - \delta_{n-1}| \leq b/n, \tag{38}$$

where a,b, and c depend only on the size of the bounded set X. He then applies the following lemma, which we state as

Application 5. A sequence of random variables $\delta_1, \delta_2, \ldots$ satisfying (36), (37), and (38) converges to 0 at a rate which depends only on a, b, and c; i.e., for every $\varepsilon > 0$ there exists an n_0 depending only on ε,a,b, and c such that

$$P\{\sup_{n \geq n_0} \delta_n \geq \varepsilon\} \leq \varepsilon$$

uniformly for all sequences (δ_n) satisfying (36), (37) and (38).

Proof. Let $z_n = \delta_n^2$, $F_n = B(\delta_1, \ldots, \delta_n)$ $(F_0 = \{\emptyset, \Omega\})$. Then by (36), (37), and (38), on

$$\{\delta_{n-1} > 0\}$$

$$E(z_n|F_{n-1}) = E((\delta_n - \delta_{n-1} + \delta_{n-1})^2|F_{n-1})$$

$$= E((\delta_n - \delta_{n-1})^2|F_{n-1}) + 2\delta_{n-1} E(\delta_n - \delta_{n-1}|F_{n-1}) + \delta_{n-1}^2$$

$$\leq \left(\frac{b}{n}\right)^2 + 2\delta_{n-1}(-2/n\ \delta_{n-1} + c/n^2) + \delta_{n-1}^2$$

$$\leq z_{n-1}(1 - 4/n) + \frac{2ac + b^2}{n^2}.$$

Hence by (38), with probability one

$$E(z_n|F_{n-1}) \leq (1 - 4/n)z_{n-1} + \frac{2ac + b^2}{n^2} \tag{39}$$

$(n \geq 4)$. It follows at once from Theorem 1 with

$$\xi_n = \frac{2ac + b^2}{(n+1)^2}$$

and $\zeta_n = (4/(n+1))z_n$ that

$\lim_{n \to \infty} z_n$ exists and is finite and

$$\sum_1^\infty \zeta_n = 4 \sum_1^\infty z_n / n+1 < \infty \quad \text{a.s.};$$

hence $\lim_{n \to \infty} z_n = 0$ a.s. To go farther and prove the required uniformity, note that from (37) and (39) we obtain by iteration for $n_0 \geq m \geq 4$

$$Ez_{n_0} \leq a^2 \prod_{k=m+1}^{n_0} (1 - 4/k) + (2ac + b^2) \sum_{k=m+1}^{n_0} k^{-2} \quad . \tag{40}$$

Now $\Pi\ (1 - 4/k)$ diverges and $\Sigma\ k^{-2}$ converges. Hence for any $\epsilon > 0$ there exists $m = m(a,b,c,\epsilon)$ so large that

$$(2ac + b^2) \sum_{m+1}^\infty k^{-2} < \epsilon^2/4$$

and $n_0 = n_0(a,b,c,\epsilon) \geq m$ such that

$$a^2 \prod_{m+1}^{n_0} (1 - 4/k) < \epsilon^2/4.$$

Then by (12) and (40)

$$P\{ \sup_{n \geq n_0} z_n \geq \epsilon\} \leq \epsilon^{-1}[Ez_{n_0} + (2ac + b^2) \sum_{k=n_0+1}^\infty k^{-2}]$$

$$\leq \epsilon^{-1}[\epsilon^2/2 + \epsilon^2/4] < \epsilon.$$

6. Cluster Analysis

In [8] MacQueen describes a method for partitioning an N-dimensional population into k sets on the basis of a sample. Let $y_1, y_2, \ldots$ be independent random vectors in N-dimensional Euclidean space each having a continuous distribution p. For a given k-tuple $(x_1, \ldots, x_k)$, $x_i \in E_n$, $i = 1,2, \ldots, k$, we define a <u>minimum distance partition</u> $S(x) = (S_1(x), \ldots, S_k(x))$ of E_N by

$$S_1(x) = T_1(x),\ S_2(x) = T_2(x) \cap S_1^c(x),\ \ldots,$$
$$S_k(x) = T_k(x) \cap S_1^c(x) \cap \ldots \cap S_{k-1}^c(x),$$

where

$$T_i(x) = \{\xi : \xi \in E_N,\ \|\xi - x_i\| \le \|\xi - x_j\|,\ j = 1,2,\ \ldots,k\}.$$

($\| \ \|$ denotes the usual Euclidean distance.) Sample $\underline{k}$-<u>means</u>

$$x^n = (x_1^n, \ldots x_k^n)$$

with associated integer weights $(w_1^n, \ldots, w_k^n)$ are defined by

$$x_i^1 = y_i,\quad w_i^1 = 1 \quad (i = 1,2, \ldots, k)$$

and for $n = 1,2, \ldots$, if $y_{k+n} \in S_i(x^n)$

$$x_i^{n+1} = \frac{w_i^n x_i^n + y_{k+n}}{w_i^n + 1},\quad x_j^{n+1} = x_j^n \quad \text{for } j \ne i$$
$$w_1^{n+1} = w_i^n + 1,\quad w_j^{n+1} = w_j^n \quad \text{for } j \ne i. \tag{41}$$

For each k-tuple $x = (x_1, \ldots, x_k)$ define

$$W(x) = \sum_1^k \int_{S_i(x)} \|y - x_i\|^2 \, dp(y)$$

and

$$V(x) = \sum_1^k \int_{S_i(x)} \|y - u_i(x)\|^2 \, dp(y),$$

where

$$u_i(x) = \int_{S_i(x)} y\,dp(y)/p(S_i(x)) \quad \text{if } p(S_i(x)) > 0$$

and $= x_i$ if $p(S_i(x)) = 0$. For ease of notation let

$S_i^n = S_i(x^n)$, $p_i^n = p(S_i^n)$, and $u_i^n = u_i(x^n)$. MacQueen's main result is

Application 6. If there exists a closed, bounded convex set $R \subset E_N$ such that $p(R) = 1$, if $p(A) > 0$ for every open subset $A \subset R$, and if p is absolutely continuous, then

$$\lim_{n \to \infty} W(x^n) \text{ exists with probability one} \tag{42}$$

and equals $V(x)$ for some x in the class of points x^* having the property that $x_i^* \neq x_j^*$ $(i \neq j)$ and $x_i^* = u_i(x^*)$ $(i = 1,2, \ldots ,k)$. Moreover with probability one

$$\lim_{m \to \infty} \sum_{n=1}^{m} \left(\sum_{i=1}^{k} p_i^n \|x_i^n - u_i^n\| \right) /m = 0. \tag{43}$$

Using only the hypothesis that R is bounded, we shall prove (42) and (43). For the remainder of the proof of Application 6, which is straightforward, see [8].

Let $F_n = B(z_1, \ldots ,z_{k+n-1})$. From

$$W(x^{n+1}) = \sum_{i=1}^{k} \int_{S_i^{n+1}} \|y - x_i^{n+1}\|^2 \, dp(y) \leq$$

$$\sum_{i=1}^{k} \int_{S_i^n} \|y - x_i^{n+1}\|^2 \, dp(y), \text{ (41)},$$

and the relation

$$\int_A \|y - x\|^2 \, dp(y) = \int_A \|y - u\|^2 \, dp(y) + p(A)\|x - u\|^2,$$

if $\int_A (y - u) \, dp(y) = 0$, it follows by straightforward computation that

$$E[W(x^{n+1})|F_n] \leq W(x^n) + \sum_{i=1}^{k} \sigma_{ni}^2 (p_i^n)^2/(w_i^n + 1)^2 \tag{44}$$

$$- \sum_{i=1}^{k} \|x_i^n - u_i^n\|^2 \, (p_i^n)^2 \, (2w_i^n + 1)/(w_i^n + 1)^2,$$

where $\sigma_{ni}^2 = \int_{S_i^n} \|y - u_i^n\|^2 \, dp(y)/p_i^n$ is uniformly bounded.

Let $z_n = W(x^n)$,

$$\xi_n = \sum_{i=1}^{k} \sigma_{ni}^2 \, (p_i^n)^2/(w_i^n + 1)^2,$$

and

$$\zeta_n = \sum_{i=1}^{k} \|x_i^n - u_i^n\|^2 (p_i^n)^2 (2w_i^n + 1)/(w_i^n + 1)^2.$$

Now

$$w_i^{n+1} = 1 + \sum_{j=1}^{n} I_i^j \text{, where } I_i^j = I_{\{y_{k+j} \in S_i^j\}} \text{ and hence}$$

$E(I_i^j | F_j) = p_i^j$. Hence by Corollary 2 and Lemma 1 of Section 3

$$\sum_{n=1}^{\infty} \frac{(p_i^n)^2}{(w_i^n + 1)^2} \leq \sum_{n=1}^{\infty} \frac{p_i^n}{(p_i^1 + \cdots + p_i^n)^2} \cdot \left(\frac{p_i^1 + \cdots + p_i^n}{w_i^{n+1}} \right)^2 < \infty \text{ a.s.}$$

$(i = 1,2, \ldots ,k)$. It follows that $\sum \xi_n < \infty$ a.s. and hence by Theorem 1 (42) holds and $\sum \zeta_n < \infty$ a.s. Since $w_i^n \leq n$, we have

$$\sum_{n=1}^{m} (p_i^n)^2 \|x_i^n - u_i^n\|^2/m \to 0 \text{ a.s. } (i = 1,2, \ldots ,k).$$

Hence by the Schwarz inequality

$$\frac{1}{m}\sum_{n=1}^{m} p_i^n \|x_i^n - u_i^n\| \le \left(\frac{1}{m}\sum_{n=1}^{m} (p_i^n)^2 \|x_i^n - u_i^n\|^2\right)^{\frac{1}{2}} \to 0$$

a.s. $(i = 1,2, \ldots ,k)$ which proves (43).

7. Hájek-Rényi-Chow Inequality

In the preceding sections we have emphasized the use of Theorems 1 and 2 in proving the almost sure convergence of a number of different processes. In this section we give two simple illustrative applications of the fundamental inequalities (9) and (12). (See also Section 5.)

Let $\{y_n, F_n, 1 \le n \le n_0\}$ be a non-negative submartingale and c_n $(1 \le n \le n_0)$ positive F_{n-1}-measurable random variables $(F_0 = \{\emptyset,\Omega\})$ such that $c_1 \ge c_2 \ge \ldots$. Chow's generalization [3] of the Hájek-Rényi inequality states that

$$P\{\max_{1\le n\le n_0} c_n y_n \ge 1\} \le E(c_1 y_1) + \sum_{k=2}^{n_0} E[c_k(y_k|F_{k-1}) - y_{k-1})]. \tag{45}$$

To prove (45), let $z_n = c_n y_n$. Then

$$E(z_{n+1}|F_n) = c_{n+1}(y_n + E(y_{n+1}|F_n) - y_n)$$

$$\le z_n + c_{n+1}(E(y_{n+1}|F_n) - y_n)$$

so that (1) is satisfied with $\beta_n = 0$, $\zeta_n = 0$, and

$$\xi_n = c_{n+1}(E(y_{n+1}|F_n) - y_n) \ge 0.$$

Hence by (12) with $m = 1$

$$P\{\max_{1\le k\le n_0} c_k y_k \ge 1\} \le E(z_1) + E\left(\sum_{k=1}^{n_0-1} \xi_k\right)$$

$$= E(c_1 y_1) + \sum_{k=2}^{n_0} E[c_k(E(y_k|F_{k-1}) - y_{k-1})].$$

In the special case that

$$y_n = (y + x_1 + \dots + x_n)^2, \quad c_n = (v+b_n)^{-2},$$

where $E(x_n | F_{n-1}) = 0$,

$$E(x_n^2 | F_{n-1}) = \sigma_n^2, \quad b_n = \sigma_1^2 + \dots + \sigma_n^2, \text{ and}$$

$v > 0$, the inequality (9) becomes (for $m = 0$)

$$E\left(\frac{y + x_1 + \dots + x_t}{v + b_t}\right)^2 \le (y/v)^2 + E\left(\sum_{n=0}^{t-1} \frac{\sigma_{n+1}^2}{(v + b_{n+1})^2}\right)$$

$$\le (y/v)^2 + E\left(\sum_0^\infty \int_{b_n}^{b_{n+1}} \frac{du}{(v + u)^2}\right) \le (y/v)^2 + 1/v, \tag{46}$$

an inequality of Dubins and Freedman [5]. The special case $y = 0$ has proved useful in problems of optimal stopping (cf. [4], p. 88).

References

[1] Blackwell, D. , An analogue of the minimax theorem for vector payoffs, Pac. Jour. Math., 6, (1956) 1-8.

[2] Blum, J., Approximation methods which converge with probability one, Ann. Math. Statist., 25, (1954) 382-386.

[3] Chow, Y. S., A martingale inequality and the law of large numbers, Proc. Amer. Math. Soc. 11 (1960) 107-111.

[4] Chow, Y. S., Robbins, H., and Siegmund, D. Great Expectations: The Theory of Optimal Stopping, Houghton Mifflin, Boston, 1971.

[5] Dubins, L. and Freedman, D., A sharper form of the Borel-Cantelli lemma and the strong law, Ann. Math. Statist. 36, (1965) 800-807.

[6] Dvoretzky, A., On stochastic approximation, Proc. Third Berk. Symp. Math. Statist. and Probab. Vol. I, University of California Press, Berkeley, 1956, pp. 39-55.

[7] Gladyshev, E. G., On stochastic approximation, Theory

of Prob. and Its Applications, 10 (1965) 275-278.

[8] MacQueen, J., Some methods for classification and analysis of multivariate observations, Proc. Fifth Berk. Symp. Math. Statist. and Probab., Vol. I, Univ. Calif. Press, Berkeley, 1967 , pp. 281-297.

[9] Neveu, J., Mathematical Foundations of the Calculus of Probabilities, Holden-Day, San Francisco, 1965, p. 148.

[10] Robbins, H. and Munro, S., A stochastic approximation method, Ann. Math. Statist., 22, (1951), 400-407.

[11] Schmetterer, L., Stochastic approximation, Proc. Fourth Berk. Symp. Math. Statist. and Probab., Vol. I, Univ. of Calif., Berkeley, 1961, pp. 587-609.

Reprinted from
Proc. Natl. Acad. Sci. USA
Vol. 75, No. 2, pp. 586–587, February 1978
Statistics

Adaptive design in regression and control

(iterated least squares/adaptive stochastic approximation/nonlinear regression/control theory/optimal dosage estimation)

T. L. LAI AND HERBERT ROBBINS

Department of Mathematical Statistics, Columbia University, New York, New York 10027

Contributed by Herbert Robbins, November 18, 1977

ABSTRACT When $y = M(x) + \epsilon$, where M may be nonlinear, adaptive regression designs of the levels $x_1, x_2, \ldots$ at which $y_1, y_2, \ldots$ are observed lead to asymptotically efficient estimates of the value θ of x for which $M(\theta)$ is equal to any desired value y^*. More importantly, these designs also make the "cost" of the observations, defined at the nth stage to be $\Sigma_1^n (x_i - \theta)^2$, to be of the order of log n instead of n, an obvious advantage in medical and other applications.

We shall consider a general, not necessarily linear, regression model

$$y_i = M(x_i) + \epsilon_i \quad (i = 1,2,\ldots), \qquad [1]$$

where the errors ϵ_i are independent, identically distributed random variables with mean 0 and variance σ^2. The adaptive designs for choosing the levels of x at which y is to be observed are motivated by applications of the following nature. Suppose that x_i is the dosage level of some drug and that y_i is the response value of the ith patient who turns up for treatment. Suppose also that an optimal response value y^* is known. If the function M were known, then the dosage level x could be kept at the value θ such that $M(\theta) = y^*$. But if M is unknown, how can the dosage levels $x_1, x_2, \ldots$ be chosen so as to approach the unknown θ as rapidly as possible?

We take $\Sigma_1^n (x_i - \theta)^2$ to be the (cumulative) *cost* at stage n. To fix ideas, consider the linear model $M(x) = \alpha + \beta x$ with $\beta > 0$ and α, β unknown, and suppose without loss of generality that the optimal response value y^* is 0. Then $\theta = -\alpha/\beta$, and if we assign the values $x_1, x_2, \ldots$ in advance and estimate θ by $\hat{\theta}_n = -\hat{\alpha}_n/\hat{\beta}_n$ after observing $x_1, y_1, \ldots, x_n, y_n$, where $\hat{\alpha}_n, \hat{\beta}_n$ are the usual least squares estimates of α, β, under mild regularity assumptions it can be shown that for large n, $\hat{\theta}_n$ is approximately normal with mean θ and variance

$$\frac{\sigma^2}{n\beta^2}\left[1 + \frac{(\bar{x}_n - \theta)^2}{\sum_1^n (x_i - \bar{x}_n)^2/n}\right] \qquad [2]$$

where the notation $\bar{x}_n$ for the arithmetic mean of any n quantities $x_1, \ldots, x_n$ will be used throughout the sequel. In ignorance of θ, to make the square bracket in Eq. 2 close to its minimal value 1 we would have to choose the x_i so as to make $\Sigma_1^n (x_i - \bar{x}_n)^2$, and hence *a fortiori* $\Sigma_1^n (x_i - \theta)^2$ = cost, of order larger than n. The results of this paper show that, by a suitable adaptive choice of $x_1, \ldots, x_n$, the unknown θ can be estimated with an asymptotic variance of $\sigma^2/n\beta^2$ but with a cost of order only $(\sigma^2/\beta^2) \log n$. Thus, the apparent dilemma of having to choose between a good final estimate of θ (of interest perhaps to future patients) and a small cost $\Sigma_1^n (x_i - \theta)^2$ (of interest to the current patients) is happily resolved for large n.

In an adaptive design, the choice of each level x_i depends on all the previous data values $x_1, y_1, \ldots, x_{i-1}, y_{i-1}$. In the linear model $M(x) = \alpha + \beta x$ with desired mean response 0 and $\theta = -\alpha/\beta$, the general model 1 can be written as

$$y_i = \beta(x_i - \theta) + \epsilon_i. \qquad [3]$$

A valuable hint toward the construction of an adaptive design in the general case comes from considering the special situation that arises when the true value of β in Eq. 3 is known. Then at stage n the least squares estimate of θ becomes

$$\theta_n^* = \bar{x}_n - (\bar{y}_n/\beta) = \theta - (\bar{\epsilon}_n/\beta). \qquad [4]$$

In this case, *no matter how the levels $x_1, \ldots, x_n$ are chosen*, whether fixed in advance or by an adaptive design, Eq. 4 shows that $E(\theta_n^* - \theta)^2 = \sigma^2/(n\beta^2)$ and

$$\sqrt{n}(\theta_n^* - \theta) \rightarrow N(0, \sigma^2/\beta^2) \text{ in distribution as } n \rightarrow \infty. \qquad [5]$$

Now, to minimize the cost, we want each x_n to be as close as possible to θ, and since our estimate of θ at stage n is θ_n^*, it is natural to choose the next level x_{n+1} to be precisely θ_n^*. This amounts to the following adaptive design. Let x_1 (the initial best guess of θ) be a random variable with finite second moment, and define inductively for $n = 1,2,\ldots$

$$x_{n+1} = \theta_n^* = \bar{x}_n - (\bar{y}_n/\beta). \qquad [6]$$

Then Eqs. 4 and 5 still hold, while from Eqs. 4 and 6

$$E\left(\sum_1^n (x_i - \theta)^2\right) = E(x_1 - \theta)^2 + \sum_1^{n-1} E(\theta_i^* - \theta)^2 \qquad [7]$$

$$= E(x_1 - \theta)^2 + \frac{\sigma^2}{\beta^2}\left[1 + \frac{1}{2} + \ldots + \frac{1}{n-1}\right]$$

$$= (\sigma^2/\beta^2) \log n + O(1).$$

The desirable properties 5 and 7 of the adaptive design 6 have been obtained under the restrictive assumption of a linear model 3 with β known. The following theorem shows that similar properties still hold for 6 when the regression function M is not linear but the slope $M'(\theta)$ at the unique root of the equation $M(x) = 0$ is known to be some positive constant β.

THEOREM 1. *Let β, θ, and the Borel function* M *be such that*

$$M(\theta) = 0, M'(\theta) = \beta > 0 \qquad [8]$$

$$\inf_{\delta \le |x-\theta| \le 1/\delta} [M(x)(x - \theta)] > 0 \textit{ for all } 0 < \delta < 1 \qquad [9]$$

$$|M(x)| \le c|x| + d \textit{ for some } c, d > 0 \textit{ and all } x. \qquad [10]$$

The costs of publication of this article were defrayed in part by the payment of page charges. This article must therefore be hereby marked "*advertisement*" in accordance with 18 U. S. C. §1734 solely to indicate this fact.

Let x_1 *be any random variable independent of* $\epsilon_1, \epsilon_2, \ldots$. *Define inductively* y_i *by* **1** *and* x_{n+1} *by* **6**. *Then as* $n \to \infty$

$$\sqrt{n}(x_n - \theta) \to N(0, \sigma^2/\beta^2) \text{ in distribution} \quad [11]$$

$$P(\lim x_n = \theta) = 1, \text{ and in fact}$$

$$P\left(\overline{\lim} \sqrt{\frac{n}{2 \log \log n}}(x_n - \theta) = \sigma/\beta\right) = 1 \quad [12]$$

$$P\left(\lim \left(\sum_1^n (x_i - \theta)^2/\log n\right) = \sigma^2/\beta^2\right) = 1. \quad [13]$$

Since in practice the value $M'(\theta) = \beta$ will rarely be known, it is natural to try replacing **6** by

$$x_{n+1} = \bar{x}_n - (\bar{y}_n/b_n), \quad [14]$$

where $b_n = b_n(x_1, y_1, \ldots\ldots, x_n, y_n)$ is some estimate of β based on the data observed by stage n. If b_n is a *consistent* estimate of β (no matter how *inefficient*), we may hope that the asymptotic properties **11, 12,** and **13** of *Theorem 1* will still hold. An obvious choice for b_n is the usual least squares estimate

$$\hat{\beta}_n = \frac{\sum_1^n (x_i - \bar{x}_n)(y_i - \bar{y}_n)}{\sum_1^n (x_i - \bar{x}_n)^2} \quad [15]$$

at least in the linear case $M(x) = \beta(x - \theta)$. But when the x_i are sequentially determined *random variables* it is not evident that $\hat{\beta}_n$ will in fact be a consistent estimate of β. [Even for the case of a fixed design $x_1, x_2, \ldots$ the *strong* consistency of $\hat{\beta}_n$ under the sole condition that $\Sigma_1^n (x_i - \bar{x}_n)^2 \to \infty$ has only recently been established (1).] However, a modified version of $\hat{\beta}_n$, asymptotically equal to it but somewhat more stable in its initial stages, has the desired properties. This is the content of

THEOREM 2. *Suppose that, in addition to* **8, 9,** *and* **10**

$$\underline{\lim}\, |M(x)/x| > 0 \text{ as } |x| \to \infty, \quad [16]$$

and that positive constants b and B *are known such that* $b \le \beta = M'(\theta) \le B$. *Define inductively* y_i *by* **1** *and* x_{n+1} *by* **14**, *where the* b_n *are defined as follows. Let* b_1 *be any constant between* b and B, *and let* c_n *be any sequence of positive constants such that*

$$\underline{\lim}\, c_n > 0, \overline{\lim}\, c_n < 1. \quad [17]$$

Define $s_n^2 = \Sigma_1^n (y_i - \bar{y}_n)^2/n$, *and for* $n \ge 2$ *if*

$$\sum_1^n (x_i - \bar{x}_n)^2 \le (c_n s_n^2/B^2) \cdot \log n \quad [18]$$

set $b_n = b_{n-1}$; *otherwise set* b_n *equal to* $\hat{\beta}_n$ *truncated below by* b *and above by* B. *Then* **11, 12,** and **13** *still hold and with probability* 1 *as* $n \to \infty$,

$$\lim b_n = \beta, \lim s_n^2 = \sigma^2. \quad [19]$$

If, furthermore, for some $r > 0$

$$M(x) = \beta(x - \theta) + O(|x - \theta|^{1+r}) \text{ as } x \to \theta \quad [20]$$

then

$$(\log n)^{1/2}(b_n - \beta) \to N(0, \beta^2) \text{ in distribution as } n \to \infty. \quad [21]$$

We call the design of *Theorem 2 iterated least squares* modified by using b_n instead of $\hat{\beta}_n$ in **14**. We now discuss a different design which we shall call *adaptive stochastic approximation*. Going back to **6**, it is easy to show that, for any constant $b > 0$, the two sequences

$$x_{n+1} = \bar{x}_n - (\bar{y}_n/b), \; x_{n+1} = x_n - (y_n/nb) \quad [22]$$

are identical. This suggests using, instead of **14**, the design

$$x_{n+1} = x_n - (y_n/nb_n) \quad [23]$$

where again b_n is an estimate of β. Properties of **23** are given by

THEOREM 3. *Suppose that* **8, 9,** *and* **10** *hold and that positive constants* b *and* B *are known such that* $b \le \beta = M'(\theta) \le B$. *Define inductively* y_i *by* **1** *and* x_{n+1} *by* **23**, *where* b_1 *is any constant between* b *and* B *and for* $n \ge 2$, b_n *is defined as follows:*

(i) *For* b_n *defined as in* Theorem 2, **11, 12, 13,** *and* **19** *still hold, and if* **20** *also holds then so does* **21.**

(ii) *Alternatively, simply put* b_n *equal to* $\hat{\beta}_n$ *truncated below by* b *and above by* B [*if* $\Sigma_1^n (x_i - \bar{x}_n)^2 > 0$]; *otherwise put* $b_n = b_{n-1}$). *Then the conclusions of* (i) *remain valid.*

Design **23** with $b_n = b$, some positive constant, is the original Robbins–Monro stochastic approximation procedure and has an extensive literature. In *Theorem 3* we have taken b_n to be a random variable, and (ii) says that for $b_n = \hat{\beta}_n$ truncated below and above by known positive bounds on β, procedure **23** makes b_n a strongly consistent estimator of β and that the conclusions of *Theorem 1* still hold for **23.**

Proofs of *Theorems 1, 2,* and *3* and simulation studies of several adaptive designs will be given elsewhere, along with a treatment of the case in which prior lower and upper bounds for β are not known; the latter case can be handled by a modification of b_n in **23.**

This research was supported by the National Institute of General Medical Sciences, the National Science Foundation, and the Office of Naval Research.

1. Lai, T. L. & Robbins, H. (1977) *Proc. Natl. Acad. Sci. USA* **74,** 2667–2669.

The Annals of Statistics
1979, Vol. 7, No. 6, 1196–1221

ADAPTIVE DESIGN AND STOCHASTIC APPROXIMATION[1]

BY T. L. LAI AND HERBERT ROBBINS

Columbia University

When $y = M(x) + \varepsilon$, where M may be nonlinear, adaptive stochastic approximation schemes for the choice of the levels $x_1, x_2, \cdots$ at which $y_1, y_2, \cdots$ are observed lead to asymptotically efficient estimates of the value θ of x for which $M(\theta)$ is equal to some desired value. More importantly, these schemes make the "cost" of the observations, defined at the nth stage to be $\Sigma_1^n(x_i - \theta)^2$, to be of the order of $\log n$ instead of n, an obvious advantage in many applications. A general asymptotic theory is developed which includes these adaptive designs and the classical stochastic approximation schemes as special cases. Motivated by the cost considerations, some improvements are made in the pairwise sampling stochastic approximation scheme of Venter.

1. Introduction. We shall consider the general regression model

$$y_i = M(x_i) + \varepsilon_i \qquad i = 1, 2, \cdots \tag{1.1}$$

where the errors $\varepsilon_1, \varepsilon_2, \cdots$ are i.i.d. random variables with mean zero and variance σ^2. Unless otherwise stated, the above notations and assumptions will be used throughout the sequel. We shall always assume that the regression function $M(x)$ is a Borel function satisfying the following three conditions:

$$M(\theta) = 0 \text{ for a unique } \theta \text{ and } M'(\theta) = \beta \text{ exists and is positive;} \tag{1.2}$$

$$\inf_{\delta < |x-\theta| < \delta^{-1}}\{M(x)(x-\theta)\} > 0 \qquad \text{for all} \quad 0 < \delta < 1; \tag{1.3}$$

$$|M(x)| < c|x| + d \qquad \text{for some} \quad c, d > 0 \quad \text{and all} \quad x. \tag{1.4}$$

Suppose that in (1.1) x_i is the dosage level of a drug given to the ith patient who turns up for treatment and that y_i is the response of the patient. Suppose the mean response of the patients under treatment should be at some optimal given level h. Without loss of generality, we shall (replacing y_i by $y_i - h$ if necessary) assume that $h = 0$. To achieve this mean response $h = 0$, if the unique (by (1.2)) root θ of the equation $M(\theta) = 0$ were known, then the dosage levels should all be set at θ. Since θ is usually unknown, how can the dosage levels x_i be chosen so that they approach θ rapidly? In the choice of the dosage levels x_i our primary objective here is in the treatment of the patients rather than in finding an efficient design to estimate the ideal dosage level θ. Calling $\Sigma_1^n(x_i - \theta)^2$ the *cost* of the design at stage n, we have announced in [8] that the apparent dilemma of choosing between a small cost and

Received March 1978; revised July 1978.

[1]This research was supported by the National Institutes of Health, the National Science Foundation, and the Office of Naval Research under grants 5R01-GM-16895, NSF MCS 76-09179, and N00014-75-C-0560

AMS 1970 *subject classifications.* Primary 62L20, 62K99; secondary 60F15.

Key words and phrases. Adaptive design, adaptive stochastic approximation, regression, logarithmic cost, asymptotic normality, iterated logarithm, pairwise sampling schemes, least squares.

a good estimate of θ can be resolved by using a suitable adaptive design. In the present paper and its companion papers [10], [11] and [12], we investigate the properties of the adaptive designs announced in [8] and prove the theorems that were stated without proof in [8]. We also consider some other adaptive designs and analyze their performance.

In an adaptive design, the choice of each level x_i will depend on the data so far observed, i.e., x_i is a function of $x_1, y_1, \cdots, x_{i-1}, y_{i-1}$. Consider first the simple case where $M(x)$ is linear so that

$$y_i = \beta(x_i - \theta) + \varepsilon_i \qquad i = 1, 2, \cdots. \tag{1.5}$$

Suppose that β is in fact *known*. To estimate θ at stage n, it is natural to use the least squares estimator

$$\theta_n^* = \bar{x}_n - \beta^{-1}\bar{y}_n (= \theta - \beta^{-1}\bar{\varepsilon}_n). \tag{1.6}$$

(Here and in the sequel, we use the notation $\bar{a}_n$ for the arithmetic mean $n^{-1}\Sigma_1^n a_i$ of any n numbers $a_1, \cdots, a_n$.) The last equality in (1.6) shows that *irrespective of how the levels x_i are chosen*, whether preassigned or sequentially determined,

$$E(\theta_n^* - \theta)^2 = \sigma^2/(n\beta^2), \tag{1.7}$$

and

$$n^{\frac{1}{2}}(\theta_n^* - \theta) \rightarrow_{\mathcal{L}} N(0, \sigma^2/\beta^2) \quad \text{as} \quad n \rightarrow \infty, \tag{1.8}$$

where $\rightarrow_{\mathcal{L}}$ denotes convergence in distribution. In particular, if we use the adaptive design

$$x_{i+1} = \theta_i^* = \bar{x}_i - \beta^{-1}\bar{y}_i \qquad i = 1, 2, \cdots \tag{1.9}$$

and let x_1 (= initial guess of θ) be a random variable with finite second moment, then it follows from (1.7) and (1.9) that the expected cost of the design (1.9) at stage n is of the order of $\log n$, i.e.,

$$E\{\Sigma_1^n(x_i - \theta)^2\} = (\sigma^2/\beta^2)\log n + 0(1). \tag{1.10}$$

While the desirable properties (1.8) and (1.10) for the adaptive design (1.9) have been obtained under the assumption of the linear model (1.5) with β known, the following theorem says that similar properties still hold in the general nonlinear case, provided again that $\beta(= M'(\theta))$ is known.

THEOREM 1. *Let $\varepsilon, \varepsilon_1, \varepsilon_2, \cdots$ be* i.i.d. *random variables with $E\varepsilon = 0$ and $E\varepsilon^2 = \sigma^2$. Let $M(x)$ be a Borel function satisfying* (1.2), (1.3) *and* (1.4). *Let x_1 be a random variable independent of $\varepsilon_1, \varepsilon_2, \cdots$. For $i = 1, 2, \cdots$, define inductively by* (1.1) *and x_{i+1} by* (1.9). *Then*

$$n^{\frac{1}{2}}(x_n - \theta) \rightarrow_{\mathcal{L}} N(0, \sigma^2/\beta^2) \quad \textit{as} \quad n \rightarrow \infty; \tag{1.11}$$

$$\lim_{n\rightarrow\infty} x_n = \theta \text{ a.s., } \textit{and in fact}, \tag{1.12}$$

$$\limsup_{n\rightarrow\infty}(n/2\log\log n)^{\frac{1}{2}}|x_n - \theta| = \sigma/\beta \text{ a.s.};$$

$$\lim_{n\rightarrow\infty}\{\Sigma_1^n(x_i - \theta)^2/\log n\} = \sigma^2/\beta^2 \text{ a.s.} \tag{1.13}$$

Theorem 1, which was stated without proof in [8], is a special case of Theorem 2 below. The almost sure convergence results (1.12) and (1.13) are of particular interest for the kind of applications described above. As time progresses, one obtains more and more subjects for treatment; and the choice of the dosage levels is a continuing process. Theorem 1 says that if one uses the adaptive design (1.9), then with probability 1 the levels x_n will converge to the ideal level θ, the cost $\Sigma_1^n(x_i - \theta)^2$ will eventually grow like $(\sigma^2/\beta^2)\log n$, and that x_n will still be an efficient estimator of θ in the sense of (1.11).

In practice, the slope β of the regression function at the level θ will be unknown. In ignorance of β, if one simply substitutes for β some guess b of its value in the recursion (1.9), then the following analogue of Theorem 1 holds.

THEOREM 2. *Let $\varepsilon, \varepsilon_1, \varepsilon_2, \cdots$ be* i.i.d. *with $E\varepsilon = 0$ and $E\varepsilon^2 = \sigma^2$. Let $M(x)$ be a Borel function satisfying* (1.2), (1.3) *and* (1.4). *Let x_1 be a random variable independent of $\varepsilon_1, \varepsilon_2 \cdots$. For $i = 1, 2, \cdots$, define inductively y_i by* (1.1) *and x_{i+1} by*

$$x_{i+1} = \bar{x}_i - b^{-1}\bar{y}_i, \tag{1.14}$$

where b is a positive constant.

(i) *Let $f(t) = 1/\{t(2-t)\}$ for $0 < t < 2$. If $b < 2\beta$, then*

$$n^{\frac{1}{2}}(x_n - \theta) \to_{\mathcal{L}} N(0, (\sigma^2/\beta^2)f(b/\beta)); \tag{1.15}$$

$$\limsup_{n\to\infty}(n/2\log\log n)^{\frac{1}{2}}|x_n - \theta| = (\sigma/\beta)f^{\frac{1}{2}}(b/\beta) \text{ a.s.}; \tag{1.16}$$

$$\lim_{n\to\infty}\{\Sigma_1^n(x_i - \theta)^2/\log n\} = (\sigma^2/\beta^2)f(b/\beta) \text{ a.s.} \tag{1.17}$$

(ii) *Assume that $M(x)$ further satisfies*

$$M(x) = \beta(x - \theta) + O(|x - \theta|^{1+\eta}) \text{ as } x \to \theta \text{ for some } \eta > 0. \tag{1.18}$$

If $b > 2\beta$, then there exists a random variable z such that

$$n^{\beta/b}(x_n - \theta) \to z \text{ a.s.}, \tag{1.19}$$

and, therefore,

$$\{\Sigma_1^n(x_i - \theta)^2/n^{1-(2\beta/b)}\} \to z^2/\{1 - (2\beta/b)\} \text{ a.s.} \tag{1.20}$$

(iii) *Suppose $b = 2\beta$ and that $M(x)$ further satisfies* (1.18). *Then*

$$(n/\log n)^{\frac{1}{2}}(x_n - \theta) \to_{\mathcal{L}} N(0, \sigma^2/b^2), \tag{1.21}$$

and

$$\{\Sigma_1^n(x_i - \theta)^2/(\log n)^2\} \to_{\mathcal{L}} (\sigma^2/b^2)\textstyle\int_0^1 w^2(t)\,dt, \tag{1.22}$$

where $w(t)$, $t \geq 0$, is the standard Wiener process.

In Section 2 we shall prove a more general result that contains Theorem 2 as a special case. Theorem 2 says that if in (1.14) $b < 2\beta$ then the cost $\Sigma_1^n(x_i - \theta)^2$ grows like a constant times $\log n$ and that $n^{\frac{1}{2}}(x_n - \theta)$ is asymptotically normal. The factor $f(b/\beta)$ in (1.15)–(1.17) has its minimum value 1 for $b = \beta$, and

$f(b/\beta) = (1 - r^2)^{-1}$ if $b = (1 \pm r)\beta$ $(0 < r < 1)$. Thus, even if our guess b of β has a relative error of 50%, the variance of the asymptotic distribution of $n^{\frac{1}{2}}(x_n - \theta)$ and the asymptotic cost $\Sigma_1^n(x_i - \theta)^2$ for the adaptive design (1.14) only exceed the corresponding minimum values for $b = \beta$ by a factor of $\frac{4}{3}$. On the other hand, if $b > 2\beta$, the cost is of a much larger order of magnitude and the rate of convergence of x_n to θ is much slower.

The adaptive design (1.14) sticks to an initial guess b of β, and its asymptotic performance is unsatisfactory when b exceeds 2β. Instead of adhering to an initial guess of β, it is natural to consider the possibility of estimating β from the data already observed and using that estimate in the choice of the next x-value. This means replacing the recursion (1.14) by

$$x_{i+1} = \bar{x}_i - b_i^{-1}\bar{y}_i \tag{1.23}$$

where $b_i = b_i(x_1, y_1, \cdots, x_i, y_i)$ is some estimate of β based on the data already observed. Adaptive designs of the type (1.23) will be discussed in [11], where we shall show that *by a suitable choice of* b_i *the desirable asymptotic properties* (1.11), (1.12), *and* (1.13) *still hold for the design* (1.23).

In this paper we consider an alternative way of modifying the adaptive design (1.9) when β is not known. We first note another way of expressing the recursive scheme (1.9), or more generally (1.14), in the following lemma.

LEMMA 1. *Let* $\{x_i, i \geq 1\}$ *and* $\{y_i, i \geq 1\}$ *be two sequences of real numbers. For any constant* c *and positive integer* n, *the following two statements are equivalent*:

$$x_{i+1} = \bar{x}_i - c\bar{y}_i \quad \textit{for all} \quad i = 1, \cdots, n; \tag{1.24}$$

$$x_{i+1} = x_i - cy_i/i \quad \textit{for all} \quad i = 1, \cdots, n. \tag{1.25}$$

Lemma 1 is easily proved by induction on n. Now, the recursion (1.25) with $c > 0$ is a special case of the general *stochastic approximation scheme*

$$x_{i+1} = x_i - c_i y_i \qquad i = 1, 2, \cdots \tag{1.26}$$

introduced by Robbins and Monro [15], where $\{c_i\}$ is an arbitrary sequence of positive constants such that

$$\Sigma_1^\infty c_i = \infty \quad \text{and} \quad \Sigma_1^\infty c_i^2 < \infty. \tag{1.27}$$

For the regression model (1.1), under the assumptions on the errors ϵ_i and on the regression function $M(x)$ described in the first paragraph, it is known (cf. [1], [15]) that (1.27) is a sufficient condition for the x_n generated by the stochastic approximation scheme (1.26) to converge to θ in mean square and with probability 1. It is also known (cf. [3], [16]) that if $c_i = (ib)^{-1}$, where b is a positive constant $< 2\beta$, then the x_n generated by (1.26) has an asymptotically normal distribution as given by (1.15). Therefore, by the equivalence in Lemma 1, the asymptotic normality (1.15) in Theorem 2 follows immediately. It is also known that if the c_i are of a larger order of magnitude than i^{-1}, then x_n may converge to θ in distribution at a rate much slower than that of (1.15). In particular, Chung [3] has considered

$c_i = i^{-(1-\delta)}$ for certain positive values of $\delta < \frac{1}{2}$ and has shown under some restrictive assumptions that

$$n^{(1-\delta)/2}(x_n - \theta) \to_{\mathcal{L}} N(0, \sigma^2/(2\beta)). \tag{1.28}$$

He has also noted that an asymptotically optimal choice of c_i for the stochastic approximation scheme (1.26) is $c_i = (i\beta)^{-1}$, at least for the linear case $M(x) = \beta(x - \theta)$ (cf. Sections 6 and 7 of [3]).

In ignorance of β, it is natural to try using $c_i = (ib_i)^{-1}$ in (1.26), where $b_i = b_i(x_1, y_1, \cdots, x_i, y_i)$ is some estimate of β based on the data already observed. Of course, we want b_i to be a strongly consistent estimator of β so that hopefully the asymptotic properties (1.11), (1.12) and (1.13) will be preserved. We shall call any adaptive design

$$x_{i+1} = x_i - y_i/(ib_i), \tag{1.29}$$

where b_i is a strongly consistent estimator of $\beta(= M'(\theta))$, an *adaptive stochastic approximation scheme*, and in Section 2 we shall show that adaptive stochastic approximation schemes have the desirable properties (1.11), (1.12) and (1.13) of Theorem 1. More generally, if b_i in (1.29) converges to some positive constant b with probability 1, we shall call (1.29) a *quasi-adaptive stochastic approximation scheme*. An asymptotic theory will be developed in Section 2 for quasi-adaptive stochastic approximation schemes, and these general results not only include Theorem 2, and therefore Theorem 1 as well, as special cases, but also establish the desired asymptotic properties (1.11), (1.12) and (1.13) for adaptive stochastic approximation schemes. In Sections 3 and 4 we shall describe two different methods of constructing adaptive stochastic approximation schemes and apply the results of Section 2 to the analysis of these procedures.

2. Asymptotic properties of quasi-adaptive stochastic approximation schemes. Throughout this section the following notations will be used. Let $\varepsilon, \varepsilon_1, \cdots$ be i.i.d. random variables with $E\varepsilon = 0$. (Although we shall often also assume that $E\varepsilon^2 < \infty$, there are certain places where we can relax this assumption.) Let $M(x)$ be a Borel function satisfying (1.2), (1.3) and (1.4). Let x_1 be a random variable independent of $\varepsilon_1, \varepsilon_2, \cdots$ Let $\mathcal{F}_0$ denote the σ-field generated by x_1, and for $k \geq 1$ let $\mathcal{F}_k$ denote the σ-field generated by $x_1, \varepsilon_1, \cdots, \varepsilon_k$. For $i = 1, 2, \cdots$, let $y_i = M(x_i) + \varepsilon_i$, where $\{x_i\}$ is a stochastic approximation scheme defined by

$$x_{i+1} = x_i - y_i/(ib_i), \tag{2.1}$$

and $\{b_i\}$ is a sequence of positive random variables.

The following representation theorem, stated without proof in [9], is a very useful tool for analyzing quasi-adaptive stochastic approximation schemes.

THEOREM 3. *Let b be a positive constant and let $\{b_n\}$ be a sequence of positive random variables such that* $\lim_{n\to\infty} b_n = b$ a.s.

(i) *For the stochastic approximation scheme* (2.1), *if* $\lim_{n\to\infty} x_n = \theta$ a.s., *then the following representation holds*:

(2.2) $$x_{n+1} = \theta + (n^{-\beta/b}/\tau_n)\{\Sigma_1^n \delta_k \varepsilon_k + \rho_0\},$$

where ρ_0, τ_k *and* δ_k *are random variables having the following properties*:

(2.3) $$\tau_k > 0 \quad \text{and} \quad \delta_k = -k^{(\beta/b)-1}\tau_k/b_k, \qquad k \geqslant 1;$$

(2.4) $$P[\tau_{n+1} - \tau_n = o(\tau_n/n) \quad \text{as} \quad n \to \infty] = 1.$$

(ii) *Suppose that* b_n *is* $\mathfrak{F}_{n-1}$*-measurable for all* $n \geqslant 1$, *and assume either that* $E(|\varepsilon|\log|\varepsilon|) < \infty$ *or that* ε *is symmetric. Then for the stochastic approximation scheme* (2.1), $\lim_{n\to\infty} x_n = \theta$ a.s., *and in the representation* (2.2) *we further obtain that*

(2.5) τ_k *and therefore* δ_k *also are* $\mathfrak{F}_{k-1}$*-measurable for all* $k \geqslant 1$.

(iii) *Assume that* $E|\varepsilon|^r < \infty$ *for some* $r > 1$, *that* $M(x)$ *also satisfies* (1.18), *and that*

(2.6) $$P[b_n - b = o(n^{-\lambda}) \text{ as } n \to \infty] = 1 \text{ for some positive constant } \lambda.$$

Then for the stochastic approximation scheme (2.1), $\lim_{n\to\infty} x_n = \theta$ a.s., *and in the representation* (2.2), *the random variables* τ_k *and* δ_k *satisfy* (2.3) *and*

(2.7) $$P[\tau_{n+1}/\tau_n = 1 + o(n^{-(1+p)}) \text{ as } n \to \infty] = 1 \text{ for some positive constant } p.$$

Consequently,

(2.8)
$$\lim_{n\to\infty} \tau_n = \tau \text{ exists and is positive a.s., and } P[\tau_n = \tau + o(n^{-p}) \text{ as } n \to \infty] = 1.$$

Moreover, if b_n *is* $\mathfrak{F}_{n-1}$*-measurable for all* $n \geqslant 1$, *then* (2.5) *also holds.*

REMARKS. In the particular case $b_n = b$ for all n, and under the stronger moment condition $E\varepsilon^2 < \infty$, similar representation results have been obtained by Major and Révész [13], Kersting [7], and Gaposhkin and Krasulina [6]. Gaposhkin and Krasulina [6] have obtained the representation (2.2) for this particular case and established the properties (2.3), (2.5) and

(2.9) $$P[\{\tau_n\} \text{ is slowly varying as } n \to \infty] = 1.$$

(A sequence $\{L(n)\}$ is said to be *slowly varying* as $n \to \infty$ if $L([cn])/L(n) \to 1$ for all $c > 0$. If $\{L(n)\}$ is slowly varying, then the sequence $\{n^\alpha L(n)\}$ is said to be *regularly varying* with exponent α.) By Theorem 4 of [2], (2.4) implies (2.9); in fact, a sequence $\{L(n)\}$ of positive numbers is slowly varying if and only if there exists a sequence $\{c_n\}$ of positive numbers such that $L(n)/c_n \to 1$ and $c_{n+1} - c_n = o(c_n/n)$. The property (2.4) is a very useful tool for studying the limiting behavior of the stochastic approximation scheme (2.1) and of the cost $\Sigma_1^n(x_i - \theta)^2$ for the design. It enables us to reduce the problem to that of the martingale $\Sigma_1^n \varepsilon_i/b_i$ (when b_i is $\mathfrak{F}_{i-1}$-measurable for all i) via a partial summation technique (cf. [9]).

The representation considered by Major and Révész [13] is somewhat different from (2.2). They assume that $M(x) = \beta(x - \theta) + U(x)$ where $U(x) = O((x - \theta)^2)$

as $x \to \theta$ (i.e., (1.18) holds with $\eta = 1$) and obtain the representation

$$x_{n+1} = \theta - n^{-\beta/b}\{\Sigma_1^n k^{(\beta/b)-1}(1 + 0(k^{-1}))(\varepsilon_k + U(x_k)) + \rho_0\}$$

for the case where $b_n = b$ for all n and under the assumption $E\varepsilon^2 < \infty$. Again, for this special case $b_n = b$, and under the assumptions $E\varepsilon^2 < \infty$ and (1.18), Kersting [7] recently showed that if $b < 2\beta$ then

$$x_{n+1} = \theta - n^{-\beta/b}\Sigma_1^n k^{(\beta/b)-1}\varepsilon_k + \rho_n$$

where the ρ_n are random variables such that $n^\lambda \rho_n \to 0$ a.s. for some $\frac{1}{2} < \lambda < \beta/b$. The methods of Kersting and of Major and Révész depend very heavily on the assumption that $b_n = b$ for all n. The following proof of Theorem 3 is based on a generalization of the argument of Gaposhkin and Krasulina.

Proof of Theorem 3. Suppose that $\lim_{n\to\infty} x_n = \theta$ a.s. Without loss of generality, we can assume that $\theta = 0$. Therefore, in view of (1.2),

(2.10) $$M(x_n) = (\beta + \xi_n)x_n, \qquad \text{where} \quad \xi_n \to 0 \text{ a.s.}$$

Hence by (2.1),

$$x_{n+1} = (1 - n^{-1}d_n)x_n - \varepsilon_n/(nb_n), \qquad \text{where} \quad d_n = (\beta + \xi_n)/b_n.$$

It then follows that

(2.11) $$x_{n+1} = \beta_{m-1,n}x_m - \Sigma_{k=m}^n \beta_{kn}\varepsilon_k/(kb_k),$$

where

$$\beta_{kn} = \Pi_{j=k+1}^n(1 - j^{-1}d_j),$$
$$k = 0, 1, \cdots, n-1; \ \beta_{nn} = 1.$$

Clearly $d_n \to \beta/b$ a.s. Therefore, for almost all ω, if k is sufficiently large, say $k > k_0(\omega)$, then

(2.12) $$\beta_{kn} = \gamma_n\gamma_k^{-1} \qquad \text{for} \quad n > k,$$

where

(2.13) $$\gamma_n = \Pi_{j=1}^n\left(\max\{1 - j^{-1}d_j, \tfrac{1}{2}\}\right).$$

Since $d_n \to \beta/b$ a.s., it is easy to see from (2.13) that with probability 1, γ_n is regularly varying with exponent $-\beta/b$ (cf. [2]). Let $\tau_n = (n^{\beta/b}\gamma_n)^{-1}$. Then τ_n is slowly varying with probability 1. To show that τ_n satisfies (2.4), we note that with probability 1,

$$\tau_n/\tau_{n+1} = (1 + n^{-1})^{\beta/b}\{1 - n^{-1}(\beta/b + o(1))\} = 1 + o(n^{-1}).$$

From (2.11) and (2.12), we obtain that

$$x_{n+1} = -(n^{\beta/b}/\tau_n)\Sigma_1^n k^{(\beta/b)-1}\tau_k\varepsilon_k/b_k$$
$$+ \{(n^{-\beta/b}/\tau_n)\Sigma_1^{k_0}k^{(\beta/b)-1}\tau_k\varepsilon_k/b_k + \beta_{k_0,n}x_{k_0+1}\}.$$

Hence (2.2) holds.

To prove (ii), suppose that b_n is $\mathfrak{F}_{n-1}$-measurable for all $n \geq 1$, and assume either that $E(|\varepsilon|\log|\varepsilon|) < \infty$ or that ε is symmetric. Then by Theorem 4 of [9], $\lim_{n\to\infty} x_n = 0$ a.s. Moreover, since b_n is $\mathfrak{F}_{n-1}$-measurable for all n, so is ξ_n (as defined by (2.10)). Hence $d_n (= (\beta + \xi_n)/b_n)$ and therefore γ_n (as defined by (2.13)) also are $\mathfrak{F}_{n-1}$-measurable, and (2.5) follows.

To prove (iii), suppose that $E|\varepsilon|^r < \infty$ for some $r > 1$ and that $\{b_n\}$ satisfies (2.6). Then

$$\Sigma_1^n (ib_i)^{-1}\varepsilon_i = \Sigma_1^n (ib)^{-1}\varepsilon_i + \Sigma_1^n (ibb_i)^{-1}(b - b_i)\varepsilon_i.$$

Since $E|\varepsilon|^r < \infty$, $\Sigma_1^n i^{-1}\varepsilon_i$ converges a.s. by the three-series theorem. Moreover, since $E(\Sigma_1^\infty i^{-1-\lambda}|\varepsilon_i|) < \infty$ and so $\Sigma_1^\infty i^{-1-\lambda}|\varepsilon_i| < \infty$ a.s., therefore by (2.6), with probability 1

$$\Sigma_1^\infty (ib_i)^{-1}|b - b_i|\,|\varepsilon_i| = \Sigma_1^\infty o(i^{-1-\lambda}|\varepsilon_i|) < \infty.$$

Hence $\Sigma_1^n (ib_i)^{-1}\varepsilon_i$ converges a.s. Therefore, using the same argument as in [1] (see also Lemma 5 in Section 3 below), it can be shown that $\lim_{n\to\infty} x_n = 0$ a.s., and so the representation (2.2) holds.

We note that by (2.3),

$$(2.14)\quad \Sigma_1^n \delta_k \varepsilon_k = -b^{-1}\left(\Sigma_1^n k^{(\beta/b)-1}\tau_k \varepsilon_k\right) - \left\{\Sigma_1^n k^{(\beta/b)-1}\tau_k\left(b_k^{-1} - b^{-1}\right)\varepsilon_k\right\}$$
$$= -b^{-1}U_{1n} - U_{2n}, \text{ say.}$$

Take $0 < \rho < \lambda$ and $q > 0$ such that $\beta/b > q$ and $\rho > q$. Then by (2.6) and (2.9), $\tau_n|b_n^{-1} - b^{-1}| = o(n^{-\rho})$ a.s., and therefore with probability 1

$$(2.15)\quad |U_{2n}| \leq n^{(\beta/b)-q}\Sigma_1^n o\left(k^{q-1-\rho}|\varepsilon_k|\right) = 0\left(n^{(\beta/b)-q}\right),$$

since $\rho > q$ implies that $\Sigma_1^\infty k^{q-1-\rho}|\varepsilon_k| < \infty$ a.s. Without loss of generality, we can assume that $r < 2$ and $\beta/b > 1 - r^{-1}$. Let $S_n = \Sigma_1^n \varepsilon_i$. Then $n^{-1/r}S_n \to 0$ a.s., and by (2.4),

$$k^{(\beta/b)-1}\tau_k - (k+1)^{(\beta/b)-1}\tau_{k+1} \sim -(b^{-1}\beta - 1)k^{(\beta/b)-2}\tau_k \text{ a.s.}$$

Therefore, in view of (2.9), with probability 1

$$(2.16)\quad U_{1n} = n^{(\beta/b)-1}\tau_n S_n + \Sigma_1^{n-1}\left\{k^{(\beta/b)-1}\tau_k - (k+1)^{(\beta/b)-1}\tau_{k+1}\right\}S_k$$
$$= 0\left(n^{(\beta/b)-1+r^{-1}}\tau_n\right).$$

Let $0 < \zeta < \min\{q, 1 - r^{-1}\}$. Then from (2.2), (2.9), (2.14), (2.15) and (2.16), it follows that $\lim_{n\to\infty} n^\zeta x_n = 0$ a.s.

Assume that $M(x)$ also satisfies (1.18). Then with η given by (1.18) and ξ_n defined by (2.10), since $n^\zeta x_n \to 0$ a.s., we can write

$$(2.17)\quad \xi_n = \xi_n' n^{-\zeta\eta}, \quad \text{where } \xi_n' \to 0 \text{ a.s.}$$

Without loss of generality, we shall assume that $\eta \leq 1$, and so $\zeta\eta < 1$. Let

$p = \min\{\zeta\eta, \lambda\}$, where λ is given by (2.6). Then by (2.6) and (2.17),

$$d_n(= (\beta + \xi_n)/b_n) = \beta/b + \xi_n'' n^{-p}, \quad \text{where } \xi_n'' \to 0 \text{ a.s.} \tag{2.18}$$

From (2.13) and (2.18), it is easy to see that (2.7) and (2.8) hold. □

We now make use of the above representation theorem to obtain the following generalization of Theorem 2 to quasi-adaptive stochastic approximation schemes.

THEOREM 4. *Assume that* $E\varepsilon^2 = \sigma^2 < \infty$, *and suppose that* $\{b_n\}$ *is a sequence of positive random variables such that* b_n *is* $\mathcal{F}_{n-1}$*-measurable for all* $n \geq 1$ *and* $\lim_{n\to\infty} b_n = b$ a.s., *where* b *is a positive constant. Let* $\{x_n\}$ *be the quasi-adaptive stochastic approximation scheme defined by* (2.1).

(i) *Let* $b < 2\beta$. *Then* (1.15), (1.16) *and* (1.17) *hold.*

(ii) *Let* $b > 2\beta$. *If* $M(x)$ *further satisfies* (1.18) *and* $\{b_n\}$ *further satisfies* (2.6), *then there exists a random variable* z *such that* (1.19) *and* (1.20) *hold.*

(iii) *Let* $b = 2\beta$. *If* $M(x)$ *further satisfies* (1.18) *and* $\{b_n\}$ *further satisfies* (2.6), *then* (1.21) *and* (1.22) *hold.*

(iv) *Suppose* $b < 2\beta$. *Let* b_n^* *be a sequence of positive random variables (not necessarily* $\mathcal{F}_{n-1}$*-measurable) such that*

$$P\left[b_n^* - b_n = o\left(n^{-\frac{1}{2}} v_n\right) \text{ as } n \to \infty\right] = 1, \tag{2.19}$$

where $v_1, v_2, \cdots$ *are* i.i.d. *positive random variables such that* $Ev_1^2 < \infty$. *Let* $x_1^* = x_1$ *and for* $i = 1, 2, \cdots$, *define* $y_i^* = M(x_i^*) + \varepsilon_i$ *and* $x_{i+1}^* = x_i^* - y_i^*/(ib_i^*)$. *Then* (1.15) *and* (1.16) *still hold with* x_i^* *in place of* x_i. *If condition* (2.19) *is strengthened to*

$$P\left[b_n^* - b_n = o\left((n \log\log n)^{-\frac{1}{2}} v_n\right) \text{ as } n \to \infty\right] = 1 \tag{2.20}$$

and there exists a positive integer m *such that* b_n^* *is* $\mathcal{F}_{n+m}$*-measurable for all* $n \geq 1$, *then* (1.17) *still holds with* x_i^* *in place of* x_i.

REMARK. The first three parts of this theorem deal with quasi-adaptive stochastic approximation schemes whose estimate b_n of β at the nth stage is based only on $(x_1, y_1, \cdots, x_{n-1}, y_{n-1})$. Although y_n is also observed, it is not used to estimate β, for otherwise b_n would not be $\mathcal{F}_{n-1}$-measurable. The requirement that b_n be $\mathcal{F}_{n-1}$-measurable gives a martingale structure to the sum $\Sigma_1^n \delta_k \varepsilon_k$ in the representation (2.2), and our proof of Theorem 4(i)–(iii) depends on this martingale structure. On the other hand, since y_n has also been observed, it seems artificial not to use it in b_n simply because this would destroy the expedient martingale property. In this connection, Theorem 4(iv) is of particular interest. It implies that given an estimator $b_n^*(= b_n^*(x_1, y_1, \cdots, x_n, y_n))$, if b_n^* is close to $b_n = b_{n-1}^*$ in the sense of (2.20), then the desired conclusions still hold, at least in the important case $b < 2\beta$. We shall see in Section 3 and [10] that most "reasonable" estimators b_n^* for the present problem satisfy the approximation property (2.20) with $b_n = b_{n-1}^*$. Note that Theorem 2 is the special case of Theorem 4 with $b_n = b$ for all n. We shall need the following three lemmas in the proof of Theorem 4.

LEMMA 2. *Let $z_1, z_2, \cdots$ be* i.i.d. *random variables with $E|z_1| < \infty$.*

(i) $\lim_{n\to\infty}(\Sigma_1^n i^{-1} z_i)/(\log n) = Ez_1$ a.s.

(ii) *Let $\{r_n\}$ be a sequence of random variables such that $P[r_n = o(n^{-\rho})$ as $n \to \infty] = 1$ for some $0 < \rho \leqslant 1$. Then, with probability* 1,

$$(2.21)\qquad \Sigma_1^n r_i z_i = o(n^{1-\rho}) \quad \text{if } \rho < 1, \qquad = o(\log n) \quad \text{if } \rho = 1.$$

PROOF. To prove (i), let $S_n = \Sigma_1^n z_i$. We note that

$$(2.22)\qquad \Sigma_1^n i^{-1} z_i = \Sigma_1^{n-1}(i^{-1} - (i+1)^{-1})S_i + n^{-1}S_n.$$

As $i \to \infty$, $i^{-1} - (i+1)^{-1} \sim i^{-2}$ and $i^{-1}S_i \to Ez_1$ a.s. Hence it follows from (2.22) that $\Sigma_1^n i^{-1} z_i \sim (Ez_1)\log n$ a.s.

To prove (ii), let $S_n' = \Sigma_1^n |z_i|$ and note that

$$\Sigma_1^n i^{-\rho}|z_i| = \Sigma_1^{n-1}(i^{-\rho} - (i+1)^{-\rho})S_i' + n^{-\rho}S_n'.$$

As $i \to \infty$, $i^{-\rho} - (i+1)^{-\rho} \sim \rho i^{-(\rho+1)}$, $i^{-1}S_i' \to E|z_1|$ a.s. and $|r_i| = o(i^{-\rho})$ a.s. Hence (2.21) follows. ▯

LEMMA 3. *Let $z_1, z_2, \cdots$ be* i.i.d. *random variables such that $Ez_1 = 0$ and $Ez_1^2 = \sigma^2 < \infty$. Let $\mathcal{G}_0 \subset \mathcal{G}_1 \subset \cdots$ be an increasing sequence of σ-fields such that z_i is $\mathcal{G}_i$-measurable and is independent of $\mathcal{G}_{i-1}$ for all $i \geqslant 1$. Let $u_1, u_2, \cdots$ be a sequence of random variables such that u_i is $\mathcal{G}_{i-1}$-measurable for all $i \geqslant 1$ and* $\lim_{n\to\infty} u_n = A$ a.s. *for some constant A. Then, redefining the random variables on a new probability space if necessary, there exists a standard Wiener process $w(t)$, $t \geqslant 0$, such that*

$$(2.23)\qquad \max_{m\leqslant n}|\Sigma_1^m k^{-\frac{1}{2}} u_k z_k - A\sigma w(\log m)|/(\log n)^{\frac{1}{2}} \to_P 0 \quad \text{as } n \to \infty.$$

PROOF. Write

$$(2.24)\qquad \Sigma_1^m k^{-\frac{1}{2}} u_k z_k = A\Sigma_1^m k^{-\frac{1}{2}} z_k + \Sigma_1^m k^{-\frac{1}{2}}(u_k - A)z_k.$$

Redefining the random variables on a new probability space if necessary, there exists a standard Wiener process $w(t)$, $t \geqslant 0$, such that

$$(2.25)\qquad \max_{m\leqslant n}|\Sigma_1^m k^{-\frac{1}{2}} z_k - \sigma w(\log m)|/(\log n)^{\frac{1}{2}} \to_P 0 \quad \text{as } n \to \infty,$$

(cf. [5]). Let $\tilde{u}_i = (u_i - A)I_{[|u_i - A| \leqslant 1]}$. Since $\lim_{n\to\infty} u_n = A$ a.s. and (2.25) holds, it remains to show that

$$(2.26)\qquad \max_{m\leqslant n}|\Sigma_1^m k^{-\frac{1}{2}} \tilde{u}_i z_i|/(\log n)^{\frac{1}{2}} \to_P 0.$$

Noting that $E(\Sigma_1^n k^{-\frac{1}{2}}\tilde{u}_i z_i)^2 = \sigma^2 \Sigma_1^n k^{-1} E\tilde{u}_i^2 = o(\log n)$, (2.26) follows easily from the martingale inequality. ▯

LEMMA 4. *With the same notations and assumptions as in Lemma 3, suppose that $\{\tau_n\}$ is a sequence of positive random variables satisfying (2.7) and therefore (2.8) as well. Then, redefining the random variables on a new probability space if necessary, we* *have as* $n \to \infty$

$$\max_{m<n} |\Sigma_1^m k^{-\frac{1}{2}} \tau_k u_k z_k - A\sigma\tau w(\log m)| / (\log n)^{\frac{1}{2}} \to_P 0. \tag{2.27}$$

PROOF. Let $S(m) = \Sigma_1^m k^{-\frac{1}{2}} u_k z_k$. By partial summation,

$$\Sigma_1^m k^{-\frac{1}{2}} \tau_k u_k z_k = \Sigma_1^{m-1}(\tau_k - \tau_{k+1}) S(k) + (\tau_m - \tau) S(m) + \tau S(m). \tag{2.28}$$

By (2.7), (2.8) and Lemma 3, for $1 \leq m < n$,

$$\begin{aligned} &\Sigma_1^{m-1}|\tau_k - \tau_{k+1}|\,|S(k)| + |\tau_m - \tau|\,|S(m)| \\ &\quad= \Sigma_1^{m-1} 0(k^{-(1+p)})\{|w(\log k) + \Delta_n(k)|\} + 0(m^{-p})\{|w(\log m) + \Delta_n(m)|\}, \end{aligned} \tag{2.29}$$

where $\Delta_n(m)$ are random variables such that

$$\max_{m<n} |\Delta_n(m)| / (\log n)^{\frac{1}{2}} \to_P 0. \tag{2.30}$$

From (2.28), (2.29), (2.30) and Lemma 3, (2.27) follows immediately. □

PROOF OF THEOREM 4(i). By a partial summation technique, we have shown in [9] that the central limit theorem (1.15) and the law of the iterated logarithm (1.16) follow from the representation in Theorem 3 and certain martingale limit theorems. We now prove the asymptotic behavior (1.17) of the cost $\Sigma_1^n (x_i - \theta)^2$. By Theorem 3(ii),

$$\Sigma_1^n (x_i - \theta)^2 = (x_1 - \theta)^2 + \Sigma_{i=1}^{n-1} i^{-2\beta/b} \tau_i^{-2} \{\Sigma_{k=1}^i \delta_k \varepsilon_k + \rho_0\}^2, \tag{2.31}$$

where ρ_0, τ_k, and δ_k are random variables satisfying (2.3), (2.4) and (2.5). Let $a = 2\beta/b$ (> 1). Define

$$S_n = \Sigma_1^n \delta_k \varepsilon_k, \quad \tilde{S}_n = S_n + \rho_0, \quad a(n) = \Sigma_{i=n}^\infty i^{-a} \sim n^{-a+1}/(a-1). \tag{2.32}$$

We note that

(2.33)

$$\begin{aligned} \Sigma_1^n i^{-a} \tau_i^{-2} \tilde{S}_i^2 &= \Sigma_1^n \tau_i^{-2} \tilde{S}_i^2 (a(i) - a(i+1)) \\ &= \Sigma_2^n a(i)(\tau_i^{-2} - \tau_{i-1}^{-2}) \tilde{S}_i^2 + \Sigma_2^n a(i) \tau_{i-1}^{-2} (\tilde{S}_i^2 - \tilde{S}_{i-1}^2) - a(n+1) \tau_n^{-2} \tilde{S}_n^2 \\ &\quad + a(1) \tau_1^{-2} \tilde{S}_1^2. \end{aligned}$$

By (2.4) and (2.9), with probability 1,

(2.34)

$$\begin{aligned} \Sigma_2^n a(i) |\tau_i^{-2} - \tau_{i-1}^{-2}| \tilde{S}_i^2 &= (2 + o(1)) \Sigma_2^n a(i) |\tau_i - \tau_{i-1}| \tau_i^{-3} \tilde{S}_i^2 + 0(1) \\ &= o(\Sigma_1^n (a(i)/i) \tau_i^{-2} \tilde{S}_i^2) + 0(1) = o(\Sigma_1^n i^{-a} \tau_i^{-2} \tilde{S}_i^2) + 0(1). \end{aligned}$$

(We add the 0(1) term in (2.34) because we have not yet shown that $\Sigma_1^n i^{-a} \tau_i^{-2} \tilde{S}_i^2 \to \infty$ with probability 1.) Obviously,

$$\Sigma_2^n a(i) \tau_{i-1}^{-2} (\tilde{S}_i^2 - \tilde{S}_{i-1}^2) = \Sigma_2^n a(i) \tau_{i-1}^{-2} (\delta_i^2 \varepsilon_i^2 + 2\delta_i \varepsilon_i S_{i-1} + 2\delta_i \varepsilon_i \rho_0). \tag{2.35}$$

Since the summands are nonnegative, it follows from (2.3), (2.9) and (2.32) that with probability 1

$$(2.36) \qquad \Sigma_2^n a(i)\tau_{i-1}^{-2}\delta_i^2\varepsilon_i^2 \sim (\Sigma_2^n i^{-1}\varepsilon_i^2)/\{b^2(a-1)\}$$
$$\sim (\sigma^2 \log n)/\{b(2\beta - b)\}.$$

The last relation above follows from Lemma 2(i) since $E\varepsilon^2 = \sigma^2 < \infty$. We note that $a(i)\tau_{i-1}^{-2}\delta_i S_{i-1} \to 0$ a.s. by (1.16), (2.2) and (2.9). Moreover, $\{\Sigma_{i=2}^n a(i)\tau_{i-1}^{-2}\delta_i S_{i-1}\varepsilon_i, \mathscr{F}_n, n \geq 2\}$ is a martingale transform. Therefore, using a standard truncation argument like that used in the proof of Lemma 3 (to ensure finite expectations) and the strong law for martingales (cf. [14], page 150), we obtain that with probability 1

$$(2.37) \quad \Sigma_2^n a(i)\tau_{i-1}^{-2}\delta_i\varepsilon_i S_{i-1} = o(\Sigma_2^n a^2(i)\tau_{i-1}^{-4}\delta_i^2 S_{i-1}^2) + 0(1)$$
$$= o(\Sigma_1^{n-1} i^{-a}\tau_i^{-2}S_i^2) + 0(1) = o(\Sigma_1^{n-1} i^{-a}\tau_i^{-2}\tilde{S}_i^2) + 0(1).$$

The 0(1) term in (2.37) indicates that $\Sigma_2^n a(i)\tau_{i-1}^{-2}\delta_i\varepsilon_i S_{i-1}$ converges a.s. on $[\Sigma_1^\infty i^{-a}\tau_i^{-2}S_i^2 < \infty]$. To see the last relation in (2.37), we note that $S_i^2 \leq 2(\tilde{S}_i^2 + \rho_0^2)$ and that $\Sigma_1^\infty i^{-a}\tau_i^{-2} < \infty$ a.s. since $a > 1$ and (2.9) holds. Since

$$\Sigma_2^\infty a^2(i)\tau_{i-1}^{-4}\delta_i^2 = \Sigma_2^\infty 0(i^{-a}\tau_i^{-2}) < \infty \text{ a.s.},$$

it follows from the (local) martingale convergence theorem ([14], page 148) and a standard truncation argument like that used in the proof of Lemma 3 (to ensure finite expectations) that

$$(2.38) \qquad \Sigma_2^n a(i)\tau_{i-1}^{-2}\delta_i\varepsilon_i \qquad \text{converges a.s.}$$

In view of the law of the iterated logarithm (1.16) and the representation (2.2), we obtain that

$$(2.39) \qquad (n^{-\beta/b}/\tau_n)\tilde{S}_n = 0(n^{-\frac{1}{2}}(\log\log n)^{\frac{1}{2}}) \text{ a.s.}$$

Therefore, with probability 1,

$$(2.40) \qquad a(n+1)\tau_n^{-2}\tilde{S}_n^2 = o(\log n).$$

From the relations (2.33)–(2.40), the desired conclusion (1.17) for $\Sigma_1^n(x_i - \theta)^2$ follows. □

To better understand the partial summation technique in the preceding proof of (1.17), consider the special case $b = \beta$ and $M(x) = \beta(x - \theta)$. In this case, as indicated in Section 1, $x_{i+1} = \theta - \beta^{-1}\bar{\varepsilon}_i$, and so (1.17) reduces to the following interesting corollary on the fluctuation behavior of sample means.

COROLLARY 1. *Let* $\varepsilon, \varepsilon_1, \cdots$ *be* i.i.d. *random variables with* $E\varepsilon = 0$ *and* $E\varepsilon^2 = \sigma^2$. *Then*

$$\lim_{n\to\infty}(\Sigma_1^n\bar{\varepsilon}_i^2)/(\log n) = \sigma^2 \text{ a.s.}$$

To analyze $\Sigma_1^n\bar{\varepsilon}_i^2 = \Sigma_1^n i^{-2}(\Sigma_1^i\varepsilon_j)^2$, partial summation is the natural method.

PROOF OF THEOREM 4(ii). Since $b > 2\beta$, $\Sigma_1^\infty \delta_i^2 < \infty$ a.s. Hence the martingale transform $\Sigma_1^n \delta_k \varepsilon_k$ converges a.s. as $n \to \infty$. Therefore, by Theorem 3(iii),

$$\tau n^{\beta/b}(x_n - \theta) \to \rho_0 + \Sigma_1^\infty \delta_k \varepsilon_k \text{ a.s.}$$

Hence (1.19) holds with $z = \tau^{-1}\{\rho_0 + \Sigma_1^\infty \delta_k \varepsilon_k\}$. The relation (1.20) is an immediate consequence of (1.19). □

PROOF OF THEOREM 4(iii). We note that since $b = 2\beta$, $\delta_k = -k^{-\frac{1}{2}}\tau_k / b_k$. Let $\tilde{S}_n = \rho_0 + \Sigma_1^n \delta_k \varepsilon_k$ as before. Since $\lim_{k\to\infty} b_k = b$ a.s. and τ_k satisfies (2.7) and (2.8) by Theorem 3(iii), Lemma 4 implies that, redefining the random variables on a new probability space if necessary, there exists a standard Wiener process $w(t)$, $t \geqslant 0$, such that

$$\max_{k \leqslant n} |\tau^{-1}\tilde{S}_k - (\sigma/b)w(\log k)| / (\log n)^{\frac{1}{2}} \to_p 0. \tag{2.41}$$

From (2.2) and (2.41), the asymptotic normality result (1.21) follows immediately. We note that with probability 1,

$$\begin{aligned} \Sigma_1^n k^{-1} w^2(\log k) &= \textstyle\int_1^n t^{-1} w^2(\log t)\, dt + o(\log n) \\ &= \textstyle\int_0^{\log n} w^2(s)\, ds + o(\log n), \qquad \text{setting } s = \log t. \end{aligned} \tag{2.42}$$

Since $\int_0^{\log n} w^2(s)\, ds$ has the same distribution as $(\log n)^2 \int_0^1 w^2(t)\, dt$, it then follows from (2.41) and (2.42) that

$$\left(\tau^{-2}\Sigma_1^{n-1} k^{-1} \tilde{S}_k^2\right) / (\log n)^2 \to_{\mathcal{L}} (\sigma^2/b^2) \textstyle\int_0^1 w^2(t)\, dt. \tag{2.43}$$

The desired conclusion (1.22) then follows from (2.43) and the fact that

$$\Sigma_1^n (x_i - \theta)^2 = (x_1 - \theta)^2 + \Sigma_1^{n-1}(k^{-1}/\tau_k^2)\tilde{S}_k^2 \sim \tau^{-2}\Sigma_1^{n-1} k^{-1} \tilde{S}_k^2 \text{ a.s.} \qquad \square \tag{2.44}$$

PROOF OF THEOREM 4(iv). Assume that (2.19) holds. We note that

$$\Sigma_1^n (ib_i^*)^{-1} \varepsilon_i = \Sigma_1^n (ib_i)^{-1} \varepsilon_i + \Sigma_1^n (ib_i b_i^*)^{-1}(b_i - b_i^*)\varepsilon_i.$$

Since b_i is $\mathcal{F}_{i-1}$-measurable and $\Sigma_1^\infty (ib)^{-2} < \infty$ a.s., the martingale transform $\Sigma_1^n (ib_i)^{-1}\varepsilon_i$ converges a.s. Since $E\{\Sigma_1^\infty i^{-\frac{3}{2}} v_i |\varepsilon_i|\} < \infty$ and so $\Sigma_1^\infty i^{-\frac{3}{2}} v_i |\varepsilon_i| < \infty$ a.s., therefore by (2.19), with probability 1,

$$\Sigma_1^\infty (ib_i b_i^*)^{-1} |b_i - b_i^*|\, |\varepsilon_i| = \Sigma_1^\infty o\left(i^{-\frac{3}{2}} v_i |\varepsilon_i|\right) < \infty.$$

Hence $\Sigma_1^n (ib_i^*)^{-1} \varepsilon_i$ converges a.s. Therefore, using the same argument as in [1] (see also Lemma 5 in Section 3 below), it can be shown that $\lim_{n\to\infty} x_n = \theta$ a.s. Hence by Theorem 3(i),

$$x_{n+1}^* = \theta + \left(n^{-\beta/b}/\tau_n^*\right)\{\Sigma_1^n \delta_k^* \varepsilon_k + \rho_0^*\}, \tag{2.45}$$

where τ_k^* and δ_k^* satisfy (2.3) and (2.4).

Let $Z^*(n) = \Sigma_1^n \varepsilon_k / b_k^*$ and $Z(n) = \Sigma_1^n \varepsilon_k / b_k$. Then by (2.19), with probability 1,

$$|Z^*(n) - Z(n)| \leqslant \Sigma_1^n o\left(k^{-\frac{1}{2}} v_k |\varepsilon_k|\right) = o\left(n^{\frac{1}{2}}\right), \tag{2.46}$$

since $E\varepsilon^2 < \infty$ and $Ev_1^2 < \infty$ (see Lemma 2(ii)). Set $Z^*(0) = 0$ and $Z^*(t) = Z^*(n)$

for $n-1<t\leqslant n$, $n=1,2,\cdots$. From (2.46) and Theorem 2 of [9], it then follows that, redefining the random variables on a new probability space if necessary, there exist standard Wiener processes $w(t)$ and $w^*(t)$, $t\geqslant 0$, such that

$$\max_{0\leqslant t\leqslant 1}|r^{-\frac{1}{2}}Z^*(rt)-(\sigma/b)w(t)|\to_P 0 \quad \text{as} \quad r\to\infty, \tag{2.47}$$

and

$$\lim_{t\to\infty}|Z^*(t)-(\sigma/b)w^*(t)|/(t\log\log t)^{\frac{1}{2}}=0 \text{ a.s.} \tag{2.48}$$

Let $\alpha=\beta/b-1$. We note that

$$\Sigma_1^n\delta_k^*\varepsilon_k=\Sigma_1^{n-1}(k^\alpha\tau_k^*-(k+1)^\alpha\tau_{k+1}^*)Z^*(k)+n^\alpha\tau_n Z^*(n), \tag{2.49}$$

and

$$k^\alpha\tau_k^*-(k+1)^\alpha\tau_{k+1}^*\sim-\alpha k^{\alpha-1}\tau_k^* \text{ a.s.} \tag{2.50}$$

by (2.4). From (2.47)–(2.50) it is not hard to show that (1.15) and (1.16) still hold with x_i^* in place of x_i (see the proof of Theorem 7 of [9]).

Now assume that the stronger condition (2.20) holds in place of (2.19), and that b_n^* is $\mathfrak{F}_{n+m}$-measurable for all $n\geqslant 1$. We shall show that (1.17) still holds with x_i^* in place of x_i. Let $a=2\beta/b$ and $a(n)=\Sigma_{i=n}^\infty i^{-a}$ as before and set $S_n^*=\Sigma_1^n\delta_k^*\varepsilon_k$, $\tilde{S}_n^*=S_n^*+\rho_0^*$. Clearly the relations (2.33)–(2.36) and (2.39) – (2.40) still hold with τ_i^*, δ_i^*, S_i^* and $\tilde{S}_i^*$ in place of τ_i, δ_i, S_i and $\tilde{S}_i$. Hence we need only show that in analogy with (2.37) and (2.38), with probability 1,

$$\Sigma_2^n a(i)\tau_{i-1}^{*-2}\delta_i^*S_{i-1}^*\varepsilon_i=o(\Sigma_1^{n-1}i^{-a}\tau_i^{*-2}S_i^{*2})+o(\log n), \tag{2.51}$$

and

$$\Sigma_2^n a(i)\tau_{i-1}^{*-2}\delta_i^*\varepsilon_i \quad \text{converges.} \tag{2.52}$$

To prove (2.51) and (2.52), we note that τ_n^* and therefore S_n^* also are $\mathfrak{F}_{n+m}$-measurable, since b_n^* is $\mathfrak{F}_{n+m}$-measurable for all $n\geqslant m$ (see the proof of Theorem 3). By (2.4), with probability 1,

(2.53)

$$\tau_i^*=\tau_{i-m-1}^*(1+o(i^{-1})) \quad \text{and} \quad \tau_{i-1}^*=\tau_{i-m-1}^*(1+o(i^{-1})) \quad \text{as} \quad i\to\infty.$$

As in (2.39), we have

$$S_n^*=0\big(\tau_n^*n^{(\beta/b)-\frac{1}{2}}(\log\log n)^{\frac{1}{2}}\big) \text{ a.s.,} \tag{2.54}$$

and therefore with probability 1

$$\Sigma_2^\infty a(i)i^{-1}\tau_{i-m-1}^{*-1}\big(i^{(\beta/b)-1}/b_i^*\big)|S_{i-1}^*|\,|\varepsilon_i| = \Sigma_2^\infty 0\big(i^{-\frac{3}{2}}(\log\log i)^{\frac{1}{2}}\big)\cdot|\varepsilon_i|<\infty. \tag{2.55}$$

By (2.20), with probability 1,

$$b_n^*=b_n+o\big((n\log\log n)^{-\frac{1}{2}}r_n\big). \tag{2.56}$$

and therefore using (2.54),

$$(2.57)\quad \Sigma_2^n a(i)\tau_i^{*-1} i^{(\beta/b)-1}|b_i^* - b_i|\,|S_i^*|\,|\varepsilon_i| = \Sigma_2^n o\big(i^{-1}v_i|\varepsilon_i|\big) = o(\log n), \qquad \text{by Lemma 2.}$$

For $i > m + 1$, let $U_i = -\tau_{i-m-1}^*\Sigma_{j=i-m}^{i-1} j^{(\beta/b)-1}\varepsilon_j/b_j$ and write

$$(2.58)\qquad S_{i-1}^* = S_{i-m-1}^* + U_i + R_i.$$

From (2.53) and (2.56), it follows that with probability 1

$$(2.59)\qquad R_i = o\Big(\Big\{i^{(\beta/b)-\frac{3}{2}}(\log\log i)^{-\frac{1}{2}}\tau_i^*\Sigma_{j=i-m}^{i-1}\big(|\varepsilon_j| + |\varepsilon_j v_j|\big)\Big\}\Big),$$

and therefore

$$(2.60)\qquad \Sigma_2^\infty a(i)\tau_i^{*-1} i^{(\beta/b)-1}|R_i|\,|\varepsilon_i| = \Sigma_{i=2}^\infty o\Big(i^{-\frac{3}{2}}|\varepsilon_i|\Sigma_{j=i-m}^{i-1}\big\{|\varepsilon_j| + |\varepsilon_j v_j|\big\}\Big) < \infty.$$

From (2.53) and (2.55)–(2.59), to prove (2.51), it suffices to show that with probability 1

$$(2.61)\quad \Sigma_{i=m+2}^n a(i)\tau_{i-m-1}^{*-1}\big(i^{(\beta/b)-1}/b_i\big)\big(S_{i-m-1}^* + U_i\big)\varepsilon_i = o\big(\Sigma_1^{n-1} i^{-a}\tau_i^{*-2}S_i^{*2}\big) + o(\log n).$$

Since τ_{i-m-1}^*, S_{i-m-1}^*, b_i and U_i are $\mathfrak{F}_{i-1}$-measurable, the left-hand side of (2.61) forms a martingale transform, so using the strong law for martingales as in (2.37),

$$(2.62)\quad \Sigma_{i=m+2}^n a(i)\tau_{i-m-1}^{*-1}\big(i^{(\beta/b)-1}/b_i\big)\big(S_{i-m-1}^* + U_i\big)\varepsilon_i = o\big(\Sigma_{m+2}^n a^2(i)\tau_i^{*-2} i^{2\beta/b-2}\big(S_{i-m-1}^* + U_i\big)^2\big) + 0(1) \text{ a.s.}$$

Since $\lim n^{-\frac{1}{2}}\varepsilon_n = 0$ a.s. and $\lim n^{-\frac{1}{2}}v_n = 0$ a.s., we obtain from (2.54), (2.58) and (2.59) that with probability 1

$$S_{i-1}^{*2} - \big(S_{i-m-1}^* + U_i\big)^2 = 2S_{i-1}^* R_i - R_i^2 = o\big(\tau_i^{*2} i^{2(\beta/b-1)+1}\big),$$

and therefore

$$(2.63)\qquad \Sigma_{i=m+2}^n a^2(i)\tau_i^{*-2} i^{2\beta/b-2}\big|S_{i-1}^{*2} - \big(S_{i-m-1}^* + U_i\big)^2\big| = o(\log n).$$

From (2.62) and (2.63), (2.61) follows as desired.

Making use of the fact that $\beta/b > \frac{1}{2}$ and an argument as in (2.53), (2.55), (2.56) and (2.57), to show that (2.52) holds with probability 1 we need only prove that

$$(2.64)\qquad \Sigma_{i=m+2}^n a(i)\tau_{i-m-1}^{*-1} i^{(\beta/b)-1}\varepsilon_i/b_i \qquad \text{converges a.s.}$$

Since τ_{i-m-1}^* and b_i are $\mathfrak{F}_{i-1}$-measurable, (2.64) follows easily from the (local) martingale convergence theorem as in (2.38). □

In the preceding proof of Theorem 4 (see also the proof of Theorem 7 of [9]), the representation given by Theorem 3 is the only property of the stochastic approximation scheme (2.1) that we have used. This suggests the following more general theorem which we shall need in Section 3.

Theorem 5. *Assume that* $E\varepsilon^2 = \sigma^2 < \infty$ *and suppose that* $\{b_n\}$ *is a sequence of positive random variables such that* b_n *is* $\mathcal{F}_{n-1}$*-measurable for all* $n \geq 1$ *and* $\lim_{n\to\infty} b_n = b$ a.s., *where* b *is a positive constant. Let* $\{\tau_n\}$ *be a sequence of positive random variables such that* (2.4) *holds and* τ_n *is* $\mathcal{F}_{n-1}$*-measurable for all* $n \geq 1$. *Let* X_n *and* ρ_n *be random variables such that for* $n \geq 1$

$$X_{n+1} = \theta - (n^{-\beta/b}/\tau_n)\{\Sigma_1^n k^{(\beta/b)-1}\tau_k\varepsilon_k/b_k + \rho_n\}, \tag{2.65}$$

where $\beta > 0$ *and* θ *are constants.*

(i) *Let* $b < 2\beta$ *and assume that*

$$\rho_n = o\left(\tau_n n^{(\beta/b)-\frac{1}{2}}\right) \text{ a.s.} \tag{2.66}$$

Then (1.15), (1.16) *and* (1.17) *still hold with* X_n *in place of* x_n.

(ii) *Let* $b > 2\beta$ *and assume that*

$$\rho_n \text{ converges a.s. to some random variable } \rho_0. \tag{2.67}$$

Suppose furthermore that τ_n *converges* a.s. *to a positive random variable* τ *as* $n \to \infty$. *Then there exists a random variable* z *such that* (1.19) *and* (1.20) *hold with* X_n *in place of* x_n.

(iii) *Let* $b = 2\beta$ *and assume that*

$$\rho_n = o\left((\log n)^{\frac{1}{2}}\right) \text{ a.s.} \tag{2.68}$$

Suppose furthermore that $\{\tau_n\}$ *satisfies the stronger assumption* (2.7) *instead of* (2.4). *Then* (1.21) *and* (1.22) *still hold with* X_n *in place of* x_n.

(iv) *Let* $b < 2\beta$. *Let* τ_n^* *and* b_n^* *be positive random variables (not necessarily* $\mathcal{F}_{n-1}$*-measurable) such that* τ_n^* *satisfies* (2.4) *(with* τ_n^* *in place of* τ_n*) and* b_n^* *satisfies the approximation property* (2.19) *for some sequence* $\{v_n\}$ *of positive* i.i.d. *random variables with* $Ev_1^2 < \infty$. *Let* $\{\rho_n\}$ *be a sequence of random variables such that* (2.66) *holds with* τ_n^* *in place of* τ_n. *Suppose for* $n \geq 1$ *that*

$$X_{n+1}^* = \theta - (n^{-\beta/b}/\tau_n^*)\{\Sigma_1^n k^{(\beta/b)-1}\tau_k^*\varepsilon_k/b_k^* + \rho_n\}. \tag{2.69}$$

Then (1.15) *and* (1.16) *still hold with* X_n^* *in place of* x_n. *If condition* (2.19) *is strengthened to* (2.20) *and there exists a positive integer* m *such that* b_n^* *and* τ_n^* *are* $\mathcal{F}_{n+m}$*-measurable for all* $n \geq 1$, *then* (1.17) *still holds with* X_i^* *in place of* x_i.

Proof. We shall only consider Part (i) of the theorem, since the argument for the other parts is similar. In the proof of Theorem 4(i), we have actually established that Theorem 5(i) holds for the special case $\rho_n = 0$ for all n. This in turn obviously implies that the central limit theorem (1.15) and the law of the iterated logarithm (1.16) also hold for the more general case where $\{\rho_n\}$ satisfies (2.66). Let

$$S_n = \Sigma_1^n k^{(\beta/b)-1}\tau_k\varepsilon_k/b_k.$$

For the special case $\rho_n = 0$ for all n, the relation (1.17) can be written as

$$\Sigma_1^n S_j^2/\left(j^{2\beta/b}\tau_j^2\right) \sim \sigma^2(\log n)/\{b(2\beta - b)\} \text{ a.s.} \tag{2.70}$$

We now show that (2.70) implies that (1.17) also holds in the more general case where $\{\rho_n\}$ satisfies (2.66). By (2.66) there exist positive random variables η_j such that

$$\eta_j \to \infty \text{ a.s.} \quad \text{and} \quad \eta_j(|\rho_j| + 1) = o\left(\tau_j j^{(\beta/b)-\frac{1}{2}}\right) \text{ a.s.,} \tag{2.71}$$

and therefore

$$\Sigma_1^n \eta_j^2(|\rho_j| + 1)^2 / \left(j^{2\beta/b}\tau_j^2\right) = o(\log n) \text{ a.s.} \tag{2.72}$$

From (2.72), it follows that with probability 1

$$\begin{aligned} \Sigma_1^n S_j^2 / \left(j^{2\beta/b}\tau_j^2\right) &= \Sigma_1^n S_j^2 I_{[|S_j| > \eta_j(|\rho_j|+1)]} / \left(j^{2\beta/b}\tau_j^2\right) + o(\log n) \\ &= (1 + o(1))\Sigma_1^n (S_j + \rho_j)^2 I_{[|S_j| > \eta_j(|\rho_j|+1)]} / \left(j^{2\beta/b}\tau_j^2\right) \\ &\quad + o(\log n) \qquad (\text{since } \eta_j \to \infty \text{ a.s.}) \\ &= (1 + o(1))\Sigma_1^n (S_j + \rho_j)^2 / \left(j^{2\beta/b}\tau_j^2\right) + o(\log n). \end{aligned} \tag{2.73}$$

From (2.70) and (2.73), it then follows that (1.17) also holds under the assumption (2.66). □

3. Venter's design and some modifications. In [17] Venter proposed a modification of the Robbins-Monro stochastic approximation scheme (1.26) to obtain successive estimates of the unknown slope β which have the desired property of converging to β with probability 1. Venter's design requires that at the mth stage ($m = 1, 2, \cdots$) two observations y_m' and y_m'' be taken, at levels $x_m' = x_m - a_m$ and $x_m'' = x_m + a_m$, where $\{a_m\}$ is a sequence of positive constants such that

$$a_m \sim am^{-\gamma} \text{ for some constants } a > 0 \text{ and } \tfrac{1}{4} < \gamma < \tfrac{1}{2}, \tag{3.1}$$

and x_m is the mth approximation to θ, defined recursively by

$$\begin{aligned} x_1 &= \text{initial guess of } \theta, \\ x_{i+1} &= x_i - y_i / (ib_i). \end{aligned} \tag{3.2}$$

The quantity y_i in (3.2) estimates the (unobserved) response at the level x_i, and is defined by

$$y_i = \tfrac{1}{2}(y_i' + y_i''). \tag{3.3}$$

Assuming that positive constants b and B are known such that

$$b < \beta < B, \tag{3.4}$$

Venter defines the slope estimate b_i in (3.2) by

$$b_i = b \vee \left\{ B \wedge i^{-1}\Sigma_{j=1}^i (y_j'' - y_j') / (2a_j) \right\}, \tag{3.5}$$

where the symbols $\vee$ and $\wedge$ denote maximum and minimum respectively.

We note that for $j = 1, 2, \cdots$,

$$y_j' = M(x_j - a_j) + \varepsilon_j', \qquad y_j'' = M(x_j + a_j) + \varepsilon_j'', \tag{3.6}$$

where the errors $\varepsilon_1', \varepsilon_2', \cdots, \varepsilon_1'', \varepsilon_2'', \cdots$ are i.i.d. with mean 0 and variance σ^2. In particular, for the linear case $M(x) = \beta(x - \theta)$, (3.6) implies that

(3.7) $$y_j'' - y_j' = 2\beta a_j + (\varepsilon_j'' - \varepsilon_j'),$$

which depends only on β but not on θ. Since $\gamma < \frac{1}{2}$, $\Sigma_1^\infty j^{-2(1-\gamma)} < \infty$, and therefore in view of (3.1) it easily follows that

(3.8) $$m^{-1}\Sigma_1^m(\varepsilon_j'' - \varepsilon_j')/a_j \to 0 \text{ a.s.}$$

Hence b_i defined by (3.5) is strongly consistent, at least in the linear case. This argument was extended by Venter [17] to general regression functions $M(x)$ which satisfy (1.2)–(1.4) and

(3.9)

$$\sup_{|y|<A}|M(x + y) - M(x)| \leqslant A^* \quad \text{for some} \quad A, A^* > 0 \quad \text{and all} \quad x;$$

(3.10) $M(x)$ is k-times continuously differentiable in some neighborhood of θ, where $k \geqslant 2$ is an integer satisfying $k\beta > \gamma B$.

At stage m, Venter's scheme has taken $n = 2m$ observations. Let $\tilde{\theta}_n = x_{m+1}$ be the estimate of θ and $\tilde{\beta}_n = b_m$ be the estimate of β given by (3.2) and (3.5) respectively, and let

(3.11) $$C_n = \Sigma_1^m(x_i' - \theta)^2 + \Sigma_1^m(x_i'' - \theta)^2$$

be the cost of these n observations. The following theorem, which can be proved by using Theorem 5, shows that although $\tilde{\theta}_n$ approaches θ at the asymptotically optimal rate given in (1.11) and (1.12) of Theorem 1, the cost C_n incurred by Venter's scheme is of a much larger order of magnitude than the logarithmic cost in (1.13) of Theorem 1. As to the conditions of the theorem, we are able to relax Venter's assumptions on $M(x)$ and also to remove the assumption (3.4) on prior knowledge of bounds for β.

THEOREM 6. *Assume that $M(x)$ satisfies* (1.2)–(1.4) *and* (3.10) *with $k = 2$. Let a_m be a sequence of positive constants satisfying* (3.1) *and let $\xi_m > \zeta_m$ be two sequences of positive constants such that*

(3.12a) $$\limsup_{m\to\infty}\zeta_m < \beta < \liminf_{m\to\infty}\xi_m,$$

(3.12b) $$\Sigma_1^\infty(i\zeta_i)^{-2} < \infty,$$

and

(3.12c) $$\Sigma_1^\infty(i\xi_i)^{-1} = \infty.$$

Let $\varepsilon_1', \varepsilon_2', \cdots, \varepsilon_1'', \varepsilon_2'', \cdots$ be i.i.d. *with mean* 0 *and variance σ^2.*

(i) *Define x_i, y_i, y_i', y_i'' by* (3.2), (3.3) *and* (3.6), *and b_i by*

(3.13) $$b_i = \zeta_i \vee \{\xi_i \wedge (i-1)^{-1}\Sigma_{j=1}^{i-1}(y_j'' - y_j')/(2a_j)\}, \quad i \geqslant 2, b_1 = \zeta_1.$$

Let $n = 2m$, $\tilde{\theta}_n = x_{m+1}$, $\tilde{\beta}_n = b_m$ *and define* C_n *as in* (3.11). *Then as* $n \to \infty$,

$$\tilde{\theta}_n \to \theta \text{ and } \tilde{\beta}_n \to \beta \text{ a.s.,} \tag{3.14}$$

$$\limsup (n/2 \log\log n)^{\frac{1}{2}} |\tilde{\theta}_n - \theta| = \sigma/\beta \text{ a.s.,} \tag{3.15}$$

$$n^{\frac{1}{2}}(\tilde{\theta}_n - \theta) \to_{\mathcal{L}} N(0, \sigma^2/\beta^2), \tag{3.16}$$

$$C_n/n^{1-2\gamma} \to 4^\gamma a^2/(1-2\gamma) \text{ a.s.,} \tag{3.17}$$

$$n^{\frac{1}{2}-\gamma}(\tilde{\beta}_n - \beta) \to_{\mathcal{L}} N(0, \sigma^2/\{4^\gamma a^2(1+2\gamma)\}), \tag{3.18}$$

where $a > 0$ *and* $\frac{1}{4} < \gamma < \frac{1}{2}$ *are given by the condition* (3.1) *on the sequence* $\{a_m\}$.

(ii) *Assume in place of* (3.12*b*) *the stronger condition*

$$\Sigma_1^\infty i^{-2+\gamma} \zeta_i^{-2} < \infty. \tag{3.19}$$

Suppose that in (i) *we replace* b_i *as defined in* (3.13) *by*

$$b_i^* = \zeta_i \vee \{\xi_i \wedge i^{-1}\Sigma_{j=1}^i (y_j'' - y_j')/(2a_j)\}, \qquad i \geq 1. \tag{3.20}$$

Then the relations (3.14)–(3.18) *still hold.*

REMARKS. (a) Venter [17] proved (3.14), (3.16) and (3.18) in Theorem 6(ii) for the special case $\zeta_m = b < \beta < B = \xi_m$ under the more restrictive smoothness conditions (3.9) and (3.10) with $k \geq 2$ such that $k\beta > \gamma B$. The asymptotic behavior (3.17) of the cost C_n, however, has not been considered in the literature.

(b) Dropping Venter's assumption (3.4) on prior knowledge of bounds for β, we can choose $\zeta_m \to 0$ and $\xi_m \to \infty$ such that (3.12c) and (3.19) hold. Obviously the condition (3.12a) is then also satisfied.

(c) For the case $M''(\theta) \neq 0$, Venter has shown that the constant γ in condition (3.1) has to be chosen $> \frac{1}{4}$ and that (3.16) actually fails to hold if $\gamma = \frac{1}{4}$ (see Theorem 3 of [17]).

(d) Fabian [4] has proved (3.14) and (3.16) in Theorem 6(i) for the special case $\zeta_m = c_1 m^{-\alpha}$ and $\xi_m = c_2 \log(m+1)$ with $0 < c_1 < c_2$ and $0 < \alpha < \frac{1}{2}$. His proof is simpler than that of Venter and does not require Venter's smoothness conditions and the assumption (3.4) on prior bounds for β. However, his method requires that the last summand $(y_i'' - y_i')/(2a_i)$ be dropped in b_i^*, and therefore he considers b_i instead of b_i^*. His argument depends heavily on the fact that b_i is $\mathcal{F}_{i-1}$-measurable, where

$$\mathcal{F}_m = \mathcal{B}(x_1, \varepsilon_1', \varepsilon_1'', \cdots, \varepsilon_m', \varepsilon_m'') \qquad \mathcal{F}_0 = \mathcal{B}(x_1). \tag{3.21}$$

Our argument is different from those of Venter and Fabian and works for both b_n and b_n^*.

The following lemma will be used in the proofs of Theorems 6 and 7.

Lemma 5. *Let $M(x)$ be a Borel function satisfying* (1.4), *and assume that* (1.3) *holds for some real number θ. Let x_n, u_n, v_n, t_n, and t_n' be random variables such that*

$$x_{n+1} = x_n - v_n\{M(x_n + t_n) + M(x_n + t_n')\} + u_n, \qquad n = 1, 2, \cdots, \tag{3.22}$$

$$\lim_{n\to\infty} t_n = \lim_{n\to\infty} t_n' = 0 \text{ a.s.}, \tag{3.23}$$

$$v_n > 0 \text{ and } \lim_{n\to\infty} v_n = 0 \text{ a.s.}, \tag{3.24}$$

and

$$\Sigma_1^N u_n \text{ converges a.s. as } N \to \infty. \tag{3.25}$$

Then x_n converges a.s. *to some random variable. If, furthermore,*

$$\Sigma_1^\infty v_n = \infty \text{ a.s.}, \tag{3.26}$$

then $\lim_{n\to\infty} x_n = \theta$ a.s.

Proof. From (3.22) and (3.25), it follows that

$$x_{N+1} + \Sigma_1^N v_n\{M(x_n + t_n) + M(x_n + t_n')\} \text{ converges a.s. as } N \to \infty. \tag{3.27}$$

When $t_n = t_n' = 0$ for all n, the rest of the proof is exactly like that of Blum in Lemma 3 and Theorem 1 of [1]. An obvious modification of Blum's argument extends to the more general case where t_n and t_n' satisfy (3.22). □

We now proceed to prove Theorem 6. We shall only prove Part (ii) of the theorem in detail. The proof of Part (i) is similar and is, in fact, simpler, and we shall comment on it after the proof of Part (ii).

Proof of Theorem 6(ii). We shall first prove that

$$\lim_{n\to\infty} x_n = \theta \text{ a.s.} \tag{3.28}$$

In view of (3.2), (3.3) and (3.6), $\{x_n\}$ satisfies (3.22) with $t_n = -t_n' = a_n$, $v_n = \frac{1}{2}(nb_n^*)^{-1}$, and $u_n = -\frac{1}{2}(\varepsilon_n' + \varepsilon_n'')/(nb_n^*)$. Therefore, by Lemma 5, to prove (3.28) it suffices to show that

$$\Sigma_1^N(\varepsilon_n' + \varepsilon_n'')/(nb_n^*) \text{ converges a.s.} \tag{3.29}$$

Since b_i^* is $\mathcal{F}_i$-measurable, x_{i+1} is also $\mathcal{F}_i$-measurable by (3.2). Therefore, although b_i^* is not $\mathcal{F}_{i-1}$-measurable, a slight modification of it gives the $\mathcal{F}_{i-1}$-measurable random variable

$$b_i' = \zeta_i \vee \left\{\xi_i \wedge \frac{1}{i}\left[\frac{M(x_i + a_i) - M(x_i - a_i)}{2a_i} + \Sigma_{j=1}^{i-1}\frac{y_j'' - y_j'}{2a_j}\right]\right\}. \tag{3.30}$$

We note that with probability 1

$$\begin{aligned}(nb_n^*)^{-1} - (nb_n')^{-1} &= (nb_n^* b_n')^{-1}(b_n' - b_n^*)\\ &= O\big(|\varepsilon_n'' - \varepsilon_n'|/(n^2\zeta_n^2 a_n)\big).\end{aligned} \tag{3.31}$$

Since $a_n \sim an^{-\gamma}$, it follows from (3.19) that

$$\Sigma_1^\infty E\{(|\varepsilon_n''|^2 + |\varepsilon_n'|^2)/(n^2\zeta_n^2 a_n)\} < \infty, \tag{3.32}$$

and therefore $\Sigma_1^\infty\{|\varepsilon_n' + \varepsilon_n''|\,|\varepsilon_n' - \varepsilon_n''|/(n^2\zeta_n^2 a_n)\} < \infty$ a.s. Hence, by (3.31),

$$\Sigma_1^N(\varepsilon_n' + \varepsilon_n'')\{(nb_n^*)^{-1} - (nb_n')^{-1}\} \text{ converges a.s. as } N \to \infty. \tag{3.33}$$

Since b_i' is $\mathcal{F}_{i-1}$-measurable, $\{\Sigma_1^n(\varepsilon_i' + \varepsilon_i'')/(ib_i'), \mathcal{F}_n, n \geq 1\}$ is a martingale. Moreover, since $b_n' \geq \zeta_n$ and $\Sigma_1^\infty(n\zeta_n)^{-2} < \infty$, $\Sigma_1^\infty E\{(\varepsilon_n' + \varepsilon_n'')^2/(nb_n')^2\} < \infty$. Therefore, by the martingale convergence theorem,

$$\Sigma_1^N(\varepsilon_n' + \varepsilon_n'')/(nb_n') \text{ converges a.s. as } N \to \infty. \tag{3.34}$$

From (3.33) and (3.34), the desired conclusion (3.29) follows and so (3.28) holds.

Using (3.10) (with $k = 1$) and (3.28), we obtain that, with probability 1,

$$M(x_j + a_j) - M(x_j - a_j) = 2a_j(\beta + o(1)) \quad \text{as } j \to \infty. \tag{3.35}$$

From (3.6), (3.8) and (3.35), it follows easily that $b_m^* \to \beta$ a.s. Hence (3.14) holds.

Let $\varepsilon_i = \frac{1}{2}(\varepsilon_i' + \varepsilon_i'')$. Then $E\varepsilon_i = 0$ and $E\varepsilon_i^2 = \frac{1}{2}\sigma^2$. Since (3.10) holds with $k = 2$ and $x_i \to \theta$ a.s., we can apply Taylor's expansion to two terms to obtain that

$$\tfrac{1}{2}\{M(x_i + a_i) + M(x_i - a_i)\} = (\beta + \eta_i)x_i + \omega_i, \tag{3.36}$$

where η_i and ω_i are $\mathcal{B}(x_i)$-measurable random variables such that

$$\eta_i \to 0 \quad \text{and} \quad \omega_i = O(a_i^2) \text{ a.s.} \tag{3.37}$$

We note that by (3.3) and (3.36),

$$\begin{aligned} y_i &= \tfrac{1}{2}\{M(x_i + a_i) + M(x_i - a_i)\} + \varepsilon_i \\ &= (\beta + \eta_i)x_i + (\varepsilon_i + \omega_i). \end{aligned} \tag{3.38}$$

Therefore, using exactly the same argument as in the proof of Theorem 3, and noting that $b_n^* \to \beta$ a.s., it can be shown that

$$\begin{aligned} x_{n+1} &= \theta - (n\tau_n^*)^{-1}\{\Sigma_1^n\tau_k^*(\varepsilon_k + \omega_k)/b_k^* + \rho_0\} \\ &= \theta - (n\tau_n^*)^{-1}\{\Sigma_1^n\tau_k^*\varepsilon_k/b_k^* + \rho_n\}, \end{aligned} \tag{3.39}$$

where ρ_0 and τ_n^* are random variables such that τ_n^* satisfies (2.4) and is positive and $\mathcal{F}_n$-measurable (since b_n^* is $\mathcal{F}_n$-measurable), and

$$\begin{aligned} \rho_n &= \rho_0 + \Sigma_1^n\tau_k^*\omega_k/b_k^* \\ &= O(\tau_n^* n^{1-2\gamma}) \text{ a.s.} \qquad \text{by (3.37).} \end{aligned} \tag{3.40}$$

Since $1 - 2\gamma < \frac{1}{2}$, (3.40) implies that $\{\rho_n\}$ satisfies (2.66) (with $b = \beta$). Moreover, b_n' is $\mathcal{F}_{n-1}$-measurable and, with probability 1,

$$b_n^* - b_n' = n^{-1}(\varepsilon_n'' - \varepsilon_n')/(2a_n) = O(n^{-(1-\gamma)}(1 + |\varepsilon_n'| + |\varepsilon_n''|)) \text{ as } n \to \infty. \tag{3.41}$$

Hence $\{x_m\}$ admits a representation of the type in Theorem 5(iv). Therefore, by

Theorem 5(iv), as $m \to \infty$

(3.42) $$m^{\frac{1}{2}}(x_m - \theta) \to_{\mathcal{L}} N\big(0, \tfrac{1}{2}\sigma^2/\beta^2\big),$$

(3.43) $$\limsup(m/2 \log\log m)^{\frac{1}{2}}|x_m - \theta| = \sigma/\big(2^{\frac{1}{2}}\beta\big) \text{ a.s.},$$

(3.44) $$\Sigma_1^m(x_i - \theta)^2/\log m \to \tfrac{1}{2}\sigma^2/\beta^2 \text{ a.s.}$$

From (3.42) and (3.43), (3.15) and (3.16) follow immediately. Since $C_n = 2\{\Sigma_1^m(x_i - \theta)^2 + \Sigma_1^m a_i^2\}$ and (3.44) holds, while

$$\Sigma_1^m a_i^2 \sim a^2\big(\tfrac{1}{2}n\big)^{1-2\gamma}/(1-2\gamma),$$

(3.17) follows.

Using (3.10) (with $k = 2$) together with (3.43) and the fact that $a_j > j^{-\frac{1}{2}}(\log\log j)^{\frac{1}{2}}$ for all large j, we can sharpen (3.35) to

(3.45)
$$M(x_j + a_j) - M(x_j - a_j) = 2a_j M'(x_j + r_j a_j) \quad (\text{where } |r_j| \leq 1)$$
$$= 2a_j\big\{\beta + 0(r_j a_j + x_j - \theta)\big\} = 2a_j(\beta + 0(a_j)) \text{ a.s.}$$

Hence, with probability 1,

(3.46) $$m^{-1}\Sigma_{j=1}^m(M(x_j + a_j) - M(x_j - a_j))/(2a_j) - \beta$$
$$= m^{-1}\Sigma_{j=1}^m 0(a_j) = 0(m^{-\gamma}) = o\big(m^{-(\frac{1}{2}-\gamma)}\big) \qquad \text{since } \gamma > \tfrac{1}{4}.$$

Since $E(\varepsilon_1'' - \varepsilon_1')^2 = 2\sigma^2$ and $\Sigma_1^m a_i^{-2} \sim a^{-2}m^{1+2\gamma}/(1+2\gamma)$, we obtain by the Feller-Lindeberg central limit theorem that

(3.47) $$(2m)^{\frac{1}{2}-\gamma}m^{-1}\Sigma_{j=1}^m(\varepsilon_j'' - \varepsilon_j')/(2a_j) \to_{\mathcal{L}} N\big(0, \sigma^2/\{4^\gamma a^2(1+2\gamma)\}\big).$$

From (3.46) and (3.47), (3.18) follows immediately. □

Proof of Theorem 6(i). Here b_i is $\mathcal{F}_{i-1}$-measurable and, therefore, the convergence of $\Sigma_1^N(\varepsilon_n' + \varepsilon_n'')/(nb_n)$ follows from the fact that $\Sigma_1^\infty(n\zeta_n)^{-2} < \infty$ and the martingale convergence theorem. The rest of the proof is the same as that of Theorem 6(ii), except that Theorem 5(i) can be used instead of Theorem 5(iv). □

A close examination of the preceding proof suggests that in order to reduce the cost C_n to the desired logarithmic order of magnitude we should choose the sequence $\{a_m\}$ such that

(3.48) $$\Sigma_1^m a_i^2 = o(\log m) \qquad \text{as} \quad n \to \infty.$$

This means that γ in condition (3.1) has to be chosen $\geq \frac{1}{2}$ instead. However, even with $\gamma = \frac{1}{2}$ and a_m modified to be of the form $m^{-\frac{1}{2}}(1 + \log m)^{-\delta}$ ($\delta > 0$) so that (3.48) holds, the relation (3.8) is no longer true, and such an a_m is too small for the b_i (or b_i^*) defined by (3.13) (or (3.20)) to be strongly consistent. In order to be able to choose a_m satisfying (3.48) instead of (3.1), we shall use another estimator b_m of

β at stage m. Let $\{a_m\}$ be any sequence of positive constants such that

(3.49) $$a_j = 0(j^{-\gamma}) \quad \text{for some} \quad \gamma > \tfrac{1}{4} \quad \text{but} \quad \Sigma_1^\infty a_j^2 = \infty.$$

Then, by the strong law,

(3.50) $$\Sigma_1^m a_j(\varepsilon_j'' - \varepsilon_j')/\left(\Sigma_1^m a_j^2\right) \to 0 \text{ a.s.}$$

Considering the linear case $M(x) = \beta(x - \theta)$, we note, in view of (3.7) and (3.50), that

(3.51) $$\tfrac{1}{2}\Sigma_1^m a_j(y_j'' - y_j')/\left(\Sigma_1^m a_j^2\right) \to \beta \text{ a.s.}$$

Thus, at least in the linear case, (3.51) gives a strongly consistent estimate of β under the minimal assumption that $\Sigma_1^m a_j^2 \to \infty$, no matter how slow the convergence may be. In the following theorem, we shall prove that this modification of Venter's scheme yields the desired growth rates for both C_n and $n^{\frac{1}{2}}(\tilde{\theta}_n - \theta)$, even in the nonlinear case.

THEOREM 7. (i) *Suppose that in Theorem* 6(i) *we replace the assumption* (3.1) *on the sequence* $\{a_m\}$ *by the weaker assumption* (3.49) *and also replace the definition* (3.13) *of* b_i *by*

(3.52) $$b_i = \zeta_i \vee \left\{\xi_i \wedge \tfrac{1}{2}\Sigma_{j=1}^{i-1} a_j(y_j'' - y_j')/\left(\Sigma_{j=1}^{i-1} a_j^2\right)\right\}, \qquad i \geq 2,\ b_1 = \zeta_1.$$

Then the relations (3.14), (3.15) *and* (3.16) *still hold. Moreover, instead of* (3.18), *we have*

(3.53) $$\left(\Sigma_{j=1}^m a_j^2\right)^{\frac{1}{2}}(b_m - \beta) \to_{\mathcal{L}} N\left(0, \tfrac{1}{2}\sigma^2\right).$$

If $\{a_m\}$ *further satisfies* (3.48), *then instead of* (3.17), *we have*

(3.54) $$C_n/\log n \to \sigma^2/\beta^2 \text{ a.s.}$$

(ii) *Let* $\{a_n\}$ *be a sequence of positive constants such that*

(3.55) $$a_m \sim am^{-\frac{1}{2}}(\log m)^{-\delta} \quad \textit{for some} \quad a > 0 \quad \textit{and} \quad 0 < \delta < \tfrac{1}{2}.$$

Then conditions (3.48) *and* (3.49) *are satisfied. Suppose that in Theorem* 6(i) *we replace the assumption* (3.1) *by* (3.55) *and also replace* b_i *as defined in* (3.13) *by*

(3.56) $$b_i^* = \zeta_i \vee \left\{\xi_i \wedge \tfrac{1}{2}\Sigma_{j=1}^{i} a_j(y_j'' - y_j')/\left(\Sigma_{j=1}^{i} a_j^2\right)\right\}.$$

Moreover, in place of (3.12b), *we assume that*

(3.57) $$\Sigma_2^\infty i^{-3/2}\zeta_i^{-2}(\log i)^{\delta - 1} < \infty.$$

Then (3.14), (3.15), (3.16), (3.53) *and* (3.54) *still hold.*

REMARK. While b_i, as defined by (3.52), is $\mathcal{F}_{i-1}$-measurable, b_i^* in (3.56) is not. Let $a_m \sim am^{-\gamma}(\log m)^{-\delta}$, where a and γ are positive constants and $\delta > 0$. Then

(3.48) fails to hold if $\gamma < \frac{1}{2}$. Obviously,

$$\text{(3.49) holds} \Leftrightarrow \tfrac{1}{4} < \gamma \leq \tfrac{1}{2} \quad \text{and} \quad \delta \leq \tfrac{1}{2} \quad \text{when} \quad \gamma = \tfrac{1}{2}.$$

It is easy to see that the conditions (3.48) and (3.49) are both satisfied for the case $\gamma = \frac{1}{2}$ and $0 < \delta \leq \frac{1}{2}$.

PROOF OF THEOREM 7(i). As in Theorem 6(i), it is easy to show, using Lemma 5, that $x_m \to \theta$ a.s. Therefore (3.35) again holds. By (3.6), (3.35) and (3.50), we still have $b_m \to \beta$ a.s. Hence, as in Theorem 6(i), (3.15), (3.16), (3.43) and (3.44) still hold. Therefore, if condition (3.48) is also satisfied, then by (3.44), with probability 1,

$$C_n = 2\{\Sigma_1^m(x_i - \theta)^2 + \Sigma_1^m a_i^2\} \sim (\sigma^2/\beta^2)\log m,$$

and so (3.54) holds.

We note that, as in (3.45), with probability 1,

$$M(x_j + a_j) - M(x_j - a_j) = 2a_j\{\beta + 0(a_j) + 0(|x_j - \theta|)\}. \tag{3.58}$$

By (3.49), $a_j = o(j^{-\rho})$ for some $\rho > \frac{1}{4}$. Therefore, by the Schwarz inequality,

$$\Sigma_1^m a_j^3 = \Sigma_1^m o(a_j j^{-2\rho}) \leq \left(\Sigma_1^m o(a_j^2)\right)^{\frac{1}{2}}\left(\Sigma_1^m j^{-4\rho}\right)^{\frac{1}{2}} = o\left(\left(\Sigma_1^m a_j^2\right)^{\frac{1}{2}}\right). \tag{3.59}$$

Moreover, by (3.43) and the Schwarz inequality, with probability 1,

$$\Sigma_1^m a_j^2|x_j - \theta| = \Sigma_1^m o\left(a_j j^{-\frac{1}{2}-\rho}(\log\log j)^{\frac{1}{2}}\right) = o\left(\left(\Sigma_1^m a_j^2\right)^{\frac{1}{2}}\right). \tag{3.60}$$

From (3.58)–(3.60), it follows that, with probability 1,

$$\left\{\tfrac{1}{2}\Sigma_{j=1}^m a_j\left(M(x_j + a_j) - M(x_j - a_j)\right)/\left(\Sigma_{j=1}^m a_j^2\right)\right\} - \beta = o\left(\left(\Sigma_{j=1}^m a_j^2\right)^{-\frac{1}{2}}\right). \tag{3.61}$$

By the Feller-Lindeberg central limit theorem,

$$\left(\Sigma_{j=1}^m a_j^2\right)^{-\frac{1}{2}}\Sigma_{j=1}^m \tfrac{1}{2}a_j(\varepsilon_j'' - \varepsilon_j') \to N\left(0, \tfrac{1}{2}\sigma^2\right). \tag{3.62}$$

From (3.61) and (3.62), (3.53) follows immediately. □

PROOF OF THEOREM 7(ii). Define

(3.63)

$$\tilde{b}_i = \zeta_i \vee \left\{\xi_i \wedge \tfrac{1}{2}\left[a_i(M(x_i + a_i) - M(x_i - a_i)) + \Sigma_{j=1}^{i-1}a_j(y_j'' - y_j')\right]/\left(\Sigma_{j=1}^i a_j^2\right)\right\}.$$

Then $\tilde{b}_i$ is $\mathcal{F}_{i-1}$-measurable and, by (3.55),

$$(nb_n^*)^{-1} - (n\tilde{b}_n)^{-1} = 0\left(|\varepsilon_n'' - \varepsilon_n'|/\left\{n^{\frac{3}{2}}\zeta_n^2(\log n)^{1-\delta}\right\}\right) \text{ a.s.} \tag{3.64}$$

Making use of (3.57) and using the same argument as in (3.32)–(3.34), it then follows that $\Sigma_1^N(\varepsilon_n'' + \varepsilon_n')/(nb_n^*)$ converges a.s. as $N \to \infty$. Hence, by Lemma 5, $x_m \to \theta$ a.s., and so, as in the proof of Theorem 7(i), $b_m^* \to \beta$ a.s. Therefore, by (3.55), with probability 1, for all large i

$$b_i^* - \tilde{b}_i = \tfrac{1}{2}a_i(\varepsilon_i'' - \varepsilon_i')/\left(\Sigma_{j=1}^i a_j^2\right) = o\left(i^{-\frac{1}{2}}(\log i)^{-\delta}(1 + |\varepsilon_i'| + |\varepsilon_i''|)\right).$$

By using the same argument as in the proof of Theorem 6(ii), it follows that (3.15), (3.16), (3.43) and (3.44) still hold. From (3.44) and (3.48), (3.54) follows. Moreover, (3.53) can be obtained by the same argument as in Theorem 7(i). □

4. Adaptive stochastic approximation schemes using least squares estimates of β. In the designs of Section 3, the main reason for choosing two levels x'_m and x''_m (instead of x_m) at the mth stage is to be able to estimate β in a consistent way by using the differences $y''_j - y'_j$ $(j = 1, 2, \cdots)$. As pointed out in (3.7), for the linear case $M(x) = \beta(x - \theta)$, these differences $y''_j - y'_j$ depend only on β and not on θ. It is natural to ask whether consistent estimates b_n of β can be found for the adaptive Robbins-Monro scheme

(4.1) $$x_{n+1} = x_n - y_n/(nb_n).$$

An obvious choice for b_n is the usual least squares estimate

(4.2) $$\hat{\beta}_n = \Sigma_1^n(x_i - \bar{x}_n)(y_i - \bar{y}_n)/\Sigma_1^n(x_i - \bar{x}_n)^2,$$

where we set $\hat{\beta}_n$ equal to some positive constant b when $\Sigma_1^n(x_i - \bar{x}_n)^2 = 0$. In the linear case $M(x) = \beta(x - \theta)$, the strong consistency of $\hat{\beta}_n$ is equivalent to

(4.3) $$\Sigma_1^n(x_i - \bar{x}_n)\varepsilon_i/\Sigma_1^n(x_i - \bar{x}_n)^2 \to 0 \text{ a.s.}$$

By Theorem 4, if $\hat{\beta}_n$ is indeed strongly consistent and $\sigma \neq 0$, then with probability 1, $n^{\frac{1}{2}}(\bar{x}_n - \theta) = O((\log\log n)^{\frac{1}{2}})$, $\Sigma_1^n(x_i - \theta)^2 \sim (\sigma^2/\beta^2)\log n$, and therefore

(4.4) $$\Sigma_1^n(x_i - \bar{x}_n)\varepsilon_i = \Sigma_1^n(x_i - \theta)\varepsilon_i + o\big(\Sigma_1^n(x_i - \theta)^2\big),$$

(4.5) $$\Sigma_1^n(x_i - \bar{x}_n)^2 = \Sigma_1^n(x_i - \theta)^2 - n(\bar{x}_n - \theta)^2 \sim \Sigma_1^n(x_i - \theta)^2.$$

Thus if $\hat{\beta}_n$ is indeed strongly consistent and $\sigma \neq 0$, then the sequence $\{x_n\}$ will behave nicely in the sense of (4.4) and (4.5), which by the strong law for martingales in turn imply (4.3) and hence the strong consistency of $\hat{\beta}_n$. This suggests that a natural way of proving the strong consistency of $\hat{\beta}_n$ is to show that the design levels x_n satisfy (4.4) and (4.5). In the case where upper and lower bounds B and b (> 0) for β are known, we are able to elaborate this idea to prove the strong consistency of

(4.6) $$b_n = b \vee \big(B \wedge \hat{\beta}_n\big)$$

and thereby to obtain from Theorem 4 the following

THEOREM 8. *Let $\varepsilon, \varepsilon_1, \cdots$ be* i.i.d. *random variables with $E\varepsilon = 0$ and $0 < E\varepsilon^2 = \sigma^2 < \infty$. Let $M(x)$ be a Borel function satisfying* (1.2)–(1.4), *and assume that $M(x)$ is continuously differentiable in some open neighborhood of θ. Let b, B be positive constants such that $b < \beta < B$. Let x_1 be a random variable independent of $\varepsilon_1, \varepsilon_2, \cdots$, and define inductively y_n, x_n, $\hat{\beta}_n$ and b_n by* (1.1), (4.1), (4.2) *and* (4.6). *Then* $\lim_{n\to\infty} b_n = \beta$ a.s., *and* (1.11), (1.12), (1.13) *still hold. If $M(x)$ further satisfies* (1.18), *then*

(4.7) $$(\log n)^{\frac{1}{2}}(b_n - \beta) \to_{\mathcal{L}} N(0, \beta^2).$$

The details of the proof of Theorem 8 are given in [10]. It is interesting to compare the asymptotic distribution (4.7) of b_n with the corresponding result (3.53) for the pairwise sampling scheme of Theorem 7. For this pairwise sampling scheme, which also satisfies (1.11)–(1.13), the relation (3.53) says that the consistent estimator b_m defined by (3.52) is asymptotically normal with variance $\frac{1}{2}\sigma^2/(\Sigma_1^m a_i^2)$, which is of a larger order of magnitude than $(\log m)^{-1}$ in view of (3.48). Thus the adaptive stochastic approximation scheme of Theorem 8 uses an asymptotically more efficient estimator of β than the pairwise sampling scheme of Theorem 7. Some simulation studies comparing the performance of these two kinds of adaptive stochastic approximation procedures for moderate sample sizes will be described in [12]. While Theorem 8 assumes that prior upper and lower bounds for β are known, we are able to remove this assumption by a modification of b_n in (4.6) and of the argument used. The details are given in [10].

REFERENCES

[1] Blum, J. R. (1954). Approximation methods which converge with probability one. *Ann. Math. Statist.* **25** 382–386.

[2] Bojanic, R. and Seneta, E. (1973). A unified theory of regularly varying sequences. *Math. Z.* **134** 91–106.

[3] Chung, K. L. (1954). On a stochastic approximation method. *Ann. Math. Statist.* **25** 463–483.

[4] Fabian, V. (1968). On asymptotic normality in stochastic approximation. *Ann. Math. Statist.* **39** 1327–1332.

[5] Freedman, D. (1971). *Brownian Motion and Diffusion*. Holden-Day, San Francisco.

[6] Gaposhkin, V. F. and Krasulina, T. P. (1974). On the law of the iterated logarithm in stochastic approximation processes. *Theor. Probability Appl.* **19** 844–850.

[7] Kersting, G. (1977). Almost sure approximation of the Robbins-Monro process by sums of independent random variables. *Ann. Probability* **5** 954–965.

[8] Lai, T. L. and Robbins, H. (1978a). Adaptive design in regression and control. *Proc. Nat. Acad. Sci. U.S.A.* **75** 586–587.

[9] Lai, T. L. and Robbins, H. (1978b). Limit theorems for weighted sums and stochastic approximation processes. *Proc. Nat. Acad. Sci. U.S.A.* **75** 1068–1070.

[10] Lai, T. L. and Robbins, H. (1978c). Consistency and asymptotic efficiency of slope estimates in adaptive stochastic approximation. Unpublished manuscript.

[11] Lai, T. L. and Robbins, H. (1978d). Iterated least squares and multiperiod control. Unpublished manuscript.

[12] Lai, T. L. and Robbins, H. (1978e). Adaptive design and estimation of the root of a regression function. Unpublished manuscript.

[13] Major, P. and Révész, P. (1973). A limit theorem for the Robbins-Monro approximation. *Z. Wahrscheinlichkeitstheorie und Verw. Gebiete* **27** 79–86.

[14] Neveu, J. (1965). *Mathematical Foundations of the Calculus of Probability*. Holden-Day, San Francisco.

[15] Robbins, H. and Monro, S. (1951). A stochastic approximation method. *Ann. Math. Statist.* **22** 400–407.

[16] Sacks, J. (1958). Asymptotic distribution of stochastic approximation procedures. *Ann. Math. Statist.* **29** 375–405.

[17] Venter, J. (1967). An extension of the Robbins-Monro procedure. *Ann. Math. Statist.* **38** 181–190.

Department of Mathematical Statistics
Columbia University
Mathematics Building
New York, New York 10027

Part Two

SEQUENTIAL EXPERIMENTATION AND ANALYSIS

B. Adaptive Allocation of Treatments in Sequential Experiments

The well-known "multi-armed bandit problem" in the statistics and engineering literature, which is prototypical of a wide variety of adaptive control and design problems, was first formulated and studied by Robbins [28]*. Let A, B denote two statistical populations with finite means μ_A, μ_B. Robbins studied in [28] how one should draw a sample $x_1, \ldots, x_n$ from the two populations in order to achieve the greatest possible expected value of the average $n^{-1}S_n = n^{-1}(x_1 + \cdots + x_n)$ as $n \to \infty$. Four years after this pioneering paper, Robbins [42]* introduced the so-called "bandit problem with finite memory," in which the decision at every stage can depend on the information from no more than m previous stages.

These seminal papers by Robbins initiated an active area of research leading to an extensive literature on the subject. Major developments in the finite-memory problem include the works of Isbell (1959), Smith and Pyke (1965), Samuels (1968), and Cover and Hellman (1970). Some important contributions to the multi-armed bandit problem are the works of Bradt, Johnson, and Karlin (1956), Bellman (1956), Vogel (1960), Feldman (1962), Chernoff and Ray (1965), Fabius and Van Zwet (1970), Berry (1972), Fox (1974), Gittins and Jones (1974), Gittins (1979), Berry and Fristedt (1979), Whittle (1980), Bather (1980, 1981), and Kelly (1981).

An extension from the finite number of populations in the multi-armed bandit problem to the case of infinitely many populations was introduced in the paper [57] with Mallows in 1963. In recent years Robbins returned to the multi-armed bandit problem. In a series of joint papers [129]*, [130], and [131] with Lai, asymptotically optimal adaptive allocation rules are developed. In particular, if π_j $(j=1, \ldots, k)$ are populations with means μ_j

*Papers reprinted in this book are marked here by an asterisk when they first appear.

and densities $f(x;\theta_j)$ with respect to some measure ν, it is shown that

$$n \max_j \mu_j - E_{\boldsymbol{\theta}} S_n \geq (C(\boldsymbol{\theta}) + o(1)) \log n,$$

for all adaptive allocation rules satisfying certain assumptions, where $\boldsymbol{\theta} = (\theta_1, \ldots, \theta_k)$, and $C(\boldsymbol{\theta})$ is a constant involving μ_i and $I(\theta_i, \theta_j) = E_{\theta_i}[\log(f(X;\theta_i)/f(X;\theta_j))]$. Simple adaptive allocation rules that attain this asymptotic lower bound are also constructed.

Another important contribution of Robbins is in the related area of treatment allocation in sequential clinical trials. Here one wants to reach a decision about the relative merits of two treatments on the basis of an experiment which allocates the "inferior" treatment as infrequently as possible. A principal result of [89]* and [96]* is that in a comparison of the means of two normal populations with a common known variance, if one restricts consideration to procedures which are invariant under a common change of location of the data, then the error probabilities are to a considerable extent independent of the sampling rule, which consequently can be selected with a view to minimizing the expected number of allocations of the inferior treatment. Various allocation rules have been studied, primarily by simulation in [96], Louis (1975), and Hayre (1979).

References

Bather, J. A. (1980). Randomized allocation of treatments in sequential trials. *Adv. Appl. Probab.* **12**, 174–182.

Bather, J. A. (1981). Randomized allocation of treatments in sequential experiments. *J. Roy. Statist. Soc. Ser. B* **43**, 265–292.

Bellman, R. E. (1956). A problem in the sequential design of experiments. *Sankhyā Ser. A* **16**, 221–229.

Berry, D. A. (1972). A Bernoulli two-armed bandit. *Ann. Math. Statist.* **43**, 871–897.

Berry, D. A. and Fristedt, B. (1979). Bernoulli one-armed bandits—arbitrary discount sequence. *Ann. Statist.* **7**, 1086–1105.

Bradt, R. N., Johnson, S. M. and Karlin, S. (1956). On sequential designs for maximizing the sum of *n* observations. *Ann. Math. Statist.* **27**, 1060–1074.

Chernoff, H. and Ray, S. N. (1965). A Bayes sequential sampling plan. *Ann. Math. Statist.* **36**, 1387–1407.

Cover, T. M. and Hellman, M. E. (1970). Two-armed bandit problems with time-invariant finite memory. *IEEE Trans. Information Theory*, **IT-14**, 185–195.

Fabius, J. and Van Zwet, W. R. (1970). Some remarks on the two-armed bandit. *Ann. Math. Statist.* **41**, 1906–1916.

Feldman, D. (1962). Contributions to the two-armed bandit problem. *Ann. Math. Statist.* **33**, 847–856.

Fox, B. L. (1974). Finite horizon behavior of policies for two-armed bandits. *J. Amer. Statist. Assoc.* **69**, 963–965.

Gittins, J. C. (1979). Bandit processes and dynamic allocation indices. *J. Roy. Statist. Soc. Ser. B* **41**, 148–177.

B. ADAPTIVE ALLOCATION OF TREATMENTS

Hayre, L. S. (1979). Two population sequential tests with three hypotheses, *Biometrika* **66**, 465–474.

Isbell, J. R. (1959). On a problem of Robbins. *Ann. Math. Statist.* **30**, 606–610.

Kelly, F. P. (1981). Multi-armed bandits with discount factor near one: the Bernoulli case. *Ann. Statist.* **9**, 987–1001.

Louis, T. A. (1975). Optimal allocation in sequential tests comparing the means of two Gaussian populations, *Biometrika* **62**, 359–369.

Samuels, S. M. (1968), Randomized rules for the two-armed bandit with finite memory. *Ann. Math. Statist.* **39**, 2103–2107.

Smith, C. V. and Pyke, R. (1965). The Robbins–Isbell two-armed bandit problem with finite memory. *Ann. Math. Statist.* **36**, 1375–1386.

Vogel, W. (1960). An asymptotic minimax theorem for the two-armed bandit problem. *Ann. Math. Statist.* **31**, 444–451.

Whittle, P. (1980), Multi-armed bandits and the Gittins index. *J. Roy. Statist. Soc. Ser. B* **42**, 143–149.

Reprinted from the
BULLETIN OF THE AMERICAN MATHEMATICAL SOCIETY
Vol. 58, No. 5, pp. 527-535
September, 1952

SOME ASPECTS OF THE SEQUENTIAL DESIGN OF EXPERIMENTS

HERBERT ROBBINS

1. Introduction. Until recently, statistical theory has been restricted to the design and analysis of sampling experiments in which the size and composition of the samples are completely determined before the experimentation begins. The reasons for this are partly historical, dating back to the time when the statistician was consulted, if at all, only after the experiment was over, and partly intrinsic in the mathematical difficulty of working with anything but a fixed number of independent random variables. A major advance now appears to be in the making with the creation of a theory of the *sequential design* of experiments, in which the size and composition of the samples are not fixed in advance but are functions of the observations themselves.

The first important departure from fixed sample size came in the field of industrial quality control, with the double sampling inspection method of Dodge and Romig [1]. Here there is only one population to be sampled, and the question at issue is whether the proportion of defectives in a lot exceeds a given level. A preliminary sample of n_1 objects is drawn from the lot and the number x of defectives noted. If x is less than a fixed value a the lot is accepted without further sampling, if x is greater than a fixed value b $(a<b)$ the lot is rejected without further sampling, but if $a \leqq x \leqq b$ then a second sample, of size n_2, is drawn, and the decision to accept or reject the lot is made on the basis of the number of defectives in the total sample of n_1+n_2 objects. The total sample size n is thus a random variable with two values, n_1 and n_1+n_2, and the value of n is stochastically dependent on the observations. A logical extension of the idea of double sampling came during World War II with the development, chiefly by Wald, of sequential analysis [2], in which the observations are made one by one and the decision to terminate sampling and to accept or reject the lot (or, more generally, to accept or reject whatever statistical "null hypothesis" is being tested) can come at any stage. The total sample size n now becomes a random variable capable in principle of assuming infinitely many values, although in practice a finite upper limit on n is usually set. The advantage of sequential

An address delivered before the Auburn, Alabama, meeting of the Society, November 23, 1951, by invitation of the Committee to Select Hour Speakers for Southeastern Sectional Meetings; received by the editors December 10, 1951.

over fixed-size sampling lies in the fact that in some circumstances the judicious choice of a sequential plan can bring about a considerable reduction in the average sample size necessary to reduce the probability of erroneous decision to a desired low level. The theory of sequential analysis is still very incomplete, and much work remains to be done before optimum sequential methods become available for treating the standard problems of statistics.

The introduction of sequential methods of sampling freed statistics from the restriction to samples of fixed size. However, it is not only the sample *size* that is involved in the efficient design of an experiment. Most statistical problems met with in practice involve more than one population, and in dealing with such problems we must specify *which population* is to be sampled at each stage. An example will serve to clarify this point. Suppose we are dealing with two normally distributed populations with unknown means μ_1, μ_2 and variances σ_1^2, σ_2^2, and that we wish to estimate the value of the difference $\mu_1-\mu_2$. In order to concentrate on the point at issue we shall suppose that the total sample size, n, is fixed. There remains the question of how the n observations are to be divided between the two populations. If $\bar{x}_1$, $\bar{x}_2$ denote the means of samples of sizes n_1, n_2 from the two populations, then $\bar{x}_1-\bar{x}_2$ is an unbiased estimator of $\mu_1-\mu_2$, with variance $\sigma^2=(\sigma_1^2/n_1)+(\sigma_2^2/n_2)$. For fixed $n=n_1+n_2$, σ^2 is a minimum when $n_1/n_2=\sigma_1/\sigma_2$. If the latter ratio is known in advance, all is well. If this ratio is not known, but if the sampling can be done in two stages, then it would be reasonable to draw preliminary samples of some size m from each of the two populations and to use the values so obtained to estimate σ_1/σ_2; the remainder of the $n-2m$ observations could then be allocated to the two populations in accordance with the sample estimate of σ_1/σ_2. The question then becomes, what is the best choice for m? If m is small, no accurate estimate of σ_1/σ_2 can be made. If m is large, then the remaining $n-2m$ observations may be too few to permit full utilization of the approximate knowledge of σ_1/σ_2. (This kind of dilemma is characteristic of all sequential design problems.) More generally, we could consider schemes in which the observations are made one by one, with the decision as to which population each observation should come from being allowed to depend on all the previous observations; the total sample size n could be fixed or could be a random variable dependent on the observations.

Despite the total absence of theory, a notable pioneering venture in the spirit of sequential design was carried out in 1938 by Mahalanobis [3] to determine the acreage under jute in Bengal. Preliminary surveys were made on a small scale to estimate the values of certain parameters, a knowledge of which was essential to the efficient design of a subsequent large scale census. In a subsequent publication [4] Mahalanobis called attention to the desirability of revising the design of any experiment as data accumulates. The question, of course, is how best to do this.

We are indebted to Wald for the first significant contribution to the theory of sequential design. His book [5] states the problem in full generality and gives the outline of a general inductive method of solution. The probability problems involved are formidable, since dependent probabilities occur in all their complexity, and explicit recipes are not yet available for handling problems of practical interest. Nevertheless, enough is visible to justify a prediction that future results in the theory of sequential design will be of the greatest importance to mathematical statistics and to science as a whole.

In what follows we shall discuss a few simple problems in sequential design which are now under investigation and which are different from those usually met with in statistical literature. Optimum solutions to these problems are not known. Still, it is often better to have reasonably good solutions of the proper problems than optimum solutions of the wrong problems. In the present state of statistical theory this principle applies with particular force to problems in sequential design.

2. A problem of two populations. Let A and B denote two statistical populations (coins, urns, manufacturing processes, varieties of seed, treatments, etc.) specified respectively by univariate cumulative distribution functions $F(x)$ and $G(x)$ which are known only to belong to some class D. We shall suppose that the expectations

$$(1) \qquad \alpha = \int_{-\infty}^{\infty} x dF(x), \qquad \beta = \int_{-\infty}^{\infty} x dG(x)$$

exist. *How should we draw a sample $x_1, x_2, \cdots, x_n$ from the two populations if our object is to achieve the greatest possible expected value of the sum $S_n = x_1 + \cdots + x_n$?*

For example, let A and B denote two coins of unknown bias, and suppose that we are allowed to make n tosses, with the promise of getting $1 for each head but nothing for tails. If $x_i = 1$ or 0 according as heads or tails occurs on the ith toss, then S_n denotes the total sum which we are to receive, and α and β ($0 \leqq \alpha, \beta \leqq 1$) are the respective probabilities of obtaining heads on a single toss of coins A and B.

As a general intuitive principle, whenever we feel pretty sure from the results of previous observations that one of the two numbers α, β is the greater, we shall want to devote more of our future observations to that population. Note that there is no terminal decision to make; that is, we are not interested in estimating $\alpha-\beta$ or in testing the hypothesis that $\alpha=\beta$, etc., after the sample is drawn. The whole problem lies in deciding *how to draw the sample*. There certainly exist practical situations in which the present problem represents more nearly what one wants to solve than would any formulation in terms of testing hypotheses, estimating parameters, or making terminal decisions. In fact, the problem represents in a simplified way the general question of how we learn—or should learn—from past experience. A reasonably good solution of the problem must therefore be found if mathematical statistics is to provide a guide to what has been called by Neyman [6] inductive behavior.

To begin with we shall consider the special case already mentioned in which A and B are coins and the unknowns α and β are the respective probabilities of obtaining heads ($x_i=1$) on a single trial. Let us take as an example of a possible sampling rule the following.

Rule R_1. For the first toss choose A or B at random. Then, for $i=1, 2, \cdots$, if the ith toss results in heads, stick to the same coin for the $(i+1)$th toss, while if the ith toss results in tails, switch to the other coin for the $(i+1)$th toss.

What are the operating characteristics of the rule R_1? The successive tosses represent the evolution of a simple Markov chain with four states, (A, H), (A, T), (B, H), (B, T), and with transition probabilities which are easily written down; for example, the probability of a transition from (A, H) on the ith toss to (A, T) on the $(i+1)$th toss is $1-\alpha$. Let p_i denote the probability of obtaining heads on the ith toss. To avoid trivialities we shall suppose that α and β are not both 0 or both 1; then $|\alpha+\beta-1|<1$. It is easy to show that

$$p_{i+1} = (\alpha + \beta - 1)p_i + (\alpha + \beta - 2\alpha\beta), \tag{2}$$

from which it follows that

$$p_i = (\alpha + \beta - 1)^{i-1}\left[p_1 - \frac{\alpha + \beta - 2\alpha\beta}{2 - (\alpha + \beta)}\right] + \frac{\alpha + \beta - 2\alpha\beta}{2 - (\alpha + \beta)}, \tag{3}$$

and hence that

$$\lim_{i\to\infty} p_i = \frac{\alpha + \beta - 2\alpha\beta}{2 - (\alpha + \beta)} = \gamma + \frac{\delta^2}{1 - \gamma}, \tag{4}$$

where we have set

(5) $$\gamma = \frac{\alpha + \beta}{2}, \qquad \delta = \frac{|\alpha - \beta|}{2}.$$

It follows that in using the rule R_1,

(6) $$\lim_{n\to\infty} E\left(\frac{S_n}{n}\right) = \lim_{n\to\infty}\left(\frac{p_1 + \cdots + p_n}{n}\right) = \gamma + \frac{\delta^2}{1-\gamma}.$$

Now, if we *knew* which of the two numbers α, β is the greater, then by using the corresponding coin exclusively we could achieve the result

(7) $$E\left(\frac{S_n}{n}\right) = \max(\alpha, \beta) = \gamma + \delta.$$

Hence it is natural to take the difference

(8) $$L(A, B, R_1) = (\gamma + \delta) - \left(\gamma + \frac{\delta^2}{1-\gamma}\right) = \delta\left[1 - \frac{\delta}{1-\gamma}\right] \geqq 0$$

as a measure of the asymptotic *loss* per toss, by a person who uses R_1, *due to ignorance of the true state of affairs.* It is easy to show that $L(A, B, R_1)$ has its maximum value, $M_1 = 3 - 2^{3/2} \cong .172$, when $\alpha = 0$ and $\beta = 2 - 2^{1/2} \cong .586$ or vice versa. Thus a person using R_1 will, for large n, lose on the average at most 17.2 cents per toss due to ignorance of which is the better coin. On the other hand, consider the rule R_0 which consists in choosing one of the two coins A, B at random and then sticking to it, come what may (or in tossing the two coins alternately). The corresponding quantity $L(A, B, R_0)$ is easily seen to have the value $(\gamma+\delta) - \gamma = \delta$, which has its maximum, $M_0 = 1/2$, when $\alpha = 0$ and $\beta = 1$ or vice versa. Clearly, R_1 is considerably better than R_0.

The rule R_0 makes the choice of the coin for the ith toss independent of the results of previous tosses, while R_1 makes this choice depend on the result of the $(i-1)$th toss only. For the most general rule R the choice of the coin for the ith toss will depend on the results $x_1, \cdots, x_{i-1}$ of all the previous tosses. For any such rule R let

(9) $$L_n(A, B, R) = \max(\alpha, \beta) - E\left(\frac{S_n}{n}\right)$$

where E denotes expectation computed on the basis of α, β, and R, and let

(10) $$M_n(R) = \max_{(\alpha,\beta)} [L_n(A, B, R)],$$

$$\phi(n) = \min_{(\mathrm{R})} [M_n(\mathrm{R})]. \tag{11}$$

It would be interesting to know the value of $\phi(n)$ and the explicit description of any "minimax" rule R for which the value $\phi(n)$ is attained.

A much simpler problem is: do there exist rules R such that

$$\lim_{n\to\infty} L_n(A, B, \mathrm{R}) = 0 \qquad \text{for every } A, B? \tag{12}$$

We shall see in the next paragraph that the answer is yes, not only in the case of the coins but for any two populations.

Returning to the general case in which A and B are arbitrary statistical populations for which the values (1) exist, consider the sampling rule $\overline{\mathrm{R}}$ defined as follows: let

$$\begin{aligned} 1 = a_1 < a_2 < \cdots < a_n < \cdots, \\ 2 = b_1 < b_2 < \cdots < b_n < \cdots \end{aligned} \tag{13}$$

be two fixed, disjoint, increasing sequences of positive integers of density 0; that is, such that the proportion of the integers $1, 2, \cdots, n$ which are either a's or b's tends to 0 as $n \to \infty$. We define inductively: if the integer i is one of the a's, take the ith observation, x_i, from population A, if i is one of the b's, take x_i from B, and if i is neither one of the a's nor one of the b's, take x_i from A or B according as the arithmetic mean of all previous observations from A exceeds or does not exceed the arithmetic mean of all previous observations from B. It can be shown to follow from the strong law of large numbers that with probability 1,

$$\lim_{n\to\infty} \frac{S_n}{n} = \max(\alpha, \beta). \tag{14}$$

This in turn can be shown to imply the relation

$$\lim_{n\to\infty} \mathrm{E}\left(\frac{S_n}{n}\right) = \max(\alpha, \beta), \tag{15}$$

so that

$$\lim_{n\to\infty} L_n(A, B, \overline{\mathrm{R}}) = \max(\alpha, \beta) - \lim_{n\to\infty} \mathrm{E}\left(\frac{S_n}{n}\right) = 0 \tag{16}$$

for any A, B such that α, β exist.

3. Some other problems of sequential design. The problem of

§2 can be generalized in various ways. For one thing, we can let the total sample size n be a random variable, either independent of the observations or dependent on them. As an example of the latter case, suppose in the problem of the two coins that we have to pay a fixed amount c for the privilege of making each toss. We may then decide to stop tossing whenever it seems pretty certain that $\max(\alpha, \beta) < c$; this amounts to a special case of a problem of three populations. We can even consider the case of a continuum of populations. Suppose we can apply a certain treatment to some plant or animal at any intensity θ in some interval, and let $F(x, \theta)$ be the cumulative distribution function of the response x to the treatment of intensity θ. The expected value

$$\alpha(\theta) = \int_{-\infty}^{\infty} x dF(x, \theta), \tag{17}$$

the "regression" of x on θ, is assumed to be unknown. Let $\{\theta_i\}$ denote any sequence of θ values, chosen sequentially by the experimenter, and let $\{x_i\}$ denote the corresponding sequence of responses, so that each x_i has the distribution $\Pr[x_i \leq x] = F(x, \theta_i)$. (I) Suppose $\alpha(\theta)$ has a unique maximum at some unknown point θ_0. How should the experimenter choose the sequence $\{\theta_i\}$ in order to maximize the expected value of the sum $S_n = x_1 + \cdots + x_n$ or, alternatively, in order to estimate the value of θ_0? (II) Suppose $\alpha(\theta)$ is an increasing function of θ which takes on a given value α_0 at some unknown point θ_0. How should the experimenter choose the sequence $\{\theta_i\}$ in order to estimate the value of θ_0? Problem I is the problem of the experimental determination of the maximum of a function when the observations are subject to a random error; Problem II is fundamental in sensitivity testing and bioassay.

It is clear that in both of these problems the choice of each θ_i should be made to depend on the responses $x_1, \cdots, x_{i-1}$ at the previous levels $\theta_1, \cdots, \theta_{i-1}$ of the treatment, so that we are dealing with problems of sequential design. The non-sequential study of Problem I was initiated by Hotelling [7] (see also [8]), but no sequential theory has yet been published. Problem II has been considered by Robbins and Monro [9]; their method is as follows. Let $\{a_n\}$ be a sequence of positive constants such that

$$\sum_1^{\infty} a_n^2 < \infty, \qquad \sum_1^{\infty} a_n = \infty, \tag{18}$$

let θ_1 be arbitrary, and set

$$\theta_{n+1} = \theta_n + a_n(\alpha_0 - x_n) \qquad (n = 1, 2, \cdots). \tag{19}$$

Then under certain mild restrictions on $F(x, \theta)$ it can be shown that

$$\lim_{n\to\infty} \theta_n = \theta_0 \quad \text{in probability.} \tag{20}$$

In this and other problems, any sequential design with reasonably good properties is likely to find an appreciative audience. This will encourage the use of random sampling methods to find empirical approximations to the operating characteristics of sequential designs when a full mathematical solution is difficult. An empirical study of the rapidity of convergence in (20) has been made by Teichroew [10].

4. The problem of optional stopping. To fix the ideas, let x be normally distributed with unknown mean θ and unit variance. Suppose we wish to test the null hypothesis, H_0, that $\theta = 0$ against the alternative, H_1, that $\theta > 0$. The standard statistical test based on a sample of fixed size n is the following. Let $S_n = x_1 + \cdots + x_n$ and reject H_0 in favor of H_1 if and only if

$$S_n > \alpha n^{1/2}, \tag{21}$$

where α is some constant. The probability of rejecting H_0 when it is true will then be

$$\epsilon(\alpha) = 1 - \Phi(\alpha), \tag{22}$$

where we have set

$$\Phi(x) = \frac{1}{(2\pi)^{1/2}} \int_{-\infty}^{x} e^{-t^2/2} dt, \tag{23}$$

and by choosing α large we can make $\epsilon(\alpha)$ as small as we please. For example, if $\alpha = 3.09$ then $\epsilon(\alpha) \cong .001$.

Suppose now that H_0 is true but that an unscrupulous experimenter wishes to get an unwary statistician to reject it. If the sample size n has not been agreed upon in advance the experimenter could adopt the technique of stopping the sampling as soon as the inequality (21) is verified. The law of the iterated logarithm of probability theory implies that with probability 1 the inequality (21) will hold for infinitely many values of n if the sampling continues indefinitely, no matter how large the value of α. Hence the experimenter is "sure" to come eventually to a value of n for which (21) holds, and by stopping the experiment at this point he will cause the statistician to reject H_0 even though it is true. This fact immediately vitiates the use of (21) as a test of H_0 if there is any possibility that optional stopping may be involved.

The simplest way for the statistician to guard against the effect of optional stopping is to insist that the size of the sample be fixed in advance of the experimentation. Such a restriction would often be too rigid for practical use. The statistician might therefore content himself with setting limits $n_1 \leqq n \leqq n_2$ on the sample size which will be flexible enough to meet the contingencies of experimentation but narrow enough to eliminate the worst effects of optional stopping. To this end the statistician would like to know the value of the function

$$g(n_1, n_2, \alpha) = \Pr \left[S_n > \alpha n^{1/2} \text{ for some } n_1, \leqq n \leqq n_2\right], \tag{24}$$

where the x_i are independent and normal (0, 1). It is quite easy to establish the inequality

$$g(n_1, n_2, \alpha) < \frac{1 - \Phi(\alpha)}{1 - \Phi\left(\alpha \cdot \frac{\lambda^{1/2} - 1}{(\lambda - 1)^{1/2}}\right)}, \text{ where } \lambda = \frac{n_2}{n_1}, \tag{25}$$

which is useful when λ is not too large, and sharper inequalities can no doubt be devised.

The problem of optional stopping has received little attention in statistical theory. (See, however, [11], especially pp. 286–292.) One need not assume that the experimenter is consciously trying to deceive the statistician—the two are often the same person—to recognize the desirability of devising methods of statistical analysis which would be relatively insensitive to the effects of optional stopping.

Bibliography

1. H. F. Dodge and H. G. Romig, *A method of sampling inspection*, Bell System Technical Journal vol. 8 (1929) pp. 613–631, and *Single sampling and double sampling inspection tables*, ibid. vol. 20 (1941) pp. 1–61.
2. A. Wald, *Sequential analysis*, New York, Wiley, 1947.
3. P. C. Mahalanobis, *A sample survey of the acreage under jute in Bengal with discussion of planning of experiments*, Snakhyā vol. 4 (1940) pp. 511–531.
4. ———, *On large-scale sample surveys*, Philos. Trans. Roy. Soc. London, Ser. B vol. 231 (1944) pp. 329–451.
5. A. Wald, *Statistical decision functions*, New York, Wiley, 1950.
6. J. Neyman, *First course in probability and statistics*, New York, Holt, 1950.
7. H. Hotelling, *Experimental determination of the maximum of a function*, Ann. of Math. Statist. vol. 12 (1941) pp. 20–45.
8. M. Friedman and L. J. Savage, *Planning experiments seeking maxima*, Chap. 3 of *Selected techniques of statistical analysis*, New York, McGraw-Hill, 1947.
9. H. Robbins and S. Monro, *A stochastic approximation method*, Ann. of Math. Statist. vol. 22 (1951) pp. 400–407.
10. D. Teichroew, unpublished.
11. W. Feller, *Statistical aspects of ESP*, Journal of Parapsychology vol. 4 (1940) pp. 271–298.

University of North Carolina

Reprinted from the Proceedings of the NATIONAL ACADEMY OF SCIENCES,
Vol. 42, No. 12, pp. 920–923. December, 1956.

A SEQUENTIAL DECISION PROBLEM WITH A FINITE MEMORY*

BY HERBERT ROBBINS

COLUMBIA UNIVERSITY

Communicated by Paul A. Smith, October 1, 1956

1. *Summary.*—We consider the problem of successively choosing one of two ways of action, each of which may lead to success or failure, in such a way as to maximize the long-run proportion of successes obtained, the choice each time being based on the results of a fixed number of the previous trials.

2. *Introduction.*—An experimenter has two coins, coin 1 and coin 2, with respective probabilities of coming up heads equal to $p_1 = 1 - q_1$ and $p_2 = 1 - q_2$, the values of which are unknown to him. He wishes to carry out an infinite sequence of tosses, at each toss using either coin 1 or coin 2, in such a way as to maximize the long-run proportion of heads obtained. His problem is to find a rule for deciding at each stage, on the basis of the results of the previous tosses, whether to use coin 1 or coin 2 for the next toss.

If he knew at the outset which coin had the larger p-value, he would of course use it exclusively, irrespective of the results of the tosses, and by doing so would know that with probability 1

$$\lim_{n \to \infty} \frac{\text{number of heads in first } n \text{ tosses}}{n} = \max (p_1, p_2). \tag{1}$$

We have shown elsewhere[1] that even without a knowledge of the p-values there exist rules for carrying out the sequence of tosses such that equation (1) holds for all p_1, p_2. However, any rule with this property must clearly be such that the decision as to which coin to use for the nth toss depends on the results of all the first $n - 1$ tosses; in other words, the rule requires a "memory" of unlimited length. We shall be interested here in seeing what can be done with rules requiring only a *finite* memory: a rule will be said to be of type r if the decision as to which coin to use for the nth toss depends only on the results of tosses $n - r, n - r + 1, \ldots, n - 1$. (By the "result" of a toss we mean both which coin was used and which face came up.) We shall exhibit a rule R_r, of type r, for which with probability 1

$$\lim_{n \to \infty} \frac{\text{number of heads in first } n \text{ tosses}}{n} = \frac{p_1 q_2^r + p_2 q_1^r}{q_1^r + q_2^r}. \tag{2}$$

Note that as $r \to \infty$ the right-hand side of equation (2) steadily increases and tends to the right-hand side of equation (1) as limit.

It would be interesting to know whether our rule R_r is "best" in the sense that the right-hand side of equation (2) is at least as great, for all p_1, p_2, as the corresponding function for any other rule of type r for which the left-hand side of equation (2) exists and is a symmetric function of p_1, p_2. We do not know whether this is true.

3. *Definition of the Rule R_r and Proof of the Theorem.*—We recall[2] that if a single coin, with probability $q = 1 - p$ of obtaining tails on each toss, is tossed repeatedly until the first run of r consecutive tails occurs, then the expected number of tosses required is

$$\frac{1 - q^r}{pq^r}.$$

Now suppose the coin is tossed repeatedly with the following stopping rule: stop if the first toss is tails, otherwise continue tossing until the first run of r consecutive tails occurs. Then the expected number of tosses required will be

$$q \cdot 1 + p \cdot \left(1 + \frac{1 - q^r}{pq^r}\right) = \frac{1}{q^r}. \tag{3}$$

We now prove the following:

THEOREM. *Define the rule R_r as follows: start tossing with coin 1. Stop if the first toss is tails, otherwise continue tossing until the first run of r successive tails occurs and then stop. This defines the first block of tosses with coin 1. Now start tossing with coin 2 and apply the same rule, obtaining the first block of tosses with coin 2. Then start again with coin 1 and apply the same rule, obtaining the second block of tosses with coin 1, and so on indefinitely, thus generating an infinite sequence of tosses consisting of alternate blocks of tosses with coins 1 and 2.*

With rule R_r so defined, we assert that equatian (2) hold with probability 1.

Proof: Let $x_i(y_i)$ denote the length of the ith block of tosses with coin 1 (2), $i = 1, 2, \ldots$. The process of tossing generates with probability 1 an infinite sequence of independent random variables

$$x_1, y_1, x_2, y_2, \ldots, x_n, y_n, \ldots. \tag{4}$$

The proportion of times that coin 1 is used during the first $2n$ *blocks* of tosses is then

$$\frac{x_1 + \ldots + x_n}{x_1 + \ldots + x_n + y_1 + \ldots + y_n} = \frac{(x_1 + \ldots + x_n)/n}{(x_1 + \ldots + x_n)/n + (y_1 + \ldots + y_n)/n}, \tag{5}$$

and, by the strong law of large numbers and equation (3), this tends with probability 1 to the limit

$$\frac{1/q_1^r}{1/q_1^r + 1/q_2^r} = \frac{q_2^r}{q_1^r + q_2^r}. \tag{6}$$

Now let u_n denote the proportion of heads obtained in the first $2n$ blocks of tosses; then

u_n = (proportion of times coin 1 is used in first $2n$ blocks of tosses)·(proportion of times heads occurs among these tosses with coin 1) + a similar product for coin 2, (7)

so that with probability 1

$$\lim_{n \to \infty} u_n = \frac{q_2^r}{q_1^r + q_2^r} \cdot p_1 + \frac{q_1^r}{q_1^r + q_2^r} \cdot p_2 = \frac{p_1 q_2^r + p_2 q_1^r}{q_1^r + q_2^r}. \tag{8}$$

Now let n be any positive integer, and define the random integer $N = N(n)$ so that

$$x_1 + y_1 + \ldots + x_N + y_N \leq n < x_1 + y_1 + \ldots + x_{N+1} + y_{N+1}. \tag{9}$$

Denoting by w_n the number of heads obtained among the first n tosses, we have

number of heads in first $2N$ blocks $\leq w_n \leq$ number of heads in first $2N$ + 2 blocks, (10)

and from (9) and (10) it follows that

$$\frac{u_N(x_1 + y_1 + \ldots + x_N + y_N)}{x_1 + y_1 + \ldots + x_{N+1} + y_{N+1}} \leq \frac{w_n}{n} \leq u_{N+1} \frac{(x_1 + y_1 + \ldots + x_{N+1} + y_{N+1})}{x_1 + y_1 + \ldots + x_N + y_N}. \tag{11}$$

But since $(x_{N+1} + y_{N+1})/N$ tends to 0 with probability 1 as n, and therefore N, becomes infinite, it follows that with probability 1

$$\lim_{n\to\infty} \frac{x_1 + y_1 + \ldots + x_{N+1} + y_{N+1}}{x_1 + y_1 + \ldots + x_N + y_N} = 1,$$

and hence with probability 1, from relations (8) and (11),

$$\lim_{n\to\infty} w_n/n = \lim_{N\to\infty} u_N = (p_1 q_2^r + p_2 q_1^r)/(q_1^r + q_2^r), \tag{12}$$

which was to be proved.

The author is indebted to John W. Tukey for a helpful remark which simplified the preceding proof.

* Work sponsored by the Office of Scientific Research of the Air Force, Contract No. AF18(600)-442, Project No. R-345-20-7.

[1] H. Robbins, "Some Aspects of the Sequential Design of Experiments," *Bull. Am. Math. Soc.*, 58, 529–532, 1952.

[2] W. Feller, *An Introduction to Probability Theory and Its Applications*, 1, (New York, 1950), 266.

Proc. Natl. Acad. Sci. USA
Vol. 81, pp. 1284–1286, February 1984
Statistics

Optimal sequential sampling from two populations

(two-armed bandit/adaptive allocation)

T. L. LAI AND HERBERT ROBBINS

Department of Statistics, Columbia University, New York, NY 10027

Contributed by Herbert Robbins, October 31, 1983

ABSTRACT Given two statistical populations with unknown means, we consider the problem of sampling $x_1, x_2, \ldots$ sequentially from these populations so as to achieve the greatest possible expected value of the sum $S_n = x_1 + \ldots + x_n$. In particular, for normal populations, we obtain the optimal rule and study its properties when the average of the two population means is assumed known, and exhibit an asymptotically optimal rule without assuming any prior knowledge about the population means.

Introduction

Let A and B denote two statistical populations with means $\mu_A \neq \mu_B$. How should we sample $x_1, x_2, \ldots$ sequentially from these populations if our objective is to achieve the greatest possible expected value of the sum $S_n = x_1 + \ldots + x_n$ as $n \to \infty$? At each stage the choice of A or B is allowed to depend on the previous observations. Let $f(g)$ denote the density, with respect to some measure ν, of the population with the larger (smaller) mean, and let $T_n(f)$ ($T_n(g)$) denote the number of observations taken from this population through stage n, so that $T_n(f) + T_n(g) = n$. Since

$$ES_n = \left(\int_{-\infty}^{\infty} xf(x)\,d\nu(x)\right)ET_n(f) + \left(\int_{-\infty}^{\infty} xg(x)\,d\nu(x)\right)ET_n(g)$$

$$= n \max(\mu_A, \mu_B) - |\mu_A - \mu_B| ET_n(g),$$

the problem of maximizing ES_n is equivalent to that of minimizing the expected sample size $ET_n(g)$ from the inferior population.

The Optimal Sampling Rule When f and g Are Known

Suppose that we know f and g. Let ψ_A and ψ_B denote the density functions of the populations A and B, respectively. At each stage it seems natural to test the hypothesis H_0: $(\psi_A, \psi_B) = (f,g)$ versus the alternative H_1: $(\psi_A, \psi_B) = (g,f)$ on the basis of the previous observations, and to sample from A if H_0 is (tentatively) accepted and to sample from B otherwise. This suggests the following rule, denoted by ρ^*. Let $y_1, y_2, \ldots$ denote successive observations from A, and let $z_1, z_2, \ldots$ denote successive observations from B. At the first stage choose A or B with probability ½ each. Let

$$L_n = \left\{\prod_{i=1}^{T(n,A)} f(y_i) \prod_{j=1}^{T(n,B)} g(z_j)\right\} \Big/ \left\{\prod_{i=1}^{T(n,A)} g(y_i) \prod_{j=1}^{T(n,B)} f(z_j)\right\},$$

where $T(n,A)$ and $T(n,B)$ denote the number of observations through stage n from A and B, respectively, and sample at stage $n + 1$ from A or B according as $L_n > 1$ or $L_n < 1$, choosing A or B with probability ½ when $L_n = 1$. Note that L_n is the usual likelihood ratio statistic in favor of H_0 (versus H_1).

For ρ^*, $E_0T_n(g) = E_1T_n(g)$ by symmetry, and as shown by Feldman (1), $E_iT_n(g)$ is bounded in n. Feldman (1) also showed that ρ^* is optimal, in the sense of minimizing (over the class of all allocation rules)

$$E_0T_n(g) + E_1T_n(g)$$

for every $n \leq \infty$. Hence, ρ^* is Bayes with respect to the prior distribution of equal probabilities for H_0 and H_1. We now compute $E_iT_\infty(g)$ for this optimal rule in two cases of interest.

Example 1. Suppose that f and g are normal densities with unit variance and means $\theta_1 > \theta_2$, and let $\theta = (\theta_1 + \theta_2)/2$. Then, $\log L_n = (\theta_1 - \theta_2)W_n$, where

$$W_n = \sum_{i=1}^{T(n,A)} (y_i - \theta) + \sum_{j=1}^{T(n,B)} (\theta - z_j). \quad [1]$$

The rule ρ^*, therefore, samples at stage $n + 1$ from A or B according as $W_n > 0$ or $W_n < 0$. Under H_0, $y_1 - \theta, y_2 - \theta, \ldots, \theta - z_1, \theta - z_2, \ldots$ are i.i.d. (independent, identically distributed) normal random variables with unit variance and mean $\delta = (\theta_1 - \theta_2)/2$, and therefore,

$$E_0T_n(g) = ½ + \sum_{i=1}^{n-1} P_0\{W_i < 0\}$$

$$= ½ + \sum_{i=1}^{n-1} \Phi(-\delta\sqrt{i}),$$

where Φ denotes the standard normal distribution function. Hence, for ρ^*

$$E_0T_\infty(g) = E_1T_\infty(g) = h(\delta), \text{ where } h(\delta) = ½ + \sum_1^\infty \Phi(-\delta\sqrt{i}).$$

Clearly, $h(\delta) \to ½$ as $\delta \to \infty$, while as $\delta \to 0$,

$$h(\delta) \sim \int_0^\infty \Phi(-\delta\sqrt{t})dt = 2\delta^{-2}\int_0^\infty u\Phi(-u)du = ½\delta^{-2}. \quad [2]$$

Some numerical values of $h(\delta)$ are shown in Table 1.

Table 1. Numerical values of $h(\delta)$

δ	3	2	1	0.7	0.5	0.3	0.2	0.1	0.05
$h(\delta)$	0.50	0.52	0.83	1.32	2.28	5.82	12.8	50.2	200.2
$\delta^2 h(\delta)$	4.51	2.09	0.83	0.65	0.57	0.52	0.51	0.502	0.5006

The publication costs of this article were defrayed in part by page charge payment. This article must therefore be hereby marked "*advertisement*" in accordance with 18 U.S.C. §1734 solely to indicate this fact.

Example 2. Suppose that f and g are Bernoulli densities (with respect to counting measure) such that

$$f(1) = g(0) = p, \quad f(0) = g(1) = q,$$

where $\frac{1}{2} < p < 1$ and $q = 1 - p$. Then, $\log L_n = (\log p/q)W_n'$, where

$$W_n' = \sum_{i=1}^{T(n,A)} (2y_i - 1) + \sum_{j=1}^{T(n,B)} (1 - 2z_i). \quad [3]$$

Under H_1, $2y_1 - 1, 2y_2 - 1, \ldots, 1 - 2z_1, 1 - 2z_2, \ldots$ are i.i.d. random variables assuming the values 1 (with probability q) and -1 (with probability p). For $k = 0, \pm 1, \ldots$, let $N(k)$ denote the number of n such that $W_n' = k$. The renewal measure of the simple random walk $\{W_n'\}$ is given by

$$1 + E_1N(0) = 1/(p - q) = E_1N(k) \text{ for } k = -1, -2, \ldots,$$
$$E_1N(k) = (q/p)^k(1 + E_1N(0)) \text{ for } k = 1, 2, \ldots \quad [4]$$

noting that (cf. ref. 2)

$$P_1\{W_n' = k \text{ for some } n \geq 1\} = 1 \text{ if } k < 0,$$
$$= (q/p)^k \text{ if } k > 0.$$

Since at stage $n + 1$ the rule ρ^* samples from A or B according as $W_n' > 0$ or $W_n' < 0$ and chooses A or B with probability $\frac{1}{2}$ each if $W_n' = 0$, it follows from Eq. 4 that

$$E_1T_\infty(g) = \tfrac{1}{2}(1 + E_1N(0)) + \sum_{k=1}^{\infty} E_1N(k) = \frac{1}{2(p - q)^2}. \quad [5]$$

Therefore,

$$E_0T_\infty(g) = E_1T_\infty(g) = \tfrac{1}{2}\delta^{-2}, \text{ where } \delta = p - q,$$

in agreement with the asymptotic result (see Eq. 2).

When $L_n = 1$, the posterior probability in favor of H_0 is the same as that in favor of H_1, so the Bayes rule can choose A or B arbitrarily at stage $n + 1$ (cf. ref. 1). In particular, the following modification of ρ^* is also Bayes. At the first stage, choose A or B with probability $\frac{1}{2}$. If A is chosen, continue sampling from A until stage $n(1) = \inf\{n: L_n < 1\}$, and sample from B at stage $n(1) + 1$. Continue sampling from B until stage $n(2) = \inf\{n > n(1): L_n \geq 1\}$, and then switch back to A until stage $n(3) = \inf\{n > n(2): L_n < 1\}$, etc. If B is chosen at the first stage, continue sampling from B until stage $n'(1) = \inf\{n: L_n > 1\}$, and then sample from A until stage $n'(2) = \inf\{n > n'(1): L_n \leq 1\}$, etc. This rule will be denoted by $\hat{\rho}$. By symmetry, $E_0T_n(g) = E_1T_n(g)$ for both $\hat{\rho}$ and ρ^*. Since $\hat{\rho}$ and ρ^* are Bayes rules, they have the same value of $E_iT_\infty(g)$.

In *Examples 1* and *2*, under both H_0 and H_1, $\{\log L_n\}$ is a random walk whose distribution does not depend on the sampling rule, and this enabled us to obtain explicit formulas for $E_iT_\infty(g)$. In general, in analogy with Eqs. 1 and 3, we have

$$\log L_n = \sum_{i=1}^{T(n,A)} h(y_i) - \sum_{j=1}^{T(n,B)} h(z_i), \quad [6]$$

where $h = \log(f/g)$. However, $h(y_i)$ and $-h(z_i)$ need no longer have the same distribution under H_0 or H_1. An upper bound for $E_0T_\infty(g) = E_1T_\infty(g)$ under the optimal rule $\hat{\rho}$ is obtained by considering a simpler rule ρ, as described below.

At the first stage, choose A or B with probability $\frac{1}{2}$ each. If A is chosen, continue sampling from A until stage $\tau_1 = \inf\{n: \Sigma_1^n h(y_i) < 0\}$ ($\inf \varnothing = \infty$), and then sample from B at stage $\tau_1 + 1$. Continue sampling from B up to σ_1 observations, where $\sigma_1 = \inf\{n: \Sigma_1^n h(z_i) < 0\}$, and switch to A at stage $\tau_1 + \sigma_1 + 1$. Proceeding inductively in this way, the rule ρ is defined by "switching times" at stages $\tau_1 + 1$, $\tau_1 + \sigma_1 + 1$, $\tau_1 + \sigma_1 + \tau_2 + 1, \ldots$, where

$$\tau_{j+1} = \inf\left\{n: \sum_{i=\tau_j+1}^{\tau_j+n} h(y_i) < 0\right\}, \quad \sigma_{j+1} = \inf\left\{n: \sum_{i=\sigma_j+1}^{\sigma_j+n} h(z_i) < 0\right\}.$$

If B is chosen at the first stage, then the switching times of ρ occur at stages $\sigma_1 + 1$, $\sigma_1 + \tau_1 + 1$, $\sigma_1 + \tau_1 + \sigma_2 + 1, \ldots$. For the Bernoulli distributions of *Example 2*, since $h(y_i)$ and $h(z_i)$ assume the values $\pm\log p/q$, $\log L_{n(1)} = \log L_{n(3)} = \ldots = -\log p/q$, $\log L_{n(2)} = \log L_{n(4)} = \ldots = 0$, $\log L_{n'(1)} = \log L_{n'(3)} = \ldots = \log p/q$, and $\log L_{n'(2)} = \log L_{n'(4)} = \ldots = 0$. Consequently, the optimal rule $\hat{\rho}$ in this case is the same as the rule ρ. The rule ρ ignores the effect of "overshoots" in the Bayes rule $\hat{\rho}$ at stages $n(1), n(2), \ldots, n'(1), n'(2), \ldots$. By analyzing these overshoots, we can modify the argument to obtain a lower bound for $E_0T_\infty(g) = E_1T_\infty(g)$ under the optimal rule $\hat{\rho}$. This is the content of *Theorem 1*.

THEOREM 1. *Let* $u, u_1, u_2, \ldots$ *be i.i.d. random variables with density* f, *and let* $v, v_1, v_2, \ldots$ *be i.i.d. random variables with density* g. *Let* $h = \log(f/g)$. *Then* $Eh(u) > 0$ *and* $Eh(v) < 0$. *Let* $U_m = \Sigma_1^m h(u_i)$, $V_m = \Sigma_1^m h(v_i)$. *Define*

$$e_v = E(\inf\{m \geq 1: V_m < 0\}),$$

$$p_U(a) = P\{U_m < a \text{ for some } m \geq 1\}, \quad p_U = p_U(0),$$

$$\pi_V(a) = \inf_{t \leq 0} P\{a + t < h(v) < t \mid h(v) < t\}.$$

Then for the optimal rule ρ^*,

$$\frac{1 + p_U}{2(1 - p_U)} e_v \geq E_0T_\infty(g) = E_1T_\infty(g)$$

$$\geq \frac{1 + p_U(a)\pi_V(a)}{2(1 - p_U(a)\pi_V(a))} e_v$$

for every $a < 0$.

Proof: To derive the upper bound, note that for the suboptimal rule ρ,

$$E_0T_\infty(g) = \tfrac{1}{2}\left\{E_0\sigma_1 + \int_{\{\sigma_1+\tau_1<\infty\}} \sigma_2 dP_0 + \ldots\right\}$$

$$+ \tfrac{1}{2}\left\{\int_{\{\tau_1<\infty\}} \sigma_1 dP_0 + \ldots\right\}$$

$$= \tfrac{1}{2}e_v + (p_U + p_U^2 + \ldots)e_v = \frac{1 + p_U}{2(1 - p_U)} e_v.$$

Since $E_0T_\infty(g) = E_1T_\infty(g)$ for both the rule ρ and the Bayes rule $\hat{\rho}$, the upper bound is established.

To derive the lower bound, suppose that H_0 is true. Then, $u_i = y_i$, $v_i = z_i$. Define $\sigma_1 = \inf\{n: V_n < 0\}$ as before, and let $R_1 = V_{\sigma_1}(<0)$. Define $\tau_1^* = \inf\{n: U_n - R_1 \leq 0\}$ and let $r_1 = U_{\tau_1^*} - R_1$ (≤ 0). Letting $V_{m,n} = \Sigma_{i=m+1}^{m+n} h(v_i)$ and $U_{m,n} = \Sigma_{i=m+1}^{m+n} h(u_i)$, define $\sigma_2^* = \inf\{n: V_{\sigma_1,n} - r_1 < 0\}$, $R_2 = V_{\sigma_1,\sigma_2} - r_1(\leq 0)$, $\tau_2^* = \inf\{n: U_{\tau_1^*,n} - R_2 \leq 0\}$, etc. We note that for any $a < 0$

$E_0[T_\infty(g)|\text{B is chosen at the first stage}]$

$$= E_0\sigma_1 + \int_{\{\sigma_1+\tau_1^*<\infty\}} \sigma_2^* \, dP_0 + \int_{\{\sigma_1+\tau_1^*+\sigma_2^*+\tau_2^*<\infty\}} \sigma_3^* \, dP_0 + \ldots$$

$$\geq e_v + P_0\{R_1 \geq a\}\, P_0\{U_n \leq a \text{ for some } n\}\, e_v$$

$$+ P_0\{R_1 \geq a\}\, P_0\{R_2 \geq a\}\, P_0^2\{U_n \leq a \text{ for some } n\}\, e_v + \ldots$$

$$\geq e_V + \pi_V(a)p_U(a)e_V + \pi_V^2(a)p_U^2(a)e_V + \ldots$$

A similar argument also shows that for any $a \leq 0$

$E_0[T_\infty(g)|\text{A is chosen at the first stage}]$

$$\geq p_U e_V + p_U\pi_V(a)p_U(a)e_V + p_U\pi_V^2(a)p_U^2(a) + \ldots$$

Hence, $E_0T_\infty(g) \geq \tfrac{1}{2} e_v + \{\pi_V(a)p_U(a) + \pi_V^2(a)p_U^2(a) + \ldots\} \times e_V$, and the lower bound is established.

For the Bernoulli distributions of *Example 2*, $e_V = (p - q)^{-1}$, $p_U = q/p$ (cf. ref. 2), and therefore the upper bound of *Theorem 1* reduces to

$$\frac{1 + p_U}{2(1 - p_V)}\, e_V = \frac{1}{2(p - q)^2},$$

which is in agreement with Eq. 5.

Normal Populations with Unknown Means

Suppose that A and B are normal populations with unit variance and means $\mu_A \neq \mu_B$. If $\bar\theta = (\mu_A + \mu_B)/2$ is known, then we can use the rule ρ^*, which chooses A or B at stage $n + 1$ according as $W_n > 0$ or $W_n < 0$, where

$$W_n = \sum_{i=1}^{T(n,A)} (y_i - \bar\theta) + \sum_{j=1}^{T(n,B)} (\bar\theta - z_j).$$

When $\bar\theta$ is unknown, we can take one observation from A and one from B at the first two stages and then try estimating $\bar\theta$ at stage n (≥ 2) by either

$$\hat\theta_n = \tfrac{1}{2}(\bar y_{T(n,A)} + \bar z_{T(n,B)})$$

(where $\bar a_n$ denotes the arithmetic mean of n numbers $a_1, \ldots, a_n$), or

$$\theta_n^* = \left(\sum_{i=1}^{T(n,A)} y_i + \sum_{j=1}^{T(n,B)} z_j\right)\Big/ n.$$

If we replace $\bar\theta$ by $\hat\theta_n$ in W_n, we get

$$\sum_{i=1}^{T(n,A)} (y_i - \hat\theta_n) + \sum_{j=1}^{T(n,B)} (\hat\theta_n - z_j) = \frac{n}{2}(\bar y_{T(n,A)} - \bar z_{T(n,B)}).$$

If we replace $\bar\theta$ by θ_n^* in W_n, we get

$$\sum_{i=1}^{T(n,A)} (y_i - \theta_n^*) + \sum_{j=1}^{T(n,B)} (\theta_n^* - z_j)$$

$$= \frac{2}{n} T(n,A)T(n,B)(\bar y_{T(n,A)} - \bar z_{T(n,B)}).$$

Hence replacing $\bar\theta$ in W_n by $\hat\theta_n$ or θ_n^* in the rule ρ^* leads to the "sample from the leader" rule. At stage $n + 1$ sample from A or B according as $\bar y_{T(n,A)} > \bar z_{T(n,B)}$ or $\bar y_{T(n,A)} < \bar z_{T(n,B)}$.

The difficulty with the "sample from the leader" rule is that we may have sampled too little from the apparently inferior population to get a reliable estimate of its mean, and we may thereby miss the actually superior population. In fact, the expected sample size from the inferior population is of order n: for $\mu_B > \mu_A$

$$E_{\mu_A,\mu_B}\, T(n,A) > (n - 1)p, \text{ where}$$

$$p = P_{\mu_A,\mu_B}\{\bar y_i > z_1 \text{ for all } i \geq 1\} > 0.$$

Some insight into how much we need to sample from the apparently inferior population is provided by studying the finite-horizon problem, where a preassigned total of N observations are to be taken, and we know one of the means, say $\mu_B = \mu^*$, but not the other. In this case, since sampling from B does not add any information about μ_B, we need only *sample for information* from A. However, because our objective is to maximize ES_N, we should stop sampling from A once we are reasonably confident that $\mu_A < \mu^*$, and then switch to B, whereupon no more information is gained about μ_A or μ_B. Hence, we can reduce the *allocation* problem to a *stopping* problem involving A alone, restricting ourselves to rules that sample from A until some stopping time $\tau_N(\leq N)$, whereupon H: $\mu_A > \mu^*$ is rejected in favor of K: $\mu_A < \mu^*$ and the remaining $N - \tau_N$ observations are taken from B (cf. ref. 3). Because $\{\bar y_i\}$ is a sequence of sufficient statistics, we can also restrict ourselves to stopping rules of the form

$$\tau_N = \inf\{i \leq N: \bar y_i - \mu^* \leq -a_{Ni}\}(\inf \varnothing = N), \qquad [7]$$

where a_{Ni} ($i = 1, \ldots, N$) are positive constants.

We now consider the case where both μ_A and μ_B are unknown and there is no preassigned horizon N. Start with one observation from A and one from B. At the conclusion of stage $n \geq 2$, if $T(n,B) > T(n,A)$, then $T(n,B) > n/2$, so the amount of information from B is relatively adequate, while there is more need to sample for information from A. Replacing N by n and μ^* by $\bar z_{T(n,B)}$ in Eq. 7, analogy with Eq. 7 suggests sampling at stage $n + 1$ from A if

$$\bar y_{T(n,A)} > \bar z_{T(n,B)} - a_{n,T(n,A)},$$

and sampling from B otherwise. Likewise, in the case $T(n,A) > T(n,B)$, we sample from B at stage $n + 1$ if

$$\bar z_{T(n,B)} > \bar y_{T(n,A)} - a_{n,T(n,B)},$$

and we sample from A otherwise. Finally, in the case $T(n,A) = T(n,B) = n/2$, we sample from B only if $\bar z_{T(n,B)} \geq \bar y_{T(n,A)}$. This sampling rule will be denoted by ρ_0.

If the constants a_{ni} are such that for every fixed i, a_{ni} is nondecreasing in $n \geq i$, and if there exist $\varepsilon_n \to 0$ for which

$$|a_{ni} - (\log n)/i| \leq \varepsilon_n(\log n)^{1/2}/i^{1/2} \text{ for all } i \leq n, \qquad [8]$$

then it can be shown for the rule ρ_0 that as $n \to \infty$

$$E_{\mu_A,\mu_B}T(n,A) \sim (2 \log n)/(\mu_A - \mu_B)^2 \text{ if } \mu_B > \mu_A,$$

$$E_{\mu_A,\mu_B}T(n,B) \sim (2 \log n)/(\mu_A - \mu_B)^2 \text{ if } \mu_A > \mu_B. \qquad [9]$$

Moreover, the rule ρ_0 is *asymptotically optimal* in the sense that $\liminf E_{\mu_A,\mu_B}T_n(g)/\log n \geq 2(\mu_A - \mu_B)^{-2}$ for every rule ρ such that $E_{\mu_A,\mu_B}\bar T_n(g) = O(\log n)$ at all parameter values $\mu_A \neq \mu_B$. The proof of these assertions is given in ref. 4.

This research was supported by the National Science Foundation and the National Institutes of Health.

1. Feldman, D. (1962) *Ann. Math. Stat.* **33**, 847–856.
2. Feller, W. (1966) *An Introduction to Probability Theory and Its Applications* (Wiley, New York), Vol. 1, 2nd Ed.
3. Bradt, R. N., Johnson, S. M. & Karlin, S. (1956) *Ann. Math. Stat.* **27**, 1060–1070.
4. Lai, T. L. & Robbins, H. (1984) *Adv. Appl. Math.*, in press.

Reprinted from
Proc. Nat. Acad. Sci. USA
Vol. 69, No. 10, pp. 2993–2994, October 1972

Reducing the Number of Inferior Treatments in Clinical Trials

(theoretical/statistical/experimental design/sequential trials)

B. J. FLEHINGER*, T. A. LOUIS, HERBERT ROBBINS, AND B. H. SINGER

Columbia University, New York, N.Y. 10027, and * IBM Research Center, Yorktown Heights, New York 10598

Communicated by J. Neyman, August 10, 1972

ABSTRACT **In clinical trials comparing two treatments, one would often like to control the probability of erroneous decision while minimizing not the *total sample size* but *the number of patients given the inferior treatment.* To do this obviously requires that one use a data-dependent allocation rule for the two treatments rather than the conventional equal sample size scheme, whether fixed or sequential. We show here how this may be done in the case of deciding which of two normally distributed treatment effects has the greater mean, when the variances are assumed to be equal and known. Similar methods can be used under other hypotheses on the underlying probability distributions, and will provide a considerable increase in flexibility in the design of sequential clinical trials.**

A class of allocation rules that do not affect the error probabilities

Let $x_1,x_2,\ldots$ and $y_1,y_2,\ldots$ be independent random variables, the x's being $N(\mu_1,\sigma^2)$ and the y's $N(\mu_2,\sigma^2)$ with σ known. We wish to decide whether μ_1 is greater or less than μ_2 by sequential observation of the x's and y's.

As shown in the *Appendix*, the usual formulation of sequential analysis suggests that to decide whether μ_1 is greater or less than μ_2 we use a sequential likelihood ratio test, patterned after that of Wald in the one-population case, by choosing two positive constants A and B and agreeing to continue sampling as long as the sample means $\bar{x}_m$ and $\bar{y}_n$ satisfy

$$-A < z_{m,n} = \frac{4mn}{(m+n)\sigma}(\bar{y}_n - \bar{x}_m) < B, \qquad (1)$$

stopping with sample sizes (M,N) = first (m,n) such that (1) fails to hold, and deciding that $\mu_1 > \mu_2$ or $\mu_1 < \mu_2$ according to whether $z_{M,N} \leq -A$ or $\geq B$.

It remains to specify the allocation rule under which we decide at each stage whether to observe the next x or the next y. We shall require only that (*i*) if $x_1,\ldots,x_m,y_1,\ldots,y_n$ have been observed, the decision whether to observe next x_{m+1} or y_{n+1} shall depend only on the values of m,n, and $\bar{y}_n - \bar{x}_m$, and (*ii*) as sampling proceeds both m and n become infinite. The usual arguments then show that if for an arbitrary θ and some $\delta > 0$, H_0 denotes the hypothesis that $\mu_1 = \theta + \delta\sigma$ and $\mu_2 = \theta - \delta\sigma$ and H_1 that $\mu_1 = \theta - \delta\sigma$ and $\mu_2 = \theta + \delta\sigma$, then

$$P_0(\text{error}) < e^{-\delta B}[1 - P_1(\text{error})],$$

and similarly

$$P_1(\text{error}) < e^{-\delta A}[1 - P_0(\text{error})];$$

the inequalities in both cases being approximate equalities and due to overshoot of the stopping boundaries. Thus, in practice if we wish to make

$$P_0(\text{error}) \leq \epsilon_0,\ P_1(\text{error}) \leq \epsilon_1,$$

we may put

$$A = \frac{1}{\delta}\cdot\log\left(\frac{1-\epsilon_0}{\epsilon_1}\right),\ B = \frac{1}{\delta}\cdot\log\left(\frac{1-\epsilon_1}{\epsilon_0}\right).$$

In particular, when we choose $\epsilon_0 = \epsilon_1 = \epsilon$ then $A = B$, and in this case since $P_0(\text{error}) = P_1(\text{error})$ we have for any $B > 0$ the relation

$$P(\text{error}) < \frac{1}{1 + e^{\delta B}} \text{ when } |\mu_1 - \mu_2| = 2\delta\sigma. \qquad (2)$$

In the special case of *pairwise* sampling with $m = n$ and $A = B$ the continuation region (1) is the interval

$$-B < \frac{2}{\sigma}\sum_1^n (y_i - x_i) < B$$

and the Wald approximation for the expected sample size gives

$$E(M) = E(N) \cong \frac{B(e^{\delta B} - 1)}{4\delta(e^{\delta B} + 1)} \text{ when } |\mu_1 - \mu_2| = 2\delta\sigma. \qquad (3)$$

But by using one of the more general allocation rules described above we can reduce $E_0(N) = E_1(M)$ to a value considerably below that of (3), at the expense, of course, of increasing the expected *total* sample size. Details of some heuristically optimal allocation rules for decreasing the expected size of the sample receiving the inferior treatment below the pairwise value (3) will be given elsewhere. We remark that for all allocation rules of the type described above, it can be shown that the symmetric continuation region (1) with $A = B$ not only gives the error probability bound (2) but also the approximation

$$E\left(\frac{MN}{M+N}\right) \cong \frac{B(e^{\delta B} - 1)}{8\delta(e^{\delta B} + 1)} \text{ when } |\mu_1 - \mu_2| = 2\delta\sigma, \qquad (4)$$

which is a generalization of (3).

The present note was suggested by, and gives the mathematical explanation for, recent results [1, 2] that show by computer simulation that data-dependent allocation rules allow one to reduce the expected size of the sample receiving the inferior treatment, and also provide a simpler means of proving certain results [3] concerning the continuous time problem of testing whether the drift of one Wiener process is greater or less than that of another. The interesting results [4] of Sobel and Weiss for the case of two binomial populations can also be obtained by the methods of this note.

Appendix

For an arbitrary θ and a $\delta > 0$ let H_0 denote the hypothesis that $\mu_1 = \theta + \delta\sigma$ and $\mu_2 = \theta - \delta\sigma$ and H_1 the hypothesis that $\mu_1 = \theta - \delta\sigma$ and $\mu_2 = \theta + \delta\sigma$. The likelihood ratio of $(x_1, \ldots, x_m, y_1, \ldots, y_n)$ under H_1 to that under H_0 is then

$$\frac{\prod_1^m \varphi\left(\frac{x_i - \theta + \delta\sigma}{\sigma}\right) \prod_1^n \varphi\left(\frac{y_i - \theta - \delta\sigma}{\sigma}\right)}{\prod_1^m \varphi\left(\frac{x_i - \theta - \delta\sigma}{\sigma}\right) \prod_1^n \varphi\left(\frac{y_i - \theta + \delta\sigma}{\sigma}\right)}.$$

where $\varphi(x) = (2\pi)^{-1/2} \exp(-x^2/2)$. Introducing the new variables

$$u_1 = x_1,\ u_i = x_i - x_1\ (i \geq 2),\ v_i = y_i - x_1\ (i \geq 1)$$

and integrating out u_1, we see that the likelihood ratio of $(u_2, \ldots, u_m, v_1, \ldots, v_n)$ under H_1 to that under H_0 depends only on δ and not on θ; in fact, a little algebra shows that it is given by the expression

$$L_{m,n} = \frac{g_1(u_2, \ldots, u_m, v_1, \ldots, v_n)}{g_0(u_2, \ldots, u_m, v_1, \ldots, v_n)}$$

$$= \exp\left\{\frac{4\delta mn}{(m+n)\sigma} (\bar{y}_n - \bar{x}_m)\right\}.$$

It is of theoretical interest that under H_0 as defined above both

$$\exp\left\{\frac{4mn\delta}{(m+n)\sigma} (\bar{y}_n - \bar{x}_m)\right\}$$

and

$$\frac{mn}{m+n}\left\{\frac{(\bar{y}_n - \bar{x}_m)}{\sigma} + 2\delta\right\}$$

are two-dimensional time-parameter martingales with respect to the σ-algebras $\mathfrak{F}_{m,n}$ generated not by $x_1, \ldots, x_m, y_1, \ldots, y_n$ but by all *differences* of these variables. This forms the probabilistic basis of (2) and (4), and points the way of generalization to other parametric hypothesis testing problems, including the normal case with unknown variance.

This research was supported, in part, by PHS Grant 5-R01-GM-16895-03.

1. Flehinger, B. J. & Louis, T. A., "Sequential medical trials with data dependent treatment allocation," *Proceedings Sixth Berkeley Symposium on Probability and Statistics* (to appear).
2. Flehinger, B. J. & Louis, T. A. (1971) "Sequential treatment allocation in clinical trials," *Biometrika* **58**, 419–426.
3. Louis, T. A. (1972) "Two Population Hypothesis Testing with Data-Dependent Allocation," Ph.D. Dissertation, Columbia University.
4. Sobel, M. & Weiss, G. H. (1970) "Play-the-winner sampling for selecting the better of two binomial populations," *Biometrika* **57**, 357–365.

Sequential Tests Involving Two Populations

HERBERT ROBBINS and DAVID O. SIEGMUND*

For testing which of two normally distributed treatments with a common variance is "superior" (has a larger mean response), we give a class of sequential probability ratio tests whose error probability functions are essentially independent of the sampling rule used. It is shown that the expected number of observations on the "inferior" treatment must be at least one-half that required by pairwise sampling, and a class of sampling rules is given which for some parameter values and significance levels almost attains this theoretical bound.

1. INTRODUCTION AND SUMMARY

Let $x_1, x_2, \cdots, x_m, \cdots$ and $y_1, y_2, \cdots, y_n \cdots$ be independent normally distributed random variables with expectations $Ex_m = \mu_1$, $Ey_n = \mu_2$ $(m, n = 1, 2, \cdots)$ and unit variance. A sequential decision procedure for testing the hypothesis H_0: $\mu_1 > \mu_2$ against the alternative H_1: $\mu_1 < \mu_2$ consists of a terminal decision rule for choosing H_0 or H_1, a stopping rule for deciding when sampling should be terminated, and a sampling rule for deciding (prior to stopping) whether the next observation should be an x or a y. As is pointed out in [4] and explained in the following, it is possible to give a stopping rule and terminal decision rule of sequential probability ratio test type for testing H_0 versus H_1, for which the probabilities of error are essentially independent of the sampling rule used. Consequently, if we assume that the x's represent the responses of a sequence of patients to treatment A and the y's responses to treatment B and if a large expected response is desirable we may ask which sampling rule to use in order to minimize *the expected number of patients assigned to the inferior treatment* (B when H_0 is true and A when H_1 is true) during the course of the experiment. It is the purpose of this paper to present Monte Carlo and theoretical analyses of some possible sampling rules.

In Section 2 we re-derive and amplify somewhat the results of [4]. Section 3 describes Monte Carlo studies of several different sampling rules. If we reparameterize our problem in terms of $\delta = \mu_2 - \mu_1$ and $\theta = (\mu_1 + \mu_2)/2$, it is instructive to consider the simplified problem in which θ is assumed to be known. In Section 4 we give the solution of this problem and show its relation to the original problem. Section 5 gives some asymptotic results.

A more complete summary of our results is contained in Section 3, which may be read independently of Section 2.

* Herbert Robbins and David O. Siegmund are professors, Department of Mathematical Statistics, Columbia University, New York, N.Y. 10027. Research for this article was supported by PHS Grant 5-R01-GM-16895-03.

2. SEQUENTIAL PROBABILITY RATIO TEST WITH ERROR PROBABILITIES INDEPENDENT OF THE SAMPLING RULE

Let $x_1, x_2, \cdots x_m, \cdots, y_1, y_2, \cdots y_n, \cdots$ be as in Section 1, and let

$$\delta = \mu_2 - \mu_1,$$
$$\theta = (\mu_1 + \mu_2)/2,$$
$$u_m = x_m - x_1, \quad v_n = y_n - x_1 \quad (m, n = 1, 2, \cdots).$$

The u's and v's are jointly normally distributed (but no longer independent) with expectations $Eu_m = 0$, $Ev_n = \delta$ not involving θ. Given $\delta^* > 0$, a test of the simple hypothesis H_0^*: $\delta = -\delta^*$ against the simple alternative H_1^*: $\delta = \delta^*$ based on $u_1, u_2, \cdots, u_m, v_1, \cdots, v_n$ may be given in terms of $L_{m,n}$, the likelihood ratio of the observations under the two simple hypotheses. By direct computation we find that

$$L_{m,n} = \exp\left(2\delta^* \cdot \frac{1}{m+n}\left(n \textstyle\sum_1^m u_i - m \sum_1^n v_j\right)\right),$$

which can be rewritten in terms of the original variables as

$$L_{m,n} = \exp\left(2\delta^* \cdot \frac{mn}{m+n}(\bar{y}_n - \bar{x}_m)\right),$$

where we have put $\bar{x}_m = m^{-1} \sum_1^m x_i$, $\bar{y}_n = n^{-1} \sum_1^n y_j$.

To test H_0^* versus H_1^* sequentially, we therefore choose $0 < B < 1 < A$ and for some sampling rule which, given $u_1, \cdots, u_m, v_1, \cdots, v_n$, tells us whether to observe next u_{m+1} or v_{n+1}, let (M, N) = first (m, n) such that

$$L_{m,n} \notin (B, A) \tag{2.1}$$

and when $M + N < \infty$ accept H_0^* or H_1^* according as $L_{M,N} \le B$ or $\ge A$. (In what follows we shall consider only sampling rules for which $M + N < \infty$ with probability one, although this must be checked in each case.)

Putting

$$A = \exp(2\delta^* a), \quad B^{-1} = \exp(2\delta^* b) \tag{2.2}$$

we may rewrite (2.1) in terms of

$$z_{m,n} = \frac{mn}{m+n}(\bar{y}_n - \bar{x}_m) \tag{2.3}$$

as (M, N) = first (m, n) such that

$$z_{m,n} \notin (-b, a). \tag{2.4}$$

Reprinted from: © **Journal of the American Statistical Association**
March 1974, Volume 69, Number 345
Theory and Methods Section
Pages 132–139

This test is a sequential probability ratio test, and by the arguments of Wald [8, p. 41]

$$P_{-\delta^*}(\text{accept } H_1^*) < A^{-1}[1 - P_{\delta^*}(\text{accept } H_0^*)] \quad (2.5)$$

and

$$P_{\delta^*}(\text{accept } H_0^*) < B[1 - P_{-\delta^*}(\text{accept } H_1^*)], \quad (2.6)$$

where we write P_δ to denote probability as a function of the parameter δ. The inequalities in (2.5) and (2.6) are due to overshoot of the stopping boundaries and may be regarded as approximate equalities. Then the simultaneous solution of (2.5) and (2.6) gives Wald's well-known approximations to the error probabilities of the sequential probability ratio test. We may use this test of the simple hypotheses H_0^* versus H_1^* as a test of the original composite hypotheses H_0 versus H_1, and then using (2.4) we obtain in place of the inequalities (2.5) and (2.6) the more general results

$$P_{-\delta}(\text{accept } H_1) < e^{-2\delta a}[1 - P_\delta(\text{accept } H_0)] \quad (\delta > 0) \quad (2.7)$$

and

$$P_\delta(\text{accept } H_0) < e^{-2\delta b}[1 - P_{-\delta}(\text{accept } H_1)] \quad (\delta > 0). \quad (2.8)$$

Again there is approximate equality, so that for $\delta > 0$

$$P_\delta(\text{accept } H_0) \cong e^{-2\delta b}(e^{2\delta a} - 1)/(e^{2\delta a} - e^{-2\delta b}), \quad (2.9)$$

$$P_{-\delta}(\text{accept } H_1) \cong (1 - e^{-2\delta b})/(e^{2\delta a} - e^{-2\delta b}). \quad (2.10)$$

The importance of the preceding result is that to the extent that there is equality in (2.9) and (2.10), the error probabilities of this test do not depend on the specific sampling rule used, provided only that our decision at each stage depends only on the u's and v's; i.e., on the differences $x_i - y_j$ and not on the x's and y's themselves. Hence we may try to find a sampling rule which in some sense minimizes $E_\delta M$ for $\delta > 0$ and $E_\delta(N)$ for $\delta < 0$.

In particular, if $a = b$ and if our sampling rule is symmetric in the x's and y's, so that $P_\delta(\text{accept } H_0) = P_{-\delta}(\text{accept } H_1)$, then from (2.7) or (2.8) we obtain

$$P_\delta(\text{error}) < 1/(1 + e^{2b\delta}) \quad (\delta \neq 0), \quad (2.11)$$

and again there is approximate equality.

To study in more detail the behavior of $z_{m,n}$, we begin by recalling that a Brownian motion process with drift δ (per unit time) is a family of random variables $X(t)$, $0 \leq t < \infty$, such that

1. $X(0) = 0$;
2. For any $0 < t_1 < \cdots < t_n < \infty$, the random variables $X(t_1), \cdots, X(t_n)$ are jointly normally distributed;
3. For $0 < s < t < \infty$, $E_\delta[X(t)] = t\delta$ and $\text{Cov}_\delta[X(t), X(s)] = \min(t, s)$;
4. $X(t)$ is a continuous function of t.

Lemma 1: For any sequence of pairs (m, n) of positive integers which is non-decreasing in each coordinate, the random sequences $\{z_{m,n}\}$ and $\{X(mn/(m + n))\}$ have the same joint distribution.

Proof: Since the $z_{m,n}$ are linear functions of jointly normal random variables, they are jointly normally distributed. Also, obviously

$$E_\delta z_{m,n} = \frac{mn}{m + n}\delta \quad \text{and} \quad \text{Var}_\delta z_{m,n} = mn/(m + n).$$

Hence, it suffices to prove that

$$\text{Cov}_\delta(z_{m+j,n+k}, z_{m,n}) = mn/(m + n) \quad (j \geq 0, k \geq 0).$$

But this follows immediately by writing

$$z_{m+j,n+k} = z_{m,n} + \frac{(m + j)\cdot\sum_{n+1}^{n+k} y_i - (n + k)\cdot\sum_{m+1}^{m+j} x_i}{m + n + j + k} + \frac{(nj - mk)(\sum_1^n y_i + \sum_1^m x_i)}{(m + n)(m + n + j + k)}$$

and taking expectations:

$$\begin{aligned}\text{Cov}_\delta(z_{m+j,n+k}, z_{m,n}) &= E_0(z_{m+j,n+k}, z_{m,n}) \\ &= E_0(z_{m,n}^2) + E_0[(z_{m+j,n+k} - z_{m,n})z_{m,n}] \\ &= E_0 z_{m,n}^2 = mn/(m + n)\end{aligned}$$

As a consequence, if $\mathfrak{F}_{m,n}$ denotes the σ-algebra generated by $u_1, \cdots, u_m, v_1, \cdots, v_n$, or equivalently $z_{i,j}(1 \leq i \leq m, 1 \leq j \leq n)$,

$$\begin{aligned}E_\delta\left(z_{m+1,n} - \frac{(m + 1)n\delta}{m + n + 1}\,\middle|\,\mathfrak{F}_{m,n}\right) &= z_{m,n} - \frac{mn\delta}{m + n} \\ &= E_\delta\left(z_{m,n+1} - \frac{m(n + 1)\delta}{m + n + 1}\,\middle|\,\mathfrak{F}_{m,n}\right),\end{aligned}$$

so the family $\{z_{m,n} - mn\delta/(m+n), \mathfrak{F}_{m,n}; m, n = 1, 2, \cdots\}$ is a martingale in two-dimensional time. For statistical purposes we use a sampling rule which, given $u_1, \cdots, u_m, v_1, \cdots, v_n$, tells us to observe next the variable x_{m+1} or y_{n+1}, so that the observed process $\{z_{m,n} - mn\delta/(m+n), \mathfrak{F}_{m,n}\}$ is a martingale in one-dimensional time. It follows (cf. [2, p. 302]) that if

$$E_\delta\left(\frac{MN}{M + N}\right) < \infty, \quad (2.12)$$

then

$$E_\delta\left(z_{M,N} - \frac{MN\delta}{M + N}\right) = 0, \quad (2.13)$$

so that for $a = b$, treating the inequality in (2.11) as an approximate equality, we have for all $\delta > 0$

$$\begin{aligned}\delta E_\delta\left(\frac{MN}{M + N}\right) &= E_\delta(z_{M,N}) \\ &\cong bP_\delta\{z_{M,N} > b\} - bP_\delta\{z_{M,N} < -b\} \\ &\cong b\left(\frac{e^{2b\delta} - 1}{e^{2b\delta} + 1}\right).\end{aligned}$$

A similar argument applies when $\delta < 0$, so that

$$E_\delta\left(\frac{MN}{M + N}\right) \cong \frac{b}{\delta}\left(\frac{e^{2b\delta} - 1}{e^{2b\delta} + 1}\right) \quad (2.14)$$

for all $\delta \neq 0$. For the case $\delta = 0$, one may use the martingale $\{z_{m,n}^2 - mn/(m+n);\ \mathfrak{F}_{m,n}\}$ and the fact that $P\{z_{M,N} > b\} = \frac{1}{2}$ to obtain

$$E_0\left(\frac{MN}{M+N}\right) \cong b^2.$$

Since the right side of (2.14) converges to b^2 as $\delta \to 0$, we shall agree to define the right side of (2.14) to be b^2 when $\delta = 0$, and then (2.14) holds for all δ.

Three important conclusions follow at once from (2.14):

1. For the important special case of *pairwise* sampling, $M = N$, so

$$E_\delta N = E_\delta M \cong \frac{2b}{\delta}\left(\frac{e^{2b\delta}-1}{e^{2b\delta}+1}\right) \quad (2.15)$$

and

$$E_\delta(M+N) \cong \frac{4b}{\delta}\left(\frac{e^{2b\delta}-1}{e^{2b\delta}+1}\right). \quad (2.16)$$

2. Since $\min(M, N) \geq MN/(M+N)$, we have for *any* sampling rule

$$\min(E_\delta M, E_\delta N) \geq E_\delta\left(\frac{MN}{M+N}\right), \quad (2.17)$$

which by (2.14) and (2.15) shows that the expected number of observations on each population must exceed (approximately) one-half the expected number under *pairwise* sampling. Moreover, equality would be obtained in the inequality $M \geq MN/(M+N)$ only if $N = \infty$, so we see that achieving approximate equality in (2.17) can be accomplished (if at all) only by making $\max(E_\delta M, E_\delta N)$ extremely large.

3. Since $x(1-x) \leq \frac{1}{4}$ for $0 \leq x \leq 1$, with equality only if $x = \frac{1}{2}$, we have

$$E_\delta(M+N) \geq 4 \cdot E_\delta\left(\frac{MN}{M+N}\right) \quad (2.18)$$

with equality only if $P_\delta\{M = N\} = 1$. Hence, in conjunction with (2.14) and (2.15), we see that the expected *total* sample size is (approximately) minimized by *pairwise* sampling.

3. MONTE CARLO RESULTS

To test H_0 versus H_1 sequentially, we put

$$z_{m,n} = \left(m \textstyle\sum_1^n y_i - n \sum_1^m x_j\right)/(m+n)$$

and, for some fixed $b > 0$, stop sampling with (M, N) = first (m, n) such that $z_{m,n} \notin (-b, b)$, and accept H_0 or H_1 according as $z_{M,N} \leq -b$ or $\geq b$. Having observed $x_1, \cdots, x_m$ and $y_1, \cdots, y_n$, the next variable to be observed (x_{m+1} or y_{n+1}) is selected according to some rule which (1) is symmetric in the x's and y's, and (2) depends on $x_1, \cdots, x_m, y_1, \cdots, y_n$ only through the pairwise differences $x_i - y_j$ $(1 \leq i \leq m, 1 \leq j \leq n)$. Under these conditions it was shown in Section 2 that if

$$P_\delta\{M + N < \infty\} = 1 \quad (\delta = \mu_2 - \mu_1),$$

then

$$P_\delta(\text{error}) < 1/(1 + e^{2b\delta}) \quad (\delta \neq 0) \quad (3.1)$$

and there is approximate equality in (3.1). Also, if $E_\delta(MN/(M+N)) < \infty$, then

$$E_\delta\left(\frac{MN}{M+N}\right) \cong \frac{b}{\delta}\left(\frac{e^{2b\delta}-1}{e^{2b\delta}+1}\right). \quad (3.2)$$

(The right side of (3.2) is defined to be b^2 by continuity at $\delta = 0$). Moreover, the expected *total* sample size $E_\delta(M+N)$ is minimized by *pairwise* sampling, for which $E_\delta(M+N) \cong 4 \cdot$[right side of (3.2)], and min $[E_\delta(M), E_\delta(N)]$ is at least as large as $\frac{1}{2}$ the common value of $E_\delta M$ and $E_\delta N$ achieved by pairwise sampling, which equals approximately $2 \cdot$[right side of (3.2)].

It seems intuitively clear that to minimize the expected number of observations with the smaller mean, when $z_{m,n}$ is close to zero, our sampling rule should approximate pairwise sampling; but when $z_{m,n}$ is close to one of the stopping boundaries $+b$ or $-b$ it should take most of the future observations from one population—the y population if $z_{m,n}$ is close to b and the x population if $z_{m,n}$ is close to $-b$. A simple class of rules having these qualitative properties is the following. Choose $c \geq b$. Observe x_1, y_1. For $m, n = 1, 2, \cdots$, having observed $x_1, \cdots, x_m$ and $y_1, \cdots, y_n$, let the next observation be y_{n+1} if

$$\frac{n-m}{m+n} \leq \frac{z_{m,n}}{c}; \quad (3.3)$$

otherwise observe x_{m+1}.

Lemma 2: For the sampling rule given by (3.3), for each δ there exist $p_\delta < 1$, $c > 0$ such that

$$P_\delta\{M + N > \nu + 1\} \leq cp_\delta^\nu \quad (\nu = 1, 2, \cdots).$$

In particular, $E_\delta(M+N)^k < \infty$ for all $k = 1, 2, \cdots$, and the approximations of Section 2 apply.

We defer the proof of Lemma 2 until the end of this section.

Tables 1 and 2 give Monte Carlo results for 1,000 trials for pairwise sampling (corresponding to (3.3) with $c = +\infty$) and for sampling rule (3.3). Where applicable we have also given the Wald approximations derived in Section 2. We have used the values $b = 6$ and $b = 9$, which for $\delta = 0.25$ give error probabilities of approximately .05 and .01, respectively. Since our procedures are symmetric, we consider only non-negative δ.

Qualitatively, these values are as one expects. Increasing the value of c increases the value of $E_\delta(M)$ for large δ but decreases it for small δ. The magnitude of these changes may be somewhat surprising. Also note that the marginally smaller value of $E_\delta(M)$ for large δ obtained by setting $c = b$ results in a considerably larger value of $E_\delta(N)$ for such δ. Slight variations on the sampling rule (3.3) give somewhat smaller values of $E_\delta(M)$ for small δ without considerable changes for large δ. For example, the preceding results suggest letting c be an increasing function of the number of observations already made. If we set $b = 6$ and

$$\begin{aligned} c &= c\,(mn/(m+n)) \\ &= 6 + 0.1\,mn/(m+n), \end{aligned} \quad (3.4)$$

and use the sampling rule (3.3) for this function c, we find by Monte Carlo that $E_0(M) \cong 85$, $E_{.25}(M) \cong 36$, $E_{.5}(M) \cong 16$.

1. Monte Carlo Results for 1,000 Trials for Pair-Wise Sampling and for Sampling Rule 3.3

$$b = 6 \begin{cases} c = 6 = b \text{ (left entry)} \\ c = 7.2 = (1.2)b \text{ (right entry)} \end{cases}$$

δ	P_δ(error) WA[a]	P_δ(error) MC[b]		$E_\delta(M) = E_\delta(N)$ Pairwise WA	$E_\delta(M) = E_\delta(N)$ Pairwise MC		$E_\delta(M)$ MC		$E_\delta(N)$ MC		$E_\delta\left(\frac{MN}{M+N}\right)$ MC	
0				72	84	84	106	95	110	96	38	38
.05	.35	.35	.32	70	80	78	86	80	123	101	36	36
.10	.23	.22	.25	64	73	72	75	66	122	100	34	32
.17	.11	.12	.10	54	63	61	57	48	128	100	30	28
.25	.05	.04	.03	43	47	48	34	36	112	90	21	23
.375	.01	.008	.01	31	34	33	22	22	105	75	16	16
.5	.002	.004	.000	24	25	25	16	16	93	62	12	12
.75	.000	.000	.000	15	17.4	17.6	9.6	10.5	74.6	47.3	8.2	8.4
1	.000	.000	.000	12	13.1	13.3	7.2	7.7	63.9	37.5	6.3	6.3
2	.000	.000	.000	6	6.8	6.8	3.7	3.9	35.0	19.3	3.2	3.2

[a] WA = Wald approximation.
[b] MC = Monte Carlo.

In summary, by using the sampling rule (3.3) we obtain values of $E_\delta(M)$ which are less than ten percent larger than those for pairwise sampling for δ near 0 and about half as large as those for pairwise sampling (the theoretical lower bound) for large values of δ. For intermediate values of δ; i.e., δ for which $.01 \leq P_\delta(\text{error}) \leq .25$ a reduction over pairwise sampling takes place. We do not know whether much further reduction is possible.

In Section 4 we obtain a second lower bound for $E_\delta(M)$ which for some values of δ is larger than the one-half of pairwise sampling lower bound given in Section 2.

Proof of Lemma 2: Let $\mathfrak{F}_{m,n} = B(u_1, \cdots, u_m, v_1, \cdots, v_n)$. It suffices to prove that for some $0 < p_\delta < 1$ and for all $\nu = m + n > 1$

$$P_\delta\{M + N = \nu + 1 \mid \mathfrak{F}_{m,n}\} \geq 1 - p_\delta \text{ a.e. on } \{M + N > \nu\}. \quad (3.5)$$

On the event

$$\{z_{m,n} \in (c(n - m)/\nu, b)\}$$

on which according to (3.3) we observe y_{n+1}, we have

$$P_\delta\{z_{m,n+1} \geq b \mid \mathfrak{F}_{m,n}\} \geq P_\delta\{z_{m,n+1} - z_{m,n} \geq 2bm/\nu \mid \mathfrak{F}_{m,n}\}. \quad (3.6)$$

According to Lemma 1 the conditional distribution of $z_{m,n+1} - z_{m,n}$ is normal with mean $\delta m^2/\nu(\nu + 1)$ and variance $m^2/\nu(\nu + 1)$, so that the right side of (3.6) equals $1 - \Phi\{2b(1 + \nu^{-1})^{\frac{1}{2}} - \delta m/[\nu(\nu + 1)]^{\frac{1}{2}}\}$ (Φ denotes the standard normal distribution function), which is bounded away from zero in ν. A similar argument applies on $\{z_{m,n} \in (-b, c(n - m)/\nu)\}$ on which x_{m+1} is observed. This proves (3.5) and hence the lemma.

4. THE CASE $(\mu_1 + \mu_2)/2$ KNOWN

Assuming that $\theta = (\mu_1 + \mu_2)/2$ is known to the experimenter simplifies considerably the original problem. One might hope that the optimal solution to this problem, which may be obtained explicitly, would provide insight

2. Monte Carlo Results for 1,000 Trials for Pair-Wise Sampling and for Sampling Rule 3.3

$$b = 9 \begin{cases} c = 9 = b \text{ (left entry)} \\ c = 10.8 = (1.2)b \text{ (right entry)} \end{cases}$$

δ	P_δ(error) WA[a]	P_δ(error) MC[b]		$E_\delta(M) = E_\delta(N)$ Pairwise WA	$E_\delta(M) = E_\delta(N)$ Pairwise MC		$E_\delta(M)$ MC		$E_\delta(N)$ MC		$E_\delta\left(\frac{MN}{M+N}\right)$ MC	
0				162	179	175	244	203	255	208	86	82
.05	.29	.31	.28	152	170	169	189	165	267	231	77	78
.10	.14	.13	.14	129	141	141	130	124	284	229	65	67
.17	.045	.04	.04	96	99	101	75	80	286	207	47	51
.25	.011	.01	.006	70	74	73	50	50	256	174	36	36
.375	.001	.000	.001	48	50	50	29	31	221	136	24	24
.5	.000	.000	.000	36	38	37	21	22	202	111	19	18.1
.75	.000	.000	.000	24	25	25.3	13.5	14.6	162	83	12.2	12.3
1	.000	.000	.000	18	19.5	19.2	10.2	10.8	134	65	9.3	9.2
2				9	9.7	9.6	5.1	5.5	74	32	4.7	4.6

[a] WA = Wald approximation.
[b] MC = Monte Carlo.

into sampling rules for the original problem. While this does not seem to be the case, the value of $E_\delta(M)(\delta > 0)$ using the optimal sampling rule in the simplified problem provides us with a lower bound for $E_\delta(M)(\delta > 0)$ in the original problem, which for small δ is better than the one-half of pairwise lower bound given in Section 2.

Without loss of generality we may assume that $\theta = 0$. Then the likelihood ratio of $(x_1, \cdots, x_m, y_1, \cdots, y_n)$ for testing the simple hypothesis $H_0^*: \delta = -\delta^*$ vs. $H_1^*: \delta = \delta^*$ is

$$L_{m,n} = \exp\left[\delta^*\left(\sum_1^n y_i - \sum_1^m x_i\right)\right]. \quad (4.1)$$

If we use a symmetric sequential probability ratio test stopping rule, then Wald's arguments show that if for

$$(M, N) = \text{first } (m, n) \text{ such that } \sum_1^n y_i - \sum_1^m x_i \notin (-2b, 2b) \quad (4.2)$$

we reject $H_0(H_0^*)$ if and only if

$$\sum_1^N y_i - \sum_1^M x_i \geq 2b, \quad (4.3)$$

we obtain the error probability approximation

$$P_\delta(\text{error}) < 1/(1 + e^{2b\delta}) \quad (\delta \neq 0), \quad (4.4)$$

which is the same as (2.11). Using Wald's lemma we also obtain

$$E_\delta(M + N) \cong \frac{4b}{\delta}\left(\frac{e^{2b\delta} - 1}{e^{2b\delta} + 1}\right), \quad (4.5)$$

and these approximations do not depend on the sampling rule, which now may be based on the x's and y's themselves—not just on their pairwise differences.

From the form of the likelihood ratio (4.1), it is evident that an optimal sampling rule is to observe either x_1 or y_1 with probability $\frac{1}{2}$ each, and for $m + n \geq 1$ to observe y_{n+1} next if

$$\sum_1^n y_i \geq \sum_1^m x_i; \quad (4.6)$$

otherwise observe x_{m+1}. If we assume that the penalty for an incorrect terminal decision is a, the cost for observing an $x(y)$ is c if $H_1^*(H_0^*)$ is true and 0 if $H_0^*(H_1^*)$ is true, and the prior probability that H_0^* is true is $\frac{1}{2}$, it may be shown that for some $b > 0$ the decision procedure given by (4.2), (4.3), and (4.6) is a Bayes test of H_0^* versus H_1^*. In fact, standard heuristic arguments in the single population case (cf. [5, p. 105]) suggest that having observed $x_1, \cdots, x_m$ and $y_1, \cdots, y_n$ the risks r_i associated with the four possible actions: stop and accept $H_0^*(r_0)$, stop and accept $H_1^*(r_1)$, observe $x_{m+1}(r_2)$, or observe $y_{n+1}(r_3)$ are as graphed as a function of the posterior probability $\pi_{m,n} = \pi$.

Hence, to proceed optimally one should stop and accept H_0^* if $\pi_{m,n} \geq 1 - \pi'$, stop and accept H_1^* if $\pi_{m,n} \leq \pi'$, sample x_{m+1} if $\frac{1}{2} < \pi_{m,n} < 1 - \pi'$, and sample y_{n+1} if $\pi' < \pi_{m,n} \leq \frac{1}{2}$. It is easily seen that $\pi_{m,n} - \frac{1}{2}$ is a decreasing symmetric function of $\sum_1^n y_i - \sum_1^m x_i$, and hence for some $b > 0$ an optimal Bayes procedure is given by (4.2), (4.3), and (4.6). It may then be shown

Optimal Bayes Procedure

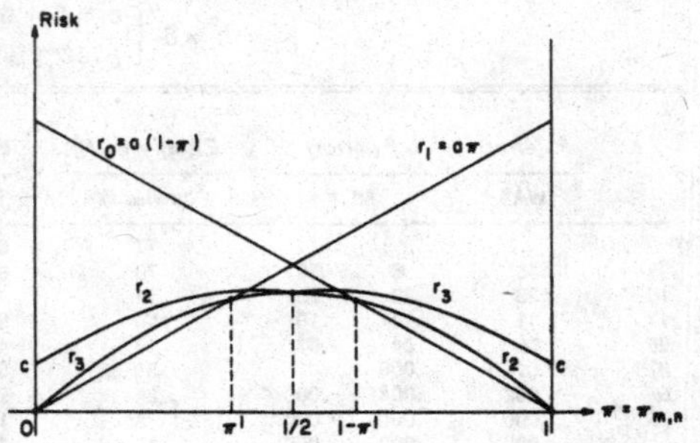

by a variational argument that to each $0 < b < \infty$ there correspond values of a and c such that the decision procedure defined by (4.2), (4.3), and (4.6) is a Bayes test. Hence any other test which is symmetric in the x's and y's and has no larger probability of error must on the average take at least as many observations on the population having the smaller mean. Since this optimal procedure does not depend on δ^*, its optimality property holds uniformly in δ for testing the hypothesis that $\delta < 0$ against the alternative that $\delta > 0$.

It follows that the value of $E_\delta(M)$ for the decision procedure defined by (4.2), (4.3), and (4.6) is (approximately) a lower bound for $E_\delta(M)$ for the sampling rules of Section 3 in which θ was unknown. For δ near 0 this lower bound is greater than the one-half of pairwise bound given in Section 2 and 3. To compute an analytic approximation to $E_\delta(M)$, observe that if $\xi_1, \xi_2, \cdots$ are independent and normally distributed with mean $\delta/2$ and unit variance, then as m and n increase the joint distribution of the sequence $\sum_1^n y_i - \sum_1^m x_i$ is the same as that of $\sum_1^{m+n} \xi_k$ $(m, n = 1, 2, \cdots)$, regardless of which sampling rule is used. Then the number of observations from the x population after the first observation (which is x_1 or y_1 with probability $\frac{1}{2}$ each) has the same distribution as the number of integers k for which $\sum_1^k \xi_i < 0$ prior to the first ν for which $\sum_1^\nu \xi_i \notin (-2b, 2b)$. Letting $X(t)$ denote a Brownian motion process with drift $\delta/2$ per unit time and $T = \inf\{t: X(t) \notin (-2b, 2b)\}$, we consider approximating $E_\delta(M)$ for the sampling rule (4.6) by

$$E(\text{measure } \{t: 0 < t < T, X(t) < 0\}), \quad (4.7)$$

which may be computed exactly. First observe that for $\delta > 0$

$$\begin{aligned} E(\text{meas } \{t: 0 < t < T, X(t) < 0\}) &= E(\text{meas } \{t: 0 < t < \infty, X(t) < 0\}) \\ -E(\text{meas } \{t: t \geq T, X(t) < 0\} \mid X(T) &= -2b) \cdot P\{X(T) = -2b\} \quad (4.8) \\ -E(\text{meas } \{t: t \geq T, X(t) < 0\} \mid X(T) &= 2b) \cdot P\{X(T) = 2b\}. \end{aligned}$$

Now a simple computation shows that for $\delta > 0$

$$E(\text{meas } \{t: 0 < t < \infty, X(t) < 0\}) = 2/\delta^2, \quad (4.9)$$

and since the approximations derived earlier are equalities for the Brownian motion process, we have from (4.4)

$$P\{X(T) = -2b\} = 1/(1 + e^{2b\delta}) = 1 - P\{X(T) = 2b\}. \quad (4.10)$$

Now, the first conditional expectation in (4.8) is just the sum of the expected length of time it takes for $X(t)$ to travel from $-2b$ to 0, which by an application of Wald's lemma is seen to be $4b/\delta$, and (4.9). Furthermore, the second conditional expectation in (4.8) is the product of the probability that starting from $2b$, $X(t)$ ever reaches zero, which is known (cf. [7, p. 1412]) to be $\exp(-2b\delta)$, and (4.9). Hence from (4.8), (4.9), and (4.10), we obtain for $\delta > 0$

$$E(\text{meas } \{t: 0 < t < T, X(t) < 0\}) = \frac{2}{\delta^2}\left(1 - \frac{2}{1 + e^{2b\delta}}\right) - \frac{4b}{\delta(1 + e^{2b\delta})}. \quad (4.11)$$

By symmetry, for $\delta = 0$, $E(\text{meas } \{t: 0 < t < T, X(t) < 0\}) = \frac{1}{2}ET = 2b^2 = \lim_{\delta \to 0}$ [right side of (4.11)]. Table 3 gives the value of (4.11) with Monte Carlo estimates of $E_\delta M$ based on 1000 trials for $b = 6$ and 9 and different values of δ.

3. Sample Sizes for Simplified Problem

δ	$b = 6$ (35)	$b = 6$ MC	$b = 9$ (35)	$b = 9$ MC
.05	63	70	129	153
.10	52	59	92	97
.17	37	38	54	56
.25	24.4	25.2	30	33

As we have remarked previously, these results apply to our original problem to give a lower bound for $E_\delta(M)$ which for small δ is greater than the one-half of pairwise lower bound given in Section 2. For example, for $b = 6$, $c = 7.2$, and $\delta = .17$, according to our Monte Carlo results $E_\delta(M) \cong 48$, compared to a lower bound of 38 given in Table 3. The one-half of pairwise lower bound is about 31.

5. ASYMPTOTIC THEORY

In this section we give a theorem on the behavior of $E_\delta(M)$ and $E_\delta(N)$ as $b \to \infty$ for the sampling rule (3.3). The results are very crude in the sense that similar asymptotic properties hold for a large class of sampling rules, many of which seem to be distinctly inferior to (3.3) for small values of b.

Theorem 1: For the sampling rule (3.3), as $b \to \infty$: (*i*) if for some $\eta > 0$, $c = b(1 + \eta)$, then for each $\delta > 0$

$$E_\delta M \sim \delta^{-1}b\,\frac{2(1+\eta)}{2+\eta}, \quad E_\delta N \sim \delta^{-1}b\,\frac{2(1+\eta)}{\eta}, \quad (5.1)$$

(*ii*) if $c = b[1 + \eta(m + n)]$, where $\eta(x) \to 0$ and $x\eta(x) \to \infty$ as $x \to \infty$, then for each $\delta > 0$

$$E_\delta(M) \sim \delta^{-1}b. \quad (5.2)$$

Proof: We shall prove only the first part of (5.1). The proofs of the second part of (5.1) and of (5.2) are similar.

Let $T = M + N$, and if ν is the total number of observations already made, let M_ν and N_ν denote the number of x's and the number of y's observed, respectively. Then $M_\nu + N_\nu \equiv \nu$ and $M = M_T$, $N = N_T$. Let

$$\lambda = \eta/2(1 + \eta) \quad (5.3)$$

Then

$$\{T > \nu, M_\nu \le \lambda\nu\} \cdot\subset \{M_{\nu+1} > M_\nu\}, \quad (5.4)$$

for if $M_\nu \le \lambda\nu$ and $M_{\nu+1} = M_\nu$, then by (5.3)

$$z_{M_\nu,N_\nu} \ge c\left(\frac{N_\nu - M_\nu}{\nu}\right) \ge b(1 + \eta)(1 - 2\lambda) = b$$

and hence $T \le \nu$.

We now record without proof several consequences of (5.4) and standard arguments:

$$\lim_{b\to\infty} M/(M + N) \ge \lambda \quad \text{a.s.}; \quad (5.5)$$

$$\lim_{b\to\infty} b^{-1}MN/(M + N) = 1/\delta \quad \text{a.s.}; \quad (5.6)$$

$$\lim_{b\to\infty} b^{-1}E_\delta[MN/(M + N)] = 1/\delta; \quad (5.7)$$

$$\lim_{b\to\infty} \sup b^{-1}N \le \delta^{-1}2(1 + \eta)/\eta \quad \text{a.s.}; \quad (5.8)$$

$$\lim_{b\to\infty} \inf b^{-1}M \ge \delta^{-1}2(1 + \eta)/(2 + \eta) \quad \text{a.s.} \quad (5.9)$$

To complete the proof it suffices to show that $b^{-1}M$ converges in probability to the right side of (5.9) and is uniformly integrable. Let $\epsilon > 0$. Since $mn/(m + n)$ is increasing in each of its arguments

$$P\left\{M \ge \frac{b}{\delta}\frac{2(1+\eta)}{2+\eta}(1 + 2\epsilon),\ N \ge \frac{b}{\delta}\frac{2(1+\eta)}{\eta}(1 + \epsilon)\right\} \le P_\delta\left\{\frac{MN}{M+N} \ge \frac{b}{\delta}(1 + \epsilon)\right\} \to 0 \quad (5.10)$$

by (5.6). Let $\nu_0 = [b/\delta]$. Then

$$\left\{M \ge \frac{b}{\delta}\frac{2(1+\eta)}{2+\eta}(1 + 2\epsilon),\ N < \frac{b}{\delta}\frac{2(1+\eta)}{\eta}(1 + \epsilon)\right\} \subset \bigcup_{\nu=\nu_0}^{\infty}\left\{T > \nu,\ M_\nu \ge \frac{b}{\delta}\frac{2(1+\eta)}{2+\eta}(1 + \epsilon),\ N_\nu < \frac{b}{\delta}\frac{2(1+\eta)}{\eta}(1 + \epsilon),\ M_{\nu+1} > M_\nu\right\}. \quad (5.11)$$

If $M_{\nu+1} > M_\nu$, then by (3.3)

$$\bar{y}_{N_\nu} - \bar{x}_{M_\nu} < b(1 + \eta)(1/M_\nu - 1/N_\nu), \quad (5.12)$$

and since the right side of (5.12) is decreasing in M_ν and increasing in N_ν, on each event on the right side of (5.11) we have

$$\bar{y}_{N_\nu} - \bar{x}_{M_\nu} < b(1+\eta)\left(\frac{\delta(2+\eta)-\delta\eta}{b2(1+\eta)(1+\epsilon)}\right) = \frac{\delta}{1+\epsilon}.$$

The proof of (5.4) also shows that

$$\{T > \nu, M_{\nu+1} > M_\nu\} \subset \{N_\nu \geq \lambda\nu\}, \quad (5.13)$$

and hence the event (5.11) implies

$$\bigcup_{\nu=\nu_0}^{\infty} \left\{T > \nu, \bar{y}_{N_\nu} - \bar{x}_{M_\nu} - \delta < -\frac{\delta\epsilon}{2}, M_\nu \geq \nu_0, N_\nu \geq \lambda\nu_0\right\}, \quad (5.14)$$

which by Lemma 1 and (7) of [7] has probability at most

$$2\left\{1 - \Phi\left[\frac{\delta\epsilon}{2}(\nu_0\lambda/2)^{\frac{1}{2}}\right]\right\} \to 0 \quad \text{as} \quad b \to \infty.$$

Hence, $b^{-1}M \xrightarrow{P} \delta^{-1}2(1+\eta)/(2+\eta)$ as $b \to \infty$. To prove that $b^{-1}M$ is uniformly integrable, first observe that

$$M_\nu N_\nu/\nu = \sum_{k=2}^{\nu} \Delta_k,$$

where

$$\Delta_2 = \tfrac{1}{2}$$

and for $k \geq 1$

$$\Delta_{k+1} = \frac{M_{k+1}N_{k+1}}{k+1} - \frac{M_k N_k}{k}$$

$$= \begin{cases} M_k^2/k(k+1) & \text{if} \quad N_{k+1} > N_k \\ N_k^2/k(k+1) & \text{if} \quad M_{k+1} > M_k. \end{cases} \quad (5.15)$$

Hence, for any event A,

$$\int_A \frac{MN}{M+N} dP_\delta = \sum_{\nu=1}^{\infty} \int_{A\{T>\nu\}} \Delta_{\nu+1} dP_\delta$$

$$\geq \sum_{\nu=1}^{\infty} \int_{A\{T>\nu, M_{\nu+1}>M_\nu\}} \Delta_{\nu+1} dP_\delta. \quad (5.16)$$

From (5.13), (5.15), and (5.16), we obtain

$$\int_A \left(\frac{MN}{M+N}\right) dP_\delta \geq \lambda^2/2 \sum_{\nu=1}^{\infty} P_\delta(A\{T > \nu, M_{\nu+1} > M_\nu\})$$

$$= \lambda^2/2\left[\int_A M dP_\delta - P_\delta(A)\right]. \quad (5.17)$$

It follows from (5.6) and (5.7) that $b^{-1}[MN/(M+N)]$ is uniformly integrable, and hence by (5.17) that $b^{-1}M$ is also. This completes the proof.

Hopefully, Theorem 1 gives some insight into the behavior of the sampling rule (3.3) and may provide some guidance in selecting values of $c = b(1+\eta)$. For the values of b and c used in Table 2, the asymptotic approximation to $E_\delta M$ given by (5.1) provides a crude approximation to the Monte Carlo estimate for values of δ for which $P_\delta(\text{error}) \cong 0$. Presumably, (5.1) provides a better approximation for larger values of c. For $b = 22.5$ and $c = 27$, corresponding to $P_\delta(\text{error}) \cong .01$ for $\delta = .10$, we obtain from (5.1)

$$E_\delta M \cong 66, 50, \text{and } 33$$

for $\delta = .375$, .5, and .75, compared with 200 repetition Monte Carlo estimates of 68, 52, and 33. The approximation provided by (5.1) to $E_\delta(N)$ is very bad in all the cases we have investigated. Fortunately, the Monte Carlo estimate of $E_\delta(N)$ is always much smaller than its asymptotic value.

6. REMARKS

We have considered only stopping rules of the type of the sequential probability ratio test, and have tried to select a sampling rule which makes the average number of observations made on the inferior population as small as possible. In the single population problem other stopping rules have been proposed (see, e.g., [1]) and might be studied in the two population problem as well. In some cases, if $|\delta|$ is small, it may be desirable to permit the experimenter a third decision: that there is no significant difference between the two populations. The two population problem in this context has been studied by Flehinger and Louis [3], who use a class of sampling rules which are different from ours and do not seem to perform as well as ours for the sequential probability ratio test stopping rule.

In some cases the very large values of $E_\delta N$ required in order to make $E_\delta M$ as small as possible ($\delta > 0$) may be regarded as too high a price to pay for what has been achieved. It would be interesting to consider minimizing $\gamma E_\delta(M) + (1-\gamma)E_\delta(N)(\delta > 0)$ for a given value γ. For $\gamma = \frac{1}{2}$ pairwise sampling is optimal, while $\gamma = 1$ corresponds to minimizing $E_\delta M$ without regard to $E_\delta N$.

At the other extreme from problems we have considered are those in which the sampling rule is determined to some extent by the nature of the investigation and is not subject to experimental control. The results of Section 2 assure us that even in these cases Wald's error probability approximations for the sequential probability ratio test are valid.

We have made considerable use of the invariance of our problem under addition of a constant to all of the variables. An important problem in which no invariance exists is that of two binomial populations. The case of two negative exponential populations exhibits scale invariance and some of the preceding considerations apply, but this problem seems to be considerably more difficult than the normal case.

A more delicate asymptotic analysis arises if $b \to \infty$ and δ varies with b in such a way that $b\delta$ approaches a non-zero finite limit. (If $b\delta \to \infty$ the situation is essentially no different from the one we have discussed.) We have been unable to carry out the required computations to develop this asymptotic theory.

The problem of this article is also of interest in the context of tests with probability of error bounded away from one-half (cf. [6]), and many of the results carry over in a straightforward manner.

[*Received February 1973. Revised July 1973.*]

REFERENCES

[1] Anderson, T.W., "A Modification of the Sequential Probability Ratio Test to Reduce the Sample Size," *Annals of Mathematical Statistics*, 31 (March 1960), 165–97.

[2] Doob, J., *Stochastic Processes*, New York: John Wiley & Sons, Inc., 1953.

[3] Flehinger, B. and Louis, T., "Sequential Medical Trials with Data Dependent Treatment Allocation," *Proceedings of the 6th Berkeley Symposium on Mathematical Statistics and Probability*, Vol. 4, 1972, 43–52.

[4] ———, Louis, T., Robbins, H. and Singer, B., "Reducing the Number of Inferior Treatments in Clinical Trials," *Proceedings of the National Academy of Sciences*, 69 (October 1972), 2993–4.

[5] Lehmann, E., *Testing Statistical Hypotheses*, New York: John Wiley & Sons, Inc., 1959.

[6] Robbins, H., "Statistical Methods Related to the Law of the Iterated Logarithm," *Annals of Mathematical Statistics*, 41 (October 1970), 1397–409.

[7] ——— and Siegmund, D., "Boundary Crossing Probabilities for the Wiener Process and Partial Sums," *Annals of Mathematical Statistics*, 41 (October 1970), 1410–29.

[8] Wald, A., *Sequential Analysis*, New York: John Wiley & Sons, Inc., 1947.

Part Two

SEQUENTIAL EXPERIMENTATION AND ANALYSIS

C. Sequential Estimation and Testing

The fundamental contributions [44]*, in 1959, and [64]* in 1965 (written jointly with Y. S. Chow), paved the way for a large number of subsequent papers on sequential estimation. The early stage of this line of research was marked by the pioneering works of Stein (1945,1949) on fixed width confidence intervals, Stein and Wald (1947), and the papers of Anscombe (1952,1953) and Cox (1952). Some interesting contributions to sequential estimation following [44] and [64] are the papers of Starr (1966a,b) on sequential point and interval estimation; the papers of Gleser (1965) and Albert (1966) on fixed width confidence regions for regression parameters; the work of Sen and Ghosh (1971,1981) on extensions to rank-order and U-statistics, the papers of Starr and Woodroofe (1969,1972), Woodroofe (1977), Vardi (1979a,b), Chow and Martinsek (1982) on bounded regret in sequential estimation, and Lai and Siegmund (1983) on fixed accuracy estimation in autoregressive models; and the papers [73], [98], [99] on sampling and stopping rules in certain sequential estimation problems.

The paper [68]* with Darling on finding the size of a finite population provides a novel application of sequential methodology to estimation problems. Another novel application is the paper [84] on estimating an integer mean.

The paper [51]* with Moriguti on testing the parameter of a Bernoulli distribution and Chernoff's (1961) paper on testing the mean of a normal distribution represent pioneering contributions to the subject of Bayes sequential tests without an indifference zone. Chernoff's (1970) monograph summarizes developments in the subject during the 1960's. One such development is Anscombe's (1963) decision-theoretic model for determining the stopping rule of a clinical trial to test which of the two treatments is better. The recent papers [116]* and [126] provide a class of stopping rules that

*Papers reprinted in this book are marked here by an asterisk when they first appear.

have nearly optimal frequentist and Bayesian properties. Another recent development is the paper by Chernoff and Petkau (1981) on Bayes stopping rules.

References

Albert, A. A. (1966). Fixed size confidence ellipsoids for linear regression parameters. *Ann. Math. Statist.* **37**, 1602–1630.

Anscombe, F. J. (1952). Large sample theory of sequential estimation. *Proc. Cambridge Philos. Soc.* **48**, 600–607.

Anscombe, F. J. (1953). Sequential estimation. *J. Roy. Statist. Soc. Ser. B* **15**, 1–29.

Anscombe, F. J. (1963). Sequential medical trials. *J. Amer. Statist. Assoc.* **58**, 365–383.

Chernoff, H. (1961). Sequential tests for the mean of a normal distribution. *Proc. Fourth Berkeley Symp. Math. Statist. Probab.* **1**, 79–95. Univ. Calif. Press.

Chernoff, H. (1972). *Sequential Analysis and Optimal Design*. Society for Industrial and Applied Mathematics, Philadelphia.

Chernoff, H. and Petkau, J. (1981). Sequential medical trials involving paired data. *Biometrika* **68**, 119–132.

Chow, Y. S. and Martinsek, A. (1982). Bounded regret of a sequential procedure for estimation of the mean. *Ann. Statist.* **10**, 909–914.

Cox, D. R. (1952). Estimation by double sampling. *Biometrika* **39**, 217–227.

Gleser, L. J. (1965). On the asymptotic theory of fixed-size sequential confidence bounds for linear regression parameters. *Ann. Math. Statist.* **36**, 463–467.

Lai, T. L. and Siegmund, D. (1983). Fixed accuracy estimation of an autoregressive parameter. *Ann. Statist.* **11**, 478–485.

Sen, P. K. and Ghosh, M. (1971). On bounded length sequential confidence intervals based on one-sample rank-order statistics. *Ann. Math. Statist.* **42**, 189–203.

Sen, P. K. and Ghosh, M. (1981). Sequential point estimation of estimable parameters based on U-statistics. *Sahkhyā Ser. A* **43**, 331–344.

Starr, N. (1966a). The performance of a sequential procedure for the fixed-width interval estimation of the mean. *Ann. Math. Statist.* **37**, 36–50.

Starr, N. (1966b). On the asymptotic efficiency of a sequential procedure for estimating the mean. *Ann. Math. Statist.* **37**, 1173–1185.

Starr, N. and Woodroofe, M. B. (1969). Remarks on sequential point estimation. *Proc. Natl. Acad. Sci. USA* **63**, 285–288.

Starr, N. and Woodroofe, M. B. (1972). Further remarks on sequential estimation: the exponential case. *Ann. Math. Statist.* **43**, 1147–1154.

Stein, C. (1945). A two-sample test for a linear hypothesis whose power is independent of the variance. *Ann. Math. Statist.* **16**, 243–258.

Stein, C. (1949). Some problems in sequential estimation. *Econometrica* **17**, 77–78.

Stein, C. and Wald, A. (1947). Sequential confidence intervals for the mean of a normal distribution with known variance. *Ann. Math. Statist.* **18**, 427–433.

Vardi, Y. (1979a). Asymptotic optimality of certain sequential estimators. *Ann. Statist.* **7**, 1034–1039.

Vardi, Y. (1979b). Asymptotic optimal sequential estimation: the Poisson case. *Ann. Statist.* **7**, 1040–1051.

Woodroofe, M. B. (1977). Second order approximations for sequential point and interval estimation. *Ann. Statist.* **5**, 984–995.

Reprinted from
Probability and Statistics
235–245 (1959)

SEQUENTIAL ESTIMATION OF THE MEAN OF A NORMAL POPULATION

HERBERT ROBBINS[1,2]
Columbia University

We consider the problem of estimating the unknown mean μ of a normal population when the variance σ^2 of the population is also unknown. As estimator we shall use the sample mean $\bar{x}_n$ of a sample of size n, so that the problem is simply that of choosing n. (In particular we shall consider two rules for determining n sequentially.) As a measure of the loss incurred when a sample of size n is taken we choose the quantity

$$L = A \cdot |\bar{x}_n - \mu| + n, \tag{1}$$

where A is some positive known constant; this involves the assumption that the cost of sampling is proportional to the sample size and that the unit of cost is chosen so that each observation costs one unit. (The use of the absolute value of the error of estimate, rather than its square, as the loss due to error of estimate is somewhat unconventional but in the present case causes no mathematical difficulties.)

We begin by asking the following question: What fixed sample size n minimizes the expected loss $E(L)$? Writing

$$L = A \cdot \left| \sqrt{n} \cdot \frac{\bar{x}_n - \mu}{\sigma} \right| \frac{\sigma}{\sqrt{n}} + n \tag{2}$$

and observing that the expression within the absolute value sign is, for samples of fixed size n, normal (0,1), so that

[1] This research was supported by the United States Air Force through the Air Force Office of Scientific Research of the Air Research and Development Command, under Contract No. AF 18(600)–442. Reproduction in whole or in part is permitted for any purpose of the United States Government.

[2] Presented at the September, 1957, meeting of the Institute of Mathematical Statistics in Atlantic City. At about the same time there appeared an article by W. D. Ray, "Sequential confidence intervals for the mean of a normal population with unknown variance," *Journal of the Royal Statistical Society*, Series B, Vol. 19 (1957), pp. 133–143, which is closely related to the present paper.

$$E\left|\sqrt{n}\cdot\frac{\bar{x}_n-\mu}{\sigma}\right|=\frac{1}{\sqrt{2\pi}}\int_{-\infty}^{\infty}|y|\,e^{\frac{y^2}{2}}\,dy=\sqrt{\frac{2}{\pi}}, \tag{3}$$

we have *for samples of fixed size n* the expected loss

$$E_n(L)=A\sqrt{\frac{2}{\pi}}\,\frac{\sigma}{\sqrt{n}}+n. \tag{4}$$

To find the value of n which minimizes this we regard n as a continuous variable, differentiate, and set the derivative equal to 0, obtaining the relation

$$A\sqrt{\frac{2}{\pi}}\,\sigma\left(-\frac{1}{2\,n^{\frac{3}{2}}}\right)+1=0, \tag{5}$$

which yields for the minimizing value of n the expression

$$n_0=\left(\frac{A\,\sigma}{\sqrt{2\pi}}\right)^{\frac{2}{3}}. \tag{6}$$

For this value of n (which need not be an integer) the value of (4) becomes

$$E_{n_0}(L)=A\sqrt{\frac{2}{\pi}}\,\sigma\left(\frac{A\,\sigma}{\sqrt{2\pi}}\right)^{-\frac{1}{3}}+\left(\frac{A\,\sigma}{\sqrt{2\pi}}\right)^{\frac{2}{3}}=\left(\frac{A\,\sigma}{\sqrt{2\pi}}\right)^{\frac{2}{3}}(2+1)=3\,n_0, \tag{7}$$

so that the loss due to the error of estimate is twice the cost of sampling.

All this presupposes that we know σ, so that we can determine n_0 by (6). What if we do not? It is then reasonable to consider the following sequential sampling rule R_1: Sample sequentially, obtaining values $x_1, x_2, \ldots$. Compute the successive sample means and variances,

$$\bar{x}_n=\frac{1}{n}\sum_{i=1}^{n}x_i,\quad s_n^2=\frac{1}{n-1}\sum_{i=1}^{n}(x_i-\bar{x}_n)^2, \tag{8}$$

and *stop sampling with the first* $n\geqslant 2$ (or, if we want to be conservative, $\geqslant 3$ or any other fixed integer) for which, in analogy with (6),

$$n\geqslant\left(\frac{A\,s_n}{\sqrt{2\pi}}\right)^{\frac{2}{3}}; \tag{9}$$

i.e. for which

$$\sum_{i=1}^{n}(x_i-\bar{x}_n)^2\leqslant\frac{2\pi}{A^2}(n-1)\,n^3. \tag{10}$$

(Since the right-hand side of (9) tends to the constant n_0 with probability 1 it is clear that the probability of an infinite sample is 0.)

In using the rule R_1 the sample size will be a random variable N. We must therefore compute $E(L)$ anew and compare it with the value $3n_0$ given by (7) to see how much we lose through ignorance of σ. To do this we make a few preliminary remarks in the form of lemmas.

LEMMA 1. For any fixed n, $\bar{x}_n$ is independent of the whole set of random variables $s_2^2, s_3^2, \ldots, s_n^2$ and hence

$$Pr\left[\sqrt{n}\cdot\frac{\bar{x}_n-\mu}{\sigma}\leqslant t\,\middle|\,s_2^2, s_3^2, \ldots, s_n^2\right]=\Phi(t), \tag{11}$$

where Φ is the normal (0,1) distribution function.

Proof. Define

$$u_i=\frac{x_i-\mu}{\sigma} \qquad (i=1, \ldots, n),$$

$$y_i=\frac{u_1+\cdots+u_i-i\,u_{i+1}}{\sqrt{i(i+1)}} \qquad (i=1, \ldots, n-1), \tag{12}$$

$$y_n=\frac{u_1+\cdots+u_n}{\sqrt{n}}.$$

The u_i are independent normal (0,1), and since the transformation to the y_i is orthogonal, the same is true of the y_i. But it is easily seen that

$$\sqrt{n}\cdot\frac{\bar{x}_n-\mu}{\sigma}=y_n,$$

$$s_i^2=\sigma^2\left(\frac{y_1^2+\cdots+y_{i-1}^2}{i-1}\right) \qquad (i=2, \ldots, n), \tag{13}$$

which proves the lemma. At the same time we have proved

LEMMA 2. The joint distribution of the sequence of random variables $\{s_n^2\}$ $(n=2, 3, \ldots)$ is the same as that of the sequence $\left\{\frac{\sigma^2}{n-1}(y_1^2+\cdots+y_{n-1}^2)\right\}$, where the y_i are independent normal (0,1).

Next we assert

LEMMA 3. Let R be any sequential sampling rule (which terminates with probability 1) in which the sample size $N=2, 3, \ldots$ is determined uniquely in such a way that $N=n$ if and only if the point $(s_2^2, \ldots, s_n^2)$ belongs to some set S_{n-1} of $(n-1)$ space. Then for any $n=2, 3, \ldots,$

$$Pr\left[\sqrt{N}\cdot\frac{\bar{x}_N-\mu}{\sigma}\leqslant t \mid N=n\right]=\Phi(t), \tag{14}$$

and hence the unconditional distribution of $\sqrt{N}\cdot\dfrac{\bar{x}_N-\mu}{\sigma}$ is also normal (0,1).

Proof. By Lemma 1,

$$Pr\left[\sqrt{N}\cdot\frac{\bar{x}_N-\mu}{\sigma}\leqslant t \mid N=n\right]$$

$$=Pr\left[(s_2^2,\ldots,s_n^2)\in S_{n-1}\quad\text{and}\quad\sqrt{n}\cdot\frac{\bar{x}_n-\mu}{\sigma}\leqslant t\right]\Big/Pr[N=n]$$

$$=Pr\left[(s_2^2,\ldots,s_n^2)\in S_{n-1}\right]\cdot Pr\left[\sqrt{n}\cdot\frac{\bar{x}_n-\mu}{\sigma}\leqslant t\right]\Big/Pr[N=n]$$

$$=\Phi(t).$$

Finally, we have

Lemma 4. For the particular sampling rule R_1 the probability distribution of the sample size N may be obtained as follows. Let $\{y_n\}$ $(n=1,2,\ldots)$ be independent normal (0,1) and let $N=n$ whenever n is the first integer $\geqslant 2$ such that

$$y_1^2+\cdots+y_{n-1}^2\leqslant\frac{(n-1)\,n^3}{n_0^3};\quad n_0^3=\frac{A^2\sigma^2}{2\pi}. \tag{15}$$

Proof. From (10), (6), and Lemma 2.

Returning now to our original problem we see by Lemma 3 that for the sampling rule R_1 the formulas for L and $E(L)$ become

$$L=A\left|\sqrt{N}\cdot\frac{\bar{x}_N-\mu}{\sigma}\right|\frac{\sigma}{\sqrt{N}}+N,$$

and

$$E_{R_1}(L)=\sum_{n=2}^{\infty}Pr[N=n]\cdot E(L\mid N=n) \tag{16}$$

$$=\sum_{n=2}^{\infty}Pr[N=n]\cdot\left(A\sqrt{\frac{2}{\pi}}\cdot\frac{\sigma}{\sqrt{n}}+n\right)$$

$$=A\sqrt{\frac{2}{\pi}}\,\sigma E(N^{-\frac{1}{2}})+E(N).$$

$$=2n_0^{3/2}\cdot E(N^{-\frac{1}{2}})+E(N),$$

which is to be compared with (7). To do this we can compute $E(N^{-\frac{1}{2}})$ and $E(N)$ by determining the probability distribution of N according to Lemma 4. The computational problem involved is considerable. Monte Carlo methods may be used to determine the distribution of N empirically; we shall give some numerical results of this sort in a moment.

Of course we hope that in using R_1, $E(N)$ will be nearly equal to n_0, irrespective of σ. It is in fact easy to see that for n_0 not too small

$$Pr[N \leqslant n_0] \geqslant \tfrac{1}{2}, \tag{17}$$

so that at least half the time the random sample size N will be no larger than the optimal fixed sample size n_0. To establish (17) we note that for any fixed integer n,

$$Pr[N \leqslant n] \geqslant Pr\left[y_1^2 + \cdots + y_{n-1}^2 \leqslant \frac{(n-1)\,n^3}{n_0^3}\right]. \tag{18}$$

Hence for $n = n_0$ we have

$$Pr[N \leqslant n_0] \geqslant Pr[y_1^2 + \cdots + y_{n_0-1}^2 \leqslant n_0 - 1] = Pr[\chi^2_{n_0-1} \leqslant n_0 - 1] \cong \tfrac{1}{2}. \tag{19}$$

The following results were obtained by Dr. Esther Seiden using an IBM 704 computer to generate random numbers. For each value of n_0 a thousand sequences $\{y_n\}$ of random normal deviates were used to compute average values E^*N and $E^*N^{-\frac{1}{2}}$.

n_0	2	4	8	16	32	64
E^*N	2.34	3.66	7.39	15.49	31.87	63.84
$E^*N^{-\frac{1}{2}}$	.665	.552	.397	.269	.180	.127
$2n_0^{\frac{3}{2}}E^*N^{-\frac{1}{2}}$	3.762	8.83	17.97	34.43	65.16	130.05
$E^*_{R_1}(L)$	6.10	12.49	25.36	49.92	97.03	193.89
$E_{n_0}(L)$	6.00	12.00	24.00	48.00	96.00	192.00

The difference $E_{R_1}(L) - E_{n_0}(L)$ is a measure of our "regret" in using R_1 as compared with the fixed sample size n_0 which we would use if we knew σ. The computations above show that the regret is small over the indicated range of values of n_0. What the minimax regret sampling rule would be we do not know.

In order to obtain an exact mathematical formula for the probabilities

$Pr[N = n]$ of our sequential sampling rule we shall find it very convenient to modify R_1 slightly by sampling *two at a time*, thus obtaining a new rule R_2 which we define as follows.

Sample sequentially, obtaining values $x_1, x_2, \ldots$, and stop with the first $n = 2m + 1$ ($m = 1, 2, 3, \ldots$) for which, in analogy with (10),

$$\sum_{i=1}^{2m+1} (x_i - \bar{x}_{2m+1})^2 \leqslant \frac{2\pi}{A^2} \cdot (2m)(2m+1)^3. \tag{20}$$

Again introducing a sequence $\{y_n\}$ of independent $N(0,1)$ random variables, we see that for the rule R_2 the probability that $N = 2m + 1$ is the same as the probability that m is the first positive integer for which

$$y_1^2 + \cdots + y_{2m}^2 \leqslant \frac{2\pi}{A^2 \sigma^2} \cdot (2m)(2m+1)^2. \tag{21}$$

Now $v = y_1^2 + y_2^2$ has the chi-square distribution with 2 degrees of freedom, with probability density function

$$\tfrac{1}{2} e^{-\frac{v}{2}} \quad (v > 0). \tag{22}$$

Hence

$$z = \tfrac{1}{2} v = \tfrac{1}{2}(y_1^2 + y_2^2) \tag{23}$$

has the negative exponential density function

$$g(z) = e^{-z} \quad (z > 0) \tag{24}$$

Thus if we introduce a sequence $\{z_n\}$ of independent random variables with the common density function (24) we can assert that *for the rule R_2 the probability that $N = 2m + 1$ is the same as the probability that m is the first positive integer for which*

$$z_1 + z_2 + \cdots + z_m \leqslant \frac{m(2m+1)^3}{n_0^3}; \quad n_0^3 = \frac{A^2 \sigma^2}{2\pi}. \tag{25}$$

The latter probability can be evaluated numerically as follows, by a method for which the author is indebted to J. E. Moyal.

Problem. Let $a_2 < a_3 < \cdots$ be any increasing sequence of positive constants $\left(\text{in the application to (25) } a_m \text{ will be } \frac{(m-1)(2m-1)^3}{n_0^3}\right)$, let $\{z_n\}$ be independent with density (24), and let $p_m = Pr[N = 2m + 1]$ be the probability that m is the first positive integer for which

$$z_1 + \cdots + z_m \leqslant a_{m+1}. \tag{26}$$

It is required to evaluate the probabilities p_m for $m = 1, 2, \ldots$.

To do this we define for $x \geqslant 0$,

$$S_m = z_1 + \cdots + z_m, \tag{27}$$

$$G_1(x) = Pr[S_1 \leqslant x] = 1 - e^{-x}, \tag{28}$$

and for $m = 2, 3, \ldots$

$$G_m(x) = Pr[S_1 > a_2,\ S_2 > a_3,\ \ldots,\ S_{m-1} > a_m;\ S_m \leqslant x]. \tag{29}$$

Then for $m = 1, 2, \ldots$

$$p_m + p_{m+1} + \cdots = Pr[S_1 > a_2,\ S_2 > a_3,\ \ldots,\ S_{m-1} > a_m] = G_m(\infty), \tag{30}$$

and hence

$$p_m = G_m(\infty) - G_{m+1}(\infty). \tag{31}$$

Now it is easily seen that

$$G_m(x) = \int_{a_m}^{x} G_1(x-y)\, dG_{m-1}(y) \quad \text{for} \quad x \geqslant a_m \quad (m = 2, 3, \ldots), \tag{32}$$

and hence, setting $g_m(x) = G_m'(x)$,

$$\left.\begin{aligned} &g_1(x) = e^{-x} \quad \text{for} \quad x \geqslant 0, \\ &g_m(x) = \int_{a_m}^{x} e^{-(x-y)} g_{m-1}(y)\, dy \quad (m = 2, 3, \ldots) \quad \text{for} \quad x \geqslant a_m. \end{aligned}\right\} \tag{33}$$

Now define

$$h_m(x) = e^{x} g_m(x) \quad (m = 1, 2, \ldots). \tag{34}$$

Then (33) becomes

$$\left.\begin{aligned} &h_1(x) = 1 \quad \text{for} \quad x \geqslant 0, \\ &h_m(x) = \int_{a_m}^{x} h_{m-1}(y)\, dy \quad (m = 2, 3, \ldots) \quad \text{for} \quad x \geqslant a_m. \end{aligned}\right\} \tag{35}$$

Clearly, $h_m(x)$ is a polynomial in x of degree $m-1$, so we can write

$$h_m(x) = \sum_{j=1}^{m-1} \frac{(x-a_m)^j}{j!} \cdot h_m^{(j)}(a_m). \tag{36}$$

16 – 585193

But from (35) the jth derivate of $h_m(x)$ is

$$h_m^{(j)}(x) = h_{m-j}(x) \quad \text{for} \quad x \geqslant a_m, \tag{37}$$

so that we have

$$h_m(x) = \sum_{j=1}^{m-1} \frac{(x-a_m)^j}{j!} \cdot h_{m-j}(a_m), \tag{38}$$

and in particular,

$$h_m(a_n) = \sum_{j=1}^{m-1} \frac{(a_n-a_m)^j}{j!} \cdot h_{m-j}(a_m) \quad \begin{pmatrix} m=2,\, 3,\, \ldots \\ n=m+1,\, m+2,\, \ldots \end{pmatrix}. \tag{39}$$

The values $h_m(a_n)$ can be computed recursively from (39), recalling that $h_1 = 1$. We then have

$$G_m(x) = \int_{a_m}^{x} g_m(y)\,dy = \sum_{j=1}^{m-1} h_{m-j}(a_m) \int_{a_m}^{x} e^{-y} \frac{(y-a_m)^j}{j!}\,dy$$

$$= \sum_{j=1}^{m-1} h_{m-j}(a_m) e^{-a_m} \cdot \int_0^{x-a_n} e^{-t} \frac{t^j}{j!}\,dt, \tag{40}$$

so that

$$G_m(\infty) = e^{-a_m} \cdot \sum_{j=1}^{m-1} h_{m-j}(a_m). \tag{41}$$

Letting

$$c_m = p_m + p_{m+1} + \cdots = G_m(\infty), \tag{42}$$

we have from (41)

$$c_m = e^{-a_m} \cdot \sum_{j=1}^{m-1} h_{m-j}(a_m) \quad (m=2,\, 3,\, \ldots); \quad c_1 = 1. \tag{43}$$

Also,

$$p_m = c_m - c_{m+1} = Pr[N = 2m+1], \tag{44}$$

and

$$E(N) = \sum_{m=1}^{\infty} (2m+1) p_m = 1 + 2 \sum_{m=1}^{\infty} m p_m = 1 + 2 \sum_{1}^{\infty} c_m. \tag{45}$$

(Of course

$$E(N^\alpha) = \sum_{m=1}^{\infty} (2m+1)^\alpha p_m$$

for any α.) As a check on our computations we observe that

$$\sum_{m=1}^{\infty} p_m = 1 \quad \text{and} \quad c_m \downarrow 0 \quad \text{as} \quad m \to \infty. \tag{46}$$

To recapitulate: in order to solve our Problem we define $h_1 = 1$ and recursively compute

$$h_m(a_n) = \sum_{j=1}^{m-1} \frac{(a_n - a_m)^j}{j!} h_{m-j}(a_m) \tag{47}$$

for $m = 2, 3, \ldots$ and $n = m+1, m+2, \ldots$, and then compute

$$c_m = e^{-a_m} \sum_{j=1}^{m-1} h_{m-j}(a_m) \quad (m = 2, 3, \ldots); \quad c_1 = 1. \tag{48}$$

Then for $m = 1, 2, \ldots$

$$Pr[N = 2m+1] = p_m = c_m - c_{m+1}. \tag{49}$$

In particular, for the sampling rule R_2 we set

$$a_m = \frac{(m-1)(2m-1)^3}{n_0^3}; \quad n_0^3 = \frac{A^2 \sigma^2}{2\pi}, \tag{50}$$

and recall that we *hope* that the sample size N will be close to n_0 most of the time.

We present herewith the results of this computation for the two cases $n_0^3 = 100$ and $n_0^3 = 1000$, corresponding to optimal fixed sample sizes of $n_0 = 4.6416$ and $n_0 = 10$. The computations were performed by Mr. Simon Stiassny of the Department of Statistics at Berkeley.

Performance of R_2

$n_0 = 4.6416$ $(n_0^3 = 100)$	
$n = 2m+1$	$p_m = Pr[N = n]$
3	0.236621
5	0.498244
7	0.263140
9	0.001995
	1.000000

$$EN = 5.061018$$

$$E(N^{-\frac{1}{2}}) = 0.459557$$

$$E_{R_2}(L) = 2n_0^{\frac{3}{2}} \cdot E(N^{-\frac{1}{2}}) + EN = 14.2522$$

$$(E_{n_0}(L) = 13.9248)$$

$n_0 = 10$	
$n = 2m+1$	$p_m = Pr[N=n]$
3	0.026639
5	0.020887
7	0.066520
9	0.244740
11	0.442514
13	0.189534
15	0.009149
17	0.000017
	1.000000

$$EN = 10.321772$$
$$E(N^{-\frac{1}{2}}) = 0.319800$$
$$E_{R_*}(L) = 30.5477$$
$$(E_{n_*}(L) = 30)$$

For larger values of n_0 the labor of hand computing becomes prohibitive. Through the courtesy of Professor Paul Minton of the Statistical Laboratory of Oklahoma State University the case $n_0 = 30$ was computed on a Univac 1103 by Mrs. Jean Richmond, with the following result.

As an application of the present theory consider the problem of sampling inspection from the following point of view. Suppose a lot of A items is submitted by the producer to the consumer who must decide by sampling not whether to accept or reject the lot but rather *how much* to pay for it. Assume that each of the A items has a fair value which can be determined exactly by inspection but that inspection of a single item costs one unit, so that to inspect the whole lot would cost A units, a sum which may be appreciable compared to the total value of the lot. Under these circumstances the consumer may take a sample of n items from the lot, where n is small compared to A, and ascertain the values $x_1, \ldots, x_n$ of the items sampled. Assuming the sample to be characteristic of the lot, a fair price for the lot would then be $\bar{x}_n \cdot A$, in the sense that the expected value of this random variable is equal to the true value of the lot. Now if from the point of view of society as a whole we regard overpayment or underpayment as equally undesirable, we may then say that the total loss due to sampling is measured by

$n_0 = 30$	
$n = 2m + 1$	$p_m = Pr[N = n]$
⋮	⋮
15	.0000
17	.0001
19	.0005
21	.0024
23	.0104
25	.0386
27	.1117
29	.2268
31	.2927
33	.2169
35	.0829
37	.0147
39	.0011
41	.0000
⋮	⋮
	.9988

$EN = 30.5856$

$E(N^{-\frac{1}{2}}) = 0.1817$

$E_{R_2}(L) = 90.3$

$(E_{n_0}(L) = 90)$

$$L = |\bar{x}_n \cdot A - \mu \cdot A| + n = A \cdot |\bar{x}_n - \mu| + n$$

where u denotes the mean of the A items in the lot. If A is large and if the values in the lot are approximately normally distributed about μ then the preceding theory applies. It would seem that for some purposes this procedure of evaluation should be preferable to the usual acceptance-rejection decision. (Such a suggestion has been made by H. Steinhaus.) In the extreme case in which the items have values 1 or 0 only, according as they are acceptable or defective, the assumption of a normally distributed lot is of course strongly violated and the rules R_1 or R_2 would be inapplicable. It would be interesting to see whether desirable sequential sampling schemes exist for the latter case.

Reprinted from THE ANNALS OF MATHEMATICAL STATISTICS
Vol. 36, No. 2, April, 1965

ON THE ASYMPTOTIC THEORY OF FIXED-WIDTH SEQUENTIAL CONFIDENCE INTERVALS FOR THE MEAN

BY Y. S. CHOW[1] AND HERBERT ROBBINS[2]

Purdue University and Columbia University

1. Introduction. Let $x_1, x_2, \cdots$ be a sequence of independent observations from some population. We want to find a confidence interval of prescribed width $2d$ and prescribed coverage probability α for the unknown mean μ of the population. If the variance σ^2 of the population is known, and if d is small compared to σ^2, this can be done as follows. For any $n \geqq 1$ define

$$\bar{x}_n = n^{-1}\sum_1^n x_i, \qquad I_n = [\bar{x}_n - d, \bar{x}_n + d],$$

and choose a to satisfy

$$(2\pi)^{-\frac{1}{2}}\int_{-a}^{a} e^{-u^2/2}\, du = \alpha.$$

Then for a sample size n determined by

$$(1) \qquad n = \text{smallest integer} \geqq (a^2\sigma^2)/d^2,$$

the interval I_n has coverage probability

$$P(\mu \in I_n) = P(\sqrt{n}|\bar{x}_n - \mu|/\sigma \leqq d\sqrt{n}/\sigma).$$

Since (1) implies that $\lim_{d\to 0} (d^2n)/(a^2\sigma^2) = 1$, it follows from the central limit theorem that

$$\lim_{d\to 0} P(\mu \in I_n) = (2\pi)^{-\frac{1}{2}}\int_{-a}^{a} e^{-u^2/2}\, du = \alpha.$$

We shall be concerned with the case in which the nature of the population, and hence σ^2, is unknown, so that no fixed sample size method is available. Define

$$(2) \qquad v_n = n^{-1}\sum_1^n (x_i - \bar{x}_n)^2 + n^{-1} \qquad (n \geqq 1),$$

let $a_1, a_2, \cdots$ be any sequence of positive constants such that $\lim_{n\to\infty} a_n = a$, and define

$$(3) \qquad N = \text{smallest } k \geqq 1 \quad \text{such that} \quad v_k \leqq (d^2k)/a_k^2.$$

The object of the present note is to prove the following

THEOREM. *Under the sole assumption that* $0 < \sigma^2 < \infty$,

---

Received 5 October 1964.

[1] Research supported by the Office of Naval Research under Contract No. Nonr-1100(26).

[2] Research supported by the Office of Naval Research under Contract No. Nonr-266(59), Project No. 042-205.

(4) $$\lim_{d\to 0} (d^2N)/(a^2\sigma^2) = 1 \quad a.s.,$$

(5) $$\lim_{d\to 0} P(\mu \varepsilon I_N) = \alpha \qquad (\textit{asymptotic "consistency"}),$$

(6) $$\lim_{d\to 0} (d^2EN)/(a^2\sigma^2) = 1. \qquad (\textit{asymptotic "efficiency"}).$$

REMARKS.

1. In case the distribution function of the x_i is continuous, Definition (2) can be replaced by, e.g.,

(7) $$v_n = n^{-1}\sum_1^n (x_i - \bar{x}_n)^2.$$

2. As will become evident from the proof, N in (3) could be defined as the smallest (or the smallest odd, etc.) integer $\geqq n_0$ such that the indicated inequality holds, where n_0 is any fixed positive integer.

2. Proof of the theorem.

LEMMA 1. *Let* y_n $(n = 1, 2, \cdots)$ *be any sequence of random variables such that* $y_n > 0$ *a.s.*, $\lim_{n\to\infty} y_n = 1$ *a.s., let* $f(n)$ *be any sequence of constants such that*

$$f(n) > 0, \qquad \lim_{n\to\infty} f(n) = \infty, \qquad \lim_{n\to\infty} f(n)/f(n-1) = 1,$$

and for each $t > 0$ *define*

(8) $$N = N(t) = \textit{smallest } k \geqq 1 \textit{ such that } y_k \leqq f(k)/t.$$

Then N *is well-defined and non-decreasing as a function of* t,

(9) $$\lim_{t\to\infty} N = \infty \quad a.s., \qquad \lim_{t\to\infty} EN = \infty,$$

and

(10) $$\lim_{t\to\infty} f(N)/t = 1 \quad a.s.$$

PROOF. (9) is easily verified. To prove (10) we observe that for $N > 1$, $y_N \leqq f(N)/t < [f(N)/f(N-1)]y_{N-1}$, whence (10) follows as $t \to \infty$.

LEMMA 2. *If the conditions of Lemma* 1 *hold and if also* $E(\sup_n y_n) < \infty$, *then*

(11) $$\lim_{t\to\infty} Ef(N)/t = 1.$$

PROOF. Let $z = \sup_n y_n$; then $Ez < \infty$. Choose m such that $f(n)/f(n-1) \leqq 2$, $(n > m)$. Then for $N > m$

$$f(N)/t = [f(N)f(N-1)]/[f(N-1)t] < 2y_{N-1} < 2z.$$

Hence for $t \geqq 1$,

(12) $$f(N)/t \leqq 2z + f(1) + \cdots + f(m).$$

(11) follows from (10), (12), and Lebesgue's dominated convergence theorem.

PROOF OF (4) AND (5). Set

(13) $$y_n = v_n/\sigma^2 = (1/n\sigma^2)(\sum_1^n (x_i - \bar{x}_n)^2 + 1),$$

(14) $$f(n) = (na^2)/a_n^2, \qquad t = (a^2\sigma^2)/d^2;$$

then (3) can be written as

$$N = N(t) = \text{smallest } k \geqq 1 \quad \text{such that} \quad y_k \leqq f(k)/t.$$

By Lemma 1,

$$(15) \qquad 1 = \lim_{t\to\infty} f(N)/t = \lim_{d\to 0} (d^2N)/(a^2\sigma^2) \quad \text{a.s.,}$$

which proves (4). Now

$$P(\mu \,\varepsilon\, I_N) = P(|x_1 + \cdots + x_N - N\mu|/\sigma\sqrt{N} \leqq d\sqrt{N}/\sigma).$$

By (15), $d\sqrt{N}/\sigma \to a$ and $N/t \to 1$ in probability as $t \to \infty$; it follows from a result of Anscombe [1] that as $t \to \infty$,

$$(x_1 + \cdots + x_N - N\mu)/\sigma\sqrt{N} \sim N(0, 1).$$

Hence

$$\lim_{t\to\infty} P(\mu \,\varepsilon\, I_N) = (2\pi)^{-\frac{1}{2}}\int_{-a}^{a} e^{-u^2/2}\, du = \alpha,$$

which proves (5).

It remains to prove (6). This is an immediate consequence of Lemma 2 whenever the distribution of the x_i is such that

$$(16) \qquad E\{\sup_n (n^{-1}\textstyle\sum_1^n (x_i - \bar{x}_n)^2)\} < \infty,$$

for then

$$(17) \qquad \lim_{t\to\infty} [Ef(N)]/t = 1,$$

and from the fact that the function $f(n)$ defined by (14) is $n + o(n)$ it follows from (17) that

$$1 = \lim_{t\to\infty} EN/t = \lim_{d\to 0} (d^2EN)/(a^2\sigma^2).$$

For (16) to hold it would suffice for the fourth moment of the x_i to be finite; however, we shall in the following prove that (6) holds without such a restriction. For this we need

Lemma 3. *If the conditions of Lemma 1 hold, if* $\lim_{n\to\infty} f(n)/n = 1$, *if for* N *defined by* (8),

$$(18) \qquad EN < \infty \ (\text{all } t > 0), \qquad \limsup_{t\to\infty} E(Ny_N)/EN \leqq 1,$$

and if there exists a sequence of constants $g(n)$ *such that*

$$g(n) > 0, \qquad \lim_{n\to\infty} g(n) = 1, \qquad y_n \geqq g(n)y_{n-1},$$

then

$$(19) \qquad \lim_{t\to\infty} EN/t = 1.$$

Proof. For any $0 < \epsilon < 1$ choose m so that

$$f(n-1) \geqq (1-\epsilon)f(n)$$

$$f(n-1) \geqq (1-\epsilon)n \qquad \text{for } n \geqq m$$

$$g(n) \geqq 1-\epsilon$$

and $E(Ny_N) \leqq (1+\epsilon)EN$ for $t \geqq m$. On the set $A = \{N \geqq m\}$ it follows that

$$[(1-\epsilon)^2/t]N^2 = (1-\epsilon)N\cdot(1-\epsilon)N/t \leqq g(N)Nf(N-1)/t$$
$$< g(N)Ny_{N-1} \leqq Ny_N .$$

Hence

$$[(1-\epsilon)^2/t](\textstyle\int_A N)^2 \leqq [(1-\epsilon)^2/t]\int_A N^2 \leqq \int_A Ny_N \leqq E(Ny_N),$$
$$[(1-\epsilon)^2/t]\textstyle\int_A N \leqq E(Ny_N)/\int_A N,$$
$$[(1-\epsilon)^2/t](EN-m) \leqq E(Ny_N)/(EN-m).$$

From (9) and (18) it follows that

$$(1-\epsilon)^2 \limsup_{t\to\infty} EN/t \leqq \limsup_{t\to\infty} E(Ny_N)/(EN) \leqq 1,$$

so that

$$\limsup_{t\to\infty} EN/t \leqq 1. \tag{20}$$

Now let $y_n' = \min(1, y_n)$. Then

$$0 < y_n' \leqq 1, \qquad y_n' \leqq y_n, \qquad \lim_{n\to\infty} y_n' = 1 \quad \text{a.s.}$$

Define

$$N' = N'(t) = \text{smallest } k \geqq 1 \quad \text{such that} \quad y_k' \leqq f(k)/t.$$

From Lemma 2, since $\sup_n (y_n') \leqq 1$,

$$1 = \lim_{t\to\infty} [Ef(N)]/t = \lim_{t\to\infty} (EN')/t.$$

But since $y_n' \leqq y_n$, $N' \leqq N$, and hence $EN' \leqq EN$. Thus

$$\liminf_{t\to\infty} (EN)/t \geqq \liminf_{t\to\infty} (EN')/t = 1,$$

which, with (20), proves (19).

PROOF OF (6). Fix $t > 0$, choose m such that $f(n)/t \geqq 1 (n \geqq m)$, choose $\delta > 0$ such that $(n-1)f(n-1) \geqq \delta n^2 (n \geqq 2)$, and define for any $r \geqq m$, $M = \min(N, r)$. By Wald's theorem for cumulative sums,

$$E(\textstyle\sum_1^M (x_i - \mu)^2) = EM\cdot E(x_i-\mu)^2 = EM\cdot\sigma^2.$$

Hence by (13),

$$E(My_M) = (1/\sigma^2)E(\textstyle\sum_1^M (x_i - \bar{x}_M)^2 + 1) \tag{21}$$
$$\leqq (1/\sigma^2)E(\textstyle\sum_1^M (x_i-\mu)^2 + 1) = EM + (1/\sigma^2).$$

Put $g(n) = (n-1)/n$, $(n \geqq 2)$; then

$y_n \geqq (1/n\sigma^2)\sum_1^{n-1}(x_i - \bar{x}_{n-1})^2 + (1/n\sigma^2) = [(n-1)/n]y_{n-1} = g(n)y_{n-1}.$
Hence

$$E(My_M) \geqq \int_{\{N>r\}} ry_r + \int_{\{N\leqq r\}} Ny_N \geqq [rf(r)/t]P(N > r) + \int_{\{2\leqq N\leqq r\}} Ny_N$$
$$\geqq rP(N > r) + \int_{\{2\leqq N\leqq r\}} [Ng(N)f(N-1)]/t$$
$$\geqq rP(N > r) + (\delta/t)\int_{\{2\leqq N\leqq r\}} N^2.$$

Hence by (21),

$$\int_{\{N\leqq r\}} N \geqq (\delta/t)\int_{\{2\leqq N\leqq r\}} N^2 - (1/\sigma^2) \geqq (\delta/t)(\int_{\{2\leqq N\leqq r\}} N)^2 - (1/\sigma^2),$$

and letting $r \to \infty$ it follows that

$$EN = \lim_{r\to\infty} \int_{\{N\leqq r\}} N < \infty,$$

which is the first part of (18). Again by Wald's theorem,

$$E(Ny_N) \leqq EN + (1/\sigma^2),$$

so by (9),

$$\limsup_{t\to\infty} [E(Ny_N)]/(EN) \leqq 1,$$

which is the second part of (18). All the conditions of Lemma 3 therefore hold, and hence

$$1 = \lim_{t\to\infty} EN/t = \lim_{d\to 0} (d^2EN)/(a^2\sigma^2),$$

which is (6). This completes the proof of the theorem of Section 1. As to Remark 1 following the theorem, it is clear that the only purpose of the term n^{-1} in (2) is to ensure that $y_n = v_n/\sigma^2 > 0$ a.s., this fact having been used in the proof of Lemma 1 to guarantee that $N \to \infty$ a.s. as $t \to \infty$. If the distribution function of the x_i is continuous the definition (7) is equally good, the only change being that the term $1/\sigma^2$ in the proof of (6) disappears.

The method used in this note is a modification of that used in [3] to prove the elementary renewal theorem. The theorem in this note has been proved when the x_i are $N(\mu, \sigma^2)$ by Stein [6], Anscombe [1], [2], and Gleser, Robbins, and Starr [4]. Some numerical computations for a slightly modified procedure have been made by Ray [5] who, apparently misled by having considered too few values of d, doubts the validity of (5) in his case. Extensive numerical computations in the $N(\mu, \sigma^2)$ case have been made by Starr and will soon be available. They indicate, for example, that for $\alpha = .95$ the lower bound for all $d > 0$ of $P(\bar{x}_N - d \leqq \mu \leqq \bar{x}_N + d)$, where N is the smallest odd integer $k \geqq 3$ such that

$$(k-1)^{-1}\sum_1^k (x_i - \bar{x}_k)^2 \leqq (d^2k)/a_k^2,$$

is about .929 if the values a_k are taken from the t-distribution with $(k-1)$ degrees of freedom.

REFERENCES

[1] Anscombe, F. J. (1952). Large sample theory of sequential estimation. *Proc. Cambridge Philos. Soc.* **48** 600–607.

[2] Anscombe, F. J. (1953). Sequential estimation. *J. Roy. Stat. Soc.* Ser. B **15** 1–21.

[3] Doob, J. L. (1948). Renewal theory from the point of view of the theory of probability. *Trans. Amer. Math. Soc.* **63** 422–438.

[4] Gleser, L. J., Robbins, H., and Starr, N. (1964). Some asymptotic properties of fixed-width sequential confidence intervals for the mean of a normal population with unknown variance. Report on National Science Foundation Grant NSF-GP-2074, Department of Mathematical Statistics, Columbia University.

[5] Ray, W. D. (1957). Sequential confidence intervals for the mean of a normal distribution with unknown variance. *J. Roy. Stat. Soc.* Ser. B **19** 133–143.

[6] Stein, C. (1949). Some problems in sequential estimation. *Econometrica* **17** 77–78.

Reprinted from The Annals of Mathematical Statistics
Vol. 38, No. 5, October 1967

FINDING THE SIZE OF A FINITE POPULATION[1]

By D. A. Darling and Herbert Robbins[2]

University of California, Berkeley

1. **Introduction.** There are many fixed sample size methods for estimating the unknown size N of a finite population by tagging the elements of a first sample and then counting the number of tagged elements of a second sample [1, p. 43]. Less well known are sequential methods [2], which have the advantage that the total sample size automatically adjusts itself to the unknown N to ensure a desired accuracy of the estimate. All such procedures only provide estimates for which the relative error is likely to be small. Suppose, however, that we want $P_N(\text{estimate} = N) \geq 1 - \alpha$, say, no matter what the value $N = 1, 2, \cdots$. How can this be done?

If we take as our estimate of N the number of distinct elements actually observed, the problem is one of finding a stopping rule such that the probability of having observed all N elements by the time sampling stops is $\geq 1 - \alpha$ for all $N \geq 1$. A concept of asymptotic efficiency may be introduced by comparing as $N \to \infty$ the expected sample size for any such rule with the fixed sample size necessary to observe all N elements with probability α. We give a procedure which is asymptotically efficient in this sense. We do not discuss the problem of finding a procedure which minimizes the Bayes expectation of the sample size for a given a priori distribution of N.

Before going on, the reader is invited to consider the following problem. Sample one element at a time with replacement, tagging each element observed so that it can be recognized if it appears again. Choose some large integer m, and stop sampling when for the first time a run of m consecutive tagged elements occurs; estimate N to be the total number of distinct elements observed. Can we choose m so large that $P_N(\text{estimate} = N) \geq .99$ for all $N = 1, 2, \cdots$? (Answer at end of paper.)

2. **A procedure based on individual waiting times.** An urn contains N white balls $(N = 1, 2, \cdots)$. We repeatedly draw a ball at random, observe its color, and replace it by a black ball. Eventually all N white balls will have been drawn and the urn will contain only black balls. The probability that this will occur at or before the nth draw is [3, pp. 92–93]

$$(1)\qquad P_{N,n} = \sum_{i=0}^{N} (-1)^i \binom{N}{i} (1 - i/N)^n,$$

and if $N, n \to \infty$ in such a way that $Ne^{-n/N} \to \lambda$, $0 < \lambda < \infty$, then

$$(2)\qquad P_{N,n} \to e^{-\lambda}.$$

For any fixed $0 < \alpha < 1$ the smallest $n = n(N, \alpha)$ such that $P_{N,n} \geq \alpha$ can be found from (1) by trial and error. For large N this is tedious, but it follows from

Received 7 December 1966.

[1] Supported in part by National Science Foundation Grant GP-[illegible].

[2] Visiting Miller Research Professor.

Reprinted from THE ANNALS OF MATHEMATICAL STATISTICS
Vol. 38, No. 5, October, 1967

FINDING THE SIZE OF A FINITE POPULATION

BY D. A. DARLING[1] AND HERBERT ROBBINS[2]

University of California, Berkeley

1. Introduction. There are fixed sample size methods for estimating the unknown size N of a finite population by tagging the elements of a first sample and then counting the number of tagged elements of a second sample [1, p. 43]. Less well known are sequential methods [2], which have the advantage that the total sample size automatically adjusts itself to the unknown N to assure a desired accuracy of the estimate. All these methods only provide estimates for which the *relative* error is likely to be small. Suppose, however, that we want P_N (estimate $= N) \geqq \alpha = .99$, say, no matter what the value $N = 1, 2, \cdots$. How can this be done?

If we take as our estimate of N the number of distinct elements actually observed, the problem is one of finding a stopping rule such that the probability of having observed all N elements by the time we stop is $\geqq \alpha$ for all $N \geqq 1$. A concept of asymptotic efficiency may be introduced by comparing as $N \to \infty$ the expected sample size for any such rule with the fixed sample size necessary to observe all N elements with probability α. We give a procedure which is asymptotically efficient in this sense. We do not discuss the problem of finding a procedure which minimizes the Bayes expectation of the sample size for a given prior distribution of N.

Before going on, the reader is invited to consider the following problem. Sample one element at a time with replacement, tagging each element observed so that it can be recognized if it appears again. Choose some large integer M, and stop sampling when for the first time a run of M consecutive tagged elements occurs; estimate N to be the total number of distinct elements observed. Can we choose M so large that P_N (estimate $= N) \geqq .99$ for all $N = 1, 2, \cdots$? (Answer at end of paper.)

2. A procedure based on individual waiting times. An urn contains N white balls ($N = 1, 2, \cdots$). We repeatedly draw a ball at random, observe its color, and replace it by a black ball. Eventually all N white balls will have been drawn and the urn will contain only black balls. The probability that this will occur at or before the nth draw is [1, pp. 92–93]

$$P_{N,n} = \sum_{i=0}^{N} (-1)^i \binom{N}{i} (1 - i/N)^n, \tag{1}$$

and if $N, n \to \infty$ so that $Ne^{-n/N} \to \lambda$, $0 < \lambda < \infty$, then

$$P_{N,n} \to e^{-\lambda}. \tag{2}$$

For any fixed $0 < \alpha < 1$ the smallest $n = n(N, \alpha)$ such that $P_{N,n} \geqq \alpha$ can be found from (1) by trial and error. For large N this is tedious, but it follows from

Received 7 December 1966.

[1] Supported in part by National Science Foundation Grant GP 6549.

[2] Visiting Miller Research Professor.

(2) that $n = N \log N + cN + o(N)$, where c is determined by the equation $e^{-e^{-c}} = \alpha$. Equivalently, if we denote by Y_N the number of the draw on which the last (Nth) white ball is drawn, then for any $-\infty < c < \infty$,

$$(3) \qquad P_N((Y_N - N \log N)/N \leqq c) \to e^{-e^{-c}} \quad \text{as} \quad N \to \infty.$$

Suppose now that N is unknown and that we wish to find a rule for deciding when to stop drawing such that for a given $0 < \alpha < 1$,

$$(4) \quad P_N(\text{all } N \text{ white balls drawn by the time we stop}) \geqq \alpha \quad (N \geqq 1).$$

No fixed sample size will do this, so we must look for a sequential procedure.

Let Y_n denote the number of the draw on which the nth white ball appears ($n = 1, \cdots, N$); thus $Y_1 \equiv 1$, and we put $Y_0 \equiv 0$, $Y_{N+1} \equiv \infty$ by convention. Define the waiting times

$$(5) \qquad X_n = Y_{n+1} - Y_n \qquad (n = 0, \cdots, N),$$

so $X_0 \equiv 1$ and $X_N \equiv \infty$. The random variables $X_1, \cdots, X_{N-1}$ are independent, with the geometric distributions

$$(6) \qquad P_N(X_n > j) = (n/N)^j \qquad (j = 0, 1, \cdots),$$

and

$$(7) \qquad E_N(X_n) = N/(N-n), \qquad \text{Var}_N(X_n) = nN/(N-n)^2.$$

Let (b_n) be any sequence of positive integers, and define B_n to be the event $X_n > b_n$ ($n = 1, \cdots, N$). Since $X_N \equiv \infty$, B_N is certain. Let J = first $n \geqq 1$ such that B_n occurs; then $1 \leqq J \leqq N$. Suppose we agree to stop drawing as soon as B_J occurs. All N white balls will have been drawn by the time we stop if and only if $J = N$. Hence the left hand side of (4) equals

$$(8) \quad P_N(J = N) = P_N\left(\textstyle\bigcap_1^{N-1} B_n'\right) = \textstyle\prod_1^{N-1} P(X_n \leqq b_n) = \prod_{n=1}^{N-1}\{1 - (n/N)^{b_n}\}.$$

It is clear that we can satisfy (4) by choosing the b_n properly. One way is the following. Define b_1^* = smallest integer b_1 such that

$$1 - (\tfrac{1}{2})b_1 \geqq \alpha.$$

Then (4) holds for $N = 2$ (and for $N = 1$ no matter what the sequence (b_n)). If $b_1^*, \cdots, b_{j-1}^*$ have been defined, set b_j^* = smallest integer b_j such that

$$\textstyle\prod_{n=1}^{j-1}\{1 - (n/(j+1))^{b_n^*}\} \cdot \{1 - (j/(j+1))^{b_j}\} \geqq \alpha.$$

The sequence (b_n^*) gives a "step-wise minimal" solution of the inequalities

$$(9) \qquad P_N(J = N) = \textstyle\prod_{n=1}^{N-1}\{1 - (n/N)^{b_n}\} \geqq \alpha \qquad (N = 2, 3, \cdots).$$

(It is not *uniformly* minimal, since if (b_n) satisfies (9) then we can increase $b_1, \cdots, b_{j-1}$ sufficiently so that a smaller b_j will still satisfy (9).) Although the b_n^* are not given by an explicit formula they can be computed numerically for any given α.

Instead of doing this we shall find a lower bound for the left hand side of (9) for the particular sequence

$$b_n = [cn] + 1 \qquad (n = 1, 2, \cdots), \tag{10}$$

where c is a suitable positive constant. We shall show that if c is chosen large enough then (9) holds, and shall find the limit of $P_N(J = N)$ as $N \to \infty$.

For the sequence (10) let β be any constant $0 < \beta < 1$. Then

$$\begin{aligned} (11)\quad P_N(J = N) \\ &= \prod_{1 \leqq n \leqq \beta N} \{1 - (n/N)^{b_n}\} \cdot \prod_{\beta N < n \leqq N-1} \{1 - (n/N)^{b_n}\} = Q_N \cdot R_N, \end{aligned}$$

where

$$Q_N \geqq \prod_{1 \leqq n \leqq \beta N} (1 - \beta^{cn}) \geqq \prod_{n=1}^{\infty} (1 - \beta^{cn}), \tag{12}$$

and since $\log(1 - x) \leqq -x$,

$$\begin{aligned} R_N &= \prod_{1 \leqq i < (1-\beta)N} \{1 - (1 - i/N)^{[c(N-i)]+1}\} \\ &\geqq \prod_{1 \leqq i < (1-\beta)N} \{1 - (1 - i/N)^{c(N-i)}\} \\ (13)\quad &= \prod_{1 \leqq i < (1-\beta)N} \{1 - e^{c(N-i)\log(1-i/N)}\} \\ &\geqq \prod_{1 \leqq i < (1-\beta)N} \{1 - e^{-ci(1-i/N)}\} \\ &\geqq \prod_{1 \leqq i < (1-\beta)N} \{1 - e^{-ci\beta}\} \geqq \prod_{i=1}^{\infty} (1 - e^{-ci\beta}). \end{aligned}$$

Defining for $0 \leqq x < 1$ the function

$$\varphi(x) = \prod_1^{\infty} (1 - x^n) \geqq 1 - \sum_1^{\infty} x^n = 1 - x/(1 - x) \to 1 \quad \text{as} \quad x \to 0, \tag{14}$$

we have the uniform lower bound

$$P_N(J = N) \geqq \varphi(\beta^c) \cdot \varphi(e^{-c\beta}) \qquad (N = 2, 3, \cdots). \tag{15}$$

In particular, if we choose β to be the root $\beta_0 = 0.56 \cdots$ of the equation

$$\beta = e^{-\beta}, \tag{16}$$

then

$$P_N(J = N) \geqq \varphi^2(e^{-c\beta_0}). \tag{17}$$

Choosing c to satisfy $\varphi^2(e^{-c\beta_0}) = \alpha$ it follows that (9) holds.

We can improve (15) somewhat for large N. Write (13) in the form

$$\log R_N = \sum_{i=1}^{\infty} a_{i,N}, \tag{18}$$

where

$$\begin{aligned} (19)\quad a_{i,N} &= \log \{1 - (1 - i/N)^{[c(N-i)]+1}\} \quad \text{for} \quad 1 \leqq i < (1 - \beta)N, \\ &= 0 \qquad\qquad \text{for } i \geqq (1 - \beta)N. \end{aligned}$$

For any fixed $i = 1, 2, \cdots$

$$\lim_{N \to \infty} a_{i,N} = \log(1 - e^{-ci}) = a_i, \text{ say}, \tag{20}$$

and for $1 \leqq i < (1-\beta)N$ we have as in (13),

$$(21)\quad 0 \geqq a_{i,N} \geqq \log\{1-(1-i/N)^{c(N-i)}\} \geqq \log(i - e^{-ci(1-i/N)}) \geqq \log(1-e^{-ci\beta}),$$

so this holds for *all* i, N, and

$$(22)\quad \textstyle\sum_{i=1}^{\infty} \log(1-e^{-ci\beta}) = \log\varphi(e^{-c\beta}) > -\infty.$$

By the dominated convergence theorem,

$$(23)\quad \lim_{N\to\infty} R_N = \lim_{N\to\infty} \exp\left[\textstyle\sum_{i=1}^{\infty} a_{i,N}\right] = \exp\left[\textstyle\sum_{i=1}^{\infty} a_i\right] = \textstyle\prod_{i=1}^{\infty}(1-e^{-ci}) = \varphi(e^{-c}).$$

Since by (12), $\varphi(\beta^c)R_N \leqq P_N(J = N) \leqq R_N$, it follows that

$$\varphi(\beta^c)\varphi(e^{-c}) \leqq \liminf_{N\to\infty} P_N(J=N) \leqq \limsup_{N\to\infty} P_N(J=N) \leqq \varphi(e^{-c}),$$

and since β can be arbitrarily near 0, and $\varphi(\beta^c) \to 1$ as $\beta \to 0$,

$$(24)\quad \lim_{N\to\infty} P_N(J=N) = \varphi(e^{-c}).$$

Thus from (17),

$$(25)\quad \varphi^2(e^{-c\beta_0}) \leqq P_N(J=N) \to \varphi(e^{-c}) \quad \text{as} \quad N\to\infty.$$

For any $\epsilon > 0$, if we increase the first $j = j(\epsilon)$ terms of (10) we can clearly strengthen (25) to read

$$(26)\quad \varphi(e^{-c}) - \epsilon \leqq P_N(J=N) \to \varphi(e^{-c}) \quad \text{as} \quad N\to\infty.$$

To see how efficient this procedure is, let us look at its sample size

$$(27)\quad S = X_0 + \cdots + X_{J-1} + b_J \leqq X_0 + \cdots + X_{N-1} + b_N.$$

From (7),

$$(28)\quad E_N(S) \leqq N(1 + \tfrac{1}{2} + \cdots + 1/N) + cN + 1,$$

which is somewhat greater than the fixed sample size $n = [N\log N + cN]$ for which we have seen that $P_{N,n} \to e^{-e^{-c}}$. Now

$$0 < \varphi(e^{-c}) < 1 - e^{-c} < e^{-e^{-c}} < 1,$$

so (26) shows that for $N\to\infty$ the probability of having drawn all the white balls by the sequential procedure is somewhat less than the corresponding probability for the fixed sample size $n = [N\log N + cN]$. Of course, the latter procedure requires a knowledge of N; moreover, the ratio of the two error probabilities is small for large values of c;

$$(29)\quad (1 - e^{-e^{-c}})/(1-\varphi(e^{-c})) \to 1 \quad \text{as} \quad c\to\infty.$$

Nevertheless, the fact remains that the sequential procedure is somewhat inefficient for any fixed c and large values of N. We therefore ask, is there any sequential procedure such that for fixed $-\infty < c < \infty$,

(30) Sample size $\leqq N \log N + cN$,

$$P_N \text{ (all } N \text{ white balls drawn by the time we stop)} \to e^{-e^{-c}} \text{ as } N \to \infty ?$$

An affirmative answer is given in the next section.

3. An asymptotically efficient procedure based on cumulative waiting times. We modify the sequential procedure of the previous section as follows. Let (a_n) be a sequence of positive constants, and define A_n to be the event that $Y_{n+1} > a_n$ $(n = 1, \cdots, N)$. Since $Y_{N+1} \equiv \infty$, A_N is certain. Let I = first $n \geqq 1$ such that A_n occurs. Then $1 \leqq I \leqq N$. We agree to stop as soon as A_I occurs. Then (cf. (8))

$$\begin{aligned} P_N \text{ (all } N \text{ white balls drawn by the time we stop)} &= P_N(I = N) \\ (31) \qquad &= P_N(\textstyle\bigcap_{n=1}^{N-1} A_n') = P_N(\bigcap_{n=1}^{N-1} (Y_{n+1} \leqq a_n)) \\ &= P_E(\textstyle\bigcap_{n=1}^{N-1} (X_0 + \cdots + X_n \leqq a_n)). \end{aligned}$$

As before, we could define a "step-wise minimal" sequence (a_n^*) such that the expression (31) is $\geqq \alpha$ for every $N \geqq 2$, but since the events A_n are not independent the explicit computation of the a_n^* is difficult. Instead, as before, we shall choose an explicit sequence (a_n) and estimate the value of (31) when N is large. Our sequence is the following. Let c be any finite constant, let n^* = smallest n such that $n \geqq e^{1-c}$, and define

$$\begin{aligned} (32) \qquad a_n &= n + 1 \quad \text{for} \quad n = 1, \cdots, n^* - 1, \\ &= n \log n + cn \quad \text{for} \quad n \geqq n^*. \end{aligned}$$

It is clear that for this procedure the sample size is always $\leqq a_N$ $(= N \log N + cN$ for $N \geqq n^*)$. And we shall prove that $P_N(I = N) \to e^{-e^{-c}}$ as $N \to \infty$. (It will be seen from the proof that as in (26) for any $\epsilon > 0$ we can increase the first $j = j(\epsilon)$ values of (a_n) so as to make

$$e^{-e^{-c}} - \epsilon \leqq P_N(I = N) \to e^{-e^{-c}} \quad \text{as} \quad N \to \infty.)$$

It is easy to check that

$$\begin{aligned} (33) \qquad a_n - a_{n-1} &\geqq 1 \quad \text{for all} \quad n \geqq 2, \\ &\geqq \log n + c \quad \text{for} \quad n \geqq n^*, \end{aligned}$$

and that the random variables

$$(34) \qquad w_n = X_{N-n}/N - 1/n \qquad (n = 1, \cdots, N)$$

are independent with

$$(35) \qquad E_N(w_n) = 0, \qquad \text{Var}_N(w_n) = 1/n^2 - 1/Nn < 1/n^2.$$

By Kolmogorov's inequality, for any $d > 0$,

(36) $P_N(w_1 + \cdots + w_n \leqq -d \text{ for some } n = 1, \cdots, N)$
$$\leqq (\textstyle\sum_1^\infty 1/n^2)/d^2 = \pi^2/6d^2.$$

We have

(37) $A_n' = (X_0 + \cdots + X_n \leqq a_n) = (w_{N-n} + \cdots + w_N \leqq b_n),$
$$b_n = a_n/N - ((N-n)^{-1} + \cdots + 1/N).$$

And for $n \geqq 2$,

(38) $$A'_{N-1} \cap (w_1 + \cdots + w_n \geqq b_{N-1} - b_{N-n}) \subset A'_{N-n}.$$

Now for $N \geqq n + n^*$,

$$b_{N-1} - b_{N-n} = ((a_{N-1} - a_{N-n})/N) + (1 + \tfrac{1}{2} + \cdots + 1/(n-1))$$

(39) $$= (n-1) \log N/N - (1 + \tfrac{1}{2} + \cdots + 1/(n-1)) + c(n-1)/N + f(1 - n/N) - f(1 - 1/N),$$

where

(40) $$0 < f(x) = -x \log x < e^{-1} \quad \text{for} \quad 0 < x < 1.$$

Hence for $N \geqq n + n^*$,

(41) $$b_{N-1} - b_{N-n} \leqq n \log N/N - \log n + \beta \quad (\beta = |c| + e^{-1}).$$

Put

(42) $$p = d + \beta + 1, \qquad k = e^p.$$

Then as $N \to \infty$,

$$k \log N/N - \log k + \beta \to -d - 1,$$

$$(N(1 - p/\log N)/N) \log N - \log (N(1 - p/\log N)) + \beta \to -d - 1,$$

and for $N \geqq N_d$,

(43) $$b_{N-1} - b_{N-n} \leqq -d \quad \text{for} \quad k \leqq n \leqq N(1 - p/\log N).$$

Hence from (38) and (36),

$$P_N(A'_{N-n} \quad \text{for all} \quad 1 \leqq n \leqq N(1 - p/\log N))$$

(44) $$\geqq P_N(A'_{N-1} \cap \cdots \cap A'_{N-k}) - P_N(w_1 + \cdots + w_n \leqq -d \text{ for some } n = 1, \cdots, N)$$
$$\geqq P_N(A'_{N-1} \cap \cdots \cap A'_{N-k}) - \pi^2/6d^2.$$

But from (3), as $N \to \infty$

(45) $P_N(A'_{N-1} \cap \cdots \cap A'_{N-k}) \geqq P_N(Y_N \leqq a_{N-k})$
$$= P_N((Y_N - N \log N)/N \leqq ((a_{N-k} - N \log N)/N) \to e^{-e^{-c}},$$

since as $N \to \infty$

$$\begin{aligned}(a_{N-k} - N \log N)/N &= ((N - k) \log (N - k) - c(N - k) - N \log N)/N \\ &= (1 - k/N) \log (1 - k/N) + (1 - k/N) \log N \\ &\quad - \log N - c(1 - k/N) \to -c.\end{aligned}$$

Hence

$$(46)\quad \liminf_{N\to\infty} P_N(A'_{N-n} \text{ for all } 1 \leqq n \leqq N(1 - p/\log N)) \geqq e^{-e^{-c}} - \pi^2/6d^2.$$

We shall show in a moment that

$$(47)\quad \lim_{N\to\infty} P_N(A_n' \text{ for all } 1 \leqq n \leqq pN/\log N) = 1.$$

It will then follow from (44) and (45) that

$$(48)\quad \liminf_{N\to\infty} P_N(A_n' \text{ for all } 1 \leqq n \leqq N - 1) \geqq e^{-e^{-c}} - \pi^2/6d^2.$$

Since d can be arbitrarily large, (48) holds without the last term. But

$$(49)\quad P_N(A_n' \text{ for all } 1 \leqq n \leqq N - 1) \leqq P_N(A'_{N-1}) \to e^{-e^{-c}}$$

by (45) for $k = 1$. Hence

$$(50)\quad \lim_{N\to\infty} P_N(A_n' \text{ for all } 1 \leqq n \leqq N - 1) = \lim_{N\to\infty} P_N(I = N) = e^{-e^{-c}}.$$

It remains only to prove (47). Now setting $a_0 = 1$

$$\begin{aligned}P_N(A_n' \text{ for all } 1 \leqq n \leqq pN/\log N) &= P_N\Big(\textstyle\bigcap_{1\leqq n\leqq pN/\log N} (X_0 + \cdots + X_n \leqq a_n)\Big) \\ &\geqq P_N\Big(\textstyle\bigcap_{1\leqq n\leqq (pN/\log N)} (X_n \leqq a_n - a_{n-1})\Big) \\ &= \textstyle\prod_{1\leqq n\leqq pN/\log N} \{1 - (n/N)^{a_n - a_{n-1}}\} \\ &\geqq 1 - \textstyle\sum_{1\leqq n\leqq pN/\log N} (n/N)^{a_n-a_{n-1}},\end{aligned}$$

and by (33), as $N \to \infty$

$$\textstyle\sum_{1\leqq n\leqq N^{\frac{1}{2}}/\log N} (n/N)^{a_n-a_{n-1}} \leqq \sum_{1\leqq n\leqq N^{\frac{1}{2}}/\log N} (n/N) \leqq 1/(\log N)^2 \to 0,$$

$$\begin{aligned}\textstyle\sum_{N^{\frac{1}{2}}/\log N<n\leqq pN/\log N} (i/N)^{a_n-a_{n-1}} &\leqq \textstyle\sum_{N^{\frac{1}{2}}/\log N<n\leqq pN/\log N} (p/\log N)^{\log n+c} \\ &\leqq (p/\log N)^c \textstyle\sum_{n>N^{\frac{1}{2}}/\log N} n^{\log p-\log\log N} \\ &\leqq (p/\log N)^c \cdot \textstyle\int_{N^{\frac{1}{2}}}^{\infty} (dx/x^2) \to 0,\end{aligned}$$

which completes the proof of (47).

The answer to the question in Section 1 is no; for any M,

$$\lim_{N\to\infty} P_N(\text{estimate} = N) = 0.$$

REFERENCES

[1] FELLER, W. (1957). *Introduction to Probability Theory and its Applications.* **1** (2nd ed.) Wiley, N. Y.

[2] GOODMAN, L. A. (1953). Sequential Sampling Tagging for Population Size Problems. *Ann. Math. Statist.* 56–69.

Rep. Stat. Appl. Res., JUSE
Vol. 9, No. 2, 1962

A-Section

A BAYES TEST OF "$p\leq\frac{1}{2}$" versus "$p>\frac{1}{2}$"*,**

By Sigeiti MORIGUTI AND Herbert ROBBINS

Columbia University, Department of Mathematical Statistics

1. Introduction.

Consider the following *sequential decision problem*:—An urn contains white and black balls, the proportion p of black balls being unknown, $0\leq p\leq 1$. A single ball is to be drawn at random from the urn tomorrow and we must guess its color; if our guess is correct we are to receive a reward of $\$A$, and if incorrect we are to receive nothing. In order to obtain some information about p we are allowed to draw as many balls as we like from the urn today, one at a time and with replacement, observing the color of each ball drawn, but for each such ball drawn we must pay \$1. Problem: how many balls should we observe today and what color should we then predict for the ball to be drawn tomorrow?

Let D denote any sequential decision procedure. This will consist of a rule for determining sequentially the (random) sample size N, and then, on the basis of the sample, for guessing "white" or "black". Let $\delta=\delta(p)$ denote the probability that D will lead to guessing "white", $0\leq\delta\leq 1$. Then our *expected gain* in using D is

$$g_p(D)=A[\delta(1-p)+(1-\delta)p]-E_p(N)\,. \tag{1.1}$$

For fixed p this is clearly maximized by choosing D to consist of not sampling at all (i.e. setting $N=0$) and putting $\delta=1$ or 0 according as $p\leq 1/2$ or $p>1/2$; i.e. guessing "white" if $p\leq 1/2$ and "black" if $p>1/2$. Calling this procedure D_p we have, for every $0\leq p\leq 1$ and every D,

$$g_p(D)\leq g_p(D_p)=A\max(p,1-p)\,. \tag{1.2}$$

The difference

$$r_p(D)=A\max(p,1-p)-g_p(D)\geq 0 \tag{1.3}$$

is our *regret* in using D; we have from (1.1) and (1.3)

$$r_p(D)=\begin{cases}2A(1/2-p)(1-\delta)+E_p(N) & \text{if } p\leq 1/2\,,\\ 2A(p-1/2)\delta+E_p(N) & \text{if } p>1/2\,.\end{cases} \tag{1.4}$$

We shall now formulate our problem in terms of hypothesis testing. Let x be a random variable with two values 0, 1 and with

$$P(x=1)=p\,,\qquad P(x=0)=1-p\,, \tag{1.5}$$

where $0\leq p\leq 1$ is an unknown parameter. Let

$$H_0:\ p\leq 1/2 \qquad\text{and}\qquad H_1:\ p>1/2$$

be the (composite) null hypothesis and alternative, and let a_i, $i=0,1$, be two possible

* This research was sponsored by the Office of Naval Research under Contract Number Nonr-266 (33), Project Number NR 042-034, and Contract Number Nonr-266 (59), Project Number NR 042-205. Reproduction in whole or in part is permitted for any purpose of the United States Government.

** Received 9 October, 1961.

actions of which a_i is appropriate when H_i is true. We assume that the loss incurred in taking action a_i where p is the true value of the parameter is given by

$$(1.6)\qquad L(a_0, p)=\begin{cases} 0 & \text{if } p\leq 1/2, \\ 2A(p-1/2) & \text{if } p>1/2; \end{cases}$$

$$(1.7)\qquad L(a_1, p)=\begin{cases} 0 & \text{if } p>1/2, \\ 2A(1/2-p) & \text{if } p\leq 1/2; \end{cases}$$

where A is some positive constant. (It is customary in hypothesis testing to take

$$(1.6)'\qquad L'(a_0, p)=\begin{cases} 0 & \text{if } p\leq 1/2, \\ c_0 & \text{if } p>1/2; \end{cases}$$

$$(1.7)'\qquad L'(a_1, p)=\begin{cases} 0 & \text{if } p>1/2, \\ c_1 & \text{if } p\leq 1/2; \end{cases}$$

where c_0, c_1 are positive constants. In many cases the loss schedule (1.6), (1.7) is more appropriate than (1.6)', (1.7)'. Suppose we can observe a sequence of independent random variables $x_1, x_2, \cdots$, each with the distribution of x, and that each observation costs \$1. We are required to stop at some point and take action a_0 or a_1. If D denotes any decision procedure and if N denotes the sample size and $\delta=\delta(p)$ the probability of taking action a_0, then the *expected loss* in using D due to incorrect decision and cost of sampling is

$$(1.8)\qquad E_p(\text{loss}\,;D)=\begin{cases} 2A(1/2-p)(1-\delta)+E_p(N) & \text{if } p\leq 1/2, \\ 2A(p-1/2)\delta+E_p(N) & \text{if } p>1/2 \end{cases}$$
$$=r_p(D).$$

Thus the expected loss in the hypothesis testing formulation equals the regret in the original urn problem.

Our task is to find a decision procedure D which in some sence minimizes (over the interval $0\leq p\leq 1$) the regret (1.3), or, alternatively, maximizes the expected gain $g_p(D)$ defined by (1.1).

We shall confine ourselves to the following Bayesian interpretation of this problem. Let

$$(1.9)\qquad f_{a,b}(p)=\frac{\Gamma(a+b)}{\Gamma(a)\Gamma(b)}p^{a-1}(1-p)^{b-1}$$

be the beta probability density function on $0\leq p\leq 1$ with parameters $a, b>0$, and for any decision procedure D let

$$(1.10)\qquad B_{a,b}(D)=\int_0^1 g_p(D)f_{a,b}(p)dp\,.$$

If there exists a D which maximizes (1.10) we shall denote it by $D_{a,b}$ and define the *value* of the ordered pair (a, b) to be

$$(1.11)\qquad V(a, b)=B_{a,b}(D_{a,b})=\max_D \int_0^1 g_p(D)f_{a,b}(p)dp\,.$$

We shall for arbitrary $A>0$

(i) show that $D_{a,b}$ exists for every $a, b>0$;
(ii) describe it constructively;
(iii) find $V(a, b)$; and

(iv) find $\delta(p, a, b)$, the probability of guessing "white", and $n(p, a, b)$, the expected sample size, in using $D_{a,b}$ when p is the proportion of black balls in the urn.

2. Lemmas and the theorem.

Lemma 1. Suppose $a+b>8A^3$. Then $D_{a,b}$ exists and consists of taking no observations and guessing "white" or "black" according as $a\leq b$ or $a>b$; moreover

$$V(a, b)=A\frac{\max(a, b)}{a+b}. \tag{2.1}$$

Proof. We remark that if p is a random variable with density (1.9) on $0\leq p\leq 1$ then

$$E(p)=\int_0^1 pf_{a,b}(p)dp=\frac{a}{a+b}, \quad E(p^2)=\frac{a(a+1)}{(a+b)(a+b+1)}, \tag{2.2}$$

and hence

$$\sigma^2=E(p^2)-E^2(p)=\frac{ab}{(a+b)^2(a+b+1)}<\frac{ab}{(a+b)^3}<\frac{1}{a+b}. \tag{2.3}$$

It follows from Chebyshev's inequality that

$$P\left[\left|p-\frac{a}{a+b}\right|\geq\frac{1}{2A}\right]\leq 4A^2\sigma^2<\frac{4A^2}{a+b}<\frac{1}{2A}. \tag{2.4}$$

Now consider any decision procedure D which involves taking at least one observation. Then $E_p(N)\geq 1$ for all $0\leq p\leq 1$, and hence from (1.1),

$$g_p(D)\leq A\max(p, 1-p)-1.$$

$$B_{a,b}(D)=\int_0^1 g_p(D)f_{a,b}(p)dp$$

$$\leq A\int_0^1 \max(p, 1-p)f_{a,b}(p)dp-1$$

$$=A\int_{\left|p-\frac{a}{a+b}\right|\geq\frac{1}{2A}} \max(p, 1-p)f_{a,b}(p)dp+A\int_{\left|p-\frac{a}{a+b}\right|<\frac{1}{2A}} \max(p, 1-p)f_{a,b}(p)dp-1.$$

But if $\left|p-\frac{a}{a+b}\right|<\frac{1}{2A}$ then $\frac{a}{a+b}-\frac{1}{2A}<p<\frac{a}{a+b}+\frac{1}{2A}$ so that

$$\max(p, 1-p)\leq\max\left(\frac{a}{a+b}+\frac{1}{2A}, 1-\frac{a}{a+b}+\frac{1}{2A}\right)=\frac{1}{2A}+\frac{\max(a, b)}{a+b}.$$

Hence from (2.4),

$$B_{a,b}(D)\leq A\int_{\left|p-\frac{a}{a+b}\right|\geq\frac{1}{2A}} 1\cdot f_{a,b}(p)dp+A\int_{\left|p-\frac{a}{a+b}\right|<\frac{1}{2A}}\left[\frac{1}{2A}+\frac{\max(a, b)}{a+b}\right]f_{a,b}(p)dp-1 \tag{2.5}$$

$$<\frac{A}{2A}+A\left[\frac{1}{2A}+\frac{\max(a, b)}{a+b}\right]-1=A\frac{\max(a, b)}{a+b}.$$

On the other hand, if we take no observations and guess "white" with probability δ and "black" with probability $1-\delta$, then for this decision procedure D^* we have

$$g_p(D^*)=A[\delta(1-p)+(1-\delta)p],$$

$$(2.6)\qquad B_{a,b}(D^*)=\int_0^1 g_p(D^*)f_{a,b}(p)dp$$
$$=A\left[\delta\frac{b}{a+b}+(1-\delta)\frac{a}{a+b}\right]$$
$$\leq A\,\frac{\max(a,b)}{a+b},$$

with equality if $\delta=1$ when $a\leq b$ and $\delta=0$ when $a>b$. This completes the proof.

LEMMA 2. If x is a random variable with distribution (1.5) and if p is a random variable with *à priori* density (1.9) then the *à posteriori* density of p given $x=1$ is $f_{a+1,b}$, and the *à posteriori* density of p given $x=0$ is $f_{a,b+1}$.

PROOF.

$$P[\alpha\leq p\leq\beta\,|\,x=1]=P[\alpha\leq p\leq\beta,\,x=1]/P[x=1]$$
$$=\int_\alpha^\beta pf_{a,b}(p)dp\Big/\int_0^1 pf_{a,b}(p)dp\,.$$

Hence the *à posteriori* density of p given $x=1$ is

$$\frac{pf_{a,b}(p)}{\int_0^1 pf_{a,b}(p)dp}=\frac{\Gamma(a+b+1)}{\Gamma(a+1)\Gamma(b)}p^a(1-p)^{b-1}=f_{a+1,b}(p)\,.$$

Similarly for $x=0$.

LEMMA 3. $D_{a,b}$ exists for every $a,b>0$, and $V(a,b)$ satisfies the relation

$$(2.7)\qquad V(a,b)=\max\left[A\,\frac{\max(a,b)}{a+b}\,,\ \frac{aV(a+1,b)+bV(a,b+1)}{a+b}-1\right].$$

PROOF. By Lemma 1 we know that $D_{a,b}$ exists for $a+b>8A^3$. Now suppose that

$$(2.8)\qquad 8A^3-1<a+b\leq 8A^3\,.$$

If we take no observations we see by (2.6) that the best we can achieve for $B_{a,b}(D)$ is $A\max(a,b)/(a+b)$. Suppose then that we take a first observation x; then

$$P(x=1)=E(p)=\frac{a}{a+b}\,,\quad P(x=0)=\frac{b}{a+b}\,.$$

If $x=1$ then we have paid \$1, and by Lemma 2 the *à posteriori* density of p is $f_{a+1,b}$. But $(a+1)+b>8A^3$, and hence by Lemma 1, $V(a+1,b)$ exists; similarly if $x=0$. We see therefore that $V(a,b)$ exists and satisfies (2.7). The proof is completed by backward induction on $a+b$.

LEMMA 4. If $a+b\geq A/2-1$ then

$$(2.9)\qquad V(a,b)=A\frac{\max(a,b)}{a+b}\,.$$

PROOF. We know by Lemma 1 that (2.9) holds for $a+b>8A^3$. It will suffice to show that if (2.9) holds for $a+b\geq\gamma$ then it must also hold for $\gamma-1\leq a+b<\gamma$ provided that $\gamma\geq A/2$.

Thus suppose that (2.9) holds for all $a+b\geq\gamma$ and that $\gamma\geq A/2$. If $\gamma-1\leq a+b<\gamma$ and $a\geq b$ then by (2.7)

— 4 —

$$(2.10)\qquad V(a,b)=\max\left[\frac{Aa}{a+b},\ \frac{aA(a+1)+bA\max(a,b+1)}{(a+b)(a+b+1)}-1\right]$$
$$\leq\max\left[\frac{Aa}{a+b},\ \frac{aA(a+1)+bA(a+1)}{(a+b)(a+b+1)}-1\right].$$

The last member will be equal to $Aa/(a+b)$ provided that we can show that

$$(a+b+1)Aa\geq aA(a+1)+bA(a+1)-(a+b)(a+b+1),$$

i.e. that

$$A\leq\frac{(a+b)(a+b+1)}{b},$$

and the last inequality does hold, since $a\geq b$, $a+b+1\geq\gamma$, $2\gamma\geq A$. Hence

$$V(a,b)\leq\frac{Aa}{a+b}=A\frac{\max(a,b)}{a+b}.$$

Clearly the inequality does not hold and we have (2.9). The same holds by symmetry if $b\geq a$. Hence (2.9) holds for all $\gamma-1\leq a+b<\gamma$, which was to be proved.

We remark parenthetically that by (2.7)

$$(2.11)\qquad V(a,a)=\max\left[\frac{A}{2},\ V(a+1,a)-1\right]\geq\max\left[\frac{A}{2},\ \frac{A(a+1)}{2a+1}-1\right]>\frac{A}{2}$$

provided that

$$\frac{A}{2}<\frac{A(a+1)}{2a+1}-1.$$

But the last inequality amounts to $2a<A/2-1$. Hence

$$(2.12)\qquad V(a,a)>\frac{A}{2}\ \text{ provided that }\ a+a<\frac{A}{2}-1.$$

Let us introduce the notation

$$(2.13)\qquad V_0(a,b)=A\frac{\max(a,b)}{a+b}.$$

Then we have proved the following main

Theorem. For any $a,b>0$,

$$(2.14)\qquad\begin{cases}V(a,b)=\max\left[V_0(a,b),\ \dfrac{aV(a+1,b)+bV(a,b+1)}{a+b}-1\right],\\ V(a,b)=V_0(a,b)\qquad\text{if}\quad a+b\geq\dfrac{A}{2}-1,\\ V(a,a)>V_0(a,a)\qquad\text{if}\quad a+a<\dfrac{A}{2}-1.\end{cases}$$

3. Computation of $V(a,b)$.

We can now compute $V(a,b)$ for all $a,b>0$ as follows. Let $a+b=u$. For $u\geq A/2-1$ we know that $V(a,b)=V_0(a,b)$. For $A/2-1\leq t<A/2-2$ we can compute $V(a,b)$ by the recursion formula (2.7), and so on backward.

The optimal decision rule $D_{a,b}$ is now characterized by the following random walk in the plane (Fig. 1): We start at (a,b). Let

$$(3.1)\qquad M(a,b)=V(a,b)-V_0(a,b)\geq0.$$

Table 1

$A=10$				
$M(1,1) = 0.7$			$V(1,1) = 5.7$	
$A=20$				
$M(1,1) = 2.3$			$V(1,1) = 12.3$	
$M(2,2) = 1.0$			$V(2,2) = 11.0$	
$M(3,3) = 0.4$			$V(3,3) = 10.4$	
$M(4,4) = 0.1$			$V(4,4) = 10.1$	
$A=50$				
$M(1,1) = 7.7$			$V(1,1) = 32.7$	
$M(2,2) = 4.0$	$M(2,1) = 0.3$		$V(2,2) = 29.0$	$V(2,1) = 33.7$
$M(3,3) = 2.6$	$M(3,2) = 0.0$		$V(3,3) = 27.6$	$V(3,2) = 30.0$
$M(4,4) = 1.8$			$V(4,4) = 26.8$	
$M(5,5) = 1.3$			$V(5,5) = 26.3$	
$M(6,6) = 0.9$			$V(6,6) = 25.9$	
$M(7,7) = 0.7$			$V(7,7) = 25.7$	
$M(8,8) = 0.5$			$V(8,8) = 25.5$	
$M(9,9) = 0.3$			$V(9,9) = 25.3$	
$M(10,10) = 0.2$			$V(10,10) = 25.2$	
$M(11,11) = 0.1$			$V(11,11) = 25.1$	
$A=100$				
$M(1,1) = 18.3$				
$M(2,2) = 11.0$	$M(2,1) = 2.7$			
$M(3,3) = 7.4$	$M(3,2) = 2.0$			
$M(4,4) = 5.3$	$M(4,3) = 1.3$			
$M(5,5) = 3.9$	$M(5,4) = 0.7$			
$M(6,6) = 2.9$	$M(6,5) = 0.3$			
$M(7,7) = 2.3$	$M(7,6) = 0.1$			
$M(8,8) = 1.9$				
$M(9,9) = 1.6$				
$M(10,10) = 1.4$				
....				
$M(15,15) = 0.6$				
....				
$M(20,20) = 0.2$				
....				
$M(24,24) = 0.0$				
$A=200$				
$M(1,1) = 40.5$				
$M(2,2) = 26.0$	$M(2,1) = 8.2$			
$M(3,3) = 18.7$	$M(3,2) = 7.0$	$M(3,1) = 0.7$		
$M(4,4) = 14.3$	$M(4,3) = 5.5$	$M(4,2) = 0.8$		
$M(5,5) = 11.3$	$M(5,4) = 4.2$	$M(5,3) = 0.6$		
$M(6,6) = 9.2$	$M(6,5) = 3.2$	$M(6,4) = 0.3$		
$M(7,7) = 7.7$	$M(7,6) = 2.5$	$M(7,5) = 0.1$		
$M(8,8) = 6.5$	$M(8,7) = 2.0$			
....	....			
$M(15,15) = 2.2$	$M(15,14) = 0.1$			
....				
$M(20,20) = 2.4$				
....				
$M(30,30) = 0.6$				
....				
$M(40,40) = 0.2$				
....				
$M(49,49) = 0.0$				

$A = 500$

a	$M(a,a)$	$M(a,a-1)$	$M(a,a-2)$	$M(a,a-3)$
1	110.5			
2	75.5	28.2		
3	57.8	26.5	6.0	
4	46.6	23.1	7.4	0.5
5	38.8	19.8	7.2	1.0
6	33.0	17.1	6.5	1.2
7	28.5	14.8	5.7	1.1
8	24.9	12.8	4.9	0.9
9	21.9	11.2	4.1	0.6
10	19.5	9.8	3.5	0.4
11	17.5	8.6	2.9	0.2
12	15.7	7.6	2.5	0.1
13	14.2	6.7	2.1	
14	12.9	5.9	1.7	
15	11.7	5.2	1.4	
...	...	...	...	
22	6.5	2.2	0.0	
...	...	...		
40	2.1	0.0		
50	1.5			
60	1.1			
80	0.6			
100	0.2			
124	0.0			

$A = 1000$

a	$M(a,a)$	$M(a,a-1)$	$M(a,a-2)$	$M(a,a-3)$	$M(a,a-4)$
1	230.3				
2	162.2	64.7			
3	127.6	63.2	17.4		
4	105.6	57.1	22.0	3.4	
5	90.2	51.1	22.5	6.0	
6	78.5	45.7	21.6	7.0	0.7
7	69.4	41.1	20.2	7.2	1.2
8	62.0	37.1	18.7	7.0	1.3
9	55.8	33.6	17.1	6.6	1.4
10	50.6	30.5	15.7	6.1	1.3
11	46.1	27.8	14.3	5.6	1.2
12	42.3	25.4	13.0	5.1	1.0
13	38.9	23.3	11.8	4.5	0.9
14	35.9	21.4	10.8	4.0	0.7
15	33.3	19.7	9.8	3.5	0.5
16	31.0	18.2	8.9	3.1	0.3
17	28.8	16.8	8.1	2.7	0.3
18	26.9	15.6	7.4	2.4	0.0
30	13.4	6.6	2.2	0.0	
47	6.3	2.1	0.0		
82	2.0	0.0			
100	1.5				
150	0.7				
200	0.2				
249	0.0				

$A=2000$

a	$M(a,a)$	$M(a,a-1)$	$M(a,a-2)$	$M(a,a-3)$	$M(a,a-4)$	$M(a,a-5)$
1	473.7					
2	340.4	141.3				
3	272.7	141.4	43.3			
4	229.8	130.9	55.6	11.9		
5	199.6	119.7	58.4	19.4	2.2	
6	176.7	109.6	57.7	23.1	5.3	
7	158.6	100.8	55.6	24.7	7.4	0.6
8	143.9	92.9	53.0	25.1	8.6	1.3
9	131.6	86.1	50.3	24.8	9.3	1.9
10	121.1	79.9	47.5	24.2	9.6	2.2
11	112.1	74.5	44.8	23.3	9.6	2.4
12	104.1	69.6	42.2	22.3	9.4	2.5
13	97.2	65.1	39.8	21.3	9.1	2.5
14	90.9	61.1	37.5	20.2	8.8	2.4
15	85.4	57.5	35.4	19.1	8.4	2.4
31	38.8	25.0	14.4	6.9	2.2	0.0
46	22.5	13.2	6.5	2.1	0.0	
66	12.5	6.2	2.0	0.0		
97	6.1	2.0	0.0			
165	2.0	0.0				
250	1.0					
350	0.4					
499	0.0					

$A=5000$

a	$M(a,a)$	$M(a,a-1)$	$M(a,a-2)$	$M(a,a-3)$	$M(a,a-4)$	$M(a,a-5)$	$M(a,a-6)$	$M(a,a-7)$	$M(a,a-8)$
1	1211.7								
2	886.1	379.4							
3	721.3	387.1	127.5						
4	617.2	365.2	166.0	42.2					
5	543.7	340.4	177.9	68.0	12.5				
6	488.1	317.4	179.6	82.0	25.4	2.6			
7	444.1	296.8	177.0	89.5	34.6	7.9			
8	408.2	278.5	172.3	93.1	41.0	12.6	1.5		
9	378.1	262.1	166.8	94.5	45.3	16.5	3.3		
10	352.4	247.5	161.0	94.4	48.0	19.5	5.1	0.2	
11	330.2	234.4	155.0	93.5	49.6	21.7	6.7	0.8	
12	310.6	222.5	149.2	91.9	50.4	23.3	8.0	1.4	
13	293.3	211.6	143.6	90.0	50.7	24.4	9.0	1.9	
14	277.7	201.7	138.2	87.9	50.6	25.2	9.9	2.3	
15	263.8	192.5	133.0	85.6	50.1	25.7	10.6	2.7	
16	251.1	184.1	128.0	83.2	49.4	25.9	11.0	3.1	0.0
17	239.5	176.3	123.3	80.8	48.6	25.9	11.4	3.4	0.2
18	228.8	169.0	118.8	78.4	47.6	25.7	11.6	3.6	0.3
19	219.1	162.3	114.5	76.0	46.5	25.5	11.7	3.8	0.4
20	210.0	156.0	110.4	73.7	45.4	25.1	11.7	3.9	0.5
25	173.2	129.6	92.8	62.8	39.5	22.4	10.8	3.8	0.5
30	146.2	109.6	78.7	53.5	33.8	19.2	9.2	3.1	0.3
35	125.4	93.9	67.3	45.7	28.7	16.1	7.5	2.3	0.0
55	75.1	54.9	38.1	24.6	14.2	6.9	2.2	0.0	
75	49.6	34.8	22.7	13.3	6.5	2.1	0.0		
98	33.1	21.8	12.9	6.4	2.1	0.0			
129	20.9	12.4	6.2	2.0	0.0				
173	12.2	6.1	2.0	0.0					
247	6.0	2.0	0.0						
415	2.0	0.0							
700	0.8								
1249	0.0								

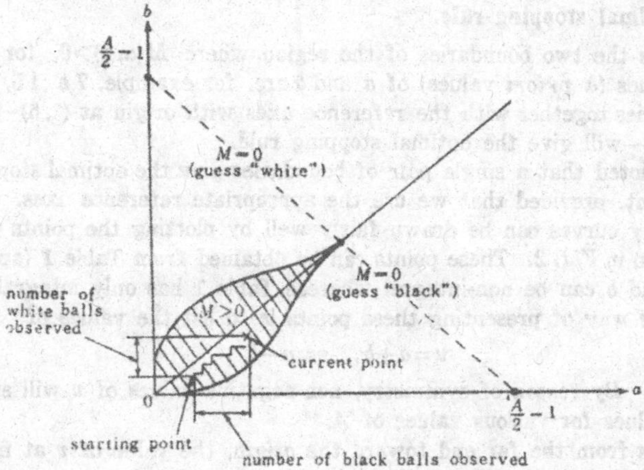

Fig. 1

If $M(a, b)=0$ we take no observations and guess "white" or "black" according as $a\leq b$ or $a>b$. If $M(a, b)>0$ we take an observation and move to $(a+1, b)$ or $(a, b+1)$ according as the ball observed is black or white, and then proceed as though we had started there. In this random walk in the (a, b)-plane the value of $a+b$ increases by 1 at each step and we ultimately reach a point for which $M(a, b)=0$, since $M(a, b)=0$ whenever $a+b>A/2-1$.

Numerical computations of $V(a, b)$ and $M(a, b)$ have been carried out on the electronic computer IBM 709 and numerical results are available for $A=10$, 20, 50, 100, 200, 500, 1000, 2000, and 5000. (Only integral values of a and b were taken, and only those values of $M(a, b)$ which are positive were printed out, together with the corresponding a, b and $V(a, b)$.) Table 1 shows a part of the results.

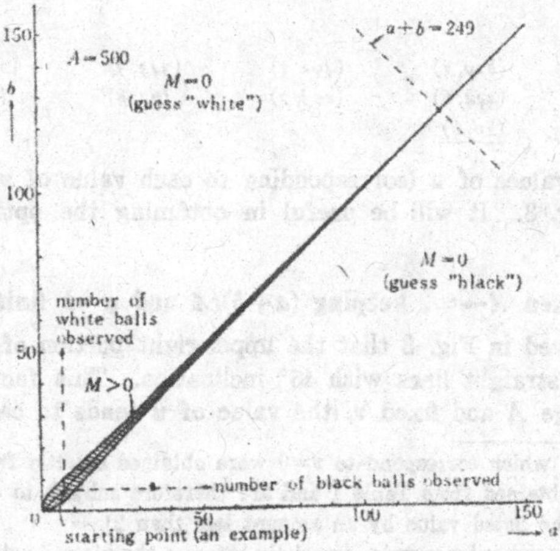

Fig. 2

— 9 —

4. The optimal stopping rule.

Fig. 2 shows the two boundaries of the region where $M(a,b)>0$, for $A=500$. If the starting values (*à priori* values) of a and b are, for example, 7 and 5, respectively, then the boundaries together with the reference axes with origin at (7, 5)——the dotted axes in Fig. 2——will give the optimal stopping rule.

It is to be noted that a single pair of boundaries give the optimal stopping rule for any starting point, provided that we use the appropriate reference axes.

The boundary curves can be drawn fairly well by plotting the points where $a-b=1,2,3,\cdots$, as seen in Fig. 2. These points can be obtained from Table 1 (approximately, since such a and b can be non-integral whereas Table 1 has only integral values of a and b). Another way of presenting these points is to list the values of

$$u=a+b,\quad v=a-b \tag{4.1}$$

for these points. By reason of symmetry, non-negative values of v will suffice. Table 2 gives these values for various values of A.

As one goes from the far end toward the origin, the value of v at first increases to a maximum, and then decreases. The points in the decreasing region are underlined in Table 2.

Table 2 $(u,v)=(a+b,a-b)$ for boundary points*

$A=100$						
	(49, 0)	(13, 1)				
$A=200$						
	(99, 0)	(89, 1)	(12, 2)			
$A=500$						
	(249, 0)	(79, 1)	(42, 2)	(21, 3)		
$A=1000$						
	(499, 0)	(163, 1)	(92, 2)	(57, 3)	(32, 4)	(8, 4)
$A=2000$						
	(999, 0)	(329, 1)	(192, 2)	(129, 3)	(88, 4)	(57, 5)
	(9, 5)					
$A=5000$						
	(249, 0)	(829, 1)	(492, 2)	(343, 3)	(254, 4)	
	(191, 5)	(144, 6)	(103, 7)	(62, 8)	(24, 8)	
	(13, 7)	(10, 6)				

If we plot the values of u (corresponding to each value of v) against A on log-log paper, we get** Fig. 3. It will be useful in obtaining the optimal decision rule for given A.

5. Behavior when $A\to\infty$, keeping $(a+b)/A$ and $a-b$ finite.

It will be observed in Fig. 3 that the upper-right portion of the diagram tends to be a set of parallel straight lines with 45° inclination. This fact on the log-log paper means that, for large A and fixed v, the value of u tends to be proportional to A. It

* Those values of u which correspond to $v=0$ were obtained directly from (2.14). All the other values of u were obtained from Table 1 and are therefore subject to error (the true value will be greater than the listed value by an amount less than 2).

** The diagram is somewhat inaccurate, especially between the plotted points. It would be desirable to have a more accurate diagram in the future.

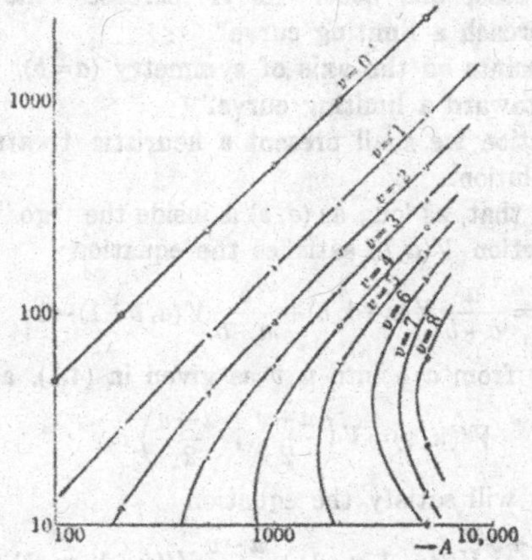

Fig. 3

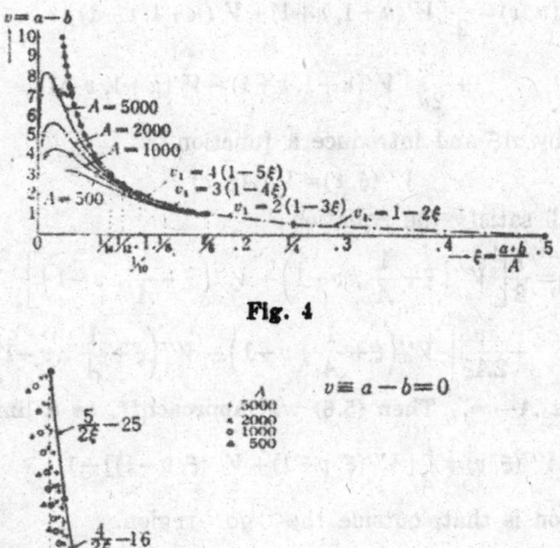

Fig. 4

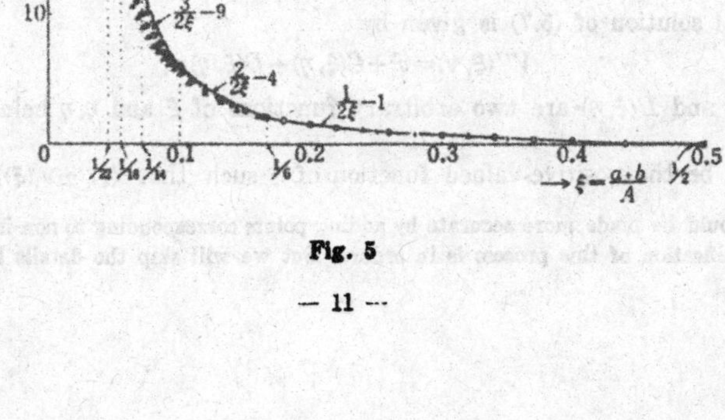

Fig. 5

suggests plotting v against $\xi \equiv u/A$. Thus we get Fig. 4, which shows the boundary points for $A=500$, 1000, 2000, and 5000. As A increases, the bonndary (in this representation) seems to approach a limiting curve*.

Fig. 5 plots, for the points on the axis of symmetry ($a=b$), $M(a, b)$ against ξ. It shows a similar tendency toward a limiting curve.

In the rest of this section we shall present a heuristic treatment of the limiting behavior of the optimal solution.

From (2.14), we know that, as long as (a, b) is inside the "go"-region—i.e. as long as $M(a, b)>0$—, the function $V(a, b)$ satisfies the equation:

(5.1) $$V(a, b)=\frac{a}{a+b}V(a+1, b)+\frac{b}{a+b}V(a, b+1)-1.$$

Now, change the variables from a, b into u, v as given in (4.1), and define

(5.2) $$V'(u, v)=V\left(\frac{u+v}{2}, \frac{u-v}{2}\right).$$

Then the function $V'(u, v)$ will satisfy the equation

(5.3) $$V'(u, v)=\frac{u+v}{2u}V'(u+1, v+1)+\frac{u-v}{2u}V'(u+1, v-1)-1,$$

i.e.

(5.4) $$V'(u, v)=\frac{1}{2}[V'(u+1, v+1)+V'(u+1, v-1)]$$
$$+\frac{v}{2u}[V'(u+1, v+1)-V'(u+1, v-1)]-1.$$

If we replace u by $A\xi$ and introduce a function

(5.5) $$V''(\xi, v)=V'(A\xi, v),$$

then this function will satisfy the equation:

(5.6) $$V''(\xi, v)=\frac{1}{2}\left[V''\left(\xi+\frac{1}{A}, v+1\right)+V''\left(\xi+\frac{1}{A}, v-1\right)\right]$$
$$+\frac{v}{2A\xi}\left[V''\left(\xi+\frac{1}{A}, v+1\right)-V''\left(\xi+\frac{1}{A}, v-1\right)\right]-1.$$

Suppose we make $A\to\infty$. Then (5.6) will approach**, as a limit,

(5.7) $$V''(\xi, v)=\frac{1}{2}[V''(\xi, v+1)+V''(\xi, v-1)]-1.$$

The boundary condition is that, outside the "go"-region.

(5.8) $$V''(\xi, v)=V_0\left(\frac{A\xi+v}{2}, \frac{A\xi-v}{2}\right)=A\frac{\max(A\xi+v, A\xi-v)}{2A\xi}=\frac{A}{2}+\frac{|v|}{2\xi}$$

A general solution of (5.7) is given by

(5.9) $$V''(\xi, v)=v^2+C(\xi, \eta)+D(\xi, \eta)v,$$

where $C(\xi, \eta)$ and $D(\xi, \eta)$ are two arbitrary functions of ξ and η, η being the residue of v mod. 1.

Let $v_1(\xi)$ be the positive-valued function of ξ such that $(\xi, \pm v_1(\xi))$ constitute the

* Fig. 4 should be made more accurate by adding points corresponding to non-integral values of v.
** Some justification of this process is in order. But we will skip the details here.

boundary of the "go"-region in the (ξ, v)-plane. The "go"-region exists only for $0<\xi<1/2$ (see Lemma 4). For a given value of ξ in this interval, let

(5.10) $$\eta_1\equiv v_1(\xi) \quad (\text{mod. } 1), \qquad 0\leq\eta_1<1.$$

We have two cases* which must be treated separately:—

Case 1: $0<\eta_1<1/2$.

In this case, there exists a positive integer m such that

(5.11) $$v_1(\xi)=\eta_1+(m-1).$$

For any η in the interval $-\eta_1<\eta<\eta_1$, and for any integer k such that $-m<k<m$, $\eta+k$ lies between $-v_1(\xi)$ and $v_1(\xi)$, and $\eta+m>v_1(\xi)$, $\eta-m<-v_1(\xi)$. Hence, the condition (5.8), together with the general solution (5.9) will give

(5.12) $$\frac{A}{2}+\frac{\eta+m}{2\xi}=(\eta+m)^2+C(\xi,\eta)+D(\xi,\eta)(\eta+m),$$

and

(5.13) $$\frac{A}{2}-\frac{\eta-m}{2\xi}=(\eta-m)^2+C(\xi,\eta)+D(\xi,\eta)(\eta-m).$$

Solving this set of equations with respect to $C(\xi,\eta)$ and $D(\xi,\eta)$, we get

(5.14) $$C(\xi,\eta)=\frac{A}{2}+(\eta^2-m^2)\left(1-\frac{1}{2m\xi}\right),$$

(5.15) $$D(\xi,\eta)=\eta\left(-2+\frac{1}{2m\xi}\right).$$

Hence, we have a solution:

(5.16) $$\begin{aligned} V''(\xi,\eta+k)&=(\eta+k)^2+\frac{A}{2}+(\eta^2-m^2)\left(1-\frac{1}{2m\xi}\right)+\eta\left(-2+\frac{1}{2m\xi}\right)(\eta+k)\\ &=\frac{A}{2}+\frac{k}{2m\xi}\eta+\left(\frac{m}{2\xi}-m^2+k^2\right),\\ &-\eta_1<\eta<\eta_1,\ k=-m+1,\ -m+2,\ \cdots,\ m-1. \end{aligned}$$

Since $(\xi, v_1(\xi))$ is the boundary, (5.16) must approach $V_0[(A\xi+v_1(\xi))/2, (A\xi-v_1(\xi))/2]$ as $\eta\to\eta_1$ with $k=m-1$. Hence,

(5.17) $$\frac{A}{2}+\frac{\eta_1+(m-1)}{2\xi}=\frac{A}{2}+\frac{m-1}{2m\xi}\eta_1+\left(\frac{m}{2\xi}-m^2+(m-1)^2\right).$$

This is equivalent to

(5.18) $$\eta_1=m-2m(2m-1)\xi.$$

The inequality $0<\eta_1<1/2$ is satisfied if and only if

(5.19) $$\frac{1}{2(2m)}<\xi<\frac{1}{2(2m-1)}.$$

From (5.11) and (5.18), we get

(5.20) $$v_1(\xi)=(2m-1)(1-2m\xi),$$

and for ξ in the interval (5.19) it holds that

(5.21) $$m-1<v_1(\xi)<m-1/2.$$

Next, for any η in the interval $\eta_1<\eta<1-\eta_1$, and for any integer k such that**

* We ignore the boundary cases such as $\eta_1=0$ or $\eta_1=1/2$, since they are trivial.

** If $m=1$, there exists no such k, and (5.16) is enough as the solution.

$-m<k<m-1$, $\eta+k$ lies between $-v_1(\xi)$ and $v_1(\xi)$, and $\eta+(m-1)>v_1(\xi)$, $\eta-m<-v_1(\xi)$. Hence we have, in place of (5.12) and (5.13),

$$\frac{A}{2}+\frac{\eta+(m-1)}{2\xi}=\{\eta+(m-1)\}^2+C(\xi,\eta)+D(\xi,\eta)\{\eta+(m-1)\}, \tag{5.22}$$

and

$$\frac{A}{2}-\frac{\eta-m}{2\xi}=(\eta-m)^2+C(\xi,\eta)+D(\xi,\eta)(\eta-m). \tag{5.23}$$

Solving these, we get

$$C(\xi,\eta)=\frac{A}{2}+\{\eta+(m-1)\}(\eta-m)\left[1-\frac{1}{(2m-1)\xi}\right], \tag{5.24}$$

$$D(\xi,\eta)=\left(\eta-\frac{1}{2}\right)\left[-2+\frac{1}{(2m-1)\xi}\right]. \tag{5.25}$$

Therefore, we have a solution:

$$\begin{aligned} V''(\xi,\eta+k)&=(\eta+k)^2+\frac{A}{2}+\{\eta^2-\eta-m(m-1)\}\left[1-\frac{1}{(2m-1)\xi}\right] \\ &\quad+\left(\eta-\frac{1}{2}\right)\left[-2+\frac{1}{(2m-1)\xi}\right](\eta+k) \\ &=\frac{A}{2}+\frac{k+\frac{1}{2}}{(2m-1)\xi}\left(\eta-\frac{1}{2}\right)+\frac{m-\frac{1}{2}}{2\xi}-\left(m-\frac{1}{2}\right)^2+\left(k+\frac{1}{2}\right)^2, \\ &\qquad \eta_1<\eta<1-\eta_1,\ k=-m+1,\ \cdots,\ m-2. \end{aligned} \tag{5.26}$$

This goes together with (5.18), (5.19), and (5.20).

Case 2: $1/2<\eta_1<1$.

In this case also, there exists a positive integer m such that (5.11) holds.

For any η in the interval $1-\eta_1<\eta<\eta_1$, and for any integer k such that $-m-1<k<m$, we have $\eta+m>v_1(\xi)$, $\eta-m-1<-v_1(\xi)$, and $-v_1(\xi)<\eta+k<v_1(\xi)$.

Similar arguments to those in the latter part of Case 1 will now give us the following results:

$$\begin{aligned} V''(\xi,\eta+k)&=\frac{A}{2}+\frac{k+\frac{1}{2}}{(2m+1)\xi}\left(\eta-\frac{1}{2}\right)+\frac{m+\frac{1}{2}}{2\xi}-\left(m+\frac{1}{2}\right)^2+\left(k+\frac{1}{2}\right)^2, \\ &\qquad 1-\eta_1<\eta<\eta_1,\ k=-m,-m+1,\ \cdots,\ m-1. \end{aligned} \tag{5.27}$$

As $\eta\to\eta_1-0$, with $k=m-1$, (5.27) must approach $V_0[(A\xi+v_1(\xi))/2, (A\xi-v_1(\xi))/2]$. Hence,

$$\frac{A}{2}+\frac{\eta_1+(m-1)}{2\xi}=\frac{A}{2}+\frac{m-\frac{1}{2}}{(2m+1)\xi}\left(\eta_1-\frac{1}{2}\right)+\frac{m+\frac{1}{2}}{2\xi}-\left(m+\frac{1}{2}\right)^2+\left(m-\frac{1}{2}\right)^2. \tag{5.28}$$

This is equivalent to

$$\eta_1=(m+1)-2m(2m+1)\xi. \tag{5.29}$$

The inequality $1/2<\eta_1<1$ is satisfied if and only if

$$\frac{1}{2(2m+1)}<\xi<\frac{1}{2(2m)}. \tag{5.30}$$

From (5.11) and (5.29), we get

(5.31) $$v_1(\xi)=2m\{1-(2m+1)\xi\},$$

and for ξ in the interval (5.30) it holds that

(5.32) $$m-\frac{1}{2}<v_1(\xi)<m.$$

Next, for any η in the interval $-1+\eta_1<\eta<1-\eta_1$, and for any integer k such that $-m<k<m$, we have $\eta+m>v_1(\xi)$, $\eta-m<-v_1(\xi)$, and $-v_1(\xi)<\eta+k<v_1(\xi)$. This will lead to the same solution (5.16), with $-1+\eta_1<\eta<1-\eta_1$ and $k=-m+1, -m+2,\cdots, m-1$. It goes in this case with (5.29), (5.30), and (5.31).

The (limiting) broken line in Fig. 4 is given by (5.20) and (5.31) with the appropriate intervals (5.19) and (5.30), respectively.

The (limiting) curve in Fig. 5 is given by (5.16) with $\eta=0$, $k=0$. (Notice that $V_0(A\xi/2, A\xi/2)=A/2$ for all ξ.)

Fig. 6 shows some of the areas and those where (5.16) is applicable (shaded areas) and those where formula (5.26) or (5.27) is applicable (white areas).

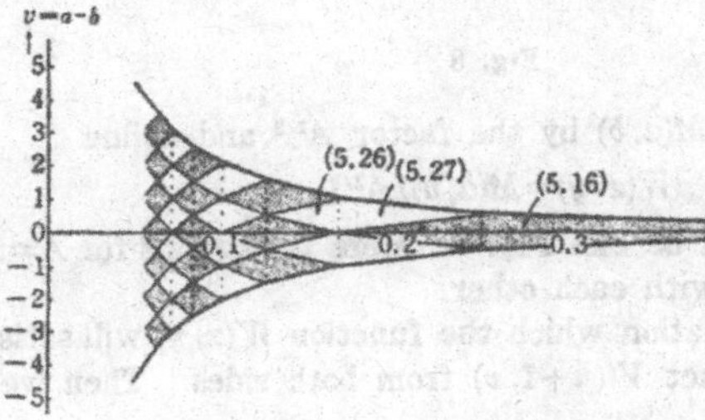

Fig. 6

6. Behavior when $A\to\infty$, keeping $(a+b)/A^{2/3}$ and $(a-b)/A^{1/3}$ finite.

Figs. 4 and 5 show considerable systematic deviations from the "theoretical" curves. As $a+b$ approaches the origin, the discrepancy becomes large, especially in Fig. 4. In fact, the boundary curve for a finite A must come down to the origin as $a+b\to 0$ (see Table 2), where as the "theoretical" curve goes up indefinitely.

In order better to treat the behavior of the optimal rule for moderate values of a and b, let us take two new variables:

(6.1) $$x=\frac{u}{A^{2/3}}=\frac{a+b}{A^{2/3}},\ y=\frac{v}{A^{1/3}}=\frac{a-b}{A^{1/3}},$$

and plot the boundary curve on (x, y)-plane. Thus we get Fig. 7.

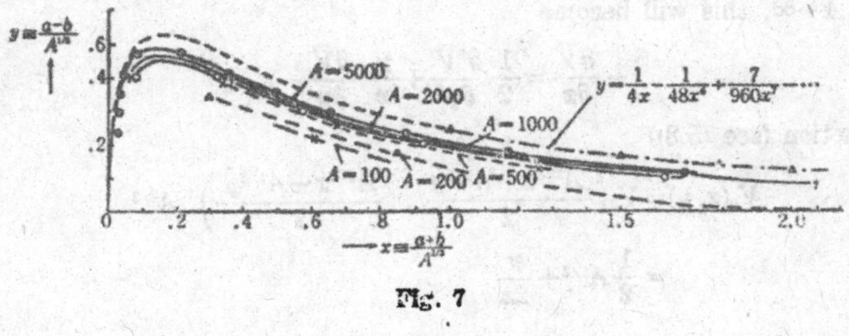

Fig. 7

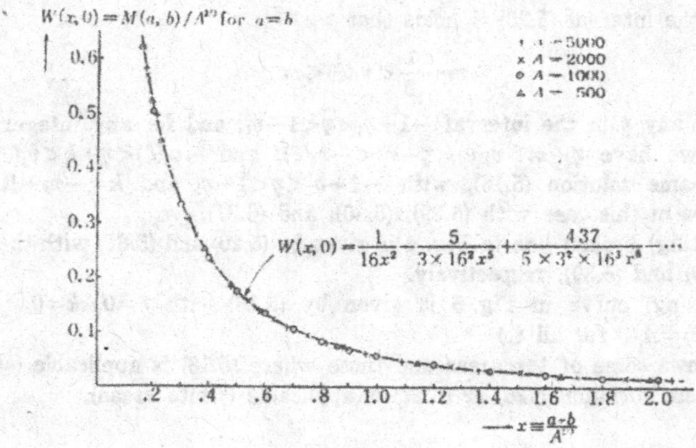

Fig. 8

If we reduce the function $M(a, b)$ by the factor $A^{2/3}$ and define

$$W(x, y) = M(a, b)/A^{2/3}, \tag{6.2}$$

then the graphs of $W(x, 0)$ will be like Fig. 8. Here the graphs for $A=500$, 1000, 2000, and 5000 paractically coincide with each other.

Now, let us derive the equation which the function $W(x, y)$ will satisfy in the limit. Take equation (5.4) and substract $V'(u+1, v)$ from both sides. Then we get

$$\begin{aligned} V'(u, v) - V'(u+1, v) = &\frac{1}{2}[V'(u+1, v+1) - 2V'(u+1, v) + V'(u+1, v-1)] \\ &+ \frac{v}{2u}[V'(u+1, v+1) - V'(u+1, v-1)] - 1. \end{aligned} \tag{6.3}$$

If we define

$$\bar{V}(x, y) = V'(A^{2/3}x, A^{1/3}y)/A^{2/3}, \tag{6.4}$$

then this function will satisfy, in place of (6.3),

$$\begin{aligned} &A^{2/3}[\bar{V}(x, y) - \bar{V}(x+A^{-2/3}, y)] \\ &= \frac{1}{2}A^{2/3}[\bar{V}(x+A^{-2/3}, y+A^{-1/3}) - 2\bar{V}(x+A^{-2/3}, y) + \bar{V}(x+A^{-2/3}, y-A^{-1/3})] \\ &\quad + \frac{y}{2x}A^{1/3}[\bar{V}(x+A^{-2/3}, y+A^{-1/3}) - \bar{V}(x+A^{-2/3}, y-A^{-1/3})] - 1. \end{aligned} \tag{6.5}$$

By making $A \to \infty$, this will become

$$-\frac{\partial \bar{V}}{\partial x} = \frac{1}{2}\frac{\partial^2 \bar{V}}{\partial y^2} + \frac{y}{x}\frac{\partial \bar{V}}{\partial y} - 1. \tag{6.6}$$

The function (see (5.8))

$$\begin{aligned} \bar{V}_0(x, y) &= V_0\left(\frac{A^{2/3}x + A^{1/3}y}{2}, \frac{A^{2/3}x - A^{1/3}y}{2}\right) A^{2/3} \\ &= \frac{1}{2}A^{1/3} + \frac{|y|}{2x} \end{aligned} \tag{6.7}$$

satisfies, in the region where $y>0$, the equation

(6.8) $$-\frac{\partial\overline{V}_0}{\partial x}=\frac{1}{2}\frac{\partial^2\overline{V}_0}{\partial y^2}+\frac{y}{x}\frac{\partial\overline{V}_0}{\partial y}.$$

Therefore, the function $W(x,y)=\overline{V}(x,y)-\overline{V}_0(x,y)$ will satisfy

(6.9) $$-\frac{\partial W}{\partial x}=\frac{1}{2}\frac{\partial^2 W}{\partial y^2}+\frac{y}{x}\frac{\partial W}{\partial y}-1$$

in the limit.

The partial differential equation (6.9) will hold in the region between the straight line $y=0$ and the boundary curve

(6.10) $y=y_1(x)$, say (see Fig. 9).

At the boundary $y=0$, symmetry will give us the condition:

(6.11) $$\left.\frac{\partial\overline{V}}{\partial y}\right|_{y=0}=0,$$

and this, together with

(6.12) $$\left.\frac{\partial\overline{V}_0}{\partial y}\right|_{y=0}=\frac{1}{2x},$$

will give us the boundary condition:

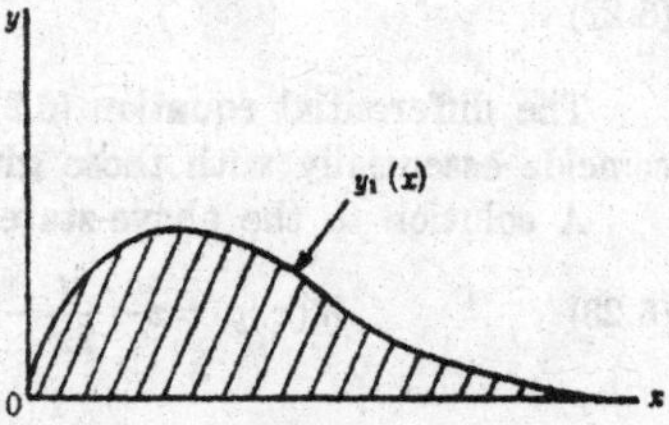

Fig. 9

(6.13) $$\left.\frac{\partial W}{\partial y}\right|_{y=0}=-\frac{1}{2x}.$$

A boundary condition at $y=y_1(x)$ is given by the continuity requirement as

(6.14) $$W|_{y=y_1(x)}=0.$$

Another boundary condition is obtained by the following arguments:—Take equation (5.4) and suppose that the point (u,v) is on the boundary, $(u+1,v+1)$ being outside, $(u+1,v-1)$ being inside the "go"-region. Then* $V'(u,v)=V_0'(u,v)$ and $V'(u+1,v+1)=V_0'(u+1,v+1)$. Let us further assume that $v-1>0$. Then substitution of $V'(u,v)=V_0'(u,v)=A/2+Av/2u$, and $V'(u+1,v+1)=V_0'(u+1,v+1)=A/2+A(v+1)/2(u+1)$ into (5.4) will give us

(6.15) $$\frac{A}{2}+\frac{Av}{2u}=\frac{1}{2}\left[\frac{A}{2}+\frac{A(v+1)}{2(u+1)}+V'(u+1,v-1)\right]$$
$$+\frac{v}{2u}\left[\frac{A}{2}+\frac{A(v+1)}{2(u+1)}-V'(u+1,v-1)\right]-1,$$

i.e.

(6.16) $$V'(u+1,v-1)=\frac{A}{2}+\frac{A(v-1)}{2(u+1)}+\frac{2u}{u-v}.$$

But

(6.17) $$V_0'(u+1,v-1)=\frac{A}{2}+\frac{A(v-1)}{2(u+1)}.$$

Therefore,

(6.18) $$V'(u+1,v-1)-V_0'(u+1,v-1)=\frac{2u}{u-v}.$$

* The function $V_0'(u,v)$ is defined as in (5.2) with $V(a,b)$ replaced by $V_0(a,b)$.

i.e.

$$W(x+A^{-2/3}, y-A^{-1/3})=A^{-2/3}\frac{2A^{2/3}x}{A^{2/3}x-A^{1/3}y}. \tag{6.19}$$

If we expand the left-hand member and neglect higher order terms, we get

$$W(x, y)-A^{-1/3}\frac{\partial W}{\partial y}+O(A^{-2/3})=O(A^{-2/3}). \tag{6.20}$$

But $W(x, y)=0$, since (x, y) is on the boundary. Hence, (6.20) implies that

$$\frac{\partial W}{\partial y}=O(A^{-1/3}). \tag{6.21}$$

Therefore, as the limit, we should expect that

$$\left.\frac{\partial W}{\partial y}\right|_{y=y_1(x)}=0. \tag{6.22}$$

The differential equation (6.9) and the boundary conditions (6.13), (6.14), and (6.22) coincide essentially with those given in Chernoff [1].

A solution to the above-stated free-boundary problem is given (for large x) by

$$\begin{aligned} W(x, y)=x-\frac{y}{2x}-xF\left(1, \frac{1}{2}; \frac{-y^2}{2x}\right) \\ +\frac{1}{16x^2}F\left(-2, \frac{1}{2}; \frac{-y^2}{2x}\right)-\frac{5}{3\times 16^2x^5}F\left(-5, \frac{1}{2}; \frac{-y^2}{2x}\right) \\ +\frac{437}{5\times 3^2\times 16^3x^8}F\left(-8, \frac{1}{2}; \frac{-y^2}{2x}\right)-\cdots, \end{aligned} \tag{6.23}$$

with

$$y_1(x)=\frac{1}{4x}-\frac{1}{48x^4}+\frac{7}{960x^7}-\cdots. \tag{6.24}$$

In (6.23), we used the notation of the confluent hypergeometric function:

$$F(\alpha, \gamma; z)=1+\frac{\alpha}{\gamma}\frac{z}{1!}+\frac{\alpha(\alpha+1)}{\gamma(\gamma+1)}\frac{z^2}{2!}+\frac{\alpha(\alpha+1)(\alpha+2)}{\gamma(\gamma+1)(\gamma+2)}\frac{z^3}{3!}+\cdots. \tag{6.25}$$

The curve (6.24) is plotted in Fig. 7 as a chain line. Its extension (dashed line) is somewhat arbitrarily drawn to indicate a plausible limit curve.

In Fig. 8, the curve (for $x\geq 1.0$) is well represented by the first three terms of

$$W(x, 0)=\frac{1}{16x^2}-\frac{5}{3\times 16^2x^5}+\frac{437}{5\times 3^2\times 16^3x^8}-\cdots, \tag{6.26}$$

which is derived from (6.23) by putting $y=0$.

A good expression of the solution in the domain $x\leq 1.0$ is desirable. Also a correction term of $y_1(x)$ (of the order of $A^{-1/3}$) would be desirable.

7. Comparison with other rules.

There are several decision rules, which are worth considering because of their simplicity.

One such rule is

R_r: *Sample until you have r balls of the same color, and bet on that color.*

For this rule,

$$\partial(p)\equiv P_p\,(\text{bet on white})=\sum_{i=0}^{r-1}\binom{r-1+i}{i}q^r p^i, \tag{7.1}$$

and therefore

$$P_p(\text{hit})=\sum_{i=0}^{r-1}\binom{r-1+i}{i}(p^{r+1}q^i+q^{r+1}p^i)\,, \tag{7.2}$$

$$E_p(N)=\sum_{i=0}^{r-1}(r+i)\binom{r-1+i}{i}(p^rq^i+q^rp^i)\,. \tag{7.3}$$

If we take the uniform distribution over (0, 1) as the *à priori* distribution of p, in other words, if we set $a=b=1$ in (1.9), then

$$\int_0^1 P_p(\text{hit})f_{1,1}(p)dp=\sum_{i=0}^{r-1}\frac{(r-1+i)!}{(r-1)!\,i!}\left[\frac{(r+1)!\,i!}{(r+2+i)!}+\frac{i!\,(r+1)!}{(r+2+i)!}\right] \tag{7.4}$$

$$=\frac{3r+1}{4r+2}=\frac{3}{4}-\frac{1}{4(2r+1)}\,,$$

$$\int_0^1 E_p(N)f_{1,1}(p)dp=\sum_{i=0}^{r-1}(r+i)\frac{(r-1+i)!}{(r-1)!\,i!}\left[\frac{r!\;i!}{(r+i+1)!}+\frac{i!\;r!}{(r+i+1)!}\right] \tag{7.5}$$

$$=2r\sum_{i=0}^{r-1}\frac{1}{r+i+1}\,.$$

Therefore the expected gain (1.10) for the rule R_r is

$$B_{1,1}(R_r)=A\left[\frac{3}{4}-\frac{1}{4(2r+1)}\right]-2r\sum_{i=0}^{r-1}\frac{1}{r+i+1}\,. \tag{7.6}$$

Since the expected gain if we knew the actual value of p would be

$$A\int_0^1 \max(p,q)dp=A\left[\int_0^{1/2} q\,dp+\int_{1/2}^1 p\,dp\right]=\frac{3}{4}A\,, \tag{7.7}$$

the regret for the rule R_r is given by

$$\frac{3}{4}A-B_{1,1}(R_r)=\frac{A}{4(2r+1)}+2r\sum_{i=0}^{r-1}\frac{1}{r+i+1}\equiv a_rA+b_r\,,\quad\text{say.} \tag{7.8}$$

Another simple rule is

R'_{2r-1}: *Sample* $2r-1$ *balls and bet on the majority.*

This is a single sampling rule with fixed sample size. A little reflection will reveal that the rule R_r is a "curtailment" of the rule R'_{2r-1}, that is, R_r is the modification of the rule R'_{2r-1} obtained by stopping as soon as the final decision becomes clear (since whichever color appears r times is certain to become the majority). Thus the decision using the rule R'_{2r-1} is exactly the same as the decision using the rule R_r; the only difference lies in the cost of sampling, which for R'_{2r-1} is simply $2r-1$. Therefore the regret for the rule R'_{2r-1} can be expressed as $a_r'A+b_r'$, with

$$a_r'=a_r=\frac{1}{4(2r+1)}\,,\qquad b_r'=2r-1\,. \tag{7.9}$$

A third rule is

R_r'': *Sample until one color is ahead of the other by* r; *and bet on the leading color.*

This is a Wald type sequential procedure, which has a pair of parallel stopping lines. Here the results on the gambler's ruin problem are applicable (see, e.g., Feller [2], pp. 313-318). Thus

$$\delta(p)\equiv P_p(\text{bet on white})=\frac{q^r}{q^r+p^r}\,, \tag{7.10}$$

$$E_p(N)=\frac{r}{q-p}\,\frac{q^r-p^r}{q^r+p^r}\,. \tag{7.11}$$

Therefore the regret can be expressed as $a_r''A+b_r''$, with

$$a_r''=\frac{3}{4}-\int_0^1 \frac{q^{r+1}+p^{r+1}}{q^r+p^r}dp\,, \tag{7.12}$$

$$b_r''=\int_0^1 \frac{r}{q-p}\,\frac{q^r-p^r}{q^r+p^r}dp\,. \tag{7.13}$$

By decomposing the fraction into partial fractions, we can evaluate (7.12) as follows:

$$\begin{aligned} a_r'' &=\frac{1}{4}-\frac{r}{2}+\frac{\pi}{2r^2}\sum_{j=1}^{r/2}\frac{\sin[(2j-1)\pi/2r]}{\cos^3[(2j-1)\pi/2r]}(r-\overline{2j-1}), \quad r:\text{ even},\\ &=\frac{1}{4}-\frac{r}{6}+\frac{\pi}{2r^2}\sum_{j=1}^{(r-1)/2}\frac{\sin[(2j-1)\pi/2r]}{\cos^3[(2j-1)\pi/2r]}(r-\overline{2j-1}), \quad r:\text{ odd}. \end{aligned} \tag{7.14}$$

Likewise,

$$\begin{aligned} b_r'' &=2\pi\sum_{j=1}^{r/4}\frac{1}{\sin[(2j-1)\pi/r]}, && \text{if } r\equiv 0 \pmod{4},\\ &=2\pi\sum_{j=1}^{(r-2)/4}\frac{1}{\sin[(2j-1)\pi/r]}+\pi, && \text{if } r\equiv 2 \pmod{4},\\ &=\frac{2\pi}{r}\sum_{j=1}^{(r-1)/2}\frac{r-(2j-1)}{\sin[(2j-1)\pi/r]}+1, && \text{if } r \text{ is odd}. \end{aligned} \tag{7.15}$$

Table 3 gives a_r'' and b_r'' for a few values of r.

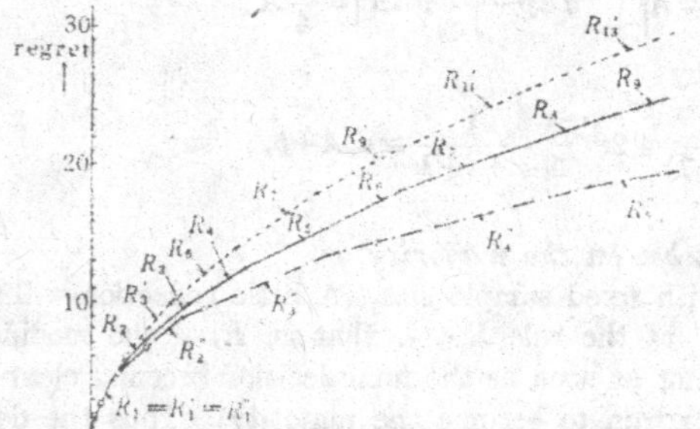

Fig. 10

Table 3 a_r'' and b_r''.

r	a_r''	b_r''
1	0.083333	1
2	0.035398	3.14159
3	0.018711	5.83680
4	0.011366	8.88577
5	$0.0^{2}75729$	12.19429
6	$0.0^{2}53839$	15.70796
7	$0.0^{2}40146$	19.39138
8	$0.0^{2}31044$	23.21963

Plotting the regret against A, we get Fig. 10. For each of the categories R_r, R'_{2r-1}, R_r'', we have a sequence of straight lines corresponding to $r=1, 2, \cdots$. For a given value of A, there exists an optimal r in each category. The regret using that optimal r is shown by the thick segment.

In Fig. 10, the regret for the optimal decision rule given in Section 3 is also shown, by a circle, for each of the values $A=10, 20, 50, \cdots$. It is interesting to note that these circles lie very close to the segments for R_r''. It means that the decision rules R_r'' based on the parallel-line stopping rule are nearly optimal (at least for $a=b=1$).

An interesting comparison can be made between the results of Section 6 and the results of [3] which gives the optimal fixed-sample-size procedure.

8. Operating characteristic curve and average sample number curve for the optimal decision rule.

Let $\delta(p, a, b)$ be the probability of guessing "white" when we use the decision rule $D_{a,b}$ and p is the actual probability of black. Then this function will satisfy the difference equation:

$$\delta(p, a, b) = p\delta(p, a+1, b) + (1-p)\delta(p, a, b+1) \tag{8.1}$$

in the "go"-region, and the boundary conditions:

$$\delta(p, a, b) = 0 \text{ in the "guess black"-region,} \tag{8.2}$$

$$\delta(p, a, b) = 1 \text{ in the "guess white"-region.} \tag{8.3}$$

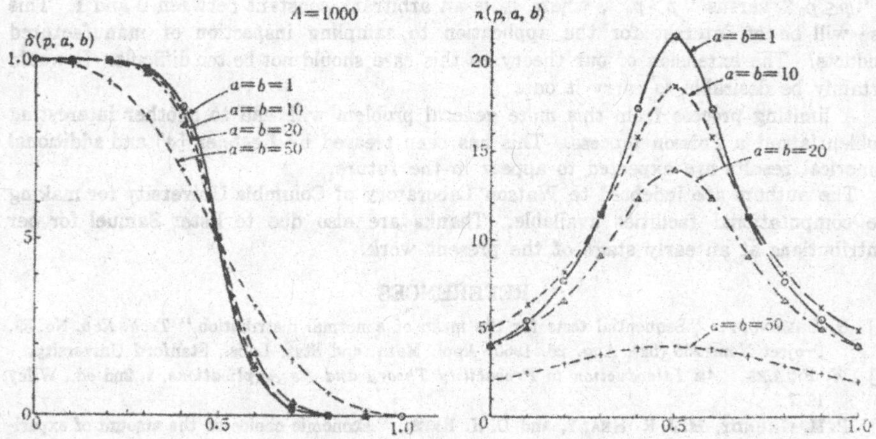

Fig. 11

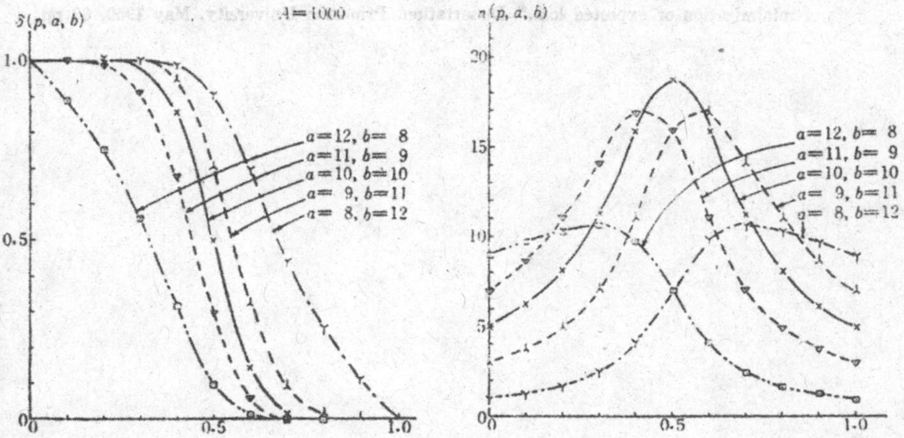

Fig. 12

Let $n(p, a, b)$ be the expected sample size when we use the decision rule $D_{a,b}$ and p is the actual probability of black. Then this function will satisfy the equation

$$n(p, a, b)=pn(p, a+1, b)+(1-p)n(p, a, b+1)+1 \tag{8.4}$$

in the "go"-region, and the boundary condition

$$n(p, a, b)=0, \text{ in both "guess white"- and "guess black"-regions.} \tag{8.5}$$

For $A=1000$, and for $p=0.1, 0.2, 0.3, 0.4$, and 0.5, the functions $\delta(p, a, b)$ and $n(p, a, b)$ have been evaluated recursively (together with $V(a, b)$ and $W(a, b)$). Part of the results are shown in Figs. 11 and 12.

9. Final remarks and acknowledgements.

A more general problem than the one discussed in the present paper is the testing of "$p \leq p_0$" versus "$p > p_0$", where p_0 is an arbitrary constant between 0 and 1. This case will be of interest for the application to sampling inspection of manufactured products. The extension of our theory to this case should not be too difficult. It would certainly be desirable to carry it out.

A limiting process from this more general problem will lead to another interesting problem about a Poisson process. This has been treated by Lechner [4] and additional numerical results are expected to appear in the future.

The authors are indebted to Watson Laboratory of Columbia University for making the computational facilities available. Thanks are also due to Ester Samuel for her contributions at an early stage of the present work.

REFERENCES

[1] H. CHERNOFF, "Sequential tests for the mean of a normal distribution," *Tech. Rep.* No. 59, Project Nonr-255 (52), Aug. 26, 1960, Appl. Math. and Stat. Labs., Stanford University.

[2] W. FELLER, *An Introduction to Probability Theory and Its Applications*, 1, 2nd ed., Wiley 1957.

[3] P. M. GRUNDY, M. J. R. HEALY, and D. H. REES, "Economic choice of the amount of experimentation," *J. Roy. Stat. Soc.*, Ser. B, 18, 1956, pp. 32–55.

[4] J. A. LECHNER, "Optimum decision rules for the comparison of two Poisson processes, based on minimization of expected loss," Dissertation, Princeton University, May 1959, 62 pp.

Proc. Natl. Acad. Sci. USA
Vol. 77, No. 6, pp. 3135–3138, June 1980
Statistics

Sequential medical trials

(stopping rules/asymptotic optimality)

T. L. LAI[†], BRUCE LEVIN[†], HERBERT ROBBINS[†], AND DAVID SIEGMUND[‡]

[†]Department of Mathematical Statistics, Columbia University, New York, New York 10027; and [‡]Department of Statistics, Stanford University, Stanford, California 94305

Contributed by Herbert Robbins, March 24, 1980

ABSTRACT A model for sequential clinical trials is discussed. Three proposed stopping rules are studied by the Monte Carlo method for small patient horizons and mathematically for large patient horizons. They are shown to be about equally effective and asymptotically optimal from both Bayesian and frequentist points of view and are markedly superior to any fixed sample size procedure.

Consider the design of a clinical trial to select the better of two treatments, A or B, for treating a specified number N of patients. The trial phase involves pairwise allocation of treatments to n pairs of patients, after which the apparently superior treatment is given to the remaining $N - 2n$ patients. We shall assume that the difference in effect of the two treatments is measured by a random variable z which is normally distributed with mean δ and variance σ^2. If $\delta > 0$, treatment A is preferred to treatment B; if $\delta < 0$, the preference is reversed; and if $\delta = 0$, neither is preferred. The problem is to decide on the number of pairs to be used for the trial phase.

Let z_i denote the difference in response between the patient receiving treatment A and the patient receiving treatment B in the ith pair on trial, and let $s_n = z_1 + \ldots + z_n$. If n pairs are put on trial and the remaining $N - 2n$ patients are given treatment A or B according to if $s_n > 0$ or $s_n < 0$, then the total number of inferior treatments is $n + (N - 2n)I(s_n < 0)$ if $\delta > 0$ and $n + (N - 2n)I(s_n > 0)$ if $\delta < 0$, where $I(\cdot)$ denotes the indicator function of the event in question. Given a stopping rule T for determining the number of pairs of patients to be put on trial, the regret $R(\delta,T)$ is defined to be the expected total difference in response between the ideal procedure, which would assign all N patients to the superior treatment, and the procedure determined by the stopping rule T; hence, $R(\delta,T)$ equals $|\delta|$ times the expected number of inferior treatments, or

$$\begin{aligned} R(\delta,T) &= \delta E_\delta\{T + (N - 2T)I(s_T < 0)\} \quad \text{if} \quad \delta > 0 \\ &= |\delta|\, E_\delta\{T + (N - 2T)I(s_T > 0)\} \quad \text{if} \quad \delta < 0. \end{aligned} \qquad [1]$$

Our primary purpose is to compare three stopping rules for minimizing Eq. 1 in the case of known σ: (*i*) the Bayes rule T_B for a flat prior on δ; (*ii*) an *ad hoc* rule T_A suggested by Anscombe (ref. 1); and (*iii*) another *ad hoc* rule T^* considered here for the first time. These rules are studied by the Monte Carlo method, and analytic approximations are obtained as $N \to \infty$. A general conclusion is that all three rules are about equally good from both a frequentist and a Bayesian point of view. We also discuss briefly the case of unknown σ.

Suppose first that σ is known and hence without loss of generality that $\sigma = 1$. Let Φ denote the distribution function and ϕ the density function of the standard normal distribution. To provide a standard of comparison for the stopping rules to be discussed below, suppose for the moment that $|\delta|$ is known and that we need only discover whether sgn δ is $+1$ or -1. Then a fixed sample size n minimizing $R(|\delta|,n)$ would be obtained by solving the equation

$$\frac{\partial}{\partial_n} R(|\delta|, n) \left(= |\delta| \frac{\partial}{\partial_n} \{n + (N - 2n)\Phi(-|\delta|n^{1/2})\}\right) = 0,$$

which defines n implicitly by

$$g(|\delta|n^{1/2}) = N/(2n), \qquad [2]$$

where

$$\begin{aligned} g(x) &= [2\Phi(x) - 1]/[x\phi(x)] + 1 \quad \text{if} \quad x > 0, \\ &= 3 \quad \text{if} \quad x = 0. \end{aligned} \qquad [3]$$

It can be shown that $g(x)$ is increasing in $x \geq 0$. For this minimizing value of n, say $n^* = n^*(|\delta|)$, one can show that for any $\delta \neq 0$

$$R(\delta,n^*) \sim (2 \log N)/|\delta| \quad \text{as} \quad N \to \infty. \qquad [4]$$

In practice $|\delta|$ will be unknown, but we can try to estimate it sequentially. Thus, Eq. 2 suggests the following procedure. Stop the trial at stage

$$T^* = \inf\{k: g(|s_k|/k^{1/2}) \geq N/(2k)\}. \qquad [5]$$

Since $g(x) \geq 3$ for all $x \geq 0$, $T^* \leq N/6$. Among other properties of T^*, we shall show that for every fixed $\delta \neq 0$, $R(\delta,T^*) \sim R(\delta,n^*)$ as $N \to \infty$. Hence, T^* is asymptotically as efficient as the optimal fixed sample size n^*, which requires knowledge of $|\delta|$.

An alternative approach in ignorance of $|\delta|$ is to assume a prior distribution G on δ and to choose a stopping rule T to minimize the integrated regret $\int_{-\infty}^{\infty} R(\delta,T)dG(\delta)$. Anscombe (ref. 1) first considered this problem with the flat prior $dG(\delta) = d\delta$. Although the integrated regret with respect to the flat prior is infinite for every stopping rule $T(\geq 1)$, the posterior regret $r(k,s)$ given that $s_k = s$ and that stopping occurs at stage k is well defined and can be shown to be

$$r(k,s) = |s| + k^{-1/2}N\{\phi(s/k^{1/2}) - k^{-1/2}|s|\,\Phi(-|s|/k^{1/2})\}. \qquad [6]$$

For fixed k, $r(k,s)$ attains its minimal value at two symmetric points, s and $-s$, defined by the equation

$$1 - \Phi(|s|/k^{1/2}) = k/N. \qquad [7]$$

Anscombe suggested using Eq. 7 to define a stopping rule

$$T_A = \inf\{k: 1 - \Phi(|s_k|/k^{1/2}) \leq k/N\}, \qquad [8]$$

which, like T^*, will be shown to be asymptotically optimal from both the Bayesian and the frequentist points of view.

The exact solution T_B to the Bayes problem considered by Anscombe can be computed by the backward induction algorithm (e.g., ref. 2, p. 50). Fig. 1 plots the stopping boundaries of T^*, T_A, and T_B for $N = 100$.

The publication costs of this article were defrayed in part by page charge payment. This article must therefore be hereby marked "*advertisement*" in accordance with 18 U. S. C. §1734 solely to indicate this fact.

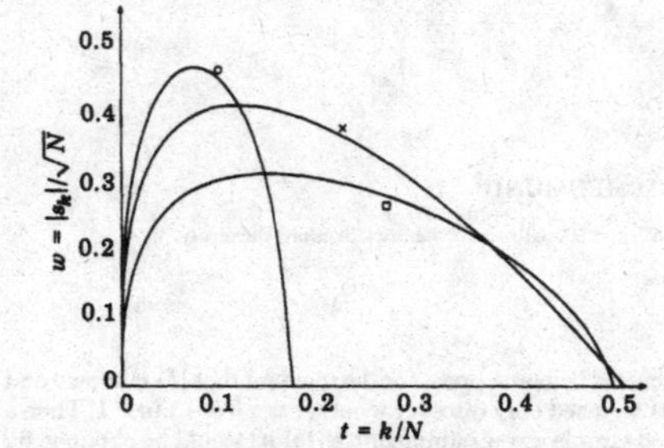

FIG. 1. The boundary T^* (○), Anscombe boundary T_A (×), and Bayes boundary T_B (□) for $N = 100$.

An asymptotically optimal class of stopping rules

Given any continuous function $f:(0, \tfrac{1}{2}] \to [0,\infty)$, define the stopping rule

$$T(f,N) = \inf\{k \le N/2: |s_k| \ge N^{1/2} f(k/N)\} (\inf\phi = N/2). \quad [9]$$

The stopping rule T^* in Eq. 5 is of this form with $f(t)$ equal to

$$f^*(t) = t^{1/2} g^{-1}(1/2t), \quad [10]$$

where g is defined in Eq. 3. Likewise Anscombe's rule T_A in Eq. 8 is also of this form with $f(t)$ equal to

$$f_A(t) = t^{1/2}\Phi^{-1}(1-t). \quad [11]$$

Let $\theta = \delta N^{1/2}$. Let $w(t)$, $t \ge 0$, denote the Wiener process with drift coefficient θ under the probability measure P_δ. Let $I_N = \{k/N: 1 \le k \le N/2\}$. Under P_δ, the sequences $\{s_k\}$ and $\{N^{1/2}w(k/N)\}$ have the same distribution, and therefore the stopping rule $T(f,N)$ has the same distribution as $N\tau_{f,N}$, where $\tau_{f,N}$ is the stopping rule defined on the Wiener process $w(t)$ by

$$\tau_{f,N} = \inf\{t \in I_N: |w(t)| \ge f(t)\} (\inf\phi = 1/2). \quad [12]$$

As $N \to \infty$, $\tau_{f,N}$ converges with probability 1 to

$$\tau(f) = \inf\left\{t \in \left(0, \frac{1}{2}\right] : |w(t)| \ge f(t)\right\}. \quad [13]$$

The regret $R(\delta; T(f,N))$ can be expressed in terms of the process $w(t)$ as

$$R(\delta; T(f,N)) = N^{1/2}\rho(\theta; \tau_{f,N}), \quad [14]$$

where we define for any stopping rule $\tau (\le 1/2)$ on $w(t)$

$$\rho(\theta;\tau) = \theta E_\delta\{\tau + (1-2\tau)I(w(\tau) \le 0)\} \text{ if } \theta > 0,$$
$$= |\theta|\, E_\delta\{\tau + (1-2\tau)I(w(\tau) \ge 0)\} \text{ if } \theta < 0.$$

In connection with the Bayes rule T_B with respect to the flat prior on δ, Chernoff and Petkau (3) recently studied the corresponding continuous time problem of minimizing the integrated risk $\int_{-\infty}^{\infty} \rho(\theta;\tau)d\theta$ among all stopping rules $\tau \le 1/2$ on the process $w(t)$. The minimizing τ is of the form $\tau(f_B)$ in Eq. 13, and they obtained the following asymptotic expansion of the Bayes optimal boundary $f_B(t)$ as $t \to 0$:

$$f_B(t) = \left\{2t\left[\log\frac{1}{t} - \frac{1}{2}\log\log\frac{1}{t} - \frac{1}{2}\log 16\pi + o(1)\right]\right\}^{1/2}. \quad [15]$$

It is interesting to compare Eq. 15 with the following asymptotic expansions of the boundaries $f^*(t)$ and $f_A(t)$ associated with the rules T^* and T_A in Eqs. 5–8: as $t \to 0$,

$$f^*(t) = \left\{2t\left[\log\frac{1}{t} + \frac{1}{2}\log\log\frac{1}{t} - \frac{1}{2}\log 4\pi + o(1)\right]\right\}^{1/2}, \quad [16]$$

$$f_A(t) = \left\{2t\left[\log\frac{1}{t} - \frac{1}{2}\log\log\frac{1}{t} - \frac{1}{2}\log 4\pi + o(1)\right]\right\}^{1/2}. \quad [17]$$

Hence, for small t the three boundaries $f^*(t)$, $f_A(t)$, and $f_B(t)$ are close to each other; and $f^*(t) > f_A(t) > f_B(t)$, as is illustrated in Fig. 1.

Let $\mathcal{C}$ denote the set of all continuous functions $f: (0,\tfrac{1}{2}] \to [0,\infty)$ satisfying the following two conditions: (i) as $t \to 0$,

$$f(t) \sim \left\{2t\log\frac{1}{t}\right\}^{1/2}; \quad [18]$$

and (ii) there exist $\eta < 3/2$ and $t_0 > 0$ such that for all $0 < t \le t_0$,

$$f(t) \ge \left\{2t\left(\log\frac{1}{t} - \eta\log\log\frac{1}{t}\right)\right\}^{1/2}. \quad [19]$$

In view of Eqs. 15–17, the boundaries f^*, f_A, and f_B all belong to $\mathcal{C}$. The following theorem shows that for every given δ, stopping rules of the form $T(f,N)$ in Eq. 9 with $f \in \mathcal{C}$ are asymptotically (as $N \to \infty$) as efficient as the optimal fixed sample size n^* which assumes $|\delta|$ known.

THEOREM 1. *Let* $f \in \mathcal{C}$. *Then for every fixed* $\delta \ne 0$,

$$R(\delta; T(f,N)) \sim (2\log N)/|\delta| \text{ as } N \to \infty. \quad [20]$$

Hence, in view of Eq. 4, for every fixed $\delta \ne 0$,

$$R(\delta; T(f,N)) \sim R(\delta; n^*).$$

Moreover, as $N \to \infty$ *and* $\delta \to 0$ *such that* $|\theta| = N^{1/2}|\delta| \to \infty$,

$$R(\delta; T(f,N)) \sim 2(\log\theta^2)/|\delta| \sim R(\delta; n^*). \quad [21]$$

The order of magnitude of the regret $R(\delta; T(f,N))$ given by [20] is asymptotically minimal in the sense of the following theorem.

THEOREM 2. *Let* $a > 1$ *and let* δ_N *be a sequence of positive constants such that* $\delta_N \to 0$ *and* $\log \delta_N^{-1} = o(\log N)$ *as* $N \to \infty$. *For every* $N \ge 2$, *let* $\mathcal{T}_N$ *denote the class of stopping rules* $T \le N/2$ *such that* $R(\delta_N; T) \le (\log N)^a/\delta_N$. *Let* $f \in \mathcal{C}$. *Then* $T(f,N) \in \mathcal{T}_N$ *for all large* N *in view of* [21], *and for every fixed* $\delta \ne 0$,

$$\inf_{T\in\mathcal{T}_N} R(\delta; T) \sim (2\log N)/|\delta| \sim R(\delta; T(f,N)) \text{ as } N \to \infty.$$

Stopping rules of the form $T(f,N)$ with $f \in \mathcal{C}$ are also asymptotically optimal from a Bayesian point of view. The following theorem shows that these rules are asymptotically Bayes with respect to *any* prior distribution on δ that has a positive continuous density in some neighborhood of the origin. Moreover, their integrated regret is of the order of $(\log N)^2$ and is therefore much smaller for large N than $\{2NG'(0) \times \int|\delta|dG(\delta)\}^{1/2}$, which can be shown to be asymptotically equivalent to the integrated regret of the Bayes rule using the optimal fixed n for a given prior G on δ.

THEOREM 3. *Let* G *be a distribution function on* $(-\infty,\infty)$ *such that* G' *is positive and continuous in some neighborhood of the origin and* $\int_{-\infty}^{\infty}/|\delta|dG(\delta) < \infty$. *Let* $f \in \mathcal{C}$. *Then as* $N \to \infty$,

$$\int_{-\infty}^{\infty} R(\delta: T(f,N))dG(\delta) \sim G'(0)(\log N)^2$$
$$\sim \inf_T \int_{-\infty}^{\infty} R(\delta; T)dG(\delta).$$

The proofs of *Theorems 1–3* will be given elsewhere.

Table 1. Risk of Anscombe's rule for various N^{a}

$\theta =$	$N = 400$			$N = 2500$			$N = 10{,}000$		
$\delta N^{1/2}$	R	P	E	R	P	E	R	P	E
0	0	0.5	0.15	0	0.5	0.14	0	0.5	0.13
1	0.36	0.31	0.14	0.37	0.33	0.13	0.37	0.33	0.12
2	0.52	0.18	0.13	0.53	0.20	0.11	0.54	0.20	0.11
3	0.56	0.09	0.11	0.56	0.11	0.10	0.57	0.12	0.09
5	0.47	0.03	0.07	0.51	0.04	0.07	0.51	0.04	0.07
10	0.36	0.002	0.03	0.36	0.005	0.03	0.38	0.01	0.03
16	0.31	0.0003	0.02	0.27	0.001	0.02	0.29	0.003	0.02
24	0.28	$0.4 \cdot 10^{-4}$	0.01	0.24	0.0002	0.01	0.23	0.0007	0.01

^{a} $\delta > 0$, $\sigma = 1$, $R = N^{-1/2} R(T,\delta)$, $P = P_\delta\{s_T < 0\}$, $E = N^{-1} E_\delta(T)$.

The case of unknown variance

In the case of known σ, writing Anscombe's stopping rule as

$$T_A = \inf \left\{k: \Phi\left(\frac{|s_k|}{k^{1/2}\sigma}\right) \geq 1 - \frac{k}{N}\right\}, \quad [22]$$

we can regard T_A as the stopping rule of a symmetric repeated significance test of $H_0: \delta < 0$ versus $H_1: \delta > 0$ with nominal significance level at stage k equal to k/N. When σ is unknown, we can estimate it from the data observed so far, and it is natural to replace σ in Eq. **22** by $\hat{\sigma}_k$, where

$$\hat{\sigma}_k^2 = \sum_{i=1}^{k} (z_i - \bar{z}_k)^2/(k-1), \quad \bar{z}_k = s_k/k.$$

For $\delta = 0$, while Φ is the distribution function of $s_k/(k^{1/2}\sigma)$ in Eq. **22**, $s_k/(k^{1/2}\hat{\sigma}_k)$ has the t distribution with $k - 1$ degrees of freedom. This suggests the following analogue of T_A for the case of unknown σ:

$$\hat{T} = \inf\left\{k \geq 2: F_{k-1}\left(\frac{|s_k|}{k^{1/2}\hat{\sigma}_k}\right) \geq 1 - \frac{k}{N}\right\}, \quad [23]$$

where F_ν denotes the distribution function of the t distribution with ν degrees of freedom. Letting $F_\nu(t) = \Phi(x_\nu(t))$ and

$$u(t;\nu) = \left\{\nu \log\left(1 + \frac{t^2}{\nu}\right)\right\}^{1/2}, \quad [24]$$

it is convenient to use in Eq. **23** the following two simple approximations to $x_\nu(t)$ due to Wallace (4):

$$u_1(t;\nu) = u(t;\nu)\left\{1 - \frac{1}{2\nu}\right\}^{1/2}; \quad [25]$$

$$u_2(t;\nu) = u(t;\nu)\left\{1 - \frac{2(1 - e^{-y^2})^{1/2}}{8\nu + 3}\right\}, \quad [26]$$

with

$$y = \frac{(0.184)(8\nu + 3)}{\nu^{1/2} u(t;\nu)}.$$

As shown by Wallace, $u(t;\nu) \geq x_\nu(t) \geq u_1(t,\nu)$, and $u_2(t;\nu)$ provides an excellent approximation to $x_\nu(t)$ over a wide range of values of ν and t. Thus instead of $\hat{T}$, it is more convenient to use the stopping rule

$$\hat{T}_i = \inf\{k \geq 2: \Phi[u_i(|s_k|/(\hat{\sigma}_k k^{1/2}); k-1)] \geq 1 - k/N\} \quad [27]$$

for $i = 1$ or 2. It may be shown analogously to *Theorems 2* and *3* that the stopping rules $\hat{T}$, $\hat{T}_1$, and $\hat{T}_2$ are asymptotically optimal as $N \to \infty$ in the case of unknown σ, from both the Bayesian and frequentist points of view.

Error probability and expected sample size approximations

Suppose again that $\sigma = 1$. Two of the three terms in the regret defined by Eq. **1** are $E_\delta(T)$, the expected number of pairs in the trial phase, and $P_\delta\{s_T < 0\}$ ($\delta > 0$), the probability of making an incorrect decision for the post-trial phase. In sequential hypothesis testing, in the absence of a specific loss function, it is customary to examine these two quantities separately in evaluating a test. Hence, it is of interest to give approximations to $E_\delta(T)$ and $P_\delta(s_T < 0)$ for $\delta > 0$.

Suppose that as $t \to 0$, for some $\eta < 3/2$ and $c > 0$ (see Eqs. **15–17**)

$$f(t) = \{2t\,(\log(1/t) - \eta \log\log(1/t) - \log c + o(1))\}^{1/2}. \quad [28]$$

By theorem 3 of ref. 5, for each $\delta \neq 0$ as $N \to \infty$

$$E_\delta\{T(f,N)\} = \delta^{-2}\{2\log(N\delta^2) - 2(1+\eta)\log\log(N\delta^2) - \log 4c^2 - 1 + K_\delta + o(1)\}, \quad [29]$$

where K_δ can be computed numerically and equals $1.166\ldots\delta + \delta^2/4 + o(\delta^2)$ as $\delta \to 0$. However, this result does not yield good numerical approximations, because for δ near 0, the factor δ^{-2} magnifies the effect of the asymptotically negligible $o(1)$ in Eq. **29**. Although difficult to justify rigorously, the ideas in ref. 5 suggest that $E_\delta\{T(f,N)\}$ should approximately equal the root x of the equation

$$x = \delta^{-2}\{2\log(N/x) - 2\eta\log\log(N/x) - \log c^2 - 1 + 2/\log(N\delta^2)\} + 1.166/\delta. \quad [30]$$

We hope that solving Eq. **30** will provide useful approximations to $E_\delta(T)$ when it is not too small and when $P_\delta\{s_T < 0\}$ is near 0. Comparisons with the Monte Carlo results in Table 1 indicate that this is the case.

It seems a fairly delicate problem to approximate $P_\delta\{s_T < 0\}$ for $\delta > 0$. However, for Brownian motion and boundaries f satisfying Eq. **28**, a modification of the argument given in ref. 6 shows that as $\theta \to \infty$

$$P_\delta\{w(\tau(f)) < 0\} \sim \frac{c(\log\theta)^{\eta - 1/2}}{2\pi^{1/2}\theta^2}. \quad [31]$$

Table 2. Regret, error probability, and expected sample size for n^*, T^*, T_A, T_B, $\hat{T}_1$, and $\hat{T}_2{}^{a}$

$\theta =$	n^*			T^*			T_A			T_B			$\hat{T}_1$			$\hat{T}_2$		
$\delta N^{1/2}$	R	P	E	R	P	E	R	P	E	R	P	E	R	P	E	R	P	E
0.5	0.22	0.43	0.17	0.21	0.39	0.13	0.22	0.42	0.18	0.22	0.42	0.10	0.22	0.41	0.14	0.22	0.42	0.13
1	0.39	0.34	0.16	0.38	0.39	0.13	0.37	0.32	0.16	0.37	0.33	0.09	0.37	0.33	0.13	0.39	0.35	0.12
2	0.61	0.22	0.15	0.53	0.20	0.12	0.53	0.17	0.15	0.56	0.23	0.09	0.56	0.21	0.12	0.57	0.21	0.12
3	0.70	0.14	0.13	0.60	0.11	0.12	0.55	0.08	0.13	0.61	0.14	0.08	0.65	0.13	0.11	0.65	0.13	0.10
4	0.73	0.10	0.11	0.58	0.06	0.10	0.54	0.03	0.11	0.61	0.09	0.07	0.61	0.08	0.08	0.65	0.09	0.08
5	0.72	0.07	0.09	0.57	0.03	0.09	0.51	0.02	0.09	0.56	0.06	0.05	0.63	0.05	0.08	0.64	0.05	0.08
6	0.71	0.05	0.08	0.50	0.01	0.08	0.47	0.01	0.07	0.50	0.04	0.05	0.61	0.04	0.06	0.62	0.04	0.06
7	0.68	0.04	0.07	0.50	0.003	0.07	0.47	0.004	0.06	0.46	0.03	0.04	0.58	0.03	0.05	0.58	0.03	0.05
8	0.66	0.03	0.06	0.51	0.001	0.06	0.44	0.002	0.05	0.43	0.02	0.04	0.51	0.01	0.05	0.56	0.02	0.05
9	0.63	0.02	0.05	0.50	0.000	0.06	0.45	0.001	0.05	0.40	0.01	0.03	0.52	0.01	0.04	0.52	0.01	0.04
10	0.60	0.02	0.04	0.50	0.000	0.05	0.41	0.001	0.04	0.37	0.008	0.03	0.52	0.01	0.04	0.52	0.01	0.04
20	0.43	0.005	0.02	0.48	0.000	0.02	0.38	0.000	0.02	0.29	0.000	0.01	0.52	0.000	0.03	0.52	0.000	0.03

^{a} $N = 100$, $\delta > 0$, $\sigma = 1$, $R = R(\delta,T)/N^{1/2}$, $P = P_\delta\{s_T < 0\}$, $E = E_\delta(T)/N$.

This suggests that for N large and $\theta = \delta N^{1/2}$ moderately large, and for, say, $f = f_A$,

$$P_\delta\{s_T < 0\} \doteq \theta^{-2}.$$

Comparison with the Monte Carlo results in Table 1 indicate that this is a good approximation for $N = 10{,}000$ and $2 \leq \theta \leq 10$. For smaller θ, [31] is not applicable. For larger θ, neglecting the excess over the boundary in using Brownian motion is too crude.

Tables 1 and 2 report the results of two Monte Carlo experiments. Table 2 gives the risk, error probability, and expected sample size of the stopping rules T^*, T_A, T_B, $\hat{T}_1$, and $\hat{T}_2$ for $\sigma = 1$ and $N = 100$ and compares these quantities with those of the optimal fixed sample size n^*, which assumes $|\delta|$ and σ known. Table 1 gives similar results for Anscombe's rule and $N = 400$, 2500, and 10,000.

This research was supported by the National Institutes of Health, the National Science Foundation, and the Office of Naval Research.

1. Anscombe, F. J. (1963) *J. Am. Stat. Assoc.* 58, 365–383.
2. Chow, Y. S., Robbins, H. & Siegmund, D. (1971) *Great Expectations: The Theory of Optimal Stopping* (Houghton Mifflin, Boston).
3. Chernoff, H. & Petkau, A. J. (1979) *Sequential Medical Trials Involving Paired Data*, Technical Report 79-20 (Inst. Appl. Math. Stat., Univ. British Columbia).
4. Wallace, D. L. (1959) *Ann. Math. Stat.* 30, 1121–1130.
5. Lai, T. L. & Siegmund, D. (1979) *Ann. Stat.* 7, 60–76.
6. Lai, T. L. & Siegmund, D. (1977) *Ann. Stat.* 5, 946–954.

Part Two

SEQUENTIAL EXPERIMENTATION AND ANALYSIS

D. Power-One Tests and Related Boundary Crossing Probabilities

Let $x_1, x_2, \ldots$ be independent and normally distributed with mean μ and variance 1, and let $S_n = x_1 + \cdots + x_n$. The standard test of $H_0: \mu = 0$ against $H_1: \mu > 0$ rejects H_0 if $S_n \geq bn^{1/2}$ at the significance level $\alpha(b) = P_0\{S_n \geq bn^{1/2}\}$. Already in [28] Robbins discussed the problem of optional stopping. Suppose that n is not fixed in advance of the experiment, but the experiment is terminated with rejection of H_0 if for *some* value of n, say some $n_0 \leq n \leq n_1$, the event $\{S_n \geq bn^{1/2}\}$ occurs. Now the significance level is

$$P_0\{S_n \geq bn^{1/2} \text{ for some } n_0 \leq n \leq n_1\}, \tag{6}$$

which can be substantially larger than $\alpha(b)$. In fact, if $n_1 = \infty$ the probability (6) is one, so an unscrupulous statistician who wants badly to reject a true null hypothesis can do so with certainty by optional stopping.

In [70]* together with Darling, Robbins proposed the beautiful idea of a test of power one as a solution to the problem of optional stopping. There it was shown that for any $\varepsilon > 0$ by increasing $bn^{1/2}$ to a suitable c_n which increases more rapidly than $n^{1/2}$ (one particularly convenient choice has $c_n \sim (n \log n)^{1/2}$ as $n \to \infty$), it is possible to have

$$P_0\{S_n \geq c_n \text{ for some } n \geq 1\} \leq \varepsilon \tag{7}$$

and for all $\mu > 0$

$$P_\mu\{S_n \geq c_n \text{ for some } n \geq 1\} = 1. \tag{8}$$

---

*Papers reprinted in this book are marked here by an asterisk when they first appear.

Thus if one samples until the first occurrence of the event $\{S_n \geq c_n\}$, the probability of rejecting a true null hypothesis is $\leq \varepsilon$, whereas the probability of rejecting a false H_0 equals one.

Subsequent development of these ideas [72], [77], [78], [81], [82]*, [83]* involves improved and more general approximations to boundary crossing probabilities like (7) and approximations under H_1 for the expected waiting time until the first occurrence of an event $\{S_n \geq c_n\}$ ([92]*, [95], Pollak and Siegmund, 1975). Since 1976 some of the techniques developed to study tests of power one have been extended to give approximations for (6), thus effectively solving the problem posed by Robbins in 1952 (e.g. Lai and Siegmund, 1977, Woodroofe, 1976, 1978). Summaries of this research with extensive bibliographies are described by Woodroofe (1982) and Siegmund (1985).

References

Lai, T. L. and Siegmund, D. (1977). A non-linear renewal theory with applications to sequential analysis I, *Ann. Statist.* **5**, 946–954.

Pollak, M. and Siegmund, D. (1975). Approximations to the expected sample size of certain sequential tests, *Ann. Statist.* **3**, 1267–1282.

Siegmund, D. (1977). Repeated significance tests for a normal mean, *Biometrika* **64**, 177–189.

Siegmund, D. (1985). *Sequential Analysis: Tests and Confidence Intervals*. Springer-Verlag, New York.

Woodroofe, M. (1976). Frequentist properties of Bayesian sequential tests, *Biometrika* **63**, 101–110.

Woodroofe, M. (1978). Large deviations of the likelihood ratio statistic with applications to sequential testing. *Ann. Statist.* **6**, 72–84.

Woodroofe, M. (1982). *Nonlinear Renewal Theory in Sequential Analysis*, Society for Industrial and Applied Mathematics, Philadelphia.

Reprinted from the PROCEEDINGS OF THE NATIONAL ACADEMY OF SCIENCES
Vol. 57, No. 5, pp. 1188–1192. May, 1967.

ITERATED LOGARITHM INEQUALITIES*

By D. A. Darling and Herbert Robbins

UNIVERSITY OF CALIFORNIA, BERKELEY

Communicated by J. Neyman, March 10, 1967

1. *Introduction.*—Let $x, x_1, x_2, \cdots$ be a sequence of independent, identically distributed random variables with mean 0, variance 1, and moment generating function $\varphi(t) = E(e^{tx})$ finite in some neighborhood of $t = 0$, and put $s_n = x_1 + \cdots + x_n$, $\bar{x}_n = s_n/n$. For any sequence of positive constants a_n, $n \geq 1$, let

$$P_m = P(|\bar{x}_n| \geq a_n \quad \text{for some } n \geq m).$$

The law of the iterated logarithm gives conditions on the sequence a_n which guarantee that $P_m \to 0$ as $m \to \infty$, e.g. it suffices that $a_n \geq (2c \log_2 n/n)^{1/2}$ for some $c > 1$ (we write log log $n = \log_2 n$, etc.), but says nothing about the rate at which P_m tends to 0. In the present note we give explicit upper bounds for P_m as a function of m for various sequences a_n, including sequences such that $a_n \sim (2 \log_2 n/n)^{1/2}$ as $n \to \infty$, and for which $P_m = O(1/\log_2 m)$. Such bounds have interesting statistical applications, based on the following considerations.

Suppose $y_1, y_2, \cdots$ are i.i.d. with distribution depending on an unknown mean μ and with known variance σ^2. Put $x_n = (y_n - \mu)/\sigma$ and define the interval $I_n = (\bar{y}_n - \sigma a_n, \bar{y}_n + \sigma a_n)$. Then

$$P(\mu \in I_n \text{ for all } n \geq m) = P(|\bar{x}_n| < a_n \text{ for all } n \geq m) = 1 - P_m \to 1 \quad \text{as } m \to \infty. \tag{1}$$

Defining $J_n = \bigcap_m^n I_j$ for $n \geq m$, we can therefore assert with probability $\geq 1 - P_m$ that for *every* $n \geq m$ the confidence interval J_n, of length $\leq 2\sigma a_n$, contains μ (cf. ref. 1).

Example 1, Optional stopping: To test H_0: $\mu = \mu_0$ with a given type I error without fixing the sample size in advance, reject H_0 if for any desired $N \geq m$, $\mu_0 \notin I_N$. Then P_{H_0} (reject H_0) $\leq P_m$, which can be made arbitrarily small by taking m large enough. The power function P_μ(reject H_0) is bounded below by a calculable function of μ which tends rapidly to 1 as $|\mu - \mu_0| \to \infty$.

Example 2, Tests with uniformly small error probability: To test H_1: $\mu < \mu_0$ against H_2: $\mu > \mu_0$, stop sampling with N = first $n \geq m$ such that $\mu_0 \notin I_n$, and accept H_1 or H_2 according as I_N is to the left or to the right of μ_0. The error probability is then $\leq P_m$ and $E_\mu N < \infty$ for all $\mu \neq \mu_0$ (cf. ref. 3).

Example 3, Tests with zero type II error: To test H_0: $\mu = \mu_0$ against H_1: $\mu > \mu_0$, reject H_0 with N = first $n \geq m$ such that I_n is to the right of μ_0. Then P_{H_0} (reject H_0) $< P_m$ and for $\mu > \mu_0$, P_μ(reject H_0) $\equiv 1$. ($E_\mu N < \infty$ for $\mu > \mu_0$ but $P_{H_0}(N = \infty) > 0$, which may be an advantage.) The case in which σ^2 is unknown can also be treated. The applicability of bounds on the P_m of (1) goes beyond the usual statistical decision framework in which a stopping rule and single terminal action are assumed.

We proceed to derive the basic inequality (8) below. Under the assumptions of the first sentence of this section, let $z_n = e^{ts_n}/\varphi^n(t)$ for any fixed t for which

$\varphi(t) < \infty$. Then z_n is a nonnegative martingale with expected value 1. A simple martingale inequality asserts that for any positive constant b,

$$P(z_n \geqq b \text{ for some } n \geqq 1) \leqq \frac{1}{b}. \tag{2}$$

(G. Haggstrom called it to our attention and has independently considered using it to obtain simultaneous confidence intervals.)

Putting $b = e^{mt^2/2}$ for any fixed m and $t > 0$ gives

$$P\left(s_n \geqq \frac{mt}{2} + \frac{n \log \varphi(t)}{t} \text{ for some } n \geqq 1\right) \leqq e^{-mt^2/2}. \tag{3}$$

Define

$$h(t) = \frac{1}{2} + \frac{\log \varphi(t)}{t^2} \quad (\rightarrow 1 \text{ as } t \rightarrow 0); \tag{4}$$

then

$$P(\bar{x}_n \geqq t\, h(t) \text{ for some } n \geqq m) \leqq e^{-mt^2/2}. \tag{5}$$

Let $m_i \rightarrow \infty$ be any increasing sequence of positive constants and t_i any sequence of positive constants, $i \geqq 1$. From (5) we have for any integer $j \geqq 1$,

$$P(\bar{x}_n \geqq t_i h(t_i) \text{ for some } m_i \leqq n < m_{i+1}, i \geqq j) \leqq \sum_{i=j}^{\infty} e^{-m_i t_i^2/2} = Q_j, \text{ say.} \tag{6}$$

Defining the sequence of constants b_n for $n \geqq m_1$ by putting

$$b_n = t_i h(t_i) \text{ for all } n \text{ such that } m_i \leqq n < m_{i+1} \ (i \geqq 1), \tag{7}$$

we can write (6) in the form

$$P(\bar{x}_n \geqq b_n \text{ for some } n \geqq m_j) \leqq Q_j \quad (j \geqq 1). \tag{8}$$

We obtain various iterated logarithm inequalities by making different choices of the sequences m_i, t_i that enter into (8).

2. *A Special Case of* (8).—Put for $i \geqq 3$

$$m_i = \exp(i/\log i), \tag{9}$$

$$t_i = (2 \log i + 4 \log_2 i + 2 \log A)^{1/2} \cdot m_i^{-1/2}, \tag{10}$$

where A is any positive constant. Then from (6),

$$Q_j = \frac{1}{A} \sum_{i=j}^{\infty} \frac{1}{i(\log i)^2} < \frac{1}{A} \cdot \frac{1}{\log (j - 1/2)} \rightarrow 0 \text{ as } j \rightarrow \infty. \tag{11}$$

We shall now find an upper bound for the b_n of (8). A little algebra shows that for $m_i \leqq n < m_{i+1}$,

$$\log i \leqq \log_2 n + \log_3 n + \log 2,$$

$$\log_2 i \leqq \log_3 n + \log 2,$$

$$m_i > n\, e^{-1/\log_2 n} \qquad (i \geqq 3).$$

Hence from (10)

$$t_i < (2 \log_2 n + 6 \log_3 n + 6 \log 2 + 2 \log A)^{1/2} n^{-1/2} e^{(2 \log_2 n)^{-1}} = f(n), \text{ say,} \quad (12)$$

and from (7)

$$b_n \leqq f(n) h(v_n), \quad (13)$$

where v_n (= the t_i of (7)) is some constant such that

$$0 < v_n \leqq f(n) \sim (2 \log_2 n/n)^{1/2} \quad \text{as } n \to \infty. \quad (14)$$

For the normal (0,1) distribution, $h(t) \equiv 1$. For coin tossing with $p = 1/2$ and more generally whenever $\varphi(t) \leqq e^{t^2/2}$, $h(t) \leqq 1$ and we can omit the term $h(v_n)$ in (13). In any case, since $h(t) \to 1$ as $t \to 0$ and $f(n) \to 0$ as $n \to \infty$, $h(v_n) \to 1$ as $n \to \infty$. From (6), (7), and (13) we have for $j \geqq 3$

$$P(\bar{x}_n \geqq f(n) h(v_n) \text{ for some } n \geqq e^{\frac{j}{\log j}}) \leqq Q_j \quad (15)$$

where $f(n)$ is defined by (12), $h(t)$ by (4), Q_j by (11), and v_n satisfies (14). By combining (15) with the analogous inequality for $-\bar{x}_n$ we obtain

$$P(|\bar{x}_n| \geqq f(n) h(v_n) \text{ for some } n \geqq e^{\frac{j}{\log j}}) \leqq 2Q_j \quad (16)$$

where now (14) is replaced by

$$0 < |v_n| \leqq f(n). \quad (17)$$

Putting $a_n = f(n)h(v_n)$ we have an inequality for the P_m of (1) with $a_n \sim (2 \log_2 n/n)^{1/2}$ and for which $P_m = O(1/\log_2 m)$.

3. *Other Choices.*—Replacing (10) by

$$t_i = (2\, c \log i + 2 \log A)^{1/2} \cdot m_i^{-1/2} \quad (18)$$

for some $c > 1$, we find the same results (15) and (16) as before, where now

$$f(n) = (2c \log_2 n + 2c \log_3 n + 2c \log 2 + 2 \log A)^{1/2} n^{-1/2} e^{(2 \log_2 n)^{-1}} \quad (19)$$

and

$$Q_j \leqq \frac{1}{A(c-1)} \frac{1}{(j - 1/2)^{c-1}}. \quad (20)$$

A somewhat different result is obtained by putting

$$m_i = \alpha^i \quad \text{where } \alpha \text{ is any number} > 1, \quad (21)$$

and retaining (18) for t_i. We obtain from (3) for $j \geqq 1$

$$P\left(\bar{x}_n \geqq \frac{1 + \alpha D_n}{2\sqrt{\alpha}} f(n) \text{ for some } n \geqq \alpha^j\right) \leqq \frac{1}{A(c-1)} \frac{1}{(j - 1/2)^{c-1}} \quad (22)$$

where now $\quad f(n) = (2c \log_2 n - 2c \log_2 \alpha + 2 \log A)^{1/2} n^{-1/2}, \quad (23)$

and $\quad D_n = 1 + 2(h(v_n) - 1)^+ \quad \text{for } 0 < v_n < \alpha^{1/2} \frac{f(n)}{n}. \quad (24)$

(If $\varphi(t) \leqq e^{t^2/2}$, then $D_n \equiv 1$.)

4. *An Extension.*—There is an immediate extension of the preceding results to additive processes. Let $X(\tau)$ be an infinitely divisible process, $\tau \geqq 0$, with $X(0) = 0$, and let $E(X(1)) = 0$, $E^2(X(1)) = 1$, and suppose $X(1)$ has a moment generating function $\varphi(t) = E(e^{tX(1)})$ in some neighborhood of $t = 0$. Suppose $X(\tau)$ is separable. A necessary and sufficient condition for $X(\tau)$ to be such a process is that log $E(e^{tX(\tau)}) = \tau g(t)$, where

$$g(t) = \int_{-\infty}^{\infty} (e^{tx} - 1 - tx) \frac{dF(x)}{x^2}$$

and $F(x)$ is a distribution function whose moment generating function exists in some neighborhood of the origin.

Then exp $(tX(\tau) - \tau g(t))$, $\tau \geqq 0$, is a positive martingale, and an almost literal repetition of the steps leading to (15) yields for $j \geqq 3$

$$P\left(\frac{X(\tau)}{\tau} > f(\tau)h(v_\tau) \text{ for some } \tau \geqq e^{j^{\log j}}\right) \leqq Q_j \tag{25}$$

where

$$f(\tau) = \left(\frac{2 \log_2 \tau + 6 \log_3 \tau + 6 \log 2 + 2 \log A}{\tau}\right)^{1/2} e^{(2 \log_2 \tau)^{-1}},$$

$$h(t) = \frac{1}{2} + \frac{1}{t^2} g(t),$$

and $v_\tau \leqq f(\tau)$, Q_j as in (11).

We remark that in the case of Brownian motion, where $h(t) \equiv 1$, there are inequalities due to Ito and McKean[4] and Strassen,[6] the latter giving asymptotic results for $j \to \infty$ also. The bound Q_j in (25) is of the same order of magnitude as their bounds, though the theorems are not, strictly speaking, comparable.

5. *Remarks.*—An inequality for the P_m of (1) should be obtainable from Chow's inequality[2] (which generalizes (2) and ref. 5) without our subdivision of the n-axis by the points m_i. Even within the framework of the present method, a great variety of inequalities can be obtained, and by a closer analysis minor improvements in the inequalities of Sections 2 and 3 are possible; for example, by letting $m_i \to \infty$ a little more slowly than in (9) we can replace the exponential factor of (12) by something closer to 1. In general, however, sharper inequalities for large n require a larger initial sample size m.

In a subsequent issue of these PROCEEDINGS we shall give results analogous to the above for random variables for which the moment generating function does not exist; such results, based on different specializations of reference 2, are necessarily less sharp. We shall also give explicit bounds on the $E_\mu N$ of Examples 2 and 3 above.

* Supported in part by National Science Foundation grant GP-6549.

[1] Blum, J. R., and J. Rosenblatt, "On some statistical problems requiring purely sequential sampling schemes," *Ann. Inst. Stat. Math.*, **18**, 351–354 (1966).

[2] Chow, Y. S., "A martingale inequality and the law of large numbers," *Proc. Am. Math. Soc.*, **11**, 107–111 (1960).

[3] Farrell, R. H., "Asymptotic behavior of expected sample size in certain one sided tests," *Ann. Math. Stat.*, **35**, 36–72 (1964).

[4] Ito, K., and H. P. McKean, "Diffusion processes and their sample paths," in *Grundlehren der Mathematischen Wissenschaften* (Berlin: Springer, 1965), vol. 125.

[5] Rényi, A., and J. Hájek, "Generalization of an inequality of Kolmogorov," *Acta. Math. Acad. Sci. Hungar.*, **6**, 281–283 (1955).

[6] Strassen, V., "Almost sure behavior of sums of independent random variables and martingales," *Proc. Fifth Berk. Symp.* (1965), in press.

The Annals of Mathematical Statistics
1970, Vol. 41, No. 5, 1397–1409

STATISTICAL METHODS RELATED TO THE LAW OF THE ITERATED LOGARITHM[1]

BY HERBERT ROBBINS

Columbia University

1. Extension and applications of an inequality of Ville and Wald. Let $x_1, \cdots$ be a sequence of random variables with a specified joint probability distribution P. We shall give a method for obtaining probability inequalities and related limit theorems concerning the behavior of the entire sequence of x's.

We begin with a result of J. Ville ([16] page 100); cf. also A. Wald ([17] page 146).

Suppose that under P for each $n \geqq 1$ the random variables $x_1, \cdots, x_n$ have a probability density function $g_n(x_1, \cdots, x_n)$ with respect to a σ-finite measure μ_n on the Borel sets of n-space, and that P' is any other joint probability distribution of the sequence $x_1, \cdots$ such that $x_1, \cdots, x_n$ have a probability density function $g_n'(x_1, \cdots, x_n)$ with respect to the same μ_n. Define the likelihood ratio $z_n = g_n'/g_n$ when $g_n > 0$. Then for any $\varepsilon > 1$,

$$P(z_n \geqq \varepsilon \text{ for some } n \geqq 1) \leqq 1/\varepsilon. \tag{1}$$

To prove this, let $N =$ first $n \geqq 1$ such that $g_n' \geqq \varepsilon g_n$, with $N = \infty$ if no such n occurs; then $P(g_n = 0 \text{ for some } n \geqq 1) = 0$ and

$$\begin{aligned} P(z_n \geqq \varepsilon \text{ for some } n \geqq 1) &= P(N < \infty) = \textstyle\sum_1^\infty \int_{(N=n)} g_n \, d\mu_n \\ &\leqq 1/\varepsilon \textstyle\sum_1^\infty \int_{(N=n)} g_n' \, d\mu_n \\ &= 1/\varepsilon \cdot P'(N < \infty) \leqq 1/\varepsilon. \end{aligned} \tag{2}$$

As a first example, cf. ([16] page 52), suppose that the x's are i.i.d. Bernoulli random variables such that $P(x_i = 1) = p$, $P(x_i = 0) = 1-p$, $0 < p < 1$. If μ_n is counting measure on the space of vectors $(x_1, \cdots, x_n)$ then $g_n(x_1, \cdots, x_n) = P(x_1, \cdots, x_n) = p^{S_n}(1-p)^{n-S_n}$, where $S_n = x_1 + \cdots + x_n$. Take P' to be the uniform mixture of Bernoulli distributions with parameter $0 < \theta < 1$; then

$$\begin{aligned} g_n'(x_1, \cdots, x_n) = P'(x_1, \cdots, x_n) &= \textstyle\int_0^1 \theta^{S_n}(1-\theta)^{n-S_n} \, d\theta \\ &= \frac{S_n!(n-S_n)!}{(n+1)!}. \end{aligned}$$

Using the notation $b(n, p, x) = \binom{n}{x} p^x (1-p)^{n-x}$ and replacing ε by $1/\varepsilon$, (1) gives the inequality

$$P(b(n, p, S_n) \leqq \varepsilon/(n+1) \text{ for some } n \geqq 1) \leqq \varepsilon \quad (0 < \varepsilon < 1).$$

Received September 8, 1969.

[1] Invited Wald Lectures given at the annual meeting of the Institute of Mathematical Statistics New York, August 19–21, 1969; research supported by N. I. H. Grant 1-R01-GM-16895-01.

If we denote by $I_n(x, \varepsilon)$ the set of all $0 \leqq \theta \leqq 1$ such that $b(n, \theta, x) > \varepsilon/(n+1)$, then this is equivalent to the "confidence" statement

$$P(p \in I_n(S_n, \varepsilon) \text{ for every } n \geqq 1) \geqq 1-\varepsilon \qquad (0 < \varepsilon, p < 1). \tag{3}$$

As a second example, which we shall study in more detail, of the use of (1), let P_θ, $-\infty < \theta < \infty$, denote the probability under which the x's are i.i.d. $N(\theta, 1)$, and take $P = P_0$. If μ_n denotes Lebesgue measure in n-space and

$$\varphi(x) = (2\pi)^{-\frac{1}{2}} \exp(-x^2/2), \qquad \Phi(x) = \int_{-\infty}^{x} \varphi(t)\, dt, \qquad S_n = x_1 + \cdots + x_n,$$

then the joint density of $x_1, \cdots, x_n$ under P_θ is

$$g_{\theta,n}(x_1, \cdots, x_n) = \prod_1^n \varphi(x_i - \theta). \tag{4}$$

We shall take P' to be any mixture of the form $P'(\cdot) = \int_{-\infty}^{\infty} P_\theta(\cdot)\, dF(\theta)$, where F is an arbitrary probability measure on $(-\infty, \infty)$, so that

$$g_n'(x_1, \cdots, x_n) = \int_{-\infty}^{\infty} g_{\theta,n}(x_1, \cdots, x_n)\, dF(\theta)$$
$$= \int_{-\infty}^{\infty} \prod_1^n \varphi(x_i - \theta)\, dF(\theta),$$

$$z_n = g_n'/g_{0,n} = \int_{-\infty}^{\infty} \exp(\theta S_n - \tfrac{1}{2} n\theta^2)\, dF(\theta). \tag{5}$$

If we define the function

$$f(x, t) = \int_{-\infty}^{\infty} \exp(xy - \tfrac{1}{2} y^2 t)\, dF(y) \tag{6}$$

and for reasons which will become apparent in the next section replace $F(\theta)$ in (5) by $F(\theta m^{\frac{1}{2}})$, where m is an arbitrary positive constant, then (5) becomes

$$z_n = \int_{-\infty}^{\infty} \exp(\theta S_n - \tfrac{1}{2} n\theta^2)\, dF(\theta m^{\frac{1}{2}}) = \int_{-\infty}^{\infty} \exp(y S_n/m^{\frac{1}{2}} - \tfrac{1}{2} n y^2/m)\, dF(y) \tag{7}$$
$$= f(S_n/m^{\frac{1}{2}}, n/m),$$

and hence by (1) we have hat for i.i.d. $N(0, 1)$ x's,

$$P(f(S_n/m^{\frac{1}{2}}, n/m) \geqq \varepsilon \text{ for some } n \geqq 1) \leqq 1/\varepsilon \qquad (m > 0, \varepsilon > 1). \tag{8}$$

In order to see the meaning of (8) more clearly, suppose now that the probability measure F is confined to $(0, \infty)$, so that $f(x, t)$ defined by (6) is an *increasing* function of x. If we define for $t > 0$ the function $A(t, \varepsilon) =$ the (positive) solution x of the equation $f(x, t) = \varepsilon$, then

$$f(x, t \ \geqq \varepsilon \ \text{ if and only if } \ x \geqq A(t, \varepsilon), \tag{9}$$

and (8) can be written as

$$P(S_n \geqq m^{\frac{1}{2}} A(n/m, \varepsilon) \ \text{ for some } \ n \geqq 1) \leqq 1/\varepsilon \qquad (m > 0, \varepsilon > 1). \tag{10}$$

We note that (10) remains valid if instead of being i.i.d. $N(0, 1)$ the x's are any i.i.d. random variables having a moment generating function Ψ such that

$$\Psi(\theta) = E(e^{\theta x_1}) \leqq e^{\frac{1}{2}\theta^2} \qquad (0 < \theta < \infty). \tag{11}$$

To see this, let $g(x)$ be the density of x_1 with respect to some measure μ on $(-\infty, \infty)$ and let $g_\theta(x) = e^{\theta x} g(x)\Psi^{-1}(\theta)$; the argument which led to (10) now carries over *a fortiori* with $\varphi(x-\theta)$ replaced by $g_\theta(x)$. An example other than the $N(0, 1)$ case of a "subnormal" distribution satisfying (11) is the symmetric random walk distribution $P(x_1 = 1) = P(x_1 = -1) = \frac{1}{2}$, for which

$$\Psi(\theta) = (e^\theta + e^{-\theta})/2 = \sum_0^\infty \theta^{2i}/(2i)! \leqq e^{\frac{1}{2}\theta^2} \qquad (-\infty < \theta < \infty).$$

Thus S_n for the symmetric random walk also satisfies (10), and for many distributions—Bernoulli, Poisson, uniform, normal square, etc.—an appropriate linear function with expectation 0 and variance 1 will provide an x_1 satisfying (11) and hence (10) in the case where F is confined to $(0, \infty)$.

We shall now illustrate the significance of (10) by considering a few particular choices of F. We write P for any distribution under which the x's are i.i.d. and (11) holds.

EXAMPLE 1. Let F be the degenerate measure which assigns mass 1 to the point $2a > 0$. Then $f(x, t) = \exp(2ax - 2a^2 t) \geqq \varepsilon$ if and only if $x \geqq at + (\log \varepsilon)/2a$. Hence (10) yields the inequality (with $d = (\log \varepsilon)/2a$)

$$P(S_n \geqq an/m^{\frac{1}{2}} + dm^{\frac{1}{2}} \text{ for some } n \geqq 1) \leqq e^{-2ad} \qquad (a, d, m \text{ all} > 0).$$

EXAMPLE 2. Let F be defined by

$$dF(y) = (2/\pi)^{\frac{1}{2}} e^{-\frac{1}{2}y^2}\, dy \quad \text{for} \quad 0 < y < \infty; = 0 \quad \text{elsewhere.}$$

Then for $t > -1$

$$f(x, t) = (2/\pi)^{\frac{1}{2}} \int_0^\infty \exp(xy - y^2(t+1)/2\, dy = \frac{2e^{\frac{1}{2}x^2/(t+1)}}{(t+1)^{\frac{1}{2}}} \Phi\left(\frac{x}{(t+1)^{\frac{1}{2}}}\right).$$

To solve the equation $f(x, t) = \varepsilon$ for x we introduce the function

$$h(x) = x^2 + 2 \log \Phi(x) \qquad (-\infty < x < \infty),$$

which increases from $-\infty$ to ∞ as x does, and is such that $h(x) \sim x^2$, $h^{-1}(x) \sim x^{\frac{1}{2}}$ as $x \to \infty$. Then $f(x, t) = \varepsilon$ for $x = A(t, \varepsilon) = (t+1)^{\frac{1}{2}} \cdot h^{-1}(2 \log \frac{1}{2}\varepsilon + \log(t+1))$, so (10) yields (with $\varepsilon = 2e^{\frac{1}{2}a^2}\Phi(a)$) the inequality

$$(12) \quad P(S_n \geqq (n+m)^{\frac{1}{2}} \cdot h^{-1}(h(a) + \log(n/m+1)) \text{ for some } n \geqq 1) \leqq \tfrac{1}{2} e^{-\frac{1}{2}a^2}/\Phi(a) \qquad (a \text{ and } m > 0).$$

EXAMPLE 3. For any $\delta > 0$ let F be defined by

$$dF(y) = \delta \cdot \frac{dy}{y \log 1/y (\log_2 1/y)^{1+\delta}} \quad \text{for} \quad 0 < y < e^{-e};$$

$$= 0 \quad \text{elsewhere,}$$

where we write $\log(\log y) = \log_2 y$, etc. It is impossible to evaluate $f(x, t)$ explicitly, but it can be shown by some analysis [14] that as $t \to \infty$

$$(13) \qquad A(t, \varepsilon) = \left\{2t\left(\log_2 t + \left(\frac{3}{2} + \delta\right)\log_3 t + \log \frac{\varepsilon}{2\delta\pi^{\frac{1}{2}}} + o(1)\right)\right\}^{\frac{1}{2}}.$$

The inequality (10) states for i.i.d. x's satisfying (11) that $P(S_n \geqq c_n$ for some $n \geqq 1) \leqq 1/\varepsilon$, where by proper choice of the probability measure F on $(0, \infty)$ we have seen that as $n \to \infty$

$$c_n = m^{\frac{1}{2}} A(n/m, \varepsilon) = an/m^{\frac{1}{2}} + dm^{\frac{1}{2}} \sim an/m^{\frac{1}{2}} \quad \text{(Example 1)}$$

$$= (n+m)^{\frac{1}{2}} h^{-1}(h(a) + \log(n/m+1)) \sim (n \log n)^{\frac{1}{2}} \quad \text{(Example 2)}$$

$$\sim (2n \log_2 n)^{\frac{1}{2}} \quad \text{(Example 3).}$$

Other choices of F give sequences $c_n \sim n^{\beta}$ $(\frac{1}{2} < \beta < 1)$, etc. By the law of the iterated logarithm,

$$P\left(\limsup_{n\to\infty} \frac{S_n}{(2n \log_2 n)^{\frac{1}{2}}} = 1\right) = 1 \tag{14}$$

whenever the x's are i.i.d. with mean 0 and variance 1. Thus the sequences c_n of Example 3 increase about as slowly as is possible to have for i.i.d. x's with mean 0 and variance 1, $P(S_n \geqq c_n$ for some $n \geqq 1) < 1$.

Useful extensions of (10) are provided by the following remarks. The reader interested primarily in applications to statistics may proceed directly to Section 3.

(I) Returning to the general inequality (1), suppose we put

$$g_n'(x_1, \cdots, x_n) = \int_a^b g_{\theta,n}(x_1, \cdots, x_n)\, dF(\theta),$$

$$z_n = g_n'/g_n \quad \text{when} \quad g_n > 0,$$

where $P'(\cdot) = \int_a^b P_\theta(\cdot)\, dF(\theta)$ and $\{P_\theta; a < \theta < b\}$ is any family of joint probability distributions for the sequence $x_1, \cdots$ such that for each $n \geqq 1$ the random variables $x_1, \cdots, x_n$ have a probability density function $g_{\theta,n}(x_1, \cdots, x_n)$ (with respect to the same μ_n) which for fixed $x_1, \cdots, x_n$ is a Borel measurable function of θ. Then for any $\varepsilon > 0$ and $j = 1, 2, \cdots$ we shall show that

$$P(z_n \geqq \varepsilon \text{ for some } n \geqq j) \leqq P(z_j \geqq \varepsilon) + 1/\varepsilon \int_{(z_j < \varepsilon)} z_j\, dP \leqq F(a, b)/\varepsilon, \tag{15}$$

where F may now be any *σ-finite* measure on the parameter interval (a, b). (When $j = 1$ and F is a *probability* measure on (a, b) the extreme terms of (15) reduce to (1).)

PROOF. Let $N =$ first $n \geqq j$ such that $g_n' \geqq \varepsilon g_n$, with $N = \infty$ if no such n occurs; then

$$\begin{aligned}
&P(z_n \geqq \varepsilon \text{ for some } n \geqq j)\\
&\quad = P(N < \infty) = P(N = j) + \textstyle\sum_{j+1}^{\infty} \int_{(N=n)} g_n\, d\mu_n\\
&\quad \leqq P(N = j) + 1/\varepsilon \textstyle\sum_{j+1}^{\infty} \int_{(N=n)} g_n'\, d\mu_n = P(N = j) + 1/\varepsilon P'(j < N < \infty)\\
&\quad \leqq P(N = j) + 1/\varepsilon P'(N > j) = P(z_j \geqq \varepsilon) + 1/\varepsilon \int_{(g_j' < \varepsilon g_j)} g_j'\, d\mu_j\\
&\quad = P(z_j \geqq \varepsilon) + 1/\varepsilon \int_{(z_j < \varepsilon)} z_j\, dP \leqq 1/\varepsilon \int_{(z_j \geqq \varepsilon)} z_j\, dP + 1/\varepsilon \int_{(z_j < \varepsilon)} z_j\, dP \leqq 1/\varepsilon \int z_j\, dP\\
&\quad = F(a, b)/\varepsilon.
\end{aligned}$$

We remark that the first inequality of (15) holds under the assumption that $\{z_n, \mathscr{F}_n; n \geqq 1\}$ is any positive supermartingale on a probability space $(\Omega, \mathscr{F}, P)$, although the proof for that case is slightly different.

Applied to the i.i.d. $N(0, 1)$ case where F is confined to $(0, \infty)$ we obtain from (15) as an extension of (10) the result that for $\varepsilon > 0, m > 0, j = 1, 2, \cdots$ and $\tau = j/m$

$$(16) \qquad P(S_n \geqq m^{\frac{1}{2}} A(n/m, \varepsilon) \text{ for some } n \geqq j)$$
$$\leqq 1 - \Phi(A(\tau, \varepsilon)/\tau^{-\frac{1}{2}} + 1/\varepsilon \textstyle\int_0^\infty \Phi(A(\tau, \varepsilon)/\tau^{-\frac{1}{2}} y \tau^{-\frac{1}{2}} \, dF(y).$$

For example, if $dF(y) = (2/\pi)^{\frac{1}{2}} dy$ for $0 < y < \infty$ then (cf. Example 2 above)

$$f(x, t) = 2e^{\frac{1}{2}x^2/t}/t^{-\frac{1}{2}} \Phi(x/t^{\frac{1}{2}}) \qquad (t > 0),$$

and from (16) for $j = m$, $\tau = 1$ we obtain (cf. (12)) for i.i.d. $N(0, 1)$ x's

$$P(S_n \geqq n^{\frac{1}{2}} h^{-1}(h(a) + \log n/m) \text{ for some } n \geqq m) \leqq 1 - \Phi(a) + \varphi(a)(a + \varphi(a)/\Phi(a)).$$

(II) If F is a *symmetric* probability measure on $(-\infty, \infty)$ then instead of (9) we see that $f(x, t) \geqq \varepsilon$ if and only if $|x| \geqq A(t, \varepsilon)$, and (10) continues to hold with S_n replaced by $|S_n|$ when the x's are i.i.d. $N(0, 1)$, or more generally when (11) holds for all $-\infty < \theta < \infty$. For example, if $F = \Phi$ we obtain when (11) holds for all $-\infty < \theta < \infty$ that

$$(17) \qquad P(|S_n| \geqq [(n+m)(a^2 + \log(n/m + 1))]^{\frac{1}{2}} \text{ for some } n \geqq 1) \leqq e^{-\frac{1}{2}a^2}.$$

Since $(a^2 + \log t)^{\frac{1}{2}} > h^{-1}(h(a) + \log t)$ for all $t > 1$, it follows from (12) that as a one-sided version of (17) we have

$$(18) \quad P(S_n \geqq [(n+m)(a^2 + \log(n/m + 1))]^{\frac{1}{2}} \text{ for some } n \geqq 1) \leqq \tfrac{1}{2} e^{-\frac{1}{2}a^2}/\Phi(a) \qquad (a > 0).$$

Likewise, if F is any symmetric measure on $(-\infty, \infty)$ and the x's are i.i.d. $N(0, 1)$ then we obtain from (15) as an analogue of (16) that with $\tau = j/m$

$$(19) \quad P(|S_n| \geqq m^{\frac{1}{2}} A(n/m, \varepsilon) \text{ for some } n \geqq j)$$
$$\leqq 2(1 - \Phi(A(\tau, \varepsilon)/\tau^{\frac{1}{2}})) + 1/\varepsilon \textstyle\int_{-\infty}^\infty \{\Phi(A(\tau, \varepsilon)/\tau^{\frac{1}{2}} - y\tau^{\frac{1}{2}})$$
$$- \Phi(-A(\tau, \varepsilon)/\tau^{\frac{1}{2}} - y\tau^{\frac{1}{2}})\} \, dF(y).$$

From (19) with $dF(y) = (2\pi)^{-\frac{1}{2}} dy$ for $-\infty < y < \infty$ and $j = m$, $\tau = 1$ we obtain for i.i.d. $N(0, 1)$ x's that

$$(20) \quad P(|S_n| \geqq [n(a^2 + \log n/m)]^{\frac{1}{2}} \text{ for some } n \geqq m) \leqq 2(1 - \Phi(a) + a\varphi(a)) \qquad (a > 0).$$

(III) Suppose the x's are i.i.d. with an absolutely continuous density function $g(x - \theta)$ under P_θ, and that $P = P_0$. Define for $n \geqq 1$ and $-\infty < \theta < \infty$,

$y_n = x_n/|x_1|$ and

$$z_{\theta,n}(x_1, \cdots, x_n) = \frac{\int_0^\infty \prod_1^n \left\{\frac{1}{\sigma} g\left(\frac{x_i}{\sigma} - \theta\right)\right\} \frac{d\sigma}{\sigma}}{\int_0^\infty \prod_1^n \left\{\frac{1}{\sigma} g\left(\frac{x_i}{\sigma}\right)\right\} \frac{d\sigma}{\sigma}}.$$

Then $z_{\theta,n}(x_1, \cdots, x_n) = z_{\theta,n}(y_1, \cdots, y_n)$, and if the joint density of $y_1, \cdots, y_n$ under P_θ is denoted by $p_\theta(y_1, \cdots, y_n)$, then it is easy to see that

$$z_{\theta,n}(y_1, \cdots, y_n) = \frac{p_\theta(y_1, \cdots, y_n)}{p_0(y_1, \cdots, y_n)}.$$

Hence if F is any measure on $(-\infty, \infty)$ and if we put

$$z_n(x_1, \cdots, x_n) = \int_{-\infty}^{\infty} z_{\theta,n}\, dF(\theta),$$

then (15) holds. For example, if $g = \varphi$ and we put $dF(\theta) = (m/2\pi)^{\frac{1}{2}}\, d\theta$, $-\infty < \theta < \infty$, we obtain after some computation the following result: if the x's are i.i.d. $N(\mu, \sigma^2)$ and

$$\bar{x}_n = n^{-1}\textstyle\sum_1^n x_i, \qquad v_n^{\,2} = n^{-1}\sum_1^n (x_i - \bar{x}_n)^2 \qquad \text{then}$$

$$(21) \qquad P(|\bar{x}_n - \mu| \geqq v_n[(tn)^{n^{-1}} - 1]^{\frac{1}{2}} \text{ for some } n \geqq m)$$
$$\leqq 2(1 - F_{m-1}(a) + a f_{m-1}(a)) \qquad (a \geqq 0,\ m = 1, 2, \cdots)$$

where $t = m^{-1}(1 + a^2/(m-1))^m$ and f_m, F_m denote the Student t density and distribution function with m degrees of freedom. Note that for large n, m

$$(tn)^{n^{-1}} - 1 \cong e^{n^{-1}\log(tn)} - 1 \cong n^{-1}\log(tn)$$
$$\cong n^{-1}\log(nm^{-1}(1 + a^2/(m-1))^m) \cong n^{-1}(a^2 + \log n/m)$$

and compare (21) with (20). The use of the scale-invariant measure $d\sigma/\sigma$ was suggested to the author by Robert Berk; cf. ([10] page 250).

2. The Wiener process and a limit theorem. Let $w(t)$ denote a standard Wiener process for $t \geqq 0$ with $w(0) = 0$. If $x_1, \cdots$ are independent $N(0, 1)$, the two sequences

$$(22) \qquad (S_1/m^{\frac{1}{2}}, S_2/m^{\frac{1}{2}}, \cdots)$$
$$(w(1/m), w(2/m), \cdots)$$

have the same joint distribution for any $m > 0$. This suggests that (10), (16), and (19) should become equalities for the Wiener process; e.g., in the case of (10) that if F is a probability measure on $(0, \infty)$ and if $f(x, 0) < \infty$ for all x, then

$$(23) \qquad P(w(t) \geqq A(t, \varepsilon) \text{ for some } t \geqq 0) = 1/\varepsilon \qquad (\varepsilon > 1).$$

To see heuristically why this should be so we remark that in (2) the only strict inequality was the replacement of $1/z_n$ by $1/\varepsilon$ on the set $(N = n)$. If m is large, the "overshoot" will be stochastically small, and the behavior of the first sequence of

(22) will be about the same as that of the continuous process $w(t)$. However, instead of trying to prove that (23) holds by letting $m \to \infty$ in (10), it is more convenient to prove (23) directly from the fact that

$$z(t) = \int_0^\infty \exp\left(w(t)y - \tfrac{1}{2}ty^2\right) dF(y) \qquad (t \geqq 0) \tag{24}$$

is a positive martingale with continuous sample paths for which $z(0) = 1$. It is easy to show that $z(t) \to 0$ in probability as $t \to \infty$, and from this it follows almost immediately by stopping $z(t)$ the first time it is $\geqq \varepsilon$ that

$$P(z(t) \geqq \varepsilon \text{ for some } t \geqq 0) = 1/\varepsilon \qquad (\varepsilon > 1),$$

which is equivalent to (23). The examples of F which we have already considered thus give exact boundary crossing probabilities for the Wiener process:

$$P(w(t) \geqq at + d \text{ for some } t \geqq 0) = e^{-2ad} \qquad (a > 0, d > 0),$$

$$P(w(t) \geqq (t+1)^{\frac{1}{2}} \cdot h^{-1}(h(a) + \log(t+1)) \text{ for some } t \geqq 0) = \tfrac{1}{2}e^{-\frac{1}{2}a^2}/\Phi(a) \qquad (a > 0)$$

$$P(w(t) \geqq A(t, \varepsilon) \text{ for some } t \geqq 0) = 1/\varepsilon \qquad (\varepsilon > 1),$$

for the function $A(t, \varepsilon)$ of Example 3 for which the asymptotic expression (13) holds,

$$P(|w(t)| \geqq [(t+1)(a^2 + \log(t+1))]^{\frac{1}{2}} \text{ for some } t \geqq 0) = e^{-\frac{1}{2}a^2} \qquad (a > 0),$$

etc. Only in the case of a linear boundary have such formulas been available up to now.

The heuristic argument which suggested the truth of (23) suggests further, because of the central limit theorem, that the limit relations

$$\begin{aligned} &\lim_{m\to\infty} P(S_n \geqq m^{\frac{1}{2}} A(n/m, \varepsilon) \text{ for some } n \geqq 0) \\ &\quad = P(w(t) \geqq A(t, \varepsilon) \text{ for some } t \geqq 0) = 1/\varepsilon \qquad (\varepsilon > 1), \end{aligned}$$

$$\begin{aligned} &\lim_{m\to\infty} P(S_n \geqq m^{\frac{1}{2}} A(n/m, \varepsilon) \text{ for some } n \geqq \tau m) \\ &\quad = P(w(t) \geqq A(t, \varepsilon) \text{ for some } t \geqq \tau) \\ &\quad = 1 - \Phi(A(\tau, \varepsilon)/\tau^{\frac{1}{2}}) + 1/\varepsilon \int_0^\infty \Phi(A(\tau, \varepsilon)/\tau^{\frac{1}{2}} - y\tau^{\frac{1}{2}})\, dF(y) \qquad (\varepsilon > 0, \tau > 0) \end{aligned} \tag{25}$$

together with analogous relations for $|S_n|$ and $|w(t)|$ should hold *whenever the* x's *are* i.i.d. *with mean* 0 *and variance* 1, normal or not. This is true under some mild assumption about the behavior of the function $A(t, \varepsilon)$ as $t \to \infty$; it is sufficient to assume that $A(t, \varepsilon)/t^{\frac{1}{2}}$ is ultimately non-decreasing. Hence the inequalities previously obtained for every finite $m > 0$ when (11) holds now become limit theorems as $m \to \infty$ with no parametric assumptions about the x's. A full discussion of this is given in [14]. The statistical significance of limit theorems such as (25) will be discussed in the following sections.

3. Confidence sequences and tests with uniformly small error probability for the mean of a normal distribution with known variance. Let $x_1, \cdots$ be independent

$N(\theta, 1)$ where θ is an unknown parameter, $-\infty < \theta < \infty$. Define the intervals $I_n = ((S_n - c_n)/n, (S_n + c_n)/n)$ $(n \geqq 1)$, where $S_n = x_1 + \cdots + x_n$ and c_n is any sequence of positive constants such that $c_n/n \to 0$ as $n \to \infty$. Then

$$P_\theta(\theta \in I_n \text{ for every } n \geqq 1) = P_0(|S_n| < c_n \text{ for every } n \geqq 1)$$
$$= 1 - P_0(|S_n| \geqq c_n \text{ for some } n \geqq 1).$$

We saw for example in (17) that if

$$c_n = [(n+m)(a^2 + \log(n/m+1))]^{\frac{1}{2}} \qquad (m > 0) \tag{26}$$

then $P_0(|S_n| \geqq c_n$ for some $n \geqq 1) \leqq e^{-\frac{1}{2}a^2}$. Hence $P_\theta(\theta \in \bigcap_1^\infty I_n) \geqq 1 - e^{-\frac{1}{2}a^2}$, which can be made as near 1 as we please by choosing a sufficiently large; e.g., for

$$a^2 \cong 6, \qquad 1 - e^{-\frac{1}{2}a^2} = .95. \tag{27}$$

Thus for $a^2 = 6$ and any $m > 0$, the sequence I_n with c_n defined by (26) forms a "confidence sequence" for an unknown θ with coverage probability $\geqq .95$ (cf. [17] pages 153–156). As $m \to \infty$ the coverage probability $\to .95$ by (25).

Choosing for example $m = 1$, the half-width of I_n is

$$\frac{c_n}{n} = \left[\frac{n+1}{n^2}(6 + \log(n+1))\right]^{\frac{1}{2}} \sim \left[\frac{\log n}{n}\right]^{\frac{1}{2}} \to 0 \qquad \text{as } n \to \infty. \tag{28}$$

Of course, for any *fixed n*

$$P_\theta((S_n - 1.96n^{\frac{1}{2}})/n < \theta < (S_n + 1.96n^{\frac{1}{2}})/n) = .95, \tag{29}$$

a 95% confidence interval for θ of half-width $1.96/n^{\frac{1}{2}}$. However, by (14)

$$P_\theta((S_n - 1.96n^{\frac{1}{2}})/n < \theta < (S_n + 1.96n^{\frac{1}{2}})/n \text{ for every } n \geqq 1) = 0,$$

and this remains true if 1.96 is replaced by any constant, no matter how large.

The advantage of the confidence sequence I_n compared to a fixed sample size confidence interval is that it allows us to "follow" the unknown θ throughout the whole sequence $x_1, \cdots$ with an interval I_n whose length shrinks to 0 as the sample size increases, in such a way that with probability $\geqq .95$ the interval I_n contains θ at every stage. (This is also true of the smaller intervals $J_n = \bigcap_1^n I_k \subset I_n$, although for some n it might happen that $J_n = \emptyset$.) The validity of the relation $P_\theta(\theta \in I_n) \geqq .95$ is therefore unaffected by the possibility that n may be a random variable dependent on the whole sequence $x_1, \cdots$; in other words, the confidence level .95 is unaffected by any kind of optional stopping which could vitiate (29).

The disadvantage of using the sequence I_n is evident from a comparison of the numerical value of (28) with $1.96/n^{\frac{1}{2}}$.

If we wish to test $H^-: \theta < 0$ versus $H^+: \theta > 0$ ($\theta = 0$ being excluded) we can define the stopping time $N =$ first $n \geqq 1$ such that $|S_n| \geqq c_n$ and accept H^+ or H^- according as $S_N \geqq c_N$ or $S_N \leqq -c_N$. Since $S_n/n \to \theta \neq 0$ under H^- or H^+, while $c_n/n \to 0$, it follows that $P_\theta(N < \infty) = 1$ for $\theta \neq 0$, while if $\theta > 0$,

$$P_\theta(\text{accept } H^-) = P_\theta(S_n \leqq -c_n \text{ before } S_n \geqq c_n)$$
$$< P_0(S_n \leqq -c_n \text{ before } S_n \geqq c_n) = \tfrac{1}{2}P_0(|S_n| \geqq c_n \text{ for some } n \geqq 1)$$
$$\leqq \tfrac{1}{2}e^{-\frac{1}{2}a^2},$$

and similarly, P_θ (accept H^+) $\leqq \frac{1}{2}e^{-\frac{1}{2}a^2}$ for any $\theta < 0$. Thus the error probability of this test is uniformly $\leqq \frac{1}{2}e^{-\frac{1}{2}a^2}$ for all $\theta \neq 0$. (Exactly the same argument holds if the x's have the distribution $P(x_i = 1) = p = 1 - P(x_i = -1)$, with $\theta = 2p - 1$.)

Of course, $P_0(N < \infty) \leqq e^{-\frac{1}{2}a^2}$, so the test will rarely terminate when $\theta = 0$. The expected sample size $E_\theta(N)$ is, however, finite for every $\theta \neq 0$, approaching ∞ as $\theta \to 0$ and 1 as $|\theta| \to \infty$.

4. Tests with power 1. Again, let $x_1, \cdots$ be independent $N(\theta, 1)$ but now let $H_0: \theta \leqq 0$, $H_1: \theta > 0$ be the hypotheses which are to be tested. Put

$$(30) \qquad \begin{aligned} N &= \text{first } n \geqq 1 \text{ such that } S_n \geqq c_n \\ &= \infty \text{ if no such } n \text{ occurs,} \end{aligned}$$

and agree when $N < \infty$ to stop sampling with x_N and reject H_0 in favor of H_1; if $N = \infty$, continue sampling indefinitely and do not reject H_0.

For $\theta \leqq 0$ we have

$$\begin{aligned} P_\theta(\text{reject } H_0) &= P_\theta(N < \infty) \leqq P_0(N < \infty) \\ &= P_0(S_n \geqq c_n \text{ for some } n \geqq 1). \end{aligned}$$

If we are using the c_n sequence (26) we have by (18)

$$(31) \qquad P_0(S_n \geqq c_n \text{ for some } n \geqq 1) < \tfrac{1}{2}e^{-\frac{1}{2}a^2}/\Phi(a)$$

(if we were to use for c_n the smaller sequence of (12), (31) would still hold and with approximate equality for large m). Hence the type I error probability of the test has the upper bound $P_\theta(\text{reject } H_0) < \frac{1}{2}e^{-\frac{1}{2}a^2}/\Phi(a)$ for all $\theta \leqq 0$, while the type II error probability is

$$P_\theta(\text{not reject } H_0) = P_\theta(S_n < c_n \text{ for all } n \geqq 1) \equiv 0 \qquad \text{for all } \theta > 0,$$

since $c_n/n \to 0$ and $S_n/n \to \theta > 0$ as $n \to \infty$. Thus the test has power 1 against the alternative $\theta > 0$.

Of course the test will rarely terminate when $\theta \leqq 0$. Some people may consider this intolerable, but that is an unreasonable attitude in many practical situations.

Concerning the expected sample size $E_\theta(N)$ when $\theta > 0$, it can be shown that for *any* stopping rule N of the sequence $x_1, \cdots$ the inequality

$$(32) \qquad E_\theta(N) \geqq -2 \log P_0(N < \infty)/\theta^2$$

must hold for every $\theta > 0$. Thus, if we are willing to tolerate an N for which $P_0(N < \infty) = .05$, then necessarily $E_\theta(N) \geqq 6/\theta^2$ for every $\theta > 0$; however, no such N will minimize $E_\theta(N)$ uniformly for all $\theta > 0$. For the N given by (30), if like (26) the function c_t is concave for $t \geqq 1$ it can be shown [5] that

$$E_\theta(N) \leqq \frac{c_{E_\theta(N)}}{\theta} + \frac{\varphi(\theta)}{\theta\Phi(\theta)} + 1, \tag{33}$$

which gives an implicit upper bound for $E_\theta(N)$ as a function of $\theta > 0$. For (26) with $m = 1$ and $a^2 = 9$, for example, we obtain from (32) and (33) the bounds

$$1040 < E_{.1}(N) < 1800$$
$$10.4 < E_1(N) < 15$$
$$2.6 < E_2(N) < 5$$
$$\vdots$$

More precise estimates of $E_\theta(N)$ could be obtained from Monte Carlo methods which, for obvious reasons, are not directly applicable to estimating the Type I error, for which we have the upper bound (31) of .0056 for $a^2 = 9$.

Other examples of the methods described above in testing, selection, and ranking procedures are indicated in the references. In the next two sections we shall discuss some non-parametric "open-ended" procedures.

5. Confidence sequences for the median. Let $z_1, \cdots$ be i.i.d. with $P(z_i \leqq M) = \frac{1}{2}$ and let $z_1^{(n)} \leqq z_2^{(n)} \leqq \cdots$ denote the ordered values $z_1, \cdots, z_n$. The usual confidence interval for M for a single value of n is based on the normal approximation to the binomial distribution which gives the relation

$$P(z_{a_1}^{(n)} \leqq M \leqq z_{a_2}^{(n)}) \cong 2\Phi(a) - 1 \qquad \text{for large } n$$

where

$$a_1 = a_1(n) = \text{largest integer} \leqq \tfrac{1}{2}(n - an^{\frac{1}{2}}),$$
$$a_2 = a_2(n) = \text{smallest integer} \geqq \tfrac{1}{2}(n + an^{\frac{1}{2}}).$$

To construct a confidence sequence for M, let

$$x_i = \begin{cases} 1 & \text{if } z_i \leqq M, \\ -1 & \text{if } z_i > M, \end{cases} \qquad S_n = x_1 + \cdots + x_n,$$

and let c_n be some sequence of positive constants. Define

$$b_1 = b_1(n) = \text{largest integer} \leqq \tfrac{1}{2}(n - c_n),$$
$$b_2 = b_2(n) = \text{smallest integer} \geqq \tfrac{1}{2}(n + c_n).$$

Then

$$P(z_{b_1}^{(n)} \leqq M \leqq z_{b_2}^{(n)} \text{ for every } n \geqq m)$$
$$\geqq 1 - P(|S_n| \geqq c_n \text{ for some } n \geqq m).$$

Using for example the sequence $c_n = [n(a^2 + \log n/m)]^{\frac{1}{2}}$ for which (20) gives the approximation for large m

$$P(|S_n| > c_n \text{ for some } n \geqq m) \cong 2(1-\Phi(a)+a\varphi(a)),$$

we have for large m

$$P(z_{b_1}^{(n)} \leqq M \leqq z_{b_2}^{(n)} \text{ for every } n \geqq m) \cong 2\Phi(a)-1-2a\varphi(a)$$
$$\equiv H(a), \text{ say,}$$
where

a	$2\Phi(a)-1$	$H(a)$
2	.9546	.7386
2.5	.9876	.9001
2.8	.9948	.9506
3	.9974	.9710
3.5	.9996	.9933

and $b_1(m) = a_1(m)$, $b_2(m) = a_2(m)$. Remember that by the law of the iterated logarithm

$$P(z_{a_1}^{(n)} \leqq M \leqq z_{a_2}^{(n)} \text{ for every } n \geqq m) = 0 \qquad (m = 1,2, \cdots).$$

6. Kolmogorov–Smirnov tests with power 1. For any two distribution functions G, H write

$$D^+(G,H) = \sup_{-\infty<t<\infty}(G(t)-H(t)), \quad D(G,H) = \sup_{-\infty<t<\infty}|G(t)-H(t)|.$$

Let $x_1, \cdots$ be i.i.d. with df $F_x(t) = P(x_i \leqq t)$ and let $y_1, \cdots$ be i.i.d. with df $F_y(t)$, the x's and y's being independent. Denote by $F_x{}^n(t)$, $F_y{}^n(t)$ the sample df's of $x_1, \cdots, x_n$ and $y_1, \cdots, y_n$.

Consider the hypothesis

$$H_0: F_x(t) \leqq F_y(t) \qquad \text{for every } -\infty < t < \infty.$$

To test H_0 define

$$N = \text{first integer } n \geqq m \text{ such that } D^+(F_x{}^n, F_y{}^n) \geqq f(n)/n$$
$$= \infty \text{ if no such } n \text{ occurs,}$$

where $f(n)$ is some positive sequence such that $f(n)/n \to 0$ as $n \to \infty$. If H_0 is false, and $D^+(F_x, F_y) = d > 0$, then by the Glivenko–Cantelli theorem, as $n \to \infty$ $D^+(F_x{}^n, F_y{}^n) \to d$ with probability 1, so that $P(N < \infty) = 1$. Hence if we agree to reject H_0 as soon as we observe that $N < \infty$, while if $N = \infty$ we do not reject H_0, then the test certainly has power 1 when H_0 is false. It remains to consider the Type I error probability.

It can be shown that when H_0 is true, no matter what F_x and F_y may be in other respects, the inequality

$$P\left(D^+(F_x{}^n, F_y{}^n) \geqq \frac{r}{n}\right) \leqq \frac{(n!)^2}{(n-r)!(n+r)!} \leqq e^{-(r^2/n+1)} \qquad (r = 0, 1, \cdots, n)$$

always holds, and hence the crude inequality

$$(34)\qquad P(N < \infty) \leqq \sum_{n=m}^{\infty} P\left(D^+(F_x^n, F_y^n) \geqq \frac{f(n)}{n}\right) \leqq \sum_{n=m}^{\infty} e^{-[f^2(n)/n+1]}.$$

holds. Choosing $f(n)$ to be $\sim [(1+\varepsilon)n \log n]^{\frac{1}{2}}$ will suffice to make the series converge and hence will guarantee an arbitrarily small Type I error probability when H_0 is true if m is chosen sufficiently large. For example, if $f(n) = [(n+1)(\log 4 + 2\log n)]^{\frac{1}{2}}$, $m = 6$ then $P(N < \infty) < .05$ whenever H_0 is true.

The law of the iterated logarithm for the sequence $D^+(F_x^n, F_y^n)$ shows that taking $f(n) \sim [(1+\varepsilon)n \log_2 n]^{\frac{1}{2}}$ would also suffice to ensure an arbitrarily small value of $P(N < \infty)$ under H_0 for large m. Upper bounds for $P(N < \infty)$ in such cases have recently been obtained by Richard Stanley (unpublished).

Concerning the value of EN when H_0 is false, it can be shown that it is always finite and that the inequality

$$EN \leqq g(d - m/g(d)), \qquad d = D^+(F_x, F_y)$$

holds, where $g(x)$ is the function inverse to $f(x)/x$.

Similar tests are available for various other non-parametric hypotheses such as

$H_1: F_x = F_y$

$H_2: F_x \leqq F;$ $\qquad F$ an arbitrary specified df,

$H_3: F_x \in \mathscr{F};$ $\qquad \mathscr{F}$ any class of df's closed under the D metric (e.g., the set $N(\mu, \sigma^2)$ with $-\infty < \mu < \infty$, $0 \leqq \sigma^2 < \infty$).

In each case the power is 1 and the expected sample size is finite under any alternative, while an arbitrarily small upper bound for the type I error can be guaranteed. The currently available bounds are crude, however, being based on inequalities similar to (34).

7. Concluding remark. The ideas involved here seem to be a natural extension (or contraction) of Wald's sequential analysis; cf. also [1], [7], and [9]. My own work has been done in collaboration with D. A. Darling in the first instance and later with D. Siegmund, to whom I wish to express my deep appreciation.

REFERENCES

[1] Blum, J. R. and Rosenblatt, J. (1966). On some statistical problems requiring purely sequential sampling schemes. *Ann. Inst. Statist. Math.* **18** 351–354.

[2] Darling, D. A. and Robbins, H. (1967a). Iterated logarithm inequalities. *Proc. Nat. Acad. Sci.* **57** 1188–1192.

[3] Darling, D. A. and Robbins, H. (1967b). Inequalities for the sequence of sample means. *Proc. Nat. Acad. Sci.* **57** 1577–1580.

[4] Darling, D. A. and Robbins, H. (1967c). Confidence sequences for mean, variance, and median. *Proc. Nat. Acad. Sci.* **58** 66–68.

[5] Darling, D. A. and Robbins, H. (1968a). Some further remarks on inequalities for sample sums. *Proc. Nat. Acad. Sci.* **60** 1175–1182.

[6] DARLING, D. A. and ROBBINS, H. (1968b). Some nonparametric sequential tests with power 1. *Proc. Nat. Acad. Sci.* **61** 804–809.

[7] FABIAN, V. (1956). A decision function. *Czechoslovak Math. J.* **6** 31–41.

[8] FARRELL, R. H. (1962). Bounded length confidence intervals for the zero of a regression function. *Ann. Math. Statist.* **33** 237–247.

[9] FARRELL, R. H. (1964). Asymptotic behavior of expected sample size in certain one sided tests. *Ann. Math. Statist.* **35** 36–72.

[10] LEHMANN, E. (1959). *Testing Statistical Hypotheses*. Wiley, New York.

[11] ROBBINS, H. and SIEGMUND, D. (1968). Iterated logarithm inequalities and related statistical procedures. *Mathematics of the Decision Sciences* **2**. American Mathematical Society, Providence 267–279.

[12] ROBBINS, H. and SIEGMUND, D. (1969a). Probability distributions related to the law of the iterated logarithm. *Proc. Nat. Acad. Sci.* **62** 11–13.

[13] ROBBINS, H. and SIEGMUND, D. (1969b). Boundary crossing probabilities for the Wiener process. Stanford Univ. Department of Statistics Tech. Report No. 4.

[14] ROBBINS, H. and SIEGMUND, D. (1970). Boundary crossing probabilities for the Wiener process and sample sums. *Ann. Math. Statist.* **41** 1410–1429.

[15] ROBBINS, H., SIEGMUND, D. and WENDEL, J. (1968). The limiting distribution of the last time $S_n \geqq n\varepsilon$. *Proc. Nat. Acad. Sci.* **61** 1228–1230.

[16] VILLE, J. (1939). *Étude Critique de la Notion de Collectif*. Gauthier-Villars, Paris.

[17] WALD, A. (1947). *Sequential Analysis*. Wiley, New York.

The Annals of Mathematical Statistics
1970, Vol. 41, No. 5, 1410–1429

BOUNDARY CROSSING PROBABILITIES FOR THE WIENER PROCESS AND SAMPLE SUMS[1]

BY HERBERT ROBBINS AND DAVID SIEGMUND

Columbia University; Stanford University and Columbia University

1. Introduction and summary. Let $W(t)$ denote a standard Wiener process for $0 \leqq t < \infty$. We compute the probability that $W(t) \geqq g(t)$ for some $t \geqq \tau > 0$ (or for some $t > 0$) for a certain class of functions $g(t)$, including functions which are $\sim (2t \log\log t)^{\frac{1}{2}}$ as $t \to \infty$. We also prove an invariance theorem which states that this probability is the limit as $m \to \infty$ of the probability that $S_n \geqq m^{\frac{1}{2}} g(n/m)$ for some $n \geqq \tau m$ (or for some $n \geqq 1$), where S_n is the nth partial sum of any sequence $x_1, x_2, \cdots$ of independent and identically distributed (i.i.d.) random variables with mean 0 and variance 1.

The main results were announced in [19]. Some aspects of the invariance theorem were considered independently by Müller [14], who also studied the rate of convergence to the limiting distribution. Statistical applications of these ideas are indicated in [3] and [18].

In Section 2 we state the general theorems and give several examples. Sections 3–5 are devoted to the proof of these results. In Section 6 we indicate the applicability of our methods to stochastic processes other than the Wiener process. Of particular interest in this regard is the analogue of Theorem 1 for Bessel diffusion processes. Section 7 raises questions which will be treated in a subsequent paper.

2. Statement of Theorems and examples. Let F denote any measure on $(0, \infty)$ which is finite on bounded intervals, and define for $-\infty < x < \infty$, $-\infty < t < \infty$, $0 < \varepsilon < \infty$

$$0 < f(x, t) = \int_0^\infty \exp(xy - y^2 t/2)\, dF(y) \leqq \infty,$$

$$-\infty \leqq A(t, \varepsilon) = \inf\{x : f(x, t) \geqq \varepsilon\} < \infty.$$

It is easily seen that

$$x < A(t, \varepsilon) \Rightarrow f(x, t) < \varepsilon, \qquad f(x, t) < \varepsilon \Rightarrow x \leqq A(t, \varepsilon), \tag{1}$$

and that if for some b, h $f(b, h) < \infty$ then for each $t > h$ the equation $f(x, t) = \varepsilon$ has the unique solution $x = A(t, \varepsilon)$. The function $A(t, \varepsilon)$ is continuous and increasing in t for $h \leqq t < \infty$, and for $t > h$ $f(x, t) \geqq \varepsilon$ if and only if $x \geqq A(t, \varepsilon)$. Set

$$\varphi(x) = (2\pi)^{-\frac{1}{2}} \exp(-x^2/2), \qquad \Phi(x) = \int_{-\infty}^x \varphi(y)\, dy.$$

THEOREM 1. (i) *For any* b, h, ε *such that* $f(b, h) < \varepsilon$,

$$P\{W(t) \geqq A(t+h, \varepsilon) - b \text{ for some } t > 0\} = f(b, h)/\varepsilon. \tag{2}$$

(ii) *For any* b, h, ε *and* $\tau > 0$,

Received October 14, 1969.

[1] Research supported by NSF Grant GP-8985, ONR Grant N000 14-67-A-0108-0018, and NIH Grant 1-R01-GM-16895-01.

(3) $P\{W(t) \geqq A(t+h, \varepsilon)-b \text{ for some } t \geqq \tau\}$

$$= 1-\Phi\left(\frac{A(\tau+h, \varepsilon)-b}{\tau^{\frac{1}{2}}}\right)+\varepsilon^{-1}\int_0^\infty \exp(by-y^2h/2)\Phi\left(\frac{A(\tau+h, \varepsilon)-b}{\tau^{\frac{1}{2}}}-y\tau^{\frac{1}{2}}\right)dF(y).$$

THEOREM 2. (i) *Suppose that $g(t)$ is continuous for $t \geqq \tau > 0$, that $t^{-\frac{1}{2}}g(t)$ is ultimately non-decreasing as $t \to \infty$, and that*

$$\int_\tau^\infty \frac{g(t)}{t^{3/2}}\exp(-g^2(t)/2t)\,dt < \infty. \tag{4}$$

Then

$$\lim_{m\to\infty} P\{S_n \geqq m^{\frac{1}{2}}g(n/m) \text{ for some } n \geqq \tau m\} = P\{W(t) \geqq g(t) \text{ for some } t \geqq \tau\}, \tag{5}$$

where $S_n = x_1 + \cdots + x_n$ and the x_i are any i.i.d. *random variables having mean* 0 *and variance* 1.

(ii) *Suppose that in addition to the hypotheses of* (i) *g is continuous for $t > 0$, that $t^{-\frac{1}{2}}g(t)$ is non-increasing for t sufficiently small, and that*

$$\int_{0+}^1 \frac{g(t)}{t^{3/2}}\exp(-g^2(t)/2t)\,dt < \infty.$$

Then (5) *continues to hold with $n \geqq \tau m$ replaced by $n \geqq 1$ and $t \geqq \tau$ by $t > 0$.*

REMARKS ON THEOREM 2. (a) The same relations are valid if S_n, $W(t)$ are replaced by $|S_n|$, $|W(t)|$.

(b) Instead of assuming that the continuous function g satisfies the indicated growth conditions, it is sufficient to assume that it majorizes some function which does.

(c) If $g(t)$ is continuous for $0 < t \leqq \tau < \infty$ and the growth conditions of (ii) hold for t sufficiently small, then

$$\lim_{m\to\infty} P\{S_n \geqq m^{\frac{1}{2}}g(n/m) \text{ for some } 1 \leqq n \leqq \tau m\} = P\{W(t) \geqq g(t) \text{ for some } 0 < t \leqq \tau\}.$$

In discussing the following examples, we use the fact (cf. [2], page 266) that the process $W^*(t)$ defined by

$$W^*(t) = tW(t^{-1}) \qquad (t > 0), \quad W^*(0) \equiv 0 \tag{6}$$

is also a standard Wiener process.

EXAMPLE 1. Let F be the degenerate measure which puts unit mass at some point $2a > 0$, and let $\varepsilon = 1$. Then $A(t, \varepsilon) = at$ and (3) with $h = 0$ (together with (6)) and

Theorem 2 give the result that

$$\begin{aligned}
&\lim_{m\to\infty} P\{\max_{n\geqq\tau m}(S_n - an/m^{\frac{1}{2}}) \geqq bm^{\frac{1}{2}}\} \\
&\quad = \lim_{m\to\infty} P\{\max_{1\leqq n\leqq m/\tau}(S_n - bn/m^{\frac{1}{2}}) \geqq am^{\frac{1}{2}}\} \\
&\quad = P\{\max_{t\geqq\tau}(W(t) - at) \geqq b\} \\
&\quad = P\{\max_{0<t\leqq\tau^{-1}} W((t) - bt) \geqq a\} \\
&\quad = 1 - \Phi(b/\tau^{\frac{1}{2}} + a\tau^{\frac{1}{2}}) + \exp(-2ab)\Phi(b/\tau^{\frac{1}{2}} - a\tau^{\frac{1}{2}}) \\
&\qquad\qquad (\tau > 0,\ a > 0,\ -\infty < b < \infty).
\end{aligned}$$

The special case $b = 0$ yields the relation

$$\begin{aligned}
&\lim_{m\to\infty} P\{\max_{n\geqq\tau m} S_n/n \geqq a/m^{\frac{1}{2}}\} \\
&\quad = \lim_{m\to\infty} P\{\max_{1\leqq n\leqq m/\tau} S_n \geqq am^{\frac{1}{2}}\} \qquad (7)\\
&\quad = P\{\max_{t\geqq\tau} W(t)/t \geqq a\} \\
&\quad = P\{\max_{0<t\leqq\tau^{-1}} W(t) \geqq a\} = 2(1 - \Phi(a\tau^{\frac{1}{2}})) \qquad (\tau > 0,\ a > 0).
\end{aligned}$$

For any $\varepsilon > 0$ define as in [20]

$$M = M(\varepsilon) = \sup\{n : S_n \geqq n\varepsilon\}.$$

By the strong law of large numbers, $P\{M < \infty\} = 1$. Since $\{M \geqq m\} = \{\max_{n\geqq m} S_n/n \geqq \varepsilon\}$, letting $m \to \infty$, $\varepsilon \to 0$ in such a way that $\varepsilon m^{\frac{1}{2}} = a > 0$, we obtain from (7) the result

$$\lim_{\varepsilon\to 0} P\{\varepsilon^2 M \geqq a^2\} = \lim_{m\to\infty} P\{m^{\frac{1}{2}} \max_{n\geqq m} S_n/n \geqq a\} = 2(1 - \Phi(a));$$

i.e., as $\varepsilon \to 0$ the random variable $\varepsilon^2 M$ converges in law to the chi-square distribution with one degree of freedom.

Using (2) instead of (3) we obtain from Theorem 1 and Theorem 2

$$\begin{aligned}
&\lim_{m\to\infty} P\{S_n \geqq an/m^{\frac{1}{2}} + bm^{\frac{1}{2}} \text{ for some } n \geqq 1\} \\
&\quad = P\{\max_{t>0}(W(t) - at) \geqq b\} = \exp(-2ab) \qquad (a > 0,\ b > 0).
\end{aligned}$$

Equation (2) with $h = 1$ also gives certain probabilities associated with the "tied down" Wiener process:

$$\begin{aligned}
&P\{\max_{0<t\leqq 1} W(t) \geqq a \mid W(1) = b\} \\
&\quad = P\{\max_{t\geqq 1} W(t)/t \geqq a \mid W(1) = b\} \\
&\quad = P\left\{\max_{t>0} \frac{b + W(t)}{1 + t} \geqq a\right\} = \exp(-2a(a - b)) \qquad (a > 0,\ b < a).
\end{aligned}$$

EXAMPLE 2. Let $dF(y) = (2/\pi)^{\frac{1}{2}}\, dy/y^{\gamma}$ for $0 < y < \infty$, $\gamma < 1$. Then

$$A(t, \varepsilon) = t^{\frac{1}{2}}\alpha^{-1}((1 - \gamma)\log t + 2\log\tfrac{1}{2}\varepsilon),$$

where we have set

$$\alpha(x) = x^2 + 2\log \textstyle\int_0^\infty \varphi(y-x)\,dy/y^\gamma \sim x^2 \quad \text{as} \quad x \to \infty,$$

and (2) with $b = 0$ and Theorem 2 imply that

$$\begin{aligned} &\lim_{m\to\infty} P\{S_m \geqq (n+hm)^{\frac{1}{2}}\alpha^{-1}((1-\gamma)\log(n/m+h)+2\log\varepsilon/2) \text{ for some } n \geqq 1\} \\ (8) \qquad &= P\{W(t) \geqq (t+h)^{\frac{1}{2}}\alpha^{-1}((1-\gamma)\log(t+h)+2\log\varepsilon/2) \text{ for some } t > 0\} \\ &= \frac{\Gamma\left(\frac{1-\gamma}{2}\right)}{2^{(\gamma+2)/2}\pi^{\frac{1}{2}}h^{(1-\gamma)/2}\varepsilon}, \qquad \left(h > 0,\ \varepsilon > \frac{\Gamma\left(\frac{1-\gamma}{2}\right)}{2^{(\gamma+2)/2}\pi^{\frac{1}{2}}h^{(1-\gamma)/2}}\right). \end{aligned}$$

For $\gamma = 0$, $\alpha(x) = x^2 + 2\log\Phi(x)$, and the right-hand side of (8) becomes $(2h^{\frac{1}{2}}\varepsilon)^{-1}$.

Setting $b = h = 0$, $\varepsilon = 2\exp(\frac{1}{2}\alpha(a))$, we obtain from (3) and Theorem 2 for any $\tau > 0$ the result

$$\begin{aligned} &\lim_{m\to\infty} P\{S_n \geqq n^{\frac{1}{2}}\alpha^{-1}((1-\gamma)\log n/m+\alpha(a)) \text{ for some } n \geqq \tau m\} \\ (9) \quad &= P\{W(t) \geqq t^{\frac{1}{2}}\alpha^{-1}((1-\gamma)\log t+\alpha(a)) \text{ for some } t \geqq \tau\} \\ &= 1-\Phi(\alpha^{-1}((1-\gamma)\log\tau+\alpha(a)))+\varphi(a)\frac{\displaystyle\int_0^\infty \Phi(\alpha^{-1}((1-\gamma)\log\tau+\alpha(a))-y)\frac{dy}{y^\gamma}}{\displaystyle\tau^{(1-\gamma)/2}\int_0^\infty \varphi(a-y)\frac{dy}{y^\gamma}}. \end{aligned}$$

For $\tau = 1$ the right-hand side of (9) simplifies to

$$1-\Phi(a)+\frac{\displaystyle\varphi(a)\int_0^\infty \Phi(a-y)\frac{dy}{y^\gamma}}{\displaystyle\int_0^\infty \varphi(a-y)\frac{dy}{y^\gamma}},$$

which for $\gamma = 0$ becomes

$$1-\Phi(a)+\varphi(a)\left(a+\frac{\varphi(a)}{\Phi(a)}\right).$$

Finally, for $\tau = h = 0$, $b > 0$, $\varepsilon > (2/\pi)^{\frac{1}{2}}\Gamma(1-\gamma)/b^{1-\gamma}$, we have

$$\begin{aligned} &\lim_{m\to\infty} P\{S_n \geqq n^{\frac{1}{2}}\alpha^{-1}((1-\gamma)\log n/m+2\log\varepsilon/2)+b \text{ for some } n \geqq 1\} \\ &\qquad = P\{W(t) \geqq t^{\frac{1}{2}}\alpha^{-1}((1-\gamma)\log t+2\log\varepsilon/2)+b \text{ for some } t > 0\} \\ &\qquad = (2/\pi)^{\frac{1}{2}}\Gamma(1-\gamma)/(\varepsilon b^{1-\gamma}). \end{aligned}$$

EXAMPLE 3. Theorem 1 generalizes in an obvious manner to the case in which F is a measure on $(-\infty, \infty)$ which assigns measure 0 to $\{0\}$. For $dF = dy/(2\pi)^{\frac{1}{2}}$,

$b=0$, $\varepsilon=e^{\frac{1}{2}a^2}$ the results are particularly elegant:

$$\lim_{m\to\infty} P\{|S_n| \geqq (n(a^2+\log n/m))^{\frac{1}{2}} \text{ for some } n \geqq \tau m\}$$
$$= P\{|W(t)| \geqq (t(a^2+\log t))^{\frac{1}{2}} \text{ for some } t \geqq \tau\}$$
$$= 2[1-\Phi((a^2+\log\tau)^{\frac{1}{2}})+((a^2+\log\tau)/\tau)^{\frac{1}{2}}\varphi(a)], \quad (\tau > e^{-a^2});$$
$$\lim_{m\to\infty} P\{|S_n| \geqq [(n+hm)(a^2+\log(n/m+h))]^{\frac{1}{2}} \text{ for some } n \geqq 1\}$$
$$= P\{|W(t)| \geqq [(t+h)(a^2+\log(t+h))]^{\frac{1}{2}} \text{ for some } t>0\}.$$
$$= h^{-\frac{1}{2}}e^{-\frac{1}{2}a^2} \qquad (h > e^{-a^2}).$$

EXAMPLE 4. For $\delta > 0$ let

$$dF(y) = dy/y(\log 1/y)(\log_2 1/y)\cdots(\log_n 1/y)^{1+\delta} \quad \text{for} \quad 0<y<1/e_n$$
$$= 0 \text{ elsewhere},$$

where we write $\log_2 x = \log(\log x)$, $e_2 = e^e$, etc. It will be shown in Section 4 that as $t\to\infty$, $A(t,\varepsilon) \sim (2t\log_2 t)^{\frac{1}{2}}$; in fact for $n \geqq 3$

$$A(t,\varepsilon) = [2t(\log_2 t + 3/2\log_3 t + \textstyle\sum_{k=4}^n \log_k t + (1+\delta)\log_{n+1} t + \log\frac{1}{2}\varepsilon/\pi^{\frac{1}{2}} + o(1))]^{\frac{1}{2}} \tag{10}$$

while for $n=2$,

$$A(t,\varepsilon) = [2t(\log_2 t + (3/2+\delta)\log_3 t + \log \varepsilon/2\pi^{\frac{1}{2}} + o(1))]^{\frac{1}{2}},$$

so Theorem 1 and Theorem 2 give a deeper content to the "easy" half of the law of the iterated logarithm. For example, for $b=h=0$ we have

$$\lim_{m\to\infty} P\{S_n \geqq m^{\frac{1}{2}}A(n/m,\varepsilon) \text{ for some } n \geqq 1\}$$
$$= P\{W(t) \geqq A(t,\varepsilon) \text{ for some } t>0\} = 1/(\delta\varepsilon) \quad (\varepsilon > 1/\delta).$$

EXAMPLE 5. For $\delta > 0$ let

$$dF(y) = dy/y(\log y)(\log_2 y)\cdots(\log_n y)^{1+\delta} \quad \text{for} \quad y > e_n$$
$$= 0 \quad \text{elsewhere}.$$

In this case $f(0,0)=\delta^{-1}$, but $f(x,0)=\infty$ for each $x>0$, so that $A(0,\varepsilon)=0$ for each $\varepsilon > \delta^{-1}$. An argument similar to that leading to (10) shows that for any $\varepsilon > \delta^{-1}$, as $t\to 0$,

$$A(t,\varepsilon) = [2t(\log_2 t^{-1} + 3/2\log_3 t^{-1} + \textstyle\sum_4^n \log_k t^{-1} + (1+\delta)\log_{n+1} t^{-1} + \log\frac{1}{2}(\varepsilon-\delta^{-1})/\pi^{\frac{1}{2}} + o(1))].$$

EXAMPLE 6. It will be shown in Section 4 that if

$$dF(y) = \tfrac{1}{2}(3/\pi)^{\frac{1}{2}}\exp(-16/27y^{-1})\,dy \qquad (0<y<\infty),$$

then as $t \to \infty$

$$A(t, \varepsilon) = t^{\frac{1}{3}} + 4^{-1} t^{\frac{1}{3}} (\tfrac{1}{2} \log t + \log \varepsilon + o(1)). \tag{11}$$

More generally, if

$$dF(y) = \gamma \exp(-\alpha y^{-\beta})\, dy,$$

for any $\beta > 0$ and appropriate values $\alpha = \alpha(\beta) > 0$ and $\gamma = \gamma(\beta) > 0$, then

$$A(t, \varepsilon) \sim t^{(1+\beta)/(2+\beta)}$$

and an expansion similar to (11) may be obtained. We omit the details.

3. Proof of Theorem 1. The proof of Theorem 1 is an application of Lemma 1 below to certain martingales defined in terms of the function $f(x, t)$. Since (cf., e.g., [21]) $\{\exp(yW(t) - \frac{1}{2}y^2 t, \mathscr{B}(W(s), s \leqq t), t \geqq 0\}$ is a martingale for each fixed y, it follows from Fubini's theorem that for any real numbers b and h

$$\{z(t), \mathscr{F}(t), t \geqq 0\} = \{f(b + W(t), t + h), \mathscr{B}(W(s), s \leqq t), t \geqq 0\}$$

is also a martingale except that $Ez(t)$ may be ∞. Although the definition of a martingale usually includes the assumption that $E|z(t)| < \infty$ (cf. [16], page 131), the proof of Lemma 1 does not actually require this hypothesis, and our departure from customary usage permits applications such as Example 2 of Section 2.

LEMMA 1. *Let ε be any positive constant and $\{z(t), \mathscr{F}(t), t \geqq \tau\}$ a nonnegative martingale. If $z(t)$ has continuous sample paths on $\{z(\tau) < \varepsilon\}$ and converges to 0 in probability on $\{\sup_{t > \tau} z(t) < \varepsilon\}$, then*

$$P\{\sup_{t > \tau} z(t) \geqq \varepsilon \mid \mathscr{F}(\tau)\} = \varepsilon^{-1} z(\tau) \quad \textit{on} \quad \{z(\tau) < \varepsilon\}. \tag{12}$$

PROOF. Define $T = \inf\{t : t \geqq \tau, z(t) \geqq \varepsilon\}$, where the inf of the empty set is taken to be $+\infty$. It is well known (e.g. [16], page 142) that $\{z(T \wedge t), \mathscr{F}(t), t \geqq \tau\}$ is a martingale. Hence for any $A \in \mathscr{F}(\tau)$ and $t \geqq \tau$,

$$\begin{aligned} \int_{A\{z(\tau) < \varepsilon\}} z(\tau)\, dP &= \int_{A\{z(\tau) < \varepsilon\}} z(T \wedge t)\, dP \\ &= \varepsilon P(A\{z(\tau) < \varepsilon, T \leqq t\}) + \int_{A\{z(\tau) < \varepsilon, T > t\}} z(t)\, dP. \end{aligned}$$

Since $I_{\{T > t\}} z(t) \leqq \varepsilon I_{\{t < T < \infty\}} + I_{\{T = \infty\}} z(t) \leqq \varepsilon$ and converges to 0 in probability as $t \to \infty$, we have by the dominated convergence theorem

$$\begin{aligned} \int_{A\{z(\tau) < \varepsilon\}} z(\tau)\, dP &= \varepsilon P(A\{z(\tau) < \varepsilon, T < \infty\}) \\ &= \varepsilon \int_{A\{z(\tau) < \varepsilon\}} P\{T < \infty \mid \mathscr{F}(\tau)\}\, dP, \end{aligned}$$

which proves (12).

From (12) it follows directly that

$$P\{\sup_{t \geqq \tau} z(t) \geqq \varepsilon\} = P\{z(\tau) \geqq \varepsilon\} + \varepsilon^{-1} \int_{\{z(\tau) < \varepsilon\}} z(\tau)\, dP. \tag{13}$$

LEMMA 2. *If $f(x, \tau) < \infty$ for some x, τ, then for any b, h $f(b + W(t), h + t) \to 0$ in probability as $t \to \infty$.*

PROOF. By replacing τ by $\tau+1$ if necessary, we may assume without loss of generality that $f(x,\tau)<\infty$ for all x. Suppose first that $h \geqq \tau$. For any $c>0$

$$f(b+ct^{\frac{1}{2}}, h+t) = (\varphi(c))^{-1}\int_0^\infty \varphi(c-yt^{\frac{1}{2}})\exp(by-\tfrac{1}{2}hy^2)\,dF(y)\to 0$$

as $t\to\infty$ by the dominated convergence theorem. Hence for any $\varepsilon>0$, for all t sufficiently large

$$P\{f(b+W(t), h+t)\geqq \varepsilon\} \leqq P\{W(t)\geqq ct^{\frac{1}{2}}\} = 1-\Phi(c),$$

which can be made arbitrarily small by taking c sufficiently large. Now suppose that $h<\tau$. Then for any $t>\tau-h$ and $\varepsilon>0$

$$\begin{aligned}P\{f(b+W(t), h+t)\geqq\varepsilon\}\\ &= \int_{-\infty}^{\infty} P\{f(b+(\tau-h)^{\frac{1}{2}}x+W(t-\tau+h), \tau+(t-\tau+h))\geqq\varepsilon\}\cdot\varphi(x)\,dx\to 0\end{aligned}$$

as $t\to\infty$ by the first part of the proof and the dominated convergence theorem.

PROOF OF THEOREM 1. (ii) Let b, h, ε be arbitrary and $\tau>0$. We may assume that $f(x,\tau+h)<\infty$ for some x, since otherwise the theorem is trivially true. For each $t\geqq\tau$ let $z(t) = f(b+W(t), t+h)$, $\mathscr{F}(t)=\mathscr{B}(W(s), s\leqq t)$, and set

$$B_1=\{W(\tau)<A(\tau+h,\varepsilon)-b\},\qquad B_2=\{f(b+W(\tau),\tau+h)<\varepsilon\}.$$

If $x<A(\tau+h,\varepsilon)$ then for some $\delta>0$ $x+\delta<A(\tau+h,\varepsilon)$ and hence $f(x+\delta,\tau+h)<\varepsilon$. Thus from the continuity of the sample paths of $W(t)$ and the dominated convergence theorem it follows that the martingale $\{z(t),\mathscr{F}(t), t\geqq\tau\}$ has continuous sample paths on B_1. But $B_1=B_2$ a.s. by (1) and the continuity of the distribution function of $W(\tau)$, and hence by Lemma 2 and (12)

$$\begin{aligned}&P\{W(t)\geqq A(t+h,\varepsilon)-b \text{ for some } t\geqq\tau\}\\ &\quad= 1-P(B_1)+\int_{B_1}P\{W(t)\geqq A(t+h,\varepsilon)-b \text{ for some } t>\tau\,|\,\mathscr{F}(\tau)\}\,dP\\ &\quad= 1-\Phi\left(\frac{A(\tau+h,\varepsilon)-b}{\tau^{\frac{1}{2}}}\right)+\int_{B_2}P\{\sup\nolimits_{t>\tau}f(b+W(t),t+h)\geqq\varepsilon\,|\,\mathscr{F}(\tau)\}\,dP\\ &\quad= 1-\Phi\left(\frac{A(\tau+h,\varepsilon)-b}{\tau^{\frac{1}{2}}}\right)+\varepsilon^{-1}\int_{B_1}f(b+W(\tau),\tau+h)\,dP\\ &\quad= 1-\Phi\left(\frac{A(\tau+h,\varepsilon)-b}{\tau^{\frac{1}{2}}}\right)+\varepsilon^{-1}\int_0^\infty \exp(by-\tfrac{1}{2}y^2h)\Phi\left(\frac{A(\tau+h,\varepsilon)-b}{\tau^{\frac{1}{2}}}-y\tau^{\frac{1}{2}}\right)dF(y).\end{aligned}$$

(i) By absorbing the factor $\exp(by-\frac{1}{2}y^2h)$ into the measure F and dividing by a constant, we may without loss of generality assume that $b=h=0$ and $f(0,0)=1<\varepsilon$. Now (3) may be written

$$\begin{aligned}(14)\qquad &P\{f(W(t),t)\geqq\varepsilon \quad\text{for some}\quad t\geqq\tau\}\\ &\qquad= P\{f(W(\tau),\tau)\geqq\varepsilon\}+\varepsilon^{-1}\int_{\{f(W(\tau),\tau)<\varepsilon\}}f(W(\tau),\tau)\,dP,\end{aligned}$$

valid for any $\tau > 0$. An argument similar to that of Lemma 2 shows that $f(W(\tau), \tau) \to 1$ in probability as $\tau \to 0$, and hence (2) follows from (14) and the dominated convergence theorem on letting $\tau \to 0$.

REMARKS. (a) Although (2) is a special case of (12), we chose to prove it as the limit of (3). The reason for this chicanery is that under the sole assumption that $f(0, 0) = 1$ it is not immediately obvious that the martingale $\{f(W(t), t), \mathscr{F}(t), t \geqq 0\}$ has sample paths which are continuous from the right at $t = 0$, a condition that is required in order that (12) apply. (That this is in fact the case follows from our argument that $f(W(t), t) \to 1 = f(0, 0)$ in probability as $t \to 0$ and the martingale convergence theorem, which asserts that this convergence takes place with probability one.)

(b) An additional argument shows that if $\varepsilon \leqq f(b, h) \leqq \infty$, then the probability on the left-hand side of (2) is 1. Hence we obtain instead of (2) the completely general statement that *for any* b, h, ε

$$(2') \qquad P\{W(t) \geqq A(t+h, \varepsilon) - b \text{ for some } t > 0) = \min(1, f(b, h)/\varepsilon).$$

(c) Part (ii) of Theorem 1 remains valid if we replace "for some $t \geqq \tau$" by "for some $t > \tau$". In contrast, part (i) with "for some $t \geqq 0$" replacing "for some $t > 0$" is false when $A(h, \varepsilon) = b$ and $f(b, h) < \varepsilon$ (see Example 5 with $b = h = 0$ and $\varepsilon > \delta^{-1}$).

(d) We shall say that a function $\psi: [\tau, \infty) \to (-\infty, \infty)$ has the property (*) if for every $\tau' > \tau$ such that $\psi(\tau) < \psi(\tau')$ and every $c \in (\psi(\tau), \psi(\tau'))$ there exists a smallest $t \in (\tau, \tau')$ such that $\psi(t) = c$. It is clear that Lemma 1 still holds if "continuous sample paths" is replaced by "sample paths having the property (*)". This remark will be used toward the end of Section 6.

4. Asymptotic expansions for $A(t, \varepsilon)$. In this section we obtain asymptotic expansions for the functions $A(t, \varepsilon)$ associated with the measures F of Examples 4, 5 and 6 of Section 2.

Suppose first that F is as in Example 4 with $n = 2$. (The case of general n requires only minor modifications.) Let $f = F'$, and for fixed $\varepsilon > 0$ let $B = B(t) = A(t, \varepsilon)/t^{\frac{1}{2}}$ be defined by the equation

$$\varepsilon = \int_0^\infty \exp(Byt^{\frac{1}{2}} - \tfrac{1}{2}y^2 t) f(y)\, dy = (\varphi(B))^{-1} \int_0^\infty \varphi(yt^{\frac{1}{2}} - B) f(y)\, dy.$$

It is easily verified that $B \to \infty$, $B = o(t^{\frac{1}{2}})$ as $t \to \infty$. Let $\gamma > 1$. Since f is decreasing in $(0, \varepsilon')$ for some $\varepsilon' > 0$ we have for all t sufficiently large

$$\begin{aligned}\varepsilon &\geqq \frac{1}{\varphi(B)} \int_0^{\gamma B/t^{1/2}} \varphi(yt^{\frac{1}{2}} - B) f(y)\, dy \\ &\geqq \frac{f(\gamma B/t^{\frac{1}{2}})}{\varphi(B)} \int_0^{\gamma B/t^{1/2}} \varphi(yt^{\frac{1}{2}} - B)\, dy \\ &= \frac{\Phi((\gamma-1)B) - \Phi(-B)}{\gamma B \varphi(B) \log t^{\frac{1}{2}}/\gamma B (\log_2 t^{\frac{1}{2}}/\gamma B)^{1+\delta}}.\end{aligned}$$

Letting $t \to \infty$, then $\gamma \to 1$, we obtain

$$\limsup_{t\to\infty} \frac{1}{B\varphi(B)\log t^{\frac{1}{2}}/B(\log_2 t^{\frac{1}{2}}/B)^{1+\delta}} \leqq \varepsilon. \tag{15}$$

Now let $0 < \alpha < \gamma < 1$. Then for t sufficiently large

$$\begin{aligned}
\varepsilon &= \frac{1}{\varphi(B)}\int_0^\infty \varphi(yt^{\frac{1}{2}}-B)f(y)\,dy = \frac{1}{\varphi(B)}\left(\int_0^{\alpha B/t^{1/2}} + \int_{\alpha B/t^{1/2}}^{\gamma B/t^{1/2}} + \int_{\gamma B/t^{1/2}}^{\infty}\right)\varphi(yt^{\frac{1}{2}}-B)f(y)\,dy \\
&\leqq \frac{1}{\varphi(B)}\left[\varphi((\alpha-1)B)F(\alpha B/t^{\frac{1}{2}}) + f(\alpha B/t^{\frac{1}{2}})\int_{-\infty}^{\gamma B/t^{1/2}} \varphi(yt^{\frac{1}{2}}-B)\,dy\right. \\
&\qquad \left. + f(\gamma B/t^{\frac{1}{2}})\int_{\gamma B/t^{1/2}}^{\infty} \varphi(yt^{\frac{1}{2}}-B)\,dy\right] \\
&\leqq \frac{1}{\varphi(B)}\left[\frac{\varphi((\alpha-1)B)}{\delta(\log_2 t^{\frac{1}{2}}/\alpha B)^{\delta}} + f(\alpha B/t^{\frac{1}{2}})(1/t^{\frac{1}{2}})\Phi((\gamma-1)B) + f(\gamma B/t^{\frac{1}{2}})(1/t^{\frac{1}{2}})(1-\Phi((\gamma-1)B)\right].
\end{aligned}$$

It follows easily from (15) that $B^2 = O(\log_2 t)$. Hence, setting $\alpha = (\log_2 t)^{-1}$ and using the inequality

$$\Phi(x) \leqq |x|^{-1}\varphi(x) \qquad (x < 0),$$

we obtain

$$\varepsilon \leqq o(1) + \frac{(\log_2 t)\exp\left[(\frac{1}{2}(B^2-(1-\gamma)^2B^2))\right]}{(1-\gamma)B^2\log t^{\frac{1}{2}}/B(\log_2 t^{\frac{1}{2}}/B)^{1+\delta}} + \frac{(2\pi)^{\frac{1}{2}}\, e^{\frac{1}{2}B^2}}{\gamma B\log t^{\frac{1}{2}}/B(\log_2 t^{\frac{1}{2}}/B)^{1+\delta}}. \tag{16}$$

From (16) we have

$$\varepsilon \leqq \text{const.}\ \frac{(\log_2 t)\, e^{\frac{1}{2}B^2}}{\log t^{\frac{1}{2}}/B} + o(1),$$

from which it follows that $\log_2 t = O(B^2)$. Hence from (15) the second term on the right-hand side of (16) is $o(1)$ as $t \to \infty$ and thus

$$\frac{(2\pi)^{\frac{1}{2}}\, e^{\frac{1}{2}B^2}}{B\log t^{\frac{1}{2}}/B(\log_2 t^{\frac{1}{2}}/B)^{1+\delta}} \geqq \gamma\varepsilon + o(1).$$

Letting $t \to \infty$, then $\gamma \to 1$, we have

$$\liminf_{t\to\infty} \frac{(2\pi)^{\frac{1}{2}}\, e^{\frac{1}{2}B^2}}{B\log t^{\frac{1}{2}}/B(\log_2 t^{\frac{1}{2}}/B)^{1+\delta}} \geqq \varepsilon. \tag{17}$$

From (15) and (17) it follows that

$$\begin{aligned}
B^2 = 2\log_2 t^{\frac{1}{2}}/B + \log B^2 + 2(1+\delta)\log_3 t^{\frac{1}{2}}/B \\
+ 2\log\varepsilon - \log 2\pi + o(1) \quad \text{as} \quad t \to \infty.
\end{aligned} \tag{18}$$

Now $\log_2 t^{\frac{1}{2}}/B = \log(\frac{1}{2}\log t - \log B) = \log(\frac{1}{2}\log t(1+o(1))) = \log_2 t - \log 2 + o(1)$, and it follows from (18) that

$$B^2 \sim 2\log_2 t.$$

Hence $\log B^2 = \log_3 t + \log 2 + o(1)$, and thus as $t \to \infty$ (18) can be simplified to

$$B^2 = 2[\log_2 t + (3/2+\delta)\log_3 t + \log \tfrac{1}{2}\varepsilon\pi^{-\frac{1}{2}} + o(1)],$$

which is equivalent to (10).

The expansion as $t \to 0$ given in Example 5 may be obtained by a similar argument, with the following important difference. Whereas the behavior of F near 0 completely determines the asymptotic behavior (as described by (10)) of $A(t, \varepsilon)$ in Example 4, and we rightfully expect the behavior of F near infinity to play an analogous role in Example 5; nevertheless, in this case the measure that F assigns to bounded sets cannot be neglected. To be precise, for all K sufficiently large, we have by the dominated convergence theorem as $t \to 0$

$$\int_0^K \exp(Ay - \tfrac{1}{2}y^2 t)\, dF(y) \to \frac{1}{\delta}\left(1 - \frac{1}{(\log_n K)^\delta}\right).$$

An argument similar to that leading to (15) allows us to infer that $B \equiv t^{-\frac{1}{2}}A = O((\log_2 t^{-1})^{\frac{1}{2}})$ and hence

$$\frac{1}{\delta} \leq \lim_{t\to 0} \frac{1}{\varphi(B)} \int_0^{(t^{1/2}\log_2 t^{-1})^{-1}} \varphi(yt^{\frac{1}{2}} - B)\, dF(y)$$

$$\leq \limsup_{t\to 0} \left[\frac{\varphi\left(\dfrac{1}{\log_2 t^{-1}} - B\right)}{\varphi(B)}\right] \cdot \frac{1}{\delta} = \frac{1}{\delta}.$$

The remainder of the argument follows as before.

Next, let F be as in Example 6, and for a given $\varepsilon > 0$ let $A = A(t, \varepsilon)$ be defined by the equation

$$f(A, t) = \varepsilon.$$

For $x > 0$ let

$$h(y) = xy - y^2 t/2 - \alpha/y \qquad (\alpha = 16/27), \tag{19}$$

so that

$$h'(y) = x - yt + \alpha/y^2 \tag{20}$$

and

$$h''(y) = -t - 2\alpha/y^3. \tag{21}$$

Fix $\delta > 0$ and let $y^* = ct^{-\frac{1}{3}}$, $x = bt^{\frac{2}{3}}$, where $b = b(t)$ and $c = c(t)$ satisfy

$$h'(y^*) = 0 \tag{22}$$

and

$$h(y^*) = \log(\varepsilon(1+\delta)t^{\frac{1}{2}}). \tag{23}$$

From (19), (20), (22), and (23), we obtain

$$b = c - \alpha/c^2 \tag{24}$$

and

$$(b - c/2 - \alpha/c^2)ct^{\frac{1}{3}} = \log(\varepsilon(1+\delta)t^{\frac{1}{2}}). \tag{25}$$

It follows from (24) and (25) that

$$c = 4/3 + \tfrac{1}{2}t^{-\frac{1}{3}}\log(\varepsilon(1+\delta)t^{\frac{1}{2}}) + o(t^{-\frac{1}{3}}). \tag{26}$$

For any $\xi \geqq y^* - t^{-5/12}$, we have from (21)

$$h''(\xi) \geqq -t - \frac{2\alpha}{(y^* - t^{-5/12})^3} = -t\left(1 + \frac{2\alpha}{(c - t^{-1/12})^3}\right). \tag{27}$$

Since by (22) $h(y) = h(y^*) + \frac{1}{2}(y - y^*)^2 h''(\xi)$ for some ξ in the interval between y and y^*, we have by (27) and (23)

$$\begin{aligned} f(bt^{\frac{2}{3}}, t) &\geqq \tfrac{1}{2}(3/\pi)^{\frac{1}{2}} \int_{y^* - t^{-5/12}}^{\infty} e^{h(y)}\, dy \\ &\geqq \tfrac{1}{2}(3t/\pi)^{\frac{1}{2}}\varepsilon(1+\delta) \int_{-t^{-5/12}}^{\infty} \exp\left[-\tfrac{1}{2}y^2 t\left(1 + \frac{2\alpha}{(c - t^{-1/12})^3}\right)\right] dy \\ &\geqq \frac{\varepsilon(1+\delta)(3/2)^{\frac{1}{2}}}{\left(1 + \dfrac{2\alpha}{(c - t^{-1/12})^3}\right)^{\frac{1}{2}}}\,[1 - \Phi(-t^{1/12})]. \end{aligned}$$

Hence from (26) we obtain

$$\liminf_{t\to\infty} f(bt^{\frac{2}{3}}, t) \geqq \varepsilon(1+\delta),$$

and it follows that

$$A(t) \leqq bt^{\frac{2}{3}} \tag{28}$$

for all sufficiently large t. After some calculation we see by (24), (26), and (28) that

$$A(t, \varepsilon) \leqq t^{\frac{2}{3}} + 4^{-1}t^{\frac{1}{3}}\log(\varepsilon(1+\delta)t^{\frac{1}{2}}) + o(t^{\frac{1}{3}}),$$

and since δ is arbitrary

$$A(t, \varepsilon) \leqq t^{\frac{2}{3}} + 4^{-1}t^{\frac{1}{3}}(\tfrac{1}{2}\log t + \log \varepsilon + o(1)).$$

A similar argument proves the reverse inequality.

5. Proof of Theorem 2. For any $0 < \tau < c < \infty$ we have

$$\begin{aligned} &P\{S_n \geqq m^{\frac{1}{2}}g(n/m) \text{ for some } \tau m \leqq n \leqq cm\} \\ &\leqq P\{S_n \geqq m^{\frac{1}{2}}g(n/m) \text{ for some } n \geqq \tau m\} \\ &\leqq P\{S_n \geqq m^{\frac{1}{2}}g(n/m) \text{ for some } \tau m \leqq n \leqq cm\} \\ &\quad + P\{S_n \geqq m^{\frac{1}{2}}g(n/m) \text{ for some } n > cm\}. \end{aligned} \tag{29}$$

Part (i) follows from (29) and Lemma 4 and Lemma 5 below by first letting $m \to \infty$ and then letting $c \to \infty$. The proof of (ii) is similar and is omitted.

LEMMA 3. *For any* $0 < \tau < c < \infty$

$$P\{\max_{\tau \leqq t \leqq c}(W(t) - g(t)) = 0\} = 0.$$

PROOF. This result has been obtained by Ylvisaker [21]. It also follows from Theorem 7 of Doob [4] and the strong Markov property.

LEMMA 4. *For any* $0 < \tau < c < \infty$

$$\lim_{m \to \infty} P\{S_n \geqq m^{\frac{1}{2}} g(n/m) \text{ for some } \tau m \leqq n \leqq cm\}$$
$$= P\{W(t) \geqq g(t) \text{ for some } \tau \leqq t \leqq c\}.$$

PROOF. This result is easily deduced from Donsker's invariance principle, Lemma 3, and (for example) Theorem 4.1 of [1]. Alternatively, given Lemma 3, it may be proved in an elementary way by the method of Erdös and Kac [6].

Let $\psi(t) = t^{-\frac{1}{2}} g(t)$, and assume that ψ is ultimately non-decreasing and that (4) holds, or, what is more convenient, that

(30) $$\sum_1^\infty \frac{\psi(n)}{n} e^{-\frac{1}{2}\psi^2(n)} < \infty.$$

By passing to $\min(\psi(n), 2(\log_2 n)^{\frac{1}{2}})$, we may assume without loss of generality that

(31) $$\psi^2(n) \leqq 4 \log_2 n.$$

From the eventual monotonicity of ψ and (30) it follows that for all sufficiently large n

(32) $$\psi^2(n) \geqq 2 \log_2 n.$$

In fact, we have

$$\sum_{n^{1/2}}^{n} \frac{\psi(k)\exp(-\frac{1}{2}\psi^2(k))}{k} \geqq \frac{\psi(n)}{\log n} \sum_{n^{1/2}}^{n} 1/k \geqq \tfrac{1}{4}\psi(n) \to \infty$$

along any subsequence of integers n for which $\psi^2(n) < 2 \log_2 n$. It follows from (30) (31), and (32) that if ν_k denotes $\exp(k/\log k)$, then

(33) $$\sum_2^\infty \frac{1}{\psi(\nu_k)} \exp(-\tfrac{1}{2}\psi^2(\nu_k)) < \infty.$$

LEMMA 5. *Suppose that $\psi(t)$ is eventually non-decreasing as $t \to \infty$ and satisfies* (30) (*and hence by the preceding remarks* (31)–(33) *as well*). *Then*

$$\liminf_{c \to \infty} \limsup_{m \to \infty} P\{S_n \geqq n^{\frac{1}{2}} \psi(n/m) \text{ for some } n > cm\} = 0.$$

PROOF. We shall use the following notation:

$$n_k = \exp[k/\log(k+\log c)], \qquad \bar{n}_k = cmn_k \qquad (k = 0, 1, \cdots),$$

$K_1, K_2, \cdots$ numerical constants not depending on c nor m (provided c is sufficiently large),

$$U(x) = \max(1, \log x), \qquad U_2(x) = U(U(x)),$$
$$H(x) = P\{x_k \leqq x\}.$$

Let $\Psi = \{\psi^*: \psi^* \text{ is eventually non-decreasing and satisfies (30)–(33)}\}$, and define

$$a_n = a_n(m) = (nU_2(n/m))^{\frac{1}{2}}$$
$$x_n' = x_n I_{\{x_n \leqq a_n\}}, \qquad x_n'' = x_n - x_n',$$
$$S_n' = \textstyle\sum_1^n x_k', \qquad S_n'' = \sum_1^n x_k''.$$

Then

$$P\{S_n \geqq n^{\frac{1}{2}}\psi(n/m) \text{ for some } n > cm\}$$
$$\leqq P\left\{S_n' \geqq n^{\frac{1}{2}}\left(\psi(n/m) - \frac{1}{U_2^{\frac{1}{2}}(n/m)}\right) \text{ for some } n > cm\right\}$$
$$+ P\{S_n'' \geqq (n/U_2(n/m))^{\frac{1}{2}} \text{ for some } n > cm\}$$
$$= p_1 + p_2, \qquad \text{say.}$$

It will be shown in Lemma 6 below that $p_2 \to 0$ as $m \to \infty$. Hence to complete the proof it suffices to show that

(34) $$\liminf_{c\to\infty} \limsup_{m\to\infty} p_1 = 0.$$

Define $\psi_1 = \psi - 2/\psi$. It is easily verified that $\psi_1 \in \Psi$. Furthermore, by (31)

(35) $$p_1 \leqq \textstyle\sum_{k=0}^{\infty} P\{\max_{\bar{n}_k \leqq n < \bar{n}_{k+1}} S_n' \geqq \bar{n}_k^{\frac{1}{2}}\psi_1(\bar{n}_k/m)\}.$$

For each $n = 1, 2, \cdots$ define

$$x_i^{(n)} = x_i I_{\{|x_i| \leqq a_n\}} \qquad (i = 1, 2, \cdots n).$$

Then $x_1^{(n)}, \cdots, x_n^{(n)}$ are i.i.d., and since $x_i^{(n)} \geqq x_i'$ $(i = 1, 2, \cdots, n)$ we have

(36) $$P\{\max_{\bar{n}_k \leqq n < \bar{n}_{k+1}} S_n' \geqq \bar{n}_k^{\frac{1}{2}}\psi_1(cn_k)\} \leqq P\{\max_{\bar{n}_k \leqq n < \bar{n}_{k+1}} S_n^{(\bar{n}_{k+1})} \geqq \bar{n}_k^{\frac{1}{2}}\psi_1(cn_k)\}$$

for all $k = 0, 1, \cdots$. Since $n|Ex_1^{(n)}| \leqq n\int_{|x|>a_n}|x|\,dH \leqq (n/U_2(n/m))^{\frac{1}{2}}$, we may as above define a function $\psi_2 \in \Psi$ such that the right-hand side of (36) is majorized by

$$P\{\max_{\bar{n}_k \leqq n < \bar{n}_{k+1}} (S_n^{(\bar{n}_{k+1})} - ES_n^{(\bar{n}_{k+1})}) \geqq \bar{n}_k^{\frac{1}{2}}\psi_2(cn_k)\},$$

which by Lemma 7 below is in turn

(37) $$\leqq 2P\{S_{\bar{n}_{k+1}}^{(\bar{n}_{k+1})} - ES_{\bar{n}_{k+1}}^{(\bar{n}_{k+1})} \geqq \bar{n}_k^{\frac{1}{2}}(\psi_2(cn_k) - [2(n_{k+1}/n_k - 1)]^{\frac{1}{2}})\}.$$

It is easily verified that

$$\frac{n_{k+1}}{n_k}-1 \leqq \exp\left(\frac{1}{\log(k+\log c)}\right)-1 \leqq \frac{K_1}{\log(k+\log c)},$$

and hence by (31) that

$$\psi_2(cn_k)-[2(n_{k+1}/n_k-1)]^{\frac{1}{2}} \geqq \psi_2(cn_k)-K_2/\psi_2(cn_k).$$

Letting $\psi_3 = \psi_2 - K_2/\psi_2$, we see that $\psi_3 \in \Psi$, and from (35)–(37) we obtain

$$(38) \qquad p_1 \leqq \textstyle\sum_{k=0}^{\infty} P\{S_{\bar{n}_{k+1}}^{(\bar{n}_{k+1})} - ES_{\bar{n}_{k+1}}^{(\bar{n}_{k+1})} \geqq \bar{n}_k^{\frac{1}{2}}\psi_3(cn_k)\}.$$

Since $\operatorname{Var} x_1^{(n)} \leqq 1$, it follows from (31) and Lemma 8 below that

$$(39) \quad P\{S_{\bar{n}_{k+1}}^{(\bar{n}_{k+1})} - ES_{\bar{n}_{k+1}}^{(\bar{n}_{k+1})} \geqq \bar{n}_k^{\frac{1}{2}}\psi_3(cn_k)\}$$
$$\leqq 1-\Phi((n_k/n_{k+1})^{\frac{1}{2}}\psi_3(cn_k))+K_3\,\bar{n}_{k+1}^{-\frac{1}{2}}(U_2(cn_{k+1}))^{-\frac{3}{2}}\textstyle\int_{|x|\leqq a_{\bar{n}_{k+1}}}|x|^3\,dH.$$

By Lemma 9 below the series of which the second term on the right-hand side of (39) is the kth summand converges to 0 as $m \to \infty$. The series

$$\textstyle\sum_{k=0}^{\infty}(1-\Phi((n_k/n_{k+1})^{\frac{1}{2}}\psi_3(cn_k)))$$

does not depend on m, and by Lemma 10 below it converges to 0 as $c \to \infty$ through the values $\exp(i/\log i)$. This proves (34) and hence the lemma.

LEMMA 6.

$$\lim_{m\to\infty} P\{S_n'' \geqq (n/U_2(n/m))^{\frac{1}{2}} \text{ for some } n \geqq m\} = 0.$$

PROOF. By the Markov inequality

$$P\{S_n'' \geqq (n/U_2(n/m))^{\frac{1}{2}} \text{ for some } n \geqq m\} \leqq E\left[\max_{n\geqq m}\frac{S_n''}{(n/U_2(n/m))^{\frac{1}{2}}}\right]$$
$$\leqq E(S_m''/m^{\frac{1}{2}})+E(\textstyle\sum_{k=m+1}^{\infty} x_k''(U_2(k/m)/k)^{\frac{1}{2}}).$$

Now

$$\begin{aligned}\textstyle\sum_{k=m}^{\infty}(U_2(k/m)/k)^{\frac{1}{2}}Ex_k'' &= \textstyle\sum_{k=m}^{\infty}(U_2(k/m)/k)^{\frac{1}{2}}\sum_{i=k}^{\infty}\int_{a_i<x\leqq a_{i+1}}x\,dH\\ &= \textstyle\sum_{i=m}^{\infty}\sum_{k=m}^{i}(U_2(k/m)/k)^{\frac{1}{2}}\int_{a_i<x\leqq a_{i+1}}x\,dH\\ &\leqq K_4\textstyle\sum_{i=m}^{\infty}(iU_2(i/m))^{\frac{1}{2}}\int_{a_i<x\leqq a_{i+1}}x\,dH\\ &\leqq K_4\textstyle\int_{x>a_m}x^2\,dH \to 0 \quad \text{as} \quad m\to\infty.\end{aligned}$$

Also

$$\begin{aligned}m^{-\frac{1}{2}}ES_m'' &\leqq m^{-\frac{1}{2}}\textstyle\sum_{n=1}^{m}\sum_{k=n}^{\infty}\int_{a_k<x\leqq a_{k+1}}x\,dH\\ &\leqq m^{-\frac{1}{2}}\textstyle\sum_{k=1}^{\infty}\min(k,m)\int_{a_k<x\leqq a_{k+1}}x\,dH\\ &\leqq m^{-\frac{1}{2}}\textstyle\sum_{k=1}^{m}k\int_{a_k<x\leqq a_{k+1}}x\,dH+m^{\frac{1}{2}}\int_{x>a_{m+1}}x\,dH\\ &\leqq m^{-\frac{1}{2}}(\textstyle\sum_{k=1}^{[\varepsilon m]}+\sum_{k=[\varepsilon m]+1}^{m})k^{\frac{1}{2}}\int_{a_k<x\leqq a_{k+1}}x^2\,dH+\int_{x>a_{m+1}}x^2\,dH\\ &\leqq \varepsilon^{\frac{1}{2}}Ex_1^2+\textstyle\int_{x>a_{[\varepsilon m]}}x^2\,dH \to 0,\end{aligned}$$

as first $m\to\infty$, then $\varepsilon\to 0$. This completes the proof.

LEMMA 7. *Let $z_1, z_2, \cdots$ be independent random variables with $Ez_n = 0$, $Ez_n^2 \leqq \sigma^2$ $(n = 1, 2, \cdots)$. For any $a > 0$, $n = 1, 2, \cdots, m = 1, 2, \cdots, n-1$,*

$$P\{\max_{m \leqq k \leqq n} \sum_{i=1}^{k} z_i \geqq a\} \leqq 2P\{\sum_{i=1}^{n} z_i \geqq a - \sigma(2(n-m))^{\frac{1}{2}}\}.$$

PROOF. This result is well known when $m = 0$. The proof given, for example, by Lamperti ([12], page 45) works for general m as well.

The following result was proved by Nagaev [15].

LEMMA 8. *Let $z_1, z_2, \cdots$ be* i.i.d. *with $Ez_1 = 0$, $Ez_1^2 = \sigma^2$, $E|z_1|^3 = \beta < \infty$. There exists a universal constant L such that*

$$\left|P\left\{\sum_{1}^{n} z_i \leqq x\sigma n^{\frac{1}{2}}\right\} - \Phi(x)\right| \leqq \frac{L\beta}{n^{\frac{1}{2}}\sigma^3(1+|x|^3)}.$$

LEMMA 9.

$$\lim_{m\to\infty} \sum_{k=1}^{\infty} \bar{n}_k^{-\frac{1}{2}}(U_2(cn_k))^{-\frac{1}{2}} \int_{|x| \leqq a_{n_k}} |x|^3 \, dH(x) = 0.$$

PROOF. In the following proof $K_1', K_2', \cdots$ denote constants which may depend on the fixed value of c, but not on m. Let

$$Q(m) = \sum_{k=1}^{\infty} \bar{n}_k^{-\frac{1}{2}}(U_2(cn_k))^{-\frac{1}{2}} \int_{|x| \leqq a_{n_k}} |x|^3 \, dH(x).$$

Then

$$Q(m) \leqq m^{-\frac{1}{2}} \int_{0<|x|<\infty} |x|^3 \left(\sum_{k \geqq 1; cn_k U_2(cn_k) \geqq x^2/m} (cn_k)^{-\frac{1}{2}}(U_2(cn_k))^{-\frac{1}{2}}\right) dH(x). \tag{40}$$

Letting $g(z)$ denote $c \exp(z/\log(z+\log c))$, making the change of variable $u = g(z)$, and letting R denote the inverse of the function $x \to xU_2(x)$, we see that the series appearing in (40) is majorized by

$$K_1' \int_{u \geqq \max(c, R(x^2/m))} u^{-\frac{3}{2}}(U_2(u))^{-\frac{1}{2}} \, du.$$

Hence

$$\begin{aligned} Q(m) &\leqq K_1' m^{-\frac{1}{2}} \left(\int_{|x| \leqq K_2' m^{1/2}} |x|^3 \, dH + \int_{|x| > K_2' m^{1/2}} |x|^3 \left(\int_{R(x^2/m)}^{\infty} u^{-\frac{3}{2}}(U_2(u))^{-\frac{1}{2}} \, du\right) dH(x)\right) \\ &= K_1'(Q_1 + Q_2), \qquad \text{say.} \end{aligned}$$

Now

$$Q_1 \leqq \varepsilon + K_2' \int_{\varepsilon m^{1/2} < |x| \leqq K_2' m^{1/2}} |x|^2 \, dH(x) \to 0$$

as first $m \to \infty$, then $\varepsilon \to 0$. Also since $R(x) \sim x/U_2(x)$, for all $x \geqq K_2' m^{\frac{1}{2}}$ we have

$$\int_{R(x^2/m)}^{\infty} u^{-\frac{3}{2}}(U_2(u))^{-\frac{1}{2}} \, du \leqq K_4'(m^{\frac{1}{2}}/|x|),$$

and thus

$$Q_2 \leqq \int_{|x| > K_2' m^{1/2}} |x|^2 \, dH \to 0$$

as $m \to \infty$.

LEMMA 10. *If* $c_i = \exp(i/\log i)$, *then for any* $\psi \in \Psi$,

$$\lim_{i\to\infty} \sum_{k=0}^{\infty} (1-\Phi((n_k/n_{k+1})^{\frac{1}{2}}\psi(c_i n_k))) = 0.$$

PROOF. Since $1-\Phi(x) \leqq Kx^{-1}e^{-\frac{1}{2}x^2}$ $(x \geqq 1)$, it suffices to show that

$$\sum_{k=0}^{\infty} \frac{1}{\psi(c_i n_k)} \exp\left(-\tfrac{1}{2}\frac{n_k}{n_{k+1}}\psi^2(c_i n_k)\right) \to 0$$

as $i \to \infty$. It is easily seen from (31) that $(1-n_k/n_{k+1})\psi^2(cn_k)$ is bounded in c and k. Hence

$$\frac{1}{\psi(cn_k)} \exp\left(-\tfrac{1}{2}\frac{n_k}{n_{k+1}}\psi^2(cn_k)\right) \leqq \frac{K_7}{\psi(cn_k)} \exp(-\tfrac{1}{2}\psi^2(cn_k)).$$

Set $c_i = \exp(i/\log i)$ and note that for large i

$$c_i n_k = \exp\left(\frac{i}{\log i} + \frac{k}{\log(k+i/\log i)}\right) \geqq \exp\left(\frac{i+k}{\log(i+k)}\right),$$

and hence by the monotonicity of ψ and by (33)

$$\sum_{k=0}^{\infty} \frac{1}{\psi(c_i n_k)} \exp(-\tfrac{1}{2}\psi^2(c_i n_k)) \leqq K_8 \sum_{k=0}^{\infty} \frac{1}{\psi(n_{i+k})} \exp(-\tfrac{1}{2}\psi^2(n_{i+k})) \to 0 \text{ as } i \to \infty.$$

REMARK. The original law of the iterated logarithm for i.i.d. random variables with mean 0 and variance 1 states that

$$P\{\limsup_{n\to\infty} S_n/(2n \log_2 n)^{\frac{1}{2}} = 1\} = 1.$$

It was proved by Khintchin [10] in the Bernoulli case and by Hartman and Wintner [8], who relied heavily on the results of Kolmogorov [11], in the general case. The more difficult problem of deciding for an arbitrary ultimately non-decreasing function ψ whether

$$P\{S_n < n^{\frac{1}{2}}\psi(n) \text{ for all sufficiently large } n\} \tag{41}$$

is 0 or 1 was posed by P. Lévy and has been studied by several authors.

Erdös [5] proved that in the symmetric Bernoulli case the probability (41) is 1 or 0 according as

$$\int^{\infty} \psi/t \exp(-\tfrac{1}{2}\psi^2/2)\,dt \tag{42}$$

converges or diverges. In the case of a standard Wiener process Itô and McKean [9] give a simple proof that the convergence of (42) implies that

$$P\{W(t) < t^{\frac{1}{2}}\psi(t) \text{ for all sufficiently large } t\} \tag{41'}$$

equals 1 and, following Motoo [13], prove that (41′) equals 0 if (42) diverges. In the context of the first boundary value problem for the heat equation this result had been discovered earlier by Petrovski [17].

The relation between (41) and (42) for general sums of i.i.d. random variables is not so adequately treated in the literature. *It follows from Lemma* 5 *that if* (42) *converges the probability* (41) *is* 1. This conclusion is implicit in a paper by Feller [7], but we are unable to justify the steps in his argument.

It may be worth noting that if a function $g(t)$ satisfying some mild regularity conditions is such that (4) holds, then for any $\delta > 0$ we can find a finite measure F on $(0, \infty)$ and an $\varepsilon > 0$ such that $g(t) \geqq A(t, \varepsilon)$ for all sufficiently large t and

$$P\{W(t) \geqq A(t, \varepsilon) \text{ for some } t > 0\} = f(0, 0)/\varepsilon < \delta.$$

6. Other stochastic processes. The idea underlying Theorem 1 is applicable to stochastic processes other than the Wiener process. For example, let $R(t)$, $t \geqq 0$, denote the distance of 3-dimensional Brownian motion from the origin. It may be checked by direct calculation that

$$\frac{\sinh yR(t)}{yR(t)} \exp(-\tfrac{1}{2}y^2 t) \qquad (t > 0) \tag{43}$$

is a martingale for each fixed $y > 0$. Hence if we define

$$f_1(x, t) = x^{-1} \int_0^\infty y^{-1} \sinh xy \exp(-\tfrac{1}{2}y^2 t)\, dF(y),$$

where F is any measure on $(0, \infty)$ such that $f(x, 1) < \infty$ for all $x > 0$, then

$$f_1(R(t), t) \qquad (t \geqq 1)$$

is a martingale. As in Section 2, let $A_1(t, \varepsilon)$ denote the solution of

$$f_1(x, t) = \varepsilon \qquad (0 < \varepsilon,\ 1 \leqq t < \infty).$$

Since $y^{-1} \sinh y \leqq e^y$ for all $y \geqq 0$, an argument similar to the proof of Lemma 2 shows that $f_1(R(t), t) \to 0$ in probability as $t \to \infty$. Hence from Lemma 1 we obtain

THEOREM 3. *For any* $a > 0$ *and* $\varepsilon = f(a, 1)$

$$P\{R(t) \geqq A_1(t, \varepsilon) \text{ for some } t \geqq 1\} = 2(1-\Phi(a)) + f(a, 1)^{-1} \int_0^\infty [\Phi(a-y) - \Phi(a+y) + 2\Phi(y) - 1]\, dF(y).$$

For the measure F of Example 4 of Section 2 it may be shown by methods similar to those of Section 4 that as $t \to \infty$

$$A_1(t,\varepsilon) = [2t(\log_2 t + 5/2 \log_3 t + \textstyle\sum_4^n \log_k t + (1+\delta)\log_{n+1} t + \log 2\varepsilon/\pi^{\frac{1}{2}} + o(1))]^{\frac{1}{2}}. \tag{44}$$

(To see that this is the "right" result, compare (44) with equation (14) of [9], page 163.)

Again, let $X(t)$ be a one-sided stable process of index $\frac{1}{2}$, i.e., let $X(t)$ be a process having stationary independent increments and Laplace transform

$$Ee^{-\lambda X(t)} = \exp(-(2\lambda)^{\frac{1}{2}} t) \qquad (t \geqq 0,\ \lambda \geqq 0). \tag{45}$$

Without loss of generality we may assume that the sample paths of $X(t)$ are non-decreasing, right-continuous, and increase only by jumps (e.g. [2], page 317). It follows from (45) that for each $y > 0$

$$\{\exp(-\tfrac{1}{2}y^2X(t)+yt),\ t \geqq 0\}$$

is a martingale. Let

$$f_2(x, t) = \int_0^\infty \exp(-\tfrac{1}{2}y^2x+yt)\,dF(y)$$

for any measure F on $(0, \infty)$ such that $f_2(0, 1) < \infty$. Then for any $x \geqq 0$, $f_2(x+X(t), t)$ $(t \geqq 1)$ is a martingale, and by the sample path properties of $X(t)$ and the dominated convergence theorem it is easy to see that the sample paths of $f(x+X(t), t)$ $(t \geqq 1)$ have the property (*) (see Remark (d) at the end of Section 3).

From (45) it follows that for each $t > 0$, $t^{-2}X(t)$ and $X(1)$ have the same distribution. Moreover, for any $x \geqq 0$ and $\delta > 0$

$$f_2(x+\delta t^2, 1+t) = (\varphi(\delta^{-\frac{1}{2}}))^{-1}\int_0^\infty \varphi(\delta^{\frac{1}{2}}yt-\delta^{-\frac{1}{2}})\exp(-\tfrac{1}{2}xy^2+y)\,dF(y) \to 0$$

as $t \to \infty$ by the dominated convergence theorem, and it follows by the argument of Lemma 2 that $f_2(x+X(t), t) \to 0$ in probability as $t \to \infty$. Letting $A_2(t, \varepsilon)$ denote the solution of the equation $f_2(x, t) = \varepsilon$, we obtain

THEOREM 4. *For any $a > 0$ and $\varepsilon = f_2(a, 1)$*

$$P\{X(t) \leqq A_2(t, \varepsilon) \text{ for some } t \geqq 1\}$$

$$= 2\left(1-\Phi\left(\frac{1}{a^{\frac{1}{2}}}\right)\right) + \frac{\varphi\left(\frac{1}{a^{\frac{1}{2}}}\right)\int_0^\infty\left[\Phi\left(ya^{\frac{1}{2}}+\frac{1}{a^{\frac{1}{2}}}\right)+e^{-2y}\Phi\left(ya^{\frac{1}{2}}-\frac{1}{a^{\frac{1}{2}}}\right)\right]dF(y)}{\int_0^\infty \varphi\left(ya^{\frac{1}{2}}-\frac{1}{a^{\frac{1}{2}}}\right)dF(y)}.$$

For the measure F of Example 4 of Section 2, since $f_2(x, t) = f(t, x)$, we obtain by inversion of (10) that

$$A_2(t, a) = t^2/2(\log_2 t + 3/2\log_3 t + \textstyle\sum_{k=4}^n \log_k t + (1+\delta)\log_{n+1} t + \log\varepsilon/\pi^{\frac{1}{2}} + o(1))$$

as $t \to \infty$.

7. Final remarks. The examples of Section 6 were chosen for computational simplicity. A closer look at them suggests many questions which will be treated in a subsequent publication. (a) What is the origin of the martingale (43), and how can analogous martingales be found for other diffusion processes? (b) Since $R(t) \to \infty$ with probability one as $t \to \infty$, there exist functions $g(t)$ which are $o(t^{\frac{1}{2}})$ such that

$$P\{R(t) \leqq g(t) \text{ for some } t \geqq 1\} < 1.$$

Does our method permit us to calculate these probabilities exactly?

(c) For the Wiener process itself, if for some $0 < \alpha \leqq \beta < \infty$, F attributes positive measure to the interval $[\alpha, \beta]$, then

$$f(x, t) \geqq \exp(x\alpha - \beta^2 t/2)F[\alpha, \beta],$$

so $f(g(t), t) \to \infty$ if $g(t)/t \to \infty$. Hence the method of Theorem 1 can only generate boundaries $g(t)$ which are $O(t)$. What is the class of martingales suitable for computing $P\{W(t) \geqq g(t)$ for some $t \geqq 1\}$ for *arbitrary continuous functions* g, and what is the relation of the class of martingales we have obtained to this much larger class? This question is closely connected with the study of certain boundary value problems for partial differential equations involving the generator $\partial/\partial t + \frac{1}{2}\partial^2/\partial x^2$ of the space-time Wiener process. We shall briefly indicate the nature of the connection.

Suppose for simplicity that $f(b, 0) < \infty$ for all b. Equation (3) implies that for any $h \geqq 0$, $-\infty < b < \infty$, and any $\varepsilon > f(b, h)$,

$$P\{W(t) \geqq A(t, \varepsilon) \text{ for some } t \geqq h \mid W(h) = b\} = f(b, h)/\varepsilon. \tag{46}$$

If we let $P(b, h)$ denote the left-hand side of (46) we have

$$\begin{aligned} P(b, h) &= f(b, h)/\varepsilon \qquad (h \geqq 0,\ b < A(h, \varepsilon)) \\ &= 1 \qquad\qquad\quad (h \geqq 0,\ b \geqq A(h, \varepsilon)). \end{aligned} \tag{47}$$

It follows from (47) that

$$\frac{\partial P}{\partial h} + \frac{1}{2}\frac{\partial^2 P}{\partial b^2} = 0 \qquad (h \geqq 0,\ b < g(h)), \tag{48}$$

where we have put $g(t) = A(t, \varepsilon)$.

Now let $g(t)$ be any positive, continuous, and increasing function of $t \geqq 0$, not necessarily of the form $A(t, \varepsilon)$. Then the left-hand side of (46) still defines a function $P(b, h)$ for $h \geqq 0$ and $-\infty < b < \infty$ which is 1 for $b \geqq g(h)$, and we may ask whether (48) continues to hold. Conversely, if P is any function defined for $h \geqq 0$ and $b \leqq g(h)$ which satisfies (48) and is 1 for $b = g(h)$, we may ask whether it is necessarily equal to the left-hand side of (46).

We are indebted to F. J. Anscombe and H. P. McKean, Jr. for several valuable suggestions.

REFERENCES

[1] Billingsley, P. (1968). *Weak Convergence of Probability Measures*. Wiley, New York.
[2] Breiman, L. (1968). *Probability*. Addison–Wesley, Reading.
[3] Darling, D. and Robbins, H. (1967). Iterated logarithm inequalities. *Proc. Nat. Acad. Sci. USA* **57** 1188–1192.
[4] Doob, J. L. (1955). A probabilistic approach to the heat equation. *Trans. Amer. Math. Soc.* **80** 216–280.
[5] Erdös, P. (1942). On the law of the iterated logarithm. *Ann. of Math.* **43** 419–436.
[6] Erdös, P. and Kac, M. (1946). On certain limit theorems of the theory of probability. *Bull. Amer. Math. Soc.* **52** 292–302.
[7] Feller, W. (1946). The law of the iterated logarithm for identically distributed random variables. *Ann. of Math.* **47** 631–638.
[8] Hartman, P. and Wintner, A. (1941). On the law of the iterated logarithm. *Amer. J. Math.* **63** 169–176.
[9] Itô, K. and McKean, H. P., Jr. (1965). *Diffusion Processes and Their Sample Paths*. Springer–Verlag, Berlin.

[10] KHINTCHIN, A. (1924). Über einen Satz der Wahrscheinlichkeitsrechnung. *Fund. Math.* **6** 9–20.

[11] KOLMOGOROV, A. (1929). Über das Gesetz des Iterierten Logarithmus. *Math. Ann.* **101** 126–135.

[12] LAMPERTI, J. (1966). *Probability*. W. A. Benjamin, Inc. New York.

[13] MOTOO, M. (1959). Proof of the law of the iterated logarithm through diffusion equation. *Ann. Inst. Statist. Math.* **10** 21–28.

[14] MÜLLER, D. W. (1968). Verteilungs Invarianzprinzipien für das Gesetz der grossen Zahlen. *Z. Wahrscheinlichkeitstheorie und Verw. Gebiete* **10** 173–192.

[15] NAGAEV, S. V. (1965). Some limit theorems for large deviations. *Theor. Probability Appl.* **10** 214–235.

[16] NEVEU, J. (1966). *Mathematical Foundations of Probability Theory*. Holden–Day, San Francisco.

[17] PETROVSKI, I. (1935). Zur ersten Randwertaufgabe der Wärmeleitungsgleichung. *Compositio Math.* **1** 383–419.

[18] ROBBINS, H. and SIEGMUND, D. (1968). Iterated logarithm inequalities and related statistical procedures. *Mathematics of the Decision Sciences*, **2** American Mathematical Society Providence, 267–279.

[19] ROBBINS, H. and SIEGMUND, D. (1969). Probability distributions related to the law of the iterated logarithm, *Proc. Nat. Acad. Sci. USA* **62** 11–13.

[20] ROBBINS, H., SIEGMUND, D., and WENDEL, J. (1968). The limiting distribution of the last time $S_n \geqq n\varepsilon$. *Proc. Nat. Acad. Sci. USA*. **61** 1228–1230.

[21] VILLE, J. (1939). Étude critique de la Notion de Collectif. Gauthier–Villars, Paris.

[22] YLVISAKER, D. (1968). A note on the absence of tangencies in Gaussian sample paths. *Ann. Math. Statist.* **39** 261–262.

A CLASS OF STOPPING RULES FOR TESTING PARAMETRIC HYPOTHESES

HERBERT ROBBINS and DAVID SIEGMUND
COLUMBIA UNIVERSITY and HEBREW UNIVERSITY

Let $f_\theta(x)$, $\theta \in \Omega$, be a one parameter family of probability densities with respect to some σ-finite measure μ on the Borel sets of the line. Denote by P_θ the probability measure under which random variables $x_1, x_2, \cdots$ are independent with the common probability density $f_\theta(x)$. Let θ_0 be an arbitrary fixed element of Ω and ε any constant between 0 and 1. We are interested in finding stopping rules N for the sequence $x_1, x_2, \cdots$ such that

$$P_\theta(N < \infty) \leqq \varepsilon \quad \text{for every } \theta \leqq \theta_0, \tag{1}$$

and

$$P_\theta(N < \infty) = 1 \quad \text{for every } \theta > \theta_0. \tag{2}$$

Among such rules, we wish to find those which in some sense minimize $E_\theta(N)$ for all $\theta > \theta_0$.

A method of constructing rules which satisfy (1) and (2) by using mixtures of likelihood ratios was given in [3]. Here we sketch an alternative method.

Let $\theta_{n+1} = \theta_{n+1}(x_1, \cdots, x_n)$ for $n = 0, 1, 2, \cdots$, be any sequence of Borel measurable functions of the indicated variables such that

$$\theta_{n+1} \geqq \theta_0. \tag{3}$$

In particular, θ_1 is some constant $\geqq \theta_0$. Define

$$z_n = \prod_1^n \frac{f_{\theta_i}(x_i)}{f_{\theta_0}(x_i)}, \qquad n = 1, 2, \cdots, \tag{4}$$

and for any constant $b > 0$, let

$$N = \begin{cases} \text{first } n \geqq 1 \text{ such that } z_n \geqq b, \\ \infty \text{ if no such } n \text{ occurs.} \end{cases} \tag{5}$$

We shall show that under a certain very general assumption on the structure of the family $f_\theta(x)$, the inequality (1) holds at least for all $b \geqq 1/\varepsilon$.

ASSUMPTION. *For every triple* $\alpha \leqq \gamma \leqq \beta$ *in* Ω,

$$\int \frac{f_\alpha(x) f_\beta(x)}{f_\gamma(x)}\, d\mu(x) \leqq 1. \tag{6}$$

Research supported by Public Health Service Grant No. 1-R01-GM-16895-03.

37

We remark without proof that this holds for the general one parameter Koopman-Darmois-Pitman exponential family and many others.

Denote by $\mathfrak{F}_n$ the Borel field generated by $x_1, \cdots, x_n$. *Then for each fixed* $\theta \leqq \theta_0$, $\{z_n, \mathfrak{F}_n, P_\theta; n \geqq 1\}$ *is a nonnegative supermartingale sequence.* For, given any $n \geqq 1$,

$$(7)\qquad E_\theta(z_{n+1}|\mathfrak{F}_n) = z_n E_\theta\left(\frac{f_{\theta_{n+1}}(x_{n+1})}{f_{\theta_0}(x_{n+1})}\middle|\mathfrak{F}_n\right) = z_n \int \frac{f_\theta(x) f_{\theta_{n+1}}(x)}{f_{\theta_0}(x)}\, d\mu(x) \leqq z_n,$$

since by hypothesis $\theta \leqq \theta_0 \leqq \theta_{n+1}$. We can therefore apply the following.

LEMMA. *Let* $\{z_n, \mathfrak{F}_n, P; n \geqq 1\}$ *be any nonnegative supermartingale. Then for any constant* $b > 0$,

$$(8)\qquad P(z_n \geqq b \textit{ for some } n \geqq 1) \leqq P(z_1 \geqq b) + \frac{1}{b}\int_{(z_1 < b)} z_2\, dP \leqq \frac{E(z_1)}{b}.$$

PROOF. Defining N by (5), we have

$$(9)\qquad P(z_n \geqq b \text{ for some } n \geqq 1) = P(z_1 \geqq b) + P(1 < N < \infty).$$

Since z_n is a nonnegative supermartingale,

$$(10)\qquad \int_{(N>1)} z_1\, dP \geqq \int_{(N>1)} z_2\, dP = \int_{(N=2)} z_2\, dP + \int_{(N>2)} z_2\, dP \geqq \cdots$$
$$\geqq \sum_{i=2}^{n} \int_{(N=i)} z_i\, dP + \int_{(N>n)} z_n\, dP \geqq bP(1 < N \leqq n) + 0,$$

because $z_i \geqq b$ on $(N = i)$ and $z_n \geqq 0$. Since n is arbitrary,

$$(11)\qquad P(1 < N < \infty) \leqq \frac{1}{b}\int_{(z_1 < b)} z_2\, dP,$$

and hence from (9)

$$(12)\qquad P(z_n \geqq b \text{ for some } n \geqq 1) \leqq P(z_1 \geqq b) + \frac{1}{b}\int_{(z_1 < b)} z_2\, dP$$
$$\leqq \frac{1}{b}\int_{(z_1 \geqq b)} z_1\, dP + \frac{1}{b}\int_{(z_1 < b)} z_1\, dP = \frac{E(z_1)}{b},$$

which proves (8).

Applying this lemma to (4) and (5), we see that for each fixed $\theta \leqq \theta_0$,

$$(13)\qquad P_\theta(N < \infty) \leqq P_\theta(z_1 \geqq b) + \frac{1}{b}\int_{(z_1 < b)} z_2\, dP_\theta$$
$$\leqq \frac{E_\theta(z_1)}{b} = \frac{1}{b}\int \frac{f_\theta(x) f_{\theta_1}(x)}{f_{\theta_0}(x)}\, d\mu(x) \leqq \frac{1}{b},$$

and hence, as claimed above, (1) holds at least for $b \geqq 1/\varepsilon$.

As an example, suppose that under P_θ the x are $N(\theta, 1)$, so that $f_\theta(x) = \varphi(x - \theta)$, where $\varphi(x)$ is the standard normal density, and that $\theta_0 = 0$. It is easily seen that if $\theta_1 > 0$ then

$$z_n = \prod_1^n \exp\left\{\theta_i x_i - \frac{\theta_i^2}{2}\right\}, \quad E_\theta(z_1) = \exp\{\theta\theta_1\},$$

(14)

$$P_\theta(z_1 \geqq b) = \Phi\left(\theta - \frac{\log b}{\theta_1} - \frac{\theta_1}{2}\right),$$

(15)
$$\int_{(z_1<b)} z_2\, dP_\theta = \int_{-\infty}^{\log b/\theta_1+\theta_1/2} \int_{-\infty}^{\infty} z_2 \varphi(x_2-\theta)\varphi(x_1-\theta)\, dx_2\, dx_1$$
$$\leqq \exp\{\theta\theta_1\}\, \Phi\left(\frac{\log b}{\theta_1} - \frac{\theta_1}{2} - \theta\right),$$

where $\Phi(x) = \int_{-\infty}^{x} \varphi(t)\, dt$. Hence, (13) gives for any $\theta \leqq 0$, the inequality

(16)
$$P_\theta\left(\prod_1^n \exp\left\{\theta_i x_i - \frac{\theta_i^2}{2}\right\} \geqq b \text{ for some } n \geqq 1\right)$$
$$\leqq \Phi\left(\theta - \frac{\log b}{\theta_1} - \frac{\theta_1}{2}\right) + \frac{1}{b}\exp\{\theta\theta_1\}\,\Phi\left(\frac{\log b}{\theta_1} - \frac{\theta_1}{2} - \theta\right)$$
$$\leqq \frac{1}{b}\exp\{\theta\theta_1\}.$$

The middle term of (16) is increasing in θ, so

(17)
$$P_\theta\left(\prod_1^n \exp\left\{\theta_i x_i - \frac{\theta_i^2}{2}\right\} \geqq b \text{ for some } n \geqq 1\right)$$
$$\leqq \Phi\left(-\frac{\log b}{\theta_1} - \frac{\theta_1}{2}\right) + \frac{1}{b}\Phi\left(\frac{\log b}{\theta_1} - \frac{\theta_1}{2}\right) \leqq \frac{1}{b}$$

for every $\theta \leqq 0$.

We shall now suppose that in addition to the requirement that $\theta_{n+1} = \theta_{n+1}(x_1, \cdots, x_n) \geqq 0$, *the sequence θ_n converges to θ with probability 1 under P_θ for each $\theta > 0$*. For example, both

(18)
$$\theta_{n+1} = \frac{\max(0, s_n)}{n}$$

and

(19)
$$\theta_{n+1} = \frac{s_n}{n} + \frac{\varphi(s_n/\sqrt{n})}{\sqrt{n}\,\Phi(s_n/\sqrt{n})},$$

where $s_n = x_1 + \cdots + x_n$, have this desired property (equation (19) is the posterior expected value of θ given $x_1, \cdots, x_n$ when the prior distribution of θ is flat for $\theta > 0$). Thus, for large n,

(20)
$$z_n = \prod_1^n \exp\left\{\theta_i x_i - \frac{\theta_i^2}{2}\right\} \approx \prod_1^n \exp\left\{\theta x_i - \frac{\theta^2}{2}\right\} = \exp\left\{\theta s_n - \frac{n\theta^2}{2}\right\} = z_n(\theta),$$

say. Now it has been remarked elsewhere [2], and a proof based on [1], pp. 107–108, is easily given, that for any fixed $\theta > 0$,

(21)
$$N_{\theta,b} = \begin{cases} \text{first } n \geqq 1 \text{ such that } z_n(\theta) \geqq b, \\ \infty \text{ if no such } n \text{ occurs,} \end{cases}$$

is optimal in the sense that if T is any stopping rule of $x_1, x_2, \cdots$ such that

$$P_0(T < \infty) \leqq P_0(N_{\theta,b} < \infty), \tag{22}$$

then $E_\theta(N_{\theta,b}) < \infty$ and $E_\theta(T) \geqq E_\theta(N_{\theta,b})$. Thus, the N using (18) or (19) may be expected to be "almost optimal" simultaneously for all values $\theta > 0$. Monte Carlo methods will be needed to get accurate estimates of $P_0(N < \infty)$ and $E_\theta(N)$ for $\theta > 0$. We have, however, been able to find the asymptotic nature of $E_\theta(N)$ as $\theta \to 0$ or $b \to \infty$ in the normal and other cases for various choices of the θ_n sequence, and the results will be published elsewhere. For example, using (18), we can show that, for $\theta > 0$,

$$E_\theta(N) \sim P_0(N = \infty)\left(\log \frac{1}{\theta} \Big/ \theta^2\right) \qquad \text{as } \theta \to 0, \tag{23}$$

and

$$E_\theta(N) = \frac{2 \log b + \log_2 b}{\theta^2} + o(\log_2 b) \qquad \text{as } b \to \infty. \tag{24}$$

By putting

$$\theta_{n+1} = \begin{cases} \dfrac{s_n}{n} & \text{if } s_n \geqq [n(2 \log_2^+ n + 3 \log_3^+ n)]^{1/2}, \\ 0 & \text{otherwise,} \end{cases} \tag{25}$$

where $\log_2 n = \log(\log n)$, and so on, equation (23) is replaced by

$$E_\theta(N) \sim 2P_0(N = \infty) \log_2 \frac{1}{\theta} \Big/ \theta^2 \qquad \text{as } \theta \to 0, \tag{26}$$

which is optimal for $\theta \to 0$.

In evaluating $P_\theta(N < \infty)$ for $\theta \leqq 0$ with an arbitrary sequence $\theta_{n+1} = \theta_{n+1}(x_1, \cdots, x_n) \geqq 0$, $n = 0, 1, 2, \cdots$, and $b > 1$, we see that this probability is equal to

$$P_\theta\left(\prod_1^n \exp\left\{\theta_i x_i - \frac{\theta_i^2}{2}\right\} \geqq b \text{ for some } n \geqq 1\right) = \sum_{n=1}^\infty \int_{(N=n)} \exp\left\{\theta s_n - \frac{n\theta^2}{2}\right\} dP_0. \tag{27}$$

For any fixed x and n the function $f(\theta) = \exp\{\theta x - n\theta^2/2\}$ is increasing for $-\infty < \theta < x/n$. Hence if the condition

$$s_n > 0 \quad \text{whenever} \quad N = n, \qquad n = 1, 2, \cdots, \tag{28}$$

is satisfied, then $P_\theta(N < \infty)$ will be an increasing function of $\theta \leqq 0$ (as is the middle term of (16)). Recalling that

$$N = \begin{cases} \text{first } n \geqq 1 \text{ such that } \sum_1^n \left(\theta_i x_i - \dfrac{\theta_i^2}{2}\right) \geqq \log b, \\ \infty \text{ if no such } n \text{ occurs,} \end{cases} \tag{29}$$

we see that if $N = 1$, then $\theta_1 x_1 \geqq \log b + \theta_1^2/2$ so $s_1 = x_1 > 0$, while if $N = n > 1$, then

$$\sum_1^{n-1} \theta_i x_i < \log b + \frac{1}{2} \sum_1^{n-1} \theta_i^2,$$
(30)
$$\sum_1^{n} \theta_i x_i \geqq \log b + \frac{1}{2} \sum_1^{n} \theta_i^2,$$

so $\theta_n x_n > 0$, and hence $\theta_n > 0$ and $x_n > 0$. In cases (18) and (25), it follows that $s_{n-1} \geqq 0$, and hence $s_n = s_{n-1} + x_n > 0$. Thus, $P_\theta(N < \infty)$ is an increasing function of $\theta \leqq 0$ in these cases. Whether this is true for the choice (19) we do not know. Likewise, we do not know whether $P_\theta(N \leqq n)$ is an increasing function of θ for each fixed $n = 1, 2, \cdots$, even for (18) or (25). For $\theta > 0$, $P_\theta(N < \infty) = 1$ and $E_\theta(N) < \infty$ in all three cases.

In the case of a general parametric family $f_\theta(x)$, we can try to make $E_\theta(N)$ small for $\theta > \theta_0$ by choosing θ_n to converge properly to θ under P_θ for $\theta > \theta_0$, but a comparison with the methods of [3] remains to be made. The present method of sequentially estimating the true value of θ when it is $>\theta_0$ appears somewhat more natural in statistical problems.

If we do not wish to take advantage of the property (6), we can use, instead of (4),

(31)
$$z_n' = \prod_1^n \frac{f_{\theta_i}(x_i)}{h_n},$$

where $h_n = h_n(x_1, \cdots, x_n) = \sup_{\theta \leqq \theta_0} \{\prod_1^n f_\theta(x_i)\}$. The use of (31) has been independently suggested by Edward Paulson. For $\theta \leqq \theta_0$, we then have

(32)
$$P_\theta(z_n' \geqq b \text{ for some } n \geqq 1) \leqq P_\theta\left(\prod_1^n \frac{f_{\theta_i}(x_i)}{f_\theta(x_i)} \geqq b \text{ for some } n \geqq 1\right) \leqq \frac{1}{b},$$

by the lemma above. It would seem, however, that (31) should be less efficient than (4) when the assumption (6) holds.

REFERENCES

[1] Y. S. Chow, H. Robbins, and D. Siegmund, *Great Expectations: The Theory of Optimal Stopping*, Boston, Houghton-Mifflin, 1971.

[2] D. A. Darling and H. Robbins, "Some further remarks on inequalities for sample sums," *Proc. Nat. Acad. Sci.*, Vol. 60 (1968), pp. 1175–1182 (see p. 1181).

[3] H. Robbins, "Statistical methods related to the law of the iterated logarithm," *Ann. Math. Statist.*, Vol. 41 (1970), pp. 1397–1409.

Reprinted from

Proc. Sixth Berkeley Symposium Math. Statist. Prob.

4, 37–41 (1972)

BULLETIN OF THE
INSTITUTE OF MATHEMATICS
ACADEMIA SINICA
Volume 1, Number 1, June 1973

STATISTICAL TESTS OF POWER ONE AND THE INTEGRAL REPRESENTATION OF SOLUTIONS OF CERTAIN PARTIAL DIFFERENTIAL EQUATIONS

BY

H. ROBBINS AND D. SIEGMUND(1)

Abstract. Replacing a sequence of observations by a Wiener process with drift, we discuss the corresponding results for the theory of tests of power one, in particular with respect to the representation theorem for martingale functions and Widder's theorem for the heat equation.

Introduction. Let Ω be a set of real numbers and let P_θ be a probability measure under which, for each $\theta \in \Omega$, the random variables $x_1, x_2, \cdots$ are independent with a common distribution. Suppose $\theta_0 \in \Omega$ and $0 < \alpha < 1$. A size $0 < \alpha < 1$ test of power one of the statistical hypothesis $\theta \leq \theta_0$ against the alternative $\theta > \theta_0$ is defined to be a stopping rule T for the sequence $x_1, x_2, \cdots$ such that

$$P_\theta\{T < \infty\} \leq \alpha \quad \text{for every} \quad \theta \leq \theta_0 \tag{1}$$

and

$$P_\theta\{T < \infty\} = 1 \quad \text{for every} \quad \theta > \theta_0 . \tag{2}$$

Among such rules we wish to find one which in some sense minimizes $E_\theta T$ for all $\theta > \theta_0$.

The theory of tests of power one has been developed in [1], [8], [9], [10] and [11]. In Chapter I of this paper we review some of these results in their most attractive mathematical setting, in which the observations $x_1, x_2, \cdots$ are replaced by a Wiener process $X(t)$, $0 \leq t < \infty$, with drift θ (per unit time). Chapter II is devoted to exploring in more detail certain mathematical questions suggested by the results of Chapter I.

Received by the editors December 23, 1972.

(1) Communicated by Y. S. Chow.

CHAPTER I

TESTS OF POWER ONE FOR THE DRIFT OF A WIENER PROCESS

1. **Simple hypothesis vs. simple alternative.** Let $X(t)$, $0 \le t < \infty$, be a Wiener process with drift θ per unit time. To obtain a test of power one of the hypothesis $\theta \le 0$ against the alternative $\theta > 0$, assume initially that Ω, the set of possible θ values, consists of only two points, say 0 and some given value $\mu > 0$. Then a complete solution to our problem is given by the following theorem.

THEOREM 1. *Let* $b > 1$ *and define*

$$T = \inf\{t : \exp(\mu X(t) - \mu^2 t/2) \ge b\}. \tag{3}$$

Then, denoting by P_0 *the probability measure when the drift is* 0,

$$P_0\{T < \infty\} = b^{-1}, \tag{4}$$

and for any stopping rule T' *such that* $P_0\{T' < \infty\} \le P_0\{T < \infty\}$, *we have, denoting by* E_μ *the expectation when the drift is* μ,

$$E_\mu T = (2 \log b)/\mu^2 \le E_\mu T'. \tag{5}$$

To prove Theorem 1 we shall use the following notation. Let $\mathfrak{F}(t)$, $0 < t < \infty$, denote an increasing family of sub-σ-algebras of a σ-algebra $\mathfrak{F}$ of subsets of a space $\mathfrak{X}$. Assume that P and Q are two probabilities on $\mathfrak{F}$ such that the restrictions $P^{(t)}$ and $Q^{(t)}$ of P and Q to the σ-algebras $\mathfrak{F}(t)$ are mutually absolutely continuous. Let $z(t) = dQ^{(t)}/dP^{(t)}$ denote the Radon-Nikodym derivatives of these restrictions $(z^{-1}(t) = dP^{(t)}/dQ^{(t)})$.

LEMMA 1. *If* $\mathfrak{F}(t)$ *and the sample paths of* $z(t)$ *are right-continuous with probability one, then for any* $\tau \ge 0$ *and stopping rule* $T \ge \tau$

$$P\{T < \infty\} = P\{T = \tau\} + \int_{\{\tau < T < \infty\}} z^{-1}(\tau)\, dQ. \tag{6}$$

If, in addition, the sample paths of $z(t)$ *are continuous, if the stopping rule* T *is defined by*

$$T = \inf\{t : t \ge \tau,\ z(t) \ge b\},$$

and if $Q\{T < \infty\} = 1$, *then*

(7) $$P\{T<\infty\} = P\{z(\tau) \geq b\} + b^{-1} \int_{\{z(\tau)<b\}} z(\tau)\, dP.$$

Proof. The σ-algebra $\mathfrak{F}(T)$ is defined to be the class of all sets $A \in \mathfrak{F}$ such that $A\{T \leq t\} \in \mathfrak{F}(t)$ for all t. Then (6) follows at once from the obvious equation $P\{T<\infty\} = P\{T=\tau\} + P\{\tau < T < \infty\}$ and the fact that $z^{-1}(T)$ is the Radon-Nikodym derivative of the restriction of P to the trace of $\mathfrak{F}(T)$ on $\{T<\infty\}$ with respect to the corresponding restriction of Q. To see this, observe that if $A \in \mathfrak{F}(T)$, then

(8) $$\int_{A\{T<\infty\}} z^{-1}(T)\, dQ = \lim_{t\to\infty} \int_{A\{T\leq t\}} z^{-1}(T)\, dQ\,.$$

But $\{z^{-1}(t), \mathfrak{F}(t), 0 \leq t < \infty\}$ is a martingale under the measure Q and hence (cf. [6, p. 534])

$$\int_{A\{T\leq t\}} z^{-1}(T)\, dQ = \int_{A\{T\leq t\}} z^{-1}(t)\, dQ = P(A\{T \leq t\})\,,$$

which together with (8) completes the proof of (6).

The additional hypotheses of the second part of the lemma justify the following equalities:

$$\int_{\{\tau<T<\infty\}} z^{-1}(T)\, dQ = b^{-1} Q\{\tau < T < \infty\} = b^{-1} Q\{T > \tau\}$$
$$= b^{-1} \int_{\{T>\tau\}} z(\tau)\, dP\,.$$

This proves (7).

To prove Theorem 1, we apply Lemma 1 with $P = P_0$, $Q = P_\mu$, and $\mathfrak{F}(t) = S(X(s), s \leq t)$, so that $z(t) = \exp(\mu X(t) - \mu^2 t/2)$. Since by the strong law of large numbers $P_\mu\{X(t)/t \to \mu\} = 1$, it follows that $P_\mu\{T<\infty\} = 1$, and hence (4) follows at once from (7) with $\tau = 0$.

To evaluate $E_\mu T$, note that by Wald's lemma, for any $t > 0$,

(9) $$E_\mu X(T \wedge t) = \mu E_\mu (T \wedge t)\,.$$

Hence, since $X(T\wedge t) \leq (\log b)/\mu + (\mu/2)(T \wedge t)$ by the definition of T, it follows that

(10) $$\mu E_\mu(T\wedge t) \leq (\log b)/\mu + (\mu/2) E_\mu (T \wedge t)\,.$$

Letting $t \to \infty$ in (10) shows that $E_\mu T < \infty$; hence using Wald's lemma again, we may replace $T \wedge t$ by T in (9). Then in place of

(10) we have by (3)

$$\mu E_\mu T = (\log b)/\mu + (\mu/2) E_\mu T,$$

from which the first part of (5) follows.

To prove the second part of (5), we apply (6) to the stopping rule T' (setting $t=0$). Since $\mathfrak{F}(0)$ is the trivial σ-algebra, $P_0(T'=0)=0$ and hence (6) becomes

$$P_0(T' < \infty) = \int_{\{T' < \infty\}} \exp(-\mu X(T') + \mu^2 T'/2)\, dP_\mu. \tag{11}$$

Now we may assume $E_\mu T' < \infty$ (and hence $P_\mu(T' < \infty) = 1$), for otherwise the second part of (5) is trivially satisfied. Then by (11), Jensen's inequality, and Wald's lemma

$$\begin{aligned} P_0(T' < \infty) &\geq \exp(-\mu E_\mu X(T') + (\mu^2/2) E_\mu T') \\ &= \exp(-(\mu^2/2) E_\mu T'), \end{aligned}$$

so

$$E_\mu T' \geq -2 \log P_0(T' < \infty)/\mu^2. \tag{12}$$

Now by hypothesis and (4) we have $P_0(T' < \infty) \leq b^{-1}$, which with (12) completes the proof.

The main ideas of this section can be found in Wald's study [12] of the sequential probability ratio test, which is closely related to the stopping rule of Theorem 1.

2. Computing $P_0(T < \infty)$ for a class of stopping rules.

The stopping rule T of Theorem 1 does not solve the original problem of testing $\theta \leq 0$ against $\theta > 0$ when $\Omega = (-\infty, \infty)$, since it depends on which possible positive value of θ is the correct one: if we choose an arbitrary value $\mu > 0$ and use the rule which is optimal for this μ, then it is easy to see that $P_\theta\{T < \infty\} < 1$ for $0 < \theta < \mu/2$, so that (2) is not satisfied.

In this section we study a more general class of stopping rules in which the measure $Q = P_\mu$ is replaced by a mixture $Q(\cdot) = \int_0^\infty P_y(\cdot)\, dF(y)$. (For a different approach, which depends on sequentially estimating the value of θ, see [10] and [11].)

Let F denote any measure on $(0, \infty)$ which is finite on bounded intervals, and define for $-\infty < t < \infty$, $-\infty < x < \infty$, and $0 < b < \infty$

$$0 < f(t, x) = \int_0^\infty \exp (yx - y^2 t/2)\, dF(y)\,,$$
$$-\infty \leq g(t, b) = \inf \{x : f(t, x) \geq b\} < \infty\,.$$

It is easily seen that

$$x < g(t, b) \Rightarrow f(t, x) < b \quad \text{and} \quad f(t, x) < b \Rightarrow x \leq g(t, b)\,,$$

and that if $f(t_0, x_0) < \infty$ for some t_0, x_0, then for each $t > t_0$ the equation $f(t, x) = b$ has the unique solution $x = g(t, b)$. The function $g(t, b)$ is continuous and increasing in t for $t_0 \leq t < \infty$, and for $t > t_0$ we have $f(t, x) \geq b$ if and only if $x \geq g(t, b)$. Set

$$\varphi(x) = (2\pi)^{-1/2} \exp (-x^2/2)\,, \qquad \Phi(x) = \int_{-\infty}^x \varphi(y)\, dy\,.$$

THEOREM 2. (i) *For any* t_0, x_0, b *such that* $f(t_0, x_0) < b$

$$P_0\{X(t) \geq g(t, b) \ \ \textit{for some} \ \ t > t_0 \,|\, X(t_0) = x_0\} = b^{-1} f(t_0, x_0)\,. \tag{13}$$

(ii) *For any* $\tau > 0$

$$\begin{aligned} P_0\{X(t) \geq g(t, b) \ \ &\textit{for some} \ \ t > \tau\} \\ &= 1 - \Phi(g(\tau, b)/\tau^{1/2}) \\ &\quad + b^{-1} \int_0^\infty \Phi(\tau^{-1/2} g(\tau, b) - y\tau^{1/2})\, dF(y)\,. \end{aligned} \tag{14}$$

(iii) *For any* $\tau \geq 0$ *the stopping rule* T *defined by*

$$T = \inf \{t : t > \tau,\ X(t) \geq g(t, b)\} \tag{15}$$

satisfies

$$E_\theta T < \infty \quad \textit{for all} \quad \theta > (1/2) \inf \{y : F(0, y] > 0\}\,. \tag{16}$$

In particular if $F(0, y] > 0$ for all y, then the rule T defined by (15) satisfies (16) and hence (2) for all $\theta > 0$. By adjusting τ and b we can make $P_\theta(T < \infty) \leq \alpha$ for all $\theta \leq 0$ (except in the trivial case $f(t, x) \equiv +\infty$), so the class of stopping rules defined by (15) satisfies (1) and (2). A more detailed study of the behavior of $E_\theta T$ is given in §3.

Before discussing the proof of Theorem 2, we give several examples. We shall use the familiar fact that under P_0 the process $X^*(t)$ defined by

$$X^*(t) = t\, X(1/t) \quad (t > 0)\,, \qquad X^*(0) = 0\,, \tag{17}$$

is again a Wiener process with 0 drift.

EXAMPLE 1. Let F be a degenerate measure putting unit mass at $\mu>0$. Then $f(t,x)=\exp(\mu x-\mu^2 t/2)$ and $g(t,b)=(\log b)/\mu+\mu t/2$, and by Theorem 2(i) with $x_0=t_0=0$,

$$P_0\{X(t)\geq(\log b)/\mu+\mu t/2 \quad \text{for some} \quad t>0\}=b^{-1} \quad (b>1)\,, \tag{18}$$

in agreement with Theorem 1. Using (14) instead of (13) and putting $a=\mu/2$, $b=\exp(2ac)$, we obtain

$$\begin{aligned} &P_0\{X(t)\geq c+at \quad \text{for some} \quad t\geq\tau\} \\ &\quad =1-\Phi(c/\tau^{1/2}+a\tau^{1/2})+\exp(-2ac)\,\Phi(c/\tau^{1/2}-a\tau^{1/2})\,, \end{aligned}$$

which by applying the transformation (17) gives the well-known result

$$\begin{aligned} &P_0\{X(t)\geq a+ct \quad \text{for some} \quad 0<t\leq\tau\} \\ &\quad =1-\Phi(c\tau^{1/2}+a/\tau^{1/2})+\exp(-2ac)\,\Phi(c\tau^{1/2}-a/\tau^{1/2})\,. \end{aligned} \tag{19}$$

EXAMPLE 2. Let $dF(y)=2\varphi(yh^{1/2})dy/y^r$ for $0<y<\infty$, $r<1$, $h>0$. Then $g(t,b)=(t+h)^{1/2}\alpha^{-1}((1-r)\log(t+h)+2\log b/2)$, where we have set

$$\alpha(x)=x^2+2\log\int_0^\infty \varphi(y-x)\,dy/y^r\sim x^2 \qquad \text{as} \quad x\to\infty\,.$$

Theorem 2(i) with $x_0=t_0=0$ gives

$$\begin{aligned} &P_0\{X(t)\geq(t+h)^{1/2}\alpha^{-1}((1-r)\log(t+h)+2\log b/2) \quad \text{for some} \quad t>0\} \\ &\quad =\Gamma((1-r)/2)/2^{(r+2)/2}\,\pi^{1/2}\,h^{(1-r)/2}\,b\,, \end{aligned}$$

provided this expression is <1.

EXAMPLE 3. Theorem 2 generalizes in an obvious manner to the case in which F is a measure on $(-\infty,\infty)$ which assigns measure 0 to $\{0\}$. For example, for $dF=dy/(2\pi)^{1/2}$, we obtain

$$\begin{aligned} &P_0\{|X(t)|\geq(t(\log t+a^2))^{1/2} \quad \text{for some} \quad t>\tau\} \\ &\quad =2[1-\Phi((\log\tau+a^2)^{1/2})+((\log\tau+a^2)/\tau)^{1/2}\,\varphi(a)]\,, \\ &\qquad\qquad (\tau>e^{-a^2})\,. \end{aligned} \tag{20}$$

For a statistical application of this example, consider the problem of testing $\theta<0$ against $\theta>0$ ($\theta=0$ being excluded). Defining $T=\inf\{t:t>\tau,\ |X(t)|\geq(t(\log t+a^2))^{1/2}\}$ and if $T<\infty$ deciding that $\theta<0$ or >0 according as $X(T)<0$ or >0, we obtain a test for which for all $\theta\neq0$

P_θ (error) $\leq (1/2)P_0(T < \infty) = 1/2$ (right-hand side of (20))

and $E_\theta T < \infty$.

EXAMPLE 4. For $\delta > 0$ let

$$dF(y) = \delta(dy/y)(\log 1/y)(\log_2 1/y)\cdots(\log_n 1/y)^{1+\delta} \quad \text{for} \quad 0 < y < 1/e_n,$$
$$= 0 \quad \text{elsewhere},$$

where we write $\log_2 x = \log(\log x)$, $e_2 = e^e$, etc.

It was shown in [9] that as $t \to \infty$, for $n \geq 3$

$$g(t, b) = [2t(\log_2 t + (3/2)\log_3 t + \sum_4^n \log_k t + (1+\delta)\log_{n+1} t + \log b/2\pi^{1/2} + o(1))]^{1/2}, \tag{21}$$

while for $n = 2$

$$g(t, b) = [2t(\log_2 t + (3/2 + \delta)\log_3 t + \log b/2\pi^{1/2} + o(1))]^{1/2}, \tag{22}$$

and by Theorem 2

$$P_0\{X(t) \geq g(t, b) \quad \text{for some} \quad t > 0\} = b^{-1} \qquad (b > 1).$$

By the law of the iterated logarithm (cf. [4, p. 33]), if we were to replace δ by 0 on the right-hand side of (21) or (22), then $P_0\{X(t) \geq g(t, b) \text{ i.e. } t \uparrow \infty\} = 1$. Hence (21) and (22) give essentially the slowest possible rate of increase of $g(t, b)$ for large t. It will be seen in §3 that this example is of particular interest in the theory of tests of power one.

It is possible to give a proof of Theorem 2 (i) and (ii) based on Lemma 1. (In fact for the stopping rule T defined by (15), (7) becomes (14) if we put $z(t) = f(t, X(t))$.) However, for certain generalizations it is more convenient to use the following closely related lemma.

LEMMA 2. *Let b be a positive constant and $\{z(t), \mathfrak{F}(t), t \geq \tau\}$ a nonnegative martingale. If $z(t)$ has continuous sample paths on $\{z(\tau) < b\}$ and converges to 0 in probability on $\{z(t) < b \text{ for all } t \geq \tau\}$, then*

$$P\{z(t) \geq b \text{ for some } t > \tau \mid \mathfrak{F}(\tau)\} = b^{-1} z(\tau) \quad \text{on} \quad \{z(\tau) < b\} \tag{23}$$

and

(24) $$P\{z(t) \geq b \quad for\ some \quad t \geq \tau\} = P\{z(\tau) \geq b\} + b^{-1}\int_{\{z(\tau)<b\}} z(\tau)dP.$$

The new reader is referred to [9] for a proof of Lemma 2 and its application to the martingale $z(t) = f(t, X(t))$ to prove Theorem 2(i), (ii).

We now turn to the proof of Theorem 2(iii) and to simplify slightly we assume that $\tau = 0$. We may also assume that $g(0, b) \geq 0$, for otherwise $T \equiv 0$ and the theorem is trivially true. It is an easy consequence of Hölder's inequality and the definition of g that $g(\cdot, b)$ is concave on $(0\ \infty)$. Furthermore if θ^* denotes $\inf\{y : F(0, y] > 0\}$, then

(25) $$\limsup_{t\to\infty} t^{-1} g(t, b) \leq \theta^*/2.$$

In fact, if $g(t, b) \geq \theta^*(1 + \varepsilon)\, t/2$ for some $0 < \varepsilon < 1$ and infinitely many values $t_i \to \infty$, then

$$\begin{aligned} b &= \int_0^\infty \exp\,(yg(t_i, b) - y^2 t_i/2)\, dF(y) \\ &\geq \exp\,(\theta^{*2}(1+\varepsilon)\, t_i/2 - \theta^{*2}(1+\varepsilon/4)^2 t_i/2)\, F[\theta^*, \theta^* + \varepsilon/4] \to \infty \end{aligned}$$

as $i \to \infty$, a contradiction which proves (25). Theorem 2(iii) now follows immediately from

LEMMA 3. *If $g(t)$ is a nonnegative, increasing, concave function satisfying*

(26) $$\limsup_{t\to\infty} g(t)/t \leq \beta,$$

then the stopping rule T defined by

(27) $$T = \inf\{t : t > 0,\ X(t) \geq g(t)\}$$

satisfies

(28) $$E_\theta T < \infty \quad for\ all \quad \theta > \beta.$$

Proof. By (27), Wald's lemma, and Jensen's inequality, for any $t > 0$

(29) $$\theta E_\theta(T \wedge t) = E_\theta(X(T \wedge t)) \leq E_\theta\, g(T \wedge t) \leq g(E_\theta(T \wedge t)).$$

Since $E_\theta(T \wedge t) \uparrow E_\theta T \leq \infty$ as $t \uparrow \infty$, we obtain (28) from (29) and (26).

3. **Asymptotic behavior of $E_\theta T$ as $\theta \to 0$.** It is impossible to compute $E_\theta T$ exactly for the class of stopping rules discussed in §2 (except in the case where $g(t, b)$ is a linear function of t). However, since $P_0\{T = \infty\} > 0$, we have $E_\theta T \to \infty$ as $\theta \to 0$, and in many cases it is possible to find the asymptotic value of $E_\theta T$ for small positive θ.

For example, under the assumptions of Lemma 3, by Wald's lemma we may replace $T \wedge t$ by T in (29). If $\beta = 0$ and the behavior of $g(t)$ for large t is known, it may be possible to solve this inequality for $E_\theta T$ as $\theta \to 0$. If $g(t) \sim (t \log t)^{1/2}$, as in Example 2 of §2 with $r = 0$, then we obtain

$$E_\theta T \leq ((2 \log 1/\theta)/\theta^2)(1 + o(1)) \quad \text{as} \quad \theta \to 0. \tag{30}$$

If we use a stopping rule T from Example 3, for which $g(t) \sim (2t \log_2 t)^{1/2}$, the log in (30) is replaced by $\log_2$.

Sharpening this argument slightly, we obtain a better asymptotic upper bound for $E_\theta T$, which we shall see later is actually an asymptotic equality. Under the assumptions of Lemma 3, we have by Wald's lemma

$$\theta E_\theta T = E_\theta X(T) = E_\theta g(T).$$

Hence for arbitrary t_0

$$\begin{aligned} \theta E_\theta(T \mid T > t_0) P_\theta(T > t_0) &\leq g(t_0) + E_\theta(g(T) \mid T > t_0) P_\theta(T > t_0) \\ &\leq g(t_0) + g(E_\theta(T \mid T > t_0)) P_\theta(T > t_0), \end{aligned}$$

where the second inequality follows from Jensen's inequality. Hence

$$\theta E_\theta(T \mid T > t_0) \leq \frac{g(t_0)}{P_\theta(T > t_0)} + g(E_\theta(T \mid T > t_0)), \tag{31}$$

which for asymptotic purposes is about the same as (29). Hence for $i = 1$ or 2, if $g(t) \sim (it \log_i t)^{1/2}$, we obtain as before

$$E_\theta(T \mid T > t_0) \leq ((2 \log_i 1/\theta)/\theta^2)(1 + o(1)) \quad \text{as} \quad \theta \to 0. \tag{32}$$

Now for fixed t_0

$$E_\theta(T \mid T > t_0) = \int_{\{T > t_0\}} T \, dP_\theta / P_\theta\{T > t_0\} \sim E_\theta(T)/P_0\{T > t_0\} \tag{33}$$

as $\theta \to 0$. Moreover, since $dP_\theta^{(t)}/dP_0^{(t)} = \exp(\theta X(t) - \theta^2 t/2)$, we have $P_\theta\{T \leq t_0\} = \int_{\{T \leq t_0\}} \exp(\theta X(t_0) - \theta^2 t_0/2) \, dP_0$ and hence by Fatou's lemma

$$P_0\{T \le t_0\} \le \liminf_{\theta \to 0} P_\theta\{T \le t_0\},$$

or equivalently

(34) $$P_0\{T > t_0\} \ge \limsup_{\theta \to 0} P_\theta\{T > t_0\}.$$

Substituting (33) and (34) into (32) and then letting $t_0 \to \infty$, we obtain

(35) $$E_\theta T \le ((2P_0(T=\infty)\log_2 1/\theta)/\theta^2)(1+o(1)) \quad \text{as} \quad \theta \to 0.$$

The preceding argument is due to Farrell [3]. To prove that the inequality appearing in (35) is actually an equality we use the following elementary inequality (cf. [11]).

LEMMA 4. *For any stopping rule R, $\theta > 0$, $a > 0$, and real numbers $0 \le t_0 < t_1 < \infty$*

$$P_\theta\{t_0 < R \le t_1\} \le 1 - \Phi(a) + \exp(\theta^2 t_1/2 + a\theta t_1^{1/2})\, P_0\{t_0 < R \le t_1\}.$$

Proof. Since $\{t_0 < R \le t_1\} \in S(X(s), s \le t_1)$ and $dP_\theta^{(t)}/dP_0^{(t)} = \exp(\theta X(t) - \theta^2 t/2)$, we have

$$\begin{aligned} P_\theta\{t_0 < R \le t_1\} &= P_\theta\{t_0 < R \le t_1,\ X(t_1) - \theta t_1 > a t_1^{1/2}\} \\ &\quad + \int_{\{t_0 < R \le t_1,\ X(t_1) \le \theta t_1 + a t_1^{1/2}\}} \exp(\theta X(t_1) - \theta^2 t_1/2)\, dP_0 \\ &\le P_\theta\{X(t_1) - \theta t_1 > a t_1^{1/2}\} \\ &\quad + \exp(\theta^2 t_1/2 + a\theta t_1^{1/2})\, P_0\{t_0 < R = t_1\}. \end{aligned}$$

COROLLARY. *For any stopping rule R such that $P_0\{R = \infty\} > 0$,*

(36) $$\lim_{\theta \to 0} \theta^2 E_\theta R = \infty.$$

Proof. Putting t_0 arbitrary but fixed and $t_1 = K/\theta^2$ in Lemma 4, we obtain

(37) $$\begin{aligned} E_\theta R \ge t_1 P_\theta\{R > t_1\} &= t_1(P_\theta\{R > t_0\} - P_\theta\{t_0 < R \le t_1\}) \\ &\ge t_1(P_\theta\{R > t_0\} - (1 - \Phi(a)) \\ &\qquad - \exp(K/2 + aK^{1/2})\, P_0\{t_0 < R \le t_1\}). \end{aligned}$$

Since $P_\theta\{R > t_0\} = \int_{\{T > t_0\}} \exp(\theta X(t_0) - \theta^2 t_0/2)\, dP_0$, by Fatou's lemma $P_0\{R > t_0\} \le \liminf_{\theta \to 0} P_\theta\{R > t_0\}$. Hence we obtain (36) from (37) upon letting $\theta \to 0$, $t_0 \to \infty$, $a \to \infty$, and $K \to \infty$ (in the indicated order).

For a stopping rule T defined by (15), knowledge of the asymptotic behavior of $g(t)$ permits an extension of this argument giving an asymptotic lower bound for $E_\theta T$ as $\theta \to 0$. For example, for $i=1$ or 2, if $g(t) \sim (it \log_i t)^{1/2}$ as $t \to \infty$, then for any $\varepsilon > 0$

(38) $$\liminf P_\theta \{T > 2(1-\varepsilon) \log_i \theta^{-1}/\theta^2\} \geq P_0 \{T = \infty\},$$

from which it follows (since $\varepsilon > 0$ is arbitrary) that

(39) $$E_\theta T \geq 2P_0(T = \infty)(\log_i \theta^{-1}/\theta^2)(1 + o(1))$$

as $\theta \to 0$. If g is also concave and increasing, so that (36) holds, then

(40) $$E_\theta T \sim 2P_0(T = \infty) \log_i \theta^{-1}/\theta^2 .$$

We shall prove (38) with $i=1$ under the assumption $g(t) \sim (t \log t)^{1/2}$. The case $i = 2$, which is similar, was treated in detail in [11].

Fix $\varepsilon > 0$, and let t_0 be arbitrary, $t_1 = 1/\theta^2$, and $t_2 = 2(1-\varepsilon) \log \theta^{-1}/\theta^2$.
Then

$$P_\theta \{T > t_2\} = P_\theta \{T > t_0\} - P_\theta \{t_0 < T \leq t_1\} - P_\theta \{t_1 < T \leq t_2\}.$$

As in the previous argument

$$\lim_{t_0 \to \infty} \liminf_{\theta \to 0} P_\theta \{t_0 < T \leq t_1\} = 0 .$$

Hence to complete the proof it suffices to show $P_\theta \{t_1 < T \leq t_2\} \to 0$ as $\theta \to 0$. But by Lemma 4

$$P_\theta \{t_1 < T \leq r_2\} \leq 1 - \Phi(a) + t_1^{(1-\varepsilon)/2 + o(1)} P_0 \{t_1 < T \leq t_1 \log t_1\},$$

and hence, since a is arbitrary, it suffices to show that

(41) $$P_0 \{X(t) \geq g(t) \quad \text{for some} \quad t_1 < t \leq t_1 \log t_1\} = o(t_1^{-(1-\varepsilon)/2})$$

as $t_1 \to \infty$. Since $g(t) \sim (t \log t)^{1/2}$, we may assume that $g(t) \geq ((1-\varepsilon/5)t \log t)^{1/2}$ for all $t \geq t_1$. Putting $r_k = k^{5/\varepsilon}$ and applying (19) with $c=0$, we see that the left-hand side of (41) is majorized by

$$\sum_{t_1 \le r_k < t_1^{\log t_1}} P_0\{X(t) \ge ((1-\varepsilon/5)r_k \log r_k)^{1/2} \quad \text{for some} \quad r_k < t \le r_{k+1}\}$$
$$\le 2 \sum_{t_1 \le r_k < t_1^{\log t_1}} P_0\{X(r_{k+1}) \ge ((1-\varepsilon/5)r_k \log r_k)^{1/2}\}$$
$$\le 2 \sum_{t_1 \le r_k < t_1^{\log t_1}} \exp\left(-\frac{1}{2}(1-\varepsilon/5)\frac{r_k}{r_{k+1}} \log r_k\right)$$
$$\le 4 \exp(-(1/2-\varepsilon/5) \log t_1)(t_1 \log t_1)^{\varepsilon/5},$$

from which (41) follows.

We see, then, that for stopping rules T defined by (27) with well-behaved functions g, the behavior of $E_\theta T$ for θ positive and near 0 is determined by the growth of $g(t)$ as $t \to \infty$; and typically, the more slowly g grows, the smaller $E_\theta T$ for θ near 0. Since by the law of the iterated logarithm, if $P_0\{T < \infty\} < 1$, then g must grow about as fast as $(2t \log_2 t)^{1/2}$, it is natural to suspect that (40) with $i = 2$ is optimal for $E_\theta T$ as $\theta \downarrow 0$. In fact, it has been shown by Farrell [3] that for any stopping rule T such that $P_0(T < \infty) < 1$

(42) $$\limsup_{\theta \downarrow 0} \theta^2 E_\theta T/\log_2 \theta^{-1} \ge 2P_0(T = \infty).$$

A simpler proof of (42) is sketched in [11], where there is also an example to show that lim sup cannot be replaced by lim inf and, in fact, that (36) cannot be improved.

CHAPTER II

REPRESENTATION THEOREMS FOR MARTINGALE FUNCTIONS AND PARTIAL DIFFERENTIAL EQUATIONS

1. **Introduction.** Let $X(t) = w(t)$, $0 \le t < \infty$, be a Wiener process with drift 0. In this chapter the probability (expectation) of events (random variables) defined in terms of $w(\tau)$ for $\tau \ge t$ conditional on $w(t) = x$ will be denoted by $P_{(t,x)}(\cdot)(E_{(t,x)}(\cdot))$. For example, the left-hand side of (13) will now be written as

(43) $$P_{(t_0,x_0)}\{w(t) \ge g(t) \quad \text{for some} \quad t > t_0\},$$

where we have suppressed the variable b in our notation for the function g.

A nonnegative measurable function $f(t, x)$ defined for $0 < t < l \le +\infty$ and $-\infty < x < \infty$ is said to be a martingale function on $(0, l)$ (for $(w(t))$ if

$$f(t, x) = E_{(t,x)} f(\tau, w(\tau)) \qquad (0 < t < \tau < l,\ -\infty < x < \infty). \tag{44}$$

In Chapter I we found the probability (43) for a certain class of functions g by exploiting the fact that

$$f(t, x) = \int_{-\infty}^{\infty} \exp(yx - y^2 t/2)\, dF(y) \quad (0 < t < \infty,\ -\infty < x < \infty) \tag{45}$$

is a martingale function on $(0, \infty)$. (In what follows we shall assume that $f(t, x) < \infty$ for all positive t.) For example, if $\int_{-\infty}^{0+} dF = 0$ and $g(t)$ is defined for $t > 0$ as the unique solution of $f(t, g(t)) = 1$, then by Theorem 2(i)

$$P_{(t,x)}\{w(\tau) \geq g(\tau) \text{ for some } \tau > t\} = f(t, x) \qquad (0 < t < \infty,\ x < g(t)). \tag{46}$$

Now, if we denote the left-hand side of (46) by $p(t, x)$, it is easy to see, at least for smooth g, that

$$p_t + \tfrac{1}{2} p_{xx} = 0 \qquad (0 < t < \infty,\ x < g(t)) \tag{47}$$

and $p(t, g(t)) = 1$; and it may be verified by differentiation under the integral in (45) that the function $f(t, x)$ satisfies (47) for all $0 < t < \infty$, $-\infty < x < \infty$.

Hence, in order to compute probabilities like (46) for as large a class of functions g as possible, it is desirable to obtain a simple representation for a large class of solutions of (47), or alternatively for a large class of martingale functions. These considerations suggest the following theorem.

THEOREM 3. *Let f be a nonnegative continuous function defined on $(0, \infty) \times (-\infty, \infty)$. The following statements are equivalent:*

(48) $f_t + \frac{1}{2} f_{xx} = 0$ *on* $(0, \infty) \times (-\infty, \infty)$;

(49) f *has the representation* (45) *for some measure* F;

(50) f *is a martingale function on* $(0, \infty)$.

Theorem 3 is proved in the next section, which is followed by a number of remarks showing the connection of this result with two theorems of Widder [13], [14] on positive solutions of the heat equation $u_t = \frac{1}{2} u_{xx}$. In §4 we give surprising analogues of Theorem 3 for

$r(t) = (w_1^2(t) + w_2^2(t) + w_3^2(t))^{1/2}$, where w_1, w_2, and w_3 are independent Wiener processes. These results are then applied to yield boundary crossing probabilities for $r(t)$, answering a question raised in [9].

2. **Proof of Theorem 3.** We shall prove (49) ⇒ (48) ⇒ (50) ⇒ (49).

If f has the representation (45), differentiation under the integral may be justified by the dominated convergence theorem. Since for each real number y the integrand $\exp(yx - y^2t/2)$ satisfies equation (48), it follows at once that f does also.

Suppose now that (48) is satisfied. Let (t, x) be arbitrary and $\tau > t$. For $a < x < c$, let

$$T_{ac} = \inf\{t' : t' > t,\ w(t') \leq a \quad \text{or} \quad w(t') \geq c\}.$$

By Lemma 1 below

$$f(t, x) = E_{(t,x)} f(T_{ac} \wedge \tau,\ w(T_{ac} \wedge \tau)). \tag{51}$$

Hence, since $f \geq 0$, we have

$$f(t, x) \geq \int_{(T_{ac} > \tau)} f(\tau, w(\tau))\, dP_{(t,x)},$$

and letting $a \to -\infty$, $c \to \infty$,

$$f(t, x) \geq E_{(t,x)} f(\tau, w(\tau)); \tag{52}$$

in particular

$$(\tau - t)^{-1/2} \int_{-\infty}^{\infty} f(\tau, y)\, \varphi((y - x)/(\tau - t)^{1/2})\, dy < \infty \qquad (\tau > t). \tag{53}$$

To prove the equality holds in (52), it suffices by (51) to show that

$$\begin{aligned} \liminf_{c \to \infty} \int_{(T_{ac} \leq \tau,\, w(T_{ac}) = c)} f(T_{ac}, c)\, dP_{(t,x)} &= 0 \\ = \liminf_{a \to -\infty} \int_{\{T_{ac} \leq \tau,\, w(T_{ac}) = a\}} & f(T_{ac}, a)\, dP_{(t,x)}. \end{aligned} \tag{54}$$

Letting $T_c = \inf\{t' : t' > t,\ w(t') = c\}$, we have from (19) with $a = 0$ the well-known result

$$P_{(t,x)}\{T_c < \tau\} = 2(1 - \Phi(|c - x|/(\tau - t)^{1/2})) \tag{55}$$

for all $0 \leq t < \tau$, $-\infty < c$, $x < \infty$. Suppose for the moment that $f_t \leq 0$. Then by (55)

$$\int_{\{T_{ac} \leq \tau, w(T_{ac}) = c\}} f(T_{ac}; c)\, dP_{(t,x)}$$

$$\leq f(t, c)\, P_{(t,x)}\{T_c \leq \tau\} \tag{56}$$

$$= 2f(t, c) \int_c^\infty \varphi\left(\frac{y-x}{(\tau-t)^{1/2}}\right) \frac{dy}{(\tau-t)^{1/2}} .$$

From the assumption that $f_t \leq 0$ and (48) it follows that f is convex in x for each t. Hence either $f(t, c)$ remains bounded or ultimately tends monotonically to $+\infty$ as $c \to \infty$. In the first case the first part of (54) obviously holds and in the second case the right-hand side of (56) is majorized for large c by

$$2\int_c^\infty f(t, y)\varphi\left(\frac{y-x}{(\tau-t)^{1/2}}\right) \frac{dy}{(\tau-t)^{1/2}},$$

which by (53) converges to 0 as $c \to \infty$ for all $t > \tau/2$. The second integral in (54) is treated similarly, and hence

$$E_{(t,x)} f(\tau, w(\tau)) = f(t, x) \tag{57}$$

for all $\tau > 0$ and $\tau/2 < t \leq \tau$. But for $\tau/4 < t \leq \tau/2$, we have for all $s \in (\tau/2, 2t)$

$$E_{(t,x)} f(\tau, w(\tau)) = E_{(t,x)}[E_{(s,w(s))} f(\tau, w(\tau))]$$
$$= E_{(t,x)} f(s, w(s)) = f(t, x) ,$$

and hence (57) holds for all $\tau/4 < t \leq \tau$. Repeating the argument shows that (57) holds for all $0 < t < \tau$. This proves (50) if f is nonincreasing in t. By (49) we see that in fact this condition is always satisfied.

If f is an arbitrary nonnegative solution of (48), for fixed $\tau > 0$ define

$$\hat{f}(t, x) = E_{(t,x)} f(\tau, w(\tau)) \qquad (0 < t < \tau)$$

and

$$\hat{\hat{f}}(t, x) = \int_t^\tau (f(t', x) - \hat{f}(t', x))\, dt' . \tag{58}$$

(The transformation (58) is due to Widder [13].) Straightforward application of the dominated convergence theorem shows that $\hat{f}$ is continuous for $0 < t \leq \tau$ and is twice continuously differentiable and satisfies (48) for $0 < t < \tau$. By (52) $f \geq \hat{f}$ and hence $\hat{\hat{f}} \geq 0$. Since f is twice continuously differentiable, by Lemma 1, it is possible to differentiate under the integral in (58) to show that $\hat{\hat{f}}$ satisfies (48) for $0<t<\tau$. Moreover, $\hat{\hat{f}}$ is decreasing in t for $t\leq\tau$ and hence by the first part of the proof

$$\begin{aligned}\hat{\hat{f}}(t-\varepsilon, x) &= E_{(t-\varepsilon, x)}\hat{\hat{f}}(\tau-\varepsilon, w(\tau-\varepsilon)) \\ &= \int_{-\infty}^{\infty}\hat{\hat{f}}(\tau-\varepsilon, y)\,\varphi\left(\frac{y-x}{(\tau-t)^{1/2}}\right)\frac{dy}{(\tau-t)^{1/2}}\end{aligned}$$

for $0<t<\tau$, $0<\varepsilon<t$. Letting $\varepsilon\to 0$, we have by monotone convergence

$$\hat{\hat{f}}(t, x) = E_{(t,x)}\hat{\hat{f}}(\tau, w(\tau)) = 0 \qquad (0<t<\tau).$$

Hence $\hat{f}=f$, which completes the proof of (50), since τ is arbitrary.

Finally, suppose that (50) holds, i. e. that

(59) $$f(t, x) = \int_{-\infty}^{\infty}\varphi\left(\frac{y-x}{(\tau-t)^{1/2}}\right)f(\tau, y)\frac{dy}{(\tau-t)^{1/2}}$$

for $0<t<\tau$, $-\infty<x<\infty$. Dividing and multiplying by $\varphi(y/\tau^{1/2})/\tau^{1/2}$ and making the change of variables $u=y$, we obtain after some algebra

(60) $$\begin{aligned}f(t, x) &= \left(\frac{\tau}{\tau-t}\right)^{1/2}\exp(-x^2/2\tau) \\ &\cdot\int_{-\infty}^{\infty}\exp\left(\frac{\tau}{\tau-t}(xu-u^2t/2)\right)\varphi(u\tau^{1/2})\,\tau^{1/2}f(\tau, u\tau)\,du.\end{aligned}$$

Consider the family F_τ of measures defined by

$$dF_\tau(u) = \tau^{1/2}\varphi(u\tau^{1/2})f(\tau, u\tau)\,du.$$

Assume for the moment that (59) holds for $t=x=0$ and hence

(61) $$\int_{-\infty}^{\infty}dF_\tau(u) = f(0, 0) < \infty.$$

By Helly's lemma there exists a sequence $\tau_1<\tau_2<\cdots<\tau_k\to\infty$ and a measure F such that $F_{\tau_k}\to F$ weakly as $k\to\infty$, and hence from (60)

$$f(t, x) = \int_{-\infty}^{\infty}\exp(xu-u^2t/2)\,dF(u).$$

This proves (49) under the assumption (61). In the general case, a change of origin together with the preceding argument yields

$$f(t, x) = \int_{-\infty}^{\infty}\exp(xy-(y^2/2)(t-1))\,dF_1(y) \qquad (1\leq t<\infty,\ -\infty<x<\infty),$$

which together with (59) yields (49) with $dF(y) = \exp(y^2/2)\,dF_1(y)$ for all $t > 0$ and $-\infty < x < \infty$.

LEMMA 1. *Let f be a continuous solution of*

(61a) $$f_t + (1/2)f_{xx} = 0$$

on $(0, l) \times (-\infty, \infty)$ for some $l > 0$. For $t > 0$, $-\infty < a < c < \infty$, define $T_{ac} = T_{ac}(t) = \inf\{t' : t' \geq t,\ w(t') \leq a \text{ or } \geq c\}$. Then for each $0 < t < \tau < l$ and $a < x < c$

$$f(t, x) = E_{(t,x)} f(T_{ac} \wedge \tau,\ w(T_{ac} \wedge \tau))$$

and f is twice continuously differentiable.

Proof. T. L. Lai has recently proved a version of this lemma for a large class of Markov processes [5]. For sufficiently smooth f it follows easily from Itô's formula for stochastic differentials.

The outline of a simple direct proof follows. Let $0 < \varepsilon < \tau < l$ and define

$$h(t, x) = E_{(t,x)} f(T_{ac} \wedge \tau,\ w(T_{ac} \wedge \tau)) \qquad (\varepsilon < t \leq \tau,\ a < x < c)\,.$$

From the known form of the first-passage distribution of the process $(t, w(t))$ to the complement of the rectangle $R = \{(t', x') : \varepsilon < t' < \tau,\ a < x' < c\}$ as a function of its initial position (t, x) it may be verified that h is continuous on the closure of R and is twice continuously differentiable and satisfies (61a) on R. An application of a classical maximum principle (cf. [7, p. 160]) shows that $h = f$ on the closure of R.

3. Remarks and relation to Widder's theorems.

The method of proof of Theorem 3 given in §2 is not the most direct, but it has the advantage of generalizing easily to certain other real-valued processes, as we shall see in §4. On the other hand our proof of (50) does not seem to generalize to higher dimensions even for the Wiener process. The following outline of a proof of (50) can be easily adapted to prove the obvious d-dimensional generalization of Theorem 3.

First observe that to prove (50) it suffices to show that

(62) $$\int_{c_0}^{\infty} \left(\int_{(T_c < \tau)} f(T_c, c)\, dP_{(t,x)} \right) dc < \infty$$

for some $c_0 > 0$. Using (55) the inner integral in (62) may be expressed analytically in terms of the probability density function of T_c, and an easy approximation argument and (52) then prove (62).

The proof of Theorem 3 with a somewhat simpler weak convergence argument allows one to prove

THEOREM 4. *Let* $0<l<\infty$ *and let* f *be a nonnegative continuous function on* $(0, l) \times (-\infty, \infty)$. *The following statements are equivalent:*

(63) $f_t + \frac{1}{2} f_{xx} = 0$ *on* $(0, l) \times (-\infty, \infty)$;

(64) $f(t, x) = \int_{-\infty}^{\infty} \frac{1}{(l-t)^{1/2}} \varphi\left(\frac{y-x}{(l-t)^{1/2}}\right) dF(y)$ *for some measure* F;

(65) f *is a martingale function on* $(0, l)$.

By defining $u(t, x) = f(l-t, x)$ for $0<t<l$ it is easy to see (at least for twice continuously differentiable f) that Theorem 4 is equivalent to the following theorem of Widder [13] on positive solutions of the heat equation.

THEOREM 5. *Let* $0<l<\infty$ *and let* u *be a nonnegative twice continuously differentiable function on* $(0, l) \times (-\infty, \infty)$. *The following statements are equivalent:*

(66) $u_t = \frac{1}{2} u_{xx}$ *on* $(0, l) \times (-\infty, \infty)$;

(67) $u(t, x) = \int_{-\infty}^{\infty} t^{-1/2} \varphi((y-x)/t^{1/2})\, dF(y)$ *for some measure* F.

On the other hand Theorem 5 may be related directly to Theorem 3 by observing that if u satisfies $u_t = \frac{1}{2} u_{xx}$ in a domain D_u, then $f(t, x) = u(t^{-1}, xt^{-1}) / \varphi(xt^{-1/2})\, t^{1/2}$ satisfies $f_t + \frac{1}{2} f_{xx} = 0$ in $D_f = \{(t, x) : (t^{-1}, xt^{-1}) \in D_u\}$, and conversely. Probabilistically, this transformation is equivalent to (17).

In a later paper Widder [14] studied positive solutions of the heat equation on the half-line $0<x<\infty$. Slight changes in the proof of Theorem 3 allow us to obtain this result as well.

THEOREM 6. *Let* $0<l<\infty$ *and let* f *be a nonnegative continuous function on* $(0, l) \times (0, \infty)$. *The following statements are equivalent*:

(68) $f_t + \frac{1}{2} f_{xx} = 0$ *on* $(0, l) \times (0, \infty)$;

$$f(t, x) = \int_0^\infty \frac{1}{(l-t)^{1/2}} \left[\varphi\left(\frac{y-x}{(l-t)^{1/2}}\right) - \varphi\left(\frac{y+x}{(l-t)^{1/2}}\right) \right] dF_1(y)$$

$$+ \int_t^{l+} \frac{x}{(s-t)^{3/2}} \varphi\left(\frac{x}{(s-t)^{1/2}}\right) dF_2(s), \tag{69}$$

for some measures F_1 ***on*** $(0, \infty)$ ***and*** F_2 ***on*** $(0, l]$.

Widder's theorem [14] may be obtained from Theorem 6 by the change of variable $t \to l - t$.

To prove Theorem 6, if (69) holds, then by differentiation under the integral it may be shown that (68) holds. (However, in justifying this differentiation it is not legitimate to treat the two terms in the first integrand separately—as Widder does—since for some admissible measures, e. g. $dF_1(y) = y^{-1} dy$ $(0 < y < 1)$, $= 0$ $(1 \le y < \infty)$, the integrals of the two terms do not converge separately.) Now let f satisfy (68). Then using Lemma 1 and letting $c \to \infty$ while leaving $a > 0$ fixed, we obtain by the argument of Theorem 3 that

$$f(t, x) = \int_a^\infty f(\tau, y) \left[\varphi\left(\frac{y-x}{(\tau-t)^{1/2}}\right) - \varphi\left(\frac{y+x-2a}{(\tau-t)^{1/2}}\right) \right] \frac{1}{(\tau-t)^{1/2}} dy$$

$$+ \int_t^\tau f(s, a) \frac{x-a}{(s-t)^{3/2}} \varphi\left(\frac{x-a}{(s-t)^{1/2}}\right) ds \tag{70}$$

$$(0 < a < x, \ 0 < t < \tau < l),$$

where we have used the probability density function of T_a (obtained by differentiating (55)) and the well-known formula

$$P_{(t,x)}\{T_a > \tau, \ w(\tau) \in dy\} = \left[\varphi\left(\frac{y-x}{(\tau-t)^{1/2}}\right) - \varphi\left(\frac{y+x-2a}{(\tau-t)^{1/2}}\right) \right] \frac{dy}{(\tau-t)^{1/2}}$$

(which may be obtained from (18) together with (17)). Letting $a \to 0$ we obtain by monotone convergence in the first integral and a weak convergence argument in the second

$$f(t, x) = \int_0^\infty f(\tau, y) \left[\varphi\left(\frac{y-x}{(\tau-t)^{1/2}}\right) - \varphi\left(\frac{y+x}{(\tau-t)^{1/2}}\right) \right] \frac{dy}{(\tau-t)^{1/2}}$$

$$+ \int_t^\tau \frac{x}{(\tau-t)^{3/2}} \varphi\left(\frac{x}{(s-t)^{1/2}}\right) dF_2(s)$$

for some measure F_2. Now letting $\tau \to l$, we obtain (69) by weak convergence. (Details of these weak convergence arguments are provided in [14], although Widder's approach to the first one seems unnecessarily complicated.)

The method of proof of Theorem 3 applies to functions which assume both positive and negative values if we are willing to impose some growth condition on $f(t, x)$ as $|x| \to \infty$ and replace the measure F in (45) by Schwartz distributions. Since the results obtainable this way are clearly incomplete and since nonnegative functions suffice for most applications to probability, we omit the details. We remark, however, that to compute the probability that w ever leaves the region $R = \{(t, x) : -at < x < ct\}$ starting from some point $(t_0, x_0) \in R$, one may use Lemma 2 applied to the martingale function

$$f(t, x) = 1 - \sum_{k=-\infty}^{\infty} [\exp (2k(a + c)x - (2k(a + c))^2 t/2)$$
$$- \exp (2((k - 1)a + kc)x - (2(k - 1)a + 2kc)^2 t/2)],$$

which is nonnegative in R and is of the form (45) with a signed measure F which puts mass $+1$ at $2[(k - 1)a + kc]$ for $k = 0, \pm 1, \pm 2, \cdots$ and mass -1 at $2k(a + c)$ for $k = \pm 1, \pm 2, \cdots$. Using (17) one may then derive the first-passage time distribution for w from the region between two nonparallel straight lines. A slightly more sophisticated use of (17) allows one to compute the probability of hitting one of two nonparallel straight lines before the other and before time t, starting from the corresponding well-known probabilities in the case of parallel lines with 0 slope.

4. The Bessel process $r(t) = (w_1^2(t) + w_2^2(t) + w_3^2(t))^{1/2}$. It is natural to expect appropriate generalizations of Theorem 3 to hold for a large class of Markov processes. In this section we discuss a simple example for which it is possible to carry out the required computations explicitly, and for which the results are strikingly different from those for the Wiener process. We apply these results to compute boundary crossing probabilities similar to those of Chapter I, answering a question raised in [9].

Let $r(t) = (w_1^2(t) + w_2^2(t) + w_3^2(t))^{1/2}$, where $w_i(t)$, $0 \leq t < \infty$ $(i = 1, 2, 3)$, are independent Wiener processes. It is easy to see that r is a diffusion process on the interval $(0, \infty)$ with differential generator

$$\mathcal{D} u(x) = \tfrac{1}{2} u''(x) + x^{-1} u'(x) \qquad (0 < x < \infty)$$

and transition probability density function

$$g(t, x, y) = (y/xt^{1/2})\,[\varphi((y-x)/t^{1/2}) - \varphi((y+x)/t^{1/2})].$$

Let f be a nonnegative continuous function defined on $(0, \infty) \times (0, \infty)$.

THEOREM 7. *The following statements are equivalent*:

(71) $f_t + \mathcal{D} f = 0$ *on* $(0, \infty) \times (0, \infty)$ *and* $xf(t, x) \to 0$ *as* $(t, x) \to (t_0, 0+)$ $(0 < t_0 < \infty)$;

(72) $f(t, x) = \int_{0-}^{\infty} (yx)^{-1} (\sinh yx) \exp(-y^2 t/2)\, dF(y)$ *for some measure* F *on* $[0, \infty)$;

(73) f *is a martingale function for* r *on* $(0, \infty)$.

THEOREM 8. *The following statements are equivalent*:

(74) $f_t + \mathcal{D} f = 0$ *on* $(0, \infty) \times (0, \infty)$;

(75)
$$f(t, x) = \int_{0-}^{\infty} (xy)^{-1} (\sinh yx) \exp(-y^2 t/2)\, dF_1(y) + \int_t^{\infty} (s-t)^{-3/2} \varphi(x/(s-t)^{1/2})\, dF_2(s)$$
for some measure F_1 *on* $[0, \infty)$ *and* F_2 *on* $(0, \infty)$;

(76) *for each* $0 < a < x < \infty$ *and* $0 < t < \tau < \infty$ $f(t, x) = E_{(t,x)} f(T_a \wedge \tau, r(T_a \wedge \tau))$, *where* $T_a = \inf\{t' : t' > t, r(t') \le a\}$.

Note that (74) corresponds to (48) and (73) to (50), while (71) and (76) are weaker statements. Versions of Theorems 7 and 8 when f is defined on $(0, l) \times (0, \infty)$ for some finite l are easily obtained.

The proofs of Theorems 7 and 8 proceed in much the same way as for Theorem 3, beginning with an obvious extension of Lemma 1. To prove the first part of (54) with w replaced by r we first note that for some $r < 1$

$$\limsup_{c \to \infty} \sup_{0 < s < \tau - t} \int_0^c g(s, x, y)\, dy < r$$

and hence the second term on the right-hand side of the inequality

$$P_{(t,x)}\{T_c \leq \tau\} \leq P_{(t,x)}\{r(\tau) \geq c\}$$
$$+ \int_{\{T_c \leq \tau\}} P_{(t,x)}\{r(\tau) < c \mid r(t'),\ t \leq t' \leq T_c\}\, dP_{(t,x)}$$

is majorized for large c by $\gamma P_{(t,x)}\{T_c \leq \tau\}$, from which it follows that

$$P_{(t,x)}\{T_c \leq \tau\} = O(P_{(t,x)}\{r(\tau) \geq c\})$$

as $c \to \infty$. Recalling that $\mathcal{D}$ may be expressed as $(d/dm)d/ds$, where m and s are the speed measure and scale function of r (cf. [4, p. 111 ff.]), we see that if $f_t \leq 0$, then $f(t, \cdot)$ is either bounded or ultimately increasing at ∞ for each t, and hence the argument of §2 continues to hold here. Thus (76) follows from (74). The corresponding argument at 0 requires a more detailed knowledge of the first-passage time distribution from x to a $(a < x)$. Observe that $h_\lambda(x) = x^{-1}\exp(-(2\lambda)^{1/2}x)$ $(0 < x < \infty)$ is a decreasing solution of $\mathcal{D}h = \lambda h$ for each $\lambda \geq 0$, and hence $f(t, x) = h_\lambda(x)\exp(-\lambda t)$ satisfies (74). Hence (76) applies, and letting $\tau \to \infty$ we have

$$E_{(t,x)}[\exp(-\lambda(T_a - t))] = (a/x)\exp(-(2\lambda)^{1/2}(x - a)) \qquad (0 < a < x). \tag{77}$$

Setting $\lambda = 0$ we have in particular

$$P_{(t,x)}\{T_a < \infty\} = a/x \qquad (0 < a < x), \tag{78}$$

and except for the factor a/x the right-hand side of (77) is the Laplace transform of the one-sided first-passage time of the standard Wiener process (cf. [4, p. 26]). Hence from (55) and (78)

$$\begin{aligned} P_{(t,x)}\{T_a \in ds\} &= \frac{a(x-a)}{x(s-t)^{3/2}}\varphi((x-a)/(s-t)^{1/2})\, ds, \\ P_{(t,x)}\{T_a \leq \tau\} &= 2a(x)(1 - \Phi((x-a)/(\tau-t)^{1/2})). \end{aligned} \tag{79}$$

The proofs of Theorems 7 and 8 may now be completed along established lines.

Theorems 7 and 8, and in particular the appearance of the second integral in (75) appear to be similar to Theorem 6 involving solutions of $f_t + \frac{1}{2}f_{xx} = 0$ on the half-line $0 < x < \infty$. There, however, the situation is actually quite different. For Brownian motion on $(0, \infty)$ the point 0 is an accessible boundary point $(P_{(t,x)}\{T_0 < \tau\} > 0$ for

$\tau > t$), and hence if f is the restriction to $(0, \infty) \times (0, \infty)$ of a solution of (48) which does not vanish for $x = 0$, the necessity for the second integral in (69) is apparent from (51) with $a = 0$. On the other hand, for the process $r(t)$ it follows from (78) that $P_{(t,x)}\{T_0 < \infty\} = 0$ for all $0 < t$, $x < \infty$, and hence it is only by actual computation that we can decide whether equality holds in (52) with r in place of w, i.e. whether

$$\liminf_{a \to 0} \int_{(T_a < \tau)} f(T_a, a)\, dP_{(t,x)} = 0 .$$

No obvious qualitative argument suffices.

The distinguishing feature of the present case is that 0 is an entrance boundary for r whereas the boundary at $+\infty$ for r and the boundaries at $\pm\infty$ for w are natural boundaries. (For the definitions of these terms see [4, p. 108].) It has recently been shown by T. L. Lai [5] that the phenomenon of Theorems 7 and 8 is quite general, in the following sense. For an arbitrary conservative, non-singular linear diffusion with inaccessible boundaries $-\infty \leq r_0 < r_1 \leq \infty$, if h is an increasing nonnegative solution of $\mathcal{D}h = \lambda h$ for some $0 \leq \lambda < \infty$ ($\mathcal{D}$ is now the characteristic operator of the process), then the function $f(t, x) = h(x) \exp(-\lambda t)$, which satisfies $f_t + \mathcal{D}f = 0$, is a martingale function if r_1 is natural, but not if r_1 is entrance. The case of general f remains an unsolved problem.

5. **Boundary crossing probabilities for $r(t)$.** In Chapter I of this paper we studied a statistical problem in which it was desirable to compute the probability that a Wiener process $w(t)$ ever exceeds a boundary curve $g(t)$ for some $t>0$. It is interesting to see whether a similar method allows one to compute such probabilities for other stochastic processes.

In [9] we also discussed the process $r(t)$ of §4, and showed that if

$$f_1(t, x) = \int_{0+}^{\infty} (yx)^{-1} (\sinh yx) \exp(-y^2 t/2)\, dF_1(y) \tag{80}$$

and

$$g_1(t, b) = \inf \{x : f_1(t, x) \geq b\} ,$$

then results analogous to Theorem 2(i) and (ii) hold for $r(t)$. For example, if F is as in Example 4 of Chapter I, then (for $n \geq 3$)

$$g_1(t, b) = [2t(\log_2 t + (5/2)\log_3 t + \sum_4^n \log_k t + (1+\delta)\log_{n+1} t + \log(2b/\pi^{1/2}) + o(1))]^{1/2} \tag{81}$$

as $t \to \infty$. (This result should be compared with equation (14) of [4, p. 163].)

Since $P_{(t_0, x_0)}\{r(t) \to \infty\} = 1$ there exist functions $g(t) \to \infty$ as $t \to \infty$ for which

$$P_{(t_0, x_0)}\{r(t) \leq g(t) \quad \text{for some} \quad t > t_0\} < 1, \tag{82}$$

and the question was raised in [9] whether our methods allowed the evaluation of the left-hand side of (82) for some of these functions. We can now answer this question in the affirmative.

Define

$$0 < f_2(t, x) = \int_t^\infty (s-t)^{-3/2} \varphi(x/(s-t)^{1/2}) \, dF_2(s)$$

for some measure F_2 for which $f_2(t, x) < \infty$ for all $(t, x) \in (0, \infty) \times (0, \infty)$ and for $b > 0$ let $g_2(t, b)$ be the unique solution of $f_2(t, x) = b$. Since $f_2(t, \cdot)$ is decreasing, it follows that $x \leq g_2(t, b)$ if and only if $f_2(t, x) \geq b$.

THEOREM 9. *For any* $(t_0, x_0) \in (0, \infty) \times (0, \infty)$ *and* $b > f_2(t_0, x_0)$,

$$P_{(t_0, x_0)}\{r(t) \leq g_2(t, b) \quad \textit{for some} \quad t > t_0\} = f_2(t_0, x_0)/b.$$

To obtain results analogous to (21) and (81), let

$$\begin{aligned} dF_2(s) &= s^{1/2}[\log s \log_2 s \cdots (\log_n s)^{1+\delta}]^{-1} ds, \qquad \text{for } s \geq e_n, \\ &= 0, \qquad \text{elsewhere}. \end{aligned} \tag{83}$$

Then it will be shown that

$$g_2(t, b) \sim t^{1/2}/b[\log t \log_2 t \cdots (\log_n t)^{1+\delta}] \tag{84}$$

as $t \to \infty$ (cf. equation (15) of [4, p. 164]).

Proof of Theorem 9. Let

$$T = \inf\{t : t > t_0, r(t) \leq g_2(t, b)\} = \inf\{t : t > t_0, f_2(t, r(t)) \geq b\}.$$

By (76) for each a $\{f(T_a \wedge t, r(T_a \wedge t)), t_0 \leq t < \infty\}$ is a martingale and hence so is $\{f(T \wedge T_a \wedge t, r(T \wedge T_a \wedge t)), t_0 \leq t < \infty\}$. Letting $a \to 0$, since $P_{(t,x)}\{T_a = \infty\} \to 1$, by the dominated convergence theorem $\{f(T \wedge t, r(T \wedge t)), t_0 \leq t < \infty\}$ is a martingale, and hence Theorem 9 follows at once from Lemma 2, provided that

$$f_2(t, r(t)) \to 0$$

in probability as $t \to \infty$.

From the definition of f_2 it follows that

$$\int_t^\infty s^{-3/2}\, dF_2(s) < \infty \qquad (t > 0)\,. \tag{85}$$

For any $\varepsilon > 0$

$$\begin{aligned} f_2(t, \varepsilon t^{1/2}) &= \int_t^\infty (s/(s-t))^{3/2} \varphi(\varepsilon(t/(s-t))^{1/2})\, dF_2(s)/s^{3/2} \\ &= \int_t^\infty (1 + t/(s-t))^{3/2} \varphi(\varepsilon(t/(s-t))^{1/2}\, dF_2(s)/s^{3/2}\,. \end{aligned}$$

It is easy to see that the function $g(x) = (1+x)^{3/2} \exp(-\varepsilon^2 x/2)$ is bounded on $[0, \infty)$ and hence by (85)

$$f_2(t, \varepsilon t^{1/2}) \to 0 \quad \text{as} \quad t \to \infty\,.$$

Since $r(t)/t^{1/2}$ has a distribution G not depending on t, for any δ, $\varepsilon < 0$ and sufficiently large t

$$P_{(0,0)}\{f_2(t, r(t)) \geq \delta\} \leq P_{(0,0)}\{r(t) \leq \varepsilon t^{1/2}\} = G(\varepsilon)\,,$$

which can be made arbitrarily small by taking ε sufficiently small.

Proof of (84). If F_2 is given by (83) with $n = 1$ (the general case is similar), for $t > e$ we may rewrite $f_2(t, x)$ after a change of variable as

$$f_2(t, x) = \int_0^\infty \varphi(xu^{1/2})(1 + tu)^{1/2}\, du/u[\log(t + 1/u)]^{1+\delta}\,.$$

Putting $g = g_2(t, b)$, we have by definition

$$b = (2\pi)^{-1/2} \int_0^\infty \exp(-g^2 u/2)(1 + tu)^{1/2}\, du/u[\log(t + 1/u)]^{1+\delta}\,. \tag{86}$$

Making the change of variable $x = ut$, we see that

$$b \leq \int_0^\infty \exp(-g^2 x/2t)(1 + x)^{1/2}\, dx/x\, (\log t)^{1+\delta}\,,$$

from which we conclude that

(87) $$g = o(t^{1/2}) \quad \text{as} \quad t \to \infty.$$

Let $0 < \alpha < \beta < \infty$. Then

$$\begin{aligned} b &\geq (2\pi)^{-1/2} \int_{\alpha g^{-2}}^{\beta g^{-2}} \exp(-g^2u/2)(1+tu)^{1/2}\, du/u[\log(t+1/u)]^{1+\delta} \\ &\geq (t^{1/2}/g[\log(t(1+g^2/\alpha t))]^{1+\delta}) \int_{\alpha g^{-2}}^{\beta g^{-2}} \exp(-g^2u/2)g^2du/g(2\pi u)^{1/2} \\ &= (t^{1/2}/g[\log(t+o(1))]^{1+\delta}) \int_{\alpha}^{\beta} \exp(-x/2)\, dx/(2\pi x)^{1/2} \end{aligned}$$

as $t \to \infty$. Since α and β are arbitrary, by letting $\alpha \to 0$ and $\beta \to \infty$, we obtain

(88) $$b \geq (t^{1/2}/g(\log t)^{1+\delta})(1+o(1)) \quad \text{as} \quad t \to \infty.$$

To prove the reverse inequality, let $0 < \alpha < \beta < \infty$. Then from (86)

(89) $$b = \int_0^{\alpha g^{-2}} + \int_{\alpha g^{-2}}^{\beta g^{-2}} + \int_{\beta g^{-2}}^{\infty}.$$

Now

$$\begin{aligned} \int_{\beta g^{-2}}^{\infty} &\leq (t^{1/2}/g(\log t)^{1+\delta}) \int_{\beta g^{-2}}^{\infty} \exp(-g^2u/2)(g^2/t+ug^2)^{1/2}\, du/u(2\pi)^{1/2} \\ &= (t^{1/2}/g(\log t)^{1+\delta}) \int_{\beta}^{\infty} \exp(-x/2)(g^2/t+x)^{1/2}\, dx/x(2\pi)^{1/2}, \end{aligned}$$

which by (87) and bounded convergence

$$= (t^{1/2}/g(\log t)^{1+\delta}) \int_{\beta}^{\infty} \exp(-x/2)\, dx/(2\pi x)^{1/2}(1+o(1)) \quad \text{as} \quad t \to \infty.$$

Since β is arbitrary and $\int_0^{\infty} \exp(-x/2)\, dx/(2\pi x)^{1/2} = 1$, to complete the proof it suffices to show that for a suitable choice of α the first two integrals in (89) can be made arbitrarily small for β near 0. Now

$$\begin{aligned} (2\pi)^{1/2} \int_0^{\alpha g^{-2}} &\leq (t^{1/2}/g) \int_0^{\alpha} (g^2/t+x)^{1/2}\, dx/x[\log t(1+g^2/tx)]^{1+\delta} \\ &\leq (t^{1/2}/g) \int_0^{\alpha} (g^2/t+x)^{1/2}\, dx/x[\log(g^2/tx)]^{1+\delta} \\ &\leq (t^{1/2}/g)(g^2/t+\alpha)^{1/2}/\delta[\log(g^2/t\alpha)]^{\delta}. \end{aligned}$$

With $\alpha = \eta g^2/t$, as $t \to \infty$, this becomes

$$(1+\eta)^{1/2}/\delta(\log \eta^{-1})^{\delta} \to 0 \quad \text{as} \quad \eta \to 0.$$

Also

$$(2\pi)^{1/2}\int_{\alpha g^{-2}}^{\beta g^{-2}} \leq (t^{1/2}/g(\log t)^{1+\delta})\int_{\alpha}^{\beta}(g^2/t+x)^{1/2}\,dx/x\,,$$

and for $\alpha=\eta g^2/t$

$$\begin{aligned}\int_{\alpha}^{\beta}(g^2/t+x)^{1/2}\,dx/x &= \int_{\alpha}^{\beta}(g^2/tx^{1/2}+x^{1/2})^{1/2}\,dx/x^{3/4}\\ &\leq (g^2/t\alpha^{1/2}+\beta^{1/2})^{1/2}\int_0^{\beta}x^{-3/4}\,dx\\ &\leq (g/(t\eta)^{1/2}+\beta^{1/2})^{1/2}\,4\beta^{1/4}\,,\end{aligned}$$

which by (87) is majorized by $8\beta^{1/2}$ (independent of η) as $t\to\infty$. Hence by first letting $t\to\infty$, then $\eta\to 0$ and then $\beta\to 0$, we have

$$b\leq (t^{1/2}/g(\log t)^{1+\delta})(1+o(1))\,,$$

which together with (88) proves (84).

6. **Relation to the Martin Boundary.** Many of the results of this chapter, in particular Theorems 3 and 8, are formally special cases of the general Martin Boundary Theory (e. g. [2]) applied to the space-time processes $(t, w(t))$ (Theorem 3) and $(t, r(t))$ (Theorem 8). However, there are certain technical assumptions contained in [2] which appear not to hold for space-time processes. Our approach is simple and direct but has the computational disadvantage that it requires a considerable amount of knowledge of certain first-passage time distributions, whereas the general Martin Boundary Theory requires knowledge only of the Green's function—in the case of space-time processes, the transition probability of the process.

It would be interesting to obtain the results of Chapter II as an application of Martin Boundary Theory, or better yet to give for a general class of Markov processes on $(0,\infty)$, say, a complete description of the possible representations of solutions of (74) in terms of the boundary behavior of the process.

Added in proof. Professor S. Sawyer has observed that f satisfies (68) if and only if $x^{-1}f(x,t)$ satisfies (74) on $(0,\infty)\times(0,l)$ so that Theorems 6 and 8 are equivalent. He has also generalized this argument to show that the integral representations of Theorems 6 and 8 exhibit all the possibilities for arbitrary one-dimensional diffusion processes. Specifically, in representing positive space-time harmonic functions there exists an integral on the boundary for accessible (Theorem 6 boundary at 0) and entrance boundaries (Theorem 8 boundary at 0) but not for natural (Theorem 6 boundary at $+\infty$) and inaccessible exit boundaries (Theorem 8 boundary at $+\infty$). Proofs of these results will be published elsewhere.

REFERENCES

1. D. Darling and H. Robbins, *Iterated logarithm inequalities*, Proc. Nat. Acad. Sci. U.S.A. **57** (1967), 1188–1192.
2. E. B. Dynkin, *The space of exits of a Markov process*, Russian Math. Surveys **24** (1969), 89–157.
3. R. Farrell, *Asymptotic behavior of expected sample size in certain one-sided tests*, Ann. Math. Statist. **35** (1964), 36–72.
4. K. Itô and H. P. McKean, Jr., *Diffusion processes and their sample paths*, Springer-Verlag, Berlin, 1965.
5. T. L. Lai, *Confidence sequences and martingales*, Columbia University dissertation, 1971.
6. M. Loeve, *Probability theory*, 3rd ed., Van Nostrand, Princeton, N. J., 1963.
7. M. Protter and H. Weinberger, *Maximum principles in differential equations*, Prentice-Hall, Englewood Cliffs, N. J., 1967.
8. H. Robbins, *Statistical methods related to the law of the iterated logarithm*, Ann. Math. Statist. **41** (1970), 1397–1409.
9. H. Robbins and D. Siegmund, *Boundary crossing probabilities for the Wiener process and partial sums*, Ann. Math. Statist. **41** (1970), 1410–1429.
10. ———, *A class of stopping rules for testing parametric hypotheses*, Proc. Sixth Berkeley Sympos. Math. Statist. and Probability (to appear).
11. ———, *The expected sample size of some tests of power one*, Technical Report, Columbia University, 1972.
12. A. Wald, *Sequential analysis*, John Wiley, New York, 1947.
13. D. V. Widder, *Positive temperatures on an infinite rod*, Trans. Amer. Math. Soc. **55** (1944), 85–95.
14. ———, *Positive temperatures on a semi-infinite rod*, Trans. Amer. Math. Soc. **75** (1953), 510–525.

DEPARTMENT OF MATHEMATICAL STATISTICS, COLUMBIA UNIVERSITY, NEW YORK, NEW YORK 10027, U.S.A.

Part Three

PROBABILITY AND INFERENCE

We begin this part of the volume with Robbins' first paper [7]* in the field of probability and statistics. This research arose from overhearing a conversation between two senior naval officers after Robbins had enlisted in the Navy during the war. The problem which the naval officers were discussing can be described as follows. To knock out an airstrip one can drop a large number of bombs, but since the bomb impacts overlap in a random manner, it does not do much good to obliterate the same area many times. The naval officers asked how many bombs were necessary to knock out 90% of the area with probability close to one, taking into account the randomness of impact patterns. Robbins offered them a suggestion for attacking the problem. Because of his lack of appropriate security clearance, he was not recruited to pursue the classified research project. Nevertheless, his work on this problem led to the fundamental papers [7] and [10] in the field of geometric probability in addition to contributing to the military effort.

After the war, Robbins was offered the position of associate professor of mathematical statistics by the University of North Carolina at Chapel Hill. He accepted the position and spent the next six years at Chapel Hill. During this relatively short period he not only studied and developed an increasingly deep interest in probability and statistics, but also made a number of profound contributions to his new field. The earliest of these contributions is the paper [12]* with P. L. Hsu, where the notion of complete convergence is formulated and a beautiful refinement of the strong law of large numbers is established. This work motivated the classical paper of Erdős (1949), which in turn led to a long sequence of papers in the 1950's and 1960's, reaching its culmination with Baum and Katz (1965). Along this line the paper [76] with Siegmund and Wendel in 1968 relates complete conver-

*Papers reprinted in this book are marked here by an asterisk when they first appear.

gence to the last time τ of boundary crossings for sample sums and gives certain asymptotic results for the moments of τ. This work provides a new direction in the development of the subject, which was undertaken in the 1970's by Slivka and Severo (1970), Chow and Lai (1975, 1978), Lai (1974, 1976, 1977), Gut (1978), Hipp (1979) and others.

Another important contribution in probability theory at Chapel Hill was Robbins' paper [16]* with Hoeffding on the central limit theorem for dependent random variables. This paper and Bernstein's (1927) pioneering work are precursors to Rosenblatt's (1956) classical paper on the central limit theorem for strongly mixing random variables.

In addition to his seminal work in compound decision theory, stochastic approximation, and sequential design of experiments presented in Parts I and II, his other well-known contributions to mathematical statistics at Chapel Hill include the distribution of quadratic forms [18], [21]* (with Pitman), and the Chapman–Robbins inequality [27]* that generalizes the classical Cramer–Rao inequality.

Robbins spent the year 1952–53 at the Institute for Advanced Study on a Guggenheim Fellowship, and then moved to Columbia University in 1953. During this period he wrote a series of important papers [30]*, [31]*, [32]*, and [33] on occupation times and ergodic behavior of random walks and Brownian motion. One well-known contribution is the Kallianpur–Robbins law for the occupation time of two-dimensional Brownian motion [30]. Another is his contribution together with T. E. Harris to ergodic theory of Markov chains having an invariant distribution of infinite total mass [32].

In 1955 appeared Robbins' short note [38]* on Stirling's formula. The paper provides a new proof and a sharp refinement of this basic result in probability. Feller (1957) incorporated this proof in the second edition of his classic textbook. In Feller's (1957) chapter on laws of large numbers, the so-called "Petersburg game" (dating back to Daniel Bernoulli) is introduced as an example of the limiting behavior of sample sums without finite first moments. The papers [48]* (with Chow) and [50] in 1961 are Robbins' well-known contributions to the Petersburg game. A related earlier contribution is the Derman–Robbins strong law of large numbers when the first moment does not exist [40].

A major thrust of Robbins' research in probability during the 1960's is his collaborative work with Chow on optimal stopping. Although this subject had its origin in the classic paper of Arrow, Blackwell, and Girshick (1949), which was refined in certain respects by Snell (1952), Chow and Robbins showed that it was a much richer subject than the early literature suggested. Reprinted here are their first two papers [49]*, [55]* on the general theory of optimal stopping, their famous joint paper [61]* with Moriguti and Samuels on the "secretary problem," and Robbins' expository article [85]*. Other important contributions to optimal stopping are the papers [62], [66], [67], and the monograph [85] (with Chow and Siegmund), which also contains an extensive bibliography on the subject. A more recent contribution is due to Berk (1975), who used optimal stopping theory to establish that a locally optimal sequential test for one-sided problems in exponential families is a sequential probability ratio test. A closely related area of Robbins' research during this period is the theory of randomly stopped sums [56], [63]*, [65].

Two other important contributions during the 60's are the papers [60]* (with A. Dvoretzky) and [75]*, on the "parking problem" and on "estimating" the total probability of the unobserved outcomes of an experiment respectively.

In the 1970's Robbins developed a growing interest in the application of probabilistic-statistical methods in legal proceedings. His earliest work in this area is the paper [93]* (with M. O. Finkelstein, a lawyer) on election challenges. During this period, he also introduced the concept of "maximally dependent" random variables in studying extreme value theory, and wrote the joint papers [102]* and [106] with Lai on the subject. Motivated by applications to stochastic approximation and adaptive control, he worked with Lai and Wei towards a definitive solution of the strong consistency problem concerning least squares estimates in regression models, leading to the papers [105], [110]*, [111] and [119]. In recent years, his work in statistical methods for legal proceedings has been concerned primarily with discrimination suits based on sex, race, and age, and has led to the fascinating papers [125]*, [132], and [133] with B. Levin and M. O. Finkelstein.

References

Arrow, K. J., Blackwell, D., and Girshick, M. A. (1949). Bayes and minimax solutions of sequential decision problems, *Econometrica* **17**, 213–244.

Baum, L. and Katz, M. (1965). Convergence rates in the law of large numbers. *Trans. Amer. Math. Soc.* **120**, 108–123.

Berk, R. H. (1975). Locally most powerful sequential tests, *Ann. Statist.* **3**, 373–381.

Bernstein, S. (1927). Sur l'extension du théorème limite du calcul des probabilités aux sommes de quantités dépendentes. *Math. Ann.* **97**, 1–59.

Chow, Y. S. and Lai, T. L. (1975). Some one-sided theorems on the tail distribution of sample sums with applications to the last time and largest excess of boundary crossings. *Trans. Amer. Math. Soc.* **208**, 51–72.

Chow, Y. S. and Lai, T. L. (1978). Paley-type inequalities and convergence rates related to the law of large numbers and extended renewal theory.

Erdős, P. (1949). On a theorem of Hsu and Robbins. *Ann. Math. Statist.* **20**, 286–291.

Feller, W. (1957). *An Introduction to Probability Theory and Its Applications*, Vol. 1. Second edition, Wiley, New York.

Gut, A. (1978). Marcinkiewicz laws and convergence rates in the law of large numbers for random variables with multidimensional indices. *Ann. Probab.* **6**, 469–482.

Hipp, C. (1979). Convergence rates of the strong law for stationary mixing sequences. *Z. Wahrsch. Verw. Gebiete* **49**, 49–62.

Lai, T. L. (1974). Limit theorems for delayed sums. *Ann. Probab.* **2**, 432–440.
Lai, T. L. (1976). On r-quick convergence and a conjecture of Strassen. *Ann. Probab.* **4**, 612–627.
Lai, T. L. (1977). Convergence rates and r-quick versions of the strong law for stationary mixing sequences. *Ann. Probab.* **5**, 693–706.
Rosenblatt, M. (1956). A central limit theorem and a strong mixing condition. *Proc. Natl. Acad. Sci. USA* **42**, 43–47.
Slivka, J. and Severo, N. C. (1970). On the strong law of large numbers. *Proc. Amer. Math. Soc.* **24**, 729–734.
Snell, J. L. (1952). Application of martingale system theorem, *Trans. Amer. Math. Soc.* **73**, 293–312.

Reprinted from
Ann. Math. Statist.
15, 70–74 (1944)

ON THE MEASURE OF A RANDOM SET

BY H. E. ROBBINS

Post Graduate School, U. S. Naval Academy

1. Introduction. The following is perhaps the simplest non-trivial example of the type of problem to be considered in this paper. On the real number axis let N points x_i ($i = 1, 2, \cdots, N$) be chosen independently and by the same random process, so that the probability that x_i shall lie to the left of any point x is a given function of x,

$$\sigma(x) = \Pr(x_i < x). \tag{1}$$

With the points x_i as centers, N unit intervals are drawn. Let X denote the set-theoretical sum of the N intervals, and let $\mu(X)$ denote the linear measure of X. Then $\mu(X)$ will be a chance variable whose values may range from 1 to N, and whose probability distribution is completely determined by $\sigma(x)$. Let $\tau(u)$ denote the probability that $\mu(X)$ be less than u. Then by definition, the expected value of $\mu(X)$ is

$$E(\mu(X)) = \int_1^N u \, d\tau(u), \tag{2}$$

where

$$\tau(u) = \Pr(\mu(X) < u). \tag{3}$$

The problem is to transform the expression for $E(\mu(X))$ so that its value may be computed in terms of the given function $\sigma(x)$.

In order to do this, we observe that, since the x_i are independent,

$$\tau(u) = \int \cdots \int_{C(u)} d\sigma(x_1) \cdots d\sigma(x_N), \tag{4}$$

where the domain of integration $C(u)$ consists of all points $(x_1, \cdots, x_N)$ in Euclidean N-dimensional space such that the linear measure of the set-theoretical sum of N unit intervals with centers at the points x_i is less than u. Here, however, a difficulty arises. Due to the possible overlapping of the intervals, the geometrical description of the domain $C(u)$ is such as to make the explicit evaluation of the integral (4) a complicated matter.

The difficulty is even more serious in the analogous problem where instead of N unit intervals on the line we have N unit circles in the plane, with a given probability distribution for their centers (x_i, y_i). Again we seek the expected value of the measure of the set-theoretical sum of the N circles. The corresponding domain $C(u)$ in $2N$-dimensional space will now be very complicated.

It is the object of this paper to show how, in such cases as these, the expected value of $\mu(X)$ may be found without first finding the distribution function $\tau(u)$.

In fact, the theorem to be stated in (15) will in many important cases yield a comparatively simple formula for $E(\mu(X))$.

2. Expected value of $\mu(X)$. In order to state the problem in full generality, let us suppose that X is a random Lebesgue measurable subset of Euclidean n dimensional space E_n. By this we shall mean that in the space T of all possible values of X there is defined a probability measure $\rho(X)$ so that for every ρ-measurable subset S of T, the probability that X shall belong to S is given by the Lebesgue-Stieltjes integral

$$\text{Pr}\,(X \in S) = \int_T C_S(X)\, d\rho(X), \tag{5}$$

where the integrand is the characteristic function of S,

$$C_S(X) = \begin{cases} 1 & \text{for} \quad X \in S \\ 0 & \text{for} \quad X \notin S. \end{cases} \tag{6}$$

In practice, the set X will be a function of a finite number of real parameters (e.g., the coordinates of the centers of the intervals or circles considered in the Introduction), $X = X(\alpha_1, \cdots, \alpha_r) = X(\alpha)$. There will be given a probability measure $\nu(\alpha)$ in the parameter space E_r, so that α will be a vector random variable in the ordinary sense. If A is any ν-measurable subset of E_r, then by definition,

$$\text{Pr}\,(\alpha \in A) = \int_{E_r} C_A(\alpha)\, d\nu(\alpha). \tag{7}$$

Now for the set S' consisting of all X such that $X = X(\alpha)$ for α in A, we define $\rho(S') = \nu(A)$. Thus a ρ-measure is defined in the space T of X, which is the general situation considered in the preceding paragraph.

Returning to the general case described in the first paragraph of this section, we shall now prove the main theorem of this paper. To this end we define, for every point x of E_n and every set X of T, the function

$$g(x, X) = \begin{cases} 1 & \text{for} \quad x \in X \\ 0 & \text{for} \quad x \notin X. \end{cases} \tag{8}$$

Moreover, for every x in E_n we let $S(x)$ denote the set of all X in T which contain x. Then for every x in E_n we have from (6),

$$g(x, X) = C_{S(x)}(X). \tag{9}$$

Let us denote the Lebesgue measure in E_n of the set X by $\mu(X)$. Assuming that the function $g(x, X)$ is a $\mu\rho$-measurable function of the pair (x, X) in the product space[1] of E_n with T, it follows from Fubini's theorem[1] that

$$\int_{E_n \times T} g(x, X)\, d\mu\rho(x, X) = \int_{E_n} \int_T g(x, X)\, d\rho(X)\, d\mu(x). \tag{10}$$

[1] See S. Saks, *Theory of the Integral*, G. E. Stechert, N. Y., 1937, pp. 86, 87.

From (5) and (9) it follows that

$$\int_T g(x, X)\, d\rho(X) = \Pr(X \in S(x)) = \Pr(x \in X). \tag{11}$$

Again by Fubini's theorem we have

$$\int_{E_n \times T} g(x, X)\, d\mu\rho(x, X) = \int_T \int_{E_n} g(x, X)\, d\mu(x)\, d\rho(X). \tag{12}$$

But from (8),

$$\int_{E_n} g(x, X)\, d\mu(x) = \int_X d\mu(x) = \mu(X). \tag{13}$$

Now from (10), (11), (12), and (13) we have

$$\int_{E_n} \Pr(x \in X)\, d\mu(x) = \int_T \mu(X)\, d\rho(X). \tag{14}$$

But the latter integral is equal to $E(\mu(X))$. Hence we have the relation

$$E(\mu(X)) = \int_{E_n} \Pr(x \in X)\, d\mu(x). \tag{15}$$

This is our fundamental result. We may state it as a

THEOREM: *Let X be a random Lebesgue measurable subset of E_n, with measure $\mu(X)$. For any point x of E_n let $p(x) = \Pr(x \in X)$. Then, assuming that the function $g(x, X)$ defined by (8) is a measurable function of the pair (x, X), the expected value of the measure of X will be given by the Lebesgue integral of the function $p(x)$ over E_n.*

3. Higher moments of $\mu(X)$. We may generalize the result (15) to obtain similar expressions for the higher moments of $\mu(X)$. For the second moment we have the expression

$$E(\mu^2(X)) = \int_T \mu^2(X)\, d\rho(X). \tag{16}$$

Now from (13),

$$\begin{aligned} \mu^2(X) = \mu(X)\cdot\mu(X) &= \int_{E_n} g(x, X)\, d\mu(x) \cdot \int_{E_n} g(y, X)\, d\mu(y) \\ &= \int_{E_n}\int_{E_n} g(x, X)\cdot g(y, X)\, d\mu(x)\, d\mu(y). \end{aligned} \tag{17}$$

Let

$$g(x, y, X) = g(x, X)\cdot g(y, X) \begin{array}{l} = 1 \text{ if } X \text{ contains both } x \text{ and } y \\ = 0 \text{ otherwise.} \end{array} \tag{18}$$

Then from (16), (17), and (18), we have as before by Fubini's theorem,

$$
(19)\quad
\begin{aligned}
E(\mu^2(X)) &= \int_T \int_{E_n} \int_{E_n} g(x, y, X)\, d\mu(x)\, d\mu(y)\, d\rho(X) \\
&= \int_{E_n} \int_{E_n} \int_T g(x, y, X)\, d\rho(X)\, d\mu(x)\, d\mu(y).
\end{aligned}
$$

But from (5) and (18) it follows that

$$(20)\qquad \int_T g(x, y, X)\, d\rho(X) = \Pr\,(x \in X \text{ and } y \in X).$$

The latter probability may be denoted by $p(x, y)$. This function will be defined over the Cartesian product, E_{2n}, of E_n with itself. Let $\mu(x, y)$ denote Lebesgue measure in E_{2n}. Then from (19) we have

$$(21)\qquad E(\mu^2(X)) = \int_{E_{2n}} p(x, y)\, d\mu(x, y),$$

where

$$(22)\qquad p(x, y) = \Pr\,(x \in X \text{ and } y \in X).$$

The formula for the mth moment of $\mu(X)$ will clearly be

$$(23)\qquad \operatorname{Exp}\,(\mu^m(X)) = \int_{E_{mn}} p(x_1, x_2, \cdots, x_m)\, d\mu(x_1, x_2, \cdots, x_m),$$

where $\mu(x_1, x_2, \cdots, x_m)$ denotes Lebesgue measure in E_{mn} and where

$$(24)\qquad p(x_1, x_2, \cdots, x_m) = \Pr\,(x_1 \in X \text{ and } x_2 \in X \cdots \text{ and } x_m \in X).$$

In the next section we shall apply formulas (15) and (21) to a specific problem.

4. Let a, p, B be given positive numbers such that $(B + a)p \leq a$ and $a \leq B$. We shall define the random linear point set X as follows. N intervals, each of length a, are chosen independently on the number axis. The probability density function for the center of the ith interval will be assumed to be constant and equal to p/a in the interval $-a/2 \leq x \leq B + (a/2)$; it may be arbitrary outside this interval. The set X is now defined as the intersection of the fixed interval I: $0 \leq x \leq B$ with the variable set-theoretical sum of the N intervals. The hypothesis of (15) is clearly satisfied. The probability that any point x in the interval I shall be contained in the ith interval of length a is clearly $(p/a)a = p$. From this it follows that

$$(25)\qquad \Pr\,(x \in X) = p(x) = \begin{cases} 1 - (1 - p)^N \text{ for } 0 \leq x \leq B \\ 0 \text{ elsewhere.} \end{cases}$$

From (15) it follows that

$$(26)\qquad E(\mu(X)) = \int_0^B p(x)\, dx = B(1 - (1 - p)^N).$$

(The same formula holds in the case where the N intervals of length a are replaced by N circles of area a and I by a plane domain of area B, provided that for every point of the domain the probability of being contained in the ith circle is equal to a constant p. A similar remark holds for spheres in space.)

To evaluate $E(\mu^2(X))$ in the linear case we make use of the identity

$$\text{Pr}\ (A \text{ and } B) = \text{Pr}\ (A) + \text{Pr}\ (B) + \text{Pr}\ (\text{neither } A \text{ nor } B) - 1, \tag{27}$$

which holds for any two events A and B. It follows from (27) and (25) that if x and y are any two points of I, then

$$\begin{aligned} p(x, y) &= \text{Pr}\ (x \in X \text{ and } y \in X) \\ &= \text{Pr}\ (x \in X) + \text{Pr}\ (y \in X) + \text{Pr}\ (x \notin X \text{ and } y \notin X) - 1 \\ &= 1 - 2(1-p)^N + \text{Pr}\ (x \notin X \text{ and } y \notin X). \end{aligned} \tag{28}$$

Let

$$h(x, y) = \text{Pr}\ (x \notin X \text{ and } y \notin X). \tag{29}$$

Then

$$h(x, y) = \begin{cases} [1 - (p/a)2a]^N = (1-2p)^N, & \text{for } |y - x| \geq a \\ [1 - (p/a)(a + |y - x|)]^N = \left(\dfrac{a - ap - p|y - x|}{a}\right)^N, & \text{for } |y - x| < a. \end{cases} \tag{30}$$

Now from (21), (28), and (29) we have

$$\begin{aligned} E(\mu^2(X)) &= \int_0^B \int_0^B p(x, y)\, dy\, dx \\ &= \int_0^B \int_0^B [1 - 2(1-p)^N + h(x, y)]\, dy\, dx \\ &= B^2[1 - 2(1-p)^N] + 2 \int_0^B \int_x^B h(x, y)\, dy\, dx. \end{aligned} \tag{31}$$

When the latter integral is evaluated the result is

$$\begin{aligned} E(\mu^2(X)) = B^2[1 - 2(1-p)^N] + (B - a)^2(1 - 2p)^N \\ + \frac{2aB(1-p)^{N+1}}{(N+1)p} - \frac{2a(B-a)(1-2p)^{N+1}}{(N+1)p} \\ - \frac{2a^2}{(N+1)(N+2)p^2}[(1-p)^{N+2} - (1-2p)^{N+2}]. \end{aligned} \tag{32}$$

Combining this with (26), we find for the variance of $\mu(X)$ the expression

$$\begin{aligned} \sigma^2 &= E(\mu^2(X)) - [E(\mu(X))]^2 \\ &= (B - a)^2(1-2p)^N - B^2(1-p)^{2N} + \frac{2aB(1-p)^{N+1}}{(N+1)p} \\ &\quad - \frac{2a(B-a)(1-2p)^{N+1}}{(N+1)p} - \frac{2a^2}{(N+1)(N+2)p^2}[(1-p)^{N+2} - (1-2p)^{N+2}]. \end{aligned} \tag{33}$$

PROCEEDINGS
OF THE
NATIONAL ACADEMY OF SCIENCES
Volume 33 February 15, 1947 Number 2

COMPLETE CONVERGENCE AND THE LAW OF LARGE NUMBERS

By P. L. Hsu and Herbert Robbins
Department of Mathematical Statistics, University of North Carolina

Communicated January 7, 1947

1. We begin by listing some standard definitions in the theory of probability. A *probability space* is a set Ω of elements ω together with a σ-field m of subsets of Ω on which is defined a completely additive measure P such that $P(\Omega) = 1$. A real-valued P-measurable function $X = X(\omega)$ is a *random variable*, and the function $F(x) = P\{X \leq x\}$, where $\{\ \}$ denotes the set of all ω such that the relation within the braces holds, is the *distribution function* of X. The sets of a sequence $A_1, A_2, \ldots$ are *independent* if for every finite set $i_1, \ldots, i_n$ of distinct integers, $P(\prod_{r=1}^{n} A_{i_r}) = \prod_{r=1}^{n} P(A_{i_r})$, and the random variables of a sequence $X_1, X_2, \ldots$ are independent if, for every sequence $x_1, x_2, \ldots$ of real numbers, the sets $\{X_1 \leq x_1\}, \{X_2 \leq x_2\}, \ldots$ are independent.

For purposes of comparison we list the following modes in which a sequence

$$X_1, X_2, \ldots \tag{1}$$

of random variables defined on Ω may converge to 0.

(*i*) The sequence (1) converges to 0 *in probability* if for every $\epsilon > 0$,

$$\lim_{n\to\infty} P\{|X_n| > \epsilon\} = 0.$$

(*ii*) The sequence (1) converges to 0 *with probability* 1 if for every $\epsilon > 0$,

$$\lim_{n\to\infty} P\{\{|X_n| > \epsilon\} + \{|X_{n+1}| > \epsilon\} + \ldots\} = 0.$$

It is easily seen that this is equivalent to the usual condition, $P\{\lim_{n\to\infty} X_n = 0\} = 1$, and that (*ii*) implies (*i*) but not conversely.

2. We shall be concerned with a third mode of convergence, which, for want of a better name, we call *complete.*

(*iii*) The sequence (1) converges to 0 *completely* if for every $\epsilon > 0$,

$$\lim_{n \to \infty} [P\{|X_n| > \epsilon\} + P\{|X_{n+1}| > \epsilon\} + \ldots] = 0.$$

Clearly, (*iii*) implies (*ii*). The example: Ω = unit interval $0 < \omega < 1$, P = Lebesgue measure, $X_n = 1$ for $0 < \omega < \frac{1}{n}$ and 0 otherwise, shows that (*ii*) does not imply (*iii*).

Let us call two sequences of random variables $X_1, X_2, \ldots$ and $Y_1, Y_2, \ldots$, defined, respectively, on probability spaces Ω and Ω_1, *F-equivalent*, if for every n the distribution function of Y_n is identical with that of X_n. Definitions (*i*) and (*iii*) are invariant under F-equivalence, while (*ii*) is not. However, *a sequence* $X_1, X_2, \ldots$ *of random variables converges to* 0 *completely if and only if every F-equivalent sequence converges to* 0 *with probability* 1. The necessity is obvious; to prove sufficiency consider a sequence $Y_1, Y_2, \ldots$ of independent random variables F-equivalent to the given sequence. If the sequence $Y_1, Y_2, \ldots$ converges to 0 with probability 1 then for any $\epsilon > 0$, $P\{\limsup_{n \to \infty}\{|Y_n| > \epsilon\}\} = 0$. Since the sets $\{|Y| > \epsilon\}$ are independent, it follows from a theorem of Borel-Cantelli[1] that $\sum_{n=1}^{\infty} P\{|Y_n| > \epsilon\} = \sum_{n=1}^{\infty} P\{|X_n| > \epsilon\} < \infty$.

It follows from this proof that if $X_1, X_2, \ldots$ is a sequence of *independent* random variables, then definitions (*ii*) and (*iii*) are equivalent.

3. Let the random variables X_n in (1) be independent with the same distribution function $F(x) = P\{X_n \leq x\}$ and such that the expectation $E(X_n) = \int_{-\infty}^{\infty} x dF(x) = 0$. The *strong law of large numbers* for identically distributed random variables states that the sequence of random variables $Y_1, Y_2, \ldots$, where for each n

$$Y_n = (X_1 + \ldots + X_n)/n \tag{2}$$

converges to 0 with probability 1. We shall show in Theorems 1 and 2 that under the same hypotheses the sequence (2) need not converge to 0 completely, but that it will do so under the further hypothesis that $\int_{-\infty}^{\infty} x^2 d F(x) < \infty$.

4. THEOREM 1. *Let (1) be a sequence of independent random variables with the same distribution function* $F(x)$ *and such that*

$$\int_{-\infty}^{\infty} x dF(x) = 0, \quad \sigma^2 = \int_{-\infty}^{\infty} x^2 dF(x) < \infty. \tag{3}$$

Then the sequence (2) converges to 0 completely; i.e., the series

$$\sum_{n=1}^{\infty} P\{|Y_n| > \epsilon\} \tag{4}$$

converges for every $\epsilon > 0$.

Proof. We shall prove the theorem for $\epsilon = 2$. This is no restriction since we can always consider $\frac{2}{\epsilon}X_n$ instead of X_n. Moreover, we may assume that $\sigma^2 > 0$.

Let $f(t) = \int_{-\infty}^{\infty} e^{itx}dF(x)$ be the characteristic function of the distribution $F(x)$. From (3) it follows that constants α, α', α'' exist such that

$$|1 - f(t)| \leq \alpha t^2, \quad |f'(t)| \leq \alpha' t, \quad |f''(t)| \leq \alpha''. \tag{5}$$

Choose and fix a positive δ so small that for $|t| \leq 4\delta$ the following conditions (6) and (7) are satisfied:

$$|\sin {}^1/_2 t| \geq Bt, \; |(1 - f(t))^2 - 4(1 - f(t)) \sin^2 {}^1/_2 t + 4 \sin^2 {}^1/_2 t| \geq Ct^2, \tag{6}$$

where B and C are constants,

$$|f(\ )| \neq 1 \quad \text{except at } t = 0. \tag{7}$$

Let Z be a random variable distributed with the density $3(2\pi)^{-1}x^{-4}\sin^4 x$ and hence the characteristic function

$$\varphi(t) = \begin{cases} 1 - {}^3/_8 t^2 + {}^3/_{32}|t|^3, & |t| \leq 2, \\ {}^1/_{32}(4 - |t|)^2, & 2 \leq |t| \leq 4, \\ 0 & 4 \leq |t|. \end{cases} \tag{8}$$

We regard Z as independent of Y_n and use addition in this sense. Since

$$P\{|Y_n| > 2\} \leq P\left\{\left|Y_n + \frac{Z}{n\delta}\right| > 1\right\} + P\left\{\left|\frac{Z}{n\delta}\right| > 1\right\} =$$
$$P\left\{\left|\frac{Z}{n\delta}\right| \leq 1\right\} - P\left\{\left|Y_n + \frac{Z}{n\delta}\right| \leq 1\right\} + 2P\left\{\left|\frac{Z}{n\delta}\right| > 1\right\},$$

and since

$$\sum_{n=1}^{\infty} P\left\{\left|\frac{Z}{n\delta}\right| > 1\right\} \leq \frac{3}{\pi}\sum_{n=1}^{\infty}\int_{n\delta}^{\infty}\frac{dx}{x^4} = \frac{1}{\pi\delta^3}\sum_{n=1}^{\infty}\frac{1}{n^3} < \infty,$$

it is sufficient to prove that

$$\sum_{n=1}^{N-1}\left[P\left\{\left|\frac{Z}{n\delta}\right| \leq 1\right\} - P\left\{\left|Y_n + \frac{Z}{n\delta}\right| \leq 1\right\}\right] = 0(1), \tag{9}$$

where $0(1)$ always denotes a quantity bounded with respect to N.

The characteristic function $f^n\left(\frac{t}{n}\right)\varphi\left(\frac{t}{n\delta}\right)$ of $Y_n + \frac{Z}{n\delta}$ vanishes for $|t| > 4n\delta$; hence by a well-known inversion formula,[2]

$$P\left\{\left|Y_n + \frac{Z}{n\delta}\right| \leq 1\right\} = \frac{1}{\pi}\int_{-4n\delta}^{4n\delta} f^n\left(\frac{t}{n}\right)\varphi\left(\frac{t}{n\delta}\right)\frac{\sin t}{t}dt = $$
$$\frac{1}{\pi}\int_{-4\delta}^{4\delta} f^n(t)\,\varphi\left(\frac{t}{\delta}\right)\frac{\sin nt}{t}dt.$$

Also,

$$P\left\{\left|\frac{Z}{n\delta}\right| \leq 1\right\} = \frac{1}{\pi}\int_{-4\delta}^{4\delta}\varphi\left(\frac{t}{\delta}\right)\frac{\sin nt}{t}dt.$$

Hence, by subtraction the left side of (9) is equal to

$$\frac{1}{\pi}\int_{-4\delta}^{4\delta}\frac{1}{t}\varphi\left(\frac{t}{\delta}\right)\sum_{n=1}^{N-1}(1 - f^n(t))\sin nt\,dt = \frac{1}{\pi}A_N, \text{ say.} \qquad (10)$$

From now on we write f for $f(t)$. Direct computation gives the result

$$\sum_{n=1}^{N-1}(1 - f^n)\sin nt = \frac{(1-f)^2 \sin {}^1/_2Nt \sin {}^1/_2(N-1)t}{q(t)\sin {}^1/_2t}$$
$$+ \frac{(1-f)\sin t}{q(t)} - \frac{4(1-f)\sin {}^1/_2t \sin {}^1/_2Nt \sin {}^1/_2(N-1)t}{q(t)}$$
$$+ \frac{(1-f)f^N \sin Nt}{q(t)} - \frac{2(1-f^{N+1})\sin {}^1/_2t \cos (N - {}^1/_2)t}{q(t)}, \qquad (11)$$

where

$$q(t) = (f - e^{it})(f - e^{-it}) = (1-f)^2 - 4(1-f)\sin^2 {}^1/_2t + 4\sin^2 {}^1/_2t. \qquad (12)$$

By (6) we have $|q(t)| \geq Ct^2$, $|q(t)\sin {}^1/_2t| \geq C'|t|^3$, where C and C' are constants. Hence when (11) is substituted into (10) and the first inequality of (5) is used, we see that the first three terms merely contribute $0(1)$. Consequently,

$$A_N = \int_{-4\delta}^{4\delta}\varphi\left(\frac{t}{\delta}\right)\frac{(1-f)f^N \sin Nt - 2(1-f^{N+1})\sin {}^1/_2t\cos(N - {}^1/_2)t}{tq(t)}dt$$
$$+ 0(1). \qquad (13)$$

For $\delta \leq |t| \leq 4\delta$ we have, by (7), $|f| \neq 1$, hence $q(t) = |f - e^{it}|^2 \geq (1 - |f|)^2 \geq a > 0$. Therefore the part of the integral in (13) extended over the range $\delta \leq |t| \leq 4\delta$ is $0(1)$, so that (13) holds with 4δ replaced by δ in the two limits of integration. For $|t| \leq \delta$, however, $\varphi\left(\frac{t}{\delta}\right) = 1 - \frac{3t^2}{8\delta^2} + \frac{3|t|^3}{32\delta^3}$, and the terms with t^2 and $|t|^3$ are easily seen to contribute $0(1)$. Hence,

$$A_N = \int_{-\delta}^{\delta} \frac{(1-f)f^N \sin Nt - 2(1-f^{N+1}) \sin {}^1/_2 t \cos (N - {}^1/_2)t}{tq(t)} dt + 0(1). \tag{14}$$

Since

$$\left| \frac{1}{tq(t)} - \frac{1}{t^3} \right| = \left| \frac{t^2 - q(t)}{t^3 q(t)} \right| \leq k \frac{t^4}{|t|^5} = \frac{k}{|t|},$$

where k is a constant, the replacement of $tq(t)$ by t^3 in (14) will make a difference of only 0(1), so that

$$A_N = \int_{-\delta}^{\delta} \frac{(1-f)f^N \sin Nt}{t^3} dt - \int_{-\delta}^{\delta} \frac{2(1-f^{N+1}) \sin {}^1/_2 t \cos (N - {}^1/_2)t}{t^3} dt + 0(1). \tag{15}$$

Let the two integrals in (15) be denoted by I_N and J_N, respectively. We have

$$I_N = \frac{2}{N} \int_{-\delta}^{\delta} \frac{(1-f)f^N}{t^3} d(\sin^2 {}^1/_2 Nt) = 0(1) + \frac{2}{N} \int_{-\delta}^{\delta} \left\{ \frac{3(1-f)f^N}{t^4} + \frac{f^N f'}{t^3} - \frac{Nf^{N-1}(1-f)f'}{t^3} \right\} \sin^2 {}^1/_2 Nt dt.$$

Using the first two inequalities of (5) we obtain the result

$$|I_N| \leq 0(1) + (6\alpha + 2\alpha') \int_{-\infty}^{\infty} \frac{\sin^2 {}^1/_2 Nt}{Nt^2} dt + 2 \int_{-\infty}^{\infty} \frac{|1-f|\,|f'|}{|t^3|} dt = 0(1),$$

since the integral involving N is independent of N.

To deal with J_N we observe first that in J_N, $\sin {}^1/_2 t$ may be replaced by ${}^1/_2 t$ and f^{N+1} by f^N, the difference thus made being 0(1). Hence

$$J_N = \int_{-\delta}^{\delta} \frac{(1-f^N) \cos (N - {}^1/_2)t}{t^2} dt + 0(1) = \int_{-\infty}^{\infty} \frac{(1-f^N) \cos (N - {}^1/_2)t}{t^2} dt + 0(1).$$

We may replace $\cos (N - {}^1/_2)t$ by $\cos Nt$, since

$$\frac{2}{\pi} \int_{-\infty}^{\infty} \frac{1-f^N}{t^2} (\cos (N - {}^1/_2)t - \cos Nt) dt = \frac{1}{\pi} \int_{-\infty}^{\infty} (1-f^N) \frac{\sin {}^1/_4 t}{{}^1/_4 t} \frac{\sin (N - {}^1/_4)t}{t} dt = P\{|U| \leq N - {}^1/_4\} - P\{|U + NY_N| \leq N - {}^1/_4\} = 0(1),$$

where U is a random variable independent of Y_N and whose characteristic function is $4/t \sin {}^1/_4t$. Hence

$$J_N = 0(1) + \frac{1}{N}\int_{-\infty}^{\infty}\frac{1-f^N}{t^2}d\sin Nt = 0(1) + \frac{2}{N^2}\int_{-\infty}^{\infty}\left\{\frac{2(1-f^N)}{t^3} + \frac{Nf^{N-1}f'}{t^2}\right\}d\sin^2 {}^1/_2t$$

$$= 0(1) + \frac{2}{N^2}\int_{-\infty}^{\infty}\left\{\frac{6(1-f^N)}{t^4} + \frac{4Nf^{N-1}f'}{t^3} - \frac{N(N-1)f^{N-2}f'^2}{t^2} - \frac{Nf^{N-1}f''}{t^2}\right\}\sin^2 {}^1/_2Ntdt.$$

Using all the inequalities (5) we have

$$|J_N| \leq 0(1) + (12\alpha + 8\alpha' + 2\alpha'')\int_{-\infty}^{\infty}\frac{\sin^2 {}^1/_2Nt}{Nt^2}dt + 2\int_{-\infty}^{\infty}\frac{|f'|^2}{t^2}dt = 0(1)$$

The proof is now complete.

5. By following the essential steps of the proof of Theorem 1 we obtain the following theorem, the proof of which is omitted from the present communication.

THEOREM 2. *If instead of conditions* (3) *we have*

$$\int_{-\infty}^{\infty} xdF(x) = 0,\ \int_{-\infty}^{\infty} |x|^a dF(x) < \infty,\ \int_{-\infty}^{\infty} x^2dF(x) = \infty \qquad (16)$$

where a is some constant such that $\frac{1}{2}(1 + 5^{1/2}) \leq a < 2$, then the series (4) *diverges for every $\epsilon > 0$.* (Example: Let X_n be distributed with the density $|x|^{-3}$ for $|x| \geq 1$ and 0 elsewhere.)

Since the finiteness of the second integral in (16) would seem rather to favor than to oppose the convergence of (4), it may be conjectured that given the first condition of (3), the finiteness of σ^2 is not only sufficient but also necessary for the convergence of (4). We have not been able to prove this.

6. The following generalization[3] of the strong law of large numbers is an immediate consequence of Theorem 1 and the remarks in section 2.

THEOREM 3 *Let $X_r^{(n)} (n = 1, 2, \ldots, r = 1, \ldots, n)$ be an array of random variables with the same distribution function $F(x)$ and such that* (1) $\int_{-\infty}^{\infty} xdF(x) = 0,\ \int_{-\infty}^{\infty} x^2dF(x) < \infty$, *and* (2) *for each n the random variables $X_1^{(n)}, \ldots, X_n^{(n)}$ are independent. Then the sequence of random variables $Y_1, Y_2, \ldots$, where for each n, $Y_n = (X_1^{(n)} + \ldots + X_n^{(n)})/n$, converges to 0 with* *probability 1. (Note that we do not assume any relation of dependence or independence between $X_n^{(r)}$ and $X_m^{(s)}$ for $r \neq s$.)*

[1] See M. Fréchet, *Recherches théoriques modernes*, Vol. 1, Paris, 1937, p. 27.

[2] See H. Cramér, *Mathematical methods of statistics*, Princeton, 1946, p. 93.

[3] Compare F. P. Cantelli, Considerazioni sulla legge uniforme dei grandi numeri ecc., *Giornale dell 'Istituto Italiano degli Attuari*, IV (1933), pp. 331–332; also H. Cramér, Su un theorema relativo alla leggi uniforme dei grandi numeri, *Ibid.*, V (1934), pp. 1–13.

Editors note: The proof of Theorem 2 is given in handwritten notes by Hsu. It is reproduced below with minor editing.

Proof of Theorem 2. Let $\frac{1}{2}(1+\sqrt{5}) \leqslant a < 2$. (The number $\frac{1}{2}(1+\sqrt{5})$ is the positive root of $a^2 - a - 1 = 0$.) Let

$$\int x\,dF = 0, \qquad \int |x|^a\,dF < \infty, \qquad \int x^2\,dF = \infty.$$

Let $X_1, X_2, \ldots$ be a sequence of independent random variables having the same distribution $F(x)$. Let $Y_n = (1/n)(X_1 + \cdots + X_n)$, and suppose that $\sum_{n=1}^{\infty} P\{|Y_n| > \epsilon\}$ were convergent. We shall obtain a contradiction.

Consider $Wn = X_n - Z_n$, in the sense of convolution, where Z_n has the same distribution F. Then

$$E(W_n) = 0, \qquad E|W_n|^a \leqslant 2^{a-1}(E|X_n|^a + E|Z_n|^a) < \infty,$$

$$E|W_n|^2 = \lim_{A\to\infty} \int_{-A}^{A}\int_{-A}^{A} (x-z)^2\,dF(x)\,dF(z)$$

$$= 2\lim_{A\to\infty}\left\{\int_{-A}^{A} x^2\,dF \cdot \int_{-A}^{A} dF - \left(\int_{-A}^{A} x\,dF\right)^2\right\} = \infty.$$

Hence the conditions on moments remain unchanged for the W_n. Moreover, we have

$$\sum_{n=1}^{\infty} P\left\{\frac{1}{n}|W_1 + \cdots + W_n| > \epsilon\right\} \leqslant \sum_{n=1}^{\infty} P\left\{\frac{1}{n}|X_1 + \cdots + X_n| > \frac{\epsilon}{2}\right\}$$

$$+ \sum_{n=1}^{\infty} P\left\{\frac{1}{n}|Z_1 + \cdots + Z_n| > \frac{\epsilon}{2}\right\}$$

$$= 2\sum_{n=1}^{\infty} P\left\{|Y_n| > \frac{\epsilon}{2}\right\} < \infty.$$

Hence the situation in the first paragraph remains the same if we replace X_n by $X_n - Z_n$. This amounts to the additional assumption that $F(x)$ *is a symmetric distribution.*

Given a symmetric F, we have for its characteristic function

$$f(t) = \int \cos tx\, dF, \qquad 1 - f(t) = 2\int \sin^2 \frac{tx}{2}\, dF, \qquad f'(t) = -\int x \sin tx\, dF,$$

whence

$$|1 - f(t)| = 1 - f(t) \leqslant 2\int \left|\sin \frac{tx}{2}\right|^a dF \leqslant \frac{|t|^a}{2^{a-1}} \int |x|^a dF \leqslant \alpha |t|^a,$$

$$|f'(t)| \leqslant \int |x|\,|\sin tx|^{a-1} dF \leqslant |t|^{a-1} \int |x|^a dF \leqslant \beta |t|^{a-1}.$$

These inequalities,

$$|1 - f(t)| \leqslant \alpha |t|^a, \qquad |f'(t)| \leqslant \beta |t|^{a-1}, \tag{5*}$$

play the parts of inequalities (5).

It is then easy to retrace all the steps (ohne weiteres) until we arrive at (15), or, what is the same thing,

$$A_N = \int_{-\infty}^{\infty} \frac{(1-f)f^N \sin Nt}{t^3}\, dt - \int_{-\infty}^{\infty} \frac{(1-f^N)\cos Nt}{t^2}\, dt + O(1). \tag{15*}$$

Let the two integrals in (15*) be I_N and J_N. Then

$$I_N = \frac{2}{N} \int \frac{(1-f)f^N}{t^3}\, d\left(\sin^2 \frac{Nt}{2}\right)$$

$$= -\frac{2}{N} \int \sin^2 \frac{Nt}{2} \left\{ \frac{-3(1-f)f^N}{t^4} - \frac{f^N f'}{t^3} + \frac{N(1-f)f^{N-1}f'}{t^3} \right\} dt$$

$$= \frac{6}{N} \int \frac{(1-f)f^N \sin^2(Nt/2)}{t^4}\, dt + \frac{2}{N} \int \frac{f^N f' \sin^2(Nt/2)}{t^3}\, dt + O(1)$$

$$= \frac{6}{N} \int \frac{(1-f)\sin^2(Nt/2)}{t^4}\, dt + \frac{2}{N} \int \frac{f' \sin^2(Nt/2)}{t^3}\, dt$$

$$- \frac{6}{N} \int \frac{(1-f)(1-f^N)\sin^2(Nt/2)}{t^4}\, dt$$

$$- \frac{2}{N} \int \frac{(1-f^N)f' \sin^2(Nt/2)}{t^3}\, dt + O(1)$$

$$= \frac{6}{N} \int \frac{(1-f)\sin^2(Nt/2)}{t^4}\, dt + \frac{2}{N} \int \frac{f' \sin^2(Nt/2)}{t^3}\, dt + 6\theta \int \frac{(1-f)^2}{t^4}\, dt$$

$$+ 2\theta \int \frac{(1-f)|f'|}{|t|^3}\, dt + O(1)$$

$$= \frac{6}{N} \int \frac{(1-f)\sin^2(Nt/2)}{t^4}\, dt + \frac{2}{N} \int \frac{f' \sin^2(Nt/2)}{t^3}\, dt + O(1),$$

where θ denotes any quantity such that $|\theta| \leqslant 1$. We obtain these θ terms because $|1 - f^N| = 1 - f^N = (1-f)(1 + f + \cdots + f^{N-1}) \leqslant N(1-f)$.

Hence

$$I_N = \frac{6}{N}\int \frac{(1-f)\sin^2(Nt/2)}{t^4}\,dt - \frac{2}{N}\int \frac{\sin(Nt/2)}{t^3}\,d(1-f) + O(1)$$

$$= \int \frac{(1-f)\sin Nt}{t^3}\,dt + O(1).$$

Again

$$J_N = \frac{1}{N}\int \frac{1-f^N}{t^2}\,d(\sin Nt) = -\frac{1}{N}\int\left(-\frac{Nf^{n-1}f'}{t^2} - \frac{2(1-f)^N}{t^3}\right)\sin Nt\,dt$$

$$= \int \frac{f^{N-1}f'}{t^2}\sin Nt\,dt + \frac{2}{N}\int \frac{1-f^N}{t^3}\sin Nt\,dt$$

$$= \int \frac{f^{N-1}f'}{t^2}\sin Nt\,dt + \frac{4}{N^2}\int \frac{1-f}{t^3}\,d\left(\sin^2\frac{Nt}{2}\right)$$

$$= O(1) + \int \frac{f^Nf'}{t^2}\sin Nt\,dt - \frac{4}{N^2}\int\left(\sin^2\frac{Nt}{2}\right)\left(-\frac{4(1-f^N)}{t^4} - \frac{Nf^{N-1}f'}{t^3}\right)dt$$

$$= \frac{16}{N^2}\int \frac{1-f^N}{t^4}\sin^2\frac{Nt}{2}\,dt + \frac{4}{N}\int \frac{f^{N-1}f'\sin^2(Nt/2)}{t^3}\,dt$$

$$+ \int \frac{f^Nf'}{t^2}\sin Nt\,dt + O(1)$$

$$= \frac{16}{N}\int \frac{1-f}{t^4}\sin^2\frac{Nt}{2}\,dt - 16\int \frac{\sin^2(Nt/2)}{t^4}\left(\frac{1-f}{N} - \frac{1-f^N}{N^2}\right)dt$$

$$+ \frac{4}{N}\int \frac{f'\sin^2(Nt/2)}{t^3}\,dt - \frac{4}{N}\int \frac{(1-f^{N-1})f'\sin^2(Nt/2)}{t^3}\,dt$$

$$+ \int \frac{f^Nf'\sin Nt}{t^2} + O(1)$$

$$= \frac{16}{N}\int \frac{1-f}{t^4}\sin^2\frac{Nt}{2}\,dt + \frac{4}{N}\int \frac{f'\sin^2(Nt/2)}{t^3}\,dt + \int \frac{f^Nf'\sin Nt}{t^2}\,dt$$

$$+ 16\theta\int \frac{(1-f)^2}{2t^4}\,dt + 4\theta\int \frac{(1-f)|f'|}{|t|^3}\,dt + O(1)$$

$$= \frac{16}{N}\int \frac{1-f}{t^4}\sin^2\frac{Nt}{2}\,dt + \frac{4}{N}\int \frac{f'\sin^2(Nt/2)}{t^3}\,dt$$

$$+ \int \frac{f^N f'\sin Nt}{t^2}\,dt + O(1).$$

The θ terms come from the inequalities

$$\frac{1}{N}|1-f^{N-1}| \leqslant 1-f,$$

$$\left|\frac{1-f}{N} - \frac{1-f^N}{N^2}\right| = \frac{1}{N^2}(1-f)|N-(1+f+\cdots+f^{N-1})|$$

$$\leqslant \frac{1}{N^2}(1-f)\{(1-f)+(1-f^2)+\cdots+(1-f^{N-1})\}$$

$$= \frac{1}{N^2}(1-f)^2\{1+(1+f)+(1+f+f^2)+\cdots$$

$$+(1+f+\cdots+f^{N-2})\}$$

$$\leqslant \frac{1}{N^2}(1-f)^2(1+2+3+\cdots+N-1) \leqslant \tfrac{1}{2}(1-f)^2.$$

Hence

$$J_N = \frac{16}{N}\int \frac{1-f}{t^4}\sin^2\frac{Nt}{2}\,dt - \frac{4}{N}\int \frac{\sin^2(Nt/2)}{t^3}\,d(1-f)$$

$$+ \int \frac{f^N f'\sin Nt}{t^2}\,dt + O(1)$$

$$= 2\int \frac{1-f}{t^3}\sin Nt\,dt + \int \frac{f^N f'\sin Nt}{t^2}\,dt + O(1). \tag{18}$$

Substituting (17) and (18) into (15*), we have

$$A_N = -\int \frac{1-f}{t^3}\sin Nt\,dt - \int \frac{f^N f'\sin Nt}{t^2}\,dt + O(1)$$

$$= \pi B_N + C_N + O(1), \tag{19}$$

where

$$B_N = -\frac{1}{\pi}\int \frac{1-f}{t^3}\sin Nt\,dt - \frac{1}{\pi}\int \frac{f'\sin Nt}{t^2}\,dt,$$

$$C_N = \int \frac{(1-f^N)f'\sin Nt}{t^2}\,dt.$$

Now,

$$B_N = -\frac{1}{\pi}\int \frac{1-f}{t^3}\sin Nt\,dt + \frac{1}{\pi}\int \frac{\sin Nt}{t^2}\,d(1-f)$$

$$= -\frac{1}{\pi}\int \frac{1-f}{t^3}\sin Nt\,dt - \frac{1}{\pi}\int (1-f)\left(-\frac{2}{t^3}\sin Nt + \frac{N\cos Nt}{t^2}\right)dt$$

$$= \frac{1}{\pi}\int \frac{1-f}{t^3}\sin Nt\,dt - \frac{N}{\pi}\int \frac{(1-f)\cos Nt}{t^2}\,dt$$

$$= \frac{2}{\pi}\int_0^\infty \frac{1-f}{t^3}\sin Nt\,dt - \frac{2N}{\pi}\int_0^\infty \frac{(1-f)\cos Nt}{t^2}\,dt$$

$$= 2\int_{-\infty}^{\infty} dF(x)\left\{\frac{2}{\pi}\int_0^\infty \frac{\sin^2(tx/2)\sin Nt}{t^3}\,dt - \frac{2N}{\pi}\int_0^\infty \frac{\sin^2(tx/2)\cos Nt}{t^2}\,dt\right\}$$

But

$$\frac{2}{\pi}\int_0^\infty \frac{\sin^2(tx/2)\sin Nt}{t^3}\,dt = \begin{cases} \frac{N}{4}(2|x|-N), & |x|>N, \\ \frac{x^2}{4}, & |x|\leqslant N, \end{cases}$$

$$\frac{2}{\pi}\int_0^\infty \frac{\sin^2(tx/2)\cos Nt}{t^2}\,dt = \begin{cases} \tfrac{1}{2}(|x|-N), & |x|>N, \\ 0, & |x|\leqslant N. \end{cases}$$

Hence, after some reduction, we obtain

$$B_N = \frac{1}{2}\int_{|x|\leqslant N} x^2\,dF(x) + \frac{N^2}{2}\int_{|x|>N} dF(x) \to \infty \quad \text{as} \quad N\to\infty.$$

It remains to prove that $C_n = O(1)$. For this purpose we write

$$f'(t) = -\int_{|x|\leqslant N} x\sin tx\,dF - \int_{|x|>N} x\sin tx\,dF = g_N(t) + h_N(t)$$

and

$$C_N = \int \frac{(1-f^N)g_N\sin Nt}{t^2}\,dt + \int \frac{(1-f^N)h_N\sin Nt}{t^2}\,dt$$
$$= D_N + E_N.$$

We have

$$|f'| \leqslant \beta|t|^{a-1}, \qquad |g_N| \leqslant \beta|t|^{a-1}, \qquad \text{as before,} \tag{20}$$

$$|g_N'| = \left|\int_{|x|<N} x^2 \cos tx\, dF\right| \leqslant \int_{|x|<N} x^2\, dF \leqslant N^{2-a}\int |x|^a\, dF \leqslant \gamma N^{2-a}, \tag{21}$$

$$|h_N| \leqslant \int_{|x|>N} |x|\, dF \leqslant N^{1-a}\int_{|x|>N} |x|^a\, dF \leqslant \delta N^{1-a}, \tag{22}$$

$$|1-f^N| \leqslant N(1-f) \leqslant \alpha N t^a, \qquad \text{as before.} \tag{23}$$

Hence

$$D_N = \frac{2}{N}\int \frac{(1-f^N)g_N}{t^2}\, d\left(\sin^2\frac{Nt}{2}\right)$$

$$= -\frac{2}{N}\int \sin^2\frac{Nt}{2}\left\{-\frac{3}{t^3}(1-f^N)g_N - \frac{Nf^{N-1}f'g_N}{t^2} + \frac{(1-f^N)g_N'}{t^2}\right\} dt$$

$$= O(1) - \frac{2}{N}\int \frac{(1-f^N)g_N'}{t^2}\sin^2\frac{Nt}{2}\, dt$$

$$= O(1) + \frac{2\theta}{N}\gamma N^{2-a}\int \frac{1-f^N}{t^2}\, dt, \qquad \text{by (21).}$$

Also,

$$E_N = \theta\delta_N^{1-a}\int \frac{(1-f^N)}{t^2}\, dt, \qquad \text{by (22).}$$

Thus it is sufficient to prove that

$$G_N = N^{1-a}\int \frac{1-f^N}{t^2}\, dt = O(1).$$

Now

$$G_N = N^{1-a}\left\{\int_{|t|<A} \frac{1-f^N}{t^2}\, dt + \int_{|t|>A} \frac{1-f^N}{t^2}\, dt\right\}$$

$$\leqslant N^{1-a}\left\{\int_{|t|<a} \alpha N t^{a-2}\, dt + 2\int_{|t|>A} \frac{dt}{t^2}\right\}, \qquad \text{by (23),}$$

$$= 2N^{1-a}\left\{\frac{\alpha}{a-1}NA^{a-1} + \frac{2}{A}\right\}.$$

Putting $A = N^{-1/a}$, we get

$$G_N \leqslant 2\left(\frac{\alpha}{a-1} + 2\right)N^{1-a+1/a} = O(1), \qquad \text{since} \quad a \geqslant \frac{1+\sqrt{5}}{2}.$$

THE CENTRAL LIMIT THEOREM FOR DEPENDENT RANDOM VARIABLES

By Wassily Hoeffding and Herbert Robbins

Introduction. The central limit theorem has been extended to the case of dependent random variables by several authors (Bruns, Markoff, S. Bernstein, P. Lévy, Loève). The conditions under which these theorems are stated either are very restrictive or involve conditional distributions, which makes them difficult to apply. In the present paper we prove central limit theorems for sequences of dependent random variables of a certain special type which occurs frequently in mathematical statistics. The hypotheses do not involve conditional distributions.

1. **The one-dimensional case.** Let

$$X_1, X_2, \cdots \tag{1}$$

be a sequence of random variables.

Definition 1. If for some function $f(n)$ the inequality $s - r > f(n)$ implies that the two sets

$$(X_1, X_2, \cdots, X_r), \qquad (X_s, X_{s+1}, \cdots, X_n)$$

are independent, then the sequence (1) is said to be *$f(n)$-dependent.*

It follows by induction that if (1) is $f(n)$-dependent, then any number of blocks of successive terms of (1) with subscripts not greater than n are independent whenever the first subscript of each block (after the first) differs from the last subscript of the preceding block by more than $f(n)$. An important special case occurs when $f(n) = m$ is independent of n. In particular, 0-dependence is equivalent to independence. In the present paper we shall confine ourselves to the case of m-dependence, where m is any positive integer.

Let (1) be m-dependent and such that $EX_i = 0$, $EX_i^2 < \infty$ $(i = 1, 2, \cdots)$; we define

$$A_i = EX_{i+m}^2 + 2\sum_{j=1}^{m} EX_{i+m-j}X_{i+m} \qquad (i = 1, 2, \cdots).$$

Then for $s > m$

$$E(X_{i+1} + \cdots + X_{i+s})^2 = E(X_{i+1} + \cdots + X_{i+s-1})^2 + A_{i+s-m} \qquad (i = 0, 1, \cdots),$$

so that

Received March 13, 1948; in revised form, May 4, 1948. Hoeffding is a research associate, Office of Naval Research Contract N7onr-284, Task Order II.

$$E(X_{i+1} + \cdots + X_{i+s})^2 = E(X_{i+1} + \cdots + X_{i+m})^2 + \sum_{h=1}^{s-m} A_{i+h} \tag{2}$$

$$(i = 0, 1, \cdots; s > m).$$

Theorem 1. *Let (1) be an m-dependent sequence of random variables such that*
(a) $EX_i = 0, E|X_i|^3 \le R^3 < \infty$ $(i = 1, 2, \cdots)$,
(b) $\lim_{p\to\infty} p^{-1} \sum_{h=1}^{p} A_{i+h} = A$ *exists, uniformly for all* $i = 0, 1, \cdots$.
Then as $n \to \infty$ *the random variable* $n^{-\frac{1}{2}}(X_1 + \cdots + X_n)$ *has a limiting normal distribution with mean 0 and variance A.*

Proof. Choose and fix α, $0 < \alpha < 1/4$, and let $k = [n^\alpha]$, $\nu = [n/k]$, where $[a]$ denotes the largest integer $\le a$. We have

$$k \le n^\alpha, \qquad n = k\nu + r \qquad (0 \le r < k). \tag{3}$$

It is easily seen that

$$k = O(\nu^{\alpha/(1-\alpha)}) = o(\nu^{1/3}), \qquad \nu = O(n^{1-\alpha}). \tag{4}$$

Now let $S = X_1 + \cdots + X_n$, and for every n so large that $k > m$ let

$$U_i = X_{ik-k+1} + \cdots + X_{ik-m} \qquad (i = 1, 2, \cdots, \nu),$$

$$S' = U_1 + \cdots + U_\nu,$$

$$T = \sum_{i=1}^{\nu-1} (X_{ik-m+1} + \cdots + X_{ik}) + (X_{\nu k-m+1} + \cdots + X_{\nu k+r}),$$

so that $S = S' + T$. The theorem will be proved [3; 254] when it is shown that as $n \to \infty$,
(i) $n^{-\frac{1}{2}}S'$ has a limiting normal distribution with mean 0 and variance A,
(ii) $n^{-\frac{1}{2}}T \to 0$ in probability.
We proceed to prove (i). Since

$$n^{-\frac{1}{2}}S' = (\nu k/n)^{\frac{1}{2}} \cdot \nu^{-\frac{1}{2}} \sum_{i=1}^{\nu} (k^{-\frac{1}{2}}U_i),$$

and since $\nu k/n \to 1$ as $n \to \infty$, it will suffice to prove that as $n \to \infty$,
(i′) $\nu^{-\frac{1}{2}} \sum_{i=1}^{\nu} (k^{-\frac{1}{2}}U_i)$ has a limiting normal distribution with mean 0 and variance A.
From (2) we have for $k > 2m$,

$$\sum_{i=1}^{\nu} EU_i^2 = \sum_{i=1}^{\nu} E(X_{ik-k+1} + \cdots + X_{ik-m})^2$$

$$= \sum_{i=1}^{\nu} \left\{ E(X_{ik-k+1} + \cdots + X_{ik-k+m})^2 + \sum_{h=1}^{k-2m} A_{ik-k+h} \right\},$$

and from (a) we have

(5) $$EX_i^2 \leq (E \mid X_i \mid^3)^{2/3} \leq R^2, \qquad E \mid X_i X_j \mid \leq (EX_i^2 EX_j^2)^{1/2} \leq R^2.$$

Hence $E(X_{ik-k+1} + \cdots + X_{ik-k+m})^2 \leq m^2 R^2$, so that

$$\left| \nu^{-1} \sum_{i=1}^{\nu} E(k^{-\frac{1}{2}} U_i)^2 - (k\nu)^{-1}(k - 2m) \sum_{i=1}^{\nu} (k - 2m)^{-1} \sum_{h=1}^{k-2m} A_{ik-k+h} \right| \leq k^{-1} m^2 R^2.$$

It follows from (b) that

(6) $$\lim_{\nu \to \infty} \nu^{-1} \sum_{i=1}^{\nu} E(k^{-\frac{1}{2}} U_i)^2 = A \geq 0.$$

Now the random variables $\{k^{-\frac{1}{2}} U_1, \cdots, k^{-\frac{1}{2}} U_\nu\}$ are independent, with a distribution that depends on n. From the Liapounoff form of the central limit theorem proved in the Appendix to this paper, it follows that a sufficient condition for (i′) is that (6) hold and that

(7) $$\max \{E \mid k^{-\frac{1}{2}} U_1 \mid^3, \cdots, E \mid k^{-\frac{1}{2}} U_\nu \mid^3\} = o(\nu^{\frac{1}{2}}).$$

By the generalized Hölder inequality,

$$E \mid X_i X_j X_k \mid \leq (E \mid X_i \mid^3 E \mid X_j \mid^3 E \mid X_k \mid^3)^{1/3} \leq R^3,$$

so that, by (4), we have for $i = 1, 2, \cdots, \nu$,

$$E \mid k^{-\frac{1}{2}} U_i \mid^3 = k^{-3/2} E \left| \sum_{f=1}^{k-m} X_{ik-k+f} \right|^3 \leq k^{-3/2}(k - m)^3 R^3 \leq k^{3/2} R^3 = o(\nu^{\frac{1}{2}}),$$

from which (7) follows. This completes the proof of (i′) and hence of (i).

Now by (a) and (5), since the summands in T are independent for $k > 2m$, we have for sufficiently large n,

$$n^{-1} ET^2 = n^{-1} \left\{ \sum_{i=1}^{\nu-1} E(X_{ik-m+1} + \cdots + X_{ik})^2 + E(X_{\nu k-m+1} + \cdots + X_{\nu k+r})^2 \right\}$$

$$\leq n^{-1}\{(\nu - 1) m^2 R^2 + (k + m)^2 R^2\} = O(n^{-\alpha}) + O(n^{2\alpha - 1}) = o(1).$$

Relation (ii) follows from Tchebycheff's inequality. This completes the proof of Theorem 1.

Theorem 1 should be compared with Bernstein's Theorem A [2; 14]. The chief difference lies in the fact that the latter involves (and uses, p. 15, line 4 from bottom) assumptions on conditional expectations, while Theorem 1 does not. Theorem A applies to the case of $f(n)$-dependence where $f(n)$ is not necessarily a constant but is of sufficiently low order of magnitude compared to n; Theorem 1 could also be generalized in this direction.

It is interesting to observe that Theorem A actually defines what we call $f(n)$-dependence to mean that any two *terms* of the sequence (1) are independent if the difference between their subscripts is sufficiently large. In the proof, however, the stronger assumption about *blocks* is used.

DEFINITION 2. If for every $i \geq 1$, $r \geq 0$ the joint distribution of $X_i, X_{i+1}, \cdots, X_{i+r}$ does not depend on i, the sequence (1) is said to be *stationary*.

THEOREM 2. *Let (1) be a stationary and m-dependent sequence of random variables such that* $EX_1 = 0$, $E|X_1|^3 < \infty$. *Then as* $n \to \infty$ *the random variable* $n^{-\frac{1}{2}}(X_1 + \cdots + X_n)$ *has a limiting normal distribution with mean 0 and variance*

$$A = EX_1^2 + 2EX_1X_2 + 2EX_1X_3 + \cdots + 2EX_1X_{m+1}.$$

Proof. This is easily seen to be a special case of Theorem 1.

One way in which m-dependent sequences occur in practice is as follows [2; 19]. Let

$$Z_1, Z_2, \cdots \tag{8}$$

be an arbitrary sequence of independent random variables. Let $g(z_1, \cdots, z_{m+1})$ be any Borel measurable function of $m + 1$ real variables, and define

$$X_i = g(Z_i, \cdots, Z_{m+i}) \qquad (i = 1, 2, \cdots). \tag{9}$$

The sequence (9) is then m-dependent, and if (8) is stationary (that is, if the Z's are identically distributed), so is (9).

As an example, let (8) be independent and stationary, with $EZ_1 = 0$, $EZ_1^2 = 1$, $E|Z_1|^3 < \infty$. Let $g(z_1, z_2) = z_1z_2$. Then (9) is stationary and 1-dependent, $EX_1 = 0$, $EX_1^2 = 1$, $E|X_1|^3 < \infty$, and

$$A = EX_1^2 + 2EX_1X_2 = EZ_1^2Z_2^2 + 2EZ_1Z_2^2Z_3 = 1.$$

Hence by Theorem 2 as $n \to \infty$ the random variable $n^{-\frac{1}{2}}(Z_1Z_2 + Z_2Z_3 + \cdots + Z_nZ_{n+1})$ has a limiting normal distribution with mean 0 and variance 1. On the other hand, if with the same sequence (8) we take $g(z_1, z_2) = z_2 - z_1$, then

$$A = E(Z_2 - Z_1)^2 + 2E(Z_2 - Z_1)(Z_3 - Z_2) = 0,$$

whence by Theorem 2 the random variable $n^{-\frac{1}{2}}(Z_{n+1} - Z_1) \to 0$ in probability as $n \to \infty$, as is obviously the case.

2. The two-dimensional case.

Definitions 1 and 2 hold unchanged if the elements of (1) are random vectors in R_N, $N > 1$. We shall state the analogue of Theorem 2 for $N = 2$.

THEOREM 3. *Let*

$$(X_1, Y_1), \quad (X_2, Y_2), \quad \cdots \tag{10}$$

be a stationary and m-dependent sequence of random vectors in R_2 *such that* $EX_1 = EY_1 = 0$, $E|X_1|^3 < \infty$, $E|Y_1|^3 < \infty$. *Then as* $n \to \infty$ *the random vector* $n^{-\frac{1}{2}}(X_1 + \cdots + X_n, Y_1 + \cdots + Y_n)$ *has a limiting normal distribution with mean* (0, 0) *and covariance matrix*

$$\begin{pmatrix} A & B \\ B & C \end{pmatrix}, \tag{11}$$

where

$$A = EX_1^2 + 2EX_1X_2 + \cdots + 2EX_1X_{m+1},$$

(12) $$B = EX_1Y_1 + (EX_1Y_2 + EX_2Y_1) + \cdots + (EX_1Y_{m+1} + EX_{m+1}Y_1),$$

$$C = EY_1^2 + 2EY_1Y_2 + \cdots + 2EY_1Y_{m+1}.$$

Proof. It follows as in (2) that for $n \geq m$,

$$E(X_1 + \cdots + X_n)^2 = E(X_1 + \cdots + X_m)^2 + (n - m)A,$$

$$E(X_1 + \cdots + X_n)(Y_1 + \cdots + Y_n) = E(X_1 + \cdots + X_m)(Y_1 + \cdots + Y_m) + (n - m)B,$$

$$E(Y_1 + \cdots + Y_n)^2 = E(Y_1 + \cdots + Y_m)^2 + (n - m)C.$$

The rest of the proof is now similar to that of Theorem 1.

The extension of Theorem 3 to the case $N > 2$, as well as to the non-stationary case, is evident and will be left to the reader.

If V_n $(n = 1, 2, \cdots)$ and V are random vectors and if the distribution function of V_n converges [3; 83] to that of V, we say that V_n converges to V *in distribution.*

LEMMA. *Let* (P_n, Q_n) $(n = 1, 2, \cdots)$ *and* (P, Q) *be random vectors in* R_2. *If*
(a) (P_n, Q_n) *converges in distribution to* (P, Q) *as* $n \to \infty$,
(b) d_n $(n = 1, 2, \cdots)$ *is a sequence of non-zero constants such that* $\lim_{n\to\infty} d_n = 0$,
(c) $H(x, y)$ *is a function of the real variables* (x, y) *which has a total differential at* $(0, 0)$, *with*

$$H_1 = \left.\frac{\partial H(x, y)}{\partial x}\right|_{(0,0)}, \qquad H_2 = \left.\frac{\partial H(x, y)}{\partial y}\right|_{(0,0)},$$

then as $n \to \infty$ *the random variable*

$$W_n = d_n^{-1}\{H(d_nP_n, d_nQ_n) - H(0, 0)\}$$

converges in distribution to

$$H_1P + H_2Q.$$

Proof. This is a special case of a theorem of Anderson and Rubin [1; 42]. A direct proof runs as follows. We have by (c),

$$H(x, y) - H(0, 0) = H_1x + H_2y + x\epsilon_1(x, y) + y\epsilon_2(x, y),$$

where $\epsilon_i(x, y) \to 0$ as $(x, y) \to (0, 0)$. Hence

$$W_n = H_1P_n + H_2Q_n + \{P_n\epsilon_1(d_nP_n, d_nQ_n) + Q_n\epsilon_2(d_nP_n, d_nQ_n)\}.$$

It is easy to see (*e.g.* by taking characteristic functions) that $H_1P_n + H_2Q_n$ converges in distribution to $H_1P + H_2Q$ as $n \to \infty$. Hence we need only show that, *e.g.*, $P_n \epsilon_1(d_n P_n, d_n Q_n) \to 0$ in probability as $n \to \infty$. From (a) and (b) it follows that $(d_n P_n, d_n Q_n) \to (0, 0)$ in probability, and hence that $\epsilon_1(d_n P_n, d_n Q_n) \to 0$ in probability, as $n \to \infty$. Now for any $\delta > 0$, $t > 0$,

$$P[| P_n \cdot \epsilon_1(d_n P_n, d_n Q_n) | > \delta] \leq P[| \epsilon_1(d_n P_n, d_n Q_n) | > \delta/t] + P[| P_n | > t].$$

Given $\epsilon > 0$ we can by (a) choose t and n_0 so large that the last term is less than ϵ for $n \geq n_0$. The preceding term tends to 0 as $n \to \infty$. Hence the first term tends to 0 as $n \to \infty$, which completes the proof.

If (P, Q) is normally distributed with mean (0, 0) and covariance matrix (11), then W_n will have a limiting normal distribution with mean 0 and variance

$$H_1^2 A + 2H_1 H_2 B + H_2^2 C. \tag{13}$$

Hence if we set

$$P_n = n^{-\frac{1}{2}}(X_1 + \cdots + X_n), \qquad Q_n = n^{-\frac{1}{2}}(Y_1 + \cdots + Y_n),$$

where the X_i, Y_i are as in Theorem 3, then (a) will hold. Setting $d_n = n^{-\frac{1}{2}}$ we obtain the following theorem.

THEOREM 4. *Let* (10) *be a sequence of random vectors in* R_2 *satisfying the conditions of Theorem* 3, *let* $H(x, y)$ *have a total differential at* (0, 0), *and let*

$$W_n = n^{\frac{1}{2}}\left\{H\left(n^{-1} \sum_1^n X_i, n^{-1} \sum_1^n Y_i\right) - H(0, 0)\right\}.$$

Then as $n \to \infty$, W_n *has a limiting normal distribution with mean* 0 *and variance* (13), *where* A, B, C *are defined by* (12).

As an example, let (8) be independent and stationary, with $EZ_1 = 0$, $EZ_1^2 = 1$, $EZ_1^6 < \infty$. Let $g(z_1, z_2) = z_1 z_2$, $h(z_1) = z_1^2 - 1$, and set

$$X_i = g(Z_i, Z_{i+1}) = Z_i Z_{i+1}, \qquad Y_i = h(z_i) = Z_i^2 - 1 \qquad (i = 1, 2, \cdots).$$

Then (X_1, Y_1), (X_2, Y_2), $\cdots$ is stationary and 1-dependent and satisfies the conditions of Theorem 3 with $A = EX_1^2 + 2EX_1X_2 = 1$. Now let $H(x, y) = x/(y + 1)$; then $H_1 = 1$, $H_2 = 0$, $H(0, 0) = 0$, so that (13) has the value 1. It follows from Theorem 4 that as $n \to \infty$ the random variable

$$W_n = n^{\frac{1}{2}}\left\{\frac{Z_1 Z_2 + \cdots + Z_n Z_{n+1}}{Z_1^2 + \cdots + Z_n^2}\right\}$$

has a limiting normal distribution with mean 0 and variance 1.

Theorem 4 may be extended in an obvious manner to functions of three or more variables. Thus, for example, the random variable

$$n^{\frac{1}{2}}\left\{\frac{Z_1 Z_2 + \cdots + Z_n Z_{n+1} - n^{-1}(\sum_1^n Z_i)^2}{Z_1^2 + \cdots + Z_n^2 - n^{-1}(\sum_1^n Z_i)^2}\right\},$$

which occurs in the theory of serial correlation, may be shown to be asymptotically normal with mean 0 and variance 1; the details are left to the reader.

APPENDIX

THEOREM. *Given a sequence* (X_{nk}, Y_{nk}) $(n = 1, 2, \cdots; k = 1, \cdots, \nu; \nu = \nu(n); \lim_{n\to\infty} \nu = \infty)$ *of sets of random vectors in* R_2, *independent for each fixed n, with* $EX_{nk} = EY_{nk} = 0$. *Let*

$$EX_{nk}^i Y_{nk}^j = \mu_{ij}^{(nk)} \qquad (i + j = 2),$$

$$\rho_{nk}^3 = \max \{E \mid X_{nk} \mid^3, E \mid Y_{nk} \mid^3\}, \qquad \rho_n^3 = \sum_{k=1}^{\nu} \rho_{nk}^3 .$$

Assume that

(a) $$\lim_{n\to\infty} \nu^{-1} \cdot \sum_{k=1}^{\nu} \mu_{ij}^{(nk)} = \mu_{ij} \qquad (i + j = 2),$$

(b) $$\lim_{n\to\infty} \nu^{-\frac{1}{2}} \rho_n = 0.$$

Then as $n \to \infty$ *the random vector* $\nu^{-\frac{1}{2}}(X_{n1} + \cdots + X_{n\nu}, Y_{n1} + \cdots + Y_{n\nu})$ *has a limiting normal distribution with mean* $(0, 0)$ *and covariances* μ_{ij}.

Proof (see [3; 215]). We shall use the inequalities

$$E \mid X_{nk}^i Y_{nk}^j \mid \leq \rho_{nk}^{i+j} \qquad (i + j = 1, 2, 3),$$

and the fact that for sufficiently large n,

$$0 \leq \nu^{-\frac{1}{2}} \rho_{nk} \leq \nu^{-\frac{1}{2}} \rho_n < 1.$$

Choose any s, t and set $u = \max \{\mid s \mid, \mid t \mid\}$. Let $F_{nk}(x, y)$ be the distribution function of (X_{nk}, Y_{nk}). Then the characteristic function of (X_{nk}, Y_{nk}) is

$$\phi_{nk}(s, t) = \int_{-\infty}^{\infty} \int_{-\infty}^{\infty} e^{i(sx+ty)} \, dF_{nk}(x, y)$$

$$= \int_{-\infty}^{\infty} \int_{-\infty}^{\infty} [1 + i(sx + ty) - \tfrac{1}{2} (sx + ty)^2 + \theta(sx + ty)^3] \, dF_{nk}(x, y)$$

$$= 1 - \tfrac{1}{2}(\mu_{20}^{(nk)} s^2 + 2\mu_{11}^{(nk)} st + \mu_{02}^{(nk)} t^2) + \theta \rho_{nk}^3 \cdot 8u^3,$$

where θ denotes any quantity such that $\mid \theta \mid \leq 1$, so that

$$\phi_{nk}(\nu^{-\frac{1}{2}} s, \nu^{-\frac{1}{2}} t) = 1 + z_{nk},$$

where for sufficiently large n,

$$z_{nk} = - \tfrac{1}{2}\nu^{-1}(\mu_{20}^{(nk)} s^2 + 2\mu_{11}^{(nk)} st + \mu_{02}^{(nk)} t^2) + \theta \nu^{-3/2} \rho_{nk}^3 \cdot 8u^3$$

$$= \theta \nu^{-1} \rho_{nk}^2 \cdot 4u^2 + \theta \nu^{-1} \rho_{nk}^2 \cdot 8u^3 = \theta \nu^{-1} \rho_{nk}^2 (4u^2 + 8u^3).$$

Hence for sufficiently large n, $|z_{nk}| < \frac{1}{2}$, so that

$$\begin{aligned}\log \phi_{nk}(\nu^{-\frac{1}{2}}s, \nu^{-\frac{1}{2}}t) &= z_{nk} + \theta z_{nk}^2 \\ &= -\tfrac{1}{2}\nu^{-1}(\mu_{20}^{(nk)}s^2 + 2\mu_{11}^{(nk)}st + \mu_{02}^{(nk)}t^2) \\ &\quad + \theta\nu^{-3/2}\rho_{nk}^3 \cdot 8u^3 + \theta\nu^{-3/2}\rho_{nk}^3(4u^2 + 8u^3)^2.\end{aligned}$$

Thus

$$\begin{aligned}\sum_{k=1}^{\nu} \log \phi_{nk}(\nu^{-\frac{1}{2}}s, \nu^{-\frac{1}{2}}t) &= -\frac{1}{2}\left[\nu^{-1}\sum_{k=1}^{\nu}\mu_{20}^{(nk)}s^2 + 2\nu^{-1}\sum_{k=1}^{\nu}\mu_{11}^{(nk)}st \right. \\ &\quad \left. + \nu^{-1}\sum_{k=1}^{\nu}\mu_{02}^{(nk)}t^2\right] \\ &\quad + \theta\nu^{-3/2}\rho_n^3[8u^3 + (4u^2 + 8u^3)^2] \\ &= -\tfrac{1}{2}[\mu_{20}s^2 + 2\mu_{11}st + \mu_{02}t^2] + o(1),\end{aligned}$$

which implies the assertion of the theorem.

If we define $\lambda = \max\{\rho_{n1}^3, \cdots, \rho_{n\nu}^3\}$ and if $\lambda = o(\nu^{\frac{1}{2}})$, then

$$\nu^{-3/2}\rho_n^3 \leq \nu^{-3/2}\cdot\nu\cdot\lambda = o(1),$$

so that (b) holds.

References

1. T. W. Anderson and Herman Rubin, *Estimation of the parameters of a single stochastic difference equation in a complete system*, Cowles Commission Staff Papers, Statistics, January, 1947.
2. Serge Bernstein, *Sur l'extension du théorème limite du calcul des probabilités aux sommes de quantités dépendantes*, Mathematische Annalen, vol. 97(1927), pp. 1–59.
3. Harald Cramér, *Mathematical Methods of Statistics*, Princeton University Press, 1946.

Department of Mathematical Statistics,
University of North Carolina.

Reprinted from
Duke Math. Jour.
15, 733–780 (1948)

APPLICATION OF THE METHOD OF MIXTURES TO QUADRATIC FORMS IN NORMAL VARIATES

BY HERBERT ROBBINS AND E. J. G. PITMAN

Institute of Statistics, University of North Carolina

1. Summary. The method of mixtures, explained in Section 2, is applied to derive the distribution functions of a positive quadratic form in normal variates and of the ratio of two independent forms of this type.

2. The method of mixtures. If

$$F_0(x), \qquad F_1(x), \tag{1}$$

is any sequence of distribution functions, and if

$$c_0, c_1, \cdots \tag{2}$$

is any sequence of constants such that

$$c_j \geq 0 \qquad (j = 0, 1, \cdots), \qquad \Sigma c_j = 1 \tag{3}$$

(all summations will be from 0 to ∞ unless otherwise noted), then the function

$$F(x) = \Sigma c_j F_j(x) \tag{4}$$

is called a *mixture* of the sequence (1).

It is sometimes helpful to interpret $F(x)$ in the following manner. Let J, X_0, X_1, $\cdots$ be variates such that J has the distribution $P[J = j] = c_j$ $(j = 0, 1, \cdots)$ and such that X_j has the distribution function $F_j(x)$. Let X be a variate such that the conditional distribution function of X given $J = j$ is $F_j(x)$. Then the distribution function of X is

$$P[X \leq x] = \Sigma P[J = j] \cdot P[X \leq x \mid J = j] = \Sigma c_j F_j(x) = F(x).$$

This interpretation of $F(x)$ will, however, not be involved in the present paper.

The following statements are proved in [1]. If $x = (x_1, \cdots, x_n)$ is a vector variable the function $F(x)$ defined by (4) is a distribution function, and for any Borel set S,

$$\int_S dF(x) = \Sigma c_j \int_S dF_j(x). \tag{5}$$

More generally, if $g(x)$ is any Borel measurable function then

$$\int_{-\infty}^{\infty} g(x)\, dF(x) = \Sigma c_j \int_{-\infty}^{\infty} g(x)\, dF_j(x) \tag{6}$$

whenever the left hand side of (6) exists. In particular, the characteristic function

$\varphi(t)$ corresponding to $F(x)$ is

$$\varphi(t) = \Sigma c_j \varphi_j(t), \tag{7}$$

where $\varphi_j(t)$ is the characteristic function corresponding to $F_j(x)$.

If each $F_j(x)$ has a derivative $f_j(x)$ then $F(x)$ has a derivative $f(x)$ given by

$$f(x) = \Sigma c_j f_j(x), \tag{8}$$

provided that this series converges uniformly in some interval including x. Conversely, if (8) is the relation between the frequency functions and if the series is uniformly convergent in every finite interval, then the relation between the distribution functions is given by (4). In practice we deduce (4) from (8), or, using the uniqueness theorem for characteristic functions, from (7).

As regards computation, we observe that for any integers $0 \leq p_1 \leq p_2$ and for any x it follows from (3) and (4) that

$$\begin{aligned} 0 \leq F(x) - \sum_{p_1}^{p_2} c_j F_j(x) &= \sum_{0}^{p_1-1} c_j F_j(x) + \sum_{p_2+1}^{\infty} c_j F_j(x) \\ &\leq \sup_{j<p_1} \{F_j(x)\} \cdot \left(\sum_{0}^{p_1-1} c_j \right) + \sup_{j>p_2} \{F_j(x)\} \cdot \left(1 - \sum_{0}^{p_1-1} c_j - \sum_{p_1}^{p_2} c_j \right) \leq 1 - \sum_{p_1}^{p_2} c_j . \end{aligned} \tag{9}$$

The existence of these upper bounds (the last a uniform one) for the error term when the series (4) is replaced by a finite sum shows that series expansions of the mixture type (4) are especially well adapted to computational work.

For some purposes it is useful to consider series expansions of the type (4) where the c_j may be of both signs and where the series Σc_j may diverge. Both parts of (3) will, however, be satisfied in the cases considered here.

If U, V are independent variates with respective distribution functions $F(x)$, $G(x)$ we shall denote the distribution function of any Borel measurable function $H(U, V)$ by

$$H(U, V)\ (F(x), G(x)).$$

Now if $F(x)$, $G(x)$ are both mixtures,

$$F(x) = \Sigma c_j F_j(x), \qquad G(x) = \Sigma d_k G_k(x),$$

then by (5),

$$\begin{aligned} P[H(U, V) \leq x] &= \iint_{\{H(u,v) \leq x\}} dF(u)\, dG(v) \\ &= \Sigma\Sigma c_j d_k \iint_{\{H(u,v) \leq x\}} dF_j(u)\, dG_k(v), \end{aligned}$$

so that

$$H(U, V)(\Sigma c_j F_j(x), \Sigma d_k G_k(x)) = \Sigma\Sigma c_j d_k H(u, v)(F_j(x), G_k(x)). \tag{10}$$

As an application of the principles set forth in this section we shall express as series of the mixture type (4) the distribution functions of any positive quadratic form in normal variates and of the ratio of any two independent forms of this type. Special cases of the problem have been dealt with by Tang [2], Hsu [3], and many others, but the method of mixtures permits a unified and simple treatment of the general case.

3. Distribution of a positive quadratic form. We shall denote by $F_n(x)$ the chi-square distribution function with $n > 0$ degrees of freedom,

$$F_n(x) = \frac{1}{2^{\frac{1}{2}n}\cdot\Gamma(\frac{1}{2}n)}\int_0^x u^{\frac{1}{2}n-1}\cdot e^{-\frac{1}{2}u}\cdot du \qquad (x > 0), \tag{11}$$

$$= 0 \qquad (x \leq 0)$$

The corresponding characteristic function is

$$\varphi_n(t) = \int_0^\infty e^{ixt}\, dF_n(x) = (1 - 2it)^{-\frac{1}{2}n} = w^{\frac{1}{2}n}, \tag{12}$$

where we have set $w = (1 - 2it)^{-1}$. We shall denote by χ_n^2 any variate with the distribution function (11).

Let a be any constant such that $a > 0$. The characteristic function of the variate $a\cdot\chi_n^2$ is

$$(1 - 2iat)^{-\frac{1}{2}n} = [a(1 - 2it) - (a - 1)]^{-\frac{1}{2}n} = a^{-\frac{1}{2}n}\cdot w^{\frac{1}{2}n}\cdot\left(1 - \left(1 - \frac{1}{a}\right)w\right)^{-\frac{1}{2}n}. \tag{13}$$

By the binomial theorem we have for any $a > 0$,

$$a^{-\frac{1}{2}n}\left[1 - \left(1 - \frac{1}{a}\right)z\right]^{-\frac{1}{2}n} = \Sigma c_j z^j \qquad \left(|z| < \left|1 - \frac{1}{a}\right|^{-1}\right), \tag{14}$$

where

$$c_j = a^{-\frac{1}{2}n}\cdot \frac{\frac{1}{2}n(\frac{1}{2}n + 1)\cdots(\frac{1}{2}n + j - 1)}{j!}\cdot\left(1 - \frac{1}{a}\right)^j \qquad (j = 0, 1. \cdots). \tag{15}$$

For $a \geq 1$ we see from (15) that all the c_j are non-negative. Likewise for $a > \frac{1}{2}$ (and hence *à fortiori* for $a \geq 1$) we have $|1 - 1/a|^{-1} > 1$ so that (14) holds for all $|z| \leq 1$; setting $z = 1$ it follows that the sum of all the c_j is equal to 1. Hence for $a \geq 1$,

$$c_j \geq 0 \qquad (j = 0, 1, \cdots), \qquad \Sigma c_j = 1.$$

Since $|w| = |1 - 2it|^{-1} \leq 1$ for all real t it follows from (13) and (14) that for $a \geq 1$,

$$(1 - 2iat)^{-\frac{1}{2}n} = \Sigma c_j\, w^{\frac{1}{2}n+j} = \Sigma c_j(1 - 2it)^{-\frac{1}{2}n-j}$$

$$= \Sigma c_j\, \varphi_{n+2j}(t). \tag{16}$$

Hence for $a \geq 1$ the distribution function $F_n(x/a)$ of the variate $a \cdot \chi_n^2$ is a mixture of χ^2 distribution functions,

$$F_n(x/a) = \Sigma c_j F_{n+2j}(x), \tag{17}$$

where the c_j, determined by the identity (14), are the probabilities of a negative binomial distribution.

It may, in fact be proved by a direct analysis, which we omit here, that (17) holds for any $a > 0$. However, if $a < 1$ then the c_j will be of alternating sign, and if $a \leq \frac{1}{2}$ then the series Σc_j will diverge. This shows incidentally that a relation of the form (4) can hold even though the series Σc_j diverges and hence the corresponding relation (7) does not hold for $t = 0$.

THEOREM 1. *Let*

$$X = a(\chi_m^2 + a_1 \chi_{m_1}^2 + \cdots + a_r \chi_{m_r}^2),$$

where the chi-square variates are independent and $a, a_1, \cdots, a_r$ *are positive constants such that*

$$a_i \geq 1 \qquad (i = 1, \cdots, r).$$

Define constants c_j *by the identity*[1]

$$\prod_{i=1}^{r} \left\{ a_i^{-\frac{1}{2}m_i} \left[1 - \left(1 - \frac{1}{a_i} \right) z \right]^{-\frac{1}{2}m_i} \right\} = \Sigma c_j z^j \qquad (|z| \leq 1); \tag{18}$$

then obviously

$$c_j \geq 0 \qquad (j = 0, 1, \cdots), \qquad \Sigma c_j = 1.$$

Let

$$M = m + m_1 + \cdots + m_r;$$

then for every x,

$$P[X \leq x] = \Sigma c_j \cdot F_{M+2j}(x/a). \tag{19}$$

For any integers $0 \leq p_1 \leq p_2$ *and every* x,

$$\begin{aligned} 0 &\leq P[X \leq x] - \sum_{p_1}^{p_2} c_j F_{M+2j}(x/a) \\ &\leq F_M(x/a) \cdot \left(\sum_0^{p_1-1} c_j \right) + F_{M+2p_2+2}(x/a) \cdot \left(1 - \sum_0^{p_1-1} c_j - \sum_{p_1}^{p_2} c_j \right) \\ &\leq 1 - \sum_{p_1}^{p_2} c_j . \end{aligned} \tag{20}$$

PROOF. The characteristic function of X/a is, by (13) and (18),

$$\varphi(t) = w^{\frac{1}{2}M} \cdot \prod_{i=1}^{r} \left\{ a_i^{-\frac{1}{2}m_i} \left[1 - \left(1 - \frac{1}{a_i} \right) w \right]^{-\frac{1}{2}m_i} \right\} = \Sigma c_j w^{\frac{1}{2}M+j} = \Sigma c_j \varphi_{M+2j}(t)$$

[1] If $r = 0$ we regard the left hand side of (18) as having the value 1.

Hence for any y,

$$P[X/a \leq y] = \Sigma c_j F_{M+2j}(y),$$

whence (19) follows on setting $x = ay$. Finally, since $F(x)$ is a decreasing function of n for fixed x, (20) follows from (9).

It should be observed that the coefficients c_j determined by (18) can be written explicitly as the multiple Cauchy products

$$c_j = \sum_{i_1+\cdots+i_r=j} \{c_{1,i_1} \cdots c_{r,i_r}\},$$

where

$$c_{i,j} = a_i^{-\frac{1}{2}m_i} \cdot \frac{\frac{1}{2}m_i(\frac{1}{2}m_i + 1) \cdots (\frac{1}{2}m_i + j - 1)}{j!} \cdot \left(1 - \frac{1}{a_i}\right)^j$$

$$(i = 1, \cdots, r; j = 0, 1, \cdots).$$

The c_j may be computed stepwise by the relations

$$c_j^{(1)} = c_{1,j},$$

$$c_j^{(s)} = \sum_{i=0}^{j} \{c_{j-i}^{(s-1)} \cdot c_{s,i}\} \qquad (s = 2, \cdots, r),$$

$$c_j^{(r)} = c_j.$$

4. Distribution of a ratio. The ratio χ_m^2/χ_n^2 of two independent chi-square variates has the distribution function

$$(21) \qquad F_{m,n}(x) = \frac{\Gamma(\frac{1}{2}(m+n))}{\Gamma(\frac{1}{2}m)\Gamma(\frac{1}{2}n)} \int_0^x u^{\frac{1}{2}m-1}(1+u)^{-\frac{1}{2}(m+n)}\,du \qquad (x \geq 0),$$

$$= 0 \qquad (x < 0).$$

In computational work we can use the tables of the Beta distribution function

$$I_x(r, s) = \frac{\Gamma(r+s)}{\Gamma(r)\cdot\Gamma(s)} \int_0^x u^{r-1} \cdot (1-u)^{s-1} \cdot du \qquad (0 < x < 1),$$

$$= 0 \ (x \leq 0), \qquad 1 \ (x \geq 1),$$

together with the identity

$$F_{m,n}(x) = I_{x/(1+x)}(\tfrac{1}{2}m, \tfrac{1}{2}n).$$

THEOREM 2. *Let*

$$(22) \qquad X = \frac{a \cdot (\chi_m^2 + a_1\chi_{m_1}^2 + \cdots + a_r\chi_{m_r}^2)}{\chi_n^2 + b_1\chi_{n_1}^2 + \cdots + b_s\chi_{n_s}^2},$$

where the χ^2 *variates are independent and* $a, a_1, \cdots, a_r, b_1, \cdots, b_s$ *are positive constants such that*

$$a_i \geq 1, \qquad b_j \geq 1$$

$$(i = 1, \cdots, r; j = 1, \cdots, s).$$

Define constants c_j, d_k *by the identities*

$$\prod_{i=1}^{r} \left\{ a_i^{-\frac{1}{2}m_i} \cdot \left[1 - \left(1 - \frac{1}{a_i} \right) z \right]^{-\frac{1}{2}m_i} \right\} = \Sigma c_j z^j,$$

$$(|z| \leq 1)$$

$$\prod_{i=1}^{s} \left\{ b_i^{-\frac{1}{2}n_i} \cdot \left[1 - \left(1 - \frac{1}{b_i} \right) z \right]^{-\frac{1}{2}n_i} \right\} = \Sigma d_k z^k;$$

then

$$c_j \geq 0, \quad \Sigma c_j = 1, \qquad d_k \geq 0, \quad \Sigma d_k = 1.$$

Let

$$M = m + m_1 + \cdots + m_r, \qquad N = n + n_1 + \cdots + n_s;$$

then for every x,

$$P[X \leq x] = \Sigma\Sigma c_j d_k \cdot F_{M+2j, N+2k}(x/a),$$

and for any integers $0 \leq p_1 \leq p_2$, $0 \leq q_1 \leq q_2$ *and every* x,

$$0 \leq P[X \leq x] - \sum_{p_1}^{p_2} \sum_{q_1}^{q_2} c_j d_k \cdot F_{M+2j, N+2k}(x/a)$$

$$\leq \left(1 - \sum_{p_1}^{p_2} c_j \right) \cdot \left(1 - \sum_{q_1}^{q_2} d_k \right).$$

PROOF. Let U, V denote respectively numerator and denominator of (22). From Theorem 1,

$$P[U \leq x] = \Sigma c_j F_{M+2j}(x/a),$$

$$P[V \leq x] = \Sigma d_k F_{N+2k}(x).$$

Hence by (10), for every x,

$$P[X \leq x] = P[U/V \leq x] = \Sigma\Sigma c_j d_k \cdot F_{M+2j, N+2k}(x/a).$$

The rest of the theorem is obvious.

COROLLARY. *Let*

$$X = \frac{\chi_M^2}{a\chi_r^2 + b\chi_s^2},$$

where the χ^2 *variates are independent and*

$$0 < a \leq b.$$

Define

$$\alpha = a/b, \qquad N = r + s,$$

$$c_j = \alpha^{\frac{1}{2}s} \cdot \frac{\frac{1}{2}s(\frac{1}{2}s + 1) \cdots (\frac{1}{2}s + j - 1)}{j!} \cdot (1 - \alpha)^j \qquad (j = 0, 1, \cdots);$$

then

$$c_j \geq 0 \qquad (j = 0, 1, \cdots), \qquad \Sigma c_j = 1,$$

and for every x,

$$P[X \leq x] = \Sigma c_j F_{M,N+2j}(ax).$$

For any integers $0 \leq p_1 \leq p_2$ and every x,

$$0 \leq p[X > x] - \sum_{p_1}^{p_2} c_j[1 - F_{M,N+2j}(ax)]$$

$$\leq [1 - F_{M,N}(ax)] \cdot \left(\sum_0^{p_1-1} c_j\right) + [1 - F_{M,N+2p_2+2}(ax)] \cdot \left(1 - \sum_0^{p_1-1} c_j - \sum_{p_1}^{p_2} c_j\right) \leq 1 - \sum_{p_1}^{p_2} c_j. \tag{23}$$

PROOF. Except for (23) this is a special case of Theorem 2. To prove (23) we observe that

$$P[X > x] = 1 - P[X \leq x] = \Sigma c_j[1 - F_{M,N+2j}(ax)],$$

and since for fixed m and x, $F_{m,n}(x)$ is an increasing function of n, (23) follows in the same way as (9).

5. The non-central case. Let Y be normal (0, 1) and let $X = (Y + d)^2$, where d is any constant. The frequency function of X is, for $x > 0$,

$$f(x) = (2\pi x)^{-\frac{1}{2}} \cdot e^{-\frac{1}{2}(d^2+x)} \cdot (e^{dx^{\frac{1}{2}}} + e^{-dx^{\frac{1}{2}}})/2.$$

By expanding the last factor into a power series it is easily seen that

$$f(x) = \Sigma p_j \cdot f_{1+2j}(x), \tag{24}$$

where $f_n(x) = F'_n(x)$ is the chi-square frequency function with n degrees of freedom and where

$$p_j = e^{-\frac{1}{2}d^2} \cdot (\tfrac{1}{2}d^2)^j/j! \qquad (j = 0, 1, \cdots).$$

Since the identity

$$e^{-\frac{1}{2}d^2(1-z)} = \Sigma p_j z^j \qquad \text{(all } z) \tag{25}$$

holds, it follows that

$$p_j \geq 0 \qquad (j = 0, 1, \cdots), \qquad \Sigma p_j = 1.$$

The series (24) is uniformly convergent in every finite interval, so that we can write the distribution function $F(x)$ and characteristic function $\varphi(t)$ of X in the forms

$$F(x) = \Sigma p_j \cdot F_{1+2j}(x),$$

$$\varphi(t) = \Sigma p_j \cdot \varphi_{1+2j}(t) = w^{\frac{1}{2}} \cdot e^{-\frac{1}{2}d^2(1-w)},$$

where again we have set $w = (1 - 2it)^{-1}$.

Now let $Y_1, \cdots, Y_n$ be independent and normal (0, 1) variates and let

$$X = (Y_1 + d_1)^2 + \cdots + (Y_n + d_n)^2, \tag{26}$$

where the d_i are constants such that

$$d_1^2 + \cdots + d_n^2 = d^2.$$

The characteristic function of X is then

$$\varphi(t) = w^{\frac{1}{2}n} \cdot e^{-\frac{1}{2}d^2(1-w)} = \Sigma p_j w^{\frac{1}{2}n+j} = \Sigma p_j \varphi_{n+2j}(t),$$

and hence the distribution function $F(x)$ of X is again a mixture of χ^2 distribution functions,

$$F(x) = \Sigma p_j \cdot F_{n+2j}(x), \tag{27}$$

where the p_j, determined by the identity (25), are the probabilities of a Poisson distribution with parameter $\lambda = \frac{1}{2}d^2$. We shall denote the non-central chi-square variate (26) by $\chi'^2_{n,d}$.

We can now generalize Theorems 1 and 2 in a straightforward manner to cover non-central chi-square variates. We shall state only the generalization of the Corollary of Theorem 2 to the case in which the numerator is non-central.

THEOREM 3. *Let*

$$X = \frac{\chi'^2_{M,d}}{a\chi_r^2 + b\chi_s^2},$$

where the χ^2 *variates are independent and*

$$0 < a \leq b.$$

Define

$$\lambda = \tfrac{1}{2}d^2, \qquad \alpha = a/b, \qquad N = r + s$$

$$p_j = e^{-\lambda} \cdot \lambda^j / j! \qquad (j = 0, 1, \cdots),$$

$$c_k = \alpha^{\frac{1}{2}s} \cdot \frac{\frac{1}{2}s(\frac{1}{2}s + 1) \cdots (\frac{1}{2}s + k - 1)}{k!} \cdot (1 - \alpha)^k \qquad (k = 0, 1, \cdots);$$

then

$$p_j \geq 0, \quad \Sigma p_j = 1, \qquad c_k \geq 0, \quad \Sigma c_k = 1,$$

and for every x,

$$P[X \leq x] = \Sigma\Sigma p_j c_k F_{M+2j,N+2k}(ax).$$

For any integers $0 \leq g_1 \leq g_2$, $0 \leq h_1 \leq h_2$,

$$0 \leq P[X \leq x] - \sum_{g_1}^{g_2} \sum_{h_1}^{h_2} p_j c_k \cdot F_{M+2j,N+2k}(ax) \leq \left(1 - \sum_{g_1}^{g_2} p_j\right) \cdot \left(1 - \sum_{h_1}^{h_2} c_k\right).$$

REFERENCES

[1] Herbert Robbins, "Mixture of distributions," *Annals of Math. Statistics*, Vol. 19 (1948), p. 360.

[2] P. C. Tang, "The power function of the analysis of variance tests with tables and illustrations of their use," *Stat. Res. Mem.*, Vol. 2 (1938), p. 126.

[3] P. L. Hsu, "Contributions to the theory of 'Student's' t-test as applied to the problem of two samples," *ibid.*, p. 1.

Reprinted from
Ann. Math. Statist.
20, 552–560 (1949)

MINIMUM VARIANCE ESTIMATION WITHOUT REGULARITY ASSUMPTIONS

By Douglas G. Chapman[1] and Herbert Robbins

University of Washington and University of North Carolina

1. Summary and Introduction. Following the essential steps of the proof of the Cramér-Rao inequality [1, 2] but avoiding the need to transform coordinates or to differentiate under integral signs, a lower bound for the variance of estimators is obtained which is (a) free from regularity assumptions and (b) at least equal to and in some cases greater than that given by the Cramér-Rao inequality. The inequality of this paper might also be obtained from Barankin's general result[2] [3]. Only the simplest case—that of unbiased estimation of a single real parameter—is considered here but the same idea can be applied to more general problems of estimation.

2. Lower bound. Let μ be a fixed measure on Euclidean n-space X and let the random vector $x = (x_1, \cdots, x_n)$ have a probability distribution which is absolutely continuous with respect to μ, with density function $f(x, \alpha)$, where α is a real parameter belonging to some parameter set A. Define $S(\alpha)$ as follows:

$$f(x, \alpha) > 0, \qquad \text{a.e. } x \text{ in } S(\alpha),$$
$$f(x, \alpha) = 0, \qquad \text{a.e. } x \text{ in } X - S(\alpha).$$

Let $t = t(x)$ be any unbiased estimator of α, so that for every α in A,

$$\int_X tf(x, \alpha)\, d\mu = \alpha. \tag{1}$$

If $\alpha, \alpha + h (h \neq 0)$ are any two distinct values in A such that

$$S(\alpha + h) \subset S(\alpha), \tag{2}$$

then, writing S for $S(\alpha)$,

$$\int_S f(x, \alpha)\, d\mu = 1, \qquad \int_{S(\alpha+h)} f(x, \alpha + h)\, d\mu = \int_S f(x, \alpha + h)\, d\mu = 1,$$

$$\int_S tf(x, \alpha)\, d\mu = \alpha, \qquad \int_S tf(x, \alpha + h)\, d\mu = \alpha + h,$$

so that

$$\int_S [t - \alpha] \sqrt{f(x, \alpha)}\, \frac{f(x, \alpha + h) - f(x, \alpha)}{hf(x, \alpha)} \sqrt{f(x, \alpha)}\, d\mu = 1.$$

---

[1] This research was supported in part by the Office of Naval Research.

[2] But again with some additional restrictions.

581

Applying Schwarz's inequality we obtain the relation

$$\begin{aligned}1 &\le \int_S [t-\alpha]^2 f(x,\alpha)\,d\mu \cdot \int_S \left[\frac{f(x,\alpha+h)-f(x,\alpha)}{hf(x,\alpha)}\right]^2 f(x,\alpha)\,d\mu \\ &= \operatorname{Var}(t\mid\alpha)\cdot\frac{1}{h^2}\left\{\int_S\left[\frac{f(x,\alpha+h)}{f(x,\alpha)}\right]^2 f(x,\alpha)\,d\mu - 1\right\}.\end{aligned} \tag{3}$$

Let

$$J = J(\alpha, h) = \frac{1}{h^2}\left\{\left[\frac{f(x,\alpha+h)}{f(x,\alpha)}\right]^2 - 1\right\};$$

then (3) can be written in the form

$$\operatorname{Var}(t\mid\alpha) \ge \frac{1}{E(J\mid\alpha)}. \tag{4}$$

Since (4) holds whenever α, $\alpha + h$ are any two distinct elements of A satisfying (2) we obtain the fundamental inequality

$$\operatorname{Var}(t\mid\alpha) \ge \frac{1}{\inf_h E(J\mid\alpha)}, \tag{5}$$

where the infimum is taken over all $h \neq 0$ such that (2) is satisfied. It should be noted that (5) holds without any restriction on $f(x, \alpha)$ and without any restriction on t other than (1).

It is possible that $E(J\mid\alpha)$ does not exist (finitely) for any h. With the usual convention that $E(J\mid\alpha) = \infty$, in this case, (5) is still a valid, though trivial, inequality.

In applications μ will often be Lebesgue measure on X. It could equally well be a discrete measure on a countable set of points in X. Furthermore, if the set where $f(x, \alpha) > 0$ is independent of α then (2) is trivially satisfied for all $\alpha + h$ in A.

We shall have occasion to compare (5) with the Cramér-Rao inequality

$$\operatorname{Var}(t\mid\alpha) \ge \frac{1}{E(\psi^2\mid\alpha)}; \qquad \psi = \psi(\alpha) = \frac{\partial}{\partial\alpha}\ln f(x,\alpha). \tag{6}$$

This inequality is usually derived for distributions with range independent of the parameter and under certain regularity conditions on both $f(x, \alpha)$ and the unbiased estimator t.

3. Examples.

Example 1. *Unbiased estimation of the mean of a normal distribution based on a random sample of size n.* Here

$$f(x,\alpha) = (2\pi)^{-(n/2)}\sigma^{-n}e^{-(1/2\sigma^2)\Sigma_{i=1}^n(x_i-\alpha)^2},$$

where σ is a positive constant, and

$$J = \frac{1}{h^2}\{e^{-(1/\sigma^2)\Sigma_{i=1}^n[(x_i-\alpha-h)^2-(x_i-\alpha)^2]} - 1\} = \frac{n}{\sigma^2k^2}\{e^{-k^2}e^{2ku} - 1\},$$

where we have set $u = \sum_{i=1}^{n}(x_i - \alpha)/(\sigma\sqrt{n})$, $\quad k = h\sqrt{n}/\sigma \neq 0$.

When the mean is α, u is normally distributed with mean 0 and variance 1, and we find after a simple computation that

$$
\begin{aligned}
&E(J \mid \alpha) = n(e^{k^2} - 1)/(\sigma^2 k^2), \\
(7)\qquad &\inf_h E(J \mid \alpha) = \lim_{k\to 0} [n(e^{k^2} - 1)/(\sigma^2 k^2)] = n/\sigma^2 = [E(\psi^2 \mid \alpha)].
\end{aligned}
$$

Hence if t is any unbiased estimator of α it follows from (5) that

$$
(8)\qquad \mathrm{Var}(t \mid \alpha) \geq \sigma^2/n.
$$

Since the sample mean $\bar{x}$ is an unbiased estimator of α with $\mathrm{Var}(\bar{x} \mid \alpha) = \sigma^2/n$, it follows that $\bar{x}$ has minimum variance in the class of *all* unbiased estimators of α.

In this example the Cramér-Rao inequality (6) yields precisely the same bound (8).

Corresponding results hold for the unbiased estimation of the variance when the mean is known. Both (5) and (6) yield the inequality

$$
\mathrm{Var}(t \mid \alpha) \geq 2\alpha^2/n,
$$

where α is the unknown variance. The equality sign holds for

$$
t = n^{-1} \sum_{i=1}^{n} (x_i - m)^2,
$$

where m is the mean of the normal population.

Example 2. *Unbiased estimation of the standard deviation of a normal population with known mean.* Here

$$
f(x, \alpha) = (2\pi)^{-(n/2)} \alpha^{-n} e^{-(1/2\alpha^2)\sum_{i=1}^{n}(x_i - m)^2}.
$$

Setting $k = h/\alpha$ we find that for $-1 < k < \sqrt{2} - 1$, $k \neq 0$,

$$
(9)\qquad E(J \mid \alpha) = \{(1 + k)^{-n}[1 - k(2 + k)]^{-(n/2)} - 1\}/(\alpha^2 k^2).
$$

In this case also, $\lim_{k\to 0} E(J \mid \alpha) = 2n/\alpha^2 = E(\psi^2 \mid \alpha)$. But the minimum value of $E(J \mid \alpha)$ is not approached in the neighborhood of $h = k = 0$, and the inequality (5) is sharper than (6). We shall consider only the case $n = 2$. Equation (9) then becomes

$$
E(J \mid \alpha) = (p + 1)^2/[\alpha^2 p^2 (2 - p^2)],
$$

where we have set $p = 1 + k$ and $0 < p < \sqrt{2}$. We have for $p = .8393$, $1/E(J \mid \alpha) = .2698\,\alpha^2$, so that by (5)

$$
\mathrm{Var}\,(t \mid \alpha) \geq .2698\,\alpha^2 > .25\alpha^2 = \frac{1}{E(\psi^2 \mid \alpha)}.
$$

It is interesting to note that the unbiased estimator

$$t = \sqrt{\tfrac{1}{2}n}\,\frac{\Gamma(\frac{1}{2}n)}{\Gamma[\frac{1}{2}(n+1)]}\sqrt{\sum_{i=1}^{n}(x_i - m)^2 \Big/ n}$$

has variance

$$\alpha^2\left[\tfrac{1}{2}n\,\frac{\Gamma^2\frac{1}{2}n}{\Gamma^2[\frac{1}{2}(n+1)]} - 1\right],$$

which for $n = 2$ becomes

$$\alpha^2\left[\frac{4}{\pi} - 1\right] = .2732\alpha^2.$$

But it can be shown using results of Lehmann and Scheffé [4], or of Hoel [5], which were derived from Blackwell's theorem on conditional expectation [6], that no other unbiased estimator can have smaller variance than t. Thus (5) does not give the *greatest* lower bound in this case.

Various examples of the application of (5) can be given where $S(\alpha)$ is not a constant and where the Cramér-Rao formula is invalid (see for example Cramér [1], p. 485). It should be noted, however, that in many of the standard problems of this type stronger results can be obtained by other methods.

Another class of estimation problems where (5) may be applied occurs if the parameter space is discrete. Again in this case the Cramér-Rao formula does not hold. An example of this type has been given by Chapman ([7], pp. 149–150). Other applications of this type and some results related to this paper were obtained recently by Hammersley [8].

4. General comparison with the Cramér-Rao inequality. Let

$$\bar{J} = \bar{J}(\alpha, h) = \left[\frac{f(x, \alpha + h) - f(x, \alpha)}{hf(x, \alpha)}\right]^2; \tag{10}$$

then

$$E(\bar{J} \mid \alpha) = E(J \mid \alpha).$$

Hence in the fundamental inequality (5) we can replace J by $\bar{J}$. But from (10) it is clear that

$$\lim_{h\to 0} \bar{J}(\alpha, h) = \left[\frac{\partial}{\partial\alpha}\ln f(x, \alpha)\right]^2 = \psi^2(\alpha)$$

whenever the latter exists.

Assuming now the usual regularity conditions under which the Cramér-Rao lower bound is derived, that $S(\alpha)$ is independent of α and that $f(x, \alpha)$ is sufficiently regular that we may pass to the limit inside the integral sign,

$$E(\psi^2 \mid \alpha) = E\left[\lim_{h\to 0}(\bar{J} \mid \alpha)\right] = \lim_{h\to 0} E(\bar{J} \mid \alpha) \geq \inf_h E(\bar{J} \mid \alpha) = \inf_h E(J \mid \alpha), \tag{11}$$

the infimum being taken over admissible values of h. It follows that *the inequality* (5) *is at least as sharp as that given by the Cramér-Rao formula* (6).

On the other hand, when $x = (x_1, \cdots, x_n)$ is a random sample from a regular distribution, and when $E(\psi^2 \mid \alpha) < \infty$, then for any *fixed* $h \neq 0$, there exists an n_0 such that for $n > n_0$

$$E(\psi^2 \mid \alpha) \leq E(J \mid \alpha). \tag{12}$$

Without loss of generality assume $E(J \mid \alpha) < \infty$. Letting $g(t, \alpha)$ denote the density function of a single x_i and ν the one-dimensional measure which generates μ, it is easily verified that

$$E(J \mid \alpha) = \frac{1}{h^2}\left(\left[\int_X \frac{g^2(t, \alpha + h)}{g(t, \alpha)}\, d\nu\right]^n - 1\right).$$

By hypothesis, except on a set of measure 0,

$$g(t, \alpha + h) = g(t, \alpha) + h\left.\frac{\partial g}{\partial \alpha}\right|_{\alpha=\alpha(h)}; \qquad \alpha \leq \alpha(h) \leq \alpha + h.$$

Hence

$$\int_X \frac{g^2(t, \alpha + h)}{g(t, \alpha)}\, d\nu = 1 + 2h \int_X \left.\frac{\partial g}{\partial \alpha}\right|_{\alpha=\alpha(h)} d\nu + h^2 \int_X g^{-1}\left(\left.\frac{\partial g}{\partial \alpha}\right|_{\alpha=\alpha(h)}\right)^2 d\nu. \tag{13}$$

Denoting the last integral of the right hand side of (13) by $R(\alpha, h)$ and noting that the relation

$$\int_X g(t, \alpha)\, d\nu = 1$$

may be differentiated under the integral sign so that the middle term vanishes, it follows that

$$E(J \mid \alpha) = \frac{[1 + h^2 R(\alpha, h)]^n - 1}{h^2} \geq nR(\alpha, h) + \tfrac{1}{2}n(n-1)h^2 R^2(\alpha, h). \tag{14}$$

On the other hand, from (11) and (14),

$$E(\psi^2 \mid \alpha) = nR(\alpha, 0). \tag{15}$$

In order that different parameters may be distinguishable we must have

$$\left.\frac{\partial g}{\partial \alpha}\right|_{\alpha=\alpha(h)} \neq 0$$

for a set of positive measure on the t-axis, and hence $R(\alpha, h) > 0$. From this and the fact that $R(\alpha, 0)$ is independent of n, (12) follows at once, for sufficiently large n, from (14) and (15).

REFERENCES

[1] H. CRAMÉR, *Mathematical Methods of Statistics*, Princeton University Press, 1946.

[2] C. R. RAO, "Information and the accuracy attainable in the estimation of statistical parameters," *Bull. Calcutta Math. Soc.*, Vol. 37 (1945), pp. 81–91.

[3] E. W. Barankin, "Locally best unbiased estimates," *Annals of Math. Stat.*, Vol. 20 (1949), pp. 477–501. (More complete references to the general problem are given in this paper.)

[4] E. L. Lehmann and H. Scheffé, "Completeness, similar regions and unbiased estimation, Part 1," *Sankhyā*, Vol. 10 (1950), pp. 305–340.

[5] P. G. Hoel, "Conditional expectation and the efficiency of estimates," *Annals of Math. Stat.*, Vol. 22 (1951), pp. 299–301.

[6] D. Blackwell, "Conditional expectation and unbiased sequential estimation," *Annals of Math. Stat.*, Vol. 18 (1947), pp. 105–110.

[7] D. G. Chapman, "Some properties of the hypergeometric distribution with applications to zoological sample censuses," *Univ. of California Publ. Statist.*, Vol. 1, No. 7 (1951), pp. 131–160.

[8] J. M. Hammersley, "On estimating restricted parameters," *Jour. Roy. Stat. Soc.*, *Ser. B*, Vol. 12 (1950), pp. 192–229.

Reprinted from
Ann. Math Statist.
22, 581–586 (1951)

Reprinted from the Proceedings of the NATIONAL ACADEMY OF SCIENCES,
Vol. 39, No. 6, pp. 525–533. June, 1953

ERGODIC PROPERTY OF THE BROWNIAN MOTION PROCESS

BY G. KALLIANPUR AND H. ROBBINS*

THE INSTITUTE FOR ADVANCED STUDY AND UNIVERSITY OF NORTH CAROLINA

Communicated by J. von Neumann, April 16, 1953

1. Introduction.—In a previous paper ("On the Equidistribution of Sums of Independent Random Variables," to appear elsewhere; an abstract will appear in *Bull. Am. Math. Soc.*, **59**, (May, 1953); we shall refer to this paper as [1]) we considered some properties of the sequence S_n of partial sums of independent and identically distributed random variables or random vectors in two dimensions. Here we show in Theorems 1–4 that the results of [1] carry over to the Brownian motion process in one and two dimensions.

2. The Two-Dimensional Case.—Let $X(t)$ denote the Brownian motion (Wiener) process on the line: $X(0) \equiv 0$, $X(t)$ is continuous for all t with probability 1, and for any $t_0 < t_1 < \ldots < t_n$ the random variables $X(t_j) - X(t_{j-1})$, $j = 1, \ldots, n$, are independent and normally distributed with zero means and variances $t_j - t_{j-1}$. Let $V(t) = (X(t), Y(t))$ denote the Brownian motion process in the plane, the two components of $V(t)$ being independent one-dimensional Brownian motion processes. Suppose that $f(x, y)$, $g(x, y)$ are real valued functions which are bounded and summable in the plane $-\infty < x < \infty$, $-\infty < y < \infty$, and set $\bar{f} = \int\int f(x, y)\, dxdy$, $\bar{g} = \int\int g(x, y)\, dxdy$, where here and in the sequel an integral sign without limits denotes integration over $(-\infty, \infty)$. We shall prove the following two theorems for plane Brownian motion. The corresponding results for the one-dimensional case involve no essentially new arguments and will be stated without proof at the end of the paper.

THEOREM 1. *If $\bar{f} \neq 0$ then for every u,*

$$\lim_{T \to \infty} \Pr\left[\frac{2\pi}{\bar{f} \log T} \int_0^T f(V(t))\, dt \leq u\right] = G(u), \tag{2.1}$$

where $G(u) = 1 - e^{-u}$ for $u \geq 0$, $\quad = 0$ for $u < 0$.

THEOREM 2. *If $\bar{g} \neq 0$ then*

$$\lim_{T \to \infty} \frac{\int_0^T f(V(t))\, dt}{\int_0^T g(V(t))\, dt} = \frac{\bar{f}}{\bar{g}} \text{ in probability.} \tag{2.2}$$

Proofs of Theorems 1 and 2: Assume $\bar{f} \neq 0$ and $\bar{g} \neq 0$ and define

$$Z_n(f) = \frac{2\pi}{\bar{f} \log n} \int_0^n f(V(t))\, dt, \qquad W_n(f) = \frac{2\pi}{\bar{f} \log n} \sum_{j=1}^n f(V(j)),$$

with corresponding definitions for $Z_n(g)$ and $W_n(g)$. For any positive integer n, each component of $V(n)$ is the sum of n independent random variables, each distributed normally with zero mean and unit variance, since

$$X(n) = \sum_{j=1}^{n} [X(j) - X(j-1)], \qquad Y(n) = \sum_{j=1}^{n} [Y(j) - Y(j-1)].$$

A theorem in [1] gives

$$\lim_{n \to \infty} \Pr[W_n(f) \leq u] = G(u). \tag{2.3}$$

We shall later prove as Lemma 1 that

$$\lim_{n \to \infty} E[Z_n(f) - W_n(f)]^2 = 0. \tag{2.4}$$

It follows from (2.4) that $Z_n(f) - W_n(f)$ tends to zero in probability, and hence from (2.3) that $Z_n(f)$ has the same limiting distribution, $G(u)$, as $W_n(f)$. This proves (2.1) as $T \to \infty$ through integer values, and the extention to arbitrary T is immediate, proving Theorem 1.

To prove Theorem 2 we shall later prove as Lemma 2 that

$$\lim_{n \to \infty} E[Z_n(f)Z_n(g) - W_n(f)W_n(g)] = 0, \tag{2.5}$$

and we make use of the fact, proved in [1], that

$$\lim_{n \to \infty} EW_n^2(f) = \lim_{n \to \infty} EW_n^2(g) = \lim_{n \to \infty} EW_n(f)W_n(g) = \int_0^\infty u^2 dG(u) = m_2, \text{ say.} \tag{2.6}$$

From (2.4)–(2.6) it follows that

$$\lim_{n \to \infty} EZ_n^2(f) = \lim_{n \to \infty} EZ_n^2(g) = \lim_{n \to \infty} EZ_n(f)Z_n(g) = m_2,$$

$$\text{and hence } \lim_{n \to \infty} E[Z_n(f) - Z_n(g)]^2 = 0. \tag{2.7}$$

As in [1], given $\epsilon > 0$ we choose $\delta = \delta(\epsilon) > 0$ such that $\delta \leq \epsilon$, $G(\delta) \leq {}^1/_2\epsilon$, and $N = N(\epsilon)$ (by (2.1) with f replaced by g) such that

$$n \geq N \text{ implies } \Pr[Z_n(g) > \delta] \geq 1 - G(\delta) - {}^1/_2\epsilon \geq 1 - \epsilon.$$

Since $$\Pr[\{Z_n(f) - Z_n(g)\}^2 < \delta^3] \geq 1 - \frac{E[Z_n(f) - Z_n(g)]^2}{\delta^3},$$

if we choose (by (2.7)) $K = K(\epsilon)$ such that

$$n \geq K \text{ implies } E[Z_n(f) - Z_n(g)]^2 \leq \delta^4,$$

then if $n \geq \max. (N, K)$,

$$\Pr[Z_n(g) > \delta] \geq 1 - \epsilon, \qquad \Pr[\{Z_n(f) - Z_n(g)\}^2 < \delta^3] \geq 1 - \epsilon,$$

which in turn imply that

$$\Pr\left[\left\{\frac{Z_n(f)}{Z_n(g)} - 1\right\}^2 < \epsilon\right] \geq \Pr[Z_n(g) > \delta, \{Z_n(f) - Z_n(g)\}^2 < \delta^3] \geq 1 - 2\epsilon.$$

Since ϵ was arbitrary this proves (2.2) as $T \to \infty$ through integer values. Again, the extension to arbitrary T is immediate. Finally, the restriction that $\bar{f} \neq 0$ can be dropped by a simple argument and the proof of Theorem 2 is complete.

3. *Proof of the Lemmas.*—We shall prove Lemma 2 first. We have

$$Z_n(f) = \frac{2\pi}{\bar{f}\log n}\sum_{j=1}^{n}\int_{j-1}^{j} f(V(t))\,dt, \quad W_n(f) = \frac{2\pi}{\bar{f}\log n}\sum_{j=1}^{n}\int_{j-1}^{j} f(V(j))\,dt.$$

Set $D_n = E[Z_n(f)Z_n(g) - W_n(f)W_n(g)]$

$$= \frac{4\pi^2}{\bar{f}\bar{g}(\log n)^2}\sum_{j,k=1}^{n}\int_{j-1}^{j}\int_{k-1}^{k} E[f(V(t))g(V(u)) - f(V(j))g(V(k))]\,dudt$$

$$= \frac{4\pi^2}{\bar{f}\bar{g}(\log n)^2}\sum_{j,k=1}^{n} a_{jk},$$

where

$$a_{jk} = \iint_{R_{jk}} [\theta(t, u) - \theta(j, k)]\,dudt, \qquad R_{jk} = \{j - 1 < t < j;\ k - 1 < u < k\}, \qquad \theta(t, u) = E[f(V(t))g(V(u))].$$

We want to show that $D_n \to 0$ as $n \to \infty$, i.e., that

$$\lim_{n\to\infty} (\log n)^{-2}\sum_{j,k=1}^{n} a_{jk} = 0. \tag{3.1}$$

To evaluate $\theta(t, u)$ we observe that for $t_1 < t_2$ the random vector $V(t_2) - V(t_1) = (X(t_2) - X(t_1), Y(t_2) - Y(t_1))$ has the joint probability density

$$\frac{1}{2\pi(t_2 - t_1)} e^{-\frac{x^2 + y^2}{2(t_2 - t_1)}}.$$

Hence for $0 < t < u$,

$$\theta(t, u) = \frac{1}{4\pi^2 t(u - t)}\iiiint f(x, y)g(x + x', y + y')\, e^{-\frac{x^2 + y^2}{2t} - \frac{x'^2 + y'^2}{2(u - t)}}\,dxdydx'dy'$$

$$= \frac{1}{4\pi^2 t(u - t)}\iiiint f(x, y)g(\zeta, \eta)e^{-Q(t,u)}\,dxdyd\zeta d\eta,$$

where we have set

$$Q(t, u) = \frac{1}{2}\left\{\frac{x^2 + y^2}{t} + \frac{(x - \zeta)^2 + (y - \eta)^2}{u - t}\right\}.$$

In what follows C will denote any constant whose numerical value is immaterial, and $F = \sup|f(x, y)|$, $G = \sup|g(x, y)|$. Then if $(t, u)\epsilon R_{jj}$ or $\epsilon R_{j,\, j+1} (j \geq 2)$,

$$|\theta(t, u)| \leq \frac{G}{4\pi^2 t(u - t)} \int\int |f(x, y)| e^{-\frac{x^2 + y^2}{2t}} dxdy \cdot$$

$$\int\int e^{-\frac{x'^2 + y'^2}{2(u - t)}} dx'dy' \leq \frac{G}{2\pi t} \int\int |f(x\ y)| dxdy \leq \frac{C}{j - 1}. \quad (3.2)$$

Also, for $0 < t < u$,

$$|\theta(t, u)| \leq \frac{1}{4\pi^2 t(u - t)} \int\int\int\int |f(x, y) \cdot g(\zeta, \eta)|$$

$$e^{-\frac{x^2 + y^2}{2t}} dxdyd\zeta d\eta \leq \frac{F}{2\pi(u - t)} \int\int |g(\zeta, \eta)| d\zeta d\eta \leq \frac{C}{u - t}. \quad (3.3)$$

We write $\sum_{j,\, k = 1}^{n} a_{jk}$ as the sum of the following terms:

(a) $\sum_{j = 2}^{n} a_{jj}$, (b) $\sum_{k = 3}^{n} a_{1k}$, (b′) $\sum_{= 3}^{n} a_{j1}$, (c) $\sum_{j = 2}^{n - 1} a_{j,\, j+1}$,

(c′) $\sum_{k = 2}^{n - 1} a_{k+1, k}$, (d) $\sum_{2 \leq j \leq k-2 \leq n-2} a_{jk}$, (d′) $\sum_{2 \leq k \leq j-2 \leq n-2} a_{jk}$, and

(e) $a_{11} + a_{12} + a_{21}$.

To prove (3.1) it will suffice to show that each of (a), (b), (c), (d) is $o\{(\log n)^2\}$.

(a) From (3.2), if $(t, u)\epsilon R_{jj}$, $j \geq 2$, $|\theta(t, u)| \leq \frac{C}{j - 1}$.

$$\text{Also, } |\theta(j, j)| = \frac{1}{2\pi j}\left|\int\int f(x, y) g(x, y) e^{-\frac{x^2 +}{2j}} dxdy\right| \leq \frac{C}{j - 1}.$$

$$\text{Hence } \left|\sum_{j=2}^{n} a_{jj}\right| \leq C \cdot \sum_{j=2}^{n} \frac{1}{j - 1} = 0(\log n). \quad (3.4)$$

(b) From (3.3), if $(t, u)\epsilon R_{1k}$, $k \geq 3$,

$$|\theta(t, u)| \leq \frac{C}{u - t} \leq \frac{C}{k - 2}.$$

$$\text{Also, } |\theta(1, k)| \leq \frac{C}{k - 2}.$$

$$\text{Hence } \left| \sum_{k=3}^{n} a_{1k} \right| \leq C \cdot \sum_{k=3}^{n} \frac{1}{k-2} = 0(\log n). \tag{3.5}$$

(c) Again from (3.2),

$$\sum_{j=2}^{n-1} a_{j,\, j+1} = 0(\log n). \tag{3.6}$$

(d) If $(t, u) \epsilon R_{jk}$, $2 \leq j \leq k - 2 \leq n - 2$, then $Q(t, u) \geq 0$ and

$$\theta(t, u) - \theta(j, k) = \frac{1}{4\pi^2 t(u-t)} \int \ldots \int f(x, y) g(\zeta, \eta) [e^{-Q(t,u)} - e^{-Q(j,k)}] dx \ldots d\eta + \frac{1}{4\pi^2} \left[\frac{1}{t(u-t)} - \frac{1}{j(k-j)} \right] \int \ldots \int \times f(x,y) g(\zeta, \eta) e^{-Q(j,k)}\, dx \ldots d\eta = J_1 + J_2.$$

Setting
$R_a = \{|x| \leq a,\ |y| \leq a,\ |\zeta| \leq a,\ |\eta| \leq a\}$, R_a' = complement of R_a, let $\epsilon > 0$ be arbitrary and choose $a > 0$ such that

$$\int_{R_a'} \ldots \int |f(x, y) g(\zeta, \eta)|\, dx \ldots d\eta < \epsilon.$$

We can write

$$J_1 = \frac{1}{4\pi^2 t(u-t)} \int_{R_a'} \ldots \int + \frac{1}{4\pi^2 t(u-t)} \int_{R_a} \ldots \int = J_1' + J_1'',$$

where

$$|J_1'| \leq \frac{\epsilon}{4\pi^2 t(u-t)} \leq \frac{C\epsilon}{(j-1)(k-j-1)}. \tag{3.7}$$

To obtain a bound for J_1'' we observe that

$$Q(t, u) - Q(j, k) = \frac{x^2 + y^2}{2} \left(\frac{1}{t} - \frac{1}{j} \right) + \frac{(x-\zeta)^2 + (y-\eta)^2}{2} \left(\frac{1}{u-t} - \frac{1}{k-j} \right) \leq \frac{x^2 + y^2}{2} \left(\frac{1}{j-1} - \frac{1}{j} \right) + \frac{(x-\zeta)^2 + (y-\eta)^2}{2} \left(\frac{1}{k-j-1} - \frac{1}{k-j} \right) = \frac{x^2 + y^2}{2(j-1)j} + \frac{(x-\zeta)^2 + (y-\eta)^2}{2(k-j-1)(k-j)} = A(\geq 0), \text{ say.}$$

Also,

$$Q(t, u) - Q(j, k) \geq \frac{(x-\zeta)^2 + (y-\eta)^2}{2} \left(\frac{1}{k-j+1} - \frac{1}{k-j} \right) = -\frac{(x-\zeta)^2 + (y-\eta)^2}{2(k-j)(k-j+1)} \geq -A.$$

Hence

$$e^{-A} - 1 \le e^{-\{Q(t,u)-Q(j,k)\}} - 1 \le e^{A} - 1, \quad \text{and}$$

$$\left|e^{-\{Q(t,u)-Q(j,k)\}} - 1\right| \le \max.\ [e^{A} - 1, 1 - e^{-A}] = e^{A} - 1.$$

Therefore

$$|J_1''| \le \frac{1}{4\pi^2 t(u-t)} \int \underset{R_a}{\cdots} \int |f(x,y)g(\zeta,\eta)| \cdot e^{-Q(j,k)} \cdot \left|e^{-\{Q(t,u)-Q(j,k)\}} - 1\right| dx \ldots d\eta \le \frac{FG}{4\pi^2(j-1)(k-j-1)} \int \underset{R_a}{\cdots} \int (e^{A} - 1)\, dx \ldots d\eta.$$

Since $e^x \le 1 + xe^b$ for $0 \le x \le b$, we have in R_a

$$e^{A} - 1 \le A e^{\left[\frac{a^2}{(j-1)j} + \frac{4a^2}{(k-j-1)(k-j)}\right]} \le A e^{\left[\frac{a^2}{2} + 2a^2\right]},$$

and hence

$$|J_1''| \le \frac{C(a)}{(j-1)(k-j-1)}\left[\frac{1}{(j-1)j} + \frac{1}{(k-j-1)(k-j)}\right]. \quad (3.8)$$

Turning to J_2, we have

$$|J_2| \le C\left|\frac{1}{t(u-t)} - \frac{1}{j(k-j)}\right| = C\left|\frac{j(k-j) - t(u-t)}{t(u-t)j(k-j)}\right| \le C\frac{|j(k-j) - t(u-t)|}{(j-1)j(k-j-1)(k-j)}.$$

Since $1 - 2j = j(k-j) - jk + (j-1)^2 \le j(k-j) - t(u-t) \le j(k-j) - (j-1)(k-1) + j^2 = k + j - 1$,

$$|j(k-j) - t(u-t)| \le \max.\ [k + j - 1, 2j - 1] = k + j - 1 = (k-j-1) + 2(j-1) + 2.$$

Hence

$$|J_2| \le C\left[\frac{1}{(j-1)j(k-j)} + \frac{1}{j(k-j-1)(k-j)} + \frac{1}{(j-1)j(k-j-1)(k-j)}\right] \le C\left[\frac{1}{(j-1)^2(k-j-1)} + \frac{1}{(j-1)(k-j-1)^2} + \frac{1}{(j-1)^2(k-j-1)^2}\right]. \quad (3.9)$$

From (3.7)–(3.9) we obtain for $2 \le j \le k - 2 \le n - 2$,

$$|a_{jk}| \leq \frac{C\epsilon}{(j-1)(k-j-1)} + C(a)\left[\frac{1}{(j-1)^3(k-j-1)} + \frac{1}{(j-1)(k-j-1)^3}\right] + C\left[\frac{1}{(j-1)^2(k-j-1)} + \frac{1}{(j-1)(k-j-1)^2} + \frac{1}{(j-1)^2(k-j-1)^2}\right]. \qquad (3.10)$$

Now from [1],

$$\sum_{2 \leq j \leq k-2 \leq n-2} \frac{1}{(j-1)(k-j-1)} \sim (\log n)^2, \qquad (3.11)$$

while

$$\sum_{=2}^{n-2} \sum_{k=j+2}^{n} \frac{1}{(j-1)^2(k-j-1)} \leq \sum_{j=2}^{n-2} \frac{1}{(j-1)^2} \cdot \sum_{m=1}^{n} \frac{1}{m} = 0\,(\log n), \qquad (3.12)$$

$$\sum_{=2}^{n-2} \sum_{k=j+2}^{n} \frac{1}{(j-1)(k-j-1)^2} \leq \sum_{j=2}^{n-2} \frac{1}{j-1} \sum_{m=1}^{\infty} \frac{1}{m^2} = 0\,(\log n). \qquad (3.13)$$

From (3.4)–(3.6) and (3.10)–(3.13) it follows that

$$\limsup_{n \to \infty} (\log n)^{-2} \left| \sum_{j,\,k=1}^{n} a_{jk} \right| \leq C\epsilon,$$

and since ϵ is arbitrary this proves (3.1) and Lemma 2.

Turning to the proof of Lemma 1, let

$$K_n = \frac{1}{\log n}\left[\int_0^n f(V(t))\,dt - \sum_{j=1}^{n} f(V(j))\right] = \frac{1}{\log n} \sum_{j=1}^{n} \int_{j-1}^{j} \left[f(V(t)) - f(V(j))\right] dt.$$

Then $EK_n^2 = (\log n)^{-2} \cdot \sum_{,k=1}^{n} b_{jk}$,

where

$$b_{jk} = \iint_{R_{jk}} E[f(V(t)) - f(V(j))][f(V(u)) - f(V(k))]\,dtdu$$

$$= \iint_{R_{jk}} [\{\theta(t, u) - \theta(t, k)\} - \{\theta(j, u) - \theta(j, k)\}]\,dtdu$$

and where we now write

$$\theta(t, u) = Ef(V(t))f(V(u)) = \frac{1}{4\pi^2 t(u - t)} \int \ldots \int f(x, y)f(\zeta, \eta)e^{-Q(t, u)}\, dx \ldots d\eta$$

for $0 < t < u$. For $2 \leq j \leq k - 2 \leq n - 2$,

$$\theta(t, u) - \theta(t, k) = \frac{1}{4\pi^2 t(u - t)} \int \ldots \int f(x, y)f(\zeta, \eta)[e^{-Q(t, u)} - e^{-Q(t, k)}]\, dx \ldots d\eta + \frac{1}{4\pi^2 t}\left[\frac{1}{u - t} - \frac{1}{k - t}\right] \int \ldots \int f(x, y)f \times (\zeta, \eta)e^{-Q(t, k)}\, dx \ldots d\eta = J_1 + J_2.$$

In R_{jk},

$$Q(t, u) - Q(t, k) = \frac{(x - \zeta)^2 + (y - \eta)^2}{2}\left(\frac{1}{u - t} - \frac{1}{k - t}\right) = \frac{(x - \zeta)^2 + (y - \eta)^2}{2} \cdot \frac{k - u}{(u - t)(k - t)} \geq 0$$

Hence $\left|e^{-Q(t, u)} - e^{-Q(t, k)}\right| = e^{-Q(t, k)}\left|e^{-\{Q(t, u) - Q(t, k)\}} - 1\right| \leq 1 - e^{-\{Q(t, u) - Q(t, k)\}} \leq Q(t, u) - Q(t, k) \leq \frac{(x - \zeta)^2 + (y - \eta)^2}{2} \cdot \frac{k - u}{(u - t)(k - t)}$.

Thus, writing

$$J_1 = \frac{1}{4\pi^2 t(u - t)} \int \underset{R_a'}{\ldots} \int + \frac{1}{4\pi^2 t(u - t)} \int \underset{R_a}{\ldots} \int = J_1' + J_1'',$$

we have $J_1' \leq \dfrac{C\epsilon}{(j - 1)(k - j - 1)}$,

$$J_1'' \leq \frac{F^2}{8\pi^2 t(u - t)} \int_{R_a} \ldots \int [(x - \zeta)^2 + (y - \eta)^2] \times dx \ldots d\eta \cdot \frac{k - u}{(u - t)(k - t)} \leq \frac{C(a)}{(j - 1)(k - j - 1)^3},$$

$$|J_2| \leq \frac{C}{t} \cdot \frac{k - u}{(u - t)(k - t)} \leq \frac{C}{(j - 1)(k - j - 1)^2}.$$

Hence we have

$$|\theta(t, u) - \theta(t, k)| \leq \frac{C\epsilon}{(j-1)(k-j-1)} + \frac{C(a)}{(j-1)(k-j-1)^3} + \frac{C}{(j-1)(k-j-1)^2}, \quad (3.14)$$

and similarly it can be shown that the right-hand side of (3.14) is an upper bound for $|\theta(j, u) - \theta(j, k)|$. Hence, as in the proof of Lemma 2,

$$\lim_{n \to \infty} (\log n)^{-2} \sum_{2 \leq j \leq k-2 \leq n-2} b_{jk} = 0,$$

and the other sums occurring in EK_n^2 can be proved to be $o\{(\log n)^2\}$ as before, completing the proof of Lemma 1.

4. *The One-Dimensional Case.*—Let $f(x)$ and $g(x)$ be real valued functions which are bounded and summable in the line $-\infty < x < \infty$, and set $\bar{f} = \int f(x)dx, \quad \bar{g} = \int g(x)dx.$

THEOREM 3. *If $\bar{f} \neq 0$ then for every u,*

$$\lim_{T \to \infty} \Pr\left[\frac{1}{\bar{f}\sqrt{T}} \int_0^T f(X(t))\, dt \leq u\right] = H(u),$$

where

$$H(u) = \begin{cases} \sqrt{\frac{2}{\pi}} \int_0^u e^{-y^2/2}\, dy & \text{for } u \geq 0, \\ 0 & \text{for } u < 0. \end{cases}$$

THEOREM 4. *If $\bar{g} \neq 0$ then*

$$\lim_{T \to \infty} \frac{\int_0^T f(X(t))\, dt}{\int_0^T g(X(t))\, dt} = \frac{\bar{f}}{\bar{g}} \text{ in probability.}$$

* John Simon Guggenheim Memorial Fellow.

ON THE EQUIDISTRIBUTION OF SUMS OF INDEPENDENT RANDOM VARIABLES

HERBERT ROBBINS[1]

1. The main theorem. Let $X_1, X_2, \cdots$ be a sequence of independent, real-valued random variables with a common distribution function $F(x) = \Pr[X_n \leq x]$, and let $S_n = X_1 + \cdots + X_n$. We are going to show in an elementary manner that the sequence $\{S_n\}$ is "equidistributed" on the line $-\infty < x < \infty$ with respect to a certain class $\mathcal{H}$ of functions $h(x)$, in the sense that for any $h(x)$ in $\mathcal{H}$,

$$\lim_{n\to\infty} \frac{1}{n} \sum_{j=1}^{n} h(S_j) = M(h) \qquad \text{with probability 1,}$$

where the constant $M(h)$ is the mean value of $h(x)$ as defined in the theory of almost periodic functions.

Let $\phi(t) = \int e^{itx} dF(x)$ denote the characteristic function of the X's. Concerning the roots of the equation $\phi(t) = 1$ one of the three following cases must hold.

CASE 1 ("GENERAL" CASE). $\phi(t) = 1$ if and only if $t = 0$.

CASE 2 ("LATTICE" CASE). $\phi(t)$ is not identically equal to 1 but there exists a value $t_0 \neq 0$ such that $\phi(t_0) = 1$. All the mass of $F(x)$ is then concentrated at one or more of the points $x = 2k\pi/t_0$ $(k = 0, \pm 1, \cdots)$. It can be shown that there exists a largest positive number β, $0 < \beta < \infty$, such that all the mass of $F(x)$ is concentrated at one or more of the points $x = k\beta$, and the number β has the property that $\phi(t) = 1$ if and only if $t = 2k\pi/\beta$ $(k = 0, \pm 1, \cdots)$.

CASE 3 (TRIVIAL CASE). $\phi(t) = 1$ for all t. All the mass of $F(x)$ is then concentrated at $x = 0$ and all the S_n are identically 0. The distinction among these three cases is fundamental in what follows.

Let $F(x)$ be arbitrary but fixed. We define the mean value operator $M(h)$ for certain complex-valued functions $h(x)$ of a real variable x, $-\infty < x < \infty$, as follows.

CASE 1.

$$M(h) = \lim_{T\to\infty} \frac{1}{2T} \int_{-T}^{T} h(x)dx$$

provided this limit exists. In particular, for any real t we have

Presented to the Society, October 25, 1952; received by the editors September 27, 1952 and, in revised form, January 22, 1953.

[1] John Simon Guggenheim Memorial Fellow.

$$M(e^{itx}) = \begin{cases} 1 & \text{for } t = 0, \\ 0 & \text{otherwise.} \end{cases}$$

CASE 2.

$$M(h) = \lim_{N\to\infty} \frac{1}{2N+1} \sum_{j=-N}^{N} h(j\beta)$$

provided this limit exists; in particular,

$$M(e^{itx}) = \begin{cases} 1 & \text{for } t = 2k\pi/\beta \quad (k = 0, \pm 1, \cdots), \\ 0 & \text{otherwise.} \end{cases}$$

CASE 3.

$$M(h) = h(0); \; M(e^{itx}) \equiv 1.$$

We now prove the fundamental

LEMMA. *For any $F(x)$ and any real number t, the relation*

$$\lim_{n\to\infty} \frac{1}{n} \sum_{j=1}^{n} e^{itS_j} = M(e^{itx}) \tag{1.1}$$

holds with probability 1.

PROOF. For $t=0$ both sides of (1.1) are equal to 1. If $t\neq 0$ but $\phi(t) = 1$, then we are dealing with Case 2 with $t = 2k\pi/\beta$. The right-hand side of (1.1) is therefore equal to 1, and since each X_n, and hence each S_n, is an integer multiple of β, the left-hand side of (1.1) is also equal to 1. It remains to consider the case $\phi(t)\neq 1$, in which we must show that with probability 1 the left-hand side of (1.1) has the value 0. Let

$$Y_n = \frac{1}{n} \sum_{j=1}^{n} e^{itS_j}.$$

Then

$$|Y_n|^2 = Y_n\bar{Y}_n = \frac{1}{n^2} \sum_{j,k=1}^{n} e^{it(S_k - S_j)}$$
$$= \frac{1}{n} + \frac{2}{n^2} \operatorname{Re}\left[\sum_{j<k} e^{it(X_{j+1}+\cdots+X_k)} \right],$$

so that

$$E|Y_n|^2 = \frac{1}{n} + \frac{2}{n^2} \operatorname{Re}\left[\sum_{j<k} \phi^{k-j}(t) \right].$$

For any number $b \neq 1$,

$$\sum_{j<k} b^{k-j} = \sum_{k=2}^{n} \sum_{j=1}^{k-1} b^{k-j} = b \sum_{k=2}^{n} (1 + b + b^2 + \cdots + b^{k-2})$$

$$= \frac{b}{1-b} \sum_{k=2}^{n} (1 - b^{k-1}) = \frac{b(n-1)}{1-b} - \left(\frac{b}{1-b}\right)^2 (1 - b^{n-1}),$$

and hence, setting $b = \phi(t) \neq 1$, we have

$$E|Y_n|^2 = \frac{1}{n} + \frac{2(n-1)}{n^2} \operatorname{Re}\left[\frac{b}{1-b}\right]$$

$$- \frac{2}{n^2} \operatorname{Re}\left[\left(\frac{b}{1-b}\right)^2 (1 - b^{n-1})\right]$$

$$\leqq \frac{A}{n} \text{ for some finite constant } A.$$

It follows that for any $\epsilon > 0$,

$$\Pr [|Y_n| > \epsilon] \leqq \frac{E|Y_n|^2}{\epsilon^2} \leqq \frac{A}{n\epsilon^2}, \qquad \sum_{n=1}^{\infty} \Pr [|Y_{n^2}| > \epsilon] < \infty,$$

and hence, by the Borel-Cantelli lemma [1, p. 154]

$$\Pr \left[\lim_{n\to\infty} Y_{n^2} = 0\right] = 1. \tag{1.2}$$

But, by a familiar form of argument, given any positive integer n we can choose $m = m(n)$ such that

$$m^2 \leqq n < (m+1)^2;$$

then

$$Y_n = \frac{1}{n} \sum_{j=1}^{n} e^{itS_j} = \frac{m^2}{n} Y_{m^2} + \frac{1}{n} \sum_{j=m^2+1}^{n} e^{itS_j},$$

$$\left| Y_n - \frac{m^2}{n} Y_{m^2} \right| \leqq \frac{n - m^2}{n} < \frac{2m+1}{m^2} \to 0 \qquad \text{as } n \to \infty. \tag{1.3}$$

Since $m^2/n \to 1$ as $n \to \infty$ it follows from (1.2) and (1.3) that

$$\Pr \left[\lim_{n\to\infty} Y_n = 0\right] = 1,$$

which completes the proof.

Definition. For any fixed distribution function $F(x)$ let $\mathcal{H} = \mathcal{H}(F)$

denote the class of all complex-valued functions $h(x)$, $-\infty < x < \infty$, such that $M(h)$ exists and

$$(1.4) \qquad \lim_{n\to\infty} \frac{1}{n} \sum_{j=1}^{n} h(S_j) = M(h) \qquad \text{with probability 1.}$$

The class $\mathfrak{H}$ depends on the function $F(x)$ which determines the probability distribution of the sequence $\{S_n\}$ and the definition of $M(h)$. (In the trivial Case 3, $\mathfrak{H}$ consists of *all* functions $h(x)$.) However, the lemma shows that $\mathfrak{H}$ certainly contains all the functions e^{itx}. It is easy to verify that $\mathfrak{H}$ is closed under addition, multiplication by complex constants, and conjugation, and that M is a linear and monotone operator on $\mathfrak{H}$. Moreover, we can show without trouble that

(I) The uniform limit $h(x)$ of any sequence $h_n(x)$ in $\mathfrak{H}$ is also in $\mathfrak{H}$, and $M(h) = \lim_{n\to\infty} M(h_n)$. More generally,

(II) If $h(x)$ is any real-valued function such that there exist two sequences $h_n'(x)$, $h_n''(x)$ in $\mathfrak{H}$ such that

$$h_n'(x) \leqq h(x) \leqq h_n''(x); \qquad \lim_{n\to\infty} [M(h_n'') - M(h_n')] = 0,$$

then $h(x)$ is in $\mathfrak{H}$ and $M(h) = \lim_{n\to\infty} M(h_n') = \lim_{n\to\infty} M(h_n'')$.

Since we know that $\mathfrak{H}$ contains all finite sums of the form

$$(1.5) \qquad \sum_{k=1}^{r} a_k e^{it_k x}$$

where the t_k are arbitrary real numbers, it follows from (I) that $\mathfrak{H}$ contains all *almost periodic* functions (= uniform limits of sums of the form (1.5)); in particular, $\mathfrak{H}$ contains all continuous and periodic functions. From (II) it follows that $\mathfrak{H}$ contains certain discontinuous functions. For example, suppose that $h(x)$ is real-valued with period p and is Riemann integrable in the interval $0 \leqq x \leqq p$. Suppose further that Case 1 holds, or that Case 2 holds with β not a rational multiple of p. Given any integer n we can find two finite sums $h_n'(x)$, $h_n''(x)$, each of the form

$$(1.6) \qquad g(x) = \sum_{k=-r}^{r} a_k e^{i2k\pi x/p}$$

and such that

$$(1.7) \quad h_n'(x) \leqq h(x) \leqq h_n''(x); \qquad \int_0^p h_n''(x)dx - \int_0^p h_n'(x)dx < \frac{p}{n}.$$

For such sums as (1.6) we have in either of the two cases

$$M(g) = a_0 = \frac{1}{p}\int_0^p g(x)dx,$$

and hence we can write (1.7) as

$$(1.8) \qquad h_n'(x) \leqq h(x) \leqq h_n''(x); \qquad M(h_n'') - M(h_n') < \frac{1}{n}\cdot$$

It follows from (II) that $h(x)$ is in $\mathcal{H}$, and we see by a simple argument that

$$(1.9) \qquad M(h) = \frac{1}{p}\int_0^p h(x)dx.$$

(In Case 1 the relation (1.9) is a trivial consequence of the definition of $M(h)$ and the periodicity of $h(x)$. In the special case of Case 2 in which $p=1$ and all the mass of $F(x)$ is concentrated at the single point β, β irrational, (1.9) is equivalent to the famous theorem [see **2**, pp. 315–316] on the equidistribution mod 1 of the fractional parts of the sequence $n\beta$.)

We summarize our results in

THEOREM 1. *The class $\mathcal{H}$ of functions $h(x)$ such that with probability* 1

$$\lim_{n\to\infty} \frac{1}{n}\sum_{j=1}^{n} h(S_j) = M(h)$$

contains all almost periodic functions. Also, if $h(x)$ has period p and is Riemann integrable in the interval $0 \leqq x \leqq p$, then with probability 1

$$\lim_{n\to\infty} \frac{1}{n}\sum_{j=1}^{n} h(S_j) = \frac{1}{p}\int_0^p h(x)dx$$

provided Case 1 *holds, or Case* 2 *with β not a rational multiple of p.*

As an application of Theorem 1 we shall prove the following

COROLLARY. *Let $h(x)$ be bounded and such that*

$$(1.10) \qquad \lim_{|x|\to\infty} h(x) = \alpha$$

exists, $-\infty < \alpha < \infty$. Then (except in the trivial Case 3*) with probability* 1,

$$(1.11) \qquad \lim_{n\to\infty} \frac{1}{n}\sum_{j=1}^{n} h(S_j) = \alpha.$$

PROOF. It will suffice to prove the assertion (1.11) in the special case when $\alpha=0$ and $h(x)$ is real and non-negative. Suppose, then, that

$$0 \leqq h(x) \leqq B < \infty, \qquad \lim_{|x|\to\infty} h(x) = 0.$$

Given any $\epsilon>0$, choose T_0 so large that

$$h(x) \leqq \epsilon \quad \text{for} \quad x \geqq T_0.$$

Let T be any constant larger than T_0 and define

$$h^*(x) = \begin{cases} B & \text{for } \ |x| < T_0, \\ \epsilon & \text{for } \ T_0 \leqq |x| \leqq T, \\ \text{periodic with period } 2T. \end{cases}$$

We may clearly suppose that $B \geqq \epsilon$; then

$$h(x) \leqq h^*(x) \quad \text{for all } x,$$

and therefore

$$\frac{1}{n}\sum_{j=1}^{n} h(S_j) \leqq \frac{1}{n}\sum_{j=1}^{n} h^*(S_j).$$

The function $h^*(x)$ is periodic with period $2T$ and is Riemann integrable for $-T \leqq x \leqq T$. Moreover, in Case 2 we shall suppose that T is chosen so that β is not a rational multiple of $2T$. Then by Theorem 1, with probability 1,

$$0 \leqq \limsup_{n\to\infty} \frac{1}{n}\sum_{j=1}^{n} h(S_j) \leqq \lim_{n\to\infty}\frac{1}{n}\sum_{j=1}^{n} h^*(S_j) = \frac{1}{2T}\int_{-T}^{T} h^*(x)dx$$

$$\leqq \frac{2BT_0 + 2\epsilon(T - T_0)}{2T} < 2\epsilon,$$

provided that

$$T > \frac{BT_0}{\epsilon}\cdot$$

Since ϵ was arbitrary it follows that with probability 1

$$\lim_{n\to\infty}\frac{1}{n}\sum_{j=1}^{n} h(S_j) = 0,$$

which was to be proved.

We remark that any function $h(x)$ satisfying the hypotheses of the Corollary is in $\mathcal{H}$ (since $M(h)=\alpha$ and (1.11) holds with probability 1) although $h(x)$ need not even be measurable.

As a special case of the Corollary it follows that *for any finite interval* I: $a \leqq x \leqq b$,

$$\lim_{n\to\infty} \frac{\text{number of terms } S_1, \cdots, S_n \text{ in } I}{n} = 0$$

with probability 1, provided only that the X's are not identically 0.

2. Sequences equidistributed modulo I. A sequence of constants $\{a_n\}$ lying in a finite interval I: $a \leqq x \leqq b$ is said to be *equidistributed in I* if for every subinterval I': $a' \leqq x \leqq b'$ $(a \leqq a' < b' \leqq b)$,

$$\lim_{n\to\infty} \frac{\text{number of terms } a_1, \cdots, a_n \text{ which lie in } I'}{n} = \frac{b'-a'}{b-a}.$$

It follows that for any Riemann integrable function $u(x)$ defined on I,

$$\lim_{n\to\infty} \frac{1}{n} \sum_{j=1}^{n} u(a_j) = \frac{1}{b-a} \int_a^b u(x)dx.$$

For any interval I: $a \leqq x \leqq b$ and any number x, not necessarily lying in I, let $(x)_I$ denote "x modulo I"; that is, let $(x)_I$ be the unique number such that $a \leqq (x)_I < b$ and such that $x-(x)_I$ is an integer multiple of $b-a$. If a sequence of constants $\{a_n\}$ is such that the sequence $(a_n)_I$ is equidistributed in I, we shall call the sequence $\{a_n\}$ *equidistributed modulo I.*

Let $S_n = X_1 + \cdots + X_n$, the X's being independent random variables with any common distribution function for which Case 1 holds, let I: $a \leqq x \leqq b$ be any finite interval, I': $a' \leqq x \leqq b'$ any subinterval of I, and let $h(x)$ be the function defined for all x as follows:

$$h(x) = \begin{cases} 1 & \text{for } x \text{ in } I', \\ 0 & \text{for } x \text{ in } I \text{ but not in } I', \\ \text{periodic with period } b-a \text{ for all } x. \end{cases}$$

It follows from Theorem 1 that with probability 1,

$$\lim_{n\to\infty} \frac{\text{number of terms } (S_1)_I, \cdots, (S_n)_I \text{ which lie in } I'}{n} = \frac{b'-a'}{b-a}, \tag{2.1}$$

and by denumerability considerations that (2.1) holds with probability 1 simultaneously for all pairs I, I' of intervals with *rational* end points $a \leqq a' < b' \leqq b$. It then follows by approximating an arbitrary interval from within or from without by a rational interval that (2.1) holds with probability 1 simultaneously for *all* intervals I, I'. Thus we have proved

THEOREM 2. *In Case* 1, *with probability* 1 *the sequence of random variables* $\{S_n\}$ *is equidistributed modulo I simultaneously for all finite intervals I.*

3. **Poincaré's roulette theorem revisited.** A horizontal circular disk is pivoted at its center to a table. Its circumference is marked off into $2m$ arcs, alternately red and black, the red arcs of angle δ_1 and the black arcs of angle δ_2. An arrow fixed to the table points initially to a point 0 on the circumference of the disk. The disk is spun in such a way that the total angle X through which the disk turns is a random variable with a probability density $f(x)$. The probability that the disk comes to rest so that the stationary arrow points to a red point on the circumference is given by the integral

$$\int h(x)f(x)dx \tag{3.1}$$

where

$$h(x) = \begin{cases} 1 & \text{for } 0 \leqq x \leqq \delta_1 \text{ (red arc)}, \\ 0 & \text{for } \delta_1 < x < \delta_1 + \delta_2 \text{ (black arc)}, \\ \text{periodic with period } \delta_1 + \delta_2. \end{cases} \tag{3.2}$$

Poincaré [3, pp. 127–129; see also 4, 5] observed that if $f(x)$ is held fixed while $m\to\infty$ and $\delta_i\to 0$ in such a way that the ratio δ_1/δ_2 remains constant, then the integral (3.1) tends to the limit $\delta_1/(\delta_1+\delta_2)$. (Poincaré actually assumed that $f(x)$ has a bounded derivative, but this assumption is unnecessary.) Thus, despite the presence of the arbitrary density function $f(x)$, for large m the disk behaves as though it had an approximately uniform distribution of probability around its circumference.

Let $\{X_n\}$ be a sequence of independent random variables with the common probability density $f(x)$, corresponding to a succession of spins of the disk *in which the disk is returned after each spin to its initial position* with the arrow pointing to 0. According to the strong law of large numbers, the relation

$$\begin{aligned} &\lim_{n\to\infty} [\text{proportion of red stopping points among first } n \text{ trials}] \\ &= \lim_{n\to\infty} \frac{1}{n}\sum_{j=1}^{n} h(X_j) = \int h(x)f(x)dx \end{aligned} \tag{3.3}$$

will hold with probability 1.

Now suppose (as is presumably the case in practice) that the successive spins of angles $X_1, X_2, \cdots$ are applied *without touching the disk between spins.* After the nth spin the disk will have turned through a total angle S_n, and Theorem 1 shows that, in contrast to (3.3), the relation

$$\lim_{n\to\infty} [\text{proportion of red stopping points among first } n \text{ trials}]$$

$$= \lim_{n\to\infty} \frac{1}{n} \sum_{j=1}^{n} h(S_j) = \frac{\delta_1}{\delta_1 + \delta_2} \tag{3.4}$$

will hold with probability 1. In fact, (3.4) will hold even if the X's do not have a probability density, provided only that their common distribution function $F(x)$ is such that Case 1 holds or Case 2 holds with β not a rational multiple of 2π. This result is truly in the spirit of Poincaré's "method of arbitrary functions."

4. A general equidistribution problem. For a given distribution function $F(x)$ and pair of functions $h_1(x)$, $h_2(x)$, when does the limit

$$\lim_{n\to\infty} \frac{\sum_{j=1}^{n} h_1(S_j)}{\sum_{j=1}^{n} h_2(S_j)} \tag{4.1}$$

exist (either with probability 1 or in probability, etc.) and have as its value a constant which represents some natural measure of the "relative mean value" of $h_1(x)$ compared with $h_2(x)$? In the special case $h_2(x) \equiv 1$ this reduces to the problem of §1. There we demanded that the value of (4.1) for $h_2(x) \equiv 1$ be equal to the mean value $M(h_1)$ as defined in the three different cases. Similarly, we might now require that the value of (4.1) be equal to

$$\lim_{T\to\infty} \frac{\int_{-T}^{T} h_1(x)dx}{\int_{-T}^{T} h_2(x)dx}, \tag{4.2}$$

and we might confine ourselves to pairs of functions for which the limit

$$\lim \frac{\int_{T_1}^{T_2} h_1(x)dx}{\int_{T_1}^{T_2} h_2(x)dx}$$

exists uniformly as $T_2 - T_1 \to \infty$.

If the individual limits

$$\lim_{n\to\infty} \frac{1}{n} \sum_{j=1}^{n} h_i(S_j) \qquad (i = 1, 2) \tag{4.3}$$

exist and are not both 0, then (4.1) exists also, so that the results of §1 give a partial answer to the present problem. However, the case in which both the limits (4.3) exist and are 0 is often of great interest, and this is not covered by the results of §1. We shall be content here with giving three simple examples illustrative of the general problem.

EXAMPLE 1 (Periodic intervals). Let $h_1(x)$ be defined by (3.2) and let $h_2(x) = 1 - h_1(x)$. In Case 1, Theorem 1 shows that, with probability 1, (4.3) exists and equals δ_i, so that (4.1) exists and equals the value δ_1/δ_2 of (4.2).

EXAMPLE 2 (Infinite intervals). Let $h_1(x) = 1$ for $x \geq 0$ and $h_1(x) = 0$ for $x < 0$, while $h_2(x) = 1 - h_1(x)$. By a theorem of Erdös and Kac [6], if the X's have mean 0 and finite second moment, the random variables

$$Z_n = \frac{1}{n} \sum_{j=1}^{n} h_1(S_j)$$

have the continuous limiting distribution function

$$G(x) = \lim_{n\to\infty} \Pr\left[Z_n \leq x\right] = \frac{2}{\pi} \arcsin x^{1/2} \quad (0 \leq x \leq 1).$$

Since $\sum_{j=1}^{n} h_2(S_j) = n(1 - Z_n)$ it follows that neither (4.3) nor (4.1) (since the ratio in (4.1) is equal to $Z_n/(1-Z_n)$) converges to any constant with probability 1 or in probability.

EXAMPLE 3 (Finite intervals). Let $h_i(x) = 1$ on some finite interval (a_i, b_i) and $h_i(x) = 0$ elsewhere. In [7] it is shown, for example, that if $F(x)$ is not a lattice distribution function and has mean 0 and finite third absolute moment, then the limit in probability of (4.1) is $(b_1 - a_1)/(b_2 - a_2)$, and that under somewhat different conditions on $F(x)$ the limit in probability of (4.1) exists more generally wherever the $h_i(x)$ are any two bounded functions in $L^1(-\infty, \infty)$, and has as its value (4.2), which is here equal to the ratio of the integrals of the functions $h_i(x)$ over $(-\infty, \infty)$, the denominator integral being assumed $\neq 0$. Here the individual limits (4.3) exist but are both 0. A sharper and more general result was proved later in [14].

5. Extension to compact groups. By using the random ergodic theory of Ulam-von Neumann [8] and Kakutani [9], the results of §1 can be extended to the case in which the random variables X_n have values in any compact topological group.

Let G be a compact (not necessarily commutative) topological group with elements g and Haar measure μ, $\mu(G)=1$, and let $X_n(\omega)$ be a sequence of independent G-valued random variables with the same distribution defined on a probability space Ω of points ω. Put $S_n(\omega)=X_1(\omega)\cdots X_n(\omega)$, where the product is taken in the sense of group multiplication in G. From Theorem 1 of [9] it follows that for any continuous function $h(g)$ defined on G,

$$\lim_{n\to\infty}\frac{1}{n}\sum_{j=1}^{n} h(g\cdot X_1(\omega)\cdots X_j(\omega)) \tag{5.1}$$
$$= \lim_{n\to\infty}\sum_{j=1}^{n} h(g\cdot S_j(\omega)) = \bar{h}(g,\omega)$$

exists for a.e. (g,ω) in $G\times\Omega$. The functions

$$h_{n,\omega}(g) = \frac{1}{n}\sum_{j=1}^{n} h(g\cdot S_j(\omega))$$

are equi-uniformly continuous on G for any fixed ω, while from Fubini's theorem it follows that for a.e. ω the limit (5.1) exists for a.e. g. Consequently for a.e. ω the limit (5.1) exists uniformly for all g in G. In particular, setting $g=1$,

$$\lim_{n\to\infty}\frac{1}{n}\sum_{j=1}^{n} h(S_j(\omega)) = \bar{h}(\omega) \tag{5.2}$$

exists a.e. on Ω.

Let $\phi_x(g)=g\cdot x$ denote right multiplication in G. The family $\{\phi_{X(\omega)}\mid\omega\in\Omega\}$ of μ-preserving transformations is said to be ergodic on G if, for every Borel subset B of G, $\mu[\phi_{X(\omega)}(B)\Delta B]=0$ for a.e. ω implies $\mu(B)=0$ or 1, where Δ denotes symmetric difference. From Theorem 3 of [9] it follows in the ergodic case that the function $\bar{h}(g,\omega)$ of (5.1) is constant a.e. on $G\times\Omega$, and hence the function $\bar{h}(\omega)$ of (5.2) is constant a.e. on Ω.

Now it is easy to see that the family $\{\phi_{X(\omega)}\mid\omega\in\Omega\}$ is ergodic on G if $X(\omega)$ has the following property: for any null set N of Ω, the set $\{X(\omega)\mid\omega\in\Omega-N\}$ generates an everywhere dense subgroup of G. If $X(\omega)$ does not have this property on G, we replace G by the smallest closed subgroup G_0 with the property that Pr $\{\omega\mid X(\omega)\in G_0\}=1$; then $X(\omega)$ has the desired property on the group G_0. Consequently, $\bar{h}(g,\omega)$ is constant a.e. on $G_0\times\Omega$ and $\bar{h}(\omega)$ is constant a.e. on Ω. Thus

$\bar{h}(\omega)$ is constant a.e. on Ω even when the family $\{\phi_{X(\omega)} | \omega \in \Omega\}$ is not ergodic on G.

Theorem 1 of §1 for $h(x)$ almost periodic follows by taking as G the van Kampen-Weil [10, 11] compactification of the real line $-\infty < x < \infty$ with respect to the function $h(x)$. In this case $h(x)$ corresponds to a continuous function $h(g)$ on the compactification. The case in which $h(g)$ is Riemann integrable on G is treated similarly.

The results of this paper are connected to some extent with work by Lévy [12] and Kawada and Ito [13] who discussed convergence of distribution functions (instead of convergence with probability 1 of sequences of random variables) for the case of the additive group of real numbers modulo 1 and a general compact group respectively.

The author is indebted to Professor S. Kakutani for the results of this section.

6. Equidistribution for general stochastic processes. We can easily extend our equidistribution arguments beyond the case of sums of independent random variables. In fact, let $X(t)$ be any stochastic process with a continuous time parameter t, defined for all $t \geqq 0$. For any fixed real λ set

$$Y(t) = \frac{1}{t} \int_0^t e^{i\lambda X(u)} du.$$

Then

$$E | Y(t) |^2 = \frac{1}{t^2} \int_0^t \int_0^t E e^{i\lambda [X(u) - X(v)]} du dv.$$

We shall assume that for every $\lambda \neq 0$ there exists a $\delta = \delta(\lambda) > 0$ such that as $t \to \infty$,

$$E | Y(t) |^2 = O(t^{-\delta}). \tag{6.1}$$

Then just as in §1 we can show that with probability 1,

$$\lim_{n \to \infty} Y(t) = \begin{cases} 1 & \text{for } \lambda = 0, \\ 0 & \text{for } \lambda \neq 0. \end{cases}$$

It follows that the equidistribution properties which hold in Case 1 for the discrete sequence $\{S_n\}$ continue to hold in integral form for the process $X(t)$.

It is therefore of interest to ask for what processes $X(t)$ the assumption (6.1) is satisfied. We shall consider only the case of a process with stationary increments, in which the distribution of the difference $X(u)-X(v)$ depends only on the difference $u-v$. Then for fixed λ, setting

$$\phi(t) = \begin{cases} Ee^{i\lambda[X(t)-X(0)]} & \text{for } t \geqq 0, \\ \bar{\phi}(-t) & \text{for } t < 0, \end{cases}$$

we can write

$$Ee^{i\lambda[X(u)-X(v)]} = \phi(u-v),$$

and hence

$$\begin{aligned} E|Y(t)|^2 &= \frac{1}{t^2}\int_0^t\int_0^t \phi(u-v)dudv \\ &= \frac{2}{t^2}\operatorname{Re}\int_0^t\int_0^v \phi(u-v)dudv \\ &= \frac{2}{t^2}\operatorname{Re}\int_0^t\int_0^v \phi(u)dudv \qquad (6.2) \\ &= \frac{2}{t^2}\operatorname{Re}\int_0^t (t-u)\phi(u)du \\ &\leqq \frac{2}{t}\int_0^t |\phi(u)|\, du. \end{aligned}$$

Hence if for some $\delta>0$, $\phi(t)=O(t^{-\delta})$ as $t\to\infty$, then setting $\delta' = \min(\delta, 1/2)$ we have $\phi(t)=O(t^{-\delta'})$ and therefore from (6.2),

$$E|Y(t)|^2 \leqq \frac{2}{t} + \frac{2C}{t}\int_1^t u^{-\delta'}du = O(t^{-\delta'}),$$

so that (6.1) holds with δ replaced by δ'. This proves

THEOREM 3. *If $X(t)$ is a stochastic process with stationary increments, such that for every $\lambda\neq 0$ there exists a $\delta=\delta(\lambda)>0$ for which*

$$Ee^{i\lambda[X(t)-X(0)]} = O(t^{-\delta}) \quad as \quad t\to\infty,$$

then for any function $h(x)$ which is almost periodic, or periodic and Riemann integrable in its period interval, the relation

$$\lim_{t\to\infty}\frac{1}{t}\int_0^t h(X(u))du = M(h) = \lim_{A\to\infty}\frac{1}{2A}\int_{-A}^{A} h(x)dx$$

holds with probability 1.

As a particular case of Theorem 3 we mention the case of Brownian movement (Wiener-Lévy) in which the random variable $X(u)-X(v)$ is normally distributed with mean 0 and variance $|u-v|$, so that for $t\geqq 0$

$$\phi(t) = e^{-\lambda^2 t/2},$$

and

$$E|Y(t)|^2 = O(t^{-1}).$$

Bibliography

1. W. Feller, *Probability theory and its applications*, vol. 1, New York, 1950.

2. H. Weyl, *Über die Gleichverteilung von Zahlen mod. Eins*, Math. Ann. vol. 77 (1916) pp. 313–352.

3. H. Poincaré, *Calcul des probabilités*, Paris, 1896.

4. M. Fréchet, *Méthode des fonctions arbitraires, etc.*, vol. 1 (Part III, Book 2 of *Traité du calcul des probabilités* (Borel)), Paris, 1938.

5. B. Hostinský, *Méthodes générales du calcul des probabilités* (Part LII of Mémorial des sciences mathématiques), Paris, 1931.

6. P. Erdös and M. Kac, *On the number of positive sums of independent random variables*, Bull. Amer. Math. Soc. vol. 53 (1947) pp. 1011–1020.

7. G. Kallianpur and H. Robbins, *On the sequence of sums of independent random variables*, unpublished.

8. S. Ulam and J. von Neumann, *Random ergodic theorems*, Bull. Amer. Math. Soc. Abstract 51-9-165.

9. S. Kakutani, *Random ergodic theorems and Markoff processes with a stable distribution*, Proceedings of the Second Berkeley Symposium on Probability and Statistics, 1950, pp. 247–261.

10. E. R. van Kampen, *Almost periodic functions and compact groups*, Ann. of Math. (2) vol. 37 (1936) pp. 48–91.

11. A. Weil, *Sur les fonctions presque périodiques de v. Neumann*, C. R. Acad. Sci. Paris vol. 200 (1935) pp. 38–40.

12. P. Lévy, *L'addition des variables aléatoires définies sur une circonférence*, Bull. Soc. Math. France vol. 67 (1939) pp. 1–41.

13. Y. Kawada and K. Ito, *On the probability distribution on a compact group*, Proceedings of the Physico-Mathematical Society of Japan vol. 22 (1940) pp. 977–998.

14. T. Harris and H. Robbins, *Ergodic theory of Markov chains with an infinite invariant measure*, to appear in Proc. Nat. Acad. Sci. U.S.A.

University of North Carolina and
The Institute for Advanced Study

Reprinted from
Proc. Amer. Math. Soc.
4, 786–799 (1953)

Reprinted from the Proceedings of the NATIONAL ACADEMY OF SCIENCES,
Vol. 39, No. 8, pp. 860-864. August, 1953

*ERGODIC THEORY OF MARKOV CHAINS ADMITTING AN INFINITE INVARIANT MEASURE**

BY T. E. HARRIS† AND HERBERT ROBBINS

COLUMBIA UNIVERSITY AND THE INSTITUTE FOR ADVANCED STUDY

Communicated by H. Whitney, June 3, 1953

1. *Introduction.*—We consider real valued Markov chains $\ldots, x_{-1}$ $x_0, x_1, \ldots$ with a stationary transition probability function

$$H(u, B) = \text{Prob}\,(x_{n+1} \epsilon B \mid x_n = u).$$

Under certain conditions there exists a probability measure π defined on the Borel sets B and satisfying

$$\pi(B) = \int \pi(du)H(u, B). \qquad (1)$$

There is then a stationary Markov chain $\ldots x_{-1}, x_0, x_1, \ldots$ with the transition function H such that Prob $(x_n \epsilon B) = \pi(B)$, $n = 0 \pm 1, \ldots.$ Birkhoff's ergodic theorem can then be applied to this chain (cf. ref. 2).

It may be that no probability measure π exists satisfying (1). (Such is the case, for example, if for every bounded Borel set B, $\lim_{k\to\infty} H_k(u, B) = 0$, where H_k denotes the k-step transition function, $H_1 = H$.) There may, however, be an infinite measure π which satisfies (1). The purpose of this note is to show how the ergodic theorem of Hopf[3] can then be applied. A similar treatment will be given elsewhere for certain stochastic processes involving a continuous time parameter (cf. ref. 5). The present note is closely related to recent work of Robbins[7] and of Kallianpur and Robbins.[4]

2. *Definition of the Measure m.—We shall assume henceforth that* (1) *has an "admissible" solution π;* i.e., π is a measure on the real Borel sets which satisfies (1), does not vanish identically, and is finite for *bounded* Borel sets. (Current unpublished work of C. Derman shows that this is so in a variety of circumstances.)

Let F_0 be the class of real Borel sets, Ω the space of sequences x of real numbers, $x = (\ldots, x_{-1}, x_0, x_1, \ldots)$, and F the Borel extension of the cylinder sets in Ω. If $A \epsilon F$ is determined by the coordinates $x_k, x_{k+1}, \ldots$, then $Q(A|x_k = u)$ will denote the probability of A relative to the Markov chain starting with $x_k = u$, as specified by H.

A measure m may be set up on F as follows. If A is a cylinder set determined by $x_k, \ldots, x_r$, let

$$m(A) = \int \pi(du) Q(A|x_j = u), \tag{2}$$

where (1) ensures that any value of $j \leq k$ may be used indifferently in (2). It can be verified that m may be extended to F, there being no essential change in the proof of the extension theorem of Kolmogorov.[6] The extended measure m satisfies (2) whenever $A \epsilon F$ depends on $x_k, x_{k+1}, \ldots$ and $j \leq k$. Moreover, $m(A) = 0$ if and only if $Q(A|x_j = u) = 0$ for a.e. (π) u; thus $m(A) = 0$ has a meaning in terms of ordinary Markov chain probabilities. From (1) it follows that if $A \epsilon F$ and if T is the shift transformation, $(Tx)_i = x_{i+1}$, then $m(A) = m(TA)$.

3. *The Dissipative Part of Ω.*—For any $B \epsilon F_0$, let R_B be the event that $x_n \epsilon B$ infinitely often for $n = 1, 2, \ldots$ ("$x_n \epsilon B$ i.o.").

ASSUMPTION 1. *If $B \epsilon F_0$ then $Q(R_B|x_0 = u) = 1$ for* a.e. (π) *u in B.*

THEOREM 1. *If Assumption 1 holds then the dissipative part of Ω has m-measure 0 (see ref. 3 for terminology).*

Sketch of Proof: Let A be a cylinder set defined by, say, $(x_0, x_1, \ldots, x_k) \epsilon B_{k+1}$, a $(k + 1)$-dimensional Borel set, and let

$$U_n = \left\{u; \frac{1}{n+1} < Q(A|x_0 = u) \leq \frac{1}{n}\right\}, \qquad n = 1, 2, \ldots.$$

Equation (2) ensures that A differs from $\sum_{n=1}^{\infty} A\cdot(x_0 \epsilon U_n)$ by a set of

m-measure 0, and hence for a.e. $(m)x$ in A, $T^n x \epsilon A$ i.o., so that $m(A \cdot \sum_1^\infty T^{-n}A) = m(A)$. Next, let W be a "wandering" set with $m(W) < \infty$ (it suffices to consider this case) and let A be a cylinder set such that $m(W - A) + m(A - W) < \epsilon$. Then (all sums $n = 1$ to ∞)

$$\begin{aligned} 0 = m(W\cdot\textstyle\sum T^n W) &> m(A\cdot\textstyle\sum T^n W) - \epsilon \\ &= \textstyle\sum m[(T^{-n}A)\cdot W] - \epsilon \geq m(W\cdot\textstyle\sum T^{-n}A) - \epsilon > \\ &\qquad m(A\cdot\textstyle\sum T^{-n}A) - 2\epsilon = m(A) - 2\epsilon, \end{aligned}$$

and hence

$$m(W) \leq m(A) + \epsilon < 3\epsilon, \qquad m(W) = 0.$$

4. *Metric Transitivity of T.*—The following assumption is stronger than Assumption 1.

ASSUMPTION 2. *If $B \epsilon F_0$ and $\pi(B) > 0$ then $Q(R_B | x_0 = u) = 1$ for* a.e. (π) u.

THEOREM 2. *If Assumption 2 holds then the shift transformation T is metrically transitive.*

Sketch of Proof: First, let $A \epsilon F$ be such that for every $k = 1, 2, \ldots$, A depends only on $x_k, x_{k+1}, \ldots$ (call such a set a "terminal" set), and suppose $A = TA$. Assumption 2 and the invariance of A can be shown to imply that $Q(A | x_0 = u)$ is constant for a.e. $(\pi)u$, and the 0–1 law shows that the constant is 0 or 1. It follows from (2) that $m(A) = 0$ or $m(\Omega - A) = 0$.

Now let $g(x)$ be a fixed positive m-summable function which depends on a finite number of coordinates of x. Hopf's theorem implies that if $f(x)$ is m-summable then

$$\lim_{n\to\infty} \frac{\sum_0^n f(T^j x)}{\sum_0^n g(T^j x)} = \mu(f; x)$$

exists for a.e. $(m)x$. If $f(x)$ depends on only a finite number of coordinates it follows from the result above on "terminal" invariant sets that

$$\mu(f; x) = \text{const.} = \frac{\int f(x)m(dx)}{\int g(x)m(dx)}. \tag{3}$$

To prove metric transitivity it is sufficient to show that (3) holds for every m-summable f.

Now let $f(x)$ be any m-summable function and let $f_k(x)$ be a sequence of functions of a finite number of coordinates such that

$$\lim_{k\to\infty} \int |f(x) - f_k(x)| m(dx) = 0.$$

Then

$$\mu(f;\, x) = \mu(f - f_k;\, x) + \frac{\int f_k(x)m(dx)}{\int g(x)m(dx)},$$

whence

$$\left|\mu(f;\, x) - \frac{\int f(x)m(dx)}{\int g(x)m(dx)}\right| = \lim_{k\to\infty}\left|\mu(f - f_k;\, x)\right|.$$

Since for any m-summable $h(x)$, $\int \mu(h;\, x)g(x)m(dx) = \int h(x)m(dx)$, it follows that

$$\lim_{k\to\infty} \int |\mu(f - f_k;\, x)|g(x)m(dx) \leq \lim_{k\to\infty} \int \mu(|f - f_k|;\, x)g(x)m(dx) = \lim_{k\to\infty} \int |f - f_k|m(dx) = 0,$$

and since $g(x) > 0$ the desired result follows.

5. *Applications.*—In what follows $h(u)$ and $k(u)$ will be π-summable functions of a real variable u with $\int k(u)\pi(du) \neq 0$. Theorems 1 and 2 in conjunction with Hopf's theorem then imply

THEOREM 3.[8] *Under Assumption 2, for a.e.* $(\pi)x_0$,

$$\text{Prob}\left\{\lim_{n\to\infty} \frac{h(x_0) + \ldots + h(x_n)}{k(x_0) + \ldots + k(x_n)} = \frac{\int_{-\infty}^{\infty} h(u)\pi(du)}{\int_{-\infty}^{\infty} k(u)\pi(du)}\right\} = 1.$$

COROLLARY. *Under Assumption 2, if* π_1 *is another admissible solution of (1) such that* $\pi(B) = 0$ *implies* $\pi_1(B) = 0$, *then for some constant* $c > 0$, $\pi_1 = c\pi$.

Next, consider a sequence of independent random variables $y_1, y_2, \ldots$ with a common distribution function $F(y)$.

ASSUMPTION 3. *The sums* $y_1 + \ldots + y_n$ *are "interval-recurrent"; i.e., for any interval I, Prob* $(y_1 + \ldots + y_n \in I \text{ i.o.}) = 1$ *(see ref. 1).*

We state without proof

LEMMA 1. *Under Assumption 3, if* $B \in F_0$ *has positive Lebesgue measure then for a.e. a, Prob*$(a + y_1 + \ldots + y_n \in B \text{ i.o.}) = 1$. *This is true for every a if some convolution* $F^{n*}(y)$ *has an absolutely continuous component.*

Set $H(u, B) = \int_B dF(y - u)$; then π = Lebesgue measure satisfies (1), and from Theorem 3 and Lemma 1 we obtain

THEOREM 4. *Under Assumption 3, for a.e. (Lebesgue measure) a,*

$$\text{Prob}\left\{\lim_{n\to\infty} \frac{\sum_{j=1}^{n} h(a + y_1 + \ldots + y_j)}{\sum_{j=1}^{n} k(a + y_1 + \ldots + y_j)} = \frac{\int_{-\infty}^{\infty} h(u)du}{\int_{-\infty}^{\infty} k(u)du}\right\} = 1. \quad (4)$$

If some convolution $F^{n*}(y)$ has an absolutely continuous component then (4) *holds for every a.*

From Theorem 4 we can deduce the following

COROLLARY. *Under Assumption 3 (even if no $F^{n*}(y)$ has an absolutely continuous component) for every a the limit relation in* (4) *holds with probability 1 simultaneously for every two functions $h(u)$, $k(u)$ which are Riemann-integrable in a finite interval and vanish elsewhere, the integral of $k(u)$ assumed to be $\neq 0$.*

Thus the prob 1 *density* of the sequence of sums $y_1 + \ldots + y_n$ implies the prob 1 *equidistribution* of these sums. This establishes a conjecture made in ref. 4.

If the y_i have only integer values then the function $\pi(B)$ = number of integer points in B, is an admissible solution of (1). Theorem 3 then implies that under Assumption 2 (which holds whenever every integer ν is a "recurrent" value of the sums $y_1 + \ldots + y_n$), for every two functions of integers $h(\nu)$, $k(\nu)$ such that

$$\sum_{-\infty}^{\infty} |h(\nu)| < \infty, \sum_{-\infty}^{\infty} |k(\nu)| < \infty, \sum_{-\infty}^{\infty} k(\nu) \neq 0,$$

$$\text{Prob}\left\{ \lim_{n\to\infty} \frac{\sum_{j=1}^{n} h(y_1 + \ldots + y_j)}{\sum_{j=1}^{n} k(y_1 + \ldots y_j)} = \frac{\sum_{-\infty}^{\infty} h(\nu)}{\sum_{-\infty}^{\infty} k(\nu)} \right\} = 1. \qquad (5)$$

* Work sponsored by the Office of Scientific Research of the Air Force.

† On leave from the RAND Corporation.

[1] Chung, K. L , and Fuchs, W. H. J., "On the Distribution of Values of Sums of Random Variables," *Mem. Am. Math. Soc.*, **6,** (1951).

[2] Doob, J. L., *Stochastic Processes*, Wiley & Son, New York, 1953, Chap. 10.

[3] Hopf, E., *Ergodentheorie*, reprinted by Chelsea, New York, 1948, pp. 46–53.

[4] Kallianpur, G., and Robbins, H., "On the Sequence of Partial Sums of Independent Random Variables," *Duke Math. J.*, (1954).

[5] Kallianpur, G., and Robbins, H., "Ergodic Property of the Brownian Motion Process", *Proc. Natl. Acad. Sci.*, **39**, (1953), pp. 525–533.

[6] Kolmogorov, A., *Grundbegriffe der Warscheinlichkeitsrechnung*, reprinted by Chelsea, New York, 1946, pp. 24–30.

[7] Robbins, H., "On the Equidistribution of Sums of Independent Random Variables", *Proc. Am. Math. Soc.*, **4** (1953), pp. 786–799.

[8] Chung has obtained this result independently and by a different method for the case in which the X_i have only a denumerable set of values. His results will appear in *Trans. Am. Math. Soc.*

Reprinted from the AMERICAN MATHEMATICAL MONTHLY
Vol. LXII, No. 1, January, 1955

A REMARK ON STIRLING'S FORMULA

HERBERT ROBBINS, Columbia University

We shall prove Stirling's formula by showing that for $n=1, 2, \cdots$

$$n! = \sqrt{2\pi}\, n^{n+1/2} e^{-n} \cdot e^{r_n} \tag{1}$$

where r_n satisfies the double inequality

$$\frac{1}{12n+1} < r_n < \frac{1}{12n}. \tag{2}$$

The usual textbook proofs replace the first inequality in (2) by the weaker inequality

$$0 < r_n$$

or

$$\frac{1}{12n+6} < r_n.$$

Proof. Let

$$S_n = \log (n!) = \sum_{p=1}^{n-1} \log (p+1)$$

and write

$$\log (p+1) = A_p + b_p - \epsilon_p \tag{3}$$

where

$$A_p = \int_p^{p+1} \log x\, dx, \quad b_p = \tfrac{1}{2}[\log (p+1) - \log p];$$

$$\epsilon_p = \int_p^{p+1} \log x\, dx - \tfrac{1}{2}[\log (p+1) + \log p].$$

The partition (3) of $\log (p+1)$, regarded as the area of a rectangle with base $(p, p+1)$ and height $\log (p+1)$, into a curvilinear area, a triangle, and a small sliver* is suggested by the geometry of the curve $y = \log x$. Then

* Taken from G. Darmois, *Statistique Mathématique*, Paris, 1928, pp. 315–317. The only novelty of the present note is the inequality (7) which permits the first part of the estimate (2).

$$S_n = \sum_{p=1}^{n-1} (A_p + b_p - \epsilon_p) = \int_1^n \log x\, dx + \tfrac{1}{2} \log n - \sum_{p=1}^{n-1} \epsilon_p.$$

Since $\int \log x\, dx = x \log x - x$ we can write

$$S_n = (n + \tfrac{1}{2}) \log n - n + 1 - \sum_{p=1}^{n-1} \epsilon_p, \tag{4}$$

where

$$\epsilon_p = \frac{2p+1}{2} \log\left(\frac{p+1}{p}\right) - 1.$$

Using the well known series

$$\log\left(\frac{1+x}{1-x}\right) = 2\left(x + \frac{x^3}{3} + \frac{x^5}{5} + \cdots\right)$$

valid for $|x| < 1$, and setting $x = (2p+1)^{-1}$, so that $(1+x)/(1-x) = (p+1)/p$, we find that

$$\epsilon_p = \frac{1}{3(2p+1)^2} + \frac{1}{5(2p+1)^4} + \frac{1}{7(2p+1)^6} + \cdots. \tag{5}$$

We can therefore bound ϵ_p above and below:

$$\epsilon_p < \frac{1}{3(2p+1)^2}\left\{1 + \frac{1}{(2p+1)^2} + \frac{1}{(2p+1)^4} + \cdots\right\} \tag{6}$$

$$= \frac{1}{3(2p+1)^2} \cdot \frac{1}{1 - \dfrac{1}{(2p+1)^2}} = \frac{1}{12}\left(\frac{1}{p} - \frac{1}{p+1}\right),$$

$$\epsilon_p > \frac{1}{3(2p+1)^2}\left\{1 + \frac{1}{3(2p+1)^2} + \frac{1}{[3(2p+1)^2]^2} + \cdots\right\} \tag{7}$$

$$= \frac{1}{3(2p+1)^2} \cdot \frac{1}{1 - \dfrac{1}{3(2p+1)^2}} > \frac{1}{12}\left(\frac{1}{p + \dfrac{1}{12}} - \frac{1}{p + 1 + \dfrac{1}{12}}\right).$$

Now define

$$B = \sum_{p=1}^{\infty} \epsilon_p, \qquad r_n = \sum_{p=n}^{\infty} \epsilon_p, \tag{8}$$

where from (6) and (7) we have

$$\frac{1}{13} < B < \frac{1}{12}. \tag{9}$$

Then we can write (4) in the form

$$S_n = (n + \tfrac{1}{2}) \log n - n + 1 - B + r_n,$$

or, setting $C = e^{1-B}$, as

$$n! = C \cdot n^{n+1/2} e^{-n} \cdot e^{r_n},$$

where r_n is defined by (8), ϵ_p by (5), and from (6) and (7) we have

$$\frac{1}{12n+1} < r_n < \frac{1}{12n}.$$

The constant C, known from (9) to lie between $e^{11/12}$ and $e^{12/13}$, may be shown by one of the usual methods to have the value $\sqrt{2\pi}$. This completes the proof.

The preceding derivation was motivated by the geometrically suggestive partition (3). The editor has pointed out that the inequalities (6) and (7) permit the following brief proof* of (2). Let

$$u_n = n! n^{-(n+1/2)} e^n.$$

Then the series

$$\log\left(1 + \frac{1}{n}\right)^{n+1/2} = 1 + \frac{1}{3(2n+1)^2} + \frac{1}{5(2n+1)^4} + \cdots$$

together with (6) and (7) yield the inequalities

$$\exp\left\{\frac{1}{12}\left(\frac{1}{n+\frac{1}{12}} - \frac{1}{n+1+\frac{1}{12}}\right)\right\} < \frac{u_n}{u_{n+1}} = e^{-1}\left(1 + \frac{1}{n}\right)^{n+1/2}$$

$$< \exp\left\{\frac{1}{12}\left(\frac{1}{n} - \frac{1}{n+1}\right)\right\}.$$

Hence

$$v_n = u_n e^{-1/12n}$$

increases and

$$w_n = u_n e^{-1/(12n+1)}$$

decreases, while

$$v_n < w_n = v_n e^{1/12n(12n+1)}.$$

Since

$$v_1 = e^{11/12}, \qquad w_1 = e^{12/13}$$

* A modification of that attributed to Cesàro by A. Fisher, *Mathematical theory of probabilities*, New York, 1936, pp. 93–95.

it follows that

$$v_n \rightarrow C, \qquad w_n \rightarrow C, \qquad v_n < C < w_n, \qquad e^{11/12} < C < e^{12/13}.$$

Thus

$$u_n = Ce^{r_n}$$

where r_n satisfies (2).

Reprinted from the Proceedings of the NATIONAL ACADEMY OF SCIENCES
Vol. 47, No. 3, pp. 330–335. March, 1961.

ON SUMS OF INDEPENDENT RANDOM VARIABLES WITH INFINITE MOMENTS AND "FAIR" GAMES*

BY Y. S. CHOW AND HERBERT ROBBINS

INTERNATIONAL BUSINESS MACHINES CORPORATION AND COLUMBIA UNIVERSITY

Communicated by Paul A. Smith, January 25, 1961

1. *Introduction.*—At the origin of this investigation is the well-known result of Feller[1] concerning the Petersburg game, in which the player receives $\$2^i$ if heads first appears at the ith toss of an unbiased coin ($i = 1, 2, \ldots$). Since the expectation of the player's gain x is infinite, a problem arises in deciding on the proper

fee for the privilege of playing the game; Feller shows that if the player pays variable entrance fees with cumulative fee $b_n = n \log_2 n$ for the first n games, and if $s_n = x_1 + \ldots + x_n$ denotes his total gain, then the game becomes "fair" in the sense that

$$\lim_{n\to\infty} \frac{s_n}{b_n} = 1 \text{ in probability.} \tag{1}$$

For a game with *finite* and *positive* expectation $Ex = \mu$, and *constant* entrance fee μ for each game, the strong law of large numbers asserts that ratio of total gain to cumulative entrance fee tends to 1 in the strong sense,

$$P\left(\lim_{n\to\infty} \frac{s_n}{b_n} = 1\right) = 1. \tag{2}$$

It is a natural to ask whether the same holds of (1). It does not; for the Petersburg game with $b_n = n \log_2 n$,

$$P\left(\lim_{n\to\infty} \frac{s_n}{b_n} = 1\right) = 0, \tag{3}$$

an interesting example in which convergence in probability does not imply convergence with probability 1.

To show this, we note that for the Petersburg random variable x,

$$P(x > a) \geq \frac{1}{a} \qquad \text{for } a \geq 1.$$

Hence, for any constant $c > 1$ and $n \geq 2$,

$$P(x > cb_n) \geq \frac{1}{cb_n} = \frac{1}{cn \log_2 n}$$

and therefore,

$$\sum_{n=1}^{\infty} P(x > cb_n) = \infty,$$

which implies by the Borel-Cantelli lemma that

$$P\left(\frac{x_n}{b_n} > c \text{ infinitely often}\right) = 1$$

and thus,

$$P\left(\overline{\lim} \frac{x_n}{b_n} = \infty\right) = 1$$

and hence,

$$P\left(\overline{\lim} \frac{s_n}{b_n} = \infty\right) = 1, \tag{4}$$

which implies (3). Since (1) holds, it follows moreover that (3) holds for *every* sequence of constants b_n.

We shall show in the next section that (3) holds for every game in which the player's gain is such that $E|x| = \infty$ and for every sequence of constants; thus *no game with* $E|x| = \infty$ *can be made "fair" in the strong sense* (2) *even by allowing variable entrance fees.*

2. *Some General Theorems.*—In this section, $x, x_1, x_2, \ldots$ will denote any sequence of independent random variables with a common distribution, $s_n = x_1 + \ldots + x_n$, and $b_1, b_2, \ldots$ will denote any sequence of constants.

LEMMA 1. If b_n/n is positive and non-decreasing, then

$$(a)\quad \sum_1^\infty P(|x| > b_n) = \infty \quad \text{implies} \quad P\left(\overline{\lim}\, \frac{|x_n|}{b_n} = \infty\right) = 1,$$

$$(b)\quad \sum_1^\infty P(|x| > b_n) < \infty \quad \text{implies} \quad P\left(\lim \frac{x_n}{b_n} = 0\right) = 1.$$

Proof: Since

$$\frac{b_n}{n} \le \frac{b_{2n}}{2n} \le \frac{b_{2n+1}}{2n+1},$$

it follows that

$$2b_n \le b_{2n} < b_{2n+1}. \tag{5}$$

Now suppose that $\sum_1^\infty P(|x| > b_n) = \infty$; then either

$$\text{(i)}\quad \sum_1^\infty P(|x| > b_{2n}) = \infty$$

or

$$\text{(ii)}\quad \sum_1^\infty P(|x| > b_{2n+1}) = \infty.$$

If (i) holds, then by the first inequality of (5),

$$\sum_1^\infty P(|x| > 2b_n) = \infty; \tag{6}$$

while if (ii) holds, then by the second inequality of (5), (i) also holds, so that (6) holds in either case. In other words,

$$\sum_1^\infty P(|x| > b_n) = \infty \quad \text{implies} \quad \sum_1^\infty P(|x| > 2b_n) = \infty, \tag{7}$$

and hence by induction for every $k = 1, 2, \ldots,$

$$\sum_1^\infty P(|x| > b_n) = \infty \quad \text{implies} \quad \sum_1^\infty P(|x| > 2^k b_n) = \infty \tag{8}$$

provided only that b_n/n is non-decreasing.

Now to prove part (*a*) of the lemma, suppose that $\sum_1^\infty P(|x| > b_n) = \infty$. By (8) and the Borel-Cantelli lemma, it follows that

$$P\left(\frac{|x_n|}{b_n} > 2^k \text{i .o.}\right) = 1,$$

so that, since k was arbitrary,

$$P\left(\overline{\lim}\, \frac{|x_n|}{b_n} = \infty\right) = 1.$$

To prove part (*b*) of the lemma, suppose that

$$P\left(\lim \frac{x_n}{b_n} = 0\right) < 1;$$

then for some $k = 1, 2, \ldots,$

$$P\left(\frac{|x_n|}{b_n} > \frac{1}{2^k} \text{ i.o.}\right) > 0,$$

and hence by the Borel-Cantelli lemma,

$$\sum P\left(|x| > \frac{b_n}{2^k}\right) = \infty.$$

By (8), this implies that

$$\sum_1^\infty P(|x| > b_n) = \infty.$$

It follows that (*b*) holds.

LEMMA 2. If b_n/n is positive and non-decreasing, then

$$(a) \quad \sum_1^\infty P(|x| > b_n) = \infty \quad \text{implies} \quad P\left(\overline{\lim} \frac{|s_n|}{b_n} = \infty\right) = 1,$$

and if, in addition, $E|x| = \infty$, then

$$(b) \quad \sum_1^\infty P(|x| > b_n) < \infty \quad \text{implies } P\left(\lim \frac{s_n}{b_n} = 0\right) = 1.$$

Proof: To prove part (*a*), suppose that $\sum_1^\infty P(|x| > b_n) = \infty$; then by Lemma 1,

$$P\left(\overline{\lim} \frac{|x_n|}{b_n} = \infty\right) = 1. \tag{9}$$

Suppose now that
$$P\left(\overline{\lim} \frac{|s_n|}{b_n} = \infty\right) < 1. \tag{10}$$

Then for some finite constant c,

$$P\left(\overline{\lim} \frac{|s_n|}{b_n} < c\right) > 0$$

and hence,
$$P\left(\frac{|s_n|}{b_n} \text{ is bounded}\right) > 0. \tag{11}$$

But since
$$\frac{|x_n|}{b_n} = \frac{|s_n - s_{n-1}|}{b_n} \le \frac{|s_n|}{b_n} + \frac{|s_{n-1}|}{b_n} \le \frac{|s_n|}{b_n} + \frac{|s_{n-1}|}{b_{n-1}}, \tag{12}$$

it follows from (10) that

$$P\left(\frac{|x_n|}{b_n} \text{ is bounded}\right) > 0,$$

which contradicts (9). Hence (10) cannot hold, which proves (*a*).

The proof of part (*b*) of the lemma is contained in a paper of Feller,[2] to which the reader is referred.

We can combine Lemmas 1 and 2 to obtain

LEMMA 3. If b_n/n is positive and non-decreasing, then

$$(a)\quad \sum_1^\infty P(|x| > b_n) = \infty \quad \text{implies} \quad P\left(\overline{\lim}\, \frac{|x_n|}{b_n} = \infty\right) = P\left(\overline{\lim}\, \frac{|s_n|}{b_n} = \infty\right) = 1,$$

and if moreover $E\,|x\,| = \infty$, then

$$(b)\quad \sum_1^\infty P(|x| > b_n) < \infty \quad \text{implies} \quad P\left(\lim \frac{x_n}{b_n} = 0\right) = P\left(\lim \frac{s_n}{b_n} = 0\right) = 1.$$

From Lemma 3 we have at once

THEOREM 1. If b_n/n is positive and non-decreasing, then

$$(a)\quad P\left(\overline{\lim}\, \frac{|x_n|}{b_n} < \infty\right) > 0,\ E|x| = \infty \quad \text{imply} \quad P\left(\lim \frac{s_n}{b_n} = 0\right) = 1,$$

$$(b)\quad E\,|x| = \infty \quad \text{implies} \quad P\left(0 < \overline{\lim}\, \frac{|s_n|}{b_n} < \infty\right) = 0.$$

In the case of the Petersburg game with $b_n = n \log_2 n$, it follows from (1) that for some sequence of integers $n_1 < n_2 < \ldots$, we have

$$P\left(\lim_{k\to\infty} \frac{s_{n_k}}{b_{n_k}} = 1\right) = 1.$$

Define $$b_j' = b_{n_k} \text{ for } n_{k-1} < j \leq n_k.$$

Then, $$P\left(\overline{\lim}\, \frac{s_n}{b_n'} = 1\right) = 1.$$

This example shows that in Theorem 1 (*b*), the condition "b_n/n non-decreasing" cannot be replaced by "b_n non-decreasing, $b_n/n \to \infty$."

We can now prove our principal result:

THEOREM 2. If $E|x| = \infty$, then for any sequence of constants b_n, either

$$P\left(\underline{\lim} \left|\frac{s_n}{b_n}\right| = 0\right) = 1 \tag{13}$$

or

$$P\left(\overline{\lim} \left|\frac{s_n}{b_n}\right| = \infty\right) = 1. \tag{14}$$

Proof. Suppose that the theorem were false; then, there would exist a sequence of constants b_n and an x with $E|x| = \infty$ such that

$$P\left(\underline{\lim} \left|\frac{s_n}{b_n}\right| > 0\right) > 0 \quad \text{and} \quad P\left(\overline{\lim} \left|\frac{s_n}{b_n}\right| < \infty\right) > 0. \tag{15}$$

Case 1. Suppose that

$$\overline{\lim} \frac{|b_n|}{n} < \infty.$$

Then from (15),

$$P\left(\overline{\lim} \frac{|s_n|}{n} < \infty\right) > 0,$$

and hence by Lemma 2(*a*) for $b_n = n$, $\Sigma P(|x| > n) < \infty$ and hence $E|x| < \infty$. Thus Case 1 is impossible.

Case 2. Suppose that

$$\overline{\lim} \frac{|b_n|}{n} = \infty.$$

Set

$$\alpha_n = n \max \left[\frac{|b_1|}{1}, \ldots, \frac{|b_n|}{n}\right] \geq |b_n| \geq 0. \tag{16}$$

Then the sequence α_n/n is positive and non-decreasing, and there exists a sequence of integers $n_1 < n_2 < \ldots$ such that

$$\alpha_{n_k} = |b_{n_k}|. \tag{17}$$

From (15) and (16), it follows that

$$P\left(\overline{\lim} \frac{|s_n|}{\alpha_n} < \infty\right) > 0,$$

and hence by Lemma 3,

$$P\left(\lim \frac{|s_n|}{\alpha_n} = 0\right) = 1. \tag{18}$$

From (17) and (18), it follows that

$$P\left(\underline{\lim} \left|\frac{s_n}{b_n}\right| = 0\right) = 1,$$

which contradicts (15). Thus Case 2 is also impossible, and the proof of the theorem is complete.

COROLLARY. If $E|x| = \infty$, then for any sequence of constants b_n,

$$P\left(\lim_{n\to\infty} \frac{s_n}{b_n} = 1\right) = 0. \tag{19}$$

* Work sponsored in part by National Science Foundation grants at Columbia and Stanford Universities.

[1] Feller, W., *Introduction to Probability Theory and its Applications* (New York: John Wiley and Sons, Inc., 1957), Vol. 1, 2nd ed., p. 237.

[2] Feller, W., "A limit theorem for random variables with infinite moments," *Am. J. Math.*, **68**, 260 (1946).

A MARTINGALE SYSTEM THEOREM AND APPLICATIONS

Y. S. CHOW
IBM RESEARCH CENTER, YORKTOWN HEIGHTS
AND
HERBERT ROBBINS
COLUMBIA UNIVERSITY

1. Introduction

Let $(W, \mathfrak{F}, P)$ be a probability space with points $\omega \in W$ and let $(y_n, \mathfrak{F}_n)$, $n = 1, 2, \cdots$, be an *integrable stochastic sequence:* y_n is a sequence of random variables, $\mathfrak{F}_n$ is a sequence of σ-algebras with $\mathfrak{F}_n \subset \mathfrak{F}_{n+1} \subset \mathfrak{F}$, y_n is measurable with respect to $\mathfrak{F}_n$, and $E(y_n)$ exists, $-\infty \leqq E(y_n) \leqq \infty$. A random variable $s = s(\omega)$ with positive integer values is a *sampling variable* if $\{s \leqq n\} \in \mathfrak{F}_n$ and $\{s < \infty\} = W$. (We denote by $\{\cdots\}$ the set of all ω satisfying the relation in braces, and understand equalities and inequalities to hold up to sets of P-measure 0.) We shall be concerned with the problem of finding, if it exists, a sampling variable s which maximizes $E(y_s)$.

To define a sampling variable s amounts to specifying a sequence of sets $B_n \in \mathfrak{F}_n$ such that

$$(1) \qquad 0 = B_0 \subset \cdots \subset B_n \subset B_{n+1} \subset \cdots ; \bigcup_1^\infty B_n = W,$$

the sampling variable s being defined by

$$(2) \qquad \{s \leqq n\} = B_n, \qquad \{s = n\} = B_n - B_{n-1}.$$

We shall be particularly interested in the case in which the sequence $(y_n, \mathfrak{F}_n)$ is such that the sequence of sets

$$(3) \qquad B_n = \{E(y_{n+1}|\mathfrak{F}_n) \leqq y_n\}$$

satisfies (1). We shall call this the *monotone case.* In this case a sampling variable s is defined by

$$(4) \qquad \{s \leqq n\} = \{E(y_{n+1}|\mathfrak{F}_n) \leqq y_n\},$$

and s satisfies

$$(5) \qquad E(y_{n+1}|\mathfrak{F}_n) \begin{cases} > y_n, & s > n, \\ \leqq y_n, & s \leqq n. \end{cases}$$

The relations (5) will be fundamental in what follows.

This research was sponsored in part by the Office of Naval Research under Contract No. Nonr-226 (59), Project No. 042-205.

In the monotone case we have for the sampling variable s defined by (4) the following characterization:

$$s = \text{least positive integer } j \text{ such that } E(y_{j+1}|\mathfrak{F}_j) \leqq y_j. \tag{6}$$

Now even in the nonmonotone case we can always define a random variable s by (6), setting $s = \infty$ if there is no such j; let us call it the *conservative* random variable. The following statement is evident: the necessary and sufficient condition that there exists a sampling variable s satisfying (5) is that we are in the monotone case, and in this case s is the conservative random variable.

In section 3 we are going to show that *in the monotone case, under certain regularity assumptions, the conservative sampling variable s maximizes $E(y_s)$.*

2. An example

Before proceeding with the general theory we shall give a simple and instructive example of the monotone case in the form of a *sequential decision problem.*

Let $x, x_1, x_2, \cdots$ be a sequence of independent and identically distributed random variables with $E(x^+) < \infty$, where we denote $a^+ = \max(a, 0)$, $a^- = \max(-a, 0)$. We observe the sequence $x_1, x_2, \cdots$ sequentially and can stop with any $n \geqq 1$. If we stop with x_n we receive the reward $m_n = \max(x_1, \cdots, x_n)$, but the cost of taking the observations $x_1, \cdots, x_n$ is some strictly increasing function $g(n) \geqq 0$, so that our net gain in stopping with x_n is $y_n = m_n - g(n)$. The decision whether to stop with x_n or to take the next observation x_{n+1} must be a function of $x_1, \cdots, x_n$ alone. *Problem:* what stopping rule maximizes the expected value $E(y_s)$, where s is the random sample size defined by the stopping rule? We assume that the distribution function $F(u) = P\{x \leqq u\}$ is known. That $E(y_n)$ exists follows from the inequality

$$y_n^+ \leqq x_1^+ + \cdots + x_n^+, \tag{7}$$

which implies that $E(y_n^+) < \infty$.

Let $\mathfrak{F}_n$ be the σ-algebra generated by $x_1, \cdots, x_n$. Then $(y_n, \mathfrak{F}_n)$ is an integrable stochastic sequence, and we have

$$\begin{aligned} E(y_{n+1}|\mathfrak{F}_n) - y_n &= \int [m_{n+1} - m_n]\, dF(x_{n+1}) - [g(n+1) - g(n)] \\ &= \int (x - m_n)^+\, dF(x) - f(n), \end{aligned} \tag{8}$$

where we have set

$$f(n) = g(n+1) - g(n) = \text{cost of taking the } (n+1)\text{st observation.} \tag{9}$$

Since we have assumed $g(n)$ to be strictly increasing, and $f(n) > 0$, it is easily seen that there exist unique constants α_n such that

$$\int (x - \alpha_n)^+\, dF(x) = f(n), \qquad n \geqq 1. \tag{10}$$

By (8) and (10),

$$E(y_{n+1}|\mathfrak{F}_n) \begin{cases} > y_n & \text{if} \quad m_n < \alpha_n, \\ \leqq y_n & \text{if} \quad m_n \geqq \alpha_n. \end{cases} \tag{11}$$

The conservative random variable s defined by (6) is therefore

$$s = \text{least positive integer } j \text{ such that } m_j \geqq \alpha_j. \tag{12}$$

We are in the monotone case if and only if this s is a sampling variable and for every n

$$\{E(y_{n+1}|\mathfrak{F}_n) \leqq y_n\} \subset \{E(y_{n+2}|\mathfrak{F}_{n+1}) \leqq y_{n+1}\}, \tag{13}$$

that is, $m_n \geqq \alpha_n$ implies $m_{n+1} \geqq \alpha_{n+1}$, which will certainly be the case, since $m_n \leqq m_{n+1}$, if $\alpha_n \geqq \alpha_{n+1}$, that is, if $f(n)$ is *nondecreasing* and hence α_n is *nonincreasing. We shall henceforth assume this to hold.* We shall now show that in this case the conservative random variable s is in fact a sampling variable, that is, that $P\{s < \infty\} = 1$. We have

$$\{s > n\} = \{m_n < \alpha_n\}, \tag{14}$$

and hence

$$\begin{aligned} P\{s < \infty\} &= 1 - \lim_n P\{s > n\} = 1 - \lim_n P\{m_n < \alpha_n\} \\ &\geqq 1 - \lim_n P\{m_n < \alpha_1\} = 1 - \lim_n P^n\{x < \alpha_1\} = 1, \end{aligned} \tag{15}$$

since by hypothesis $f(1) > 0$ so that by (10), $P\{x < \alpha_1\} < 1$. In fact, for any $r \geqq 0$,

$$\begin{aligned} E(s^r) &= \sum_{n=1}^{\infty} n^r P\{s = n\} \leqq \sum_{n=1}^{\infty} n^r P\{s > n - 1\} \\ &\leqq 1 + \sum_{n=2}^{\infty} n^r P\{m_{n-1} < \alpha_1\} \\ &= 1 + \sum_{n=2}^{\infty} n^r P^{n-1}\{x < \alpha_1\} < \infty, \end{aligned} \tag{16}$$

so that s has finite moments of all orders.

It is of interest to consider the special case $g(n) = cn$, $0 < c < \infty$. Here $f(n) = c$ and $\alpha_n = \alpha$, where α is defined by

$$\int (x - \alpha)^+ \, dF(x) = c, \tag{17}$$

and s is the first $j \geqq 1$ for which $x_j \geqq \alpha$. Hence

$$P\{s = j\} = P\{x \geqq \alpha\} P^{j-1}\{x < \alpha\},$$

$$E(s) = \frac{1}{P\{x \geqq \alpha\}},$$

$$E(y_s) = \sum_{j=1}^{\infty} P\{s = j\} E(m_j - cj | s = j), \tag{18}$$

$$E(m_j | s = j) = E(x_j | x_1 < \alpha, \cdots, x_{j-1} < \alpha, x_j \geqq \alpha) = \frac{1}{P\{x \geqq \alpha\}} \int_{\{x \geqq \alpha\}} x \, dF(x),$$

so that

$$E(y_s) = \frac{1}{P\{x \geqq \alpha\}} \left[\int_{\{x \geqq \alpha\}} x \, dF(x) - c \right] = \frac{1}{P\{x \geqq \alpha\}} \left[\int (x - \alpha)^+ \, dF(x) - c + \alpha P\{x \geqq \alpha\} \right] = \alpha, \tag{19}$$

an elegant relation.

3. General theorems

In the following three lemmas we assume that $(y_n, \mathfrak{F}_n)$ is any integrable stochastic sequence and that s and t are any sampling variables such that $E(y_s)$ and $E(y_t)$ exist.

LEMMA 1. *If for each n,*

$$E(y_s | \mathfrak{F}_n) \geqq y_n \qquad \text{if} \quad s > n, \tag{20}$$

and

$$E(y_t | \mathfrak{F}_n) \leqq y_n \qquad \text{if} \quad s = n, \ t > n, \tag{21}$$

then

$$E(y_s) \geqq E(y_t). \tag{22}$$

Conversely, if $E(y_s)$ is finite and (22) *holds for every t, then* (20) *and* (21) *hold for every t.*

PROOF.

$$\begin{aligned} E(y_s) &= \sum_{n=1}^{\infty} \int_{\{s=n, t \leqq n\}} y_s \, dP + \sum_{n=1}^{\infty} \int_{\{s=n, t>n\}} y_n \, dP \\ &= \sum_{n=1}^{\infty} \int_{\{s \geqq n, t=n\}} y_s \, dP + \sum_{n=1}^{\infty} \int_{\{s=n, t>n\}} y_n \, dP \\ &\geqq \sum_{n=1}^{\infty} \int_{\{s \geqq n, t=n\}} y_n \, dP + \sum_{n=1}^{\infty} \int_{\{s=n, t>n\}} y_t \, dP \\ &= E(y_t). \end{aligned} \tag{23}$$

To prove the converse, for a fixed n let

$$V = \{s > n \quad \text{and} \quad E(y_s|\mathfrak{F}_n) < y_n\}; \tag{24}$$

then $V \in \mathfrak{F}_n$. Define

$$t' = \begin{cases} s, & \omega \notin V, \\ n, & \omega \in V. \end{cases} \tag{25}$$

Then t' is a sampling variable. Since $E(y_s)$ is finite, by (22) $E(y_n) < \infty$ and then $E(y_{t'})$ exists. But

$$\begin{aligned} E(y_{t'}) &= \int_{\{t'=s\}} y_{t'}\,dP + \int_V y_{t'}\,dP = \int_{\{t'=s\}} y_s\,dP + \int_V y_n\,dP \\ &\geqq \int_{\{t'=s\}} y_s\,dP + \int_V y_s\,dP = E(y_s). \end{aligned} \tag{26}$$

But by (22), $E(y_{t'}) \leqq E(y_s)$. Hence

$$\int_V y_n\,dP = \int_V y_s\,dP \tag{27}$$

and therefore $P(V) = 0$, which proves (20). To prove (21) let

$$V = \{s = n, t > n, \quad \text{and} \quad E(y_t|\mathfrak{F}_n) > y_n\}, \tag{28}$$

and define

$$t' = \begin{cases} s, & \omega \notin V, \\ t, & \omega \in V. \end{cases} \tag{29}$$

Then

$$\begin{aligned} E(y_{t'}) &= \int_{\{t'=s\}} y_{t'}\,dP + \int_V y_{t'}\,dP = \int_{\{t'=s\}} y_s\,dP + \int_V y_t\,dP \\ &\geqq \int_{\{t'=s\}} y_s\,dP + \int_V y_n\,dP = \int_{\{t'=s\}} y_s\,dP + \int_V y_s\,dP = E(y_s), \end{aligned} \tag{30}$$

and again $P(V) = 0$, which proves (21).

LEMMA 2. *If for each n,*

$$E(y_{n+1}|\mathfrak{F}_n) \geqq y_n, \qquad s > n, \tag{31}$$

and if

$$\liminf_n \int_{\{s>n\}} y_n^+\,dP = 0, \tag{32}$$

then for each n,

$$E(y_s|\mathfrak{F}_n) \geqq y_n, \qquad s \geqq n. \tag{33}$$

PROOF. (compare [2], p. 310). Let $V \in \mathfrak{F}_n$ and $U = V\{s \geqq n\}$. Then

$$\int_U y_n\,dP = \int_{V\{s=n\}} y_n\,dP + \int_{V\{s>n\}} y_n\,dP \tag{34}$$
$$\leqq \int_{V\{s=n\}} y_n\,dP + \int_{V\{s>n\}} y_{n+1}\,dP$$
$$= \int_{V\{n\leqq s\leqq n+1\}} y_s\,dP + \int_{V\{s>n+1\}} y_{n+1}\,dP$$
$$\leqq \cdots \leqq \int_{V\{n\leqq s\leqq n+r\}} y_s\,dP + \int_{V\{s>n+r\}} y_{n+r}\,dP$$
$$\leqq \int_{V\{n\leqq s\leqq n+r\}} y_s\,dP + \int_{\{s>n+r\}} y_{n+r}^+\,dP.$$

Therefore

$$\int_U y_n\,dP \leqq \int_{V\{s\geqq n\}} y_s\,dP + \liminf_n \int_{\{s>n\}} y_n^+\,dP = \int_U y_s\,dP, \tag{35}$$

which is equivalent to (33).

LEMMA 3. *If for each n,*

$$E(y_{n+1}|\mathfrak{F}_n) \leqq y_n, \qquad s \leqq n, \tag{36}$$

and if

$$\liminf_n \int_{\{t>n\}} y_n^-\,dP = 0, \tag{37}$$

then

$$E(y_t|\mathfrak{F}_n) \leqq y_n, \qquad s = n,\quad t \geqq n. \tag{38}$$

PROOF. Let $V \in \mathfrak{F}_n$ and $U = V\{s = n, t \geqq n\}$. Then

$$\int_U y_n\,dP = \int_{V\{s=n,\,t=n\}} y_n\,dP + \int_{V\{s=n,\,t>n\}} y_n\,dP \tag{39}$$
$$\geqq \int_{V\{s=n,\,t=n\}} y_n\,dP + \int_{V\{s=n,\,t>n\}} y_{n+1}\,dP$$
$$= \int_{V\{s=n,\,n\leqq t\leqq n+1\}} y_t\,dP + \int_{V\{s=n,\,t>n+1\}} y_{n+1}\,dP$$
$$\geqq \cdots \geqq \int_{V\{s=n,\,n\leqq t\leqq n+r\}} y_t\,dP + \int_{V\{s=n,\,t>n+r\}} y_{n+r}\,dP$$
$$\geqq \int_{V\{s=n,\,n\leqq t\leqq n+r\}} y_t\,dP - \int_{\{t>n+r\}} y_{n+r}^-\,dP.$$

Therefore

$$\int_U y_n\,dP \geqq \int_{V\{s=n,\,t\geqq n\}} y_t\,dP - \liminf_n \int_{\{t>n\}} y_n^-\,dP = \int_U y_t\,dP, \tag{40}$$

which is equivalent to (38).

We can now state the main result of the present paper.

THEOREM 1. *Let* $(y_n, \mathfrak{F}_n)$ *be an integrable stochastic sequence in the monotone case and let* s *be the conservative sampling variable*

$$s = \text{least positive integer } j \text{ such that } E(y_{j+1}|\mathfrak{F}_j) \leqq y_j. \tag{41}$$

Suppose that $E(y_s)$ *exists and that*

$$\liminf_n \int_{\{s>n\}} y_n^+\,dP = 0. \tag{42}$$

If t *is any sampling variable such that* $E(y_t)$ *exists and*

$$\liminf_n \int_{\{t>n\}} y_n^-\,dP = 0, \tag{43}$$

then

$$E(y_s) \geqq E(y_t). \tag{44}$$

PROOF. From lemmas 1, 2, and 3 and relations (5).

We shall now establish a lemma (see [2], p. 303) which provides sufficient conditions for (42) and (43).

LEMMA 4. *Let* $(y_n, \mathfrak{F}_n)$ *be a stochastic sequence such that* $E(y_n^+) < \infty$ *for each* $n \geqq 1$, *and let* s *be any sampling variable. If there exists a nonnegative random variable* u *such that*

$$E(su) < \infty, \tag{45}$$

and if

$$E[(y_{n+1} - y_n)^+|\mathfrak{F}_n] \leqq u, \qquad s > n, \tag{46}$$

then

$$E(y_s^+) < \infty, \qquad \lim_n \int_{\{s>n\}} y_n^+\,dP = 0. \tag{47}$$

PROOF. Define

$$z_1 = y_1^+, \qquad z_{n+1} = (y_{n+1} - y_n)^+ \quad \text{for} \quad n \geqq 1, \qquad w_n = z_1 + \cdots + z_n. \tag{48}$$

Then

$$y_n^+ \leqq w_n \tag{49}$$

(and hence $y_n^+ \leqq w_s$ if $s \geqq n$), and by (46)

$$E(z_{n+1}|\mathfrak{F}_n) \leqq u, \qquad s > n. \tag{50}$$

Hence

$$(51)\qquad \begin{aligned} E(y_s^+) &\leqq E(w_s) = \sum_{n=1}^{\infty} \int_{\{s=n\}} w_n\, dP = \sum_{n=1}^{\infty} \sum_{j=1}^{n} \int_{\{s=n\}} z_j\, dP \\ &= \sum_{j=1}^{\infty} \sum_{n=j}^{\infty} \int_{\{s=n\}} z_j\, dP = \sum_{j=1}^{\infty} \int_{\{s>j-1\}} z_j\, dP \\ &= E(y_1^+) + \sum_{j=2}^{\infty} \int_{\{s>j-1\}} E(z_j|\mathfrak{F}_{j-1})\, dP \\ &\leqq E(y_1^+) + \sum_{j=2}^{\infty} \int_{\{s>j-1\}} u\, dP = E(y_1^+) + \sum_{j=2}^{\infty} \sum_{n=j}^{\infty} \int_{\{s=n\}} u\, dP \\ &= E(y_1^+) + \sum_{n=2}^{\infty} \int_{\{s=n\}} (n-1)u\, dP = E(y_1^+) + E(su) - E(u) \\ &\leqq E(y_1^+) + E(su) < \infty, \end{aligned}$$

and hence from (49)

$$(52)\qquad \lim_n \int_{\{s>n\}} y_n^+\, dP \leqq \lim_n \int_{\{s>n\}} w_s\, dP = 0.$$

REMARK. Lemma 4 remains valid if we replace a^+ by a^- or by $|a|$ throughout.

4. Application to the sequential decision problem of section 2

Recalling the problem of section 2, let $x, x_1, x_2, \cdots$ be independent and identically distributed random variables with $E(x^+) < \infty$, $\mathfrak{F}_n$ the σ-algebra generated by $x_1, \cdots, x_n$, $g(n) \geqq 0$, $f(n) = g(n+1) - g(n) > 0$ and nondecreasing, $m_n = \max(x_1, \cdots, x_n)$, and $y_n = m_n - g(n)$. The constants α_n are defined by

$$(53)\qquad E[(x-\alpha_n)^+] = f(n)$$

and are nonincreasing; we are in the monotone case, and the conservative sampling variable s is the first $j \geqq 1$ such that $m_j \geqq \alpha_j$; thus

$$(54)\qquad \{s > n\} = \{m_n < \alpha_n\}.$$

We have shown in section 2 that

$$(55)\qquad P\{s < \infty\} = 1, \qquad E(s^r) < \infty \quad \textit{for} \quad r \geqq 0.$$

We wish to apply theorem 1. As concerns s it will suffice to show that $E(y_s^+) < \infty$ and that

$$(56)\qquad \lim_n \int_{\{s>n\}} y_n^+\, dP = 0,$$

which we shall do by using lemma 4. Let

$$(57)\qquad Y_n = m_n^+ - g(n).$$

Then

(58) $$Y_n^+ = y_n^+, \quad E(Y_n^+) = E(y_n^+) \leqq E(x_1^+ + \cdots + x_n^+) = nE(x^+) < \infty,$$

and

(59) $$\begin{aligned} E[(Y_{n+1} - Y_n)^+ | \mathfrak{F}_n] &= E\{[m_{n+1}^+ - m_n^+ - f(n)]^+ | \mathfrak{F}_n\} \\ &\leqq E[(m_{n+1}^+ - m_n^+) | \mathfrak{F}_n] \leqq E(x_{n+1}^+ | \mathfrak{F}_n) \\ &= E(x^+) < \infty. \end{aligned}$$

Hence by lemma 4, setting $u = E(x^+)$,

(60) $$E(y_s^+) = E(Y_s^+) < \infty$$

and

(61) $$\lim_n \int_{\{s>n\}} y_n^+ \, dP = \lim_n \int_{\{s>n\}} Y_n^+ \, dP = 0,$$

which were to be proved.

To establish the conditions on t of theorem 1 we assume that $Ex^- < \infty$; then since $y_n^- \leqq x_1^- + g(n)$ it follows that $E(y_n^-) < \infty$. Define a random variable u by setting

(62) $$u(\omega) = f(n) \quad \text{if} \quad t(\omega) = n.$$

Since

(63) $$(y_{n+1} - y_n)^- \leqq f(n)$$

and $f(n)$ is nondecreasing, it follows that

(64) $$E[(y_{n+1} - y_n)^- | \mathfrak{F}_n] \leqq u \quad \text{if} \quad t \geqq n.$$

We now assume that $f(n) \leqq h(n)$, where $h(n)$ is a polynomial of degree $r \geqq 0$, and that $E(t^{r+1}) < \infty$. Then

(65) $$E(tu) = \sum_{n=1}^{\infty} \int_{\{t=n\}} nf(n) \, dP \leqq \sum_{n=1}^{\infty} nh(n) P\{t = n\}.$$

Since

(66) $$E(t^{r+1}) = \sum_{n=1}^{\infty} n^{r+1} P\{t = n\} < \infty,$$

it follows that $E(tu) < \infty$. Then by the remark following lemma 4,

(67) $$E(y_t^-) < \infty \quad \text{and} \quad \lim_n \int_{\{t>n\}} y_n^- \, dP = 0,$$

and all the conditions of theorem 1 are established. Thus we have proved

THEOREM 2. *Suppose that* $E|x| < \infty$ *and that in addition to the conditions on* $g(n)$ *in the first paragraph of this section we have* $f(n) \leqq h(n)$, *where* $h(n)$ *is a polynomial of degree* $r \geqq 0$. *If* t *is any sampling variable for which* $E(t^{r+1}) < \infty$ *then* $-\infty < E(y_t) \leqq E(y_s) < \infty$, *where* s *is the conservative sampling variable defined by* (54).

If $g(n) = nc$ then $f(n) = c$ and we can take $r = 0$. Hence

COROLLARY 1. *If* $E|x| < \infty$ *and* $y_n = m_n - cn$, $0 < c < \infty$, *then if* t *is any sampling variable for which* $E(t) < \infty$, $E(y_t) \leqq E(y_s) = \alpha$ *[see (19)], where* α *is defined by* $E(x - \alpha)^+ = c$ *and* $s =$ *the first* $j \geqq 1$ *such that* $x_j \geqq \alpha$. *Thus* s *is optimal in the class of all sampling variables with finite expectations.*

To replace the condition $E(t^{r+1}) < \infty$ in theorem 2 and corollary 1 by conditions on y_t we require the following theorem which is of interest in itself. We omit the proof.

THEOREM 3. *Let* $F(u)$ *be a distribution function. Define* $G(u) = \prod_{n=1}^{\infty} F(u + n)$. *Then* $G(u)$ *is a distribution function if and only if*

$$\int_0^\infty u\, dF(u) < \infty, \tag{68}$$

and for any integer $b \geqq 1$,

$$\int_0^\infty u^b\, dG(u) < \infty \tag{69}$$

if and only if

$$\int_0^\infty u^{b+1}\, dF(u) < \infty. \tag{70}$$

COROLLARY 2. *If* $y_n = m_n - cn$, $0 < c < \infty$, *and* b *is any integer* $\geqq 1$, *then*

$$E(\sup_{n \geqq 1} y_n^+)^b < \infty \tag{71}$$

if and only if

$$E(x^+)^{b+1} < \infty. \tag{72}$$

PROOF. We can assume $c = 1$. Define

$$G(u) = P\{\sup_{n \geqq 1} y_n^+ \leqq u\}. \tag{73}$$

Then for $u \geqq 0$,

$$\begin{aligned} G(u) &= P\{x_1 \leqq u + 1, x_2 \leqq u + 2, \cdots, x_n \leqq u + n, \cdots\} \\ &= \prod_{n=1}^{\infty} F(u + n). \end{aligned} \tag{74}$$

By theorem 3,

$$E(\sup_{n \geqq 1} y_n^+)^b = \int_0^\infty u^b\, dG(u) < \infty \tag{75}$$

if and only if

$$\int_0^\infty u^{b+1}\, dF(u) = E(x^+)^{b+1} < \infty. \tag{76}$$

THEOREM 4. *Assume* $E|x| < \infty$, $E(x^+)^2 < \infty$. *If* $y_n = m_n - g(n)$ *where* $g(n)$ *is a polynomial of degree* $r \geqq 1$ *such that*

$$g(1) > 0, \tag{77}$$

$g(n+1) - g(n)$ is positive and nondecreasing, then for any sampling variable t,

$$E(y_t) \leqq E(y_s), \tag{78}$$

where s is the conservative sampling variable defined by (54).

PROOF. By theorem 2, if $E(t^r) < \infty$ then (78) holds. Hence we can assume that $E(t^r) = \infty$. Now

$$\begin{aligned} & g(1) > 0, && f(1) = g(2) - g(1) > 0, \\ & g(2) \geqq g(1) + f(1), && g(3) - g(2) \geqq f(1), \\ & g(3) \geqq g(1) + 2f(1), && \\ & \cdots\cdots && \\ & g(n) \geqq g(1) + (n-1)f(1). && \end{aligned} \tag{79}$$

Let

$$a = \frac{1}{2} \min [g(1), f(1)] > 0. \tag{80}$$

Then by (79),

$$g(n) \geqq an \quad \text{for} \quad n \geqq 1. \tag{81}$$

Let

$$\tilde{y}_n = m_n - \frac{a}{2} n. \tag{82}$$

By corollary 2, $E(\tilde{y}_t^+) < \infty$. Then since

$$y_t = \tilde{y}_t + \frac{a}{2} t - g(t) \leqq \tilde{y}_t^+ - \frac{1}{2} g(t) \tag{83}$$

we have

$$E(y_t) \leqq E(\tilde{y}_t^+) - \frac{1}{2} E[g(t)] = -\infty, \tag{84}$$

so that (78) holds in this case too.

REMARK. If in the case $g(n) = cn$ we define $\bar{y}_n = x_n - cn$, then

$$\bar{y}_n \leqq y_n, \qquad \bar{y}_s = y_s. \tag{85}$$

Hence for any sampling variable t,

$$E(\bar{y}_t) \leqq E(y_t) \leqq E(y_s) = E(\bar{y}_s), \tag{86}$$

so that s is also optimal for the stochastic sequence $(\bar{y}_n, \mathfrak{F}_n)$.

5. A result of Snell

As an application of lemmas 1 and 2, we are going to obtain Snell's result on sequential game theory [3].

LEMMA 5 (*Snell*). *Let $(y_n, \mathfrak{F}_n)$ be a stochastic sequence satisfying $y_n \geqq u$ for* *each n with* $E|u| < \infty$. *Then there exists a semimartingale* $(x_n, \mathfrak{F}_n)$ *such that for every sampling variable t and every n,*

(87) $$E(x_t|\mathfrak{F}_n| \geqq x_n \quad \text{if} \quad t \geqq n, \qquad x_n \geqq E(u|\mathfrak{F}_n),$$

(88) $$x_n = \min [y_n, E(x_{n+1}|\mathfrak{F}_n)],$$

and

(89) $$\liminf x_n = \liminf y_n.$$

We will assume the validity of this lemma, and prove the following theorem by applying lemmas 1 and 2.

THEOREM 5 (*Snell*). *Let* $(y_n, \mathfrak{F}_n)$ *and* $(x_n, \mathfrak{F}_n)$ *satisfy the conditions of lemma* 5. *For* $\epsilon \geqq 0$ *define* $s = j$ *to be the first* $j \geqq 1$ *such that* $x_j \geqq y_j - \epsilon$. *If* $\epsilon > 0$, *then*

(90) $$E(y_s) \leqq E(y_t) + \epsilon$$

for every sampling variable t. If $\epsilon = 0$ *and if* $P\{s < \infty\} = 1$, *then* (90) *still holds.*

PROOF. It is obvious that in both cases s is a sampling variable. We need to verify that $P\{s < \infty\} = 1$. If $\epsilon > 0$, by (89) this is true.

Since $(x_n, \mathfrak{F}_n)$ is a semimartingale,

(91) $$E(x_{n+1}|\mathfrak{F}_n) \geqq x_n.$$

By (88) and the definition of s,

(92) $$E(x_{n+1}|\mathfrak{F}_n) = x_n \quad \text{for} \quad s > n.$$

Since $-x_n \leqq E(-u|\mathfrak{F}_n)$ and $E|u| < \infty$, by lemma 2 and (92), we have

(93) $$E(x_s|\mathfrak{F}_n) \leqq x_n \quad \text{for} \quad s > n.$$

By (87), (93), and lemma 1, we obtain $E(x_s) \leqq E(x_t)$, and therefore, by definition of s,

(94) $$E(y_s) \leqq E(x_s) + \epsilon \leqq E(x_t) + \epsilon \leqq E(y_t) + \epsilon.$$

Thus the proof is complete.

J. MacQueen and R. G. Miller, Jr., in a recent paper [1], treat the problem of section 2 by completely different methods. Reference should also be made to a paper by C. Derman and J. Sacks [4], in which the formulation and results are very similar to those of the present paper.

REFERENCES

[1] J. MACQUEEN and R. G. MILLER, JR., "Optimal persistence policies," *J. Operations Res. Soc. America*, Vol. 8 (1960), pp. 362–380.

[2] J. L. DOOB, *Stochastic Processes*, New York, Wiley, 1953.

[3] J. L. SNELL, "Application of martingale system theorems," *Trans. Amer. Math. Soc.*, Vol. 73 (1952), pp. 293–312.

[4] C. DERMAN and J. SACKS, "Replacement of periodically inspected equipment (an optimal optional stopping rule)," to appear in *Naval Res. Logist. Quart.*, Vol. 7 (1960).

Reprinted from

Proc. Fourth Berkeley Symposium Math. Statist. Prob.

1, 93–104 (1961)

Z. Wahrscheinlichkeitstheorie 2, 33–49 (1963)

On Optimal Stopping Rules

By

Y. S. CHOW and HERBERT ROBBINS*

1. Introduction

Let $y_1, y_2, \ldots$ be a sequence of random variables with a given joint distribution. Assume that we can observe the y's sequentially but that we must stop some time, and that if we stop with y_n we will receive a payoff $x_n = f_n(y_1, \ldots, y_n)$. What stopping rule will maximize the expected value of the payoff?

In this paper we attempt to give a reasonably general theory of the existence and computation of optimal stopping rules, previously discussed to some extent in [*1*] and [*12*]. We then apply the theory to two particular cases of interest in applications. One of these belongs to the general domain of dynamic programming; the other is the problem of showing the Bayesian character of the WALD sequential probability ratio test.

2. Existence of an optimal rule

Let $(\Omega, \mathscr{F}, P)$ be a probability space with points ω, let $\mathscr{F}_1 \subset \mathscr{F}_2 \subset \ldots$ be a non-decreasing sequence of sub-σ-algebras of $\mathscr{F}$, and let $x_1, x_2, \ldots$ be a sequence of random variables defined on Ω with $E|x_n| < \infty$ and such that $x_n = x_n(\omega)$ is measurable $(\mathscr{F}_n)$. A sampling variable (s.v.) is a random variable (r.v.) $t = t(\omega)$ with values in the set of positive integers (not including $+\infty$) and such that $\{t(\omega) = n\} \in \mathscr{F}_n$ for each n, where by $\{\ldots\}$ we mean the set of all ω for which the indicated relation holds. For any s.v. t we may form the r.v. $x_t = x_{t(\omega)}(\omega)$. We shall be concerned with the problem of finding a s.v. t which maximizes the value of $E(x_t)$ in the class of all s.v.'s for which this expectation exists.

We shall use the notation $x^+ = \max(x, 0)$, $x^- = \max(-x, 0)$, so that $x = x^+ - x^-$. To simplify matters we shall suppose that $E(\sup_n x_n^+) < \infty$; then for any s.v. t, $x_t \leqq \sup_n x_n^+$, and hence $-\infty \leqq E(x_t) \leqq E(\sup_n x_n^+) < \infty$. Denoting by C the class of all s.v.'s, it follows that $E(x_t)$ exists for all $t \in C$ but may have the value $-\infty$.

In what follows we shall occasionally refer to [*1*] for the details of certain proofs.

Definition. A s.v. t is *regular* if for all $n = 1, 2, \ldots$

$$t > n \Rightarrow E(x_t | \mathscr{F}_n) > x_n. \tag{1}$$

Note that if t is any regular s.v. then

$$E(x_t) = \int_{\{t=1\}} x_t + \int_{\{t>1\}} x_t \geqq \int_{\{t=1\}} x_1 + \int_{\{t>1\}} x_1 = E(x_1) > -\infty.$$

---

* Research supported in part by National Science Foundation Grant NSF-G14146 at Columbia University.

Lemma 1. *Given any s.v. t, define*

$$t' = \text{first integer } j \geqq 1 \text{ such that } E(x_t | \mathscr{F}_j) \leqq x_j .$$

Then t′ is a s.v. and has the following properties:

(a) *t' is regular,*

(b) $t' \leqq t$,

(c) $E(x_{t'}) \geqq E(x_t)$.

Proof. If $t = n$ then $E(x_t | \mathscr{F}_n) = x_n$, so that $t' \leqq n$. Thus $t' \leqq t < \infty$, and hence (b) holds. For any $A \in \mathscr{F}_n$,

$$\int_{A\{t' \geqq n\}} x_{t'} = \sum_{j=n}^{\infty} \int_{A\{t'=j\}} x_j \geqq \sum_{j=n}^{\infty} \int_{A\{t'=j\}} E(x_t | \mathscr{F}_j) = \int_{A\{t' \geqq n\}} x_t . \tag{2}$$

Putting $n = 1$ and $A = \Omega$, (2) yields the inequality (c). Finally, from (2) and the definition of t',

$$t' > n \Rightarrow E(x_{t'} | \mathscr{F}_n) \geqq E(x_t | \mathscr{F}_n) > x_n ,$$

which proves (a).

Lemma 2. *Let $t_1, t_2, \ldots$ be any sequence of regular s.v.'s and define*

$$\tau_i = \max(t_1, \ldots, t_i), \quad \tau = \sup_i t_i = \lim_{i\to\infty} \tau_i .$$

Then the τ_i are regular s.v.'s, $\tau_1 \leqq \tau_2 \leqq \cdots$, and

$$\max(Ex_{t_1}, \ldots, Ex_{t_i}) \leqq E(x_{\tau_i}) \leqq E(x_{\tau_{i+1}}) \leqq \cdots . \tag{3}$$

Moreover, if $P(\tau < \infty) = 1$ then τ is a regular s.v. and

$$Ex_\tau \geqq \lim_{i\to\infty} E(x_{\tau_i}) \geqq \sup Ex_{t_i} . \tag{4}$$

Proof. For any $i, n = 1, 2, \ldots$ and any $A \in \mathscr{F}_n$ we have

$$\int_{A\{\tau_i \geqq n\}} x_{\tau_i} = \sum_{j=n}^{\infty} \left(\int_{A\{\tau_i = j \geqq t_{i+1}\}} x_{\tau_{i+1}} + \int_{A\{\tau_i = j < t_{i+1}\}} x_j \right) \leqq \sum_{j=n}^{\infty} \left(\int_{A\{\tau_i = j \geqq t_{i+1}\}} x_{\tau_{i+1}} + \int_{A\{\tau_i = j < t_{i+1}\}} x_{t_{i+1}} \right) = \int_{A\{\tau_i \geqq n\}} x_{\tau_{i+1}} .$$

Hence, since $\tau_1 \leqq \tau_2 \leqq \cdots$, it follows that

$$\tau_i \geqq n \Rightarrow E(x_{\tau_i} | \mathscr{F}_n) \leqq E(x_{\tau_{i+1}} | \mathscr{F}_n) \leqq E(x_{\tau_{i+2}} | \mathscr{F}_n) \leqq \cdots . \tag{5}$$

Since t_1 is regular and $\tau_1 = t_1$, it follows that

$$t_1 > n \Rightarrow x_n < E(x_{t_1} | \mathscr{F}_n) = E(x_{\tau_1} | \mathscr{F}_n) \leqq E(x_{\tau_2} | \mathscr{F}_n) \leqq \cdots .$$

By symmetry,

$$t_j > n \Rightarrow E(x_{\tau_i} | \mathscr{F}_n) > x_n , \qquad j = 1, \ldots, i,$$

and hence

$$\tau_i > n \Rightarrow E(x_{\tau_i} | \mathscr{F}_n) > x_n ,$$

so that each τ_i is regular. Setting $n = 1$ in (5) we obtain

$$E(x_{t_1} | \mathscr{F}_1) = E(x_{\tau_1} | \mathscr{F}_1) \leqq E(x_{\tau_2} | \mathscr{F}_1) \leqq \cdots , \tag{6}$$

so that

$$E(x_{t_1}) \leqq E(x_{\tau_1}) \leqq E(x_{\tau_2}) \leqq \cdots ,$$

and by symmetry

$$E(x_{t_j}) \leqq E(x_{\tau_i}), \qquad j = 1, \ldots, i,$$

which proves (3).

Turning our attention to τ we observe that since $x_\tau = \lim_{i\to\infty} x_{\tau_i}$, and since

$$E(\sup_i x_{\tau_i}) \leqq E(\sup_n x_n^+) < \infty,$$

we have by Fatou's lemma for conditional expectations [*2*, p. 348] that

$$E(x_\tau | \mathscr{F}_n) \geqq \limsup_{i\to\infty} E(x_{\tau_i} | \mathscr{F}_n). \tag{7}$$

Hence by (5) and (7),

$$\tau > n \Rightarrow \tau_i > n \quad \text{for some} \quad i \Rightarrow x_n < E(x_{\tau_i} | \mathscr{F}_n) \leqq E(x_{\tau_{i+1}} | \mathscr{F}_n) \leqq \cdots$$
$$\Rightarrow \limsup_{i\to\infty} E(x_{\tau_i} | \mathscr{F}_n) > x_n \Rightarrow E(x_\tau | \mathscr{F}_n) > x_n,$$

so that τ is regular. Finally, from (6) and (7) we have

$$E(x_\tau | \mathscr{F}_1) \geqq E(x_{\tau_i} | \mathscr{F}_1),$$

so that (4) holds.

Corollary 1. *Let t_0 be any s.v., and let $C(t_0)$ denote the class of all s.v.'s t such that $t \leqq t_0$. Then there exists a s.v. $\tau \in C(t_0)$ such that*

$$E(x_\tau) = \sup_{t\in C(t_0)} E(x_t). \tag{8}$$

Proof. Take any sequence $t_1, t_2, \ldots$ of s.v.'s in $C(t_0)$ such that

$$\sup_i E(x_{t_i}) = \sup_{t\in C(t_0)} E(x_t).$$

By Lemma 1 we may assume that the t_i are regular. Set $\tau = \sup_i t_i$; then $\tau \in C(t_0)$ and the conclusion follows from Lemma 2.

Corollary 2. *Suppose there exists a s.v. τ_0 such that*

$$E(x_{\tau_0}) = \sup_{t\in C} E(x_t). \tag{9}$$

Choose any sequence $t_1, t_2, \ldots$ of regular s.v.'s such that

$$\sup_i E(x_{t_i}) = \sup_{t\in C} E(x_t), \tag{10}$$

and set $\tau = \sup_i t_i$. Then

$$\tau \leqq \tau_0, \tag{11}$$

so that τ is a s.v., and

$$E(x_\tau) = \sup_{t\in C} E(x_t). \tag{12}$$

The s.v. τ thus defined does not depend on the particular choice of $\tau_0, t_1, t_2, \ldots$, since by (11) *and* (12) *it is the minimal s.v. τ such that* (12) *holds.*

Proof. By Lemma 1 of [*1*], $t_i \leqq \tau_0$ for each i, so that (11) holds, and (12) then follows from Lemma 2.

Lemma 3. *Assume that*

$$x_n = x_n' - x_n'' \tag{13}$$

3*

where x_n', x_n'' are measurable ($\mathcal{F}_n$) for each n, and are such that

$$E[\sup_n (x_n')^+] = B < \infty, \tag{14}$$

$$x_n'' \geqq 0, \quad \lim_{n\to\infty} x_n'' = \infty. \tag{15}$$

Let $t_1, t_2, \ldots$ be any sequence of s.v.'s such that

$$E(x_{t_i}) \geqq K > -\infty, \tag{16}$$

and set $\tau = \liminf_{i\to\infty} t_i$. Then $P(\tau < \infty) = 1$.

Proof. For any integers i and m,

$$\int_{\{t_i \geqq m\}} x_{t_i} = \int_{\{t_i \geqq m\}} (x_{t_i}' - x_{t_i}'') \leqq \int_{\{t_i \geqq m\}} (\sup_n (x_n')^+ - \inf_{j \geqq m} x_j'') \leqq B - \int_{\{t_i \geqq m\}} w_m,$$

where we have set

$$w_m = \inf_{j \geqq m} x_j''.$$

Since

$$\int_{\{t_i < m\}} x_{t_i} \leqq B,$$

we have

$$K \leqq E(x_{t_i}) \leqq 2B - \int_{\{t_i \geqq m\}} w_m.$$

Let $A_i = \{\inf_{j \geqq i} t_j \geqq m\} \subset \{t_i \geqq m\}$; then since $w_m \geqq 0$,

$$K \leqq 2B - \int_{A_i} w_m,$$

and letting $i \to \infty$ we have

$$K \leqq 2B - \int_{\{\tau \geqq m\}} w_m \leqq 2B - \int_{\{\tau = \infty\}} w_m.$$

Let $m \to \infty$; then since

$$0 \leqq w_1 \leqq w_2 \leqq \cdots \to \liminf_{n\to\infty} x_n'' = \infty,$$

it follows that

$$\int_{\{\tau = \infty\}} \infty \leqq 2B - K < \infty,$$

so that $P(\tau = \infty) = 0$.

Lemma 4. *Under the assumptions* (13), (14), (15) *of Lemma 3, there exists a s.v. τ such that*

$$E(x_\tau) = \sup_{t\in C} E(x_t). \tag{17}$$

Proof. Let $t_1, t_2, \ldots$ be any sequence of s.v.'s such that

$$\sup_i E(x_{t_i}) = \sup_{t\in C} E(x_t). \tag{18}$$

By Lemma 1 we may suppose that the t_i are regular and therefore that

$$E(x_{t_i}) \geqq E(x_1) > -\infty.$$

Set

$$\tau_i = \max(t_1, \ldots, t_i), \quad \tau = \sup_i t_i = \lim_{i\to\infty} \tau_i.$$

By Lemma 2,

$$E(x_{\tau_i}) \geqq E(x_{t_i}) \geqq E(x_1),$$

and $\tau_1 \leqq \tau_2 \leqq \cdots$. By Lemma 3, $P(\tau < \infty) = 1$. Hence by Lemma 2,

(19) $$E(x_\tau) \geqq \sup_i E(x_{t_i}),$$

and (17) follows from (18) and (19).

The main results so far may be summarized in the following theorem.

Theorem 1. *Assume that* $E(\sup_n x_n^+) < \infty$.

(i) *Choose any sequence* $t_1, t_2, \ldots$ *of regular s.v.'s such that*

(20) $$\sup_i E(x_{t_i}) = \sup_{t \in C} E(x_t)$$

(this can always be done), and define the r.v.

(21) $$\tau = \sup_i t_i.$$

Then $P(\tau < \infty) = 1$ *if and only if there exists a s.v.* τ_0 *such that*

(22) $$E(x_{\tau_0}) = \sup_{t \in C} E(x_t),$$

and τ *is then the minimal s.v. satisfying* (22).

(ii) *Assumptions* (13), (14), (15) *are sufficient to ensure that* $P(\tau < \infty) = 1$.

Proof.

(i) If $P(\tau < \infty) = 1$ then by the argument of Lemma 4,

$$E(x_\tau) = \sup_{t \in C} E(x_t).$$

And if any s.v. τ_0 exists satisfying (22), then $P(\tau < \infty) = 1$ by Corollary 2 of Lemma 2, and $\tau \leqq \tau_0$.

(ii) Follows from Lemma 4.

The main defect of Theorem 1 is that it gives no indication of how to choose a sequence of regular s.v.'s $t_1, t_2, \ldots$ satisfying (20). We now turn our attention to this problem.

3. The rules s_N and s

Let C_N denote the class of all s.v.'s t for which $t \leqq N$. We shall first show (cf. [3]) how to construct a certain regular s.v. s_N in C_N such that

(23) $$E(x_{s_N}) = \sup_{t \in C_N} E(x_t).$$

To do this we define for each $N \geqq 1$ a finite sequence of r.v.'s $\beta_1^N, \ldots, \beta_N^N$ by recursion backwards, starting with β_N^N, using the formula

(24) $$\beta_n^N = \max[x_n, E(\beta_{n+1}^N \mid \mathscr{F}_n)], \quad n = 1, \ldots, N; \quad \beta_{N+1}^N = -\infty.$$

Thus

(25) $$\beta_N^N = \max[x_N, -\infty] = x_N,$$

and β_n^N is measurable ($\mathscr{F}_n$). We now define

(26) $$s_N = \text{first } n \geqq 1 \text{ such that } \beta_n^N = x_n .$$

Note that

(27) $$\beta_{s_N}^N = x_{s_N},$$

and, since $\beta_N^N = x_N$,

(28) $$s_N \leqq N,$$

so that $s_N \in C_N$. Moreover,

(29) $$s_N > n \Rightarrow E(\beta_{n+1}^N | \mathscr{F}_n) = \beta_n^N > x_n,$$

and

(30) $$E(\beta_{n+1}^N | \mathscr{F}_n) \leqq \beta_n^N, \quad \text{all} \quad n = 1, \ldots, N.$$

From [*1*, Lemmas 1, 2, 3] applied to the finite sequence $\beta_1^N, \ldots, \beta_N^N$ it follows that s_N is regular, since

(31) $$s_N > n \Rightarrow E(x_{s_N} | \mathscr{F}_n) = E(\beta_{s_N}^N | \mathscr{F}_n) \geqq \beta_n^N > x_n,$$

and that

(32) $$E(x_{s_N}) = E(\beta_{s_N}^N) \geqq E(\beta_t^N) \geqq E(x_t) \quad \text{all} \quad t \in C_N .$$

Thus the sequence $s_1, s_2, \ldots$ has the following properties:

(33) $$s_N \text{ is regular}, \quad s_N \leqq N, \quad (23) \text{ holds},$$

and, since $C_1 \subset C_2 \subset \ldots$, it follows that

(34) $$E(x_1) = E(x_{s_1}) \leqq E(x_{s_2}) \leqq \cdots \to \lim_{N\to\infty} E(x_{s_N}).$$

It is easy to show by induction from (24) and (25) that

(35) $$x_N = \beta_N^N \leqq \beta_N^{N+1} \leqq \cdots.$$

Hence from (26) we have

(36) $$1 = s_1 \leqq s_2 \leqq \cdots,$$

and we define

(37) $$s = \sup_N s_N = \lim_{N\to\infty} s_N \leqq +\infty .$$

Lemma 5. *If* $P(s < \infty) = 1$, *then*

(38) $$E(x_s) \geqq \lim_{N\to\infty} E(x_{s_N}).$$

Proof. By (33) and Lemma 2 applied to the sequence $s_1, s_2, \ldots$.

Lemma 6. *If t is any s.v. such that*

(39) $$\liminf_{n\to\infty} \int_{\{t>n\}} x_n^- = 0$$

then

(40) $$\lim_{N\to\infty} E(x_{s_N}) \geqq E(x_t).$$

Proof. Set

(41) $$t_N = \min(t, N) \in C_N .$$

Then

(42) $$\int_{\{t \leqq N\}} x_t = E(x_{t_N}) - \int_{\{t > N\}} x_N \leqq E(x_{s_N}) - \int_{\{t > N\}} x_N \leqq E(x_{s_N}) + \int_{\{t > N\}} x_N^- .$$

Letting $N \to \infty$ it follows from (39) that (40) holds.

Corollary. *If $x_n^- \leqq c n^\alpha$ for some c, $\alpha \geqq 0$, and if $E(t^\alpha) < \infty$, then*

$$\lim_{N \to \infty} E(x_{s_N}) \geqq E(x_t) .$$

Proof. From Lemma 6 and the relation

$$\int_{\{t > n\}} x_n^- \leqq c \int_{\{t > n\}} n^\alpha \leqq c \int_{\{t > n\}} t^\alpha \to 0 .$$

Theorem 2. *Assume that*

$$x_n = x_n' - x_n'' = x_n^* - x_n^{**},$$

where all the components are measurable $(\mathscr{F}_n)$ and

(43) $$E[\sup_n (x_n')^+] = B < \infty ,$$

(44) $$0 \leqq x_1'' \leqq x_2'' \leqq \cdots, \quad \lim_{n \to \infty} x_n'' = \infty ,$$

(45) *the $(x_n^*)^-$ are uniformly integrable for all n,*

and

(46) $$x_n^{**} \leqq c x_n'' \quad \textit{for some} \quad 0 < c < \infty .$$

Then $s = \sup_N s_N$ is a s.v. and

$$E(x_s) = \sup_{t \in C} E(x_t) = \lim_{N \to \infty} E(x_{s_N}) .$$

Proof. For any s.v. t we have from (44) and (46) that for $t > N$,

$$x_N = (x_N^*)^+ - (x_N^*)^- - x_N^{**} \geqq -[(x_N^*)^- + c x_t''],$$

so that

(47) $$\int_{\{t > N\}} x_N^- \leqq \int_{\{t > N\}} [(x_N^*)^- + c x_t''] .$$

Now if $E(x_t) \neq -\infty$ then from (43)

$$E(x_t) = E(x_t') - E(x_t'') \leqq E(x_t'^+) \leqq B < \infty ,$$

so that $E(x_t')$ and hence $E(x_t'')$ is finite. From (47) and (45) it follows that (39) holds. From Lemma 3, $P(s < \infty) = 1$, and hence from Lemmas 5 and 6,

$$E(x_s) \geqq \lim_{N \to \infty} E(x_{s_N}) \geqq E(x_t) .$$

Since this is trivially true when $E(x_t) = -\infty$ the result follows.

It is of interest to express $E(x_{s_N})$ explicitly. To do this we observe that by the submartingale property (30),

$$E(x_{s_N}) = E(\beta^N_{s_N}) = \sum_{n=1}^{N} \int_{\{s_N=n\}} \beta^N_n = \sum_{n=1}^{N-1} \int_{\{s_N=n\}} \beta^N_n + \int_{\{s_N>N-1\}} \beta^N_N$$

$$\text{(48)} \qquad \le \sum_{n=1}^{N-2} \int_{\{s_N=n\}} \beta^N_n + \int_{\{s_N=N-1\}} \beta^N_{N-1} + \int_{\{s_N>N-1\}} \beta^N_{N-1}$$

$$= \sum_{n=1}^{N-2} \int_{\{s_N=n\}} \beta^N_n + \int_{\{s_N>N-2\}} \beta^N_{N-1} \le \cdots \le \int_{\{s_N>0\}} \beta^N_1 = E(\beta^N_1).$$

But since $E(\beta^N_{s_N}) \geqq E(\beta^N_1)$ it follows that

$$\text{(49)} \qquad E(x_{s_N}) = E(\beta^N_1).$$

Thus under the conditions on the x_n of Theorem 2,

$$\text{(50)} \qquad E(x_s) = \lim_{N\to\infty} E(x_{s_N}) = \lim_{N\to\infty} E(\beta^N_1).$$

From (35) the limits

$$\text{(51)} \qquad \beta_n = \lim_{N\to\infty} \beta^N_n$$

exist. By the theorem of monotone convergence for conditional expectations [2, p. 348] it follows from (35) that

$$\text{(52)} \qquad E(\beta^n_n \mid \mathscr{F}_n) \leqq E(\beta^{n+1}_n \mid \mathscr{F}_n) \leqq \cdots \to E(\beta_n \mid \mathscr{F}_n),$$

and hence from (24) that the β_n satisfy the relations

$$\text{(53)} \qquad \beta_n = \max[x_n, E(\beta_{n+1} \mid \mathscr{F}_n)], \quad n = 1, 2, \ldots.$$

Define for the moment

$$\text{(54)} \qquad s^* = \text{first } i \geqq 1 \text{ such that } x_i = \beta_i.$$

We shall show that

$$\text{(55)} \qquad s^* = \sup_N s_N = s.$$

For if $s^* = n$, then by (54) $x_i < \beta_i$ for $i = 1, \ldots, n-1$, and hence for sufficiently large N, $x_i < \beta^N_i$ for $i = 1, \ldots, n-1$, so that $s_N \geqq n$. Hence $s \geqq n$ and therefore $s \geqq s^*$. Conversely, if $s = n$ then for sufficiently large N, $s_N = n$, and hence $x_i < \beta^N_i$ for $i = 1, \ldots, n-1$ so that $x_i < \beta_i$ for $i = 1, \ldots, n-1$ and therefore $s^* \geqq n$. Thus $s^* \geqq s$.

We may now restate Theorem 2 in the following form.

Theorem 2′. *Assume that the hypotheses on the x_n of Theorem 2 are satisfied. For each $N \geqq 1$ define $\beta^N_1, \beta^N_2, \ldots, \beta^N_N$ by (24) and set*

$$\text{(56)} \qquad s = \textit{first } i \geqq 1 \textit{ such that } x_i = \beta_i = \lim_{N\to\infty} \beta^N_i.$$

Then s is a s.v. and

$$\text{(57)} \qquad E(x_s) = \lim_{N\to\infty} E(\beta^N_1) = \sup_{t\in C} E(x_t).$$

This generalizes a theorem of Arrow, Blackwell, and Girshick [3].

4. The monotone case

If the sequence of r.v.'s $x_1, x_2, \dots$ is such that for every $n = 1, 2, \dots$,

(58) $$E(x_{n+1}|\mathscr{F}_n) \leqq x_n \Rightarrow E(x_{n+2}|\mathscr{F}_{n+1}) \leqq x_{n+1},$$

we shall say that we are in the *monotone case* (to which [*1*] is devoted). In this case the calculation of the s_N defined by (26), and of $s = \sup_N s_N$, become much simpler.

Lemma 7. *In the monotone case we may compute s_N and s by the formulas*

(59) $$s_N = \min[N, \textit{ first } n \geqq 1 \textit{ such that } E(x_{n+1}|\mathscr{F}_n) \leqq x_n],$$

and

(60) $$s = \sup_N s_N = \textit{first } n \geqq 1 \textit{ such that } E(x_{n+1}|\mathscr{F}_n) \leqq x_n.$$

Proof. (a) we begin by proving that in the monotone case, for $n = 1, 2, \dots, N-1$,

(61) $$E(x_{n+1}|\mathscr{F}_n) \leqq x_n \Rightarrow E(\beta_{n+1}^N|\mathscr{F}_n) \leqq x_n.$$

For $n = N-1$ this is trivial, since $\beta_N^N = x_n$. Assume therefore that (61) is true for $n = j+1$. Then

$$E(x_{j+1}|\mathscr{F}_j) \leqq x_j \Rightarrow E(x_{j+2}|\mathscr{F}_{j+1}) \leqq x_{j+1} \Rightarrow$$
$$E(\beta_{j+2}^N|\mathscr{F}_{j+1}) \leqq x_{j+1} \Rightarrow \beta_{j+1}^N = x_{j+1} \Rightarrow$$
$$E(\beta_{j+1}^N|\mathscr{F}_j) = E(x_{j+1}|\mathscr{F}_j) \leqq x_j,$$

which establishes (61) for $n = j$.

(b) Recall that by (26),

(62) $$s_N = \text{first } n \geqq 1 \text{ such that } \beta_n^N = x_n.$$

Define for the moment

(63) $$s_N' = \min[N, \text{ first } n \geqq 1 \text{ such that } E(x_{n+1}|\mathscr{F}_n) \leqq x_n].$$

(c) Suppose $s_N' = n < N$. Then by (61), $E(\beta_{n+1}^N|\mathscr{F}_n) \leqq x_n$, so that $\beta_n^N = x_n$ and hence $s_N \leqq n = s_N'$. If $s_N' = N$ then also $s_N \leqq s_N'$. Thus $s_N \leqq s_N'$ always.

(d) Suppose $s_N = n \leqq N$. Then $E(\beta_{n+1}^N|\mathscr{F}_N) \leqq x_n$. Since $\beta_{n+1}^N \geqq x_{n+1}$ it follows that $E(x_{n+1}|\mathscr{F}_n) \leqq x_n$. Hence $s_N' \leqq n$ and therefore $s_N' \leqq s_N$.

It follows from (c) and (d) that $s_N' = s_N$, which proves (59), and (60) is immediate.

5. An example

Let $y, y_1, y_2, \dots$ be independent r.v.'s with a common distribution, let $\mathscr{F}_n$ be the σ-algebra generated by $y_1, \dots, y_n$, and let

(64) $$x_n = \max(y_1, \dots, y_n) - a_n,$$

where we assume to begin with only that the a_n are constants such that

(65) $$0 \leqq a_1 < a_2 < \cdots$$

and that $Ey^+ < \infty$. Set

(66) $$m_n = \max(y_1, \ldots, y_n), \quad b_n = a_{n+1} - a_n > 0.$$

Then

$$x_{n+1} = m_{n+1} - a_{n+1} = m_{n+1} - a_n - b_n = x_n + (y_{n+1} - m_n)^+ - b_n.$$

Hence

(67) $$E(x_{n+1} | \mathscr{F}_n) - x_n = E[(y - m_n)^+] - b_n.$$

Define constants γ_n by the relation

(68) $$E[(y - \gamma_n)^+] = b_n$$

(graphically, b_n is the area in the z, y-plane to the right of $y = \gamma_n$ and between $z = 1$ and $z = F(y)$). Then it is easy to see from (67) and (68) that

(69) $$E(x_{n+1} | \mathscr{F}_n) \leqq x \quad \text{if and only if} \quad m_n \geqq \gamma_n.$$

We are in the monotone case when

(70) $$b_1 \leqq b_2 \leqq \cdots.$$

For if (70) holds, and if $E(x_{n+1} | \mathscr{F}_n) \leqq x_n$, then by (68), $m_n \geqq \gamma_n$ and hence $m_{n+1} \geqq m_n \geqq \gamma_n \geqq \gamma_{n+1}$, so that $E(x_{n+2} | \mathscr{F}_{n+1}) \leqq x_{n+1}$. We can therefore assert that when (70) holds

(71) $$s_N = \min[N, \text{ first } n \geqq 1 \text{ such that } m_n \geqq \gamma_n],$$

and

(72) $$s = \sup_N s_N = \text{first } n \geqq 1 \text{ such that } m_n \geqq \gamma_n.$$

An example of the monotone case is given by choosing $a_n = cn^\alpha$ with $c > 0$, $\alpha \geqq 1$. When $\alpha = 1$ all the γ_n coincide and have the value γ given by

(73) $$E[(y - \gamma)^+] = c.$$

For $0 < \alpha < 1$ we are not in the monotone case and no simple evaluation of s_N and s is possible.

It is interesting to note that if we set

(74) $$\tilde{x}_n = y_n - a_n$$

instead of (64), then, setting $\mu = Ey$,

(75) $$E(x_{n+1} | \mathscr{F}_n) - \tilde{x}_n = \mu - b_n - y_n,$$

and we are never in the monotone case. However, for $a_n = cn$ we have by the above,

(76) $$\begin{aligned} s &= \text{first } n \geqq 1 \text{ such that } m_n \geqq \gamma \\ &= \text{first } n \geqq 1 \text{ such that } y_n \geqq \gamma. \end{aligned}$$

Thus

(77) $$x_s = m_s - cs = y_s - cs = \tilde{x}_s,$$

while for any s.v. t, since $\tilde{x}_n \leqq x_n$, we have

(78) $$\tilde{x}_t \leqq x_t .$$

It follows that

$$\sup_{t \in C} E(\tilde{x}_t) \leqq \sup_{t \in C} E(x_t),$$

and that if the distribution of the y_n is such that

(79) $$E(x_s) = \sup_{t \in C} E(x_t),$$

then also

$$E(\tilde{x}_s) = \sup_{t \in C} E(\tilde{x}_t).$$

We shall now investigate whether in fact (79) holds, and for this we shall use Theorem 2. Write

$$x_n = \max(y_1, \dots, y_n) - a_n = x_n' - x_n'' = x_n^* - x_n^{**}$$

where we have set

(80) $$\begin{cases} x_n' = \max(y_1, \dots, y_n) - a_n/2, & x_n'' = a_n/2, \\ x_n^* = \max(y_1, \dots, y_n), & x_n^{**} = a_n. \end{cases}$$

Assume that the constants a_n are such that

(81) $$0 \leqq a_1 \leqq a_2 \leqq \dots \to \infty .$$

Then (44) and (46) hold, and to apply Theorem 2 it will suffice to show that

(82) $$E \sup_n [\max(y_1, \dots, y_n) - a_n/2] < \infty$$

and that the r.v.'s

(83) $$[\max(y_1, \dots, y_n)]^- \text{ are uniformly integrable.}$$

The latter relation is trivial as long as $E|y| < \infty$, since

$$[\max(y_1, \dots, y_n)]^- \leqq y_1^- .$$

It remains only to verify (82).

To find conditions for the validity of (82) in the case

$$a_n = c n^\alpha, \quad c, \alpha > 0,$$

we shall need the following lemma, the proof of which will be deferred until later.

Lemma 8. *Let $w, w_1, w_2, \dots$ be independent, identically distributed, non-negative r.v.'s and for any positive constants c, α set*

$$z = \sup_n [\max(w_1, \dots, w_n) - c n^\alpha].$$

Then

(84) $$P(z < \infty) = 1 \textit{ if and only if } E(w^{1/\alpha}) < \infty,$$

and

(85) *for any* $\beta > 0$, $E(z^{1/\beta}) < \infty$ *if and only if* $E(w^{1/\alpha + 1/\beta}) < \infty$.

Now suppose that the common distribution of the y_n is such that

(86) $$E|y| < \infty, \quad E[(y^+)^{1+\alpha}] < \infty.$$

Then
$$\sup_n \left[\max(y_1, \ldots, y_n) - \frac{cn^\alpha}{2}\right] \leqq \sup_n \left[\max(y_1^+, \ldots, y_n^+) - \frac{cn^\alpha}{2}\right],$$
so that by (85) for $\beta = 1$, $w = y^+$,
$$E \sup_n \left[\max(y_1, \ldots, y_n) - \frac{cn^\alpha}{2}\right] < \infty,$$
verifying (82). Thus, *if* $a_n = cn^\alpha$ $(c, \alpha > 0)$ *and if* $E|y| < \infty$ *and* $E[(y^+)^{1+\alpha}] < \infty$, *then defining the s.v. s by* (56) *we have*
$$E(x_s) = \sup_{t \in C} E(x_t).$$

This generalizes a result of [*1*], where it was assumed that $\alpha \geqq 1$, to the more general case $\alpha > 0$. See also [*5, 6, 7*] for the case $\alpha = 1$.

A similar argument holds for the sequence
$$x_n = y_n - cn^\alpha,$$
replacing $\max(y_1, \ldots, y_n)$ by y_n in (80).

We may summarize these results in

Theorem 3. *Let $y, y_1, y_2, \ldots$ be independent and identically distributed random variables, let c, α be positive constants, and let*
$$x_n = \max(y_1, \ldots, y_n) - cn^\alpha, \quad \tilde{x}_n = y_n - cn^\alpha.$$
Then if
$$E|y| < \infty, \quad E[(y^+)^{1+\alpha}] < \infty$$
there exist s.v.'s s and $\tilde{s}$ such that
$$E(x_s) = \sup_{t \in C} E(x_t), \quad E(\tilde{x}_{\tilde{s}}) = \sup_{t \in C} E(\tilde{x}_t).$$
For $\alpha \geqq 1$,
$$s = \textit{first } n \geqq 1 \textit{ such that } \max(y_1, \ldots, y_n) \geqq \gamma_n,$$
where γ_n is defined by
$$E[(y - \gamma_n)^+] = c[(n+1)^\alpha - n^\alpha].$$

Proof of Lemma 8. If w is any r.v. with distribution function F, then $E(w) < \infty$ is equivalent to $\sum_1^\infty [1 - F(n)] < \infty$, which in turn is equivalent to the convergence of $\prod_1^\infty F(n)$. Hence $E(w^{1/\alpha}) < \infty$ if and only if $\prod_1^\infty F(n^\alpha)$ converges. Now for $u > 0$ let
$$G(u) = P(z \leqq u) = P[\bigcap_{n=1}^\infty \bigcap_{i=1}^\infty \{w_i \leqq n^\alpha + u\}] = P[\bigcap_{i=1}^\infty \bigcap_{n=i}^\infty \{w_i \leqq n^\alpha + u\}]$$
$$= P[\bigcap_{i=1}^\infty \{w_i \leqq n^\alpha + u\}] = \prod_1^\infty F(n^\alpha + u).$$
It follows that $\lim_{u \to \infty} G(u) = 1$ if and only if $\prod_1^\infty F(n^\alpha)$ converges; thus (84) holds.

To prove (85), we have $E(z) < \infty$ if and only if $\prod_1^\infty G(n)$ converges. Hence [*4*, p. 223], $E(z^{1/\beta}) < \infty$ is equivalent to

$$(87) \qquad \sum_{m=1}^{\infty} \sum_{n=n_0}^{\infty} \log F[n^\beta + m^\alpha] > -\infty \quad \text{for some } n_0 \text{ such that } \quad F(n_0^\beta) > 0 .$$

Now

$$\int_1^\infty dm \int_{n_0}^\infty \log F(n^\beta + m^\alpha)\, dn = \int_{1+n_0^\beta}^\infty \log F(u)\, du \int_{n_0^\beta}^{u-1} \frac{1}{\alpha\beta} v^{1/\beta-1}(u-v)^{1/\alpha-1} dv .$$

Hence (87) is equivalent to

$$-\infty < \int_{n_0^\beta}^\infty \log F(u)\, du \int_0^u v^{1/\beta-1}(u-v)^{1/\alpha-1} dv = B\left(\frac{1}{\alpha}, \frac{1}{\beta}\right) \int_{n_0}^\infty u^{1/\alpha+1/\beta-1} \log F(u)\, du ,$$

But $E(w^{1/\alpha+1/\beta}) < \infty$ is equivalent to

$$-\infty < \int_{n_0^{1/\alpha+1/\beta}}^\infty \log F(t^{\alpha\beta/\alpha+\beta})\, dt = \frac{\alpha+\beta}{\alpha\beta} \int_{n_0^{1/\alpha+1/\beta}}^\infty u^{1/\alpha+1/\beta-1} \log F(u)\, du ,$$

which proves (85).

6. Application to the sequential probability ratio test

The following problem in statistical decision theory has been treated in [*8*, *9*, *3*, *10*, *11*]. We shall consider it here as an illustration of our general method.

Let $y_1, y_2, \ldots$ be independent, identically distributed random variables with density function f with respect to some σ-finite measure μ on the line. It is desired to test the hypothesis $H_0 : f = f_0$ versus $H_1 : f = f_1$ where f_0 and f_1 are two specified densities. The loss due to accepting H_1 when H_0 is true is assumed to be $a > 0$ and that due to accepting H_0 when H_1 is true is $b > 0$; the cost of taking each observation y_i is unity. A sequential decision procedure (δ, N) provides for determining the sample size N and making the terminal decision δ; the expected loss for (δ, N) is

$$a\alpha_0 + E_0(N) \quad \text{when } H_0 \text{ is true},$$
$$b\alpha_1 + E_1(N) \quad \text{when } H_1 \text{ is true}$$

where

$$\alpha_0 = P_0(\text{accepting } H_1), \quad \alpha_1 = P_1(\text{accepting } H_0).$$

If there is an a priori probability π that H_0 is true (and hence probability $1-\pi$ that H_1 is true) the global "risk" for (δ, N) is given by

$$r(\pi, \delta, N) = \pi[a\alpha_0 + E_0(N)] + (1-\pi)[b\alpha_1 + E_1(N)].$$

For a given sampling variable N it is easy to determine the terminal decision rule δ which minimizes $r(\pi, \delta, N)$ for fixed values of a, b, and π. For the part of

$r(\pi, \delta, N)$ that depends on δ is (omitting symbols like $d\mu(y_1) \ldots d\mu(y_n)$)

$$\pi a \alpha_0 + (1-\pi) b \alpha_1 = \pi a \sum_{n=1}^{\infty} \int_{\{N=n,\, \text{accept } H_1\}} f_0(y_1) \ldots f_0(y_n) +$$

$$+ (1-\pi) b \sum_{n=1}^{\infty} \int_{\{N=n,\, \text{accept } H_0\}} f_1(y_1) \ldots f_1(y_n)$$

$$\geqq \sum_{n=1}^{\infty} \int_{\{N=n\}} \min[\pi a f_0(y_1) \ldots f_0(y_n), \quad (1-\pi) b f_1(y_1) \ldots f_1(y_n)]$$

$$= \sum_{n=1}^{\infty} \int_{\{N=n\}} \min[\pi_n a, (1-\pi_n) b] [\pi f_0(y_1) \ldots f_0(y_n) +$$

$$+ (1-\pi) f_1(y_1) \ldots f_1(y_n)],$$

where

$$\pi_n = \pi_n(y_1, \ldots, y_n) = \frac{\pi f_0(y_1) \ldots f_0(y_n)}{\pi f_0(y_1) \ldots f_0(y_n) + (1-\pi) f_1(y_1) \ldots f_1(y_n)}.$$

For the given sampling rule N define δ' by

$$\begin{cases} \text{accept } H_1 \text{ if } N = n \text{ and } \pi_n a \leqq (1-\pi_n) b, \\ \text{accept } H_0 \text{ if } N = n \text{ and } \pi_n a > (1-\pi_n) b. \end{cases}$$

Then

$$\pi a \alpha_0(\delta, N) + (1-\pi) b a_1(\delta, N) \geqq \pi a \alpha_0(\delta', N) + (1-\pi) b \alpha_1(\delta', N).$$

Hence to find a pair (δ, N) which for given π minimizes $r(\pi, \delta, N)$ (a "Bayes" decision procedure) amounts to solving the following problem: for given $0 < \pi < 1$ let $y_1, y_2, \ldots, y_n, \ldots$ have the joint density function for each n equal to

$$\pi f_0(y_1) \ldots f_0(y_n) + (1-\pi) f_1(y_1) \ldots f_1(y_n),$$

where f_0, f_1 are given univariate density functions. For given $a, b > 0$ let

$$h(t) = \min[at, b(1-t)] \qquad (0 \leqq t \leqq 1),$$

$$\begin{cases} \pi_0 = \pi \\ \pi_n = \pi_n(y_1, \ldots, y_n) = \dfrac{\pi f_0(y_1) \ldots f_0(y_n)}{\pi f_0(y_1) \ldots f_0(y_n) + (1-\pi) f_1(y_1) \ldots f_1(y_n)} \quad (n \geqq 1), \\ x_n = x_n(\pi_n) = -h(\pi_n) - n \qquad (n \geqq 0). \end{cases}$$

We want to find a s. v. s such that $E(x_s) =$ maximum. The problem is trivial if a or b is $\leqq 1$ since then $h(t) < 1$ and $x_0 < x_n$ for all n, so that $E(x_s) =$ max. for $s = 0$. We shall therefore assume that $a > 1$, $b > 1$.

We observe that the assumptions of Theorem 2 are satisfied by setting

$$x_n = x_n' - x_n'' = x_n^* - x_n^{**}$$

with

$$\begin{cases} x_n' = x_n^* = -h(\pi_n), & \left(0 \leqq h(\pi_n) \leqq \dfrac{ab}{a+b}\right), \\ x_n'' = x_n^{**} = n, \end{cases}$$

so that $s = \sup_N s_N$ is the desired s.v. Thus Theorem 2 guarantees the existence of a Bayes solution of our decision problem.

To find the (minimal) Bayes sampling variable s requires that we compute the quantities $\beta_0^N, \beta_1^N, \ldots, \beta_N^N$ for each $N \geqq 0$ (note that in the present context we are allowed to take no observations on the y_t and to decide in favor of H_0 or H_1 with $x_0 = -h(\pi)$). We have

$$\beta_n^N = \max[x, E(\beta_{n+1}^N|\mathscr{F}_n)], \quad n = 0, 1, \ldots, N; \quad \beta_{N+1}^N = -\infty,$$

and by Theorem 2′

$$s = \text{first} \quad n \geqq 0 \quad \text{such that} \quad x_n = \beta_n = \lim_{N\to\infty} \beta_n^N .$$

Observing that

$$\pi_{n+1} = \frac{\pi_n f_0(y_{n+1})}{\pi_n f_0(y_{n+1}) + (1-\pi_n) f_1(y_{n+1})}$$

it follows easily that

$$\beta_n^N(y_1, \ldots, y_n) = \gamma_n^N(\pi_n), \qquad n = 0, 1, \ldots, N+1,$$

where

$$\gamma_n^N(t) = \max\left\{-h(t) - n, \int_{-\infty}^{\infty} \gamma_{n+1}^N \left(\frac{t f_0(y)}{t f_0(y) + (1-t) f_1(y)}\right) \quad [t f_0(y) + (1-t) f_1(y)]\right\}$$

$$(n = 0, 1, \ldots, N); \quad \gamma_{N+1}^N(t) = -\infty .$$

Now set

$$g_n^N(t) = -\gamma_n^N(t) - n, \quad n = 0, 1, \ldots, N+1 .$$

Then

$$g_n^N(t) = \min[h(t), G_{n+1}^N(t) + 1],$$

where

$$G_n^N(t) = \int_{-\infty}^{\infty} g_n^N \frac{t f_0(y)}{t f_0(y) + (1-t) f_1(y)} [t f_0(y) + (1-t) f_1(y)]$$

for $n = 0, 1, \ldots, N$ with

$$g_{N+1}^N(t) = \infty .$$

Obviously,

$$g_n^N(t) = g_{n+1}^{N+1}(t), \quad g_n^N(t) \geqq g_n^{N+1}(t) \quad \text{for} \quad n = 0, 1, \ldots, N+1,$$

so that

$$\lim_{N\to\infty} g_n^N(t) = g_n(t) = g(t) \quad \text{exists.}$$

By the Lebesgue theorem of dominated convergence,

$$g(t) = \min[h(t), G(t) + 1]$$

where

$$G(t) = \int_{-\infty}^{\infty} g\left(\frac{t f_0(y)}{t f_0(y) + (1-t) f_1(y)}\right) [t f_0(y) + (1-t) f_1(y)] .$$

And

$$\beta_n^N(y_1, \ldots, y_n) = \gamma_n^N(\pi_n) = -g_n^N(\pi_n) - n,$$

so that

$$\beta_n = \lim_{N\to\infty} \beta_n^N = -g(\pi_n) - n$$

and hence

$$s = \text{first} \quad n \geqq 0 \quad \text{such that} \quad g(\pi_n) = h(\pi_n); \quad E(x_s) = \beta_0 = -g(\pi).$$

We shall now investigate the nature of the function $g(t)$ which characterizes s.

If a function $a(t)$ is concave for $0 \leqq t \leqq 1$ and if

$$A(t) = \int_{-\infty}^{\infty} a\left(\frac{t f_0(y)}{t f_0(y) + (1-t) f_1(y)}\right) [t f_0(y) + (1-t) f_1(y)],$$

then it is an easy exercise to show that $A(t)$ is also concave on $0 \leqq t \leqq 1$. Since $h(t)$ is concave, $g_N^N(t) = h(t)$ is concave, and hence $G_N^N(t)$ is concave. Hence by induction all the $g_n^N(t)$ and $G_n^N(t)$ are concave, as are therefore $g(t)$ and $G(t)$. Note also that

$$g(0) = G(0) = g(1) = G(1) = 0.$$

Now put

$$\alpha_1(t) = at - G(t) - 1,$$

$$\alpha_2(t) = b(1-t) - G(t) - 1,$$

$$\alpha(t) = h(t) - G(t) - 1 = \min[\alpha_1(t), \alpha_2(t)].$$

Then for $a, b > 1$,

$$\alpha_1(0) = \alpha_2(1) = -1 < 0,$$

$$\alpha_1(1) = a - 1 > 0,$$

$$\alpha_2(0) = b - 1 > 0.$$

Since $G(t)$ is concave, $G(0) = G(1) = 0$, and at is linear, there exists a unique number $\pi' = \pi'(a, b)$ such that

$$\alpha_1(t) \begin{cases} < 0 & \text{for } t < \pi' \\ = 0 & \text{for } t = \pi' \\ > 0 & \text{for } t > \pi' \end{cases} \qquad \left(\frac{1}{a} \leqq \pi' < 1\right).$$

Similarly, there exists a unique number $\pi'' = \pi''(a, b)$ such that

$$\alpha_2(t) \begin{cases} > 0 & \text{for } t < \pi'' \\ = 0 & \text{for } t = \pi'' \\ < 0 & \text{for } t > \pi'' \end{cases} \qquad \left(0 < \pi'' \leqq 1 - \frac{1}{b}\right).$$

Hence

$$\begin{aligned} s &= \text{first} \quad n \geqq 0 \quad \text{such that} \quad g(\pi_n) = h(\pi_n) \\ &= \text{first} \quad n \geqq 0 \quad \text{such that} \quad h(\pi_n) \leqq G(\pi_n) + 1 \\ &= \text{first} \quad n \geqq 0 \quad \text{such that either} \quad \alpha_1(\pi_n) \quad \text{or} \quad \alpha_2(\pi_n) \leqq 0 \\ &= \text{first} \quad n \geqq 0 \quad \text{such that} \quad \pi_n \leqq \pi' \quad \text{or} \quad \pi_n \geqq \pi''. \end{aligned}$$

If $\pi'' \leqq \pi'$ then $s \equiv 0$. If $\pi' < \pi''$ then s is the first $n \geqq 0$ for which π_n does not lies in the open interval (π', π''), and the decision procedure is a Wald sequential probability ratio test.

References

[1] Chow, Y. S., and H. Robbins: A martingale system theorem and applications. Proc. Fourth Berkeley Symposium on Math. Stat. and Prob. 1, 93–104 (1961).

[2] Loève, M.: Probability theory. Van Nostrand 1960.

[3] Arrow, K. J., D. Blackwell and M. A. Girshick: Bayes and minimax solutions of sequential decision problems. Econometrica 17, 213–244 (1949).

[4] Knopp, K.: Theory and applications of infinite series. Blackie 1928.

[5] MacQueen, J., and R. G. Hiller, Jr.: Optimal persistence policies. J. Oper. Res. Soc. America 8, 362–380 (1960).

[6] Derman, C., and J. Sacks: Replacement of periodically inspected equipment. Naval Res. Logist. Quart. 7, 597–607 (1960).

[7] Sakaguchi, M.: Dynamic programming of some sequential sampling designs. Jour. Math. Analysis and Applications 2, 446–466 (1961).

[8] Wald, A., and J. Wolfowitz: Optimum character of the sequential probability ratio test. Ann. Math. Stat. 19, 326–339 (1948).

[9] — — Bayes solutions of sequential decision problems. Ann. Math. Stat. 21, 82–99 (1950).

[10] Blackwell, D., and M. A. Girshick: Theory of games and statistical decisions. Wiley 1954.

[11] Weiss, L.: Statistical decision theory. McGraw-Hill 1960.

[12] Snell, J. L.: Application of martingale system theorems. Trans. Amer. Math. Soc. 73, 293–312 (1952).

Dept. of Mathematical Statistics, Columbia University
New York 27, N.Y. (USA)

(Received October 15, 1962)

Reprint from
ISRAEL JOURNAL OF MATHEMATICS
(Formerly: Bulletin of the Research Council of Israel, Section F)
Vol. 2, No. 2, June 1964

OPTIMAL SELECTION BASED ON RELATIVE RANK*
(the "Secretary Problem")

BY
Y. S. CHOW, S. MORIGUTI, H. ROBBINS AND S. M. SAMUELS

ABSTRACT
n rankable persons appear sequentially in random order. At the ith stage we observe the relative ranks of the first i persons to appear, and must either select the ith person, in which case the process stops, or pass on to the next stage. For that stopping rule which minimizes the expectation of the absolute rank of the person selected, it is shown that as $n \to \infty$ this tends to the value

$$\prod_{j=1}^{\infty}\left(\frac{j+2}{j}\right)^{1/j+1} \cong 3.8695.$$

1. **Introduction.** n girls apply for a certain position. If we could observe them all we could rank them absolutely with no ties, from best (rank 1) to worst (rank n). However, the girls present themselves one by one, in random order, and when the ith girl appears we can observe only her rank relative to her $i-1$ predecessors, that is, $1+$ the number of her predecessors who are better than she. We may either select the ith girl to appear, in which case the process ends, or reject her and go on to the $(i+1)$th girl; in the latter case the ith girl cannot be recalled. We must select one of the n girls. Let X denote the absolute rank of the girl selected. The values of X are $1, \cdots, n$, with probabilities determined by our selection strategy. What selection strategy (*i.e.* stopping rule) will minimize the expectation $EX =$ expected absolute rank of the girl selected?

To formulate the problem mathematically, let $x_1, \cdots, x_n$ denote a random permutation of the integers $1, \cdots, n$, all $n!$ permutations being equally likely. The integer 1 corresponds to the best girl, $\cdots, n$ to the worst. For any $i = 1, \cdots, n$ let $y_i = 1 +$ number of terms $x_1, \cdots, x_{i-1}$ which are $< x_i$ (y_i=relative rank of ith girl to appear). It is easy to see that the random variables $y_1, \cdots, y_n$ are independent, with the distribution

$$(1) \qquad P(y_i = j) = \frac{1}{i} \qquad (j = 1, \cdots, i),$$

and that

$$(2) \quad P(x_i = k \mid y_1 = j_1, \cdots, y_{i-1} = j_{i-1}, y_i = j) = P(x_i = k \mid y_i = j)$$
$$= \binom{k-1}{j-1}\binom{n-k}{i-j} \Big/ \binom{n}{i},$$

Received August 22, 1964
*Research supported by Office of Naval Research and Aerospace Research Laboratories. Reproduction in whole or in part is permitted for any purpose of the United States Government.

so that

$$E(x_i \mid y_i = j) = \sum_{k=1}^{n} kP(x_i = k \mid y_i = j) = \frac{n+1}{i+1} j. \tag{3}$$

For any stopping rule τ the expected absolute rank of the girl selected is therefore $EX = E\left(\frac{n+1}{\tau+1} y_\tau\right)$. We wish to minimize this value by optimal choice of τ.

To find an optimal τ by the usual method of backward induction we define for $i = 0, 1, \cdots, n-1$, $c_i = c_i(n) =$ minimal possible expected absolute rank of girl selected if we must confine ourselves to stopping rules τ such that $\tau \geqq i + 1$. We are trying to find the value c_0. Now

$$c_{n-1} = E\left(\frac{n+1}{n+1} y_n\right) = \frac{1}{n} \sum_{1}^{n} j = \frac{n+1}{2}, \tag{4}$$

and for $i = n-1, n-2, \cdots, 1$,

$$c_{i-1} = E\left(\min\left(\frac{n+1}{i+1} y_i, c_i\right)\right) = \frac{1}{i} \sum_{j=1}^{i} \min\left(\frac{n+1}{i+1} j, c_i\right). \tag{5}$$

These equations allow us to compute successively the values $c_{n-1}, c_{n-2}, \cdots, c_1, c_0$ and contain the implicit definition of an optimal stopping rule. Equation (5) can be rewritten more simply if we denote by $[x]$ the greatest integer $\leqq x$ and set

$$s_i = \left[\frac{i+1}{n+1} c_i\right] \qquad (i = n-1, \cdots, 1); \tag{6}$$

then (5) becomes

$$\begin{aligned} c_{i-1} &= \frac{1}{i}\left\{\frac{n+1}{i+1}(1 + 2 + \cdots + s_i) + (i - s_i) c_i\right\} \\ &= \frac{1}{i}\left\{\frac{n+1}{i+1} \cdot \frac{s_i(s_i+1)}{2} + (i - s_i) c_i\right\}. \end{aligned} \tag{7}$$

Defining $s_n = n$, an optimal stopping rule is, stop with the first $i \geqq 1$ such that $y_i \leqq s_i$; the expected absolute rank of the girl selected using this rule is c_0.

We observe from (4) and (5) that

$$c_0 \leqq c_1 \leqq \cdots \leqq c_{n-1} = \frac{n+1}{2}, \tag{8}$$

and from (6) and (8) that $s_i \leqq i$ and

$$s_1 \leqq s_2 \leqq \cdots \leqq s_{n-1} = \left[\frac{n}{2}\right]. \tag{9}$$

For example, let $n = 4$. Then from (4), (6),and(7),

$$c_3 = \frac{5}{2}, s_3 = 2, c_2 = \frac{1}{3}\left(\frac{5}{4}\cdot\frac{2.3}{2} + \frac{5}{2}\right) = \frac{25}{12}, s_2 = 1,$$

$$c_1 = \frac{1}{2}\left(\frac{5}{3}\cdot\frac{1.2}{2} + \frac{25}{12}\right) = \frac{15}{8}, \quad s_1 = 0, c_0 = c_1 = \frac{15}{8},$$

and an optimal stopping rule is given by the vector $(s_1, \cdots, s_4) = (0,1,2,4)$. The values of c_0 for $n = 10$, 100, 1000 are found by similar computation to be respectively 2.56, 3.60, 3.83.

D. V. Lindley [1] has treated this problem heuristically for large n by replacing (7) by a single differential equation. His results indicate that for $n \to \infty$, c_0 should approach a finite limit, but his method is too rough to give the value of this limit. A more adequate but still heuristic approach involves replacing (7) by an infinite sequence of differential equations, one for each value $0, 1, \cdots$ of s_i. This method indicates that $\lim c_0$ (all limits as $n \to \infty$) has the value

$$\prod_{j=1}^{\infty}\left(\frac{j+2}{j}\right)^{1/j+1} \cong 3.8695. \tag{10}$$

It is not clear how to make this heuristic argument rigorous by appealing to known theorems on the approximation of difference equations by differential equations. Instead, we shall give a direct proof that c_0 tends to the value (10).

2. The basic inequalities. We shall derive rather crude upper and lower bounds for the constants c_i which will permit the evaluation of $\lim c_0$. To this end we define the constants

$$t_i = \frac{i+1}{n+1}c_i \qquad (i = 0, 1, \cdots, n-1). \tag{1}$$

From (1.8) it follows that

$$t_0 < t_i < \cdots < t_{n-1} = \frac{n}{2} \tag{2}$$

and (1.7) becomes

$$t_{i-1} = \frac{s_i(s_i+1) + 2(i - s_i)t_i}{2(i+1)}, \qquad s_i = [t_i](i = 1, \cdots, n-1). \tag{3}$$

For fixed i set

$$t_i = s_i + \alpha, \qquad 0 \leqq \alpha < 1; \tag{4}$$

then (3) becomes

$$t_{i-1} = \frac{t_i(1 + 2i - t_i)}{2(i+1)} - \frac{\alpha(1-\alpha)}{2(i+1)}. \tag{5}$$

Put

$$T(x)=\frac{x(1+2i-x)}{2(i+1)}; \tag{6}$$

then for $x \leqq i+\frac{1}{2}$,

$$T'(x)=\frac{1+2i-2x}{2(i+1)} \geqq 0 \tag{7}$$

so $T(x)$ is increasing for $x \leqq i+\frac{1}{2}$, and by (5),

$$t_{i-1} \leqq T(t_i). \tag{8}$$

We now prove the first basic inequality.

LEMMA 1.

$$t_i \leqq \frac{2n}{n-i+3} \qquad (i=0,\cdots,n-1). \tag{9}$$

Proof. (9) is true for $i=n-1$ since by (2)

$$t_{n-1}=\frac{n}{2}=\frac{2n}{4}.$$

Assume (9) holds for some $1 \leqq i \leqq n-1$; we shall prove that it holds for $i-1$. We know by (1.8) and (1) that

$$t_{i-1}=\frac{i}{n+1}c_{i-1} \leqq \frac{i}{n+1}\cdot\frac{n+1}{2}=\frac{i}{2}. \tag{10}$$

If $\frac{i}{2} \leqq \frac{2n}{n-i+4}$ then (9) holds for $i-1$. If $\frac{i}{2}>\frac{2n}{n-i+4}$ then $i+\frac{1}{2} \geqq \frac{2n}{n-i+3}$, so by (8), since $T(x)$ is increasing for $x \leqq i+\frac{1}{2}$,

$$t_{i-1} \leqq T(t_i) \leqq T\left(\frac{2n}{n-i+3}\right)=\frac{n\{(1+2i)(n-i+3)-2n\}}{(i+1)(n-i+3)^2} \leqq \frac{2n}{n-i+4}, \tag{11}$$

since the last inequality is equivalent to

$$(n-i+3)^2+(2n-2i-1)(n-i+3)+2n \geqq 0,$$

which is true for $n-i \geqq 1$. Hence (9) holds for $i-1$ in this case also, and the lemma is proved.

COROLLARY 1.

$$c_0<8 \qquad (n=1,2,\cdots). \tag{12}$$

Proof. In (9) set $i=\left[\frac{n}{2}\right]$. Then

$$c_0 \leqq c_i=\frac{n+1}{i+1}t_i \leqq \frac{n+1}{i+1}\cdot\frac{2n}{n-i+3} \leqq \frac{2n(n+1)}{\frac{n}{2}\left(\frac{n}{2}+3\right)}<8. \tag{13}$$

We next observe that

$$t_{i-1} \geqq \frac{i}{i+1}\left\{1 - \frac{t_i}{2(i+1)}\right\} \qquad (i = 1, \cdots, n-1). \tag{14}$$

This inequality follows from (5) if we show that

$$\frac{t_i(1 + 2i - t_i)}{2(i+1)} - \frac{\alpha(1-\alpha)}{2(i+1)} \geqq \frac{i}{i+1} t_i \left\{1 - \frac{t_i}{2(i+1)}\right\},$$

which reduces to

$$t_i\left(1 - \frac{t_i}{i+1}\right) \geqq \alpha(1-\alpha) \qquad (\alpha = t_i - [t_i]), \tag{15}$$

and this is true if $t_i \geqq 1$, since $t_i \leqq \dfrac{i+1}{2}$ by (10), and if $t_i < 1$, since then $t_i = \alpha$.

We now establish the second basic inequality.

LEMMA 2.

$$t_i \geqq \frac{3(i+1)}{2(n-i+2)} \qquad (i = 0, \cdots, n-1). \tag{16}$$

Proof. (16) is true for $i = n-1$; suppose it is true for some $1 \leqq i \leqq n-1$. Define

$$T(x) = x\left\{1 - \frac{x}{2(i+1)}\right\},$$

which is increasing for $x \leqq i+1$. Since by (10) $t_i \leqq \dfrac{i+1}{2}$, we have by (14),

$$t_{i-1} \geqq \frac{i}{i+1} T(t_i) \geqq \frac{i}{i+1} T\left(\frac{3(i+1)}{2(n-i+2)}\right)$$

$$= \frac{3i(4n-4i+5)}{8(n-i+2)^2} \geqq \frac{3i}{2(n-i+3)}.$$

which is equivalent to $i \leqq n-1$. This proves the lemma.

We have seen (1.9) that for any positive integer k, if $n \geqq 2k$ then $s_{n-1} \geqq k$. We now define for each $k = 1, 2, \cdots$ and each $n \geqq 2k$,

$$i_k = \text{smallest integer } j \geqq 1 \text{ such that } s_j \geqq k. \tag{17}$$

We note that $s_{i_1 - 1} = 0$ and hence from (1.7),

$$c_0 = c_1 = \cdots = c_{i_1 - 1}. \tag{18}$$

COROLLARY 2.

(19) $$\underline{\lim}\ \frac{i_1}{n} \geqq \frac{1}{8}.$$

Proof. If $i_1 > [n/2]$ then $i_1 \geqq [n/2] + 1 > n/2$, $(i_1/n) > \frac{1}{2}$. If $i_1 \leqq [n/2]$ then by (13),

$$1 \leqq s_{i_1} \leqq t_{i_1} = \frac{i_1+1}{n+1} c_{i_1} \leqq \frac{i_1+1}{n+1} c_{[n/2]} < \frac{i_1+1}{n+1} \cdot 8.$$

(We remark that $i_1 > 1$ for $n > 2$, since if $i_1 = 1$ then $s_1 = 1$ and $c_0 = \dfrac{n+1}{2}$, which only holds for $n \leqq 2$.)

COROLLARY 3. *On every set*

(20) $$\left\{\alpha \leqq \frac{i}{n} \leqq \beta;\ 0 < \alpha < \beta < 1\right\},$$

(21) $$\lim (t_i - t_{-1}) = 0$$

uniformly.

Proof. From (14) and (9),

$$0 \leqq t_i - t_{i-1} \leqq t_i - \frac{i}{i+1} t_i \left\{1 - \frac{t_i}{2(i+1)}\right\} = \frac{t_i}{i+1} + \frac{it_i^2}{2(i+1)^2}$$

$$\leqq \frac{(1+t_i)^2}{2(i+1)} \leqq \frac{\left(1 + \dfrac{2n}{n-i}\right)^2}{2(i+1)} \leqq \frac{\left(1 + \dfrac{2}{1-\beta}\right)^2}{2\alpha} \cdot \frac{1}{n} \to 0.$$

COROLLARY 4. *For* $k = 1, 2, \cdots$ *and* $n \geqq 12k$,

(22) $$\frac{i_k}{n} \geqq 1 - \frac{2}{k},$$

(23) $$\frac{i_k}{n} \leqq 1 - \frac{1}{2k}.$$

Proof. By (9),

$$s_{i_k} \geqq k \Rightarrow t_{i_k} \geqq k \Rightarrow \frac{2n}{n-i_k} \geqq k \Rightarrow \frac{i_k}{n} \geqq 1 - \frac{2}{k},$$

which proves (22). (23) holds if $i_k \leqq \left[\dfrac{n}{2}\right]$, for then

$$\frac{i_k}{n} \leqq \frac{1}{2} \leqq 1 - \frac{1}{2k},$$

and if $i_k > \left[\frac{n}{2}\right]$ then by (16),

$$s_{i_k-1} < k \Rightarrow t_{i_k-1} < k \Rightarrow \frac{3i_k}{2(n-i_k+3)} < k \Rightarrow \frac{3i_k}{2k} < n - i_k + 3 \Rightarrow \frac{3n}{4k} < n - i_k + 3$$

$$\Rightarrow \frac{3}{4k} < 1 - \frac{i_k}{n} + \frac{3}{n} \Rightarrow \frac{i_k}{n} < 1 - \frac{3}{4k} + \frac{3}{n} \leqq 1 - \frac{1}{2k} \text{ for } n \geqq 12k.$$

COROLLARY 5.

$$\lim t_{i_k} = \lim t_{i_k-\gamma} = k \qquad (k, \gamma = 1,2,\cdots). \tag{24}$$

Proof. $t_{i_k-\gamma} < k \leqq t_{i_k}$.

Choose α, β so that $0 < \alpha < \frac{1}{8}$, $1 - \frac{1}{2k} < \beta < 1$. Then by (19) and (23),

$$\alpha < \frac{i_k - \gamma}{n} < \frac{i_k}{n} < \beta \tag{25}$$

for sufficiently large n. Hence by Corollary 3,

$$\lim (t_{i_k} - t_{i_k-\gamma}) = 0.$$

COROLLARY 6. *For* $k = 1,2,\cdots$

(26) $s_{i_k} = k$ *for sufficiently large* n,

(27) $\lim (i_{k+1} - i_k) = \infty$.

Proof. $k \leqq s_{i_k} \leqq t_{i_k}$ together with (24) proves (26).

$$\lim (t_{i_{k+1}} - t_{i_k}) = 1 \text{ and } \lim (t_{i_{k+1}} - t_{i_{k+1}-\gamma}) = 0$$

by (24), and these relations imply (27).

3. **Proof of the Theorem.** Choose and fix a positive integer k and let n by (2.26) be so large that $s_{i_k} = k$, $s_{i_{k+1}} = k+1$. For $i_k \leqq i < i_{k+1}$ define

$$v_i = t_i - \frac{k}{2}. \tag{1}$$

Substituting in (2.3) we find that

$$v_{i-1} + \frac{k}{2} = \frac{k(k+1) + 2(i-k)\left(v_i + \frac{k}{2}\right)}{2(i+1)} = \frac{k}{2} + \frac{i-k}{i+1} v_i,$$

$$v_i = \frac{i+1}{i-k} v_{i-1}. \tag{2}$$

Hence for $i_k < i < i_{k+1}$,

$$(3)\qquad v_i = \frac{i+1}{i-k} v_{i-1} = \frac{i+1}{i-k}\,\frac{i}{i-k-1}\cdots\frac{i_k+2}{i_k-k+1} v_{i_k}$$

$$= \frac{i+1}{i_k+1}\cdot\frac{i}{i_k}\cdots\frac{i+1-k}{i_k+1-k}\cdot v_{i_k} = v_{i_k}\prod_{j=1}^{k+1}\left(\frac{i+j-k}{i_k+j-k}\right),$$

and hence

$$(4)\qquad t_i = \frac{k}{2} + \left(t_{i_k} - \frac{k}{2}\right)\prod_{j=1}^{k+1}\left(\frac{i+j-k}{i_k+j-k}\right).$$

Set $i = i_{k+1} - 1$; then

$$(5)\qquad t_{i_{k+1}-1} = \frac{k}{2} + \left(t_{i_k} - \frac{k}{2}\right)\prod_{j=1}^{k+1}\left(\frac{i_{k+1}+j-k-1}{i_k+j-k}\right).$$

From (2.19) and (2.24) it follows that

$$k+1 = \frac{k}{2} + \frac{k}{2}\prod_{j=1}^{k+1}\lim\left(\frac{i_{k+1}+j-k-1}{i_k+j-k}\right) = \frac{k}{2} + \frac{k}{2}\lim\left(\frac{i_{k+1}}{i_k}\right)^{k+1}$$

and hence

$$(6)\quad \lim\frac{i_{k+1}}{i_k} = \left(\frac{k+2}{k}\right)^{1/k+1},\ \lim\frac{i_1}{i_k} = \lim\frac{i_1/n}{i_k/n} = \prod_{j=1}^{k-1}\left(\frac{j}{j+2}\right)^{1/j+1}.$$

From (2.22)

$$(7)\quad \left(1-\frac{2}{k}\right)\prod_{j=1}^{k-1}\left(\frac{j}{j+2}\right)^{1/j+1} \le \underline{\lim}\,\frac{i_1}{n} \le \overline{\lim}\,\frac{i_1}{n} \le \prod_{j=1}^{k-1}\left(\frac{j}{j+2}\right)^{1/j+1}.$$

Letting $k\to\infty$,

$$(8)\qquad \lim\frac{i_1}{n} = \prod_{j=1}^{\infty}\left(\frac{j}{j+2}\right)^{1/j+1}.$$

Now by (2.24) and (2.18),

$$1 = \lim t_{i_1-1} = \lim\left(\frac{i_1}{n+1}c_{i_1-1}\right) = \lim\left(\frac{i_1}{n}c_0\right)$$

$$(9)\qquad = \lim c_0\cdot\prod_{j=1}^{\infty}\left(\frac{j}{j+2}\right)^{1/j+1}.$$

Thus we have proved the

THEOREM. $$\lim c_0 = \prod_{j=1}^{\infty}\left(\frac{j+2}{j}\right)^{1/j+1}.$$

4. Remarks.

1. It is interesting to note that $c_0 = c_0(n)$ is a strictly increasing function of n; thus in view of the Theorem,

$$c_0(n) < 3.87 \qquad (n = 1, 2, \cdots). \tag{1}$$

A direct proof that $c_0(n)$ is strictly increasing based on the formulas of Section 1 is difficult, since there is no obvious relation between the c_i for different values of n. However, a direct probabilistic proof can be given which involves no use of the recursion formulas. Let $(s_1, \cdots, s_n, n+1)$ be any stopping rule for the case of $n+1$ girls such that $s_i < i$ for $i = 1, \cdots, n$ (any optimal stopping rule has this property). Define for $i = 1, \cdots, n$

$$t_j(i) = \begin{cases} s_j & \text{for } j = 1, \cdots, i-1, \\ s_{j+1} & \text{for } j = i, \cdots, n, \end{cases} \tag{2}$$

and

$$t_j(n+1) = \begin{cases} s_j & \text{for } j = 1, \cdots, n-1, \\ n & \text{for } j = n. \end{cases} \tag{3}$$

It is easy to see that at least one of the stopping rules defined by $(t_1(i), \cdots, t_n(i))$, $i = 1, \cdots, n+1$ must give a value of c_0 for the case n which is less than that given by $(s_1, \cdots, s_n, n+1)$ for the case $n+1$. Hence $c_0(n) < c_0(n+1)$.

2. We assumed that the n girls appear in random order, all $n!$ permutations being equally likely. The minimal expected absolute rank of the girl chosen is then $c_0 < 3.87$ for all n. Suppose now that the order in which the girls are to appear is determined by an opponent who wishes to maximize the expected absolute rank of the girl we choose. No matter what he does, by choosing at random the first, second, $\cdots$, last girl to appear we can achieve the value

$$EX = (1 + 2 + \cdots + n)/n = (n+1)/2.$$

And in fact *there exists an opponent strategy such that, no matter what stopping rule we use,* $EX = (n+1)/2$. Let $x_1 = 1$ or n, each with probability 1/2; let x_{i+1} = either the largest or the smallest of the integers remaining after $x_1, \cdots, x_i$ have been chosen, each with probability 1/2. If we define for $i = 1, \cdots, n$

$$z_i = E(x_i \mid y_1, \cdots, y_i), \quad \mathcal{B}_i = \mathcal{B}(y_1, \cdots, y_i), \tag{4}$$

then it is easy to see that $\{z_i, \mathcal{B}_i (i = 1, \cdots, n)\}$ is a martingale, so that for any stopping rule τ,

$$E(X) = E(z_\tau) = E(z_1) = (n+1)/2. \tag{5}$$

Many extensions and generalizations of the problem considered in this paper suggest themselves at once. Some further results will be presented elsewhere.

Reference

1. D. V. Lindley, *Dynamic programming and decision theory*, Applied Statistics **10** (1961), 39–51.

Purdue University
University of Tokyo
Columbia University
Purdue University

OPTIMAL STOPPING

HERBERT ROBBINS, Columbia University

The theory of probability began with efforts to calculate the odds in games of chance. In this context, optimal stopping problems concern the effect on a gambler's fortune of various possible systems for deciding when to stop playing a sequence of games. Such problems are of interest in statistics, where the experimenter must constantly ask whether the increase in information contained in further data will outweigh the cost of collecting it.

Optimal stopping theory provides a general mathematical framework in which such problems can be precisely formulated and in some cases solved completely. The examples which we shall consider here are of a simpler nature than those arising in statistics, but will serve to illustrate some of the problems that arise in the general theory.

EXAMPLE 1. A fair coin is tossed repeatedly. After each toss we have the option of stopping or going on to the next toss, our decision at each stage being allowed to depend on the outcome thus far. We must stop after some finite (but not necessarily preassigned) number of tosses, and it is agreed that if we stop after the nth toss we are to receive a reward x_n which is a given function of the outcomes of the first n tosses. When should we stop so as to maximize our expected reward?

Let us introduce random variables $y_1, y_2, \cdots$ to represent the successive tosses, the y_i being independent with the common probability distribution $P(y_i=1)=P(y_i=-1)=1/2$; $y_i=1$ denoting heads on the ith toss and $y_i=-1$ tails. The reward sequence will then consist of a sequence of functions $x_1, x_2, \cdots$, where $x_n=f_n(y_1, \cdots, y_n)$. A *stopping rule* is then a random variable t with values in the set $\{1, 2, 3, \cdots\}$ and such that the event $[t=n]$ depends solely on the values of $y_1, \cdots, y_n$ and not on future values $y_{n+1}, \cdots$. Using a stopping rule t our reward x_t will be a random variable whose expectation Ex_t measures the performance on the average of the stopping rule t. The supremum $V=\sup\{Ex_t\}$ over the class C of all possible stopping rules t for which Ex_t exists is called the *value* of the sequence $\{x_n\}$, and if a stopping rule t exists such that $Ex_t=V$, t is said to be *optimal*.

Prof. Robbins earned his Harvard Ph.D. under Hassler Whitney. He followed an assistantship under Marston Morse at the Institute and an NYU instructorship with a four-year career in the U. S. Navy. He then was Associate Prof. and Prof. at the Univ. of N. Carolina before assuming his present position at Columbia. He spent leaves at Berkeley, Minnesota, Purdue, Michigan, and I.A.S.

He has served as President of the Institute of Math. Stat. and delivered the Rietz and Wald lectures. He was a Guggenheim Fellow in 1952–53.

Since his thesis in topology, Robbins's main research has been in probability theory and mathematical statistics. His book (with Y. S. Chow and D. O. Siegmund) *Great Expectations; The Theory of Optimal Stopping* will be published shortly by Houghton Mifflin. His previous *What is Mathematics?* with Richard Courant (Oxford Univ. Press, 1941) is a landmark in mathematical exposition. *Editor.*

Thus far we have said nothing about the nature of the reward sequence $x_n = f_n(y_1, \cdots, y_n)$. To begin with, let us take

$$(1) \qquad x_n = \frac{n2^n}{n+1} \cdot \prod_1^n \left(\frac{y_i + 1}{2}\right) \qquad (n = 1, 2, \cdots)$$

and analyze the resulting situation. Equation (1) is just a symbolic way of saying that if we stop after the nth toss with all heads we are to receive $n\, 2^n/(n+1)$, while if any one of the first n tosses has been a tail we are to receive nothing. A little reflection will show that we need consider only the class of stopping rules $\{t_k\}$, $k = 1, 2, \cdots$, where $t_k = k$; i.e., t_k stops after the kth toss no matter what sequence of heads and tails has appeared. Clearly

$$Ex_{t_k} = \frac{1}{2^k} \cdot \frac{k2^k}{k+1} + \left(1 - \frac{1}{2^k}\right) \cdot 0 = \frac{k}{k+1},$$

and therefore $V = 1$ but no optimal stopping rule exists.

We remark in passing that at any stage n in which all heads have appeared, so that $x_n = n\, 2^n/(n+1)$, the conditional expected reward for making one more toss before stopping is

$$E\left(x_{n+1} \middle| x_n = \frac{n2^n}{(n+1)}\right) = \frac{1}{2} \cdot \frac{(n+1)2^{n+1}}{(n+2)} = \frac{2^n(n+1)}{(n+2)} > x_n.$$

Hence it is always "foolish" to stop with all heads. But if we do not act "foolishly" at some point we shall wait for the first tail to occur and our final reward will always be 0. Thus acting "wisely" at each stage is the worst long-range policy.

EXAMPLE 2. The same, except that now

$$(2) \qquad x_n = \frac{y_1 + \cdots + y_n}{n} \qquad (n = 1, 2, \cdots).$$

This problem is much harder than the preceding one because of the enormous family of possible stopping rules which must be considered and evaluated. A simple instance of such a rule is

$$(3) \qquad t = \begin{cases} 1 & \text{if } y_1 = 1, \text{ otherwise} \\ n & \text{if } n \text{ is the first integer such that } y_1 + \cdots + y_n = 0. \end{cases}$$

The first question we must ask is, does (3) define a legitimate stopping rule in the sense that $P(t < \infty) = 1$? In probability theory it is shown that in repeated tossings of a fair coin, for any fixed integer $k = 0, \pm 1, \pm 2, \cdots$, the probability is 1 that the difference (number of heads in first n tosses) $-$ (number of tails in first n tosses) will assume the value k infinitely often. In our notation the case $k = 0$ corresponds to the event $y_1 + \cdots + y_n = 0$, and hence $P(t < \infty) = 1$. It remains to evaluate Ex_t for (2) and (3). We have

$$Ex_t = \frac{1}{2}\cdot\frac{1}{1} + \frac{1}{2}\cdot 0 = \frac{1}{2}.$$

It follows that $V \geqq \frac{1}{2}$, and of course $V \leqq 1$, since $x_n \leqq 1$ in all cases. By trial and error we can invent other stopping rules t for which $Ex_t > \frac{1}{2}$, but we will find none for which $Ex_t \geqq .9$, say. However, it is not so easy to *prove* that $V < .9$, still less to find the exact value of V and to determine whether an optimal t exists in this example.

In cases like this it is tempting to try to "put the problem on a computer." Briefly, here is what a computer can do. Suppose we restrict ourselves to the class C_N of all possible stopping rules t which take on only values in the set $\{1, 2, \cdots, N\}$, where N is some fixed positive integer; in other words, we restrict ourselves to stopping rules which always stop after at most N tosses of the coin. Denoting by v_N the supremum of Ex_t over the class C_N, we see that in this case C_N is a finite class and therefore an optimal rule in C_N must exist. Even so, for $N = 1000$, say, the class C_N is still too large for an exhaustive analysis. At this point the general theory of optimal stopping comes to our aid. Whenever the number of stages in the problem is bounded (even if the random variables involved have continuous rather than discrete distributions) there is always an optimal rule and a more or less constructive algorithm for finding it (cf. (22)). In the present problem a computer can actually be programmed to find v_N quite quickly for all N up to a few thousand. By definition the sequence v_N is nondecreasing, and therefore $v = \lim_{N\to\infty} v_N$ exists. But the values of v_N produced by a computer for successive values of N display no obvious pattern, and it is impossible even to guess the exact value of v from computer evidence, *still less to decide whether* $V = v$ *or* $V > v$.

In the present problem it has been proved [6, 11] that an optimal t does exist and that $v = V$, but the exact description of t and the value of V are not known.

EXAMPLE 3. As in Example 1 but now with

$$x_n = \min(1, y_1 + \cdots + y_n) - n/(n+1) \qquad (n \geqq 1).$$

Consider the stopping rule

$$t = \text{first } n \geqq 1 \text{ such that } y_1 + \cdots + y_n = 1. \tag{4}$$

That $P(t < \infty) = 1$ follows as in Example 2, and since $n/(n+1) < 1$,

$$Ex_t = 1 - E\left(\frac{t}{t+1}\right) > 0$$

(the exact value of Ex_t would require some probability theory to compute). A little thought will show that t is in fact optimal for this example and hence that $V = Ex_t > 0$.

On the other hand, since the y_i are independent and identically distributed with $Ey_i = \frac{1}{2}\cdot 1 + \frac{1}{2}\cdot(-1) = 0$, Wald's lemma (see below) shows that if, unlike (4), t is any stopping rule for which $Et < \infty$, and hence in particular if $t \in C_N$ for some $N = 1, 2, \cdots$, then $E(y_1 + \cdots + y_t) = 0$, so that

$$Ex_t \leqq E(y_1 + \cdots + y_t) - E\left(\frac{t}{t+1}\right) \leqq -\frac{1}{2}.$$

Hence, $v_N \leqq -\frac{1}{2}$ for all $N \geqq 1$, and hence $v = \lim_{N\to\infty} v_N \leqq -1/2$, while as we have seen $V > 0$.

A slight variation on this example is given by

EXAMPLE 4. Let $y_1, y_2, \cdots$ be independent (but not identically distributed) random variables such that

$$P(y_i = 1 - a_i) = P(y_i = -1 - a_i) = 1/2, \quad \text{with } a_i = 1/i(i+1), \tag{5}$$

and let

$$x_n = y_1 + \cdots + y_n \quad (n \geqq 1) \tag{6}$$

be the reward if we stop at the nth stage. Thus x_n represents the net gain of a gambler who plays a succession of unfavorable games,

$$Ey_i = \tfrac{1}{2}(1 - a_i) + \tfrac{1}{2}(-1 - a_i) = -a_i < 0,$$

and stops after the nth. It might seem that whatever stopping rule t he might use, his expected net gain after stopping, Ex_t, would be <0. However, let

$$t = \text{first } n \geqq 1 \text{ such that } \sum_1^n (y_i + a_i) = k, \tag{7}$$

where k is any preassigned positive integer. The argument in Example 2 again shows that $P(t < \infty) = 1$, while by (6) and (7)

$$Ex_t = E\left(\sum_1^t y_i\right) = k - E\left(\sum_1^t a_i\right) = k - E\left(\sum_1^t \left(\frac{1}{i} - \frac{1}{i+1}\right)\right)$$

$$= k - E\left(\frac{t}{t+1}\right) > k - 1.$$

Since k can be as large as we please, we see that $V = +\infty$ for this example. (The reader may decide whether there is an optimal t in this case, i.e., one for which $Ex_t = +\infty$.) We remark that although the one-step conditional expected reward $E(x_{n+1} \mid x_n) = x_n + E(y_{n+1})$ is always *less* than the present reward x_n, the use of a proper stopping rule makes the game profitable.

This example suggests the following question. If $y_1, y_2, \cdots$ are independent *and identically distributed* with $Ey_i < 0$, does there exist a stopping rule t such that $E(\sum_1^t y_i) > 0$? We shall now show that there does not.

We shall need the strong law of large numbers of probability theory.

THEOREM. (Kolmogorov) *Let $y_1, y_2, \cdots$ be independent and identically distributed random variables with the common distribution function $F(y)=P(y_i<y)$, and put $x_n=\sum_1^n y_i$. If*

$$Ey_i = \int_0^\infty ydF(y) + \int_{-\infty}^0 ydF(y) \tag{8}$$

exists (i.e., if the two integrals in (8) are not respectively $+\infty$ and $-\infty$), denote it by μ. Then

(a) *If μ exists, $-\infty \leqq \mu \leqq \infty$, then $P(\lim_{n\to\infty}(x_n/n)=\mu)=1$.*

(b) *If for some finite constant c,*

$$P\left(\lim_{n\to\infty}\frac{x_n}{n}=c\right)=1, \tag{9}$$

then μ exists and equals c.

We remark parenthetically that (9) may hold with $c=+\infty$ or $-\infty$ even though μ does not exist.

Now let t be any stopping rule for an independent and identically distributed sequence $y_1, y_2, \cdots$ for which $\mu=Ey_i$ exists, $-\infty\leqq\mu\leqq\infty$, and consider the randomly stopped sum $x_t=\sum_1^t y_i$. For any $n\geqq 1$ let

$$\begin{aligned} S_n = (y_1+\cdots+y_{t_1}) + (y_{t_1+1}+\cdots+y_{t_1+t_2}) + \cdots \\ + (y_{t_1+\cdots+t_{n-1}+1}+\cdots+y_{t_1+\cdots+t_n}), \end{aligned}$$

where $t_1=t$, t_2 is t applied to the sequence $y_{t_1+1}, y_{t_1+2}, \cdots$, etc. It is easy to see that the n groups of terms in S_n are in fact n independent random variables with the same probability distribution as x_t, and that $t_1, t_2, \cdots$ are independent random variables with the same distribution as t. Since $\mu=Ey_i$ exists by hypothesis and since Et always exists (since $t\geqq 1$), it follows from (a) that with probability 1 as $n\to\infty$

$$\frac{S_n}{n} = \left\{\frac{y_1+\cdots+y_{t_1+\cdots+t_n}}{t_1+\cdots+t_n}\right\}\cdot\left\{\frac{t_1+\cdots+t_n}{n}\right\} \to \mu\cdot Et \tag{10}$$

provided only that the last expression is not of the form $0\cdot\infty$. Applying (a) again to $x_t=\sum_1^t y_i$ we have [15] the

COROLLARY. *If $y_1, y_2, \cdots$ are independent and identically distributed, if $\mu=Ey_i$ exists, and if t is any stopping time of the sequence $y_1, y_2, \cdots$ for which $E(\sum_1^t y_i)$ exists, then*

$$E\left(\sum_1^t y_i\right) = \mu\cdot Et \tag{11}$$

whenever the right side of (11) is not of the form $0\cdot\infty$.

From (b) and (10) we obtain

WALD'S LEMMA [18]. *If μ and Et are both finite then $E(\sum_1^t y_i)$ always exists and satisfies* (11).

Wald's Lemma admits of many generalizations. For example, if $y_1, y_2, \cdots$ are independent but not necessarily identically distributed random variables for which $E|y_i| \leqq C < \infty$ for all $i \geqq 1$, if $\mu_i = Ey_i$, and if t is any stopping rule of the sequence $y_1, y_2, \cdots$ such that $Et < \infty$, then $E(\sum_1^t y_i)$ exists and $E(\sum_1^t y_i) = E(\sum_1^t \mu_i)$. Applied to Example 4 in which $|y_i| \leqq 2$ we see that if in that example t is any stopping time of the sequence $y_1, y_2, \cdots$ such that $Et < \infty$, then

$$Ex_t = E\left(\sum_1^t \mu_i\right) = -E\left(\sum_1^t a_i\right) = -E\left(\frac{t}{t+1}\right) \leqq -\frac{1}{2}.$$

Thus to achieve $Ex_t > -1/2$ as we did with (7), we have to use a t for which $Et = \infty$. This detracts somewhat from the prospect of getting rich in the long run by using (7) or something like it.

We are now able to justify our negative answer to the question raised in the paragraph following Example 4. The Corollary shows that if $y_1, y_2, \cdots$ are independent and identically distributed with $Ey_i = \mu < 0$, and if t is any stopping rule for which $E(\sum_1^t y_i)$ exists (even though Et may be $+\infty$), then

$$E\left(\sum_1^t y_i\right) = \mu \cdot Et < 0; \tag{12}$$

the optimal stopping rule is therefore $t \equiv 1$, and $V = \mu < 0$. (We remark that there may exist t's such that $E(\sum_1^t y_i)$ does not exist, but such t's are excluded from consideration under our definition of optimality.)

Our next example is of a different character from the preceding ones in that it can be supposed to apply to certain problems in real life rather than to imaginary gambling situations.

EXAMPLE 5. Let $y_1, y_2, \cdots$ be independent random variables with a common distribution function $F(y) = P(y_i < y)$, and let

$$x_n = \max(y_1, \cdots, y_n) - cn \qquad (n \geqq 1) \tag{13}$$

where c is some positive constant. What stopping rule t maximizes Ex_t? A complete solution to this question is given by

(i) If $\int_0^\infty y\,dF(y) = \infty$ then an optimal stopping rule is

$$t = \text{first } n \geqq 1 \text{ such that } y_n \geqq b \ (b \text{ any finite constant}), \tag{14}$$

and $V = Ex_t = +\infty$, while

(ii) if $\int_0^\infty y\,dF(y) < \infty$ there exists a unique number β such that

$$\int_\beta^\infty (y - \beta)\,dF(y) = c; \tag{15}$$

an optimal stopping rule is

$$t = \text{first } n \geqq 1 \text{ such that } y_n \geqq \beta, \tag{16}$$

and $V = Ex_t = \beta$.

Proof. (i) Let $p = p(y_i \geqq b) = 1 - F(b) > 0$ and $q = 1 - p$. Then for (14)

$$P(t = n) = pq^{n-1}, \qquad P(t < \infty) = \sum_1^\infty pq^{n-1} = p \cdot \frac{1}{1-q} = 1,$$

$$Et = \sum_1^\infty npq^{n-1} = p \frac{d}{dq}\left(\sum_0^\infty q^n\right) = p \cdot \frac{1}{(1-q)^2} = \frac{1}{p}, \tag{17}$$

and

$$E \max(y_1, \cdots, y_t) = \frac{\int_b^\infty y dF(y)}{p} = \infty.$$

Hence by (17)

$$Ex_t = \infty - \frac{c}{p} = \infty,$$

so t is necessarily optimal.

(ii) Consider the function

$$\phi(b) = \int_b^\infty (y - b) dF(y),$$

which is equal to the area of the region in a y, z-plane bounded below by the curve $z = F(y)$, on the left by the line $y = b$, and above by the line $z = 1$. It is geometrically evident that there is a unique solution of (15) for any given $c > 0$, and for t defined by (16) we have as in (i)

$$Et = \frac{1}{p}, \qquad Ex_t = \left(\int_\beta^\infty y dF(y) - c\right)/p \qquad (p = 1 - F(\beta)).$$

Hence by (15)

$$Ex_t = \beta. \tag{18}$$

To prove that the t defined by (16) is optimal, let t' be any stopping rule such that $Ex_{t'}$ exists and is $> -\infty$. Choose any $b > \beta$ and observe that

$$x_n = \max(y_1, \cdots, y_n) - cn \leqq b + \sum_1^n ((y_i - b)^+ - c), \tag{19}$$

where we define $a^+ = \max(a, 0)$. The sequence of random variables $w_1, w_2, \cdots$, where $w_i = (y_i - b)^+ - c$, is independent and identically distributed with $\mu = Ew_i = \int_{-\infty}^{\infty}((y-b)^+ - c)dF(y) = \phi(b) - c < 0$ since $b > \beta$ and β satisfies (15). Now $Ex_{t'} > -\infty$ by hypothesis, so by (19)

$$E\left(\sum_1^{t'} w_i\right) \geqq Ex_{t'} - b > -\infty, \tag{20}$$

and hence by Corollary 1, whether Et' be finite or infinite,

$$E\left(\sum_1^{t'} w_i\right) = \mu Et' < 0; \tag{21}$$

thus by (20) and (21) $Ex_{t'} < b$. Since b was any constant $> \beta$ this implies that $Ex_{t'} \leqq \beta$ and hence by (18) that t is optimal and $V = \beta$.

The original proof of this result [4] along lines suggested by the general theory was considerably more complicated and was valid only under the unnecessary hypothesis that $\int_0^\infty y^2 dF(y) < \infty$. The direct proof just given is based on the trivially simple inequality (19) which reduces the problem from one involving the maximum of $y_1, \cdots, y_n$ to one involving a simple sum. This inequality was pointed out to the author by David Burdick.

We remark that the optimal stopping rule (16) in case (ii) is suggested by the one-step argument which would have led us astray in Examples 1 and 4 (and in Example 3, also for that matter). For, putting $m_n = \max(y_1, \cdots, y_n)$ we have

$$\begin{aligned} x_{n+1} &= m_{n+1} - c(n+1) = \max(m_n, y_{n+1}) - c(n+1) \\ &= m_n + (y_{n+1} - m_n)^+ - c(n+1) \\ &= x_n + (y_{n+1} - m_n)^+ - c, \end{aligned}$$

so that $E(x_{n+1} | y_1, \cdots, y_n) = x_n + E(y_{n+1} - m_n)^+ - c > x_n$ if and only if

$$\int_{m_n}^{\infty} (y - m_n)dF(y) > c;$$

i.e., if and only if $m_n < \beta$.

We remark also that if we replace (13) by $\bar{x}_n = y_n - cn$ $(n \geqq 1)$, the optimal solution remains the same, since

$$\bar{x}_n \leqq x_n \quad \text{but} \quad \bar{x}_t = x_t$$

with t defined by (14) or (16).

The general theory of optimal stopping is more than a collection of examples and *ad hoc* methods of solution. It draws heavily on martingale theory and the other apparatus of modern probability theory, to which it also contributes. Rather than sketch the general methods in a necessarily incomplete form we have preferred to present a few simple examples in the hope that the reader will be induced to consult some of the references below for a more formal presentation.

Our last example will be one in which the number of random variables is finite from the outset and hence no difficulties of principle arise concerning the existence of an optimal solution, but the explicit result is none the less surprising.

EXAMPLE 6. An employer interviews a finite number N of applicants for a position. They are Miss 1, Miss 2, $\cdots$, Miss N, in decreasing order of excellence, and they are interviewed in random order, each of the $N!$ permutations being equally likely. The rules of the game are as follows. After each interview the employer can rank the girl just interviewed relative to her predecessors but is ignorant of her absolute rank relative to the whole group of N. Thus if the first few girls in order of appearance were Misses

$$7, 3, 9, 4, \cdots$$

the employer would know successively that

the first girl has relative rank 1
the second has relative rank 1
the third has relative rank 3
the fourth has relative rank 2, etc.

The employer must hire one applicant, but if at any stage he does not hire the girl just interviewed he must dismiss her and cannot call her back later. *The object of the employer is to minimize the expected absolute rank of the girl hired.*

If t is any stopping rule for this problem and x_t denotes the absolute rank of the girl hired using t, then the possible values of x_t range from 1 (best girl hired) to N (worst hired). If t consists of always hiring the first girl to appear (or hiring at random one of the N girls) then

$$Ex_t = (N+1)/2,$$

since the first (or a randomly chosen) applicant is equally likely to have any of the absolute ranks 1, 2, $\cdots$, N. Clearly, a good strategy t, based on the information on relative ranks, should be able to do better than that. Denote

$$R_N = \inf_t \{Ex_t\} \qquad (N \geq 1),$$

where the inf is taken over all the (finite) number of possible strategies available with a group of size N. For example, when $N=1$ only one strategy is available and $R_N=1$, for $N=2$ there are only two possible strategies, equally good, and $R_2=\frac{3}{2}$, for $N=3$ there are four strategies available of which the best gives $R_3=\frac{5}{3}$, etc.

It is possible by using the general theory to devise a computer program for finding R_N for any given $N=1, 2, \cdots$. It amounts to the following: Define

$$c_{N-1} = (N+1)/2, \quad \text{and for } i = N-1, N-2, \cdots, 1$$

compute successively

$$(22)\quad \begin{cases} s_i = \left[\dfrac{i+1}{N+1}\,c_i\right], & \text{where } [x] = \text{largest integer} \leqq x, \text{ and} \\ c_{i-1} = \dfrac{1}{1}\left(\dfrac{N+1}{i+1}\cdot\dfrac{s_i(s_i+1)}{2} + (i-s_i)c_i\right). \end{cases}$$

Then $R_N = c_0$. The backward induction from c_{N-1} to c_0 is an example of the general algorithm for solving optimal stopping problems when there are only a finite number of successive stages $1, \cdots, N$ at which stopping is permitted. In the present problem, c_i represents the minimal expected absolute rank attainable by any stopping rule which must stop somewhere between the $(i+1)$th and Nth stages inclusive.

Using this procedure it is found that

$$(23)\qquad R_{10} = 2.56, \quad R_{100} = 3.60, \quad R_{1000} = 3.83.$$

(The optimal strategy in each case is also produced by the computer but we shall not bother to describe it here.) No general formula for R_N other than the recursive definition by way of (22) is available, and for each N the recursion has to be started afresh.

Inspection of (23) shows that the values R_N seem to be increasing, but surprisingly slowly. It can in fact be proved that R_N increases steadily with N and hence that the constant

$$R = \lim_{N\to\infty} R_N \leqq +\infty$$

is well defined. What is its value, finite or infinite? In [3] it is proved that

$$R = \prod_{j=1}^{\infty}\left(\frac{j+2}{j}\right)^{1/(j+1)} \cong 3.8695.$$

Thus, no matter how large N is, there exists a strategy for which the expected absolute rank of the girl hired is less than 4, even though the actual absolute rank of the girl hired may be as great as N.

If instead of arriving in random order with each permutation equally likely, the applicants are sent in by a malicious competitor of the employer who wants to force him to do badly, the competitor can use certain permutations more often than others in such a way that no matter what strategy the employer uses (even knowing the competitor's mode of randomization) he will always have expected absolute rank of applicant hired $=(N+1)/2$, the purely random expectation. This is relevant to certain real or imaginary military applications of the hiring problem.

There are many variants [12] of this problem, the most common being that in which the employer, instead of trying to minimize the expected absolute rank of the girl hired, tries to maximize the probability of hiring the best girl, regardless of whom he hires when he does not get the best. For this case it has long been known that a probability $\cong 1/e$ of hiring the best can be attained by the proper strategy. The problem as we have put it is, of course, more difficult and perhaps more interesting.

References

1. K. J. Arrow, D. Blackwell, and M. A. Girshick, Bayes and minimax solutions of sequential decision problems, Econometrica, 17 (1949) 213–244.

2. L. Breiman, Stopping-rule problems, Applied Combinatorial Mathematics (ed. E. F. Beckenbach), Wiley, New York, 1964, pp. 284–319.

3. Y. S. Chow, S. Moriguti, H. Robbins, and S. M. Samuels, Optimal selection based on relative rank, Israel J. Math., 2 (1964) 81–90. The problem was stated by A. Cayley, "Mathematical questions and their solutions," Educational Times, 22 (1875) 18–19.

4. Y. S. Chow and H. Robbins, A martingale system theorem and applications, Proc. Fourth Berkeley Symp. on Math., Stat. and Prob., 1 (1960) 93–104.

5. ——— and ———, On optimal stopping rules, Zeitsch. f. Wahr., 2 (1963) 33–49.

6. ——— and ———, On optimal stopping rules for s_n/n, Illinois J. Math., 9 (1965) 444–454.

7. ——— and ———, On values associated with a stochastic sequence, Proc. Fifth Berkeley Symp. on Math. Stat. and Prob., 1 (1965) 427–440. A class of optimal stopping problems, *ibid.*, 419–426.

8. ———, H. Robbins, and D. O. Siegmund, Great Expectations: The Theory of Optimal Stopping, Houghton Mifflin, Boston, to appear in 1970.

9. ———, H. Robbins, and H. Teicher, Moments of randomly stopped sums, Ann. Math. Statist., 36 (1965) 789–799.

10. J. L. Doob. Stochastic Processes, Wiley, New York, 1953.

11. A. Dvoretzky, Existence and properties of certain optimal stopping rules, Proc. Fifth Berkeley Symp. on Math. Stat. and Prob., 1 (1965).

12. J. P. Gilbert and F. Mosteller, Recognizing the maximum of a sequence, J. Amer. Statist. Ass., 61 (1966) 35–73.

13. G. W. Haggstrom, Optimal stopping and experimental design, Ann. Math. Statist., 37 (1966) 7–29.

14. ———, Optimal sequential procedures when more than one stop is required, *ibid.*, 38 (1967) 1618–1626.

15. H. Robbins and E. Samuel, An extension of a lemma of Wald, J. Appl. Prob., 3 (1966) 272–273.

16. D. O. Siegmund, Some problems in the theory of optimal stopping rules, Ann. Math. Statist., 38 (1967) 1627–1640.

17. J. L. Snell, Application of martingale system theorems, Trans. Amer. Math. Soc., 73 (1952) 293–312.

18. A. Wald, On cumulative sums of random variables, Ann. Math. Statist., 15 (1944) 283–296.

Reprinted from the American Mathematical Monthly
Vol. 77, No. 4, April, 1970

Reprinted from THE ANNALS OF MATHEMATICAL STATISTICS
Vol. 36, No. 3, June, 1965

MOMENTS OF RANDOMLY STOPPED SUMS

BY Y. S. CHOW, HERBERT ROBBINS, AND HENRY TEICHER

Purdue University, Columbia University, and Purdue University

1. Introduction. Let $(\Omega, \mathfrak{F}, P)$ be a probability space, let $x_1, x_2, \cdots$ be a sequence of random variables on Ω, and let $\mathfrak{F}_n$ be the σ-algebra generated by $x_1, \cdots, x_n$, with $\mathfrak{F}_0 = (\phi, \Omega)$. A *stopping variable* (of the sequence $x_1, x_2, \cdots$) is a random variable t on Ω with positive integer values such that the event $[t = n] \varepsilon \mathfrak{F}_n$ for every $n \geqq 1$. Let $S_n = \sum_1^n x_i$; then $S_t = S_{t(\omega)}(\omega) = \sum_1^t x_i$ is a randomly stopped sum. *We shall always assume that*

$$E|x_n| < \infty, \qquad E(x_{n+1} \mid \mathfrak{F}_n) = 0, \qquad (n \geqq 1). \tag{1}$$

The moments of S_t have been investigated since the advent of Sequential Analysis, beginning with Wald [9], whose theorem states that for *independent, identically distributed* (iid) x_i *with* $Ex_i = 0$, $Et < \infty$ implies that $ES_t = 0$. For higher moments of S_t, the known results [1, 3, 4, 5, 10] are not entirely satisfactory. We shall obtain theorems for ES_t^r ($r = 2, 3, 4$); the case $r = 2$ is of special interest in applications. For iid x_i with $Ex_i = 0$ and $Ex_i^2 = \sigma^2 < \infty$, we shall show that $Et < \infty$ implies $ES_t^2 = \sigma^2 Et$.

2. The second moment. It follows from assumption (1) that $(S_n, \mathfrak{F}_n; n \geqq 1)$ is a *martingale*; i.e., that

$$E|S_n| < \infty, \qquad E(S_{n+1} \mid \mathfrak{F}_n) = S_n \qquad (n \geqq 1). \tag{2}$$

The following well-known fact ([3], p. 302) will be stated as

LEMMA 1. *Let* $(S_n, \mathfrak{F}_n; n \geqq 1)$ *be a martingale and let t be any stopping variable such that*

$$E|S_t| < \infty, \qquad \liminf \int_{[t>n]} |S_n| = 0; \tag{3}$$

then

$$E(S_t \mid \mathfrak{F}_n) = S_n \quad if \quad t \geqq n \qquad (n \geqq 1), \tag{4}$$

and hence $ES_t = ES_1$.

LEMMA 2. *If* $E \sum_1^t |x_i| < \infty$, *then* (3) *holds.*

PROOF. $|S_t| \leqq \sum_1^t |x_i|$, so that $E|S_t| < \infty$, and

$$\lim \int_{[t>n]} |S_n| \leqq \lim \int_{[t>n]} \sum_1^t |x_i| = 0.$$

In the remainder of this section we shall suppose, in addition to (1) that

$$Ex_n^2 < \infty \qquad (n \geqq 1) \tag{5}$$

and we define for $n \geqq 1$

Received 31 December 1964.

(6) $$Z_n = S_n^2 - \sum_1^n x_i^2.$$

The sequence $(Z_n, \mathfrak{F}_n; n \geqq 1)$ is also a martingale, with $EZ_1 = 0$.

For any stopping variable t, let $t(n) = \min(n, t)$; then Lemma 1 applies to Z_n and $t(n)$, so that $EZ_{t(n)} = 0$, and hence

(7) $$ES_{t(n)}^2 = E\sum_1^{t(n)} x_i^2.$$

Letting $n \to \infty$ we have $S_{t(n)}^2 \to S_t^2$ and $\sum_1^{t(n)} x_i^2 \uparrow \sum_1^t x_i^2$. Hence, by Fatou's lemma and (7),

(8) $$ES_t^2 \leqq \lim ES_{t(n)}^2 = \lim E\sum_1^{t(n)} x_i^2 = E\sum_1^t x_i^2.$$

The question now arises under what circumstances equality holds in (8). (By Lemma 1 this will be the case if (3) holds with S replaced by Z, but, as we shall see, this requirement is unnecessarily stringent.) According to (8), *we need only consider the case in which $ES_t^2 < \infty$, and it will suffice to prove that*

(9) $$ES_t^2 \geqq ES_{t(n)}^2 \qquad (n \geqq 1).$$

Lemma 3. *If*

(10) $$\liminf \int_{[t>n]} |S_n| = 0,$$

then $ES_t^2 = E\sum_1^t x_i^2$.

Proof. We may suppose that $ES_t^2 < \infty$ whence, by (10) and Lemma 1, (4) holds. Hence

$$\begin{aligned} ES_t^2 &= \int_{[t \leqq n]} S_t^2 + \int_{[t>n]} (S_n + (S_t - S_n))^2 \\ &\geqq \int_{[t \leqq n]} S_t^2 + \int_{[t>n]} S_n^2 + 2\int_{[t>n]} S_n E(S_t - S_n \mid \mathfrak{F}_n) = ES_{t(n)}^2. \end{aligned}$$

Lemma 4. *If*

(11) $$\liminf \int_{[t>n]} S_n^2 < \infty,$$

then (10) *holds.*

Proof. Suppose (10) does not hold; then $\liminf \int_{[t>n]} |S_n| = \epsilon > 0$. Hence for any constant $0 < a < \infty$,

$$\liminf \int_{[t>n]} S_n^2 \geqq a \liminf \int_{[t>n, |S_n|>a]} |S_n| = a\epsilon,$$

which contradicts (11), since a may be arbitrarily large.

Lemma 5. *If* $E\sum_1^t x_i^2 < \infty$, *then* (11) *holds.*

Proof. Setting $S_0 = 0$ we have

$$\begin{aligned} \int_{[t>n]} S_n^2 &= \sum_{i=1}^n \left(\int_{[t>i]} S_i^2 - \int_{[t>i-1]} S_{i-1}^2\right) \\ &\leqq \sum_{i=1}^n \int_{[t \geqq i]} (S_i^2 - S_{i-1}^2) \leqq \sum_1^\infty \int_{[t \geqq i]} x_i^2 = E\sum_1^t x_i^2 < \infty. \end{aligned}$$

From Lemmas 1–5 we have

Theorem 1. *Let $(S_n, \mathfrak{F}_n; n \geqq 1)$ be a martingale with $ES_n^2 < \infty$ and let t be any stopping variable. Set $x_1 = S_1$, $x_{n+1} = S_{n+1} - S_n$. Then*

$$ES_t^2 \leqq E\sum_1^t x_i^2. \tag{12}$$

If any one of the four conditions

$$\liminf \int_{[t>n]} |S_n| = 0, \qquad \liminf \int_{[t>n]} S_n^2 < \infty, \tag{13}$$
$$E\sum_1^t |x_i| < \infty, \qquad E\sum_1^t x_i^2 < \infty$$

holds, then

$$ES_t^2 = E\sum_1^t x_i^2. \tag{14}$$

If either $E\sum_1^t |x_i| < \infty$ or $E\sum_1^t x_i^2 < \infty$, *then* (3) *and* (4) *hold.*

Theorem 1 generalizes (a) and (b) of Theorem II of [1]. In order to apply it, we first verify

LEMMA 6. *For any stopping variable t and any* $r > 0$,

$$E\sum_1^t |x_i|^r = E\sum_1^t E(|x_i|^r|\mathfrak{F}_{i-1}).$$

PROOF.

$$E\sum_1^t |x_i|^r = \sum_{j=1}^\infty \int_{[t=j]} \sum_{i=1}^j |x_i|^r = \sum_{i=1}^\infty \int_{[t\geqq i]} |x_i|^r$$
$$= \sum_{i=1}^\infty \int_{[t\geqq i]} E(|x_i|^r \mid \mathfrak{F}_{i-1}) = E\sum_1^t E(|x_i|^r \mid \mathfrak{F}_{i-1}).$$

For independent x_n, we have from Theorem 1 and Lemmas 1 and 6

THEOREM 2. *Let* $x_1, x_2, \cdots$ *be independent with* $Ex_n = 0$, $E|x_n| = a_n$, $Ex_n^2 = \sigma_n^2 < \infty\,(n \geqq 1)$ *and let* $S_n = \sum_1^n x_i$. *Then if t is a stopping variable, either of the two relations*

$$E\sum_1^t a_i < \infty, \qquad E\sum_1^t \sigma_i^2 < \infty \tag{15}$$

implies that $ES_t = 0$ *and*

$$ES_t^2 = E\sum_1^t x_i^2 = E\sum_1^t \sigma_i^2. \tag{16}$$

If $\sigma_n^2 = \sigma^2 < \infty$, *then* $Et < \infty$ *implies*

$$ES_t^2 = E\sum_1^t x_i^2 = \sigma^2 Et. \tag{17}$$

Some stronger sufficient conditions for (17) have been given in ([10], [1], [5], [3] (p. 351), [4]).

COROLLARY 1. *Let* $x_1, x_2, \cdots$ *be independent with* $Ex_n = 0$, $Ex_n^2 = 1$, *and define* t^*(resp. t_*) $=$ 1st $n \geqq 1$ *such that* $|S_n| > n^{\frac{1}{2}}$(resp. $<$)($= \infty$ *otherwise*). *Then* $Et^* = Et_* = \infty$.

PROOF. If $Et^* < \infty$, then t^* is a genuine stopping variable, i.e., $P(t^* < \infty) = 1$, and by the definition of t^* and (17),

$$Et^* = ES_{t^*}^2 > Et^*,$$

a contradiction; similarly for t_*.

We note that both t^* and t_* are genuine stopping variables if the x_n are, in addition, identically distributed.

The example $P[x_n = 1] = P[x_n = -1] = \frac{1}{2}$ shows that the $>$ ($<$) cannot be replaced by $\geqq$ ($\leqq$), since $Ex_n = 0$, $Ex_n^2 = 1$, and $t^* = t_* = 1$. On the other hand, if t^* is redefined as the first $n > 1$ for which $|S_n| \geqq n^{\frac{1}{2}}$, Et^* is again infinite; similarly for t_*.

Corollary 1 is a generalization of Theorem 1 of [2]. The following corollary generalizes Theorem 2 of [2].

COROLLARY 2. *Let $x_1, x_2, \cdots$ be independent with $Ex_n = 0$, $Ex_n^2 = 1$, $P[|x_n| \leqq a < \infty] = 1$. For $0 < c < 1$ and $m = 1, 2, \cdots$, define t = first $n \geqq m$ such that $|S_n| > cn^{\frac{1}{2}}$. Then $Et < \infty$.*

PROOF. For $k = m, m + 1, \cdots$, put $t' = \min(t, k)$ and $A_k = [\omega: m < t \leqq k]$. Then t' is a stopping variable and by Theorem 2

$$kP[t > k] + \int_{[t \leqq k]} t = Et' = ES_{t'}^2 = \int_{[t > k]} S_k^2 + \int_{[t \leqq k]} S_t^2$$

or

$$kP[t > k] + \int_{A_k} t \leqq c^2 kP[t > k] + \int_{A_k} (ct^{\frac{1}{2}} + a)^2 + m.$$

Hence

$$(1 - c^2)(kP[t > k] + \int_{A_k} t) \leqq 2ac \int_{A_k} t^{\frac{1}{2}} + O(1).$$

Therefore, as $k \to \infty$, $\int_{A_k} t = O(1)$ and $P[t > k] = O(k^{-1}) = o(1)$, so that t is a genuine stopping variable and $Et < \infty$.

COROLLARY 3. *If $x_1, x_2, \cdots$, are* iid *with $Ex_n = 0$, $Ex_n^2 = \sigma^2$, $P[|x_n| \leqq a < \infty] = 1$, and if $ES_t^2 < \infty$ for a stopping variable t, then $Et < \infty$ if and only if*

$$\liminf nP[t > n] = 0. \tag{18}$$

PROOF. The "only if" part is obvious. Now suppose (18) holds. Then since $\int_{[t > n]} |S_n| \leqq anP[t > n]$, the first condition of (13) holds and hence $\sigma^2 Et = ES_t^2 < \infty$, so that $Et < \infty$ if $\sigma^2 > 0$. (If $\sigma^2 = 0$, then $P[x_n = 0] = 1$ and hence t is equal a.e. to a fixed positive integer, so $Et < \infty$ in this case too.)

Applied to the case $P[x_i = 1] = P[x_i = -1] = \frac{1}{2}$, with t = first $n \geqq 1$ such that $S_t = 1$, we have by Wald's theorem $Et = \infty$, but by Corollary 3 the stronger result $\liminf nP[t > n] > 0$.

COROLLARY 4. *Let $(x_n, n \geqq 1)$ satisfy $E(x_{n+1} \mid \mathfrak{F}_n) = 0$ and let $E(x_{n+1}^2 \mid \mathfrak{F}_n) = \sigma_{n+1}^2 < \infty$ be constant for $n \geqq 0$. Then for $\epsilon > 0$,*

$$P[\max_{n \leqq m} |S_n| \geqq \epsilon] \leqq \epsilon^{-2} \sum_1^m \sigma_n^2.$$

If moreover $\sup_{n \geqq 1} |x_n| = z$ with $Ez < \infty$, then

$$P[\max_{n \leqq m} |S_n| \geqq \epsilon] \geqq 1 - [E(\epsilon + z)^2 / \sum_1^m \sigma_n^2]. \tag{19}$$

PROOF. Define t = first $n \geqq 1$ such that $|S_n| \geqq \epsilon$. Then $t' = \min(t, m)$ is a bounded stopping variable. Hence, by (14) and Lemma 6,

$$\epsilon^2 P[\max_{n \leqq m} |S_n| \geqq \epsilon] = \epsilon^2 P[t \leqq m] \leqq ES_{t'}^2 = E \sum_1^{t'} \sigma_n^2 \leqq \sum_1^m \sigma_n^2.$$

If $Ez < \infty$, then

$$E(\epsilon + z)^2 \geqq ES_{t'}^2 = E \sum_1^{t'} \sigma_n^2 \geqq \int_{[t \geqq m]} \sum_1^m \sigma_j^2 = (\sum_1^m \sigma_j^2) P[t \geqq m]$$

and (19) holds.

The first part of Corollary 4 is a special case of submartingale inequalities ([6], p. 391), and the second part generalizes slightly one of the Kolmogorov inequalities ([6], p. 235) which requires that z be constant.

3. The fourth moment. The analysis in the case of the fourth moment of S_t is somewhat easier than that of the third moment and consequently is presented first. In this section Ex_n^4 will be supposed finite and $Ex_1 = 0$. Define for $r = 1, 2, 3, 4$, and $n = 1, 2, \cdots$

$$u_{r,n} = E(x_n^r \mid \mathfrak{F}_{n-1}), \qquad U_{r,n} = \sum_1^n u_{r,j},$$
$$(20) \qquad v_{r,n} = E(|x_n|^r \mid \mathfrak{F}_{n-1}), \qquad V_{r,n} = \sum_1^n v_{r,j},$$
$$T_{r,n} = \sum_1^n |x_j|^r, \qquad T_{1,n} = T_n.$$

In these terms, Lemma 6 asserts that $ET_{r,t} = EV_{r,t}$.

LEMMA 7. *If $ES_t^2 < \infty$ and* $\liminf \int_{[t>n]} |S_n| = 0$, *then*

$$E(S_t^2 \mid \mathfrak{F}_n) \geqq S_n^2 \quad and \quad E(|S_t| \mid \mathfrak{F}_n) \geqq |S_n| \quad for \quad t > n.$$

PROOF. For any $A \in \mathfrak{F}_n$, by Lemma 1

$$\int_{A[t>n]} S_t^2 = \int_{A[t>n]} [S_n^2 + 2S_n(S_t - S_n) + (S_t - S_n)^2] \geqq \int_{A[t>n]} S_n^2.$$

Hence the first inequality of the lemma holds, and the second inequality follows immediately from Lemma 1 and the fact that $E(|S_t| \mid \mathfrak{F}_n) \geqq |E(S_t \mid \mathfrak{F}_n)|$.

THEOREM 3. *If t is a stopping variable such that* $E[t \sum_1^t E(x_j^4 \mid \mathfrak{F}_{j-1})] < \infty$, *then $ES_t^4 < \infty$ and*

$$(21) \qquad ES_t^4 = EU_{4,t} + 4ES_tU_{3,t} + 6ES_t^2U_{2,t} - 6E \sum_1^t u_{2,j}U_{2,j}.$$

PROOF. Set $Y_n = S_n^4 - 6S_n^2U_{2,n} - 4S_nU_{3,n} - U_{4,n} + 6\sum_{j=1}^n u_{2,j}U_{2,j}$ and $t' = \min(t, k)$. Since $\{Y_n, \mathfrak{F}_n; n \geqq 1\}$ is a martingale with $EY_1 = 0$, by Lemma 1,

$$ES_{t'}^4 = 6ES_{t'}^2U_{2,t'} + 4ES_{t'}U_{3,t'} + EU_{4,t'} - 6E(\sum_{j=1}^{t'} u_{2,j}U_{2,j})$$
$$\leqq 6(E^{\frac12}S_{t'}^4)(E^{\frac12}U_{2,t'}^2) + 4(E^{\frac14}S_{t'}^4)(E^{3/4}V_{3,t'}^{4/3}) + EU_{4,t'},$$

whence, if $ES_{t'}^4 > 0$,

$$(22) \qquad E^{\frac12}S_{t'}^4 \leqq 6E^{\frac12}U_{2,t'}^2 + 4(E^{3/4}V_{3,t'}^{4/3})(ES_{t'}^4)^{-\frac14} + (EU_{4,t'})(ES_{t'}^4)^{-\frac12}.$$

Now if $p > 1$, $r > 0$,

$$(23) \qquad V_{r,n} = \sum_{j=1}^n E\{|x_j|^r \mid \mathfrak{F}_{j-1}\} \leqq n^{(p-1)/p}(\sum_{j=1}^n E^p\{|x_j|^r \mid \mathfrak{F}_{j-1}\})^{1/p}$$
$$\leqq n^{(p-1)/p}(\sum_{j=1}^n E\{|x_j|^{pr} \mid \mathfrak{F}_{j-1}\})^{1/p} = n^{(p-1)/p}V_{pr,n}^{1/p}$$

and thus setting $p = 2$, $r = 2$ and then $p = \frac43$, $r = 3$,

$$(24) \qquad EU_{2,t}^2 = EV_{2,t}^2 \leqq EtV_{4,t} < \infty, \qquad EV_{3,t}^{4/3} \leqq Et^{\frac13}V_{4,t} < \infty.$$

Moreover, $EU_{4,t} \leqq E(tU_{4,t}) < \infty$ and $E(\sum_{j=1}^t u_{2,j}U_{2,j}) \leqq EU_{2,t}^2 < \infty$. Thus, the LHS of (22) is a bounded function of k, implying via Fatou's lemma that $ES_t^4 < \infty$.

Since $|Y_n| \leqq S_n^4 + 6S_n^2 U_{2,n} + 4|S_n| V_{3,n} + U_{4,n} + 6\sum_{j=1}^n u_{2,j} U_{2,j} = Y_n'$ (say), it follows from the preceding that

$$E|Y_t| \leqq EY_t' \leqq ES_t^4 + 6(E^{\frac{1}{2}}S_t^4)(E^{\frac{1}{2}}U_{2,t}^2) + 4(E^{\frac{1}{4}}S_t^4)(E^{3/4}V_{3,t}^{4/3}) + EU_{4,t} + 6EU_{2,t}^2 < \infty.$$

From (24), $ET_{2,t} = EU_{2,t} < \infty$. Thus, (8) of Section 2 and Lemmas 4 and 5 are valid, whence by Lemma 7, $E\{S_t^2 \mid \mathfrak{F}_k\} \geqq S_k^2$ for $t > k$, $k = 1, 2, \cdots$. Consequently,

$$\int_{[t>n]} S_t^4 = \int_{[t>n]} [S_n^4 + 2S_n^2(S_t^2 - S_n^2) + (S_t^2 - S_n^2)^2 \geqq \int_{[t>n]} S_n^4 + 2\int_{[t>n]} S_n^2 E\{S_t^2 - S_n^2 \mid \mathfrak{F}_n\} \geqq \int_{[t>n]} S_n^4$$

implying $\int_{[t>n]} S_n^4 = o(1)$ and concomitantly

$$\int_{[t>n]} S_n^2 U_{2,n} \leqq \left(\int_{[t>n]} S_n^4\right)^{\frac{1}{2}}\left(\int_{[t>n]} U_{2,t}^2\right)^{\frac{1}{2}} = o(1),$$

$$\int_{[t>n]} |S_n| V_{3,n} \leqq \left(\int_{[t>n]} S_n^4\right)^{\frac{1}{4}}\left(\int_{[t>n]} V_{3,t}^{4/3}\right)^{3/4} = o(1),$$

(25) $$\int_{[t>n]} U_{4,n} \leqq \int_{[t>n]} U_{4,t} = o(1),$$

$$\int_{[t>n]} \sum_{j=1}^n u_{2,j} U_{2,j} \leqq \int_{[t>n]} U_{2,n}^2 \leqq \int_{[t>n]} U_{2,t}^2 = o(1).$$

Thus, $\int_{[t>n]} |Y_n| \leqq \int_{[t>n]} Y_n' = o(1)$ and by Lemma 1, $EY_t = EY_1 = 0$.

Alternative expressions for ES_t^4 are possible as indicated in

THEOREM 4. *If* $E(t\sum_{j=1}^t E\{x_j^4 \mid \mathfrak{F}_{j-1}\}) < \infty$, *then setting* $S_0 = 0$,

$$ES_t^4 = 6E\sum_{j=1}^t S_{j-1}^2 u_{2,j} + 4E\sum_{j=1}^t S_{j-1} u_{3,j} + EU_{4,t}.$$

The proof of Theorem 4 is similar to that of Theorem 3 and will be omitted.

COROLLARY. *If* $E(tU_{4,t}) < \infty$, *then*

$$E\left(6\sum_{j=1}^t S_{j-1}^2 u_{2,j} + 4\sum_{j=2}^t S_{j-1} u_{3,j}\right) = 6ES_t^2 U_{2,t} + 4ES_t U_{3,t} - 6E\left(\sum_{j=1}^t u_{2,j} U_{2,j}\right).$$

It is intuitively clear that terms with like coefficients are equal, and indeed we have

LEMMA 8. *If* $E(tU_{4,t}) < \infty$, *then* $ES_t U_{3,t} = E(\sum_{j=2}^t S_{j-1} u_{3,j})$ *and* $E(S_t^2 U_{2,t}) = E(\sum_{j=2}^t S_{j-1}^2 u_{2,j}) + E(\sum_{j=1}^t u_{2,j} U_{2,j})$.

PROOF. It suffices to verify the first of the two relationships since the second will then follow from the corollary to Theorem 4. Suppose first that

(26) $$E\left(\sum_{j=1}^t |x_j| V_{r,j}\right) < \infty.$$

Then

$$\sum_{k=1}^\infty \int_{[t=k]} \sum_{j=1}^k x_j U_{r,j} = \sum_{j=1}^\infty \int_{[t\geqq j]} x_j U_{r,j} = \sum_{j=1}^\infty \int_{[t\geqq j]} E(x_j \mid \mathfrak{F}_{j-1}) U_{r,j} = 0,$$

whence

$$E\left(\sum_{j=1}^{t} S_{j-1}u_{r,j}\right) = \sum_{k=1}^{\infty} \int_{[t=k]} \left[\sum_{j=2}^{k} S_{j-1}u_{r,j} + \sum_{j=1}^{k} x_j U_{r,j}\right]$$

(27)
$$= \sum_{k=1}^{\infty} \int_{[t=k]} S_k U_{r,k}$$
$$= ES_t U_{r,t}.$$

Thus, if $t' = \min(t, N)$, (27) holds with t replaced by t' irrespective of (26). However,

(28)
$$ES_t U_{3,t} = \sum_{k=1}^{N} \int_{[t=k]} S_k U_{3,k} + \int_{[t>N]} S_t U_{3,t}$$
$$= ES_{t'} U_{3,t'} - \int_{[t>N]} S_N U_{3,N} + \int_{[t>N]} S_t U_{3,t},$$

and analogously

(29)
$$E\left(\sum_{j=2}^{t} S_{j-1}u_{3,j}\right) = E\left(\sum_{j=2}^{t'} S_{j-1}u_{3,j}\right) - \int_{[t>N]} \sum_{j=2}^{N} S_{j-1}u_{3,j} + \int_{[t>N]} \sum_{j=2}^{t} S_{j-1}u_{3,j}.$$

Now $E|S_t U_{3,t}| \leqq EY_t' < \infty$, and employing Lemma 7,

$$E\sum_1^t |S_{j-1}u_{3,j}| = \sum_{k=1}^{\infty} \int_{[t=k]} \sum_1^k |S_{j-1}u_{3,j}| = \sum_{j=1}^{\infty} \int_{[t\geqq j]} |S_{j-1}u_{3,j}|$$
$$\leqq \sum_1^{\infty} \int_{[t\geqq j]} |S_t u_{3,j}| \leqq E|S_t| V_{3,t} \leqq EY_t' < \infty.$$

These facts plus (25) imply that all unwanted terms of (28) and (29) are $o(1)$ and the result follows.

Identities and inequalities analogous to (27) abound and several of these will be catalogued as

LEMMA 9. $E(\sum_{n=1}^{t} S_n^2) \leqq EtS_t^2$ *under the conditions of Lemma* 7.

$$E\left(\sum_{n=1}^{t} S_n\right) = EtS_t \quad \text{if} \quad EtT_t < \infty.$$
$$E\left(\sum_{n=1}^{t} T_n\right) \leqq EtT_t \quad \text{if} \quad EtT_t < \infty.$$

PROOF.

$$E\sum_{n=1}^{t} S_n^2 = \sum_{k=1}^{\infty} \int_{[t=k]} \sum_{n=1}^{k} S_n^2$$
$$= \sum_{n=1}^{\infty} \int_{[t\geqq n]} S_n^2 \leqq \sum_{n=1}^{\infty} \int_{[t\geqq n]} E(S_t^2 \mid \mathfrak{F}_n)$$
$$= \sum_{n=1}^{\infty} \int_{[t\geqq n]} S_t^2 = \sum_{n=1}^{\infty} \sum_{k=n}^{\infty} \int_{[t=k]} S_t^2$$
$$= \sum_{k=1}^{\infty} k \int_{[t=k]} S_t^2 = EtS_t^2$$

employing Lemma 7. Similarly,

$$E\left(\sum_{n=1}^{t} T_n\right) = \sum_{n=1}^{\infty} \int_{[t\geqq n]} T_n \leqq \sum_{n=1}^{\infty} \int_{[t\geqq n]} T_t = EtT_t.$$

Finally,

$$E\left(\sum_{n=1}^{t} S_n\right) = \sum_{n=1}^{\infty} \int_{[t\geqq n]} S_n = \sum_{n=1}^{\infty} \int_{[t\geqq n]} E(S_t \mid \mathfrak{F}_n) = \sum_{n=1}^{\infty} \int_{[t\geqq n]} S_t$$
$$= EtS_t$$

in view of Lemmas 1 and 2 and the validity of interchanging the order of summation and integration.

4. The third moment. In this section $E(|x_n|^3)$ will be supposed finite and $Ex_1 = 0$. Define

$$(30)\qquad \begin{aligned} Y_n &= S_n^3 - 3S_nU_{2,n} - U_{3,n}, \\ W_n &= S_n^3 - 3\sum_{j=1}^n S_{j-1}u_{2,j} - U_{3,n}, \\ Z_n &= S_n^3 - 3\sum_{j=1}^n S_ju_{2,j} - U_{3,n}. \end{aligned}$$

It is readily checked that $(Y_n, \mathfrak{F}_n; n \geqq 1)$, $(W_n, \mathfrak{F}_n; n > 1)$, $(Z_n, \mathfrak{F}_n; n > 1)$ are all martingales and that $EY_1 = EW_1 = EZ_1 = 0$.

THEOREM 5. *If $EV_{3,t} < \infty$ and $EV_{1,t}^3 < \infty$, or equivalently if $ET_t^3 < \infty$, then $E|S_t|^3 < \infty$ and $ES_t^3 = 3E(\sum_{j=1}^t S_{j-1}u_{2,j}) + EU_{3,t}$.*

PROOF. Suppose that $EV_{3,t} < \infty$, $EV_{1,t}^3 < \infty$. (Their equivalence with $ET_t^3 < \infty$ will be deferred to Lemma 10). Then

$$(31)\qquad \begin{aligned} E|S_t|^3 &= \sum_{k=1}^\infty \int_{[t=k]} \sum_{n=1}^k (|S_n|^3 - |S_{n-1}|^3) \leqq \sum_{k=1}^\infty \sum_{n=1}^k \int_{[t=k]} (|x_n|^3 \\ &\quad + 3|S_{n-1}|x_n^2 + 3S_{n-1}^2|x_n|) \\ &\leqq 6\sum_{k=1}^\infty \sum_{n=1}^k \int_{[t=k]} (|x_n|^3 + S_{n-1}^2|x_n|) \\ &= 6[E(\sum_{n=1}^t |x_n|^3) + E(\sum_{n=1}^t S_{n-1}^2|x_n|)]. \end{aligned}$$

By Lemma 6,

$$(32)\qquad E(\sum_{n=1}^t |x_n|^3) = EV_{3,t} < \infty.$$

On the other hand, $ES_t^2 \leqq ET_t^2 \leqq 1 + ET_t^3 < \infty$ and

$$\int_{[t>k]} |S_k| \leqq \int_{[t>k]} T_k \leqq \int_{[t>k]} T_t \leqq \int_{[t>k]} (1 + T_t^3) = o(1)$$

in view of the asserted equivalence. Thus, Lemma 7 holds, whence

$$(33)\qquad \begin{aligned} &E(\sum_{n=1}^t S_{n-1}^2|x_n|) \\ &= \sum_{k=1}^\infty \sum_{n=1}^k \int_{[t=k]} S_{n-1}^2|x_n| = \sum_{n=1}^\infty \int_{[t\geqq n]} S_{n-1}^2 v_{1,n} \\ &\leqq \sum_{n=1}^\infty \int_{[t\geqq n]} E(S_t^2 \mid \mathfrak{F}_{n-1})v_{1,n} = \sum_{n=1}^\infty \int_{[t\geqq n]} S_t^2 v_{1,n} \\ &= \sum_{k=1}^\infty \sum_{n=1}^k \int_{[t=k]} S_t^2 v_{1,n} = ES_t^2 V_{1,t} \leqq (E^{2/3}|S_t|^3)(E^{\frac{1}{3}}V_{1,t}^3). \end{aligned}$$

Replace t by $t' = \min(t, k)$ in (31). Then from (32) and (33),

$$E|S_{t'}|^3 \leqq 6EV_{3,t'} + 6(E^{2/3}|S_{t'}|^3)(E^{\frac{1}{3}}V_{1,t'}^3) = O(1) + O(1)E^{2/3}|S_{t'}|^3$$

whence, by Fatou's lemma,

$$(34)\qquad E|S_t|^3 < \infty.$$

Next, (34) implies that the expectation in the LHS of (33) is finite whence

$$(35)\qquad \begin{aligned} E(\sum_{n=1}^t |S_{n-1}|u_{2,n}) &= \sum_{n=1}^\infty \int_{[t\geqq n]} |S_{n-1}|x_n^2 = E(\sum_{n=1}^t |S_{n-1}|x_n^2) \\ &\leqq E[\sum_{n=1}^t (|x_n|^3 + |S_{n-1}|^2|x_n|)] < \infty. \end{aligned}$$

Combining (34) and (35), $E|W_t| < \infty$. Since, paralleling (31),

$$\int_{[t>k]} |S_k|^3 \leqq 6 \int_{[t>k]} \sum_{n=1}^{k} (|x_n|^3 + S_{n-1}^2 |x_n|) = o(1),$$

$\int_{[t>k]} |W_k| = o(1)$ and the theorem follows from Lemma 1.

Lemma 10. *$EV_{3,t} < \infty$ and $EV_{1,t}^3 < \infty$ if and only if $ET_t^3 < \infty$.*

Proof. Suppose $EV_{3,t} < \infty$ and $EV_{1,t}^3 < \infty$. The argument of (31) with T_t replacing S_t yields

$$ET_t^3 \leqq 6 \sum_{k=1}^{\infty} \sum_{n=1}^{k} \int_{[t=k]} (|x_n|^3 + T_{n-1}^2 |x_n|).$$

The inequality of (33) also obtains with T replacing S in view of the fact that $T_t \geqq T_{n-1}$ on the set $[t \geqq n]$. Thus, analogously, $ET_{t'}^3 \leqq O(1) + O(1)E^{2/3}T_{t'}^3$, implying $ET_t^3 < \infty$.

Conversely, if $ET_t^3 < \infty$, clearly $EV_{3,t} = ET_{3,t} \leqq ET_t^3 < \infty$. Moreover,

$$\begin{aligned}
EV_{1,t'}^3 &= \sum_{j=1}^{\infty} \int_{[t'=j]} \sum_{n=1}^{j} (V_{1,n}^3 - V_{1,n-1}^3) \\
&\leqq \sum_{j=1}^{\infty} \sum_{n=1}^{j} \int_{[t'=j]} (v_{1,n}^3 + 3V_{1,n-1}^2 v_{1,n} + 3V_{1,n-1} v_{1,n}^2) \\
&\leqq O(1) + 6 \sum_{j=1}^{\infty} \sum_{n=1}^{j} \int_{[t'=j]} V_{1,n-1}^2 v_{1,n} \\
&= O(1) + 6 \sum_{n=1}^{\infty} \int_{[t' \geqq n]} |x_n| \, V_{1,n-1}^2 \\
&\leqq O(1) + 6 \sum_{n=1}^{\infty} \sum_{j=n}^{\infty} \int_{[t'=j]} |x_n| \, V_{1,t'}^2 \\
&\leqq O(1) + 6\, ET_{t'} V_{1,t'}^2 \\
&\leqq O(1) + O(1) E^{2/3} V_{1,t'}^3,
\end{aligned}$$

which implies, as earlier, that $EV_{1,t}^3 < \infty$ and completes the proof.

Theorem 6. *If $ET_t^3 < \infty$ and $Et^{\frac{1}{2}}V_{3,t} < \infty$, $ES_t^3 = 3ES_tU_{2,t} + EU_{3,t} < \infty$.*

Proof. As in Theorem 5, after setting $p = \frac{3}{2}$, $r = 2$ in (23) of Section 3 to obtain

$$E|S_tU_{2,t}| \leqq (E^{\frac{1}{3}}|S_t|^3)(E^{2/3}U_{2,t}^{3/2}) \leqq (E^{\frac{1}{3}}|S_t|^3)(E^{2/3}t^{\frac{1}{2}}V_{3,t}).$$

Corollary 1. *Under the conditions of Theorem* 6, $E(\sum_{j=1}^{t} x_j u_{2,j}) = 0$.

Proof. Analogously, $EZ_t = 0$, whence $E(W_t - Z_t) = 0$.

Corollary 2. *Under the conditions of Theorem* 6, $ES_tU_{2,t} = E(\sum_{j=1}^{t} S_{j-1}u_{2,j})$.

The single requirement $ET_t^3 < \infty$, although equivalent to the two conditions of Theorem 5, is difficult to check. The following single condition is easily seen to imply all those of Theorems 5 and 6:

$$E(t^2 V_{3,t}) < \infty, \tag{36}$$

and in addition yields

$$\begin{aligned}
ET_t^3 = {} & 3ET_t^2V_{1,t} + 3ET_t(V_{2,t} - 2\sum_{j=1}^{t} V_{1,j}v_{1,j}) + EV_{3,t} \\
& - 3E(\sum_{j=1}^{t} V_{1,j}v_{2,j}) - 3E(\sum_{j=1}^{t} V_{2,j}v_{1,j}) + 6E(\sum_{j=1}^{t} v_{1,j} \sum_{i=1}^{j} V_{1,i}v_{1,i}).
\end{aligned}$$

5. Sums of independent random variables. In this section, the random variables $x_1, x_2, \cdots$ will be supposed independent. If $Ex_n = 0$, all prior theorems are, of course, applicable but may be reformulated in especially simple terms with conditions that are susceptible of immediate verification. For example, from Theorems 3 and 6, we obtain:

THEOREM 7. *If $x_1, x_2, \cdots$ are independent with $Ex_n = 0$, $Ex_n^2 = \sigma^2$, $Ex_n^3 = \gamma$, $Ex_n^4 = \beta < \infty$ and t is a stopping rule with $Et^2 < \infty$, then $ES_t^4 < \infty$ and*

$$ES_t^4 = 6\sigma^2 EtS_t^2 + 4\gamma EtS_t + \beta Et - 3\sigma^4 Et(t+1).$$

THEOREM 8. *If $x_1, x_2, \cdots$ are independent with $Ex_n = 0$, $Ex_n^2 = \sigma^2$, $Ex_n^3 = \gamma$, $E|x_n|^3 \leqq C < \infty$, and if t is a stopping variable with $Et^3 < \infty$, then $ES_t^3 = \gamma Et + 3\sigma^2 EtS_t < \infty$.*

PROOF. According to Theorem 6 and Lemma 10, it suffices to verify that

$$EV_{3,t} \leqq E(t^{\frac{1}{2}} V_{3,t}) \leqq CEt^{3/2} < \infty,$$

$$EV_{1,t}^3 \leqq E[t(1+C)]^3 < \infty.$$

In the final theorem, the requirement of Theorem 8 that $Et^3 < \infty$ will be relaxed at the expense of increasing the moment assumptions on x_n.

THEOREM 9. *If $x_1, x_2, \cdots$ are independent with $Ex_n = 0$, $Ex_n^2 = \sigma^2$, $Ex_n^3 = \gamma$, $Ex_n^4 \leqq C < \infty$, and if t is a stopping variable with $Et^2 < \infty$, then $ES_t^3 = \gamma Et + 3\sigma^2 EtS_t$.*

PROOF. Here, the martingale Y_n of (30) simplifies to $Y_n = S_n^3 - 3\sigma^2 nS_n - n\gamma$. The theorem will follow from Lemmas 1 and 2 once it is established that $E\sum_1^t |Y_{n+1} - Y_n| = E\sum_1^t E(|Y_{n+1} - Y_n| \mid \mathfrak{F}_n) < \infty$. Now if B and D are finite constants (not necessarily the same in each appearance),

$$E(|S_{n+1}^3 - S_n^3| \mid \mathfrak{F}_n) \leqq 6E(|x_{n+1}|^3 + S_n^2 |x_{n+1}| \mid \mathfrak{F}_n) \leqq BS_n^2 + D,$$

$$E(|(n+1)S_{n+1} - nS_n| \mid \mathfrak{F}_n) = E(|S_n + (n+1)x_{n+1}| \mid \mathfrak{F}_n) \leqq S_n^2 + nD,$$

whence

$$E(|Y_{n+1} - Y_n| \mid \mathfrak{F}_n) \leqq BS_n^2 + nD.$$

Next, Lemma 9 is applicable below since (17) insures $ES_t^2 < \infty$ while Lemmas 6 and 2 guarantee (10). Consequently,

$$\begin{aligned} E\sum_1^t E(|Y_{n+1} - Y_n| \mid \mathfrak{F}_n) &\leqq B\cdot E(\textstyle\sum_1^t S_n^2) + D\cdot Et(t+1) \\ &\leqq B\cdot EtS_t^2 + D\cdot Et(t+1) \\ &\leqq B\cdot(E^{\frac{1}{2}}t^2)(E^{\frac{1}{2}}S_t^4) + D\cdot Et(t+1) < \infty. \end{aligned}$$

REFERENCES

[1] BLACKWELL, D. and GIRSHICK, M. A. (1947). A lower bound for the variance of some unbiased sequential estimates. *Ann. Math. Statist.* **18** 277–280.

[2] BLACKWELL, D. and FRIEDMAN, D. (1964). A remark on the coin tossing game. *Ann. Math. Satist.* **35** 1345–1347.

[3] DOOB, J. L. (1953). *Stochastic Processes.* Wiley, New York.

[4] JOHNSON, N. L. (1959). A proof of Wald's theorem on cumulative sums. *Ann. Math. Statist.* **30** 1245–1247.

[5] Kolmogorov, A. N. and Prohorov, Y. (1949). On sums of a random number of random terms. *Uspehi Mat. Nauk* **4** 168–172 (In Russian).
[6] Loève, M. (1960). *Probability Theory.* (2nd ed.). Van Nostrand, Princeton.
[7] Seitz, J. and Winkelbauer, K. (1953). Remarks concerning a paper of Kolmogorov and Prohorov. *Czechoslovak. Math. J.* **3** 89–91. (In Russian, with English summary.)
[8] Winkelbauer, K. (1953). Moments of cumulative sums of random variables. *Czechoslovak. Math. J.* **3** 93–108.
[9] Wald, A. (1944). On cumulative sums of random variables. *Ann. Math. Statist.* **15** 283–296.
[10] Wolfowitz, J. (1947). The efficiency of sequential estimates. *Ann. Math. Statist.* **18** 215–230.

ON THE „PARKING" PROBLEM

by

A. DVORETZKY[1] and H. ROBBINS[2]

1. Introduction. Consider the following random process in which cars of length 1 are parked on a street $[0, x]$ of length $x \geqq 1$. The first car is parked so that the position of its center is a random variable which is uniformly distributed on $\left[\frac{1}{2}, x - \frac{1}{2}\right]$. If there remains space to park another car then a second car is parked so that its center is a random variable which is uniformly distributed over the set of points in $\left[\frac{1}{2}, x - \frac{1}{2}\right]$ whose distance from the first car is $\geqq \frac{1}{2}$. If there remains an empty interval of length $\geqq 1$ on the street then a third car is parked, its center being uniformly distributed over the set of points whose distance from the cars already parked and the ends of the street is $\geqq \frac{1}{2}$. The process continues until there remains no empty interval of length $\geqq 1$. We denote by N_x the total number of cars parked and extend the definition of N_x to all $x \geqq 0$ by putting $N_x = 0$ for $0 \leqq x < 1$.

The „parking problem" is the study of the distribution of the integer-valued random variable N_x as $x \to \infty$. This problem was called to our attention by C. Derman and M. Klein in 1957. In 1958 A. Rényi [1] proved that the expectation $\mu(x) = \mathbf{E}(N_x)$ satisfies the relation

$$\mu(x) = \lambda_1 x + \lambda_1 - 1 + O(x^{-n}) \qquad (n \geqq 1) \tag{1.1}$$

(O and o refer throughout to the argument increasing to infinity); the constant λ_1 is given by

$$\lambda_1 = \int_0^\infty e^{-2\int_0^t \frac{1-e^{-u}}{u}\,du}\,dt \qquad \lambda_1 \simeq 0.748 \tag{1.2}$$

To prove (1.1) Rényi employs the Laplace transform of a certain integral equation satisfied by $\mu(x)$; using similar methods P. Ney [2] has studied the higher

[1] Hebrew University Jerusalem
[2] Columbia University New York

moments of N_x. In the present paper we show by a direct analysis of the integral equation that (1.1) can be strengthened to

$$\mu(x) = \lambda_1 x + \lambda_1 - 1 + O\left(\left(\frac{2e}{x}\right)^{x-3/2}\right) \tag{1.3}$$

and that the variance $\sigma^2(x) = \mathbf{E}(N_x - \mu(x))^2$ satisfies

$$\sigma^2(x) = \lambda_2 x + \lambda_2 + O\left(\left(\frac{4e}{x}\right)^{x-4}\right) \tag{1.4}$$

where λ_2 is some positive constant. We show moreover that the standardized random variable $Z_x = (N_x - \mu(x))/\sigma(x)$ has the limiting normal (0,1) distribution as $x \to \infty$; this is done in two ways, the first by showing that all the moments of Z_x converge to the normal moments, and the second by a direct argument using the central limit theorem for sums of independent random variables.

In Section 2 we derive the integral equations satisfied by $\mu(x)$ and quantities related to the higher moments of N_x; these equations form the basis of our study as well as those of RÉNYI and NEY. Section 3 deals with the asymptotic behaviour of the solutions of these equations; our work here is somewhat similar to that of N. G. DE BRUIJN [3]. The results of Section 3 are applied in Section 4 to the parking problem. The second proof of the asymptotic normality of Z_x is given in Section 5. Various remarks will be found in Section 6.

2. Derivation of the integral equations. For $x \geqq 0$ let $[t, t+1]$ be the random interval occupied by the first car parked on a street $[0, x+1]$ of length $x+1$. The parking process described in Section 1 is such that the number of cars which will eventually be parked to the left of the first car is independent of the number which will be parked to the right of it; moreover, the number of cars eventually parked to the left of the first car, i.e. on $[0, t]$, has the same distribution as N_t, while the number parked to the right of the first car, i.e. on $[t+1, x+1]$, has the same distribution as N_{x-t}. Hence *the conditional distribution of N_{x+1} given that the first car occupies $[t, t+1]$ is the same as the distribution of $N_t + N_{x-t} + 1$ with N_t, N_{x-t} independent*. Denoting by $|\,t$ conditioning on the event that the first car is parked at $[t, t+1]$ we therefore have

$$\mathbf{E}(N_{x+1}\,|\,t) = \mathbf{E}(N_t) + \mathbf{E}(N_{x-t}) + 1 \qquad (0 \leqq t \leqq x) \tag{2.1}$$

(here we do not use the independence of N_t and N_{x-t}). Since by hypothesis t is uniformly distributed on $[0, x]$ it follows that, setting

$$\mu(x) = \mathbf{E}(N_x)\,, \tag{2.2}$$

we have

$$\mu(x+1) = \frac{2}{x}\int_0^x \mu(t)\,dt + 1 \qquad (x > 0)\,. \tag{2.3}$$

Defining the function

$$f(x) = \mu(x) + 1 \tag{2.4}$$

we see that f satisfies the somewhat simpler equation

$$f(x+1) = \frac{2}{x}\int_0^x f(t)\,dt \qquad (x > 0). \tag{2.5}$$

Together with the initial conditions

$$f(x) = 1 \quad (0 \leqq x < 1), \quad f(1) = 2 \tag{2.6}$$

this determines $f(x)$ consecutively over the intervals $1 < x \leqq 2$, $2 < x \leqq 3$, ... Thus we find

$$f(x) = 2 \qquad (1 < x \leqq 2), \tag{2.7}$$

$$f(x) = 4 - \frac{2}{x-1} \qquad (2 < x \leqq 3), \tag{2.8}$$

$$f(x) = 8 - \frac{10}{x-1} - \frac{4}{x-1}\log(x-2) \qquad (3 < x \leqq 4), \tag{2.9}$$

at which the integration of (2.5) becomes difficult.

Using the independence of N_t and N_{x-t} we have for the function

$$\sigma^2(x) = \mathbf{D}^2(N_x) = \mathbf{E}(N_x - \mu(x))^2 \tag{2.10}$$

the relation

$$\mathbf{D}^2(N_{x+1}\,|\,t) = \sigma^2(t) + \sigma^2(x-t) \qquad (0 \leqq t \leqq x) \tag{2.11}$$

Since

$$\mathbf{D}^2(N_{x+1}) \geqq \mathbf{E}(\mathbf{D}^2(N_{x+1}\,|\,t)), \tag{2.12}$$

it follows from (2.11) that

$$\sigma^2(x+1) \geqq \frac{2}{x}\int_0^x \sigma^2(t)\,dt \qquad (x > 0). \tag{2.13}$$

Let

$$L(x) = \lambda_1 x + \lambda_1 - 1, \tag{2.14}$$

where λ_1 is a constant to be determined later, and define for $k = 0, 1, \ldots$

$$\varphi_k(x) = \mathbf{E}((N_x - L(x))^k). \tag{2.15}$$

Since

$$L(x+1) = L(t) + L(x-t) + 1, \tag{2.16}$$

we have

$$\mathbf{E}[(N_{x+1} - L(x+1))^k\,|\,t] = \mathbf{E}[\{(N_t - L(t)) + (N_{x-t} - L(x-t))\}^k], \tag{2.17}$$

and on integrating we find that

$$\varphi_k(x+1) = \frac{1}{x} \sum_{i=0}^{k} \binom{k}{i} \int_0^x \varphi_i(t)\, \varphi_{k-i}(x-t)\, dt \qquad (x > 0)\,. \tag{2.18}$$

3. The integral equations. Our results on the behaviour as $x \to \infty$ of functions satisfying certain integral equations are summarized in the following two theorems.

Theorem 1. *Let $f(x)$ be defined for $x \geqq 0$ and satisfy*

$$f(x+1) = \frac{2}{x} \int_0^x f(t)\, dt + p(x+1) \qquad (x > 0) \tag{3.1}$$

where $p(x)$ is continuous for $x > 1$ and is such that, setting

$$p_x = \sup_{x \leqq t \leqq x+1} |p(t)| \qquad (x > 1)\,, \tag{3.2}$$

we have

$$\sum_{i=2}^{\infty} \frac{p_i}{i} < \infty\,. \tag{3.3}$$

Then there exists a constant λ such that, setting

$$R_j = \frac{2j+1}{j} p_{j+1} + \frac{2(j+1)(j+3)}{j} \sum_{i=j+2}^{\infty} \frac{p_i}{i+1} \qquad (j = 1, 2, \ldots)\,, \tag{3.4}$$

we have

$$\sup_{n+1 \leqq x \leqq n+2} |f(x) - \lambda x - \lambda| \leqq \frac{2^n}{n!} \sup_{1 \leqq x \leqq 2} |f(x) - \lambda x - \lambda| + \frac{2^n}{n!} \sum_{j=1}^{n} \frac{j!}{2^j} R_j \qquad (n = 1, 2, \ldots)\,. \tag{3.5}$$

Corollary. If $\alpha > 2e$ and $f(x)$ satisfies **(3.1)** with

$$p(x) = O\left(\left(\frac{\alpha}{x}\right)^{x+\beta}\right), \tag{3.6}$$

then

$$f(x) = \lambda x + \lambda + O\left(\left(\frac{\alpha}{x}\right)^{x+\beta-1}\right). \tag{3.7}$$

The second theorem is much less precise but easier to prove.

Theorem 2. *Let $g(x)$ be defined for $x \geqq 0$ and satisfy*

$$g(x+1) = \frac{2}{x} \int_0^x g(t)\, dt + O(x^\gamma) \qquad (x > 0) \tag{3.8}$$

with $\gamma > 1$. Then

$$g(x) = O(x^\gamma)\,. \tag{3.9}$$

Corollary. *Let $g(x)$ be defined for $x \geqq 0$ and satisfy*

$$g(x+1) = \frac{2}{x}\int_0^x g(t)\,dt + Ax^\beta + O(x^\gamma) \qquad (x > 0) \tag{3.10}$$

with $\beta > \gamma > 1$. Then

$$g(x) = \frac{\beta+1}{\beta-1}Ax^\beta + O(x^{\max(\beta-1,\gamma)})\,. \tag{3.11}$$

Proof of Theorem 1. The proof is less involved and leads to a somewhat sharper error estimate if, as in the case of $\mu(x)$, the term $p(x)$ vanishes identically. For the sake of brevity, however, we shall treat the general case directly.

From (3.1) we have for positive x and y,

$$f(y+1) = \frac{2}{y}\int_0^x f(t)\,dt + \frac{2}{y}\int_x^y f(t)\,dt + p(y+1) =$$

$$= \frac{1}{y}[xf(x+1) - xp(x+1)] + \frac{2}{y}\int_x^y f(t)\,dt + p(y+1)$$

or

$$f(y+1) = \frac{x}{y}f(x+1) + \frac{2}{y}\int_x^y f(t)\,dt + p(y+1) - \frac{x}{y}p(x+1)\,. \tag{3.12}$$

Define

$$I_x = \inf_{x \leqq t \leqq x+1} \frac{f(t)}{t+1}\,, \quad S_x = \sup_{x \leqq t \leqq x+1} \frac{f(t)}{t+1} \qquad (x \geqq 0)\,. \tag{3.13}$$

Notice that $f(x) = x + 1$ satisfies (3.1) with $p \equiv 0$, and hence-that

$$y + 2 = \frac{x}{y}(x+2) + \frac{2}{y}\int_x^y (t+1)\,dt\,. \tag{3.14}$$

Subtracting (3.14) multiplied by I_x from (3.12) we have

$$\begin{aligned} f(y+1) - I_x\cdot(y+2) &= \frac{x}{y}[f(x+1) - I_x\cdot(x+2)] + \\ &+ \frac{2}{y}\int_x^y [f(t) - I_x\cdot(t+1)]\,dt + p(y+1) - \frac{x}{y}p(x+1)\,. \end{aligned} \tag{3.15}$$

Hence for $x \leqq y \leqq x+1$, in view of (3.13) and (3.2),

$$f(y+1) - I_x \cdot (y+2) \geqq 0 + 0 - p_{x+1} - p_{x+1} = -2p_{x+1}. \tag{3.16}$$

It follows that

$$I_{x+1} \geqq I_x - \frac{2p_{x+1}}{x+2} \qquad (x > 0). \tag{3.17}$$

Applying (3.17) successively with x replaced by $x+1, x+2, \ldots$ we obtain

$$I_y \geqq I_x - \Delta_x, \qquad (y \geqq x > 0). \tag{3.18}$$

where by definition

$$\Delta_x = 2 \sum_{i=1}^{\infty} \frac{p_{x+i}}{x+i+1} \qquad (x > 0). \tag{3.19}$$

In exactly the same manner we obtain the inequality

$$S_y \leqq S_x + \Delta_x \qquad (y \geqq x > 0). \tag{3.20}$$

From (3.18) we have

$$\liminf_{y \to \infty} I_y \geqq I_x - \Delta_x \qquad (x > 0). \tag{3.21}$$

Since $\Delta_x = o(1)$ by (3.3) it follows that

$$\liminf_{y \to \infty} I_y \geqq \limsup_{x \to \infty} I_x. \tag{3.22}$$

From this and (3.18) with $x = 1$ we find that

$$I_\infty = \lim_{x \to \infty} I_x \text{ exists, and } I_\infty > -\infty. \tag{3.23}$$

Similarly,

$$S_\infty = \lim_{x \to \infty} S_x \text{ exists, and } S_\infty < \infty. \tag{3.24}$$

Since $I_x \leqq S_x$ it follows that

$$-\infty < I_\infty \leqq S_\infty < \infty. \tag{3.25}$$

From (3.12) we have for $x, y > 0$

$$\begin{aligned} f(y+1) - f(x+1) &= \frac{x-y}{y} f(x+1) + \\ &+ \frac{2}{y} \int_x^y f(t)\,dt + p(y+1) - \frac{x}{y} p(x+1). \end{aligned} \tag{3.26}$$

By (3.13) and (3.25), $f(x) = O(x)$, and hence by (3.26)

$$\sup_{x \leqq y \leqq x+1} |f(y+1) - f(x+1)| = O(1) + 2p_x. \tag{3.27}$$

But this implies by (3.3) that

$$S_x - I_x = o(1) \tag{3.28}$$

and therefore that

$$I_\infty = S_\infty \neq \pm\infty . \tag{3.29}$$

We now define λ as the common value in (3.29),

$$\lambda = \lim_{x\to\infty} I_x = \lim_{x\to\infty} S_x = \lim_{x\to\infty} \frac{f(x)}{x+1} . \tag{3.30}$$

By (3.18) and (3.20),

$$I_x - \Delta_x \leqq \lambda \leqq S_x + \Delta_x \qquad (x > 0). \tag{3.31}$$

Next we observe that for every $x > 1$ there exists a number x' satisfying

$$x \leqq x' \leqq x + 1 , \quad \left| \frac{f(x')}{x'+1} - \lambda \right| \leqq \Delta_x . \tag{3.32}$$

Indeed, since by (3.1) $f(x)$ is continuous for $x > 1$ the non-existence of such an x' would imply that either

$$I_x > \lambda + \Delta_x \quad \text{or} \quad S_x < \lambda - \Delta_x , \tag{3.33}$$

contradicting (3.31). We denote by x_n a value x' satisfying (3.32) for $x = n$; thus for $n = 2, 3, \ldots$

$$|f(x_n) - \lambda(x_n + 1)| \leqq (n + 2)\,\Delta_n \qquad (n \leqq x_n \leqq n + 1) \tag{3.34}$$

Now set

$$f^*(x) = f(x) - \lambda(x + 1) . \tag{3.35}$$

Then f^* again satisfies (3.1), and applying (3.12) with $n \leqq y \leqq n + 1$ and $x = x_{n+1} - 1$ we obtain from (3.34) for $n = 1, 2, \ldots$

$$\begin{aligned} |f^*(y + 1)| \leqq{} & \frac{n+1}{n}(n+3)\,\Delta_{n+1} + \frac{2}{n} \sup_{n\leqq t\leqq n+1} |f^*(t)| + \\ & + p_{n+1} + \frac{n+1}{n} p_{n+1} \end{aligned} \tag{3.36}$$

Putting

$$T_x = \sup_{x\leqq t\leqq x+1} |f^*(t)| \qquad (x > 0) , \tag{3.37}$$

we obtain from (3.36)

$$\begin{aligned} T_{n+1} \leqq{} & \frac{2}{n} T_n + \frac{2n+1}{n} p_{n+1} + \frac{(n+1)(n+3)}{n} \Delta_{n+1} = \\ & = \frac{2}{n} T_n + R_n \qquad (n = 1, 2, \ldots) \end{aligned} \tag{3.38}$$

where R_n is defined by (3.4). Successive application of this inequality for $n = 1, 2, 3, \ldots$ yields the inequality

$$T_{n+1} \leqq \frac{2^n}{n!} T_1 + \frac{2^n}{n!}\left[\frac{1!}{2} R_1 + \frac{2!}{2^2} R_2 + \ldots + \frac{n!}{2^n} R_n\right]. \tag{3.39}$$

In view of (3.37) this is precisely (3.5), and this completes the proof of Theorem 1.

Proof of Corollary. If (3.6) holds, then by (3.4)

$$R_j = O\left(\left(\frac{\alpha}{j}\right)^{j+\beta+1}\right)$$

and hence (3.5), since $\alpha > 2e$,

$$\frac{2^n}{n!}\sum_{j=1}^{n}\frac{j!}{2^j} R_j = O\left(\left(\frac{\alpha}{n}\right)^{n+\beta+1}\right).$$

Thus by (3,5)

$$\sup_{n+1 \leqq x \leqq n+2} |f(x) - \lambda x - \lambda| = O\left(\left(\frac{2e}{n}\right)^{n+\frac{1}{2}}\right) + O\left(\left(\frac{\alpha}{n}\right)^{n+\beta+1}\right) =$$

$$= O\left(\left(\frac{\alpha}{n}\right)^{n+\beta+1}\right),$$

from which (3.7) follows.

Proof of Theorem 2. We have

$$g(x+1) = \frac{2}{x}\int_0^x g(t)\,dt + \eta(x) \qquad (x > 0),$$

where

$$\eta(x) = O(x^\gamma), \qquad \gamma > 1.$$

Choose $x_0 > 1$ and $H > 0$ such that

$$|\eta(x)| \leqq Hx^\gamma \quad \text{for} \quad x \geqq x_0 - 1,$$

$$\int_0^{x_0} |g(t)|\,dt \leqq \frac{H}{\gamma - 1}(x_0 - 1)^{\gamma+1} = \frac{\gamma+1}{\gamma-1} H \int_0^{x_0-1} t^\gamma\,dt. \tag{3.40}$$

Then for $x_0 - 1 \leqq x \leqq x_0$ we have

$$|g(x+1)| \leqq \frac{2}{x}\int_0^{x_0} |g(t)|\,dt + Hx^\gamma \leqq$$

$$\leqq \frac{2H}{x(\gamma-1)}(x_0 - 1)^{\gamma+1} + Hx^\gamma \leqq \tag{3.41}$$

$$\leqq \frac{2Hx^\gamma}{\gamma - 1} + Hx^\gamma = \frac{\gamma+1}{\gamma-1} Hx^\gamma.$$

Hence

$$\int_0^{x_0+1} |g(t)|\,dt = \int_0^{x_0} |g(t)|\,dt + \int_{x_0}^{x_0+1} |g(t)|\,dt \leq$$

$$\leq \frac{\gamma+1}{\gamma-1} H \int_1^{x_0} (t-1)^\gamma\,dt + \int_{x_0}^{x_0+1} \frac{\gamma+1}{\gamma-1} H(t-1)^\gamma\,dt =$$

$$= \frac{\gamma+1}{\gamma-1} H \int_0^{x_0} t^\gamma\,dt\,,$$

so that (3.40) holds with x_0 replaced by x_0+1. Hence by (3.41), for $x_0 \leq x \leq$ $\leq x_0 + 1$ we have

$$|g(x+1| \leq \frac{\gamma+1}{\gamma-1} H x^\gamma\,. \tag{3.42}$$

By induction, (3.42) holds for all $x \geq x_0 - 1$, which proves (3.9).

Proof of Corollary. Set

$$g^*(x) = \frac{\beta+1}{\beta-1} A x^\beta\,.$$

Then

$$g^*(x+1) = \frac{\beta+1}{\beta-1} A(x+1)^\beta = \frac{\beta+1}{\beta-1} A x^\beta + O(x^{\beta-1}) =$$

$$= \frac{2}{x} \int_0^x g^*(t)\,dt + A x^\beta + O(x^{\beta-1})\,.$$

Hence, setting

$$\bar{g}(x) = g(x) - g^*(x)$$

we have for $x > 0$,

$$\bar{g}(x+1) = g(x+1) - g^*(x+1) = \frac{2}{x} \int_0^x \bar{g}(t)\,dt + O(x^{\max(\beta-1,\gamma)}).$$

Hence by Theorem 2,

$$\bar{g}(x) = O(x^{\max(\beta-1,\gamma)})$$

which proves (3.11).

Remarks.

1. If G is the lim sup as $x \to \infty$ of $g(x)$ in (3.8) divided by x^γ, then by taking x_1 sufficiently large we have

$$\frac{|g(x)|}{x^\gamma} \leq G + \varepsilon \qquad \text{for all} \quad x \geq x_1\,.$$

Then for $x \geq x_1$,

$$|g(x+1)| \leq \frac{2}{x_1} \int_0^{x_1} |g(t)|\,dt + \frac{2}{x} \int_0^x (G+\varepsilon)\, t^\gamma\,dt + \eta(x)\,,$$

where $\eta(x)$ denotes the error term in (3.8).

Hence

$$G = \limsup_{x\to\infty} \frac{g(x+1)}{(x+1)^\gamma} \leqq 0 + \frac{2(G+\varepsilon)}{\gamma+1} + \limsup_{x\to\infty} \frac{\eta(x)}{(x+1)^\gamma}.$$

Suppose now that (3.8) holds with O replaced by o. Since $\varepsilon > 0$ was arbitrary it follows that

$$G \leqq \frac{2G}{\gamma+1},$$

and since $\gamma > 1$ it follows that $G = 0$. Hence Theorem 2 holds if O is replaced by o in both (3.8) and (3.9).

2. Theorem 1 continues to hold if (3.1) is replaced by

$$(3.1)' \qquad f(x+1) = \frac{2}{x}\int_0^x f(t)\,dt + \frac{C}{x} + p(x+1) \qquad (x > 0)$$

where C is any constant; this follows from the fact that the fundamental relation (3.12) follows from (3.1)′. Thus e.g. if $p(x+1)$ in (3.1) is of the form

$$\frac{C}{x} + O\left(\left(\frac{a}{x}\right)^{x+\beta}\right) \qquad (a > 2e)$$

then (3.7) still holds.

4. Application to the parking problem. Since $f(x) = \mu(x) + 1$ satisfies, by (2.5), the equation (3.1) with $p \equiv 0$, we have by Theorem 1 that

$$(4.1) \qquad \lim_{x\to\infty} \frac{\mu(x)}{x} = \lambda_1$$

exists, and by (3.31) for every $x > 0$,

$$(4.2) \qquad \inf_{x\leq t\leq x+1} \frac{\mu(t)+1}{t+1} = I_x \leqq \lambda_1 \leqq S_x = \sup_{x\leq t\leq x+1} \frac{\mu(t)+1}{t+1}.$$

Taking $x = 2$ we obtain easily from (2.8) that

$$(4.3) \qquad 0.66\ldots = \frac{2}{3} \leqq \lambda_1 \leqq 3 - \sqrt{5} = 0.76\ldots,$$

and (2.9) yields much narrower bounds. Since I_x and S_x approach λ_1 very rapidly it is easy to obtain extremely good approximations from (4.2) (cf. (1.2)). Since $\mu(x) = 1$ for $1 \leqq x \leqq 2$, even the crude approximation $1/2 < \lambda_1 < < 1$ yields

$$(4.4) \qquad \begin{aligned} \sup_{1\leq x\leq 2} |\mu(x) + 1 - \lambda_1 x - \lambda_1| &= \max_{1\leq x\leq 2} |2 - \lambda_1 x - \lambda_1| = \\ &= \max(|2 - 2\lambda_1|, |2 - 3\lambda_1|) < 1. \end{aligned}$$

Hence from Theorem 1 with $p \equiv 0$ we have

Theorem 3. *There exists a constant* $\lambda_1 \left(\frac{1}{2} < \lambda_1 < 1\right)$ *such that the expectation* $\mu(x)$ *of* N_x *satisfies the relation*

$$\sup_{n+1 \leq x \leq n+2} |\mu(x) + 1 - \lambda_1 x - \lambda_1| < \frac{2^n}{n!} \qquad (n = 0, 1, \ldots). \tag{4.5}$$

By Stirling's formula it follows that

$$\mu(x) - \lambda_1 x - \lambda_1 + 1 = O\left(\left(\frac{2e}{x}\right)^{x-3/2}\right). \tag{4.6}$$

We now define $L(x)$ and $\varphi_k(x)$ by (2.14) and (2.15) with λ_1 given by (4.1) Then by (2.18) with $k = 2$ we have

$$\varphi_2(x+1) = \frac{2}{x}\int_0^x \varphi_2(t)\,dt + \frac{2}{x}\int_0^x \varphi_1(t)\,\varphi_1(x-t)\,dt \qquad (x > 0) \tag{4.7}$$

But $\varphi_1(x)$ is precisely the left hand member of (4.6), and therefore

$$\sup_{0<t<x} |\varphi_1(t)\,\varphi_1(x-t)| = O\left(\left(\frac{4e}{x}\right)^{x-3}\right). \tag{4.8}$$

Thus $f(x) = \varphi_2(x)$ satisfies (3.1) with $p(x)$ estimated by (4.8). From this we deduce

Theorem 4. *There exists a constant* $\lambda_2 > 0$ *such that the variance* $\sigma^2(x)$ *of* N_x *satisfies the relation*

$$\sigma^2(x) = \lambda_2 x + \lambda_2 + O\left(\left(\frac{4e}{x}\right)^{x-4}\right). \tag{4.9}$$

Proof. $\varphi_2(x)$ satisfies (4.9) by the Corollary to Theorem 1, and

$$\sigma^2(x) - \varphi_2(x) = -(\varphi_1(x))^2,$$

which, by (4.6), is absorbed into the error term. It remains to show that $\lambda_2 > 0$. This may be done numerically from estimates obtained in the course of the proof of Theorem 1, but it is much simpler to deduce it as follows. Since $\sigma^2(x) \neq 0$ for $2 < x < 3$ it follows from (2.13) that $\sigma^2(x) > \frac{\delta}{x}$ for some $\delta > 0$. But this contradicts (4.9) unless $\lambda_2 > 0$.

We now prove a result on the central moments of N_x.

Theorem 5. For every $k = 1, 2, \ldots$ and $\varepsilon > 0$,

$$\mathbf{E}((N_x - \mu(x))^k) = c_k x^{\left[\frac{k}{2}\right]} + O\left(x^{\left[\frac{k}{2}\right]-1+\varepsilon}\right) \tag{4.10}$$

($[x]$ denotes the greatest integer $\leq x$), where the c_k are constants and

$$c_{2k} = \frac{(2k)!}{2^k\,k!}\lambda_2^k. \tag{4.11}$$

Proof. Since by (2.15) for $k = 1$

$$N_x - \mu(x) = N_x - L(x) - (\mu(x) - L(x)) = N_x - L(x) - \varphi_1(x),$$

it follows from (4.6) that (4.10) is equivalent to

$$\varphi_k(x) = c_k x^{\left[\frac{k}{2}\right]} + O\left(x^{\left[\frac{k}{2}\right]-1+\varepsilon}\right). \tag{4.12}$$

By (4.6) and (4.9), (4.10) holds for $k = 1,2$ and (4.11) holds for $k = 1$. By (2.18)

$$\varphi_3(x+1) = \frac{2}{x}\int_0^x \varphi_3(t)\,dt + \frac{6}{x}\int_0^x \varphi_1(t)\,\varphi_2(x-t)\,dt,$$

and by (4.6) and (4.9) the second integrand is $O\left(\left(\frac{C}{x}\right)^x\right)$ with a suitable C. Hence φ_3 satisfies (3.1) with p estimated as in (3.6). It follows from (3.7) that $\varphi_3(x) = c_3 x + O(1)$ and thus (4.12) holds for $k \leq 3$.

Now let $m > 3$ and assume that (4.12) holds for $k < m$. Then by (2.18),

$$\varphi_m(x+1) = \frac{2}{x}\int_0^x \varphi_m(t)\,dt + \sum_{i=1}^{m-1}\binom{m}{i}\frac{1}{x}\int_0^x \varphi_i(t)\,\varphi_{m-i}(x-t)\,dt \quad (x > 0). \tag{4.13}$$

By the induction assumption

$$\varphi_i(t)\,\varphi_{m-i}(x-t) = c_i c_{m-i}\, t^{\left[\frac{i}{2}\right]}(x-t)^{\left[\frac{(m-i)}{2}\right]} + O\left(x^{\left[\frac{i}{2}\right]+\left[\frac{m-i}{2}\right]-1+\varepsilon}\right). \tag{4.14}$$

Since

$$\frac{1}{x}\int_0^x t^{\left[\frac{i}{2}\right]}(x-t)^{\left[\frac{m-i}{2}\right]}\,dt = \frac{\left[\frac{i}{2}\right]!\left[\frac{m-i}{2}\right]!}{\left(\left[\frac{i}{2}\right]+\left[\frac{m-i}{2}\right]+1\right)!}\, x^{\left[\frac{i}{2}\right]+\left[\frac{m-i}{2}\right]} \tag{4.15}$$

and since

$$\max_{1\leq i\leq m-1}\left(\left[\frac{i}{2}\right]+\left[\frac{m-i}{2}\right]\right) = \left[\frac{m}{2}\right] \qquad \text{for } m \geq 3,$$

the sum on the right hand side of (4.13) is

$$\text{Const.}\ x^{\left[\frac{m}{2}\right]} + O\left(x^{\left[\frac{m}{2}\right]-1+\varepsilon}\right). \tag{4.16}$$

Since $\left[\frac{m}{2}\right] \geq 2$ for $m > 3$, (4.12) for $k = m$ follows from (4.13) by the Corollary of Theorem 2. Thus (4.12) holds for all $k = 1, 2, \ldots,$

By (4.13), (4.14), and (4.15) the constant in (4.16) for $m = 2k$ is

$$\sum_{j=1}^{k-1} \binom{2k}{2j} j! \, \frac{(k-j)!}{(k+1)!} c_{2j} c_{2k-2j} . \tag{4.17}$$

Assume that (4.11) holds for $c_2, c_4, \ldots, c_{2k-2}$. By (4.17) the coefficient of x^k in the equation

$$\varphi_{2k}(x+1) = \frac{2}{x} \int_0^x \varphi_{2k}(t)\, dt + cx^k + O(x^{k-1+\varepsilon})$$

is

$$\frac{(k-1)\,(2k)!}{(k+1)!\, 2^k} \lambda_2^k ,$$

so that by the Corollary of Theorem 2

$$\varphi_{2k}(x) = \frac{k+1}{k-1} \, \frac{(k-1)\,(2k)!}{(k+1)!\, 2^k} \lambda_2^k x^k + O(x^{k-1+\varepsilon}) ,$$

and hence

$$c_{2k} = \frac{(2k)!}{k!\, 2^k} \lambda_2^k ,$$

so that (4.11) holds for all $k = 1, 2, \ldots$. This completes the proof of Theorem 5.

Theorem 6. *The random variable*

$$Z_x = \frac{N_x - \mu(x)}{\sigma(x)}$$

is asymptotically normal (0,1) *as* $x \to \infty$.

Proof. By (4.10), (4.11) and (4.9) for $\varepsilon = 1/2$,

$$\mathbf{E}(Z_x^k) = \frac{c_k x^{\left[\frac{k}{2}\right]} + o\left(x^{\left[\frac{k}{2}\right]}\right)}{(\lambda_2 x + o(x))^{\frac{k}{2}}}$$

where $\lambda_2 > 0$ and

$$c_{2k} = \frac{(2k)!}{2^k\, k!} \lambda_2^k \qquad (k = 1, 2, \ldots) .$$

Hence

$$\lim_{x \to \infty} \mathbf{E}(Z_x^k) = \begin{cases} \dfrac{k!}{2^{\frac{k}{2}} \left(\frac{k}{2}\right)!} & (k \text{ even}) , \\ 0 & (k \text{ odd}) . \end{cases}$$

Since these are the moments of the normal (0,1) distribution which is uniquely determined by its moments, the theorem follows from the moment convergence theorem.

5. Another proof of the asymptotic normality. This proof will use very much less information about the moments of N_x than that of the preceding section. In fact it will be based entirely on the relation

$$\sigma^2(x) = \lambda_2 x + o(x)\,, \qquad (\lambda_2 > 0)\,. \tag{5.1}$$

We shall need two simple lemmas.

Lemma 1. *Let $\psi(x)$ be a non-negative function defined for $x \geqq 0$, bounded over finite intervals and satisfying $\psi(x) = o(x)$. Then $n = o(x)$ implies*

$$\sup \sum_{i=1}^{n} \psi(x_i) = o(x)\,, \tag{5.2}$$

the sup being taken over all sets of non-negative $x_1, \ldots, x_n$ with $x_1 + \ldots + x_n = x$.

Indeed, $\psi(x) < H + Hx$ for all $x \geqq 0$ with a suitable H. Let $\delta > 0$ be given and choose $a = a(\delta)$ so that $\psi(x) < \delta x$ for $x > a$. Divide the sum in (5.2) into two parts, one over the i with $x_i \leqq a$, the other over the remaining i. Then the first sum is $\leqq n(H + H a)$ while the second is $< \delta x$. Hence the left side of (5.2) is bounded by $2\delta x$ for large x and the lemma is established.

Lemma 2. *Given $\varepsilon > 0$, there exists $\delta = \delta(\varepsilon) > 0$ such that if $Y_0, Y_1, \ldots, Y_n$ are independent random variables satisfying*

$$\left| \sum_{i=0}^{n} \mathbf{E}(Y_i) \right| \leqq \delta\,, \tag{5.3}$$

$$\left| \sum_{i=0}^{n} \mathbf{D}^2(Y_i) - 1 \right| \leqq \delta\,, \tag{5.4}$$

$$| Y_i - \mathbf{E}(Y_i) | \leqq \delta\,, \qquad (i = 0, 1, \ldots, n), \tag{5.5}$$

then the distribution function of $\sum_{i=0}^{n} Y_i$ approximates uniformly to within ε the normal distribution with zero mean and unit variance.

It is clearly sufficient to establish the lemma with (5.3) replaced by $\mathbf{E}(Y_i) = 0$ $(i = 0, 1, \ldots, n)$ and δ replaced by 0 in (5.4). But then the lemma follows at once from the 'triangular' version of Liapounov's theorem.

We now proceed to the proof of the asymptotic normality of

$$Z_x = \frac{N_x - \mu(x)}{\sigma(x)}\,. \tag{5.6}$$

Let $n = n_x$ be a fixed non-negative integer-valued function of x defined for $x > 2$ and satisfying

$$0 \leqq n_x \leqq x/2\,, \quad n_x = o(x)\,. \tag{5.7}$$

(Eventually it will be specified further.) Consider the first $n - 1 \geqq 1$ cars parked on $[0, x]$. Denote by y_1 the distance between 0 and the leftmost car, by y_2 that between this car and the one parked second from the left, etc., by y_n the distance between the car parked to the extreme right and x. Then (see the derivation of the italicized statement in Section 2) the conditional distribution of N_x given $\underline{y} = (y_1, y_2 \ldots, y_n)$ is the same as the distribution of $n - 1 + N_{y_1} + N_{y_2} + \ldots + N_{y_n}$ with $N_{y_1}, N_{y_2}, \ldots, N_{y_n}$ independent. Therefore, the conditional distribution of Z_x given $\underline{y}$ is equal to the distribution of $\sum_{i=0}^{n} Y_i$, with the Y_i independent and defined by

$$(5.8) \qquad Y_i = \frac{N_{y_i}}{\sigma(x)} \ (i = 1, \ldots, n), \ Y_0 = \frac{n - 1 - \mu(x)}{\sigma(x)}.$$

Applying Lemma 1 with $\psi(x) = | \sigma^2(x) - \lambda_2 x |$ we deduce from (5.1) and (5.7) that

$$\sum_{i=1}^{n} | \sigma^2(y_i) - \lambda_2 y_i | = o(x), \quad \text{or} \quad \sum_{i=1}^{n} \sigma^2(y_i) = \lambda_2 x + o(x)$$

for every $\underline{y}$. Hence we obtain

$$(5.9) \qquad \mathbf{D}^2(Z_x | \underline{y}) = \sum_{i=0}^{n} \mathbf{D}^2(Y_i | \underline{y}) = 1 + o(1)$$

for the conditional variance of Z_x. Thus (5.4) holds for $Y_i = Y_i(\underline{y})$ for all sufficiently large x and all random vectors $\underline{y}$.

From

$$1 = \mathbf{E}(Z_x^2) = \mathbf{E}\{\mathbf{D}^2(Z_x | \underline{y}) + \mathbf{E}^2(Z_x | \underline{y})\}$$

and (5.9) we see that

$$(5.10) \qquad \mathbf{E}(\mathbf{E}^2(Z_x | \underline{y})) = o(1)$$

Let A_x be the event: y is such that $| \mathbf{E}(Z_x | \underline{y}) | \leq \delta$; then it follows from (5.10) that for any fixed $\delta > 0$,

$$(5.11) \qquad \lim_{x \to \infty} \mathbf{P}(A_x) = 1,$$

and $\underline{y} \in A_x$ implies that $Y_i = Y_i(\underline{y})$ satisfy (5.3).

We now specify the function $n = n_x$ by putting

$$(5.12) \qquad n = [x^{1/2} \log^2 x]$$

and let $B_x = B_x(\eta)$ denote the event

$$(5.13) \qquad \max_{i=1,\ldots,n} y_i < \eta x^{1/2}, \qquad (\eta > 0)$$

Take $k = \left[\frac{2x^{1/2}}{\eta}\right] + 1$ and divide $[0, x]$ into k equi-long intervals $I_1, \ldots, I_k$. Then if (5.13) were false it would imply that at least one of the intervals I_j $(j = 1, \ldots, k)$ is disjoint from the first $n - 1$ cars parked. The probability of this is smaller than

$$k\left(1 - \frac{1}{k}\right)^{n-1} < \left(\frac{2x^{1/2}}{\eta} + 1\right)\left(1 - \frac{\eta}{2x^{1/2}}\right)^{x^{1/2}\log^2 x - 2}$$

and, thus, tends to zero as $x \to \infty$. Hence

$$\lim_{x \to \infty} \mathbf{P}(B_x) = 1 . \tag{5.14}$$

But $Y_0(\underline{y})$ is a constant and, taking $\eta < \frac{\delta}{2} \lambda_2^{1/2}$, (5.13) implies $|Y_i(\underline{y})| < \delta$ $(i = 1, \ldots, n)$ for large x and hence that $Y_i = Y_i(\underline{y})$ satisfy (5.5).

From the above and Lemma 2 we conclude that the conditional distribution of Z_x given $A_x \cap B_x$ is asymptotically normal with zero mean and unit variance. It then follows from (5.11) and (5.14) that the same holds for the distribution of Z_x itself, and the proof is complete.

6. Remarks. 1. The parking process described in Section 1 may also be described as the process of taking independent observations on a rectangular random variable, but rejecting all those observations which differ by less than unity from any previously observed and not rejected observation. The retained observations form a finite dependent stochastic sequence and we have studied the asymptotic behaviour of the length of this sequence. It would be interesting to extend the results to other kinds of dependence, and the preceding section indicates such possibilities; however, one would have to prove some relations like (5.1) and (5.2) and we do not know how to do this under reasonably general assumptions (see, however, the next remark).

2. Returning to the parking problem, we may equivalently consider a street of unit length and cars of length $1/x$ with x tending to infinity. This suggests at once generalizing the problem by replacing the rectangular density by other probability densities. Assume e.g. that the position of the center of each parked car is a random variable whose density is constant on each half of the street but that the constants in the two halves are different. Even in this simple case it is not quite trivial to prove rigorously that the expected total number of cars parked will be approximately $\lambda_1 x$, of those parked in the left half approximately $\lambda_1 x/2$ etc. However, the technique of the end of Section 5 can be used here. This makes it possible to treat densities which are step functions etc.; we expect to study in a future paper the case of continuous densities.

3. In the uniform density case the distribution of the lengths of the empty spaces between the parked cars has been considered by G. Bánkövi [4].

4. It is natural to consider the parking problem in more dimensions. No functional equation similar to the one derived here is available, and a rigorous treatment becomes extremely difficult. In the plane one would consider, say, placing unit squares, with sides parallel to the axes, uniformly in a convex region. Such curiosities occur as lowering the expected total of squares placed while increasing the region (consider, in the (u, v) plane the regions

$$-5/4 \leq u \leq 5/4,\ 0 \leq v \leq 1 \ \text{and}\ v \geq 0,\ u + v \leq 9/4,\ v - u \leq 9/4) .$$

In the one-dimensional case it is clear from the functional equation (transformed as in (3.12)) that $\mu(x)$ is monotone, but the analogous result for homothetic regions in the plane is not evident, even if we confine ourselves to regions which are squares. Some numerical studies of the problem of placing squares in the plane have been carried out by Mrs. I. PALÁSTI, [5].

5. Differentiating (2.3) we have $x\mu'(x+1)+\mu(x+1)=2\mu(x)+1$. In view of (4.6) $\mu'(x)$ is approximated extremely closely by $\lambda_1(x+1)/(x-1)$. Higher derivatives may be treated similarly ($\mu(x)$ is, of course, n times differentiable for $x>n$). The same remarks apply to $\sigma^2(x)$ etc.

6. The estimates of the error involved in Theorem 1 can be somewhat sharpened, but this necessitates much work and we seem to have reached the point of diminishing returns. It may be more interesting to study other functional equations by the same method.

(Received January 3, 1964)

REFERENCES

[1] RÉNYI, A.: "On a one-dimensional problem concerning space-filling". *Publ. of the Math. Inst. of the Hungarian Acad. of Sciences*, **3** (1958) 109—127.

[2] NEY, P. E.: "A random interval filling problem". *Annals of Math. Stat.* **33** (1962), 702—718.

[3] DE BRUIJN, N. G.: „On some linear functional equations". *Publicationes Mathematicae* (Debrecen) **1** (1950) 129—134.

[4] BÁNKÖVI, G.: „On gaps generated by a random space filling procedure". *Publ. Math. Inst. Hung. Acad. Sci* **7** (1962) 395—407.

[5] PALÁSTI, I.: „On some random space filling problems". *Publ. Math. Inst. Hung. Acad. Sci.* **5** (1960) 353—360.

О ЗАДАЧЕ »ПАРКИРОВАНИЯ«

A. DVORETZKY и H. ROBBINS

Резюме

В работе [1] A. RÉNYI исследовал одномерную задачу о случайном. заполнении пространства (модель «паркирования»). Процедура состоит в последовательном расположении на отрезке $(0, x)$ случайным образом непересекающихся единичных отрезков. Число расположимых отрезков N_x — случайная величина.

Авторы исследуют асимптотическое поведение моментов величины N_x ((4.6), (4.9), (4.10)). Доказывается двумя способами, что величина Z_x (нормированная величина N_x) имеет асимптотически нормальное распределение с параметрами (0,1) при $x\to\infty$.

Reprinted from
Publ. Math. Inst. Hung. Acad. Sci. Ser. A
9, 209–225 (1964)

The Annals of Mathematical Statistics
1968, Vol. 39, No. 1, 256–257

ESTIMATING THE TOTAL PROBABILITY OF THE UNOBSERVED OUTCOMES OF AN EXPERIMENT[1]

BY HERBERT E. ROBBINS

Columbia University

An experiment has the possible outcomes $E_1, E_2, \cdots$ with unknown probabilities $p_1, p_2, \cdots; p_i \geqq 0, \sum_i p_i = 1$. In n independent trials suppose that E_i occurs x_i times, $i = 1, 2, \cdots$, with $\sum_i x_i = n$. Let $\varphi_i = 1$ or 0 according as $x_i = 0$ or $x_i \neq 0$. Then the random variable $u = \sum_i p_i \varphi_i$ is the sum of the probabilities of the unobserved outcomes. How can we "estimate" u? (The quotation marks appear because u is not a parameter in the usual statistical sense.)

Suppose we make one more independent trial of the same experiment, and that in the total of $n + 1$ trials E_i occurs y_i times, $i = 1, 2, \cdots$, with $\sum_i y_i = n + 1$. (Each $y_i = x_i$, except for one value of the subscript.) Let $\psi_i = 1$ or 0 according as $y_i = 1$ or $y_i \neq 1$. Consider the statistic $v = (n + 1)^{-1} \sum_i \psi_i$, which is the number of "singleton" outcomes of the $n + 1$ trials, divided by $n + 1$. In contrast to u, v is observable. The idea of using something like v to estimate u goes back to A. M. Turing according to [1], where the problem is discussed from a somewhat different point of view. We shall show that v is a good "estimator" of u in the sense that setting $w = u - v$ we have always

$$Ew = 0, \qquad Ew^2 < (n + 1)^{-1}.$$

In fact,

$$Ew = \sum_i (p_i q_i^n - (n + 1)(n + 1)^{-1} p_i q_i^n) = \sum_i 0 = 0,$$

and a little algebra shows that $(n + 1)\, Ew^2$ is equal to the expression

$$(1) \qquad \sum_i p_i q_i^n (1 + (n - 1)p_i) - \sum_{i \neq j} p_i p_j (1 - p_i - p_j)^n.$$

Hence since $1 - x \leqq e^{-x}$,

$$(n + 1) Ew^2 \leqq \sum_i p_i e^{-np_i} e^{(n-1)p_i} = \sum_i p_i e^{-p_i} < \sum_i p_i = 1.$$

In the special case in which some k of the p_i are equal to $1/k$ and all the others are 0, the expression (1) reduces to

$$(2) \qquad (1 + (n - 1)k^{-1})(1 - k^{-1})^n - (1 - k^{-1})(1 - 2k^{-1})^n.$$

Set

$$a_n = \sup (1) \text{ for all probability vectors } (p_1, p_2, \cdots),$$
$$b_n = \sup (2) \text{ for all } k = 1, 2, \cdots;$$

Received 21 August 1967.

[1] This research was sponsored by the Office of Naval Research under Contract Nonr-4259(08) at Columbia University.

then $b_n \leqq a_n \leqq 1$. Putting $k = n/\lambda$, keeping λ fixed, and letting $n \to \infty$, we have

$$(2) \to (1 + \lambda)e^{-\lambda} - e^{-2\lambda} \leqq (1 + \lambda^*)e^{-\lambda^*} - e^{-2\lambda^*} = b \cong .61,$$

where $\lambda^* \cong .85$ is the root of $\lambda = 2e^{-\lambda}$. Hence $b_n \to = b$ as $n \to \infty$. We do not know if $a_n = b_n$, $a_n \leq b$, or $a_n \to b$. In any case, the universal inequality $Ew^2 < (n + 1)^{-1}$ can certainly be improved, and can be used together with the Chebyshev inequality to obtain "confidence intervals" for u. (Similar inequalities for higher moments of w may yield shorter intervals.) There are many other statistics v based on the $n + 1$ trials such that always $E(u - v) = 0$. We do not know which is best in the sense of minimizing $E(u - v)^2$.

REFERENCE

[1] Good, I. J. (1953). On the population frequencies of species and the estimation of population parameters. *Biometrika* 40 237–264.

MATHEMATICAL PROBABILITY IN ELECTION CHALLENGES

MICHAEL O. FINKELSTEIN*
HERBERT E. ROBBINS**

Defeated candidates in primary elections sometimes challenge the results in court and collect evidence of irregularities in support of their claims that the contests should be rerun. Frequently, this evidence consists solely of proof that certain numbers of persons voted who were not qualified, with no evidence of fraud and no indication as to how such persons voted. How large must this group be before a new election should be ordered?

The New York Election Law provides that a new primary election may be ordered when the "irregularities . . . render impossible a determination as to who rightfully was . . . elected."[1] Consider a two-candidate contest in which the winner prevails by one hundred votes out of ten thousand. If there are 150 irregular voters, it is *possible* that more than 125 of them voted for the winner, so that their elimination would reverse the election. Does this possibility mean that the rightful winner cannot be determined within the meaning of the statute? The courts have answered this question with intuitive assessments of the probability that the result would be reversed if the challenged votes were removed. Thus, the Court of Appeals has articulated and applied the principle that the party attempting to impeach the results must show that the "irregularities are sufficiently large in number to establish the *probability* that the result would be changed by a shift in, or invalidation of, the questioned votes."[2]

Two polar assessments of the relevant probabilities may be illustrated by comparing *Ippolito v. Power*[3] with *De Martini v. Power*.[4] In *Ippolito*, the winner's plurality was 17 votes out of 2,827; there were 101 suspect or invalid votes. The court affirmed the lower court's ordering of a new election. Evidently relying on intuition, the court concluded that "it does not strain the probabilities to assume a likelihood that the questioned votes produced or could produce a change in the result."[5]

In *De Martini*, out of 5,250 votes, 136 were declared irregular and in-

---

* Lecturer in Law, Columbia University. A.B., 1955, J.D., 1958, Harvard University.
** Professor of Statistics, Columbia University. A.B., 1935, Ph.D., 1938, Harvard University.

1. N.Y. Election Law § 330(2) (McKinney 1964). There is no comparable statutory provision for ordering a new general election. For a comprehensive discussion of section 330 challenges, see Note, *Primary Challenges in New York: Caselaw Coleslaw v. Election Protection*, 73 Colum. L. Rev. 318 (1973).

2. Ippolito v. Power, 22 N.Y.2d 594, 597-98, 241 N.E.2d 232, 233, 294 N.Y.S.2d 209, 211 (1968) (emphasis added). This standard was quoted with approval and applied in De Martini v. Power, 27 N.Y.2d 149, 151, 262 N.E.2d 857, 858, 314 N.Y.S.2d 609, 610 (1970).

3. 22 N.Y.2d 594, 241 N.E.2d 232, 294 N.Y.S.2d 209 (1968).

4. 27 N.Y.2d 149, 262 N.E.2d 857, 314 N.Y.S.2d 609 (1970).

5. 22 N.Y.2d at 598, 241 N.E.2d at 233, 294 N.Y.S.2d at 211.

validated, no fraud being involved. The winner's plurality was 62 votes. The lower courts and Court of Appeals differed in their estimates of the relevant probability. The Supreme Court ordered a new election because "it is not beyond likelihood that the small difference of 62 votes could be altered in a new election."[6] The Appellate Division unanimously affirmed. In reversing, the Court of Appeals observed that the majority of the winner would not evaporate unless at least 99 votes—*i.e.*, at least 72.8% of the irregularities—had been cast in her favor. The court found this unlikely: "It taxes credulity to assume that, in so close a contest, such an extreme percentage of invalid votes would be cast in one direction." It concluded that "a valid determination is not rendered impossible . . . by the remote possibility of a changed result . . . ."[7]

Subjective estimates of the relevant probabilities have thus varied. There is, however, no reason to leave matters on a purely subjective basis. Using an assumption about the character of invalid voting which will be defensible in many cases, the relevant probabilities can be readily computed.

Consider all the votes cast in a primary election as balls placed in an urn: black balls which predominate are those votes cast for the winner; white balls are for the loser. A certain number of balls representing the irregular voters are then withdrawn at random from the urn, an operation which corresponds to their invalidation. What is the probability that, after the withdrawal, the number of black balls no longer exceeds the number of white? Note the key assumption that the balls are withdrawn at random, *i.e.*, that each ball has the same probability of being withdrawn. In terms of the real election situation, each voter is deemed to have the same probability of casting an invalid vote. This assumption will of course be untenable if evidence of fraud or patterns of irregular voting indicates that a disproportionate number of improper votes were cast for one candidate. But in the absence of such evidence, the assumption of random distribution of the improper votes is warranted. Whether or not mathematics is used to assess the probabilities, some implicit or explicit view as to the pattern of irregular voting seems inevitable. The assumption that each voter had an equal probability of casting an improper vote is the only neutral and non-arbitrary view that can be taken when there is no evidence to indicate that the probabilities are not equal. Thus in *Ippolito*, *De Martini*, and other cases,[8] where there was no evidence to disturb the assumption of randomness, the mathematical probability analysis depicted by the urn model is a correct expression of the intuitive probability used by the Court of Appeals in formulating the burden of proof standard for a new election.

---

6. Quoted at 27 N.Y.2d at 151, 262 N.E.2d at 857, 314 N.Y.S.2d at 610. Despite the court's casual language, the proper inquiry is whether the election result at bar was affected through irregular voting, not whether a new election would yield a different result.

7. 27 N.Y.2d at 151, 262 N.E.2d at 858, 314 N.Y.S.2d at 611.

8. *E.g.*, Posner v. Power, 18 N.Y.2d 703, 220 N.E.2d 269, 273 N.Y.S.2d 480 (1966); Santucci v. Power, 25 N.Y.2d 897, 252 N.E.2d 128, 304 N.Y.S.2d 593 (1969).

In terms of the urn model, the mathematical probability of a reversal is simply the number of combinations in which the balls representing the invalid votes may be withdrawn from the urn so as to produce a reversal, divided by the total number of combinations in which these balls may be withdrawn from the urn. If the number of votes in question was small enough, these combinations could be simply enumerated. In most practical applications, however, the combinations are far too numerous for simple counting. An approximation of this probability with sufficient accuracy for legal purposes can be obtained by using the formula

$$Z = d\sqrt{\frac{s-k}{sk}}$$

where *d* is the winner's plurality; *s* is the number of votes cast either for the winner or his challenger; and *k* is the number of invalid votes cast either for the winner or his challenger.[9] The value of *z* determines the probability of reversal; as it increases this probability declines rapidly. The following table gives some benchmarks:[10]

Value of z	*Probability of Reversal*
0.5	~~.81~~
1.0	.16
1.5	.07
2.0	.02
3.0	.001

The application of this formula demonstrates a first good reason for using mathematical methods, namely, that uneducated intuition is not a reliable guide to the true probabilities. On the facts in *Ippolito*, analysis indicates about a 5 percent chance that the election would have been reversed by the removal of the irregular votes.[11] Did the court realize that the chance was this small when

9. The derivation of this formula appears in the Appendix. The formula uses the total vote for the election, and this in general should be sufficiently accurate. If, however, votes for subdistricts are available, and if the pattern of voting varied substantially from subdistrict to subdistrict, it may be desirable to make separate computations for each subdistrict and to aggregate them as shown in the subdivision formula in the Appendix.

If there are only two candidates and no other contest, *k* will equal the number of challenged votes. If there are more than two candidates or if there are other contests (*e.g.*, a primary for another party being run simultaneously, as is frequently the case), some of the invalid votes may have been cast for the third candidate or in the other contest. To account for this effect precisely would raise mathematical complexities. Frequently, it can be demonstrated that the probability of a reversal is extremely small even if it is assumed that all the invalid votes were cast for the winner or his challenger, an assumption which clearly generates a larger probability of reversal than if a more realistic assumption were adopted. In closer cases, the probabilities fairly may be approximated by assuming that the winner and challenger combined received the same proportion of invalid votes as they did of the total vote for all candidates and contests. On this assumption *k* will be proportionately smaller than the total number of invalid votes.

10. *z* is simply a standard normal variate. A full table of values and associated probabilities may be found in most textbooks on statistics. *See, e.g.*, FREUND, MATHEMATICAL STATISTICS, Table III at 366 (1962).

11. $z = 17\sqrt{\frac{2827 - 101}{(2827)(101)}} = 1.6$. This is a value associated with a 5 percent probability of reversal.

it concluded there was a "likelihood" of reversal?[12] One cannot be sure, but 5 percent seems a rather small probability to be termed a likelihood. If so, *Ippolito* may well be wrong in the sense that the court was acting on an overestimate of the probability of reversal.[13]

In *De Martini,* although the Appellate Division found a reversal "not beyond likelihood," the mathematical test demonstrates the contrary. The chance that the winner's plurality of only 62 votes would be eroded to a tie or loss by the removal of 136 irregular votes was less than one in a million.[14] Thus the Court of Appeals, reversing, was clearly correct when it concluded that the chance of a reversal was "remote." The court's judgment on this question may have been influenced by a statistical analysis in appellant's brief which, using a weak form of mathematical estimation, demonstrated that the probability of a reversal was less than 3.7 percent.[15] If the court had this figure in mind when it characterized the probability of removal as "remote," it evidently made a mistake in *Ippolito* when it found that this probability was substantial although the true figure was only 5 percent.

The court was even more clearly mistaken in *Santucci v. Power,*[16] when it affirmed an order directing a new election on the basis of 640 irregularities and a winner's plurality of 95 votes. In citing *Ippolito,* the court evidently believed that the probability of a reversal was substantial. Mathematical analysis demonstrates, however, that this probability was in fact less than one in ten thousand.[17]

In assessing the probability of reversal required to order a new election,

---

12. No mathematical analysis was presented in the briefs for either party.

13. In *Ippolito* the court relied on its affirmance of a new election in Nodar v. Power, 18 N.Y.2d 697, 220 N.E.2d 267, 273 N.Y.S.2d 273 (1966), which it said involved "almost identical facts." But the plurality in *Nodar* was 27 votes, more than 50 percent larger than in *Ippolito*. Although the numbers in both cases were small, the difference is of some consequence. The formula applied to the *Nodar* facts is $z = 27\sqrt{\frac{1417-109}{(1417)\,(109)}} =$ 2.4, a value of z associated with a 1 percent probability of reversal. This is five times smaller than in *Ippolito* and surely should be deemed inconsequential for legal purposes.

14. $z = 62\sqrt{\frac{5250-136}{(5250)\,(136)}} = 5.2$, a value of z associated with a probability of less than one in a million.

15. Brief for Appellant, Appendix. The method used was Chebychev's inequality.

16. 25 N.Y.2d 897, 252 N.E.2d 128, 304 N.Y.S.2d 593 (1969).

17. After a proportionate reduction in the number of irregularities to account for their distribution among candidates other than the first two (in accordance with the procedure stated in note 9 *supra*) the numbers in *Santucci* were:

$$z = 95\sqrt{\frac{116{,}057-448}{(116{,}057)\,(448)}} = 3.8.$$

On the other hand, a defensible decision was made in Posner v. Power, 18 N.Y.2d 703, 220 N.E.2d 269, 273 N.Y.S.2d 480 (1966) where the court found 370 to 412 irregularities in a four-candidate contest in which the plurality of the winner was 24 votes. Assuming 412 irregular votes and allocating a proportionate number of irregularities to the two last-place candidates, the probability of a reversal was 9 percent, which would seem sufficient to justify the court's affirmance of the order directing a new election.

it should be recalled that this probability will be less than 50 percent regardless of the number of invalid votes removed. Consequently, the critical probability must be some not insubstantial figure, but short of the 50+ percent implied by a "more-likely-than-not" test.[18]

A second reason for using a mathematical approach is to prevent resort to misleadingly simple rules of thumb. As the formula shows, the probability of reversal depends principally upon the size of the plurality and the number of irregularities (and less sensitively on the total vote). Its results cannot, however, be expressed as a simple relation between these variables. For example, if the plurality is small enough, a substantial probability of reversal will exist if the number of irregularities is twice the plurality; but if the plurality is large, irregularities four or five times larger may be insufficient to cast even a shadow of doubt on the results.[19] Thus, the rule used in some cases that a new election will be ordered when the number of irregularities exceeds a certain multiple of the plurality is incorrect as a method of intuitive estimation and misleading as a method of analyzing the precedents.[20]

A third reason for using mathematical techniques is that uncertainty over intuitive estimates of probability tends to obscure significant legal issues which arise in certain cases. An examination of the recent case of *Lowenstein v. Larkin,* in which the Appellate Division unanimously set aside Congressman John Rooney's primary victory over Allard Lowenstein,[21] and was affirmed on the opinion below by the Court of Appeals,[22] serves to illustrate this point.

In ordering a new election, the Appellate Division purported to apply the rule of section 330 that the election was "characterized by such . . . irregularities as to render impossible a determination as to who rightfully was nominated." If the court meant by this that the invalid votes made the result uncertain, a judicial recall of the election was not supported by the facts recited in the court's opinion.

The court found that, among other instances of misconduct, due to errors

18. The Court of Appeals, probably unwittingly, has sometimes sounded as if it might apply such a preponderance test. *See, e.g.,* Ippolito v. Power, 22 N.Y.2d 594, 597-98, 241 N.E.2d 232, 233, 294 N.Y.S.2d 209, 211 (1968) (the irregularities "would not be sufficiently large in number to establish *the probability* that the result would be changed by a shift in, or invalidation of, the questioned votes.") (emphasis supplied).

19. A plurality of five votes with twice as many irregularities might justify a new election since the probability of reversal could be approximately 5 percent. On the other hand, with a plurality of 100 votes, even four times as many irregularities would not justify a new election since the probability of reversal would be less than one in a million.

20. *See, e.g.,* Santucci v. Power, 25 N.Y.2d 897, 252 N.E.2d 128, 304 N.Y.S.2d 593 (1969) (order directing a new election affirmed when irregularities were analyzed as being "at least six and one-half times the winning margin"); Posner v. Power, 18 N.Y.2d 703, 220 N.E.2d 269, 273 N.Y.2d 480 (1966) (order directing a new election affirmed when irregularities were analyzed as being "15 to 17 times the margin of winning votes"); Merola v. Power, 60 Misc. 2d 245, 248, 303 N.Y.S.2d 229, 232 (Sup. Ct.), *aff'd mem.,* 33 App. Div. 2d 514 (1st Dep't 1969) (new election ordered and prior cases analyzed on the basis that the irregularities exceeded three times the plurality).

21. 40 App. Div. 2d 604, 335 N.Y.S.2d 799 (2d Dep't 1972) (*mem.*).

22. 31 N.Y.2d 654, 288 N.E.2d 133, 336 N.Y.S.2d 249 (1972) (*mem.*).

by the Board of Elections, "hundreds of persons" were turned away from the polling places to which they had been assigned and that others were improperly permitted to vote because challenges were ignored or because their registrations should have been cancelled, but were not. The exact impact of these mistakes evidently was unknown, the court being able to find only that "at least" 1,920 irregular votes were cast in the election out of a total of 29,567.

The case was removed from the ordinary run by evidence of official favoritism for Rooney. Campaigning for Rooney at polling places and ignoring the challenges of Lowenstein poll watchers were among the practices the court found to have tainted the election. There was, however, nothing in the circumstances cited in the court's opinion to suggest that those improperly permitted to vote favored Rooney while those improperly excluded or inhibited from voting favored Lowenstein. On the court's findings, one could assume no more against Rooney than that the improper voters (and those inhibited or discouraged from voting) were a random sample of the total voting population.

The number of irregularities was large—at least 1,920 votes—and the court may have believed that this was sufficient to create a substantial probability of reversal, even on a random sample basis. Rooney's plurality, however, was also large—890 votes—and the number of irregular votes required to generate even a small probability of reversal rises very rapidly with the winner's plurality, probably much more rapidly than most statistically uneducated persons would suppose. In fact, given this large plurality, it would have taken more than 27,000 irregularities to create even a 2 percent probability of a reversal. Lowenstein did not claim irregularities on anything approaching this scale.[23]

The mathematical demonstration illuminates the nature of the specific findings of fact or legal conclusions that would have been required to justify a new election. If the court relied on the traditional standard—an appreciable probability of reversal—it would have had to conclude that the mistakes of the Board or its officials were not in fact neutral, but favored Rooney, in the sense that those voting improperly favored Rooney while those improperly inhibited from voting favored Lowenstein. No such findings appear in the opinion, although the evidence might have justified this conclusion.

Alternatively, the court could have taken an enlarged view of section 330, and held that serious irregularities in procedure by Board officials, particularly deviations favoring one candidate, constituted sufficient grounds for ordering a new election, even in cases in which it could not be shown that

23. In Celler v. Larkin, 71 Misc. 2d 17, 335 N.Y.S.2d 791 (Sup. Ct.), *aff'd mem.*, 40 App. Div. 2d 603, 335 N.Y.S.2d 801 (2d Dep't), *aff'd mem.*, 31 N.Y.2d 658, 288 N.E.2d 135, 336 N.Y.S.2d 251 (1972), Elizabeth Holtzman won the Democratic primary for Congress by some 600 votes out of 30,000. In the trial of Celler's challenge to her victory, Professor Robbins testified that Celler would have had to show more than 16,900 irregularities out of the 30,000 votes cast to create even a one-in-a-thousand chance of a reversal. (T. at 422-23). The courts rejected the challenge.

such errors affected, or were likely to have affected, the outcome. Since section 330 does not seem to read this broadly, and since a liberal reading should be "confined to the subject matters plainly enumerated therein,"[24] an extension of section 330 to cover such cases would be a significant development in the law meriting judicial discussion.[25]

Nothing of that sort appears in the opinion of the Appellate Division. Instead, its decision appears to rest on an undifferentiated mix of disapproval for official irregularities, and hazy or mistaken intuitive notions of probability. If the method of mathematical probability had been recognized in *Lowenstein*, the attention of the courts would have been focused on the legal issues which were critical in that case and which are likely to arise in future cases: whether there was sufficient evidence to reject the assumption of random distribution of the improper votes, and if not, whether official irregularities favoring one candidate, but without demonstrable effect on the result, are a permissible statutory ground for ordering a new election.

APPENDIX

Global analysis of challenged elections.

Suppose the winning candidate A has a votes and the losing candidate B has b votes, with a greater than b. Of the total of $s = a + b$ votes, k are removed at random. Let x denote the number of these k votes which are for A. There will be a *reversal* if after the removal there are at least as many votes for B as for A; *i.e.*, if $a - x \leqq b - (k - x)$, or

$$x \geqq (a - b + k)/2. \tag{1}$$

In a random withdrawal of k balls from $s = a + b$ balls in an urn, the number x of A balls withdrawn is a random variable with a "hypergeometric" distribution (sampling without replacement from a finite population); the mean and variance of x are given by

$$\Sigma(x) = ka/s, \qquad \mathrm{Var}(x) = kab(s-k)/s^2(s-1).$$

The standardized random variable

$$z = (x - \Sigma(x))/\sqrt{\mathrm{Var}(x)}$$

is then (replacing $s - 1$ by s for simplicity) given by the formula

$$z = \frac{sx - ka}{\sqrt{\left(kab\left(1 - \frac{k}{s}\right)\right)}};$$

z has mean 0 and variance 1. The condition (1) for reversal in terms of z is

24. Matter of Hyer, 187 Misc. 946, 948, 63 N.Y.S.2d 874, 876 (Sup. Ct. 1946).

25. Perhaps the necessary interpretation could be made by leaning heavily on the word "rightfully" in section 330. Arguably, the winner cannot claim to be rightfully selected in an election marred by serious breaches of the rules in his favor by officials charged with the duty of neutrality.

that the numerator of z be greater than or equal to $s(a-b+k)/2-ka=(s-k)(b-a)/2$; *i.e.* that z be greater than or equal to the constant

$$C=\frac{(b-a)\cdot\sqrt{(s(s-k))}}{2\sqrt{(kab)}}.$$

Now, since $a+b=s$, $ab\leqq s^2/4$, with approximate equality when a and b are nearly equal. In any case, we see that

$$C\geqq(b-a)\cdot\sqrt{\left(\frac{1}{k}-\frac{1}{s}\right)}. \tag{2}$$

Since z is approximately normally distributed, *the probability of a reversal is about equal to the probability that a standard normal random variable z will exceed the constant given by the right-hand side* of (2).

Analysis by subdistricts.

If k_i votes are removed in the ith district, an elaboration of the global analysis shows that the condition for a reversal is that an approximately standard normal random variable z will exceed the constant

$$\frac{(a-b+k)-2\left[\text{ sum over } i \text{ of }\left(\frac{k_i a_i}{a_i+b_i}\right)\right]}{2\cdot\text{sq. rt.}\left[\text{ sum over } i \text{ of }\left(\frac{k_i a_i b_i}{a_i+b_i}\right)\right]}$$

where again $k=$ [sum over i of k_i] $=$ total number of votes removed from the $a+b$ votes.

Reprinted from
Columbia Law Rev.
73, 241–248 (1973)

Reprinted from
Proc. Nat. Acad. Sci. USA
Vol. 73, No. 2, pp. 286–288, February 1976
Statistics

Maximally dependent random variables

(maxima/dependence)

T. L. LAI AND HERBERT ROBBINS

Department of Mathematical Statistics, Columbia University, New York, N.Y. 10027

Contributed by Herbert Robbins, November 14, 1975

ABSTRACT Let $X_1, \ldots, X_n$ have an arbitrary common marginal distribution function F, and let $M_n = \max(X_1, \ldots, X_n)$. It is shown that $EM_n \le m_n$, where $m_n = a_n + n\int_{a_n}^{\infty}[1 - F(x)]dx$ and $a_n = F^{-1}(1 - n^{-1})$, and that $EM_n = m_n$ when $X_1, \ldots, X_n$ are "maximally dependent"; i.e., $P(M_n > x) = \min\{1, n[1 - F(x)]\}$ for all x. Moreover, as $n \to \infty$, $a_n \sim m_n \sim m_n^*$, where $m_n^* = EM_n$ when $X_1, \ldots, X_n$ are independent, provided that $[1 - F(cx)]/[1 - F(x)] \to 0$ as $x \to \infty$ for every $c > 1$, and $E(X_1^-)^r < \infty$ for some $r > 0$. The case in which F is standard normal is considered in detail.

For any real number a it is evident that

$$M_n \le a + \sum_1^n (X_i - a)^+. \qquad [1]$$

Hence $EM_n \le h(a)$ for all a, where $h(a) = a + n\int_a^{\infty}[1 - F(x)]dx$. A simple argument shows that $h(a)$ is minimized at $a = a_n$, where

$$a_n = F^{-1}\left(1 - \frac{1}{n}\right) \qquad [2]$$

and we define $F^{-1}(t) = \inf\{x: f(x) \ge t\}$ $(\inf \phi = \infty)$, $0 \le t \le 1$. It follows that

$$EM_n \le a_n + n\int_{a_n}^{\infty}[1 - F(x)]dx = m_n, \quad \text{say.} \qquad [3]$$

To show that there exists a joint distribution of $X_1, \ldots, X_n$ with any given F as marginal such that equality is attained in [3] we begin by noting that

$$P(M_n > x) \le \min\{1, n[1 - F(x)]\} \quad \text{for all } x. \qquad [4]$$

and call $X_1, \ldots, X_n$ *maximally dependent* if

$$P(M_n > x) = \min\{1, n[1 - F(x)]\} \quad \text{for all } x. \qquad [5]$$

THEOREM 1. *Assume that* $EX_1^+ < \infty$. *Then* $EM_n = m_n$ *if and only if* $X_1, \ldots, X_n$ *are maximally dependent.*
We omit the proof.

For any n and F there exist maximally dependent $X_1, \ldots, X_n$ with F as their common marginal. (In fact, there exists a sequence $X_1, X_2, \ldots$ with F as their common marginal and such that $X_1, \ldots, X_n$ are maximally dependent for every $n = 1, 2, \ldots$.) We omit the details of the construction and instead prove the following theorem (cf. ref. 1) which extends [5] to the nonidentically distributed case when F is continuous at a_n.

THEOREM 2. *Given distribution functions* $F_1, \ldots, F_n$, *let* $\alpha = \alpha_n = \inf\{x: \Sigma_1^n[1 - F_i(x)] \le 1\}$. *Assume that* $F_i(\alpha) \neq 0$ *and* F_i *is continuous at* α *for every* i. *For* $k = 1, \ldots, n$ *let*

$$A_k = \{x_k > \alpha \text{ and } x_i \le \alpha \text{ for } i \neq k\},\ c_k = \left[\prod_{i \neq k} F_i(\alpha)\right]^{-1},$$

and define the joint distribution of $X_1, \ldots, X_n$ *by the properties*
(*i*) $P(\bigcup_1^n A_k) = 1$
(*ii*) $P(I) = c_k \cdot \prod_1^n [F_i(u_i) - F_i(v_i)]$
where $I = \{v_1 < x_1 \leqslant u_1, \ldots, v_n < x_n \leqslant u_n\}$ *is any rectangle contained in* A_k, $k = 1, \ldots, n$. *Then the distribution function of* X_i *is* F_i, $i = 1, \ldots, n$, *and*

$$P[\max(X_1, \ldots, X_n) > x] = \min\left\{1, \sum_1^n [1 - F_i(x)]\right\} \quad \text{for all } x. \qquad [6]$$

We omit the proof.

When $X_1, \ldots, X_n$ are *independent* with a common marginal F,

$$EM_n = n\int_{-\infty}^{\infty} xF^{n-1}(x)dF(x) = m_n^*, \quad \text{say.} \qquad [7]$$

The evaluation of m_n^* is often much harder than that of m_n, which by [3] is an upper bound for m_n^*. We give some examples in which m_n and m_n^* are compared.

(*a*) Uniform distribution on [0, 1]. Here $m_n = (2n - 1)/2n$, $m_n^* = n/(n + 1)$, both tending to 1 as $n \to \infty$, with $(1 - m_n)/(1 - m_n^*) \to \frac{1}{2}$.

(*b*) Exponential distribution. Here $F(x) = 1 - e^{-x}$ $(x \ge 0)$, $a_n = \log n$, $m_n = 1 + \log n$. Letting C denote Euler's constant,

$$\begin{aligned} m_n^* &= n\int_0^{\infty} x(1 - e^{-x})^{n-1}e^{-x}dx = n\sum_{k=0}^{n-1}(-1)^k\binom{n-1}{k} \times \int_0^{\infty} xe^{-(k+1)x}dx \\ &= \sum_{j=1}^{n}(-1)^{j-1}\binom{n}{j}\int_0^1 x^{j-1}dx = \int_0^1 (1 + y + \ldots + y^{n-1})dy \\ &= \log n + C + 0(n^{-1}), \end{aligned}$$

so that

$$m_n^* = m_n - (1 - C) + 0(n^{-1}) = a_n + C + 0(n^{-1}) \qquad [8]$$

(c) Standard normal distribution. For $n \geq 2$

$$a_n = \Phi^{-1}(1 - n^{-1}), \quad m_n = a_n + n\int_{a_n}^{\infty}(x - a_n)\varphi(x)dx = n\varphi(a_n). \quad [9]$$

Thus

$$a_2 = 0, \quad m_2 = (2/\pi)^{1/2}, \quad a_3 = 0.43, \quad m_3 = 1.09.$$

We shall now show that

$$m_n > a_n > (2 \log n - \log \log n - \log 4\pi)^{1/2} \quad \text{for } n \geq 5 \quad [10]$$

$$m_n < (2 \log n - \log \log n)^{1/2} \quad \text{for } n \geq 3. \quad [11]$$

To prove [10] we define ϵ_n by

$$a_n = (2 \log n - \log \log n - \log 4\pi + \epsilon_n)^{1/2}. \quad [12]$$

Since

$$1 - \Phi(x) > \varphi(x)/(x + x^{-1}) \quad \text{for } x > 0,$$

it follows that

$$a_n + a_n^{-1} > n\varphi(a_n) = (2e^{-\epsilon_n} \log n)^{1/2}, \quad [13]$$

$$a_n^2 + a_n^{-2} + 2 > 2e^{-\epsilon_n} \log n. \quad [14]$$

Assume that $\epsilon_n \leq 0$. Then from [12] and [14]

$$\frac{2 \log n - \log \log n - \log 4\pi + a_n^{-2} + 2}{2 \log n} > e^{-\epsilon_n} \geq 1,$$

which is a contradiction for $n \geq 7$ since $a_n > 1$, $\log \log n > 0.6$, and $\log \log n + \log 4\pi - a_n^{-2} - 2 > 0$. By numerical check, [10] also holds for $n = 5,6$.

To prove [11] we first check it numerically when $\log n \leq 7.5$. For larger n, since $\epsilon_n > 0$ it follows from [12] that

$$2e^{-\epsilon_n} \log n = [n\varphi(a_n)]^2 = m_n^2 > a_n^2 > 2 \log n - \log \log n - \log 4\pi > 0,$$

and hence

$$\epsilon_n < \log \{1 + (\log \log n + \log 4\pi)/(2 \log n - \log \log n - \log 4\pi)\} < (\log \log n + \log 4\pi)/(2 \log n - \log \log n - \log 4\pi) = (2 \log n)/(2 \log n - \log \log n - \log 4\pi) - 1. \quad [15]$$

From [13], $m_n = n\varphi(a_n) < a_n + a_n^{-1}$, so by [12] and [15]

$$m_n^2 < a_n^2 + a_n^{-2} + 2 < 2 \log n - \log \log n - \log 4\pi + 1 + (2 \log n + 1)/(2 \log n - \log \log n - \log 4\pi) < 2 \log n - \log \log n, \text{ for } \log n \geq 7.5.$$

From [11] it follows that if $X_1, \ldots, X_n$ are marginally standard normal, independent or not, then

$$E \max (X_1, \ldots, X_n) < (2 \log n - \log \log n)^{1/2} \quad \text{for } n \geq 3, \quad [16]$$

a fact that would be difficult to prove without using the simple inequality [3]. The inequality [16] is quite sharp for the maximally dependent case, where as $n \to \infty$ it can be shown that

$$m_n = (2 \log n)^{1/2} - \frac{1}{2}(2 \log n)^{-1/2} \times (\log \log n + \log 4\pi - 2) + O[(\log n)^{-3/2}(\log \log n)^2], \quad [17]$$

$$a_n = \{2 \log n - \log \log n - \log 4\pi + O[(\log \log n)/\log n]\}^{1/2}, \quad [18]$$

$$\mathrm{Var}\, M_n = (2 \log n)^{-1} + (2 \log n)^{-2} \log \log n + O[(\log n)^{-2}]. \quad [19]$$

In the independent case (cf. ref. 2, p. 376) we have

$$m_n^* = (2 \log n)^{1/2} - \frac{1}{2}(2 \log n)^{-1/2}(\log \log n + \log 4\pi - 2C) + O[(\log n)^{-1}] \quad [20]$$

where C is Euler's constant, and

$$\mathrm{Var}\, M_n = (\pi^2/6)(2 \log n)^{-1} + O[(\log n)^{-2} \log \log n]. \quad [21]$$

Examples (b) and (c) above are special cases of the following theorem, in which Y. S. Chow contributed an essential idea to the proof.

THEOREM 3. *If $F(x) < 1$ for all x then*

(i) As $n \to \infty$, $a_n \sim m_n \sim m_n^*$ if the following conditions hold:

$$\lim_{x \to \infty} \frac{1 - F(cx)}{1 - F(x)} = 0 \quad \text{for every } c > 1, \text{ and } \int_{-\infty}^{0} |x|^r dF(x) < \infty \text{ for some } r > 0. \quad [22]$$

(ii) If $X_1, \ldots, X_n$ are independent then [22] is equivalent to

$$M_n/a_n \xrightarrow{L_p} 1 \quad \text{for every } p > 0. \quad [23]$$

(iii) If $X_1, \ldots, X_n$ are maximally dependent then [23] is equivalent to

$$\lim_{x \to \infty} \frac{1 - F(cx)}{1 - F(x)} = 0 \quad \text{for every } c > 1. \quad [24]$$

The proof, together with that of the following theorem, will be presented elsewhere.

THEOREM 4. *Suppose $F(x) < 1$ for all x. (i) If $X_1, \ldots, X_n$ are maximally dependent then the following statements are equivalent:*

$$\lim_{x \to \infty} \frac{1 - F(x + c)}{1 - F(x)} = 0 \quad \text{for every } c > 0 \quad [25]$$

$$M_n - a_n \xrightarrow{L_p} 0 \quad \text{for every } p > 0 \quad [26]$$

$$EM_n^2 < \infty \text{ for all large } n, \text{ and } \lim_{n \to \infty} \mathrm{Var}\, M_n = 0. \quad [27]$$

Consequently, in this case

$$EM_n^p = a_n^p + o(a_n^{p-1}). \qquad [28]$$

(ii) If $X_1, \ldots, X_n$ are independent then [26] *and* [27] *are each equivalent to:*

$$F \text{ satisfies [25], and}$$
$$\int_{-\infty}^{0} |x|^r dF(x) < \infty \qquad \text{for some } r > 0. \qquad [29]$$

We shall discuss elsewhere the asymptotic distribution of M_n in the maximally dependent case, together with strong limit theorems for a sequence $X_1, X_2, \ldots$ of identically distributed random variables such that $X_1, \ldots, X_n$ are maximally dependent for every $n = 1,2, \ldots$; such sequences always exist for any given marginal F.

This research was supported by grants from the National Science Foundation and the Office of Naval Research. The grant numbers are NSF-GP-33570X and ONR-N00014-75-C-0560.

1. Mallows, C. L. (1969) *SIAM Rev.* 11, 410.
2. Cramér, H. (1946) *Mathematical Methods of Statistics* (Princeton University Press).

Reprinted from

Proc. Natl. Acad. Sci. USA
Vol. 75, No. 7, pp. 3034–3036, July 1978
Statistics

Strong consistency of least squares estimates in multiple regression

(orthogonal random variables/double array)

T. L. LAI, HERBERT ROBBINS, AND C. Z. WEI

Department of Mathematical Statistics, Columbia University, New York, New York 10027

Contributed by Herbert Robbins, May 8, 1978

ABSTRACT The strong consistency of least squares estimates in multiple regression models with independent errors is obtained under minimal assumptions on the design and weak moment conditions on the errors.

Consider the multiple regression model

$$y_i = \beta_1 x_{i1} + \ldots + \beta_p x_{ip} + \epsilon_i \quad (i = 1,2,\ldots) \qquad [1]$$

where $\epsilon_1, \epsilon_2, \ldots$ are independent random variables such that $E\epsilon_i = 0$ and $E\epsilon_i^2 = \sigma^2$ with $0 < \sigma < \infty$, and $\{x_{ij}\}$ $(i = 1,2,\ldots;$ $j = 1,\ldots,p)$ is an arbitrary double array of constants. Let X_n denote the matrix $\{x_{ij}\}_{1 \le i \le n, 1 \le j \le p}$. Let $Y_n = (y_1, \ldots, y_n)'$ and let $\beta = (\beta_1, \ldots, \beta_p)'$, where $'$ denotes transpose. For $n \ge p$, the least squares estimate b_n of the vector β based on the design matrix X_n and the response vector Y_n is given by

$$b_n = (X'_n X_n)^{-1} X'_n Y_n \qquad [2]$$

provided that $X'_n X_n$ is nonsingular. Assuming $X'_n X_n$ to be nonsingular, b_n is an unbiased estimate of β, with $\mathrm{Cov}(b_n) = \sigma^2 (X'_n X_n)^{-1}$. Hence, for b_n to converge as $n \to \infty$ to β in quadratic mean and hence in probability, it is sufficient that

$$(X'_n X_n)^{-1} \to 0 \text{ as } n \to \infty. \qquad [3]$$

([3] is also necessary for b_n to converge to β in probability; see ref. 1.)

The question whether [3] implies that b_n converges to β almost surely, however, is much harder. When the errors ϵ_i are normal, Anderson and Taylor (ref. 2) have shown that [3] implies the almost sure convergence of b_n to β. Without the assumption of normality, they have also shown that b_n converges to β almost surely under the assumption that the errors ϵ_i are identically distributed, generalized Gaussian random variables and that

$$\mathrm{tr}[(X'_n X_n)^{-1}] = o[1/(\log n)] \text{ as } n \to \infty \qquad [4]$$

(see ref. 3). The latter assumption on the design is much stronger than [3]. Earlier, Drygas (ref. 1) obtained the strong consistency of b_n under the alternative assumption that there exist positive constants $k_n \to \infty$ and a positive definite matrix Σ such that

$$\frac{1}{k_n}(X'_n X_n) \to \Sigma \text{ as } n \to \infty. \qquad [5]$$

Although this condition reduces to [3] when $p = 1$, it is much stronger than [3] when $p > 1$. Whether [3] is sufficient for the strong consistency of b_n without the assumption of normality has remained an open question. The following theorem, which only assumes the second moments of ϵ_i to be uniformly bounded, gives an affirmative answer to this problem.

THEOREM 1. *Suppose that in* [1] *the random variables* $\epsilon_1, \epsilon_2, \ldots$ *are independent*, $E\epsilon_i = 0$ *for all* i, *and* $\sup_i E\epsilon_i^2 < \infty$. *If* [3] *holds, then* $b_n \to \beta$ *with probability 1.*

Let $b_n = (b_{n1}, \ldots, b_{np})'$. Thus, b_{nj} is the least squares estimate of β_j, and its variance is $\sigma^2 c_{jj}^{(n)}$ (assuming $E\epsilon_i^2 = \sigma^2$ for all i), where

$$C_n = (c_{ij}^{(n)})_{1 \le i,j \le p} = (X'_n X_n)^{-1}. \qquad [6]$$

Theorem 2 below deals with the strong consistency of b_{nj} and implies *Theorem 1* as an immediate corollary. As the proof of *Theorem* 2 shows, the assumption of independence for the random variables ϵ_i in *Theorem 1* can be replaced by the weaker condition that the ϵ_i form a martingale difference sequence; i.e.,

$$E(\epsilon_{i+1} | \epsilon_1, \ldots, \epsilon_i) = 0 \text{ for all } i \ge 1. \qquad [7]$$

THEOREM 2. *Suppose that in* [1] *the random variables* $\epsilon_1, \epsilon_2, \ldots$ *form a martingale difference sequence such that* $E\epsilon_i = 0$ *for all* i *and* $\sup_i E\epsilon_i^2 < \infty$. *Assume that* $X'_n X_n$ *is nonsingular for some and therefore for all large* n. *For all large* n, *define* C_n *by* [6]. *Fix* $j = 1, \ldots, p$. *If* $c_{jj}^{(n)} \to 0$ *as* $n \to \infty$, *then for every* $\delta > 0$, *with probability 1,*

$$b_{nj} - \beta_j = (\{c_{jj}^{(n)} |\log c_{jj}^{(n)}|^{1+\delta}\}^{1/2}) \text{ as } n \to \infty. \qquad [8]$$

Theorem 2 generalizes an earlier result of Lai and Robbins (ref. 4) that proved for the simple linear model $y_i = \beta_1 + \beta_2 t_i + \epsilon_i$ the strong consistency of the slope estimate b_{n2} under the minimal assumption

$$\sum_1^n (t_i - \bar{t}_n)^2 \to \infty \left(\bar{t}_n = \frac{1}{n}\sum_1^n t_i\right)$$

on the design. The method of ref. 4, which is based on an embedding technique to reduce the problem to the normal case, requires the finiteness of some moment of order higher than 2 and does not extend easily to the general multiple regression model [1]. We therefore develop another method below to prove *Theorem* 2.

To prove *Theorem* 2, it suffices to consider only b_{n1} (i.e., $j = 1$). For $p \ge 2$, defining the $(p-1)$-dimensional vector

$$T_n = (x_{n2}, \ldots, x_{np})' \qquad [9]$$

and partitioning the matrix $X'_n X_n$ as

$$X'_n X_n = \begin{pmatrix} \sum_{i=1}^n x_{i1}^2 & K_n \\ K'_n & H_n \end{pmatrix} \qquad [10]$$

so that H_n is a $(p-1) \times (p-1)$ matrix, we have the following representation of b_{n1}.

LEMMA 1. *Let* $p \ge 2$. *Assume that* $X'_n X_n$ *is positive definite for* $n \ge m$ $(\ge p)$. *Define* H_n, K_n, *and* T_n *by* [9] *and* [10].

The costs of publication of this article were defrayed in part by the payment of page charges. This article must therefore be hereby marked "*advertisement*" in accordance with 18 U. S. C. §1734 solely to indicate this fact.

Then for $n \geq m$,

$$b_{n1} = \beta_1 + \frac{\sum_{i=1}^{n} (x_{i1} - K_n H_n^{-1} T_i) \epsilon_i}{\sum_{i=1}^{n} (x_{i1} - K_n H_n^{-1} T_i)^2} \quad [11]$$

Define for $n \geq m$

$$u_n = \sum_{i=1}^{n} (x_{i1} - K_n H_n^{-1} T_i) \epsilon_i, \quad [12]$$

$$w_n = u_n - u_{n-1}, \; d_n = x_{n1} - K_n H_n^{-1} T_n.$$

Then for $n > m$,

$$w_n = d_n \left\{ \epsilon_n - T'_n H_{n-1}^{-1} \left(\sum_{i=1}^{n-1} T_i \epsilon_i \right) \right\}, \quad [13]$$

and defining C_n *as in* [6], *we have*

$$1/c_{11}^{(n)} = \sum_1^n (x_{i1} - K_n H_n^{-1} T_i)^2$$

$$= (1/c_{11}^{(m)}) + \sum_{m+1}^{n} d_i^2 (1 + T_i' H_{i-1}^{-1} T_i). \quad [14]$$

Moreover, if the random variables ϵ_i *are uncorrelated and have zero mean and the same variance* σ^2 $(0 \leq \sigma < \infty)$, *then for* $l > n > m$,

$$E(w_l w_n) = 0. \quad [15]$$

The proof of *Lemma 1* is omitted; some of the details can be found in ref. 2. Let $s_n = 1/c_{11}^{(n)}$. Then s_n is nondecreasing by [14]. Suppose that $s_n \to \infty$ as $n \to \infty$. Then in view of [11], [12], [14], and the Kronecker lemma, to prove [8], it suffices to show that with probability 1

$$\sum_{i=m+1}^{n} \frac{w_i}{\{s_i |\log s_i|^{1+\delta}\}^{1/2}} \text{ converges as } n \to \infty. \quad [16]$$

Assume that $\{\epsilon_i\}$, $i \geq 1$, is a martingale difference sequence such that

$$E\epsilon_i = 0 \text{ for all } i \text{ and } \sup_i E\epsilon_i^2 < \infty. \quad [17]$$

By [14] and the integral comparison test,

$$\sum_{m+1}^{\infty} \frac{d_i^2}{s_i |\log s_i|^{1+\delta}} < \infty,$$

and hence by the martingale convergence theorem, with probability 1,

$$\sum_{i=m+1}^{n} \frac{d_i \epsilon_i}{\{s_i |\log s_i|^{1+\delta}\}^{1/2}} \text{ converges as } n \to \infty. \quad [18]$$

Therefore, in view of [13], to prove [16] it suffices to show that

$$\sum_{i=m+1}^{n} \frac{d_i T'_i H_{i-1}^{-1} \left(\sum_{j=1}^{i-1} T_j \epsilon_j \right)}{\{s_i |\log s_i|^{1+\delta}\}^{1/2}} \quad [19]$$

converges as $n \to \infty$ with probability 1. Noting that $H_n = \Sigma_1^n T_i T'_i$ and that

$$\sum_{m+1}^{\infty} \frac{d_i^2 (1 + T'_i H_{i-1}^{-1} T_i)}{s_i |\log s_i|^{1+\delta}} < \infty$$

by [14] and the integral comparison test, the desired convergence of the series [19] follows from the following

THEOREM 3. *Let* $\{\epsilon_n\}$, $n \geq 1$, *be a martingale difference sequence satisfying* [17]. *Let* k *be a positive integer. For each* $n \geq 1$, *let* T_n *be a* k*-dimensional vector of constants and let* $H_n = \Sigma_1^n T_i T'_i$. *Assume that* H_m *is positive definite for some* m (*so that* H_n *is positive definite for all* $n \geq m$). *Let* $\{a_n\}$, $n \geq 1$, *be a sequence of constants such that*

$$\sum_{m+1}^{\infty} a_i^2 (1 + T'_i H_{i-1}^{-1} T_i) < \infty. \quad [20]$$

Then with probability 1,

$$\sum_{i=m+1}^{n} a_i T'_i H_{i-1}^{-1} \left(\sum_{j=1}^{i-1} T_j \epsilon_j\right) \text{ converges as } n \to \infty. \quad [21]$$

The proof of *Theorem 3* will be given elsewhere. In the next section, we shall give a simpler proof of the strong consistency of b_{nj} under stronger assumptions on the errors ϵ_i. This alternative approach can also handle the case where $X'_n X_n$ is singular.

Generalized inverse and double arrays

Suppose that in [1] the random variables $\epsilon_1, \epsilon_2, \ldots$ are independent with $E\epsilon_i = 0$ and $E\epsilon_i^2 = \sigma^2$ for all i. The quantity u_n defined by [12] is a weighted sum $\Sigma_{i=1}^n a_{ni} \epsilon_i$ of the independent random variables ϵ_i, where the weights a_{ni} form a double array of constants. On the other hand, the relation [15] implies that u_n is also of the form $w_0 + \Sigma_{m+1}^n w_i$, where $\{w_0 (= u_m), w_{m+1}, \ldots\}$ is a sequence of mutually orthogonal random variables. We now use this dual structure of u_n to obtain a very simple proof of *Theorem 2* when the random variables ϵ_i satisfy certain fourth-order product moment assumptions. In fact, we shall prove the following more general.

THEOREM 4. *Let* $\{\epsilon_i\}$, $i \geq 1$, *be a sequence of random variables such that* $E\epsilon_i = 0$, $E\epsilon_i^4 < \infty$ *for all* i, *and*

$$E(\epsilon_i \epsilon_j) = E(\epsilon_i^3 \epsilon_j) = E(\epsilon_i^2 \epsilon_j \epsilon_k) = E(\epsilon_i \epsilon_j \epsilon_k \epsilon_l) = 0 \text{ for any distinct } i, j, k, l. \quad [22]$$

Assume that there exists a positive constant K *such that for all* i, j,

$$E(\epsilon_i^2 \epsilon_j^2) \leq K(E\epsilon_i^2)(E\epsilon_j^2). \quad [23]$$

Let $S_n = \Sigma_1^n a_{ni} \epsilon_i$, *where* $\{a_{ni}\}$, $n = 1, 2, \ldots$, $i = 1, \ldots, n$, *is a double array of constants, and let* $W_n = S_n - S_{n-1}$. *Suppose that for some positive integer* m

$$E(W_l W_n) = 0 \text{ if } l > n \geq m. \quad [24]$$

(i) *If* $\sup_n ES_n^2 = \infty$, *then letting* $v_n = ES_n^2$,

$$\lim_{n \to \infty} \frac{S_n}{v_n^{1/2} (\log v_n)^\delta} = 0 \text{ with probability 1} \quad [25]$$

for every $\delta > 1/4$.

(ii) *If* $\sup_n ES_n^2 < \infty$, *then* S_n *converges to some random variable* S *with probability 1.*

Suppose in *Theorem 2*, instead of assuming that the ϵ_i form a martingale difference sequence, we assume that they satisfy the conditions of *Theorem 4* and that they have the same variance $\sigma^2 (< \infty)$. Then, in view of *Lemma 1*, the desired conclusion [8] of *Theorem 2* is an immediate corollary of *Theorem 4*.

If $\epsilon_1, \epsilon_2, \ldots$ are independent with zero means and finite fourth moments, then clearly [22] holds and so does [23] for $i \neq j$. When $i = j$, [23] reduces to $E\epsilon_i^4 \leq K(E\epsilon_i^2)^2$, which obviously holds for some K if $\sup E\epsilon_i^4 < \infty$ and $\inf E\epsilon_i^2 > 0$. To prove *Theorem 4*, we shall use the following.

LEMMA 2. *Let* $\{S_n\}$, $n \geq 1$, *be a sequence of random variables and let* $\{c_n\}$, $n \geq 1$, *be a sequence of nonnegative constants. Suppose there exist positive constants* M *and* m *such that*

$$E(S_N - S_n)^4 \leq M(\sum_{n+1}^{N} c_i)^2 \text{ for all } N > n \geq m. \quad [26]$$

(i) *Assume that* $\Sigma_1^\infty c_i = \infty$, *and let* $\{v_n\}$, $n \geq 1$, *be a sequence of positive constants such that* $v_n \sim \Sigma_1^n c_i$. *Then* [25] *holds for every* $\delta > 1/4$.

(ii) *If* $\Sigma_1^\infty c_i < \infty$, *then* $\lim_{n\to\infty} S_n$ *exists with probability 1.*

Part (*i*) of *Lemma 2* has been established by Móricz (ref. 5, p. 309), and part (*ii*) of the lemma can be proved by a straightforward modification of Móricz's argument.

Proof of *Theorem 4*. We first make use of the double array representation $s_n = \Sigma_1^n a_{ni}\, \epsilon_i$ to show that

$$E(S_N - S_n)^4 \leq 3K\{E(S_N - S_n)^2\}^2 \text{ for all } N > n. \quad [27]$$

From the double array representation, for $N > n$,

$$S_N - S_n = \sum_{1}^{n} (a_{Ni} - a_{ni})\, \epsilon_i + \sum_{n+1}^{N} a_{Ni}\, \epsilon_i. \quad [28]$$

For fixed N and n with $N > n$, let $\theta_i = a_{Ni}$ for $N \geq i \geq n + 1$ and $\theta_i = a_{Ni} - a_{ni}$ for $1 \leq i \leq n$. Letting $\sigma_i^2 = E\epsilon_i^2$, it follows from [22] and [28] that

$$E(S_N - S_n)^2 = \sum_{1}^{N} \theta_i^2\, \sigma_i^2. \quad [29]$$

Moreover, by [22], [23], and [28],

$$E(S_N - S_n)^4 = \sum_{1}^{N} \theta_i^4\, E\epsilon_i^4 + 6 \sum_{1\leq i<j\leq N} \theta_i^2\, \theta_j^2\, E(\epsilon_i^2\, \epsilon_j^2)$$

$$\leq K\left\{\sum_{1}^{N} \theta_i^4\, \sigma_i^4 + 6 \sum_{1\leq i<j\leq N} \theta_i^2\, \sigma_i^2\, \theta_j^2\, \sigma_j^2\right\}. \quad [30]$$

From [29] and [30], [27] follows.

By the orthogonality relation [24], for $N > n \geq m$,

$$E(S_N - S_n)^2 = \sum_{n+1}^{N} EW_i^2. \quad [31]$$

To prove part (*i*) of the theorem, we note that $v_n = ES_n^2 \sim \Sigma_m^n EW_i^2$. In view of [27] and [31], the inequality [26] is satisfied with $c_i = EW_i^2$ and $M = 3K$, and therefore *Lemma 2*(*i*) gives the desired conclusion. Likewise part (*ii*) of the theorem follows from *Lemma 2*(*ii*).

Returning to the multiple regression model [1], we now consider the case where X'_nX_n may be singular. Let R^p denote the p-dimensional Euclidean space of column vectors. For $\alpha \in R^p$, $\alpha' b$ is unique for all solutions $b \in R^p$ of the normal equation $X'_nX_n\, b = X'_nY_n$ if and only if α is a linear combination of the column vectors of X'_n (see ref. 6, p. 181). Let L be the linear subspace of R^p generated by the set of vectors

$$\{(x_{i1}, \ldots, x_{ip})': i = 1,2, \ldots\}. \quad [32]$$

($L = R^p$ if X'_nX_n is nonsingular for some and therefore for all large n.) A solution to the equation $X'_nX_nb = X'_nY_n$ is

$$b = X_n^+Y_n \quad [33]$$

where X_n^+ denotes the Moore–Penrose generalized inverse of the matrix X_n (see ref. 1). This reduces to the unique solution [2] when X'_nX_n is nonsingular. Even when X'_nX_n is singular, $\alpha' X_n^+ Y_n$ is the unique least squares estimate of $\alpha'\beta$ for all large n if $\alpha \in L$. Assume that in [1] the random variables ϵ_i are uncorrelated and have a common variance $\sigma^2 > 0$. Then the variance of $\alpha' X_n^+ Y_n$ is $\sigma^2\alpha' X_n^+(X_n^+)'\alpha$. Moreover, for $\alpha \in L$, $\alpha' X_n^+Y_n$ is an unbiased estimate of $\alpha'\beta$ for all large n, and a necessary and sufficient condition for $\alpha' X_n^+ Y_n$ to converge to $\alpha'\beta$ in probability is that

$$\alpha' X_n^+(X_n^+)'\alpha \to 0 \text{ as } n \to \infty \quad [34]$$

(see ref. 1). By making use of *Theorem 4*, we obtain the strong consistency of $\alpha' X_n^+ Y_n$ under the condition [34] on the design in the following.

THEOREM 5. *Suppose that in* [1] *the random variables* $\epsilon_1, \epsilon_2, \ldots$ *satisfy* [22], $E\epsilon_i = 0$ *and* $E\epsilon_i^2 = \sigma^2$ *for all* i, *and* $\sup_i E\epsilon_i^4 < \infty$. *Let* L *be the linear subspace of* R^p *generated by the set* [32], *and let* $\alpha \neq 0$ *belong to* L. *Let* $\rho_n^2(\alpha) = \alpha' X_n^+(X_n^+)'\alpha$. *Then the sequence* $\{\rho_n(\alpha)\}$ *is eventually nonincreasing, and* $\rho_n(\alpha) > 0$ *for all large* n. *Moreover, if* $\lim \rho_n(\alpha) = 0$, *then with probability 1*

$$\alpha' X_n^+Y_n - \alpha'\beta = o(|\rho_n(\alpha)|\log \rho_n(\alpha)|^\delta|) \text{ as } n \to \infty \quad [35]$$

for every $\delta > 1/4$.

Proof. Let $\|\alpha\|$ denote $(\alpha'\alpha)^{1/2}$. Since $\alpha \in L$, for each $n \geq n_0$ (sufficiently large) there exists $z_n \in R^p$ such that $\alpha = X'_nX_n\, z_n$ (see ref. 6, p. 181). Since $\alpha \neq 0$, $X_nz_n \neq 0$. For $N \geq n_0$, define

$$a_n = (\|\alpha\|^2/\|X_nz_n\|^2)\, X_nz_n. \quad [36]$$

Then $a_n \neq 0$, and letting $a_n = (a_{n1}, \ldots, a_{nn})'$, we have

$$\alpha' X_n^+ Y_n = \alpha'\beta + \frac{\|\alpha\|^2 \sum_{1}^{n} a_{ni}\epsilon_i}{\|a_n\|^2} \quad [37]$$

(see ref. 1, p. 122). Let

$$S_n = \sum_{1}^{n} a_{ni}\, \epsilon_i,\; W_n = S_n - S_{n-1}\; (n > n_0). \quad [38]$$

Since the ϵ_i are uncorrelated and have the same variance σ^2, it follows from Lemma 3.3(a) of ref. 1 that $E(W_lW_n) = 0$ for $l > n > n_0$. Moreover, $\|a_n\|^2$ is nondecreasing in $n \geq n_0$ by Lemma 3.3(b) of ref. 1, and since

$$\rho_n^2(\alpha) = \|\alpha\|^4/\|a_n\|^2 \quad [39]$$

in view of [37], the sequence $\{\rho_n(\alpha)\}$, $n \geq n_0$, is nonincreasing and positive. Without loss of generality, we shall assume that $\sigma > 0$. Then by the Schwarz inequality, [23] holds for some positive constant K since $\sup E\epsilon_i^4 < \infty$. Moreover, if $\rho_n(\alpha) \to 0$, then $ES_n^2 = \sigma^2\|a_n\|^2 \to \infty$ by [39], and therefore the desired conclusion [35] follows from [37], [39], and *Theorem 4* (*i*).

This research was supported by the National Science Foundation, the National Institute of General Medical Sciences, and the Office of Naval Research.

1. Drygas, H. (1976) *Z. Wahrscheinlichkeitstheorie verw. Geb.* **34**, 119–127.
2. Anderson, T. W. & Taylor, J. B. (1976) *Ann. Statist.* **4**, 788–790.
3. Anderson, T. W. & Taylor, J. B. (1976) *Technical Report No. 213* (Institute for Mathematical Studies in Social Sciences, Stanford University, Stanford, CA).
4. Lai, T. L. & Robbins, H. (1977) *Proc. Natl. Acad. Sci. USA* **74**, 2667–2669.
5. Móricz, F. (1976) *Z. Wahrscheinlichkeitstheorie verw. Geb.* **35**, 299–314.
6. Rao, C. R. (1965) *Linear Statistical Inference and Its Applications* (Wiley, New York).

Statistics & Probability Letters 1 (1983) 137–139
North-Holland Publishing Company

March 1983

A Note on the 'Underadjustment Phenomenon'

Herbert Robbins
Department of Mathematical Statistics, Columbia University, Morningside Heights, New York, NY 10027, U.S.A.

Bruce Levin
Division of Biostatistics, Columbia University School of Public Health and the G. Sergievsky Center, New York, NY 10027, U.S.A.

Received August 1982; revised version received October 1982

Abstract. In recent work in the area of employment discrimination statistics we have noticed that a basic formula quantifying the degree of underadjustment bias in regression coefficients seems not to have been clearly stated and proved. We believe the underadjustment phenomenon to be of great importance in assessing the validity of multiple regression studies of wage disparities in equal employment litigation, and of observational studies involving two populations in general. Since the argument is based entirely on some simple mathematical features of the linear model, the result should be of interest in other fields of application as well.

Keywords. Regression, underadjustment, employment discrimination.

The following derivation explains in mathematical terms the underadjustment bias in any latent variable model. The main result is given by equation (9). The major points can be illustrated with three variables for employees:

Y = salary,

θ = productivity (a latent variable),

X = an observable proxy for θ.

We begin by assuming in some given population of employees a bivariate joint distribution for (θ, X) in which a linear regression of X on θ and of θ on X exists.

Assumption 1. *The joint distribution of (θ, X) is such that for some constants a, b*

$$\mathbf{E}(X \mid \theta) = a + b\theta, \tag{1}$$

and likewise $\mathbf{E}(\theta \mid X)$ is a linear function of X.

From (1) we have

$$\mathbf{E}\,X = a + b\mathbf{E}\theta, \tag{2}$$

$$\begin{aligned}\mathbf{E}(\theta X) &= \mathbf{E}\,\mathbf{E}(\theta X \mid \theta) = \mathbf{E}(a\theta + b\theta^2) \\ &= a\,\mathbf{E}\,\theta + b\,\mathbf{E}\,\theta^2,\end{aligned}$$

$$\begin{aligned}\mathrm{Cov}(\theta, X) &= \mathbf{E}(\theta X) - (\mathbf{E}\,\theta)(\mathbf{E}\,X) \\ &= a\,\mathbf{E}\,\theta + b\,\mathbf{E}\,\theta^2 - a\,\mathbf{E}\,\theta - b\,\mathbf{E}^2\,\theta \\ &= b\sigma_\theta^2,\end{aligned}$$

$$\rho = \mathrm{Corr}(\theta, X) = \frac{\mathrm{Cov}(\theta, X)}{\sigma_\theta \sigma_X} = b\frac{\sigma_\theta}{\sigma_X}. \tag{3}$$

Thus (1) can be written as

$$\begin{aligned}\mathbf{E}(X \mid \theta) &= \mathbf{E}\,X + b(\theta - \mathbf{E}\,\theta) \\ &= \mathbf{E}\,X + \rho\frac{\sigma_X}{\sigma_\theta}(\theta - \mathbf{E}\,\theta),\end{aligned}$$

and similarly

$$\begin{aligned}\mathbf{E}(\theta \mid X) &= \mathbf{E}\,\theta + \rho\frac{\sigma_\theta}{\sigma_X}(X - \mathbf{E}X) \\ &= \frac{\mathbf{E}\,X - a}{b} + \frac{\rho^2}{b}(X - \mathbf{E}\,X) \quad \text{from (2) and (3)}\end{aligned}$$

$$= \frac{(1-\rho^2)\mathbf{E}\,X + \rho^2 X - a}{b} \tag{4}$$

(we are here assuming that $b \neq 0$).

Assumption 2. *For some constants* α, β, γ, *and error term* ε,

$$Y = \alpha + \beta\theta + \gamma X + \varepsilon,$$

where $\mathbf{E}(\varepsilon \mid \theta, X) = 0$ *for all* (θ, X), *so that* $\mathbf{E}\,\varepsilon = 0$.

In words, salary depends on productivity, and may also depend explicitly on X. An important special case is $\gamma = 0$, in which mean salary does not depend on X, holding productivity constant. Assumption 2 and equation (2) imply that

$$\begin{aligned}\mathbf{E}\,Y &= \alpha + \beta\,\mathbf{E}\,\theta + \gamma\,\mathbf{E}\,X \\ &= (\alpha - a\beta/b) + (\gamma + \beta/b)\mathbf{E}X. \end{aligned} \tag{5}$$

Assumption 3. *Assumptions* 1 *and* 2 *hold for both male and female employees with the same values of* a, b, ρ, α, β *and* γ *in the two populations. However,* $\mathbf{E}\,X$, $\mathbf{E}\,\theta$, σ_X, σ_θ *are not assumed to be constant over sex.*

It now follows from (4) that

$$E_{\mathrm{m}}(\theta \mid X) - E_{\mathrm{f}}(\theta \mid X) = \frac{(1-\rho^2)}{b}(E_{\mathrm{m}}\,X - E_{\mathrm{f}}\,X), \tag{6}$$

which by Assumption 2 gives

$$E_{\mathrm{m}}(Y \mid X) - E_{\mathrm{f}}(Y \mid X) = \frac{\beta}{b}(1-\rho^2)(E_{\mathrm{m}}\,X - E_{\mathrm{f}}\,X). \tag{7}$$

From (5) we have

$$E_{\mathrm{m}}\,Y - E_{\mathrm{f}}\,Y = (\gamma + \beta/b)(E_{\mathrm{m}}\,X - E_{\mathrm{f}}\,X), \tag{8}$$

and hence

$$\boxed{E_{\mathrm{m}}(Y \mid X) - E_{\mathrm{f}}(Y \mid X) = \frac{(1-\rho^2)}{(1 + b\gamma/\beta)}(E_{\mathrm{m}}\,Y - E_{\mathrm{f}}\,Y).} \tag{9}$$

This is our basic result. In deriving it we assumed that b in (1) was not zero. However, if $b = 0$, then, by (3), $\rho = 0$ and hence $\mathbf{E}(\theta \mid X) = \mathbf{E}\,\theta$, in which case both sides of (9) are equal to $\beta(E_{\mathrm{m}}\theta - E_{\mathrm{f}}\theta)$ and (9) continues to hold.

The crux of the argument is embodied in (6), which shows that when $\rho^2 < 1$ and $E_{\mathrm{m}}\,X \neq E_{\mathrm{f}}\,X$, there will be a non-zero difference in the expected productivity of men and women for the same proxy value of X. It is this which leads to a non-zero 'sex-coefficient' $E_{\mathrm{m}}(Y \mid X) - E_{\mathrm{f}}(Y \mid X)$, expressed in (9) as the fraction $(1-\rho^2)/(1 + b\gamma/\beta)$ of the unadjusted salary difference $E_{\mathrm{m}}\,Y - E_{\mathrm{f}}\,Y$.

We emphasize that (9) continues to hold even if $a = 0$ and $b = 1$ in (1). Hence, *even if* X *is an unbiased estimator of* θ *for both men and women*, there will still be an underadjustment for productivity of the gross wage difference by regressing on X. For example, with $\rho = 0.60$, $\gamma = 0$, a sex-coefficient of almost $\frac{2}{3}$ of the unadjusted salary difference is entirely consistent with the equitable employment practice embodied in Assumptions 1, 2 and 3, even assuming unbiased estimation of θ by the proxy X.

The underadjustment phenomenon is well known, at least qualitatively, to statisticians working in the area of employment discrimination statistics. A partial list of references in this field is Wolins (1978), Roberts (1979, 1980), Birnbaum (1979), McCabe (1980), Dempster (1981) and Peterson (1981). The bias has long been discussed in the statistical literature under headings such as 'inadequately measured variables' (Stouffer, 1936), 'regression fallacies' (Thorndyke, 1942), 'errors-in-variables' (Cochran, 1957) and 'Lord's paradox' (Lord, 1960). However, neither the basic result (9) nor Assumptions 1–3 on which it rests appear to have been explicitly stated before. Roberts' papers (1979, 1980) on reverse regression contain results in a similar vein, although different from ours. Goldberger (1982) has provided an alternative model specification in which direct regression provides unbiased estimates, despite the fact that θ and X are imperfectly correlated.

Acknowledgment

We wish to thank Professor Goldberger for helpful comments on an earlier draft of this note.

References

Birnbaum, M.H. (1979), Procedures for the detection and correction of salary inequities, in: Pezzullo and Brittingham, eds., *Salary Inequity*, pp. 121–144.

Cochran, W.G. (1957), Analysis of covariance: Its nature and uses, *Biometrics* **13**, 261–281.

Dempster, A.P. (1981), Causal inference, prior knowledge, and the statistics of employment discrimination, Res. Rept. S-78, Dept. of Statistics, Harvard University.

Goldberger, A.S. (1982), Reverse regression and salary discrimination, Preliminary draft, Dept. of Economics, Univ. of Wisconsin, Madison.

Lord, F.M. (1960), Large-sample covariance analysis when the control variable is fallible, *J. Amer. Statist. Assoc.* **55**, 307–321.

McCabe, G.P. (1980), The interpretation of regression analysis results in sex and race discrimination problems, *The American Statistician* **34**, 213–215.

Peterson, D.W. (1981), Pitfalls in the use of regression analysis for the measurement of equal employment opportunity, *Internat. J. Policy Anal. and Inform. Systems* **5**, 64.

Roberts, H.V. (1979), Harris Trust and Savings Bank: An analysis of employee compensation, Rept. 7946, Center for Math. Studies in Business and Economics, University of Chicago.

Roberts, H.V. (1980), Statistical biases in the measurement of employment discrimination, in: Livernash, ed., *Comparable Worth – Issues and Alternatives* (Equal Employment Advisory Council, Washington D.C.) pp. 183–191.

Stouffer, S.A. (1936), Evaluating the effect of inadequately measured variables in partial correlation analysis, *J. Amer. Statist. Assoc.* **31**, 348–360.

Thorndyke, R.L. (1942), Regression fallacies in the matched groups experiment, *Psychometrika* **7**, 85–102.

Wolins, L. (1978), Sex differentials in salaries: Faults in analysis of covariance, *Science* **200**, 717.

PERMISSIONS

Springer-Verlag would like to thank the original publishers of Herbert Robbins' papers for granting permission to reprint specific papers in this collection. The following list contains the credit lines for those articles.

Interview Reprinted from *College Math J.* **15**, pp. 2–24.

[25] Reprinted from *Proc. Second Berkeley Symposium Math. Statist. Prob.* **1**, University of California Press, pp. 131–148.

[39] Reprinted from *Ann. Math. Statist.* **26**, pp. 37–51.

[41] Reprinted from *Proc. Third Berkeley Symposium Math. Statist. Prob.* **1**, University of California Press, pp. 157–163.

[59] Reprinted from *Ann. Math. Statist.* **35**, pp. 1–20.

[103] Reprinted from *Proc. Natl. Acad. Sci. USA* **74**, pp. 2670–2671.

[118] Reprinted from *Proc. Natl. Acad. Sci. USA* **77**, pp. 6988–6989.

[123] Reprinted from *Statistical Decision Theory and Related Topics III* **2**, pp. 251–261. Copyright © 1982, by Academic Press, Inc.

[124] Reprinted from *Ann. Statist.* **11**, pp. 713–723.

[26] Reprinted from *Ann. Math. Statist.* **22**, pp. 400–407.

[87] Reprinted from *Optimizing Methods in Statistics*, pp. 233–257. Copyright © 1971, by Academic Press, Inc.

[107] Reprinted from *Proc. Natl. Acad. Sci. USA* **75**, pp. 586–587.

[113] Reprinted from *Ann. Statist.* **7**, pp. 1196–1221.

[28] Reprinted from *Bull. Amer. Math. Soc.* **58**, pp. 527–535.

[42] Reprinted from *Proc. Natl. Acad. Sci. USA* **42**, pp. 920–923.

[129] Reprinted from *Proc. Natl. Acad. Sci. USA* **81**, pp. 1284–1286.

[89] Reprinted from *Proc. Natl. Acad. Sci. USA* **69**, pp. 2993–2994.

[96] Reprinted from *J. Amer. Statist. Assoc.* **69**, pp. 132–139.

[44] Reprinted from *Probability and Statistics*, pp. 235–245, Almqvist and Wiksells, Stockholm.

[64] Reprinted from *Ann. Math. Statist.* **36**, 457–462.
[68] Reprinted from *Ann. Math. Statist.* **38**, 1392–1398.
[51] Reprinted from *Rep. Statist. Appl. Res. JUSE* **9**, 39–60.
[116] Reprinted from *Proc. Natl. Acad. Sci. USA* **77**, pp. 3135–3138.
[70] Reprinted from *Proc. Natl. Acad. Sci. USA* **57**, pp. 1188–1192.
[82] Reprinted from *Ann. Math. Statist.* **41**, pp. 1397–1409.
[83] Reprinted from *Ann. Math. Statist.* **41**, pp. 1410–1419.
[91] Reprinted from *Proc. Sixth Berkeley Symposium Math. Statist. Prob.* **4**, pp. 37–41, University of California Press.
[92] Reprinted from *Bull. Inst. Math., Acad. Sinica*, **1**, pp. 93–120.
[7] Reprinted from *Ann. Math. Statist.* **15**, pp. 70–74; also **18**, p. 297.
[12] Reprinted from *Proc. Natl. Acad. Sci. USA* **33**, pp. 25–31.
[16] In *Duke Mathematical Journal*, **15** (1948), pp. 773–780. Copyright 1948, copyright renewed 1976 by Duke University Press.
[21] Reprinted from *Ann. Math. Statist.* **20**, pp. 552–560.
[27] Reprinted from *Ann. Math. Statist.* **22**, pp. 581–586.
[30] Reprinted from *Proc. Natl. Acad. Sci. USA* **39**, pp. 525–533.
[31] Reprinted from *Proc. Amer. Math. Soc.* **4**, pp. 786–799.
[32] Reprinted from *Proc. Natl. Acad. Sci. USA* **39**, pp. 860–864.
[38] Reprinted from *Amer. Math. Monthly* **62**, pp. 26–29.
[48] Reprinted from *Proc. Natl. Acad. Sci. USA* **47**, pp. 330–335.
[49] Reprinted from *Proc. Fourth Berkeley Symposium Math. Statist. Prob.* **1**, pp. 93–104, University of California Press.
[55] Reprinted from *Z. Wahrsch. Verw. Gebiete* **2**, pp. 33–49.
[61] Reprinted from *Israel J. Math.* **2**, pp. 81–90.
[85] Reprinted from *Amer. Math. Monthly* **77**, pp. 333–343.
[63] Reprinted from *Ann. Math. Statist.* **36**, pp. 789–799.
[60] Reprinted from *Publ. Math. Inst. Hung. Acad. Sci. Ser. A* **9**, pp. 209–225.
[75] Reprinted from *Ann. Math. Statist.* **39**, pp. 256–257.
[93] Copyright © 1973 by the Directors for the Columbia Law Review Association, Inc. All rights reserved. This article originally appeared at **73** *Colum. L. Rev.* 241–248 (1973). Reprinted by permission.
[102] Reprinted from *Proc. Natl. Acad. Sci. USA* **73**, pp. 286–288.
[110] Reprinted from *Proc. Natl. Acad. Sci. USA* **75**, pp. 3034–3036.
[125] Reprinted from *Statist. Probab. Letters* **1**, pp. 137–139, North Holland, Amsterdam.

Printed in the United States
By Bookmasters